MW00784642

Inventory and Production Management in Supply Chains

Fourth Edition

Inventory and Production Management in Supply Chains

Fourth Edition

Edward A. Silver
University of Calgary (retired), Alberta, Canada

David F. Pyke
University of San Diego, California, USA

Douglas J. Thomas
Penn State University, Pennsylvania, USA

CRC Press
Taylor & Francis Group
Boca Raton   London   New York

CRC Press is an imprint of the
Taylor & Francis Group, an **informa** business

CRC Press
Taylor & Francis Group
6000 Broken Sound Parkway NW, Suite 300
Boca Raton, FL 33487-2742

© 2017 by Taylor & Francis Group, LLC
CRC Press is an imprint of Taylor & Francis Group, an Informa business

No claim to original U.S. Government works

Printed on acid-free paper
Version Date: 20160830

International Standard Book Number-13: 978-1-4665-5861-8 (Hardback)

This book contains information obtained from authentic and highly regarded sources. Reasonable efforts have been made to publish reliable data and information, but the author and publisher cannot assume responsibility for the validity of all materials or the consequences of their use. The authors and publishers have attempted to trace the copyright holders of all material reproduced in this publication and apologize to copyright holders if permission to publish in this form has not been obtained. If any copyright material has not been acknowledged please write and let us know so we may rectify in any future reprint.

Except as permitted under U.S. Copyright Law, no part of this book may be reprinted, reproduced, transmitted, or utilized in any form by any electronic, mechanical, or other means, now known or hereafter invented, including photocopying, microfilming, and recording, or in any information storage or retrieval system, without written permission from the publishers.

For permission to photocopy or use material electronically from this work, please access www.copyright.com (http://www.copyright.com/) or contact the Copyright Clearance Center, Inc. (CCC), 222 Rosewood Drive, Danvers, MA 01923, 978-750-8400. CCC is a not-for-profit organization that provides licenses and registration for a variety of users. For organizations that have been granted a photocopy license by the CCC, a separate system of payment has been arranged.

Trademark Notice: Product or corporate names may be trademarks or registered trademarks, and are used only for identification and explanation without intent to infringe.

Library of Congress Cataloging-in-Publication Data

Names: Silver, Edward A. (Edward Allen), 1937- author. | Pyke, D. F. (David F.) author. | Silver, Edward A. (Edward Allen), 1937- Decision systems for inventory management and production and planning. | Silver, Edward A. (Edward Allen), 1937- Inventory management and production planning and scheduling.
Title: Inventory and production management in supply chains / Edward A. Silver, David F. Pyke, Douglas J. Thomas.
Description: Fourth Edition. | Boca Raton : Taylor & Francis, 2017. | Revised edition of Inventory management and production planning and scheduling. | Includes index.
Identifiers: LCCN 2016022678 | ISBN 9781466558618 (hardback : alk. paper)
Subjects: LCSH: Inventory control--Decision making. | Production planning--Decision making.
Classification: LCC HD40 .S55 2017 | DDC 658.7/87--dc23
LC record available at https://lccn.loc.gov/2016022678

Visit the Taylor & Francis Web site at
http://www.taylorandfrancis.com

and the CRC Press Web site at
http://www.crcpress.com

Edward A. Silver dedicates this work to Maxine, Michelle, Norman, and Heidi

David F. Pyke dedicates this work to Susan, James, Daniel, and Cory Ad majorem Dei gloriam

Douglas J. Thomas dedicates this work to Traci, Alison, Kate, and Maya

Contents

x ■ Contents

SECTION III SPECIAL CLASSES OF ITEMS

SECTION IV MANAGING INVENTORY ACROSS MULTIPLE LOCATIONS AND MULTIPLE FIRMS

Preface

Dramatic advances in information and communication technologies have allowed firms to be more closely connected to a broad and global network of suppliers providing components, finished goods, and services. Customers want their products quickly and reliably, even if they are coming from another continent; and they are often willing to look for other suppliers if products are frequently delivered late. As a result, inventory management and production planning and scheduling have become even more vital to competitive success.

Inventory management and production planning and scheduling have been studied in considerable depth from a theoretical perspective. Yet, the application of these theories is still somewhat limited in practice. A major gap has existed between the theoretical solutions, on the one hand, and the real-world problems, on the other.

Our primary objective in the first three editions of the book was to bridge this gap through the development of *operational* inventory management and production planning decision systems that would allow management to capitalize on readily implementable improvements to current practices. Extensive feedback from both academicians and practitioners has been very gratifying in this regard. Our primary objective is unchanged in this revised edition. In the third edition, we placed an emphasis on how computing tools, such as spreadsheets, can be used to apply the models presented in this book. While sophisticated supply chain planning software is now widely available, we still find that spreadsheets are ubiquitous in practice. As such, we have retained and expanded content on the spreadsheet application of the models in this book.

Advances in information technology have made it easier for firms to exchange information with their trading partners. Indeed, much of the academic and practitioner literature on supply chain management has focused on how firms can collaborate and share information to reduce costs in their shared supply chains. Generally speaking, these initiatives have had a positive economic impact. As discussed in Chapter 1, inventory as a percent of sales, and logistics costs as a percent of gross domestic product, have been consistently declining. Our opinion is that these savings are just the tip of the iceberg. It is now commonplace for an employee in an inventory or manufacturing planning role to (a) have powerful tools on their personal computer, and (b) have access to demand and supply-related data. This means it has never been easier to implement rigorous planning models such as those presented in this book. This is not to say that it is easy, just *easier* than before. There are new challenges to planning in modern supply chains where operations are distributed throughout the globe and across many companies. As such, we have included new material in this edition addressing how to collaborate from a planning perspective across departments within an organization and with external trading partners.

As with all editions of this book, we have drawn on the writings of literally hundreds of scholars who have extended theory and who have implemented theory in practice. We have also incorporated many helpful suggestions from our colleagues.

Where appropriate, new modeling approaches and an updated discussion of the literature have been added throughout the book. Here, we briefly highlight some of the major changes in this edition.

1. The first two chapters from the previous edition have been updated and streamlined into one chapter. This chapter highlights recent macroeconomic trends in inventory and logistics costs, and includes a discussion of how these changes affect operations and corporate strategy.
2. Chapters from the previous edition on Just-in-Time and short-range production scheduling have been updated and consolidated into a single chapter.
3. A new chapter on coordinating inventory management in the supply chain has been added. This chapter addresses collaboration and coordination within a firm through processes such as sales and operations planning, as well as external collaboration through initiatives such as collaborative planning, forecasting and replenishment, and vendor managed inventory. Contractual approaches for coordinating inventory decisions between firms are also covered.
4. The treatment of how to manage inventory for items with low-volume, erratic, or non-stationary demand has been expanded. This includes the addition of derivations and spreadsheet applications for the Poisson distribution.
5. The discussion of supply chain planning software tools including Enterprise Resource Planning (ERP) systems and advanced planning tools has been updated and expanded.
6. The treatment of managing multiple items through the use of composite exchange curves has been expanded. To accompany this material, data sets and associated problems are available as supplemental material. These data sets provide students with the opportunity to apply the concepts in the book to realistic, practical settings.

We have continued to attempt to provide a deep and rigorous treatment of the material without presenting complicated mathematics in the main text. Additional derivations have been added in this edition, but this content generally appears in an appendix. We have chosen this style of presentation deliberately so that sufficient material of interest is made available to the analytically inclined reader, while at the same time providing a meaningful text for a less analytically oriented audience.

Section I of the book presents a discussion on inventory management and production planning decisions as important components of total business strategy. The topics covered include operations strategy, the diverse nature of inventories, the complexity of production/inventory decision making, cost measurement, and an introduction to the important concept of exchange curves. A separate chapter is devoted to the determination of forecasting strategy and the selection of forecasting methods.

Section II is concerned with traditional decision systems for the inventory control of individual items. The use of approximate decision rules, *based on sound logic*, permits realistic treatment of time-varying (e.g., seasonal) demand, probabilistic demand, and the many different attitudes of management toward costing the risk of insufficient capacity in the short run.

Section III deals with special classes of items, including the most important (Class A) and the large group of low-activity items (Class C). Also discussed are procedures for dealing with items that can be maintained in inventory for only relatively short periods of time—for example, style goods and perishable items (the newsvendor problem).

Section IV addresses three types of coordination of groups of items primarily in nonproduction contexts. First, there is the situation of a family of items, at a single stocking location, that share a common supplier, a common mode of transport, or common production equipment. The second type of coordination is where a single firm is managing inventory across several of their locations in a multiechelon network. For such a setting, replenishment requests from one stocking point become part of the demand at another location. The third context is where items are managed across a network where locations in the network are managed by different firms. This setting requires new approaches to share information and align incentives among trading partners to effectively manage inventory throughout the supply chain.

Section V is concerned with decision making in a production environment. A general framework for such decision making is presented. It covers aggregate production planning, material requirements planning, enterprise resource planning systems, JIT, OPT®, and short-range production scheduling.

This book should be of interest to faculty and students in programs of business administration, industrial/systems engineering, and management sciences/operations research. Although *the presentation is geared to a foundational course in production planning, scheduling, and inventory management,* the inclusion of extensive references permits its use in advanced elective courses and as a starting point for research activities. At the same time, the book should continue to have a broad appeal to practicing analysts and managers.

Acknowledgments

The manuscript, as is often the case, had its origin in the teaching notes used by the authors for several years in courses they taught at the Amos Tuck School at Dartmouth College, The University of Auckland, Boston University, The University of Calgary, The University of Canterbury (New Zealand), The Smeal College of Business at Penn State, The Swiss Federal Polytechnique Institute, and the University of Waterloo. A large number of students (including many part-time students holding employment in industry) have provided excellent critiques.

A significant portion of the book has developed out of research supported by the Natural Sciences and Engineering Research Council of Canada, the Defence Research Board of Canada, and the Ford Foundation. Support was also provided by the Tuck School, the School of Business at the University of San Diego, and the Center for Supply Chain Research in the Smeal College of Business. The authors gratefully acknowledge all the aforementioned support. Numerous consulting assignments involving the authors have also had substantial influence on the contents of the book.

Special mention must be made of two authorities in the field who, early in our careers, encouraged us to work in the general area of inventory and production management—namely, Robert G. Brown, as a colleague at Arthur D. Little, Inc., and Morris A. Cohen, as a PhD advisor at the Wharton School, University of Pennsylvania. Also, we wish to express our sincere thanks to Rein Peterson for his contributions as coauthor of earlier editions of this book.

Many of our professional colleagues have provided helpful comments concerning our research papers, the earlier editions of the book, and the drafts of the current edition. In addition, numerous other colleagues have made available drafts of papers in the more advanced topic areas. Many of the publications of these contributors are referenced in this book. We appreciate all these important contributions. The list of all these colleagues would be too long to include, but we do wish to specifically mention Daniel Costa (Nestle S.A.), Robert Lamarre (Gestion Conseil Robert Lamarre), and David Robb (University of Auckland).

Finally, on such a major task, a special word of thanks is necessary for those individuals who have edited and proofread the many drafts of the manuscript. While a student at Penn State, Jimmy Chen (Bucknell University) carefully reviewed the manuscript, providing very helpful suggestions and edits. Lauren Bechtel, Sharon Cox, Rachel Gimuriman, and Denis Harp were extremely helpful in supporting editing and formatting of tables and figures.

Edward A. Silver
David F. Pyke
Douglas J. Thomas

Authors

Edward A. Silver is a professor emeritus of operations and supply chain management in the Haskayne School of Business at the University of Calgary. Until his retirement he held the Carma Chair at the University of Calgary. Prior to his appointment at the University of Calgary, he was a professor of management sciences in the Faculty of Engineering at the University of Waterloo. He also previously taught at Boston University and, as a visiting professor, at the Swiss Federal Polytechnique Institute (in Lausanne, Switzerland), the University of Canterbury (in Christchurch, New Zealand), and the University of Auckland (New Zealand).

A native of Montreal, Professor Silver completed a bachelor of civil engineering (applied mechanics) at McGill University and a science doctorate in operations research at the Massachusetts Institute of Technology. He is a licensed professional engineer and has been a member of a number of professional societies, including the American Production and Inventory Control Society, the Canadian Operational Research Society (of which he was the president in 1980–1981), the Institute of Industrial Engineers, the Institute for Operations Research and the Management Sciences, the International Society for Inventory Research (of which he was the president in 1994–1996), the Production and Operations Management Society, and the Operational Research Society (UK).

Professor Silver has presented seminars and talks at national and international meetings of a number of professional societies as well as educational institutions throughout North America and in parts of Europe, Asia, and New Zealand. He has published close to 170 articles in a broad range of professional journals. Dr. Silver has also served in an editorial capacity for several journals.

Professor Silver spent four years as a member of the Operations Research Group of the international consulting firm, Arthur D. Little Inc. Subsequently he has done independent consulting for a wide range of industrial and government organizations throughout North America and elsewhere. These consulting activities have addressed both tactical and strategic problems arising in the management of operations. Specific areas of application have included inventory management, supply chain management, process improvement, production planning, and logistics management. An additional important activity has been his involvement in several executive development programs and other workshops related to the inventory/production field.

Dr. Silver has received extensive professional recognition. This has included being named as a Fellow of five organizations, specifically the Institute of Industrial Engineers (1995), the Manufacturing and Services Operations Management Society (2000), the International Society for Inventory Research (2000), the Institute for Operations Research and the Management Sciences

(2003), and the Production and Operations Management Society (2010). He also was the recipient of two of the highest awards of the Canadian Operational Research Society, namely the Award of Merit (1990) and the Harold Larnder Memorial Prize (2007). Finally, he was awarded a Visiting Erskine Fellowship at the University of Canterbury, Christchurch, New Zealand (1998) and a Visiting Research Fellowship by the Japan Society for the Promotion of Science (2002).

David F. Pyke, PhD, is professor of operations and supply chain management in the School of Business at the University of San Diego (USD). He was dean of the business school from 2008 until 2015, leading the school to a significant presence in national and international business school rankings. Formerly, he was the Benjamin Ames Kimball Professor of Operations Management, and associate dean at the Amos Tuck School of Business Administration at Dartmouth College. He obtained his BA from Haverford College, MBA from Drexel University, and his MA and PhD from the Wharton School of the University of Pennsylvania. He was awarded an honorary MA from Dartmouth College in 1999.

He has taught executive programs at USD, Tuck, Wharton, and in other environments. He has taught globally at the International University of Japan, the Helsinki School of Economics, and the WHU-Otto-Beisheim-Hochschule, Vallendar, Germany. Professor Pyke has consulted for The Rand Corporation, Accenture, Corning, DHL, Nixon, Eaton, Home Depot, Lemmon Company and Markem, among others. He has also served as an expert witness on supply chain management issues for securities cases. He serves on the Board of Directors of GW Plastics, Concepts NREC, and until recently the Lwala Community Alliance, a nonprofit focused on community development in Kenya. He is an advisor to GuardHat, Inc., Analytics Ventures, and Moore Venture Partners, an operating partner of Tuckerman Capital LLC, and partner with San Diego Social Venture Partners.

Professor Pyke's research interests include operations management, supply chain management, integrated enterprise risk management, pricing, inventory systems, manufacturing in China, production management, and manufacturing strategy. He has published numerous papers in journals such as *Management Science, Operations Research, Production and Operations Management, Sloan Management Review, Journal of Operations Management, Naval Research Logistics, European Journal of Operational Research*, and *Interfaces*. He coedited a book, *Supply Chain Management: Innovations for Education*, with M. Eric Johnson, in the *POMS Series in Technology and Operations Management*. He has served on the editorial boards for *Management Science* and *Naval Research Logistics*, among others. He is a member of the Institute for Operations Research and Management Sciences, and the Production and Operations Management Society.

Douglas J. Thomas, PhD, is a professor of supply chain management in the Smeal College of Business at Pennsylvania State where he was the faculty director of the MBA program from 2011 to 2014. He also serves as chief scientist for Plan2Execute, LLC, a firm that provides supply chain software and consulting solutions in warehouse management, transportation management, and advanced planning. Doug earned his MS and PhD from Georgia Institute of Technology in industrial engineering and has a BS in operations research from Cornell University. Prior to returning to graduate school, he worked for C-Way Systems, a software company specializing in manufacturing scheduling. In addition to his years on the faculty at Penn State, Doug has had the pleasure of serving as a visiting faculty member at INSEAD (in Fontainebleau, France), the Johnson Graduate School of Management at Cornell University and The Darden School at the University of Virginia.

Doug currently teaches courses in the areas of supply chain management and quantitative modeling in MBA, executive MBA, and PhD programs. A frequent faculty leader in executive development programs, Doug has led numerous executive education sessions in Africa, Asia, Europe, and North America, including programs at Penn State, INSEAD, and Georgia Institute of Technology as well as custom programs for Accenture, DuPont, ExxonMobil, IBM, Ingersoll-Rand, Mars, Office Depot, Parker-Hannifin, Pfizer, Schlumberger, and the U.S. Marine Corps. He has testified as an expert witness and consulted for several large organizations on supply chain strategy, including Accenture, CSL Behring, Dell, ExxonMobil, and Lockheed Martin Aerospace Corporation.

His research interests include coordinating production and inventory planning across the extended enterprise and connecting decision models to logistics performance measurement. His work has appeared in several academic and practitioner journals in the areas of logistics and operations management, including *Management Science, Manufacturing and Service Operations Management*, and *Production and Operations Management*. He serves as a senior editor for *Production and Operations Management*. He is a member of the Institute for Operations Research and Management Sciences, and the Production and Operations Management Society.

THE CONTEXT AND IMPORTANCE OF INVENTORY MANAGEMENT AND PRODUCTION PLANNING

1

Since the mid-1980s, the strategic benefits of inventory management and production planning and scheduling have become obvious. The business press has highlighted the success of Japanese, European, and North American firms that have achieved unparalleled efficiency in manufacturing and distribution. In recent years, many of these firms have "raised the bar" yet again by coordinating with other firms in their supply chains. For instance, instead of responding to unknown and highly variable demand, they share information so that the variability of the demand they observe is significantly lower.

In spite of the media attention devoted to these developments, two mistakes are far too common. First, many managers assume that new levels of efficiency can be attained simply by sharing information and forming "strategic alliances" with their supply chain partners. These managers do not understand that "the devil is in the details," and that knowing what to do with the data is as important as getting the data in the first place. Developing sound inventory management and production planning and scheduling methods may seem mundane next to strategy formulation, but these methods are a critical element of long-term survival and competitive advantage. Second, many analysts assume that implementing sophisticated inventory and production methods will solve all the problems. These analysts may achieve a high level of efficiency by optimizing given the lead times and demand variability the firm observes. But they do not understand that this efficiency is insignificant compared to the efficiencies available by *changing the givens*.

In Chapter 1, we discuss the importance of inventories and production planning at an aggregate level. This includes a discussion of the role of aggregate inventories in the business cycle. We also present an overview of corporate strategy, and discuss the linkages of finance and marketing strategies with inventory management and production planning and scheduling. We then present a framework for formulating an operations strategy, and we describe how inventory management and production planning and scheduling fit in that context.

Chapter 2 presents several frameworks that will be helpful to operating managers, and to general managers. We describe various statistical properties of inventories, and we discuss how inventories can be classified. A useful framework for understanding different production processes is presented. Finally, we discuss the costs involved, and briefly note where they can be found.

Since managerial expectations about the future have such a tremendous impact on inventory management and production planning and scheduling decisions, forecasting the demand variable is given special treatment in a separate chapter. In Chapter 3, we present a number of strategies and techniques that could be adopted to cope with the unknown future.

Chapter 1

The Importance of Inventory Management and Production Planning and Scheduling

Some of the strongest and fastest-growing industries throughout the world are in the service sector. The consulting and financial services industries, in particular, hire thousands of college graduates each year. This fact has prompted many to suggest that manufacturing in developed countries has not only declined in importance, but is on a path toward extinction. It seems clear, however, that nations care deeply about manufacturing. This is particularly true in certain industries. Our observations suggest that, although they receive little attention in the press, the textile/apparel, machine tool,[*] and food industries are considered vital. Why is this so? We think it is due to at least two reasons. The first is that food and machine tools are fundamental to national security. Without an industry that manufactures and distributes food, a nation is very vulnerable if it is isolated due to a conflict; and in such situations, without machine tools, the hardware that drives the economy will grind to a halt. The second reason is employment. The textile/apparel industry employs millions of people worldwide, many in low-skill, entry-level, jobs. Without this industry a nation can face high levels of unemployment and has reduced capacity to bring people, such as immigrants, into the workforce.

Many other industries including microprocessors, computers, and automobiles are also considered vital in today's world. These industries are often the source of innovation, productivity improvements, and high-skill jobs. Consider, for example, the productivity improvements that resulted from the introduction of computer numerically controlled machine tools. So, while services have been growing in importance, we submit that manufacturing is still fundamental to the health of most modern economies. Adoption of the methods discussed in this book will ultimately help to strengthen both developed and developing economies, to the extent that they are successfully implemented in manufacturing and logistics firms. Many of the methods in this book also

---

[*] Machine tools are the machines that make machines, and therefore are the fundamental building blocks for any manufacturing operation.

apply to service industries such as banking, hospitals, hotels, restaurants, schools, and so on, where firms hold inventories of supplies. These stocks must be managed carefully, and in some cases, careful management of these stocks can be critical to the performance of the firm. Two recent studies highlight the importance of inventory management to firm performance. Hendricks and Singhal (2009) show that excess inventory announcements are followed by a negative stock market reaction, and Chen et al. (2007) relate abnormal inventory levels to poor stock returns.

Costs associated with production, inventory, and logistics are quite economically significant. Table 1.1 illustrates logistics costs in the United States. Note that the value of inventories in U.S. firms exceeded $2 trillion in 2007, and increased to close to $2.5 trillion by 2014. Total logistics costs, expressed as a percentage of gross domestic product (GDP), have slowly decreased in recent years, from 9.9% in 2000 to 8.2% in 2014. Several key factors continue to make effective inventory management challenging, including increasing product variety, shortening product lifecycles, and an increase in global sourcing. Despite these pressures, inventory levels for manufacturers have been dropping. Figure 1.1 shows inventory-to-sales ratios over time for U.S. retailers, manufacturers, and wholesalers. The figure suggests that inventory levels were decreasing up until 2009, perhaps starting to increase since then. In a rigorous examination of U.S. firms between 1981 and 2000, Chen et al. (2005) report that inventory levels dropped at an average annual rate of 2%. The majority of this improvement came from a reduction in work-in-process (WIP) inventory, with finished goods inventories not declining significantly. In a related study, Rajagopalan and Malhotra (2001) also find that raw material and WIP inventories declined for U.S. firms between 1961 and 1994.

While inventory levels as a percentage of sales have been declining, inventories have shifted downstream, closer to the customer. Figure 1.2 shows the fraction of inventory held by manufacturers compared to the fraction held by wholesalers and retailers. In 1992, manufacturers held 46% of total inventory compared to 37% in 2014. There are several factors that may be causing this shift. In recent years, firms have sourced their products from all over the globe in an effort to find cost efficiencies. Global sourcing initiatives will invariably increase lead times leading to higher inventory levels for wholesalers and retailers. A recent empirical study by Jain et al. (2013) reported that firms with more global sourcing have higher inventory levels. In addition to global sourcing initiatives, product variety offered to customers continues to increase. Firms with many products and short product lifecycles often have high gross margin and high demand uncertainty (Fisher 1997). An empirical study by Rumyantsev and Netessine (2007) confirms that higher gross margin and more demand uncertainty translate to higher inventory levels. An empirical study of U.S. retailers by Kesavan et al. (2016) observes that retailers with high inventory turnover are able to more effectively adjust purchase quantities to react to demand fluctuations, and the negative consequences of excess and shortage of inventory are far less severe for retailers with high inventory turnover.

For analysts, managers, consultants, and entrepreneurs, the opportunities to add value to manufacturing or logistics firms are enormous. On the basis of research and our own experience, it is evident that most firms do not fully understand the complexities of managing production and inventory throughout their supply chain. While there are certainly opportunities for the creative manager in the realms of finance and marketing, our focus is on the benefits that can be won by careful and competent management of the flow of goods throughout the supply chain. One of the authors regularly requires students in an inventory management class to work on consulting projects with local firms. Over the years, we have seen that in more than 90% of the cases, improved inventory or production management would lead to cost savings of at least 20%, without sacrificing customer service. This figure has been replicated in our consulting experience as well. One of our

Table 1.1 U.S. Logistics Costs ($ Billion)

Year	GDP	Value of Business Inventory (a)	Inventory Carrying Rate (%) (b)	Inventory Carrying Costs (c) = (a × b)	Transportation Costs (d)	Administrative Costs (e)	Total U.S. Logistics Costs (f) = (c + d + e)	Total Logistics Costs as a % of GDP
2000	10,280	1,525	25.3	386	594	39	1,019	9.9
2001	10,620	1,448	22.8	330	609	37	976	9.2
2002	10,980	1,495	20.7	309	582	35	926	8.4
2003	11,510	1,557	20.1	313	607	36	956	8.3
2004	12,270	1,698	20.4	346	652	39	1,037	8.5
2005	13,090	1,842	22.3	411	739	46	1,196	9.1
2006	13,860	1,994	24.0	479	809	50	1,338	9.7
2007	14,480	2,119	24.1	511	855	54	1,420	9.8
2008	14,720	2,050	21.4	439	872	52	1,363	9.3
2009	14,420	1,927	19.3	372	705	43	1,120	7.8
2010	14,960	2,130	19.2	409	770	47	1,226	8.2
2011	15,520	2,301	19.1	439	819	50	1,308	8.4
2012	16,160	2,392	19.1	457	847	52	1,356	8.4
2013	16,770	2,444	19.1	467	885	54	1,406	8.4
2014	17,420	2,496	19.1	477	901	55	1,433	8.2

Source: Council of Supply Chain Management Professionals Annual State of Logistics Report® 2015.

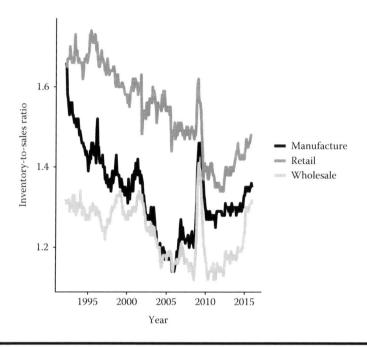

Figure 1.1   Inventory-to-sales ratios (seasonally adjusted) for U.S. retailers, manufacturers, and wholesalers. (From U.S. Census Bureau. November 2015 Manufacturing and Trade Inventories and Sales report.)

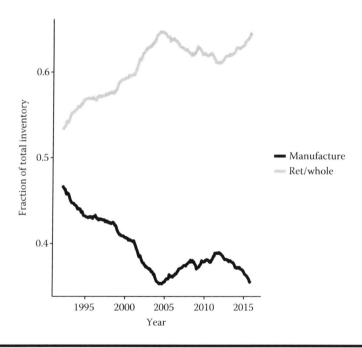

Figure 1.2   Distribution of inventory between manufacture and combined wholesale and retail (seasonally adjusted). (From U.S. Census Bureau. November 2015 Manufacturing and Trade Inventories and Sales report.)

goals in this book is to provide the reader with knowledge and skill so that he or she can bring real value to employers and, as a result, to the economy as a whole.

In Section 1.1, we discuss how business cycles affect aggregate inventory investments. We turn our attention to the connection between corporate strategy and inventory management and production planning in Section 1.2. In Section 1.3, we extend the strategy discussion to the areas of marketing and finance. Section 1.4 contains a framework for operations strategy. Section 1.5 deals with an important ingredient of decision making—namely, the specification of appropriate measures of effectiveness.

1.1 Why Aggregate Inventory Investment Fluctuates: The Business Cycle

It is an unfortunate fact that economies go through cycles–periods of expansion when employment rates are high and the general mood is one of unending prosperity, followed by periods of contraction when unemployment grows and there is a general feeling of malaise about the economy (Mack 1967; Reagan and Sheehan 1985; Blinder and Maccini 1991). As we shall see, inventories play an important role in these cycles. Many economists have tried to compile comprehensive models of the business cycle to explain all patterns of fluctuations that have occurred historically (e.g., Schmitt-Grohé and Uribe 2012). To date, no one has succeeded in building an all-purpose model. Nevertheless, it is apparent that although each cycle is somewhat different, especially with regard to its exact timing and relative magnitude, the cycles continue and there are several common underlying factors. These are illustrated in Figure 1.3.

Prior to the peak labeled in Figure 1.3, the economy is expanding and managers are optimistic about future sales. However, managers can be overzealous in their optimism. In fact, at the peak, because of overoptimistic expectations, too many products are manufactured by the economy and cannot be sold. These surplus goods increase aggregate inventories, and so producers start to decrease production levels. Eventually, the rate of sale exceeds the rate of production of goods.

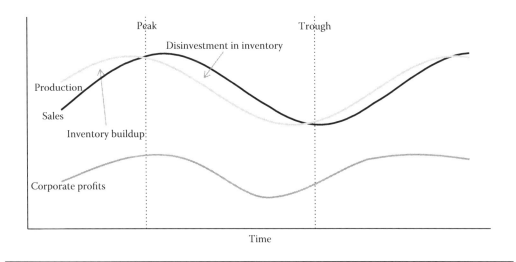

Figure 1.3   The business cycle.

The resulting disinvestment in inventories creates a recession during which prices, production, and profits fall and unemployment is prevalent. The financial crisis of 2007–2008 led to a rapid drop in sales and a subsequent buildup of inventory. This can be seen in the inventory-to-sales ratio of that time period in Figure 1.1.

After a time, an economic recovery is generated by a slowing in the rate of inventory liquidation. Some top managers, expecting that prices will recover, start to slowly expand their operations while costs are low. More and more firms slowly start to hire additional labor, purchase more raw materials, and thereby infuse more money into circulation in the economy. Consumers, with money to spend once again, start to bid up prices of available goods. Once prices of goods start to rise, more and more executives get on the bandwagon by expanding their operations and thereby accentuate the expansionary phase of the cycle. The boom that results eventually is brought to an end when costs of materials are bid up once again by competing firms, when the labor force begins to demand higher wages, and when the scarcity of money for further expansion causes the banks to raise their interest rates. (At this writing, interest rates remain quite low.) The crisis phase that follows is a period of uncertainty and hesitation on the part of consumers and business. Executives find that their warehouses are restocked with excess inventory that, once again, cannot be sold. The business cycle then is ready to repeat itself as explained above.

In Figure 1.3, the following key relationships are illustrated by the data. The peaks and troughs in corporate profits before taxes precede the peaks and troughs in production. Inventory investment lags slightly and thereby contributes to higher cyclical amplitudes in production than are really necessary.

This explanation is, of course, highly simplified, but it does illustrate the main forces, especially the role of inventories, at work during each cycle. Note that the expectations of business and consumers, as well as the ability of decision makers to react quickly and correctly to change, are important determinants of the length and severity of a cycle. Expectations about the future have been shown to depend on the following variables: the trend of recent sales and new orders, the volume of unfilled orders, price pressures, the level of inventories in the recent past, the ratio of sales to inventories (the turnover ratio to be discussed in Section 1.3.1), interest rates on business loans, the current level of employment, and the types of decision-making systems used by management. It is precisely this last point that we address in this book. We want to provide managers with sophisticated, yet understandable approaches for managing production and inventory in the supply chain, enabling them to make good decisions in a timely manner.

Recent research suggests that business cycles are less volatile than they were 50 years ago due, in part, to advances in inventory management and production planning and scheduling (Bloom et al. 2014). Many of the topics we will discuss in this book serve to more closely match production with demand. Therefore, managers can have a more timely perspective on the market, and are less surprised by economic downturns. Other models, such as the probabilistic inventory models discussed in Chapters 6 through 9, have been shown to decrease inventory volatility when applied correctly. Still, there is much work to be done, and there are many opportunities for knowledgeable and creative operations managers to help their firms weather the volatility of economic cycles.

1.2 Corporate Strategy and the Role of Top Management

Earlier in this chapter, we saw that, although management of production and inventory throughout the supply chain can be critical to the ability of a firm to achieve and maintain competitive advantage, top managers often do not recognize the importance of these issues. Operations managers, for their part, often neglect the role their activities play in the strategic direction of the firm, and even in its operations strategy. Both senior and operations managers need to understand the nature of corporate and operations strategy, and they need to understand how inventory management and production planning and supply chain issues impact other functions. In this chapter, we briefly discuss corporate strategy and the functional strategies that derive from it, including marketing, finance, and operations.

It is important to note that some firms have reorganized, or reengineered, to eliminate functional areas and replace them with *business processes*. For instance, a major multinational food and beverage firm recently reengineered in a way that redefined roles to be more responsive to the customer. A common reengineering approach is to replace the operations, logistics, and marketing functions with teams that are process focused. For instance, one team may be devoted to *generating demand*, while another focuses on *fulfilling demand*. The members of the generate demand team perform many of the tasks that traditionally have been done by the marketing department but they may carry out other functions as well, including new product development. The fulfill-demand team often looks like the manufacturing or operations department but may include other functions such as supply chain, logistics, sales, and marketing. The idea is to align the organizational structure with the processes the firm uses to satisfy its customers. Barriers between functions that created delays and tension are removed, enabling the firm to meet customer orders in a more seamless way. For our purposes, the management of inventory, production planning, and supply chains can be thought of as part of the operations function or the fulfill-demand process. In this book, we shall refer to marketing and operations functions rather than to generate and fulfill-demand processes. Nevertheless, the procedures and insights we present apply in either case.

It is not our intention to cover the topic of strategic planning in any depth. There are a number of texts on this topic (e.g., Porter 1980; Mintzberg and Quinn 1991; Grant and Jordan 2015). Instead, our objective is to show that decision making in the production and inventory areas must not be done in a vacuum, as is too often the case in practice, but rather must be coordinated with decisions in other functional areas by means of corporate strategic planning.

The key organizational role of top managers is strategic business planning. Senior management has the responsibility for defining in broad outline what needs to be done, and how and when it should be done. Top management also must act as the final arbiter of conflicts among operating divisions and has the ultimate responsibility for seeing that the general competitive environment is monitored and adapted to effectively. In most business organizations, four levels of strategy can be delineated (listed from the highest to the lowest level):

1. *Enterprise Strategy:* What role does the organization play in the economy and in society? What should be its legal form and how should it maintain its moral legitimacy?
2. *Corporate Strategy:* What set of businesses or markets should the corporation serve? How should resources be deployed among the businesses?
3. *Business Strategy:* How should the organization compete in each particular industry or product/market segment? On the basis of price, service, or what other factor?
4. *Functional Area Strategy:* At this level, the principal focus of strategy is on the maximization of resource productivity and the development of distinctive competencies.

While each level of strategy can be seen as being distinct, they must fit together to form a coherent and consistent whole. Typically, each level is constrained by the next higher one.

We will have more to say on strategic planning in Chapter 13 where a specific framework for decision making in a production environment will be developed.

1.3 The Relationship of Finance and Marketing to Inventory Management and Production Planning and Scheduling

1.3.1 Finance

Inventories have an important impact on the usual aggregate scorecards of management performance—namely, on the balance sheet and the income statement.[*] First, inventories are classified as one of the current assets of an organization. Thus, *all other things being equal*, a reduction in inventories lowers assets relative to liabilities. However, the funds freed by a reduction in inventories normally would be used to acquire other types of assets or to reduce liabilities. Such actions directly influence the so-called *current ratio*, the ratio of current assets to current liabilities, which is the most commonly used measure of liquidity.

Income statements represent the flow of revenues and expenses for a given period (e.g., 1 year). Specifically,

$$\text{Operating profit} = \text{Revenue} - \text{Operating expenses} \tag{1.1}$$

Changes in inventory levels can affect both of the terms on the right side of Equation 1.1. Sales revenue can increase if inventories are allocated among different items in an improved way. Also, because inventory carrying charges represent a significant component of operating expenses, this term can be reduced if aggregate inventory levels decrease. The labor component of operating expenses can also be reduced by more effective production scheduling and inventory control.

A primary aggregate performance measure for inventory management is inventory turnover, or *stockturns*. In this book, we define inventory turnover as

$$\text{Inventory turnover} = \frac{\text{Annual sales or usage (at cost)}}{\text{Average inventory (in \$)}} \tag{1.2}$$

Turnover can be a very useful measure, especially when comparing divisions of a firm or firms in an industry. See Gaur et al. (2005) for a detailed analysis of inventory turnover in the retail sector. From Equation 1.2, we can see that an increase in sales without a corresponding increase in inventory will increase the inventory turnover, as will a decrease in inventory without a decline in sales.

There are several dangers with turnover as a performance measure, however. Too frequently we have seen leaders of manufacturing firms respond to turnover less than the industry average by issuing an edict to increase turnover to the industry average. In one company, while competitive firms purchased some component parts, this company manufactured them because of a special competence in this type of manufacturing. WIP inventory was necessary to keep the production process flowing smoothly. To increase inventory turnover, all they would have to do is to buy more and make less. When the edict came down, the controller recalculated the decision rules for inventory ordering so that lot sizes were cut and raw material inventory would be reduced. Shipments to customers continued for a few months as warehouse inventories were depleted, but then customer service degenerated when finished goods and component inventories bottomed out. Rampant stockouts created massive, expensive expediting. The smaller lot sizes drove up the product cost, and profits plummeted. In short, strategic competitive advantage can have a significant effect on the determination of the appropriate inventory turnover number.

---

[*] Droms (1979) provides a nontechnical description of financial reports.

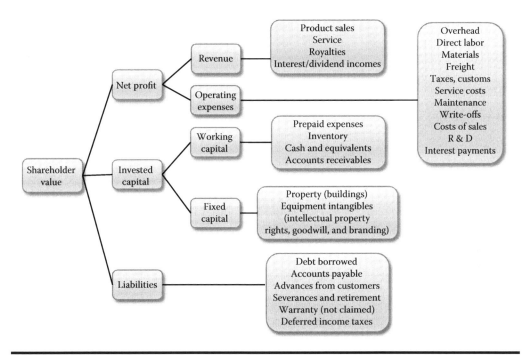

Figure 1.4   Components of shareholder value.

One of the most common measures of managerial performance is the *return on investment* (ROI), which represents the profit (after taxes) divided by the average investment (or level of assets). From the above discussion, it is clear that inventories have an important impact on the ROI (also see Problem 1.18).

Finally we note that many of these relationships can be summarized in Figure 1.4, which is due to Lee, H. and S. Whang (1997, Personal communication). Inventory is a component of working capital, and the interest payments to finance that inventory are a component of operating expenses. Improvements in inventory management, therefore, can have a significant impact on shareholder value. Likewise, improvements in production planning and scheduling can increase customer service and, therefore, product sales, and can decrease expenses related to direct labor, materials, and freight. Thus, net profit can increase, again increasing shareholder value. It is useful to consider other potential impacts of supply chain management, inventory management, and production planning and scheduling on the components of Figure 1.4 (see Problem 1.14).

1.3.2 Marketing

The most common complaint we hear from operations managers about marketing is that marketing managers simply do not understand how difficult it is to manufacture and distribute a wide variety of products. Because of low-volume production of many products, productive capacity is lost to setups, and product demand is more difficult to forecast. Workers and equipment must be more flexible, and inventories must be higher. Marketing managers prefer a wide variety of products because they are listening to customers and trying to respond to their needs and desires. High inventories are desirable because demand can be met without delay. Furthermore, sales and marketing managers may make pricing decisions, such as steep promotional discounts, that increase unit sales, but send a shock through the factory and supply chain. These fundamental conflicts can be mitigated by a process orientation as discussed in the introduction to this chapter. Even with a process orientation, however, conflicts are bound to arise. This again highlights the need for senior managers to be involved in setting policy for supply chain management, inventory management, and production planning and scheduling.

1.4 Operations Strategy

In this section, we develop a framework for operations strategy and we emphasize the place of the management of inventory, production planning, and supply chain in the broader context of the operations function.

We begin by stressing again that the operations strategy of the firm should be in concert with other functional area strategies, including marketing, finance, and human resources. Sometimes, however, one function should take precedence over the others. One firm went through a painful period of not responding to customer needs in a long-term effort to improve customer satisfaction. The operations group needed to develop the capability to manufacture high-quality items in high volume. To avoid disruptions during the improvement process, customer desires were neglected in the short term, and marketing was frustrated. In the long term, however, customers were delighted by the level of quality and by the responsiveness of the firm.

The operations strategy itself is made up of three levels: mission, objectives, and management levers. (See Figure 1.5.)

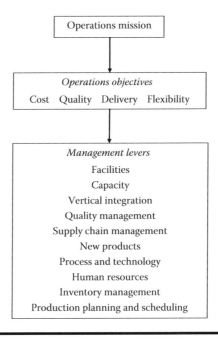

Figure 1.5   A framework for operations strategy.

1.4.1 Mission

The operations mission defines a direction for the operations function. McDonald's, for instance, uses four terms to describe its operational mission: quality, cleanliness, service, and value. For many years, the annual report has prominently highlighted these terms, most recently in the 2014 shareholder letter from the CEO. Because the mission should not change significantly over time, the mission statement is often somewhat vague. Otherwise, it is likely that it would have to be reworded frequently as changes occur in the environment, competition, or product and process technology. Employees need to know that there is a consistent direction in which their company is moving.

It is important however that the mission be attractive to excellent people. If it is a sleepy statement of direction that resembles the mission of most other firms in the industry, it will be difficult to attract the best employees, not to mention investors. Therefore, a mission statement should incorporate some of the excitement of top management and should communicate to employees, investors, and customers that this is an excellent firm.

1.4.2 Objectives

Because the mission statement is vague, it is difficult to know whether or not it has been achieved. How can one know, for instance, whether "quality" has been achieved? The second level of an operations strategy, operations objectives, provides carefully defined, measurable goals that help the firm achieve its mission. For many years, firms have used four operational objectives: cost, quality, delivery, and flexibility.

It is critical that the objectives are *defined* carefully, clearly *measurable*, and *ranked*. They must be defined carefully because these terms are often used loosely. Quality at a McDonald's restaurant is very different from quality at a five-star restaurant, which is very different from quality in a hospital operating room. It is also important that the objectives be measurable so that management knows whether they are meeting their goals or not. It is quite possible, and perhaps desirable, to use more than one measure for each objective (for instance, warranty cost and parts-per-million defective for the quality objective). The objectives should be rank ordered as well, so that managers can give priority to the correct objective when it is necessary to make tradeoffs. A manager of a high volume manufacturing line made it quite clear to one of the authors that, in his opinion, cost was more important than delivery. When questioned further, however, he noted that in fact he had occasionally gone over budget by using overtime in order to meet a due date. In other words, he said he ranked cost above delivery, but his behavior indicated that delivery was more important. Subsequent discussions with senior managers helped him understand that delivery was, in fact, primary.

In the 1970s, many people in operations thought that cost and quality were opposites, as were delivery and flexibility. For instance, one could aim for low-cost production, but then it would be impossible to achieve high quality. Likewise, if one had rapid delivery, it would be impossible to be flexible. More recent experience, however, suggests that cost and quality are complements rather than opposites. Warranty, prevention, and detection costs decrease as quality improves. Rework and congestion on the factory floor also decrease significantly, thereby reducing the cost of products. In addition, recent experience would indicate that rapid delivery of customized products can be possible. This is especially true with technology such as flexible automation, electronic data interchange, supplier web portals, radio frequency identification (RFID), and scanning technology. Most firms have instances, however, in which tradeoffs among the objectives must be made. Therefore, although combined improvements are possible, the objectives should still be ranked.

The cost objective can be considered in one of the three categories: low, competitive, or relatively unimportant. In a *low-cost* environment, such as that of certain discount retailers, the goal is to have the lowest cost products or services in the industry. Firms that aim for *competitive* costs do not necessarily strive to have the lowest cost products, but rather want to be competitive with most other firms in the industry. Some firms produce prototypes or have a unique product for which they can charge a *premium*; hence, cost becomes less important. Cost measures include dollars per unit, inventory turns, and labor hours per unit (or other productivity measures). In the United States, low cost tended to be the primary objective of manufacturing firms from the 1950s to the mid-1970s.

The quality objective rose to the fore in the mid-1970s to the mid-1980s with the advent of Japanese quality improvements and the sale of Japanese products in the United States. In particular, the automotive industry experienced the stunning effects of high-quality Japanese products. Quality can be defined by understanding which of its multiple dimensions are important. Garvin (1987) describes eight dimensions of quality: performance, conformance, reliability, durability, serviceability, features, aesthetics, and perceived quality. Quality measures include percent defective, percent returns, results from satisfaction surveys, warranty dollars, and so on.

Delivery can be defined on two dimensions, speed and reliability. For instance, some firms compete on delivering within 24 hours of the customer request. Others may take longer but will assure the customer that the products will be provided reliably within the quoted delivery time. Some companies rank delivery last among the objectives. Prototype manufacturing, for instance, sometimes involves complete customization and, therefore, may require long delivery times. Measures for delivery include percent on time, time from request to receipt, fill rate (the percent of demand met from the shelf), and so on.

The fourth objective is flexibility. Flexibility has three dimensions—volume, new product, and product mix. *Volume* flexibility is the ability to adjust for seasonal variations and fluctuations. *New product* flexibility is the speed with which new products are brought from concept to market. Automotive firms in recent years have made great strides in new product flexibility. A niche car allows a firm to jump in quickly to low-volume, profitable, markets. This flexibility is impossible if development time is eight to 10 years, as was traditionally the case with U.S. and European manufacturers. Most manufacturers have now reduced development time to roughly three-and-a-half years, a target first reached by several Japanese automotive firms. *Product mix* flexibility is the ability of a company to offer a wide range of products. This may simply mean that the catalog contains many items, or it may mean that the firm has the ability to develop customized products. Many machine tool companies produce a single product for a given customer, and then never produce that exact product again. Hence, they would rank product mix flexibility very high.

It has been argued that delivery and flexibility are the most important objectives in some industries because cost-cutting programs and quality improvement programs have "leveled the playing field" on cost and quality. These firms then compete on "time"—rapid introduction of new products and rapid delivery of existing products. The phrase *time-based competition* has been used to describe this phenomenon.

Finally, we note that the objectives are dynamic. For instance, as a new product begins full-scale production, the firm may emphasize flexibility (to design changes) and on-time delivery so that market share is not lost to competitors. Over time, as the product design stabilizes, the emphasis may change toward quality and cost.

1.4.3 Management Levers

Although the operations objectives provide measurable goals, they do not indicate how a firm should pursue those goals. The 10 management levers—facilities, capacity, vertical integration, quality management, supply chain management, new products, process and technology, human resources, inventory management, and production planning and scheduling—provide the tactical steps necessary to achieve the goals.[*]

It is interesting to note that in the late 1970s, researchers did not include quality management, supply chain relationships, new products, and human resources in the list of management levers (Wheelwright 1978). Today, these areas are considered critical to competitive success. It is also certain that new levers will be introduced in the next decade. The categories for the operations objectives, on the other hand, have not changed.

Facilities decisions concern the location and focus of factories and other facilities. Where does a company locate its factories and distribution centers to meet the needs of its customers? Is there a need for multiple facilities or is a single location sufficient? Does each facility perform all functions or is one facility focused on a particular market, a particular process, or a particular product? (See Sheu and Krajewski 1996). Many manufacturing companies have one plant that acts as a "parent plant" that is responsible for odd-ball parts and new product introduction. When products reach high-volume manufacture, they are moved to a satellite plant where efforts are focused on excellence in that particular product and/or process. It is critical to evaluate each plant manager based on the specific charter for his or her facility.

Capacity decisions involve magnitude and timing of capacity expansion. Capacity decisions interact with the facility location decisions. For instance, some firms set limits on the number of employees that may work at any one location; by remaining smaller than a given size, better communication and teamwork are possible. As demand grows, additional expansion at the same site would violate the limit, and a new plant site must be found. A Massachusetts textile manufacturer had rapidly expanding sales in Europe. When capacity in the Massachusetts factory was no longer able to meet total demand, they had to determine whether to expand in Massachusetts, elsewhere in the United States, or in Europe. The final decision was to build in eastern Germany to exploit lower transportation costs and import duties for European sales, and to take advantage of tax breaks offered by the German government.

Vertical integration concerns make/buy decisions. Some firms make components, perform final assembly, and distribute their products. Others focus only on product design and final assembly, relying on other firms to manufacture components and to distribute finished goods. Facilities, capacity, and vertical integration decisions are primarily made for the long term because they involve bricks and mortar and significant investment.

Quality management encompasses the tools, programs, and techniques used to achieve the quality goals. It is distinct from the quality objective in that the objective specifies the definition and measurable targets, and the lever specifies the means to achieve the targets. Quality management includes things such as statistical process control (SPC), Taguchi methods, quality circles, and so on. Many aspects of process improvement fall under this management lever, although process improvement leads to increased productivity as well as improved quality. It is important to note that different definitions of quality may dictate the use of different quality procedures.

---

[*] Portions of this discussion are drawn from Fine and Hax (1985).

The *supply chain management* lever focuses on relationships with suppliers and customers. Indeed, it is concerned with the entire network of firms and activities from, say, digging ore out of the ground, to getting the product into the hands of the customer. Furthermore, it is concerned with the handling of the product after it has finished its useful life. In general, supply chain management uses tools to efficiently move information, products, and funds across the entire supply chain. Chapter 12 contains an extensive discussion of supply chain management.

The *new products* category involves the procedures and organizational structures that facilitate new product introduction. The days of a focused team of engineers working without input from marketing or manufacturing are gone. Most firms now employ multifunctional teams, composed of design engineers, marketing personnel, manufacturing managers, and production line workers. The new product lever specifies the reporting relationships as well as any procedures for setting milestones in the development process.

The *process and technology* category encompasses the choice of a production process and the level of automation. The product-process matrix, to be discussed in Chapter 2, is a useful framework for analyzing process choices. Decisions about implementing new process technology are included here as well.

Human resources involves the selection, promotion, placement, reward, and motivation of the people that make the business work.

Inventory management encompasses decisions regarding purchasing, distribution, and logistics, and specifically addresses when and how much to order. This management lever is the topic of Chapters 4 through 11, although Chapter 6 alone provides a comprehensive discussion of the issues involved.

Finally, *production planning and scheduling* focuses on systems for controlling and planning production. Chapters 13 through 16 deal with specific production planning and scheduling decisions and systems.

1.4.4 General Comments

We conclude this section with several general comments. First, researchers and practitioners are continually introducing new approaches addressing some important aspect of management. Inventory management and production planning and scheduling methods that would have been considered impossible just a few years ago are commonplace today. The revolution in information technology and computer networking allows members of a supply chain to have access to manufacturers and to track consumer demand on a daily basis, and thereby more closely match production with demand. Many of these new approaches fall under the board area now widely known as *supply chain management*. The supply chain lever, of course, pertains to supply chain management, but so do inventory management (as we shall see in Chapters 11 and 12), production planning and scheduling, vertical integration, and new products. Occasionally it is necessary to introduce a new lever to focus attention on the important issues, but other times, the existing levers adequately serve the purpose. Even with new supply chain initiatives, however, inventory management and production planning and scheduling continue to be a headache for many managers. Evidence suggests that even though a massive amount of real-time data is available, many firms do not use it, either because they are overwhelmed with data, or because they do not know what to do with it. Although real-time information can significantly reduce the demand variability a firm observes, significant variability and the associated problems usually remain. Managers must understand the value of the information, and they must choose inventory policies and production planning and scheduling systems that are appropriate for their situations. One of our primary goals in this book is to communicate the basic principles and procedures in this field so that managers, consultants, and analysts can provide insight and value to their firms in this dynamic world.

Second, as a firm audits its manufacturing strategy, it is important to recognize that the policies in place for each of the 10 levers should be consistent not only with the operations objectives, but also among themselves. If there are inconsistencies, the firm should initiate action programs to mitigate any negative effects. For instance, many companies historically pursued quality improvement programs without changing worker incentives. Workers were rewarded for the volume of their output, without regard for quality. In other words, the quality objectives and policies were inconsistent with the human resources policies. Reward systems had to be rearranged to fit with the quality goals. Likewise, as we discuss inventory management and production planning and scheduling in great depth throughout the remainder of this book, keep in mind the relationship of these management levers to the other eight, and to the operations objectives. Extensive effort to reduce inventory cost may not be warranted, for instance, if the firm can charge a premium price and really needs to focus on flexibility. In a similar vein, customer service levels, which will be discussed in Chapter 6, should be established with clear knowledge of the cost and delivery objectives. These issues will be developed further in Chapter 13.

Finally, in the process of auditing a manufacturing strategy, managers should understand distinctive competencies at the detailed level of the management levers. These distinctive competencies should influence the objectives, mission, and business strategy. Thus, information flows both ways, from business strategy to management levers and from levers to business strategy.

1.5 Measures of Effectiveness for Inventory Management and Production Planning and Scheduling Decisions

Analysts in the production/inventory area have tended to concentrate on a single quantifiable measure of effectiveness, such as cost, sometimes recognizing certain constraints, such as limited space, desired customer service, and so forth. In actual fact, of course, the impact of decisions is not restricted to a single measure of effectiveness. The appropriate measures of effectiveness should relate back to the underlying objectives of management. Unfortunately, some of the following objectives are very difficult to quantify:

1. Minimizing political conflicts (in terms of the competing interests) within the organization.
2. Maintaining a high level of flexibility to cope with an uncertain future.
3. Maximizing the chance of survival of the firm or the individual manager's position within the organization.
4. Keeping at an acceptable level the amount of human effort expended in the planning and operation of a decision system.

Moreover, as will be discussed in Chapter 2, even some of the more tangible cost factors are very difficult to estimate in practice.

From the perspective of top management, the measures of effectiveness must be aggregate in nature. Management is concerned with *overall* inventory levels and customer service levels (by broad classes of items) as opposed to, for example, minimization of costs on an individual item basis. Furthermore, rather than optimization in the absolute sense, a more appealing justification for a new or modified decision system is that it simply provides significant improvements (in terms of one or more measures of effectiveness) over an existing system.

One other complexity is that certain decisions in inventory management and production planning and scheduling should not necessarily be made in isolation from decisions in other areas. As an illustration, both Schwarz (1981) and Wagner (1974) point out that there are important interactions between the design of a physical distribution system (e.g., the number, nature, and location of warehouses) and the selection of an appropriate inventory management system. (Chapter 11 will deal further with this particular issue.) Other researchers address another type of interaction, namely, between transportation (shipping) decisions and those of inventory management. (See Ernst and Pyke 1993; Henig et al. 1997.) Another interaction, vividly portrayed by the success of Japanese management systems, is between inventory management and quality assurance. Finally, interactions between pricing and inventory decisions have generated much interest in research and in practice. (See Chan et al. 2004; Fleischmann et al. 2004; Chen and Chen 2015). In this book, we do not develop mathematical models that directly deal with these large-scale interactive problems. However, in several places, further descriptive advice and references will be presented.

1.6 Summary

In this chapter, we have examined a number of strategic issues related to inventory management and production planning and scheduling. In much of the remainder of the book, our attention will be focused on specific issues of how best to design and operate an inventory management and production planning and scheduling decision system for a specific organizational context. Nevertheless, it is important not to lose sight of the broader perspective of this chapter.

Problems

1.1 Select two appropriate, recent articles from a newspaper and identify how their contents could impinge on decision making in inventory management and production planning.

1.2 Why do inflation and supply shortages tend to increase real inventories? What is the effect of a reduction in the rate of inflation on inventory investment?

1.3 Investigate and discuss the inventory control procedures at a small (owner-managed) company. When does a company become too small to have its inventories managed scientifically?

1.4 Repeat Problem 1.3 for production planning and scheduling procedures.

1.5 Repeat Problem 1.3 for a small restaurant.

1.6 Look for evidence in the newspaper or business press to support or contradict the research on business cycles cited in Section 1.1. Discuss your findings.

1.7 Find an article in the business press that indicates the position of top management toward manufacturing issues, such as inventory management or production planning and scheduling. Discuss your findings.

1.8 Consider any other course you have taken. Briefly discuss how some key concepts from it are likely to interact with decisions in the inventory management or production planning and scheduling areas.

1.9 For each of three or four specific local industries, what do you think is the approximate breakdown between raw materials, WIP, and finished goods inventories? Discuss your answer.

1.10 In Section 1.5, mention was made of an interaction between the design of a physical distribution system and the choice of an inventory management system. Briefly comment on this interaction.

1.11 Examine the interaction specified in Section 1.5 again, but now between the choice of a transportation mode and the selection of inventory decision rules. Briefly discuss your results.

1.12 Briefly comment on how inventory management might interact with each of the following functional areas of the firm:
   a. Maintenance
   b. Quality control
   c. Distribution (shipping)

1.13 A management decision system designed for providing information on how well a business is doing, relative to other businesses in the same industry segment, is appropriate for
   a. Top management
   b. Middle management
   c. Marketing management
   d. Factory management
   e. Office staff
   Explain the reasoning behind your choice.

1.14 a. What components of Figure 1.4 would change, and in what direction, if production planning and scheduling systems were improved?
   b. What components of Figure 1.4 would change, and in what direction, if inventory management systems were improved?
   c. What components of Figure 1.4 would change, and in what direction, if supply chain management systems were improved?

1.15 Discuss the potential interactions between the production planning and scheduling lever and the inventory management lever.

1.16 What new topics do you think will be included on the list of management levers in 10 years? What levers will be eliminated?

1.17 If it is true that U.S. firms focused on cost, then quality, and then on delivery and flexibility (i.e., time), what will they focus on next?

1.18 An important financial performance measure is the return on investment:

$$ROI = \frac{Profit}{Total\ assets}$$

Assume that inventories are 34% of current assets.
   a. Consider the typical company and assume that it has a ratio

$$\frac{Current\ asset}{Total\ assets} = 0.40$$

   What impact will a 25% decrease in inventory have on the ROI? What percent decrease in inventories is needed to increase the ROI by 10%?
   b. Develop an algebraic expression for the impact on ROI of an x% decrease in inventories when

$$\frac{Current\ asset}{Total\ assets} = f$$

1.19   Analyze the financial statements for the last 5 years of two companies from different industries. (E.g., a grocery chain and an electronic goods manufacturer.) Calculate their turnover ratios. Try to explain why their turnover ratios are different. Are there any trends apparent that can be explained?

1.20   Why is the turnover ratio widely used? What are its benefits and what are its drawbacks?

1.21   Blood is collected for medical purposes at various sites (such as mobile units), tested, separated into components, and shipped to a hospital blood bank. Each bank holds the components in inventory and issues them as needed to satisfy transfusion requests. There are eight major types of blood and each type has many components (such as red cells, white cells, and plasma). Each component has a different medical purpose and a different lifetime. For example, the lifetime for white cells is only 6 hours, but the lifetime for red cells is now 42 days. What type of management issues are involved in this particular type of inventory management situation? (Discuss from strategic, tactical, and operational perspectives.) Does perishability require different records than for other situations? How?

References

Blinder, A. S. and L. J. Maccini. 1991. Taking stock: A critical assessment of recent research on inventories. *Journal of Economic Perspectives 5*, 73–96.

Bloom, N., M. Floetotto, N. Jaimovich, I. Saporta Eksten, and S. Terry. 2014. Really uncertain business cycles. *US Census Bureau Center for Economic Studies Paper No. CES-WP-14-18*.

Chan, L. M., Z. M. Shen, D. Simchi-Levi, and J. L. Swann. 2004. Coordination of pricing and inventory decisions: A survey and classification. In D. Simchi-Levi, S. David Wu, and Z. M. Shen (Eds.), *Handbook of Quantitative Supply Chain Analysis*, pp. 335–392. US: Springer.

Chen, H., M. Z. Frank, and O. Q. Wu. 2005. What actually happened to the inventories of American companies between 1981 and 2000? *Management Science 51*(7), 1015–1031.

Chen, H., M. Z. Frank, and O. Q. Wu. 2007. US retail and wholesale inventory performance from 1981 to 2004. *Manufacturing and Service Operations Management 9*(4), 430–456.

Chen, M. and Z.-L. Chen. 2015. Recent developments in dynamic pricing research: Multiple products, competition, and limited demand information. *Production and Operations Management 24*(5), 704–731.

Droms, W. G. 1979. *Finance and Accounting for Nonfinancial Managers*. Reading, MA: Addison-Wesley.

Ernst, R. and D. Pyke. 1993. Optimal base stock policies and truck capacity in a two-echelon system. *Naval Research Logistics 40*, 879–903.

Fine, C. and A. Hax. 1985. Manufacturing strategy: A methodology and an illustration. *Interfaces 15*(6), 28–46.

Fisher, M. L. 1997. What is the right supply chain for your product? *Harvard Business Review 75*(2)(March/April), 105–116.

Fleischmann, M., J. M. Hall, and D. F. Pyke. 2004. Smart pricing. *MIT Sloan Management Review 45*(2), 9–13.

Garvin, D. A. 1987. Competing on the eight dimensions of quality. *Harvard Business Review* (November–December), 101–109.

Gaur, V., M. L. Fisher, and A. Raman. 2005. An econometric analysis of inventory turnover performance in retail services. *Management Science 51*(2), 181–194.

Grant, R. M. and J. J. Jordan. 2015. *Foundations of Strategy*. UK: John Wiley & Sons.

Hendricks, K. B. and V. R. Singhal. 2009. Demand–supply mismatches and stock market reaction: Evidence from excess inventory announcements. *Manufacturing and Service Operations Management 11*(3), 509–524.

Henig, M., Y. Gerchak, R. Ernst, and D. Pyke. 1997. An inventory model embedded in a supply contract. *Management Science 43*(2), 184–189.

Jain, N., K. Girotra, and S. Netessine. 2013. Managing global sourcing: Inventory performance. *Management Science 60*(5), 1202–1222.

Kesavan, S., T. Kushwaha, and V. Gaur. 2016. Do high and low inventory turnover retailers respond differently to demand shocks? *Manufacturing and Service Operations Management 18*(2), 198–215.

Mack, R. P. 1967. *Information Expectations and Inventory Fluctuations*. New York: Columbia University Press.

Mintzberg, H. and J. B. Quinn. 1991. *The Strategy Process: Concepts, Contexts, Cases*. Englewood Cliffs, NJ: Prentice-Hall.

Porter, M. E. 1980. *Competitive Strategy: Techniques for Analyzing Industries and Competitors*. New York: Free Press.

Rajagopalan, S. and A. Malhotra. 2001. Have US manufacturing inventories really decreased? An empirical study. *Manufacturing and Service Operations Management 3*(1), 14–24.

Reagan, P. and D. P. Sheehan. 1985. The stylized facts about the behavior of manufacturers' inventories and backorders over the business cycle: 1959–1980. *Journal of Monetary Economics 15*(2), 217–246.

Rumyantsev, S. and S. Netessine. 2007. What can be learned from classical inventory models? A cross-industry exploratory investigation. *Manufacturing and Service Operations Management 9*(4), 409–429.

Schmitt-Grohé, S. and M. Uribe. 2012. What's news in business cycles. *Econometrica 80*(6), 2733–2764.

Schwarz, L. 1981. Physical distribution: The analysis of inventory and location. *AIIE Transactions 13*(2), 138–150.

Sheu, C. and L. J. Krajewski. 1996. A heuristic for formulating within-plant manufacturing focus. *International Journal of Production Research 34*(11), 3165–3185.

Wagner, H. M. 1974. The design of production and inventory systems for multifacility and multiwarehouse companies. *Operations Research 22*(2), 278–291.

Wheelwright, S. 1978. Reflecting corporate strategy in manufacturing decisions. *Business Horizons* (February), 56–62.

Chapter 2

Frameworks for Inventory Management and Production Planning and Scheduling

In this chapter, we first discuss, in Sections 2.1 and 2.2, the complexity of inventory management and production planning and scheduling decisions (even without considering interactions with other functional areas, as outlined in Section 1.3 of Chapter 1). Then, in Sections 2.3 through 2.5, we present various aids for coping with the complexity. Section 2.6 discusses costs and other relevant variables. Sections 2.7 and 2.8 provide some suggestions concerning the use of mathematical models. Sections 2.9 and 2.10 are concerned with two very different approaches for estimating relevant cost factors identified earlier in the chapter. Finally, Section 2.11 provides a detailed look at the process of a major study of inventory management or production planning and scheduling systems.

2.1 The Diversity of Stock-Keeping Units

Some large manufacturing companies and military organizations stock over 500,000 distinct items in inventory. Large distributors and retailers, such as department stores, carry about 100,000 goods for sale. A typical medium-sized manufacturing concern keeps in inventory approximately 10,000 types of raw materials, parts, and finished goods.

Items produced and held in inventory can differ in many ways. They may differ in cost, weight, volume, color, or physical shape. Units may be stored in crates, in barrels, on pallets, in cardboard boxes, or loose on shelves. They may be packaged by the thousands or singly. They may be perishable because of deterioration over time, perishable through theft and pilferage, or subject to obsolescence because of style or technology. Some items are stored in dust-proof, temperature-controlled rooms, while others can lie in mud, exposed to the elements.

Demand for items also can occur in many ways. Items may be withdrawn from inventory by the thousands, by the dozen, or unit by unit. They may be substitutes for each other, so that, if one item is out of stock, the user may be willing to accept another. Items can also be complements; that is, customers will not accept one item unless another is also available. Units could be picked up by a customer, or they may have to be delivered by company-owned vehicles or shipped by rail, boat, airplane, or truck. Some customers are willing to wait for certain types of products; others expect immediate service on demand. Many customers will order more than one type of product on each purchase order submitted.

Goods also arrive for inventory by a variety of modes and in quantities that can differ from how they will eventually be demanded. Some goods arrive damaged; others differ in number or kind from that which was requisitioned (from a supplier or from an in-house production operation). Some items are unavailable because of strikes or other difficulties in-house or at a supplier's plant. Delivery of an order may take hours, weeks, or even months, and the delivery time may or may not be known in advance.

Decision making in production, inventory, and supply chain management is therefore basically a *problem of coping with large numbers and with a diversity of factors external and internal to the organization*. Given that a specific item is to be stocked at a particular location, three basic issues must be resolved:

1. How often the inventory status should be determined
2. When a replenishment order should be placed
3. How large the replenishment order should be

Moreover, the detailed daily individual item decisions must, in the aggregate, be consistent with the overall objectives of management.

2.2 The Bounded Rationality of a Human Being

An inescapable fact alluded to in the first chapter was the need to view the inventory management and production planning and scheduling problems from a systems standpoint. Before we plunge into a detailed discussion of the possible solutions to the problem as we have posed it, let us consider explicitly and realistically our expectations regarding the nature of the results we seek.

Simon (1957, pp. 196–206) has pointed out that all decision makers approach complex problems with a framework or model that simplifies the real situation. A human being's brain is simply incapable of absorbing and rationalizing all of the many relevant factors in a complex decision situation. A decision maker cannot effectively conceive of the totality of a large system without assistance. All decision makers are forced to ignore some relevant aspects of a complex problem and base their decisions on a smaller number of carefully selected factors. These selected factors always reflect decision makers' personal biases, abilities, and perceptions of the realities they think they face, as well as the decision technology that is available to them at any point in time.

The decisions of inventory management and production planning and scheduling are complex. They extend beyond the intuitive powers of most decision makers because of the many interconnected systems, both physical and conceptual, that have to be coordinated, rationalized, adapted to, or controlled. The decisions must be viewed *simultaneously* from the point of view of the individual item in its relation to other similar items, the total aggregate inventory investment, the master plan of the organization, the production–distribution systems of suppliers and customers, and the economy as a whole.

The challenge is therefore twofold. Since a decision maker's ability to cope with diversity is limited, decision systems and rules must be designed to help expand the bounds put upon that individual's ability to rationalize. However, because most decision makers probably have already developed personalized approaches to the inventory/production decision, decision systems have to be designed, and their use advocated, in the context of existing resources and managerial capabilities.

Inventory management and production planning and scheduling as practiced today, like the practice of accounting, because of the predominance of numerical data and a relatively long history, are mixtures of economically sound theory, accepted industrial practices, tested personalized approaches, and outright fallacies. A goal of this book is to influence the amount of sound theory that actually gets used, while trying to clarify existing fallacies. However, one must accept the fact that existing theory is, and will be, for some time to come, insufficient to do the whole job. There will always be room for personalized, tested-in-practice, approaches to fill the gaps in theory. Therefore, within this book, we in effect weave our own brand of personalized approaches with theory. This is why, from an intellectual point of view, we find inventory and production decisions both challenging and exciting.

2.3 Decision Aids for Managing Diverse Individual Items

The managerial aids are basically of two kinds, conceptual and physical.

2.3.1 Conceptual Aids

The following is a partial list of conceptual aids for decision making:

1. Decisions in an organization can be considered as a hierarchy—extending from long-range strategic planning, through medium-range tactical planning, to short-range operational control. Typically, different levels of management deal with the three classes of decisions. We have more to say about such a hierarchy, specifically related to production situations, in Chapter 13.
2. A related type of hierarchy can be conceptualized with respect to decision making in inventory management and production planning and scheduling. At the highest level, one chooses a particular type (or types) of control system(s). At the next level, one selects the values of specific parameters within a chosen system (e.g., the desired level of customer service). Finally, one operationalizes the system, including data collection, calculations, reporting of results, and so forth (Brown 1982).
3. The large number of physical units of inventory can be classified into a smaller number of relatively homogeneous organizational categories. Managers then manage inventories in the context of this smaller number of decision classifications (to be discussed in Section 2.4).
4. The complexity of production planning and inventory management decisions can be reduced by identifying, through analysis and empirical research, only the most important variables for explicit consideration (e.g., ordering costs and demand rates). This is an especially important essence of the operational theory we expound in this book.

2.3.2 Physical Aids

Primary physical decision aids include the following:

1. First, and foremost, decision makers can employ spreadsheets and other computer programs. Throughout this book, we comment on how spreadsheets can be used to compute many of the formulas presented; and we ask the reader to develop his or her own spreadsheets in many of the end-of-chapter problems. Occasionally, we note that the algorithms are too difficult for simple spreadsheet implementation and suggest the use of other computer programs. In either case, the computer can handle vast amounts of data faster than the manager could ever dream of doing manually. The American Production and Inventory Control Society (APICS) publishes a magazine (*APICS—The Performance Advantage*) that regularly reviews relevant software.

2. A decision maker's physical span of control can also be expanded by harnessing the *principle of management by exception.* Decision rules are used within the computer system that focus the attention of the decision maker only on unusual or important events.

2.4 Frameworks for Inventory Management

In this section, we present several frameworks designed to help reduce the complexity of managing thousands of items.

2.4.1 Functional Classifications of Inventories

We recommend six broad decision categories for controlling aggregate inventories: cycle stock, congestion stock, safety (or buffer) stock, anticipation inventories, pipeline inventories, and decoupling stock. In our opinion, senior management can and must express an opinion on how much aggregate inventory is required in each of these broad categories. While we cannot expect that such opinions will always be in the form of clear-cut, simple answers, the categorization helps managers view inventories as being controllable rather than as an ill-defined evil to be avoided at all costs.

Cycle inventories result from an attempt to order or produce in batches instead of one unit at a time. The amount of inventory on hand, at any point, that results from these batches is called cycle stock. The reasons for batch replenishments include (1) economies of scale (because of large setup costs), (2) quantity discounts in purchase price or freight cost, and (3) technological restrictions such as the fixed size of a processing tank in a chemical process. The amount of cycle stock on hand at any time depends directly on how frequently orders are placed. As we will see, this can be determined in part by senior management who can specify the desired trade-off between the cost of ordering and the cost of having cycle stock on hand. A major consideration in Chapters 4, 5, 7, and 10 will be the determination of appropriate cycle stocks.

Congestion stocks are inventories due to items competing for limited capacity. When multiple items share the same production equipment, particularly when there are significant setup times, inventories of these items build up as they wait for the equipment to become available. Chapter 16 will discuss congestion stocks in some depth.

Safety stock is the amount of inventory kept on hand, on the average, to allow for the uncertainty of demand and the uncertainty of supply in the short run. Safety stocks are not needed when the future rate of demand and the length of time it takes to get complete delivery of an order are known with certainty. The level of safety stock is controllable in the sense that this investment is directly related to the desired level of customer service (i.e., how often customer demand is met from stock). The determination of safety stock levels in response to prespecified managerial measures of customer service will be discussed in Chapters 6 through 11.

Anticipation inventory consists of stock accumulated in advance of an expected peak in sales. When demand is regularly lower than average during some parts of the year, excess inventory (above cycle and safety stock) can be built up so that, during the period of high anticipated requirements, extra demand can be serviced from stock rather than from, for example, working overtime in the plant. That is, the aggregate production rate can be stabilized through the use of anticipatory stock. Anticipation stock can also occur because of seasonality of supply. For example, tomatoes, which ripen during a rather short period, are processed into ketchup that is sold throughout the year. Likewise, climatic conditions, such as in the Arctic, can cause shipments to be restricted to certain times of the year. Anticipation stocks can also consist of inventories built up to deal with labor strikes, war crises, or any other events that are expected to result in a period of time during which the rate of supply is likely to be lower than the rate of demand. Chapter 14 will deal with the rational determination of anticipation inventories as a part of production planning.

Pipeline (or *WIP*) inventories include goods in transit (e.g., in physical pipelines, on trucks, or in railway cars) between levels of a multiechelon distribution system or between adjacent work stations in a factory. The pipeline inventory of an item between two adjacent locations is proportional to the usage rate of the item and to the transit time between the locations. Land's End, Dell Computer, and other direct-mail firms have negligible pipeline inventory in the distribution system because they deliver directly to the consumer. Pipeline inventories will be considered in Chapters 11, 15, and 16.

Decoupling stock is used in a multiechelon situation to permit the separation of decision making at the different echelons. For example, decoupling inventory allows decentralized decision making at branch warehouses without every decision at a branch having an immediate impact on, say, the central warehouse or factory. In this light, Zipkin (1995) notes that inventory is a "boundary" phenomenon. Large amounts of inventory tend to exist between firms, or between divisions within a firm. As these inventory levels are reduced in an attempt to reduce costs, inventory managers must take into account the larger system of suppliers and customers throughout the supply chain. The topics of multiechelon inventory management and supply chain management (to be discussed in Chapters 11 and 12) attempt to capture some of these system-wide effects and manage them effectively. As we dig deeply into the fundamental principles and formulas, it is essential to keep in mind the upstream and downstream effects of our decisions. Chapter 16 will include discussion related to decoupling stocks.

Note that our six functional categories were defined in order to concentrate attention on the organizational purposes of the inventories, especially with regard to control and manageability, rather than on accounting measures (e.g., raw materials, supplies, finished components, etc.). Most managers are intuitively aware of the functions that inventories in each of the six categories must play. Some of the benefits (and costs) may even be measured by existing cost-accounting systems. However, most accounting systems do not keep track of opportunity costs, such as customer disservice through lost sales or the cost of extra paperwork generated by an order that must be expedited to meet an unexpected emergency.

While it will not always be possible to precisely measure the costs and benefits of each subgroup of inventories, over the years, many experienced managers have garnered a sense of perspective regarding the effect of reducing or increasing the amount of inventory investment allocated to each of these six categories. One of our goals in this book is the development of procedures to harness this experience, along with measurable costs and benefits and analytical techniques, to yield better decisions from an overall systems viewpoint.

It is quite possible that an individual manager may not have had sufficient experience to have developed an intuitive feel for the six functional categories that we have defined above. This could especially be true in large organizations. In such cases, instead of classifying a company's total aggregate inventory investment, it may be more meaningful for a manager to deal with a smaller inventory investment, such as that of one division or strategic business unit.

Meaningful functional groupings do vary from organization to organization. But, from a theoretical and practical viewpoint, at some point, the six functional categories—cycle stock, congestion, safety, anticipation, pipeline, and decoupling stocks—need to be analyzed from a cost–benefit standpoint to provide an aggregate perspective for the control of individual items.

2.4.2 The A–B–C Classification as a Basis for Designing Individual Item Decision Models

Managerial decisions regarding inventories must ultimately be made at the level of an individual item or product. The specific unit of stock to be controlled will be called a *stock-keeping unit (or SKU), where an SKU will be defined as an item of stock that is completely specified as to function, style, size, color, and, often, location.* For example, the same-style shoes in two different sizes would constitute two different SKUs. Each combination of size and grade of steel rod in raw stock constitutes a separate SKU. An oil company must regard each segregation of crude as a separate SKU. A tire manufacturer would normally treat the exact same tire at two geographically remote locations as two distinct SKUs. Note that this classification system can result in the demand for two SKUs being highly correlated in practice, because a certain portion of customers will always be willing to substitute, say, a blue jacket for the one that is red.[*] As we will see, particularly in Chapters 10, 11, and 15, decision rules often must coordinate a number of distinct SKUs.

Close examination of a large number of actual multi-SKU inventory systems has revealed a useful statistical regularity in the usage rates of different items. (We deliberately use the broader-term usage, rather than demand, to encompass supplies, components, spare parts, etc.) Typically, somewhere on the order of 20% of the SKUs account for 80% of the total annual dollar usage. This suggests that all SKUs in a firm's inventory should not be controlled to the same extent. Figure 2.1 illustrates the typical Distribution by Value (DBV) observed in practice.[†]

A DBV curve can be developed as follows: The value v, in dollars per unit, and the annual usage (or demand) D, in specific units, of each SKU in inventory are identified. Then, the product Dv is calculated for each SKU, and the Dv values for all SKUs are ranked in descending order starting with the largest value, as in Table 2.1. Then, the corresponding values of the cumulative percent of total dollar usage and the cumulative percent of the total number of SKUs in inventory are plotted on a graph such as Figure 2.1. Experience has revealed that inventories of consumer goods will typically show a lesser concentration in the higher-value SKUs than will an inventory of industrial SKUs. Furthermore, empirically, it has been found that quite often the distribution of usage values across a population of items can be adequately represented by a lognormal distribution. Therefore, we can relatively simply estimate the aggregate effects of different inventory control policies. In the Appendix of this chapter, we show (1) the mathematical form of the distribution, (2) its moments (expected values of various powers of the variable), (3) a graphic procedure for testing the adequacy

---

[*] In practice, of course, one can define an SKU in other (e.g., less detailed) ways depending on the decision being made and the level of detail at which one wants to control inventories.

[†] Incidentally, the same DBV concept is applicable to a wide variety of other phenomena such as the distribution of (1) incomes within a population of individuals, (2) donation sizes among a group of donors, and so forth. It is also very similar to the Pareto chart used in quality management. Similar methods are used in market analysis.

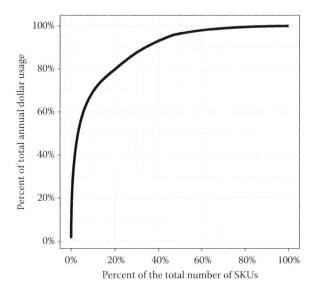

Figure 2.1 DBV of SKUs.

of the fit to a given DBV, and (4) a graphic method of estimating the two parameters of the distribution (m, the average value of Dv, and b which is a measure of the dispersion or spread of the Dv values).[*]

A table, such as Table 2.1, is one of the most valuable tools for handling the diversity of disaggregate inventories because it helps to identify the SKUs that are the most important. These SKUs will be assigned a higher priority in the allocation of management time and financial resources in any decision system we design. It is common to use three priority ratings: A (most important), B (intermediate in importance), and C (least important). The number of categories appropriate for a particular company depends on its circumstances and the degree to which it wishes to differentiate the amount of effort allocated to various groupings of SKUs. For example, one can always subdivide the DBV into further categories, such as "moderately important," etc., as long as the resulting categories receive differentiated treatment. A minimum of three categories is almost always used, and we use this number to present the basic concepts involved.

Class A items should receive the most personalized attention from management. The first 5%–10% of the SKUs, as ranked by the DBV analysis, are often designated for this most important class of items although up to 20% of the SKUs can be designated as class A in some settings. Usually these items also account for somewhere in the neighborhood of 50% or more of the total annual dollar movement ($\sum Dv$) of the population of items under consideration. In Chapters 5, 7, 9, 10, and 11, we discuss decision systems that are suitable for use with class A items.

---

[*] The parameter b is not the standard deviation of Dv. It is the standard deviation of the underlying normally distributed variable, namely $\ln(Dv)$.

Table 2.1   Sample Listing of SKUs by Descending Dollar Usage

Sequential Number	SKU I.D.	Cumulative Percent of SKU	Annual Usage Value (Dv)	Cumulative Usage	Cumulative Percent of Total Usage
1	–	0.5	$3,000	$3,000	13.3
2	–	1.0	2,600	5,600	24.9
3	–	1.5	2,300	7,900	35.1
–	–	–	–	–	–
–	–	–	–	–	–
–	–	–	–	22,498	–
199	–	99.5	2	22,500	100
200	–	100.0	0	22,500	100

Class B items are of secondary importance in relation to class A. These items, because of their *Dv* values or other considerations, rate a moderate but significant amount of attention. The largest number of SKUs fall into this category—usually more than 50% of total SKUs, accounting for most of the remaining 50% of the annual dollar usage. Some books on inventory control tend to recommend a somewhat lower portion of total SKUs for the B category. However, because of the availability of computers, we suggest that as many SKUs as possible be monitored and controlled by a computer-based system, with appropriate management-by-exception rules. In Chapters 4, 5, 6, 9, 10, and 11, we present a large number of decision systems that are suitable for use with class B items. Some of the decision rules are also useful for controlling A items, although in the case of A items, such rules are *more* apt to be overruled by managerial intervention. Furthermore, model parameters, such as costs and the estimates of demand, will be reviewed more often for A items.

Class C items are the relatively numerous remaining SKUs that make up only a minor part of the total dollar investment. For these SKUs, decision systems must be kept as simple as possible. One objective of A–B–C classification is to identify this large third group of SKUs that can potentially consume a large amount of data input and managerial time. Typically, for low-value items, most companies try to keep a relatively large number of units on hand to minimize the amount of inconvenience that could be caused by a stockout of such insignificant parts.

For C items especially, and to a lesser degree for the others, as much grouping as possible of SKUs into control groups based on similar annual usage rates, common suppliers, similar seasonal patterns, same end users, common lead times, and so on, is desirable to reduce the total number of discrete decisions that must be processed. Each control group can be designed to operate using a single order rule and monitoring system. For example, if one SKU in the group requires an order because of low inventories, most of the other items will also be ordered at the same time to save on the cost of decision making. Two bin systems, because they require a minimum of paperwork, are especially popular for controlling class C items. They will be discussed in Chapters 6 and 8. Coordinated control systems will be presented in Chapter 10.

While Figure 2.1 is typical, the precise number of members in each of the A, B, and C categories depends, of course, on how spread out the DBV curve actually is. For example, the greater the spread of the distribution, the more SKUs fall into class C.

An A–B–C classification need not be done on the basis of the DBV curve alone. Managers may shift some SKUs among categories for a number of reasons. For example, some inexpensive SKUs may be classified as "A" simply because they are crucial to the operation of the firm. Some large-volume consumer distribution centers plan the allocation of warehousing space on the basis of usage rate and cubic feet per unit. The SKUs with high usage and large cubic feet are stored closer to the retail sales counter. Similarly a distribution by (profit × volume) per SKU is sometimes used to identify the best-selling products. Items at the lower end of such an A–B–C curve become candidates for being discontinued. Krupp (1994) argues for a two-digit classification, where the first digit is based on dollar usage as in traditional A–B–C analysis, and the second is based on the number of customer transactions. Flores and Whybark (1987) also recommend using a two-dimensional classification where the first is the traditional A–B–C and the second is based on criticality.

In related research, Cohen and Ernst (1988) use a statistical technique called cluster analysis to group items across many dimensions, including criticality. Reynolds (1994) provides a classification scheme, appropriate for process industries, that helps managers focus attention on important items even if they are rarely used. Note that manufacturers and distributors often have different perspectives on the A–B–C classification: An A item for one may be a C item for the other. See for example, Lenard and Roy (1995).

2.5 A Framework for Production Planning and Scheduling

In this section, we develop an important framework, called the product-process matrix, for classifying production planning and scheduling systems. We begin by discussing the product life cycle.

2.5.1 A Key Marketing Concept: The Product Life Cycle

The lifetime sales of many branded products reveal a typical pattern of development, as shown in Figure 2.2. The length of this so-called life cycle appears to be governed by the rate of technological change, the rate of market acceptance, and the ease of competitive entry.

Each year, new clothing styles are introduced in the knowledge that their whole life cycle may last only a season (a few months). Other products, such as commercial aircraft, are expected to be competitive for decades. Some products have been known to begin a new cycle or to revert to an earlier stage as a result of the discovery of new uses, the appearance of new users, or the invention of new features. Kotler (1967) cites the example of television sales, which have exhibited a history of spurts as new sizes of screens were introduced, and the advent of color television, which put sales back to an earlier, rapid growth stage in that industry. As high-definition television becomes popular, and as television is blended with computers for Internet access, it is likely that sales will spurt again. Although the life cycle is an idealized concept, three basic generalizations seem to hold:

1. Products have a limited life. They are introduced to the market, may (or may not) pass through a strong growth phase, and eventually decline or disappear.

2. Product profits tend to follow a predictable course through the life cycle. Profits are absent in the introductory stage, tend to increase substantially in the growth stage, slow down and then stabilize in the maturity and saturation stages, and can disappear in the decline stage.
3. Products require a different marketing, production planning, inventory management, and financial strategy in each stage. The emphasis given to the different functional areas must also change.

When product life cycles are reasonably short, production processes must be designed to be reusable or multipurpose in nature. Alternatively, the amount of investment in single-purpose equipment must be recoverable before the product becomes obsolete. Moreover, the appropriate planning and scheduling systems to use for an individual product are likely to change as the relative importance of the item changes during its life cycle. In particular, special control aspects for the development and decline stages of Figure 2.2 will be discussed in Chapters 3, 5, 7, and 8.

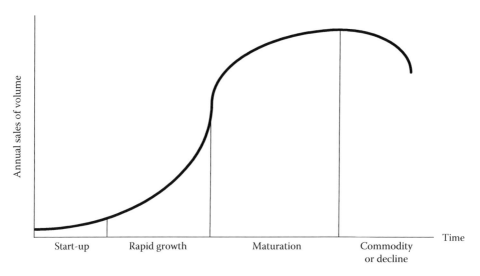

		Increasing standardization	Emergence of a "dominant design"	High standardization, commodity charactistics
Product variety	Great variety	Increasing standardization	Emergence of a "dominant design"	High standardization, commodity charactistics
Industry structure	Small competitors	Fallout and consolidation	Few large companies	"Survivors"
Form of competition	Product characteristics	Product quality and availability	Price and delivery dependability	Price
Change in growth rate	Little	Increasing rapidly	A little	Decreases rapidly, then slow, or little, if the market is renewed
Change in product design	Very great	Great	Little, unless major renewal of the product	Little
Functional areas most involved	R&D	Engineering, production	Production, marketing, and finance	Marketing, finance

Figure 2.2 The fundamental stages of product/market evolution.

2.5.2 Different Types of Production Processes

As one walks around different factories and talks to factory managers, it becomes clear that there are vast differences among different types of production processes. In this section, we distinguish among four broad classes of processes—job shop, batch flow, assembly line, and continuous process—because the differences among them have important implications for the choice of the production planning and scheduling system. Job shops typically manufacture customized products such as printed circuit boards and certain types of metal parts. A commercial printer, which must tailor each job to the customer's wishes, is also an example of a job shop. Batch flow processes manufacture, for example, apparel, machine tools, and some pharmaceuticals. Assembly products encompass, for example, automobiles, home furnishings, electrical equipment, and computers. Typical illustrations of continuous process (also known as process, continuous flow, or simply flow) products include chemicals (plastics, drugs, soaps, fertilizers, etc.), refined petroleum products, foods and beverages, and paper goods. The differences among these are vast and multidimensional. In what follows, we draw, in particular, on the work of Taylor et al. (1981) and Schmenner (1993).

The contrast in product and market characteristics can be seen in Table 2.2. We highlight just a few points, but we encourage the reader to examine the table in detail. Because of the more standardized nature of products in the process industries, there tends to be more production to stock. Assembly lines can make-to-stock or assemble-to-order, whereas job shops usually make-to-order. Batch flow processes can either make-to-order or make-to-stock depending on the products and customers.

Process industries tend to be the most capital intensive of the four, due to higher levels of automation. Also, process industries tend to have a flow-type layout that is clear but inflexible. That is, materials flow through various processing operations in a fixed routing. However, as we move to the left in this table, the flow increasingly follows numerous, different, and largely unconstrained paths. The production lines in process industries tend to be dedicated to a relatively small number of products with comparatively little flexibility to change either the rate or the nature of the output. In this environment, capacity is quite well defined by the limiting (or pacing/bottleneck) operation. Toward the left of the table, there tends to be more flexibility in both the rate and output. Moreover, both the bottleneck and the associated capacity tend to shift with the nature of the workload (which products are being produced and in what quantities).

Because of the relatively expensive equipment and the relatively low flexibility in output rate, process industries tend to run at full capacity (i.e., 3 shifts per day, 7 days per week) when they are operating. This and the flow nature of the process necessitate highly reliable equipment, which, in turn, normally requires substantial preventive maintenance. Such maintenance is usually carried out when the entire operation is shut down for an extended period of time. Moreover, much longer lead times are typically involved in changing the capacity in a process industry (partly because of environmental concerns, but also because of the nature of the plant and equipment). Job shops and batch flow industries tend to use less-expensive equipment. Maintenance can be done on an "as-needed" basis. Assembly lines are more flexible than continuous processes, but less so than job shops and batch flow lines. Often, the equipment is very inexpensive—simply hand tools; but sometimes automated assembly is used, which is very expensive. Maintenance policies, then, tend to be based on the cost of the equipment.

The number of raw materials used tends to be lower in process situations as compared with assembly; in fact, coordination of raw materials, components, and so on, as well as required labor input, is a major concern in assembly. However, there can be more natural variability in the characteristics of these raw materials in the process context (e.g., ingredients from agricultural or mineral

Table 2.2   Differences in Product and Market Characteristics

		Type of Process/Industry		
Characteristics	*Job Shop*	*Batch Flow*	*Assembly*	*Process*
Number of customers	Many	Many, but fewer	Less	Few
Number of products	Many	Fewer	Fewer still	Few
Product differentiation	Customized	Less customized	More standardized	Standardized (commodities)
Marketing characteristics	Features of the product	Quality and features	Quality and features or availability/price	Availability/price
Families of items	Little concern	Some concern	Some concern	Primary concern
Aggregation of data	Difficult	Less difficult	Less difficult	Easier
By-products	Few	Few	Few	More
Need for traceability	Little	Intermediate	Little	High
Material requirements	Difficult to predict	More predictable	Predictable	Very predictable
Control over suppliers	Low	Moderate	High	Very high
Vertical Integration	None	Very Little	Some backward, often forward	Backward and forward
Inventories				
Raw materials	Small	Moderate	Varies, frequent deliveries	Large, continuous deliveries
WIP	Large	Moderate	Small	Very small
Finished goods	None	Varies	High	Very high

(Continued)

Table 2.2 (*Continued*) Differences in Product and Market Characteristics

Characteristics	Type of Process/Industry			
	Job Shop	*Batch Flow*	*Assembly*	*Process*
QC responsibility	Direct labor	Varies	QC specialists	Process control
Production information requirements	High	Varies	Moderate	Low
Scheduling	Uncertain, frequent changes	Frequent expediting	Often established in advance	Inflexible, sequence dictated by technology
Operations challenges	Increasing labor and machine utilization, fast response, and breaking bottlenecks	Balancing stages, designing procedures, responding to diverse needs, and breaking bottlenecks	Rebalancing line, productivity improvement, adjusting staffing levels, and morale	Avoiding downtime, timing expansions, and cost minimization
End-of-period push for output	Much	Frequent	Infrequent	None (can't do anything)
Capital versus labor/material intensive	Labor	Labor and material	Material and labor	Capital
Typical factory size	Usually small	Moderate	Often large	Large
Level of automation	Low	Intermediate	Low or high	High
Number of raw materials	Often low	Low	High	Low
Material requirements	Difficult to predict	More predictable	Predictable	Very predictable
Bottlenecks	Shifting frequently	Shifting occasionally, but predictable	Generally known and stationary	Known and stationary
Speed (units/day)	Slow	Moderate	Fast	Very fast

(*Continued*)

Table 2.2 (*Continued*)   Differences in Product and Market Characteristics

Characteristics	Type of Process/Industry			
	Job Shop	*Batch Flow*	*Assembly*	*Process*
Process flow	No pattern	A few dominant patterns	Rigid flow pattern	Clear and inflexible
Type of equipment	General purpose	Combination of specialized and general purpose	Specialized, low, or high tech	Specialized, high tech
Flexibility of output	Very	Intermediate	Relatively low (except some assemble to order)	Low
Run length	Very short	Moderate	Long	Very long
Definition of Capacity	Fuzzy, often expressed in dollars	Varies	Clear, in terms of output rates	Clear, expressed in physical terms
Capacity addition	Incremental	Varies	Chunks, requires rebalancing	Mostly in chunks, requires synchronization
Nature of maintenance	As needed	As needed, or preventive when idle	As needed	Shutdown
Energy usage	Low	Low, but can be higher	Low	High
Process changes required by new products	Incremental	Often Incremental	Incremental or radical	Always radical

sources).[*] Job shops and batch flow operations may have very few inputs as well, but this can vary with the context.

Although there may be relatively few products run on a particular flow line in the process industries, the products do tend to group into families according to a natural sequence. An example would be paints, where the natural sequence of changes would be from lighter to darker colors with a major changeover (once a cycle) from the darkest to the lightest color. As a consequence, in contrast with the other three types of processes, a major concern is ascertaining the appropriate sequence and the time interval between consecutive cycles among the products. The relative similarity of items run on the same line in process industries also makes it easier to aggregate demand data, running hours, and so forth than is the case in the other three.

The flow nature of production in the process industries leads to less WIP inventories than is the case, for example, in job shops. This relative lack of buffering stock, in turn, implies a crucial need for adequate supplies of the relatively few raw materials. However, in this case, the same line of reasoning applies to high volume assembly lines. Job shops tend to have large amounts of WIP, but little finished goods inventory. Without the high levels of WIP, the output of the shop may decrease dramatically. However, because so many job shops manufacture to-order, the products are shipped immediately upon completion, and there is little inventory of finished goods.

Finally, note the operations challenges listed in Table 2.2. Managers in job shops focus on labor and machine utilization and breaking bottlenecks, because the shifting bottlenecks can idle people and equipment. A severe bottleneck can, at certain times, delay shipment of critical customer orders. Batch flow managers focus on balancing stages of production, often because there is a single bottleneck with which other work centers must coordinate. Staffing and line balancing are major concerns in assembly operations. Output tends to be very high, and different operations highly synchronized, so that small changes in the allocation of work to different stations on the line can have a significant impact on profitability. However, assembly line work is tedious, and managers find that they must focus on issues of worker morale. Finally, as is clear from the descriptions of process industries, managers in these industries must keep equipment running with a minimum of downtime.

2.5.3 The Product-Process Matrix

Recall that we are striving to divide production situations into classes that require somewhat different strategies and procedures for planning and scheduling. The product life cycle and the process characteristics are major aids in that direction. However, Hayes and Wheelwright (1979a,b) present a concept that provides us with considerable further insight. They point out that while the product life cycle is a useful framework, it focuses only on marketing. They suggest that many manufacturing processes go through a cycle of evolution, sometimes called the process cycle, beginning with job shops, making few-of-a-kind products, that are labor intensive and generally not very cost efficient, and moving to continuous processes, which are less flexible, automated, and capital intensive. Hayes and Wheelwright suggest that the product life and process life cycle stages cannot be considered separately. One cannot proceed from one level of mechanization to another, for

---

[*] Thompson 1988 notes that it is difficult to define a "unit" in process industries, so that on-hand figures in an automated inventory system may be wrong. The product differs by grade of material, potency, and quality specifications. Therefore, adjustments in traditional inventory-recording systems are necessary, including recording shelf life, date of receipt, and expiration date.

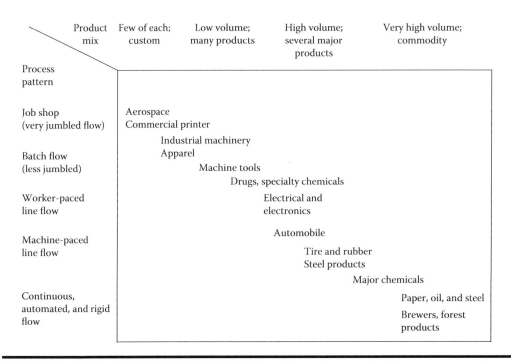

Figure 2.3   The product-process matrix.

example, without making some adjustments to the products and management decision systems involved. Nor can new products be added or others discontinued without considering the effect on the production process.

Hayes and Wheelwright encapsulate the above remarks into a graphical representation known as a product-process matrix. Rather than showing their original form of the matrix, we have portrayed in Figure 2.3 an adapted version (based partly on suggestions in Schmenner (1993), and Taylor et al. (1981)) that is more suitable for our purposes. The columns of the matrix represent the product life cycle phases, going from the great variety associated with startup products on the left-hand side, to standardized commodity products on the right-hand side. The rows represent the major types of production processes. Note that we have now divided assembly into worker-paced and machine-paced flow.[*]

A number of illustrations are shown in the figure. We see that, by and large, fabrication is in the top-left corner, process industries toward the bottom-right corner, and assembly in the middle. However, there are some exceptions. For example, drugs and specialty chemicals, which are often thought of as process industry products, are centrally located because they are produced in batch flow lines. Likewise, steel products are toward the bottom right, even though they require some fabrication.

---

[*] Spencer and Cox (1995) take a slightly different angle on the product-process matrix, defining the vertical axis as the "volume of available capacity devoted to the product line." At the top of this axis, where we have placed job shops, this value is low, whereas for continuous processes, the value is high. Repetitive manufacturing is then placed between batch and continuous, and is defined as "the production of discrete units in high volume concentration of available capacity using fixed routings." See also Silver (1996).

Hayes and Wheelwright discuss the strategic implications of nondiagonal positions. Operating in the top-right corner of the matrix can lead to high opportunity costs because the firm is attempting to meet high market demand with a process flow that is designed for low-volume flexibility. The result should be high levels of lost sales. Operating in the lower-left corner can lead to high out-of-pocket costs because the firm has invested in expensive production equipment that is rarely utilized because market demand is low. See also Safizadeh et al. (1996) who survey 144 managers and discover that most factories actually do operate close to the diagonal, and those that do not tend to exhibit poor performance.

To provide additional insight, we draw on the work of Schonberger (1983) who points out that the strategy for production planning and scheduling should depend on how easily one can associate raw material and part requirements with the schedule of end products. As we will now show, there is a direct connection between the position on the product-process matrix and the ease of the this association. In the lower right-hand corner of Figure 2.3, the association tends to be quite easy (continuous process systems). This position is, by and large, occupied by capacity-oriented process industries. As we move up to the left and pass through high-volume assembly into lower-volume assembly and batch flow, the association becomes increasingly difficult. In this region, one is dealing primarily with materials and labor-oriented fabrication and assembly industries. Finally, in the top left-hand corner, where fabrication is dominant, the association may again become easier but, because production here is primarily make-to-order, detailed, advanced planning and scheduling are like usually not feasible.

Based on these comments, and others in this section, we can introduce different production planning and scheduling systems that are appropriate for firms located at various positions on the product-process matrix. Because we deal with these systems in depth later in the book, we simply note where they will be discussed. See Figure 2.4 and Table 2.3, where we briefly describe the primary focus of each system.

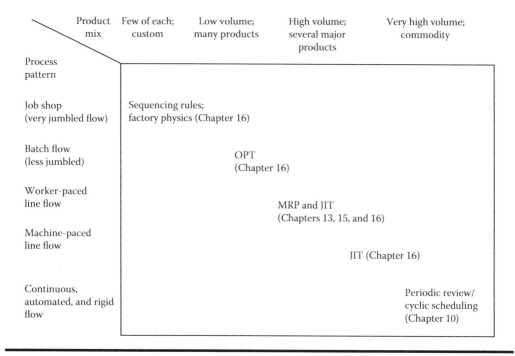

Figure 2.4 Production planning and scheduling and the product-process matrix.

Table 2.3   Production Planning and Scheduling Systems

System	Relevant Chapters of the Book	Nature of Relevant Industries	Primary Focus of the System
Sequencing rules Factory physics	16	Low-volume fabrication	Flexibility to cope with many different orders; Meeting due dates; Increasing throughput; and Predicting lead times
OPT	16	Batch; low-volume assembly	Bottleneck management
Material resources planning (MRP)	13, 15	Medium-volume assembly	Effective coordination of material and labor
JIT	16	High volume, repetitive fabrication, and assembly	Minimizing setup times and inventories; High quality
Periodic review/cyclic scheduling	10	Continuous process	Minimizing sequence-dependent setups; High-capacity utilization

Finally, we note that advanced production technology can allow a firm to move off the diagonal and still be profitable. Flexible manufacturing systems (FMS) achieve flexible, low-volume production with the efficiency of high-volume systems. Automated setups allow rapid changeovers among different products, and computer-controlled machining allows for efficient production. Thus, it is possible to operate profitably to the lower left of the diagonal.

2.6 Costs and Other Important Factors

Through empirical studies and deductive mathematical modeling, a number of factors have been identified that are important for inventory management and production planning and scheduling decisions.

2.6.1 Cost Factors

Here, we briefly describe a number of cost factors. Sections 2.9 and 2.10 will be concerned with estimating their values. The discussion of one other type of cost, the cost of changing the aggregate production rate, or work force size, will be left until Chapter 14 where, for the first time, we deal with medium-range aggregate production-planning problems in which such costs are relevant.

2.6.1.1 The Unit Value or Unit Variable Cost, v

The unit value (denoted by the symbol v) of an item is expressed in dollars per unit. For a merchant[*] it is simply the price (including freight) paid to the supplier, plus any cost incurred to make it ready for sale. It can depend, via quantity discounts, on the size of the replenishment. For producers, the unit value of an item is usually more difficult to determine. However, one thing is certain; it is seldom the conventional accounting or "book value" assigned in most organizations. The value of an item ideally should measure the actual amount of money (variable cost) that has been spent on the SKU to make it available for usage (either for fulfilling customer demand or for internal usage as a component of some other items). As we will see in Section 2.9, the determination of v is not an easy task. Nevertheless, we shall also see that most models we present are relatively robust to errors in cost estimation. A good starting point is the cost figure given by accounting, adjusted for obvious errors.

As an example, a small retailer buys ski hats for $20, and then hires a high-school student to unpack them and put them on the shelves. The cost per unit of the student's time is $1, so that the value of v should be $21. A second example is a manufacturer whose cost-accounting system shows that an item costs $4.32 per unit to manufacture. Upon further examination, however, it is clear that this item is particularly difficult to manufacture, and that a production supervisor devotes up to 30% of her time overseeing the process. The true cost of the item should reflect the value of the supervisor's time. At 30% of a $40,000 salary, and 25,000 units produced per year, the cost of the item should be increased by $(\$40,000)(0.30)/(25,000 \text{ units}) = \$0.48$ per unit.

The unit value is important for two reasons. First, the total acquisition (or production) costs per year clearly depend on its value. Second, the cost of carrying an item in inventory depends on v.

2.6.1.2 The Cost of Carrying Items in Inventory

The cost of carrying items in inventory includes the opportunity cost of the money invested, the expenses incurred in running a warehouse, handling and counting costs, the costs of special storage requirements, deterioration of stock, damage, theft, obsolescence, insurance, and taxes. The most common convention of costing is to use

$$\text{Carrying costs per year} = \bar{I}vr \qquad (2.1)$$

where $\bar{I}$ is the average inventory in units (hence $\bar{I}v$ is the average inventory expressed in dollars), and r is the carrying charge, the cost in dollars of carrying one dollar of inventory for 1 year.

By far, the largest portion of the carrying charge is made up of the opportunity costs of the capital tied up that otherwise could be used elsewhere in an organization and the opportunity costs of warehouse space claimed by inventories. Neither of these costs are measured by traditional accounting systems.

The opportunity cost of capital can be defined easily enough. It is, theoretically speaking, the ROI that could be earned on the next most attractive opportunity that cannot be taken advantage of because of a decision to invest the available funds in inventories. Unfortunately, such a marginal cost concept is difficult to implement in practice. For one thing, the next most attractive investment opportunity can change from day to day. Does this mean that the cost of the capital portion of

---

[*] We use the term merchant to denote a nonproducer.

carrying should also be changed from day to day? From a theoretical point of view, the answer is yes. In fact, one should theoretically account for the time value of money (a topic we discuss in Chapter 4).[*] In practice, such factors are difficult to administer; instead, the cost of capital is set at some level by decree and is changed only if major changes have taken place in a company's environment. For example, after due consideration by senior management, a policy is declared that "only investments which expect to earn more than a specified percent can be implemented." The cost of capital used, of course, has to depend on the degree of risk inherent in an investment. (For the same reason, banks charge a higher rate of interest on second mortgages than on the first.) As a result, in practice, the opportunity cost of capital can range from the bank's prime lending rate to 50% and higher. Small companies often have to pay the higher rate because they suffer from severe capital shortages and they lack the collateral to attract additional sources of working capital. A further complication is that inventories are often financed from a mix of sources of funds, each of which may have a different rate. The true cost of carrying the inventory should be based on a weighted average of these rates.

Inventory investment, at least in total, is usually considered to be of relatively low risk because in most cases, it can be converted into cash relatively quickly. However, the degree of risk inherent in inventory investment varies from organization to organization, and the inventory investment itself can change the firm's level of risk (Singhal 1988; Singhal and Raturi 1990; Singhal et al. 1994). Some of the most important impediments to quick conversion to cash can be obsolescence, deterioration, pilferage, and the lack of immediate demand at the normal price. Each of these factors increases the cost of capital over and above the prime rate because of a possible lack of opportunity of quick cash conversion.

The value of r is not only dependent on the relative riskiness of the SKUs, it also depends on the costs of storage that are a function of bulkiness, weight, special handling requirements, insurance, and possibly taxes. Such detailed analysis is seldom applied to all SKUs in inventory. To make the inventory decision more manageable both from a theoretical and a practical point of view, a single value of r is usually assumed to apply for most items. Note from Equation 2.1 that this assumes that more expensive items are apt to be riskier to carry and more expensive to handle or store. Such a convenient relationship doesn't always hold true for all SKUs.[†] Note again that r itself could depend on the total size of the inventory; for example, r would increase if a company had to begin using outside warehousing.

As an example, imagine a company that can borrow money from the bank at 12%. Recent investment opportunities, however, suggest that money available to the firm could earn in the neighborhood of 18%. The value of r should be based on the latter figure, perhaps inflated for the other reasons noted above, because the opportunity cost of the money tied up in inventory is 18%.

2.6.1.3 The Ordering or Setup Cost, A

The symbol A denotes the fixed cost (independent of the size of the replenishment) associated with a replenishment. For a merchant, it is called an ordering cost and it includes the cost of order

---

[*] The time value of money asserts that a dollar today is worth more than a dollar in the future because it can earn a return in the interim.

[†] In principle, one could instead use $\bar{I}h$ where h is the cost to carry one unit in inventory for a year, or a combination of $\bar{I}(vr + h)$.

forms, postage, telephone calls, authorization,[*] typing of orders, receiving, (possibly) inspection, following up on unexpected situations, and handling of vendor invoices. The production setup cost includes many of these components and other costs related to interrupted production. For example, we might include the wages of a skilled mechanic who has to adjust the production equipment for this part. Then, once a setup is completed, there often follows a period of time during which the facility is produced at lower quality or slower speed while the equipment is fine tuned and the operator adjusts to the new part. This is normally called the "learning effect" and the resulting costs are considered to be part of the setup cost because they are the result of a decision to place an order.[†] Finally, notice that during the setup and learning period, opportunity costs are, in effect, incurred because production time on the equipment is being lost during which some other item could be manufactured. Hence, the latter opportunity cost is only incurred if the production facility in question is being operated at capacity.

Several of these factors can become quite complicated. Consider for a moment the wages of the skilled mechanic who performs the setup. If this person is paid only when setting up a machine, the wages are clearly a part of the setup cost. What happens, however, if the person is on salary, so that whether or not the machine is set up, the wages are still paid? Should these wages be a part of the setup cost? The answer is the subject of much debate in operations and cost accounting. Our answer is that it depends on the use of the mechanic's time when he is not setting up the machine, and on whether a long-term or short-term perspective is taken. If he would be involved in other activities, including setting up other machines, when he is not setting up the machine for this part, the cost should be included. In other words, there is an opportunity cost for his time. Most factories we have visited are in this situation. People are busy; when they are not working on a particular setup, they are working on other things. There is another side to the story, however. If we decide not to set up the machine for this part, we do not generate actual dollar savings in the short term. The wages are still paid. So, in the short term, the cost of the mechanic's time is fixed, and therefore, one could argue that it should *not* be a part of the setup cost. Let us assume that the mechanic's time is used for other activities. The key then is whether a long-term or short-term perspective is taken. A long-term view suggests that the wages should be included because in the long term, this person could be laid off, so that the decision to set up infrequently affects the money the firm pays. (Kaplan (1988) argues that virtually all costs are variable in the long term, because people can be fired, plants closed, and so on.) A short-term view suggests that the wages should not be included. The final issue, then, is whether a long-term or short-term decision is being made. In most cases, the setup cost, A, will be applied to repeated decisions over a long period of time—say a year. We would then argue that the wages of this mechanic should be included in A. In other words, we want to include all costs that are relevant to the decision. To borrow a phrase: "If what you do does not affect the cost, the cost should not affect what you do." This type of reasoning also applies to many of the other factors listed in the previous paragraph. Further discussion on this topic is in Section 2.9.

Clearly, one could spend months trying to nail down the A value precisely. Often, it is far more beneficial to change the underlying processes that determine A. For example, the widespread adoption of information technology that facilitates computer-to-computer transactions has substantially

---

[*] The level of authorization may depend on the dollar size of the replenishment; for example, special authorization from a senior executive may be needed if the order exceeds $10,000.

[†] This learning effect often implicitly assumes that the setup process is completely forgotten between production runs.

lowered fixed order setup costs. We shall comment further on the effects of such technologies throughout the book.

2.6.1.4 The Costs of Insufficient Capacity in the Short Run

These costs could also be called the costs of avoiding stockouts and the costs incurred when stockouts take place. In the case of a producer, they include the expenses that result from changing over equipment to run emergency orders and the attendant costs of expediting, rescheduling, split lots, and so forth. For a merchant, they include emergency shipments or substitution of a less-profitable item. All these costs can be estimated reasonably well. However, there are also costs, which are much more nebulous, that can result from not servicing customer demand. Will the customer be willing to wait while the item is backordered, or is the sale lost for good? How much goodwill is lost as a result of the poor service? Will the customer ever return? Will the customer's colleagues be told of the disservice?[*] Such questions can, in principle, be answered empirically through an actual study for only a limited number of SKUs. For most items, the risks and costs inherent in disservice have to remain a matter of educated, considered opinion, not unlike the determination of the risks inherent in carrying inventories. In Chapter 6, we examine a number of different methods of modeling the costs of disservice. In every case, however, it is important to explicitly consider the customer's perspective.

2.6.1.5 System Control Costs

System control costs are those associated with the operation of the particular decision system selected. These include the costs of data acquisition, data storage and maintenance, and computation. In addition, there are less-tangible costs of human interpretation of results, training, alienation of employees, and so on. Although difficult to quantify, this category of "costs" may be crucial in the choice of one decision system over another.

2.6.2 Other Key Variables

Table 2.4[†] provides a rather extensive listing of potentially important variables. We elaborate on just three factors.[‡]

2.6.2.1 Replenishment Lead Time, L

A stockout can only occur during periods when the inventory on hand is "low." Our decision about when an order should be placed will always be based on how low the inventory should be allowed to be depleted before the order arrives. The idea is to place an order early enough so that the expected number of units demanded during a replenishment lead time will not result in a stockout very often. We define the *replenishment lead time*, as the time that elapses from the moment at which it

---

[*] See Reichheld and Sasser Jr. (1990) who discuss the extremely high cost of losing a customer in service industries. See also Chang and Niland (1967) and Oral et al. (1972).

[†] Based on A. Saipe, Partner, Thorne Stevenson & Kellogg, "Managing Distribution Inventories," Executive Development Programme, York University, Toronto, Canada, September 12, 1982.

[‡] See Silver (1981), Barancsi and Chikán (1990), Porteus (1990), and Prasad (1994) for other classification schemes, as well as overviews of inventory theory. An interesting aside is that Edgeworth (1888) is the earliest reference we know of a probabilistic inventory theory.

Table 2.4   Inventory Planning Decision Variables

Service requirements • Customer expectations • Competitive practices • Customer promise time required • Order completeness required • Ability to influence and control customers • Special requirements for large customers	*Customer-ordering characteristics* • Order timing • Order size • Advanced information for large orders • Extent of open or standing orders • Delay in order processing
Demand patterns • Variability • Seasonality • Extent of deals and promotions • Ability to forecast • Any dependent demand? • Any substitutable products?	*Supply situation* • Lead times • Reliability • Flexibility • Ability to expedite • Minimum orders • Discounts (volume, freight) • Availability • Production versus nonproduction
Cost factors • Stockout (pipeline vs. customer) • Carrying costs • Expediting • Write-offs • Space • Spoilage, etc.	*Nature of the product* • Consumable • Perishable • Recoverable/repairable
	Other issues • A–B–C pattern • Timing and quality of information • Number of stocking locations • Who bears the cost of inventory?

is decided to place an order, until it is physically on the shelf ready to satisfy customers' demands. The symbol L will be used to denote the replenishment lead time. It is convenient to think of the lead time as being made up of five distinct components:[*]

1. Administrative time at the stocking point (order preparation time): the time that elapses from the moment at which it is decided to place the order until it is actually transmitted from the stocking point.
2. Transit time to the supplier: this may be negligible if the order is placed electronically or by telephone, but transit time can be several days if a mailing system is used.

---

[*] The discussion deals specifically with the case of an external (purchased) replenishment but is also conceptually applicable for internal (production) replenishments.

3. Time at the supplier: if L is variable, this time is responsible for most of the variability. Its duration is materially influenced by the availability of stock at the supplier when the order arrives.
4. Transit time back to the stocking point.
5. Time from order receipt until it is available on the shelf: this time is often wrongly neglected. Certain activities, such as inspection and cataloging, can take a significant amount of time.

We will see that the variability of demand over L can be the result of several causes. If demand varies in a reasonably predictable manner depending on the time of the year, we say it is seasonal. If it is essentially unpredictable, we label the variability as random. Demand over lead time also varies because the time between ordering and receipt of goods is seldom constant. Strikes, inclement weather, or supplier production problems can delay delivery. During periods of low sales, lead times can turn out to be longer than expected because the supplier is accumulating orders to take advantage of the efficiency of longer production runs. When sales are high, longer lead times can result because high demand causes backlogs in the supplier's plant. Finally, in periods of short supply, it may be impossible to acquire desired quantities of items, even after any lead time.

2.6.2.2 Production versus Nonproduction

Decisions in a production context are inherently more complicated than those in nonproduction situations. There are capacity constraints at work centers as well as an interdependency of demand among finished products and their components. Gregory et al. (1983) note that the production and inventory planning and control procedures for a firm should depend on (1) whether production is make-to-stock or make-to-order (which, in turn, depends on the relation between customer promise time and production lead time) and (2) whether purchasing is for known production or anticipated production (a function of the purchasing lead time). Chapters 13 through 16 (and to some extent, Chapters 10 and 11) will address production situations.

2.6.2.3 Demand Pattern

Besides the factors listed in Table 2.4, we mention the importance of the stage in the product life cycle (Section 2.5.1). Different control procedures are appropriate for new, mature, and declining items. The nature of the item can also influence the demand pattern; for example, the demand for spare parts is likely to be less predictable than the requirements for components of an internally produced item.

2.7 Three Types of Modeling Strategies

Now that we have identified the costs and other key variables, we comment on three types of modeling strategies. First, however, we wish to emphasize that the results of any mathematical analysis must be made consistent with the overall corporate strategy (Section 1.2) and must be tempered by the behavioral and political realities of the organization under study. In particular, a proper problem diagnosis is often more important than the subsequent analysis (see Lockett 1981) and, in some cases, corporate strategy and the behavioral or political circumstances may rule out the meaningful use of any mathematical model.

There are three types of strategies that involve some modelings:

1. Detailed modeling and analytic selection of the values of a limited number of decision variables.
2. Broader-scope modeling, with less attempt at optimization.
3. Minimization of inventories with very little in the way of associated mathematical models.

2.7.1 Detailed Modeling and Analytic Selection of the Values of a Limited Number of Decision Variables

The strategy here is to develop a mathematical model that permits the selection of the values of a limited set of variables so that some reasonable measure of effectiveness can be optimized. A classical example (to be treated in Chapter 4) is the economic order quantity, which, under certain assumptions, minimizes the total of ordering and inventory carrying costs per unit time. In general, a mathematical model may permit a deductive (closed-form) solution, an iterative solution (such as in the Simplex Method of linear programming), or a solution by some form of trial-and-error procedure (such as in the use of a simulation model).

2.7.2 Broader-Scope Modeling with Less Optimization

Here, the strategy is to attempt to develop a more realistic model of the particular situation. However, the added realism often prevents any clearly defined optimization; in fact, there may not even be a mathematically stated objective function. One strives for a feasible solution that one hopes will provide reasonable performance. This is the philosophy underlying Material Requirements Planning (to be discussed in Chapters 13 and 15).

2.7.3 Minimization of Inventories with Little Modeling

Here, the strategy is to attempt to minimize inventories without the help of mathematical models. The Just-in-Time (JIT) and Optimized Production Technology (OPT) philosophies are fall into this category. They strive for elimination of wastes (including inventories) and for continuous improvement. These powerful philosophies, to be discussed in Chapter 16, have since been supplemented with mathematical models that help managers refine and explain their operation. However, the philosophies are not grounded in models, but in an approach to doing business.

2.8 The Art of Modeling

Whichever modeling strategy is chosen, one must attempt to incorporate the important factors, yet keep the model as simple as possible. Incorrect modeling can lead to costly, erroneous decisions. As models become more elaborate, they require more managerial time, take more time to design and maintain, are more subject to problems of personnel turnover, and require more advocation with top management and everyone else in an organization (i.e., system control costs are increased). There is considerable art in developing an appropriate model; thus, a completely prescriptive approach is not possible. There are some excellent references (see, e.g., Little 1970; Ackoff 1981; Hillier and Lieberman 1990; Powell 1995a,b; Powell and Batt 2011) on the general topic of modeling. In particular, Little recommends that decision models should be understandable to the decision maker (particularly the underlying assumptions), complete, evolutionary, easy to control, easy to communicate with, robust (i.e., insensitive to errors in input data, etc.), and adaptive. We now list a number of suggestions that analysts may find helpful in modeling complex production/inventory situations:

1. As discussed in Chapter 1, the measures of effectiveness used in a model must be consistent with the objectives of the organization.
2. One is usually not looking for absolute optimization, but instead for significant improvements over current operations. Thus, we advocate the development and use of the so-called heuristic decision rules. These are procedures, based on sound logic, that are designed to yield reasonable, not necessarily mathematically optimal, answers to complex real problems (see Zanakis 1989; Reeves 1993; Powell and Batt 2011).
3. A model should permit results to be presented in a form suitable for management review. Therefore, results should be aggregated appropriately and displayed graphically if possible.
4. One should start with as simple a model as possible, only adding complexities as necessary. The initial, simpler versions of the model, if nothing else, provide useful insights into the form of the solution. We have found it helpful in multi-item problems to first consider the simplest cases of two items, then three items, and so on. In the same vein, it is worthwhile to analyze the special case of a zero lead time before generalizing to a nonzero lead time. We recall the incident of an overzealous mathematician who was employed by a consulting firm. Given a problem to solve involving just two items, he proceeded to wipe out the study budget by first solving the problem for the general case of n items and then substituting $n = 2$ in the very last step. In the same vein, one should introduce probabilistic elements with extreme caution for at least three reasons: (i) the substantially increased data requirements, (ii) the resulting computational complexities, and, perhaps most important, (iii) the severe conceptual difficulties that many practicing managers have with probabilistic reasoning. Throughout the book, we first deal with deterministic situations, and then modify the decision rules for probabilistic environments. Moreover, results are often insensitive to the particular probability distribution used for a specific random variable having a given mean and variance (see Section 6.7.14 of Chapter 6); thus, a mathematically convenient distribution can be used (e.g., the normal distribution described in Appendix II at the end of the book).
5. In most cases in the book, we advocate modeling that leads to analytic decision rules that can be implemented through the use of formulas and spreadsheets. However, particularly when there are dynamic or sequential[*] effects with uncertainty present (e.g., forecast errors), it may not be possible to analytically derive (through deductive mathematical reasoning) which of two or more alternative courses of action is best to use in a particular decision situation. In such a case, one can turn to the use of simulation. A simulation may be used, which still involves a model of the system. However, now, instead of using deductive mathematical reasoning, one instead, through the model, simulates through time the behavior of the real system under each alternative of interest. Conceptually, the approach is identical with that of using a physical simulation model of a prototype system—for example, the use of a small-scale hydraulic model of a series of reservoirs on a river system. More basically, prior to

---

[*] An example of a sequential effect would be the following: if A occurs before B, then C results; but if B occurs before A, then D takes place.

even considering the possible courses of action, simulation can be used to simply obtain a better understanding of the system under study (perhaps via the development of a descriptive analytic model of the system). References on simulation include Fishman (1978) and Kelton and Law (2000).

6. Where it is known *a priori* that the solution to a problem will possess a certain property, this information should be used, if possible, to simplify the modeling or the solution process. For example, in many cases, a very costly event (such as a stockout) will only occur on an infrequent basis. By neglecting these infrequent occurrences, we can simplify the modeling and obtain a straightforward trial solution of the problem. Once this trial solution is found of the above, the assumption is tested. If it turns out to be reasonable, as is often the case, the trial solution becomes the final solution. (See White 1969.)

7. When facing a new problem, one should at least attempt to show an equivalence with a different problem for which a solution method is already known (see Naddor 1978; Nahmias 1981).

2.9 Explicit Measurement of Costs

It often comes as a surprise to technically trained individuals that cost accountants and managers usually cannot determine exactly the costs of some of the variables they specify in their models. That cost measurement is, in practice, a problem that has not been conclusively solved, is evident from the fact that a number of alternative cost-accounting systems are in use. Kaplan (1988) points out that there are three fundamental purposes for cost-accounting systems: valuing the inventory for financial statements, providing feedback to production managers, and measuring the cost of individual SKUs. The choice of a cost system should be consistent with its purpose, and many firms should have more than one system.

One basic problem arises because it is not possible and often is not economical to trace all costs (variable, semivariable, or fixed) to each and every individual SKU in the inventory. An allocation process that distributes, often arbitrarily, fixed or overhead costs across all units is inevitable. (Recall the discussion of the ordering cost *A*, in Section 2.6.1.) Rosen (1974) categorizes some of the many possible cost systems, and we have summarized these in Table 2.2.

In Table 2.5, column 1, process costing is defined as an accounting method whereby all costs are collected by cost centers such as the paint shop, the warehouse, etc. After a predetermined collection period, each SKU that passed through the process will get allocated a share of the total costs incurred in the cost center. For example, all chairs painted in a particular month could be allocated exactly the same painting cost (which could be different the next month).

Alternatively, *job order costing* could be used. Under this accounting method, a particular order for chairs would be tracked as it progressed through the shop, and all costs incurred by this particular order would be recorded. Thereby, the "cost" of producing the same chair under either of the two costing systems could be very different. The reason for having these two different systems in practice is simple. The job order costing system is more expensive in terms of the amount of bookkeeping required, but it provides information for more detailed cost control than process costing. (Note that the accountant has to select between decision systems for collecting and keeping track of costs just like we must design inventory/production decision systems that have the desired level of sophistication.)

In addition to selecting between bookkeeping systems, the accountant must decide on the basis of valuation. There are four choices listed in column 2 of Table 2.5. Under actual costing, the

Table 2.5 Costing and Control Alternatives

(1)	(2)	(3)
Basic Characteristics of the System	**Which Valuation?**	**Include versus Exclude**
	Actual?	Fixed manufacturing
	Predetermined?	Overhead in inventory cost?
	Standard?	
A	A	A
Process-Costing Methods	Actual direct material	Full absorption costing (includes a fixed portion of manufacturing overhead)
	Actual direct labor	
	Actual manufacturing overhead	
B	B	B
Job order costing methods	Actual direct material	Direct variable costing (does not include a fixed portion of manufacturing overhead)
	Actual direct labor	
	Predetermined manufacturing overhead	
	C	
	Actual direct material	
	Predetermined direct labor, overhead	
	D	
	Standard direct material	
	Standard direct labor	
	Standard manufacturing overhead	

product cost v is determined on the basis of accumulated actual cost directly incurred during a given period. Materials, labor, and overhead costs are accumulated, which are and then are spread over all units produced during the period. Note that, as a result, the cost of a given item may vary from one period to another.

Under *standard costing* the materials, labor, and factory overhead are charged against each production unit in accordance with a standard (hourly) rate regardless of how much actual effort (time) it takes. Any deficit (negative variance) between total standard costs charged and actual costs incurred over a costing period (say a year) is charged as a separate expense item on the income statement and does not end up as a part of the cost of an item, v. Alternatively, surpluses (positive variances), that is, where production takes less time than the standard, can artificially inflate v unless all items produced are revalued at the end of an accounting year. The possibility of a difference between standard and actual costs results from the fact that standards may have been set for different purposes or they may have been set at ideal, long-run, or perceived "normal" rates that turn out, after the fact, to be unattainable. Also, existing standards may not have been adjusted to account for process improvements and learning.

The term *predetermined* in Table 2.5 refers to the allocation of costs to an SKU based on short-term budgets as opposed to engineered standards. Predetermined cost allocations are usually somewhat arbitrary and may not reflect the amount of expense (as could be determined from a careful engineering study) incurred by an SKU. Predetermined costing has the virtue of ensuring that, from an accounting point of view, all costs incurred over a costing period get allocated somewhere.

Cost-accounting systems are further differentiated in column 3 of Table 2.5 by whether they utilize absorption costing or direct costing. Under *absorption costing* (also called full costing), little distinction is made between fixed and variable manufacturing costs. All overhead costs are charged according to some formula, such as the percent of total direct labor hours used by the product being manufactured. In spite of the fact that direct labor now represents as little as 5% of the total product cost, many firms still allocate overhead on that basis. Activity-based costing (also denoted by ABC) attempts to allocate the actual amount of each overhead resource spent on each SKU. Imagine a firm that produces three products, each of which uses about the same number of direct labor hours. Product X2Z, however, requires a significant amount of effort by supervisors, quality assurance, and expediters because the production processes are not well suited to it. Allocation of overhead on the basis of direct labor hours will make it appear that the other two products are less profitable than they really are; and product X2Z's cost will include a smaller portion of overhead than it should. Managers armed with this information may decide to drop a product that is actually very profitable, and to keep a product that actually is losing money. ABC systems have grown in popularity because of these very issues. See also Cooper and Kaplan (1987), Shank and Govindarajan (1988), and Karmarkar and Pitbladdo (1994).

The concept of *direct costing* involves the classification of costs into fixed and variable elements. Costs that are a function of time rather than volume are classified as fixed and are not charged to individual SKUs. For example, all fixed manufacturing expenses, executive and supervisory salaries, and general office expenses are usually considered to be fixed since they are not directly a function of volume. However, note that some costs are in fact semivariable; they vary only with large fluctuations in volume. A doubling of the production rate, for example, could make it necessary to hire extra supervisory personnel.

Semivariable costs (Figure 2.5) in practice are often handled by assuming that volume can be predicted with sufficient accuracy so that, over a relevant range, costs can be viewed as being only of two kinds: variable and fixed. This is not always a simple matter.

It should be clear that the first step in any attempt at explicit cost measurement has to be a determination of what assumptions are made by the existing cost-accounting system within a firm. As Rosen (1974) points out, a small business needing information for external reporting and bookkeeping purposes may choose from Table 2.5 a system (one selection from each of the three columns) that is A–A–A because it may be the least costly from a bookkeeping standpoint. A similar company wishing to postpone tax payments may choose A–A–B, and so forth. Again, in spite of the resistance from financial managers, it may be desirable to use more than one system.

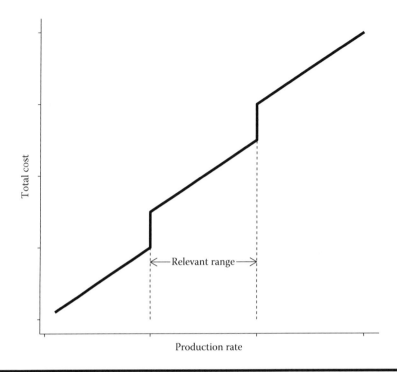

Figure 2.5 Semivariable costs.

Note that while the above description of costing systems provides many options, the most important costs of all, from an inventory-planning standpoint, do not appear at all. Inventory and production-planning decisions require the *opportunity costs*. These costs do not appear in conventional accounting records. Accountants are primarily concerned with the recording of *historical costs*, whereas decision makers must anticipate future costs so that they can be avoided if possible. Existing cost data in standard accounting records, therefore, may not be relevant to decision making or at best have to be recast to be useful.

To a decision maker, *relevant costs* include those that will be incurred in the *future* if a particular action is chosen. They also include costs that could be avoided if a particular action is not taken. Therefore, the overall overhead costs are only relevant if they can be affected by an inventory or production-planning decision. Allocated (predetermined) overhead, based on some ad hoc formula that is useful for financial reporting purposes, generally is not relevant for our purposes. One must therefore carefully examine the costing procedures of an organization from the point of view of their relevancy to the decision at hand.

2.10 Implicit Cost Measurement and Exchange Curves

Suppose we examine carefully the existing accounting systems of a firm and determine the particular species with which we are dealing. Suppose further that we modify the historical cost data to the best of our ability to reflect all relevant costs (including opportunity costs). Then, finally, suppose we use the most advanced techniques of inventory control to determine the appropriate inventory investment, based on our adjusted costs. What if this "appropriate" total investment in inventories that we propose turns out to be larger than the top management is willing to accept?

Two possible conclusions could be drawn. First, one could argue that we did not correctly determine the relevant costs and that we did not properly capture the opportunity costs. Second, one could argue that the top management is wrong. But what to do? The key lies in realizing that the inventory-planning decision deals with the design of an entire system consisting of an ordering function, a warehousing system, and the servicing of customer demand—all to top management specification. One cannot focus on an individual SKU and ask: What is the marginal cost of stocking this or that individual item without considering its impact on other SKUs or on the system as a whole? Brown (1967, pp. 29–31) argues that there is no "correct" value for r, the cost of carrying inventories, in the explicit measurement sense used by accounting. Instead, the carrying charge r, he says, is a top management policy variable that can be changed from time to time to meet the changing environment. The only "correct" value of r, according to Brown, is the one that results in a total system where aggregate investment, total number of orders per year, and the overall customer-service level are in agreement with the corporate, operations, and marketing strategy. For example, the specification of a low r value would generate a system with relatively large inventory investment, good customer service, and low order replenishment expenses. Alternatively, higher values of r would encourage carrying of less inventories, poorer customer service, and higher ordering costs.

There is of course, no reason, from a theoretical point of view, why all costs (including A, v, and the cost of disservice) could not be considered as policy variables. In practice, this is in effect what is often done, at least partially, by many inventory consultants. An attempt is first made to measure all costs explicitly to provide some baseline data ("ballpark" estimates). Then, the resulting inventory decision system is modified to conform to the aggregate specifications. Only cost estimates that achieve the aggregate specifications are ultimately used during implementation. Such an approach is feasible in practice partly because most of the decision models for inventory management and production planning and scheduling are relatively insensitive to errors in cost measurement. In Chapters 4 and 6, we describe in detail a methodology called exchange curves, for designing inventory decision systems using cost information and policy variables specified by the top management. See also the discussion in Lenard and Roy (1995). For now, we just give a brief overview illustration for the decision of choosing order quantities for a population of items.

For the economic order quantity decision rule (to be discussed in Chapter 4), it turns out that the order quantity of an individual item depends on the ratio A/r. As A/r increases, the order quantity increases; hence, the average inventory level of the item increases and the number of replenishments per year decreases. These same general effects apply for all items in the population under consideration. Therefore, as A/r increases, the total average inventory (in dollars) increases and the total number of replenishments per year decreases. Thus, an aggregate curve is traced as A/r is varied (as shown in Figure 2.6). If the management selects a desired aggregate operating point on the curve, this implies a value of A/r. The use of this A/r value in the economic order quantity formula for each item will give the desired aggregate operating characteristics.

2.11 The Phases of a Major Study of an Inventory Management or Production Planning and Scheduling System

We have now outlined some of the important costs and other factors for inventory and production decisions. The final part of this chapter describes the process of a major (internal or external)

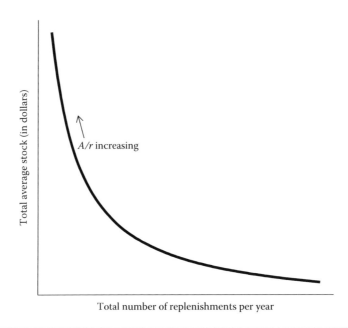

Figure 2.6 Example of an exchange curve.

consulting study. In this process, we strongly recommend building decision systems as a process of organizational intervention, whereby a model builder attempts to improve on the quality of decisions being made. By intervening, the consultant accepts a great responsibility—that of ensuring that the daily routines in the client's organization are carried on, at least as well as before, during the period of transition to the newly designed decision procedures and thereafter. Acceptance of such a responsibility implies that implementation considerations have to dominate all phases of the design of an operational decision system. In particular, successfully implemented innovations require careful planning, and once in place, they must be maintained, evaluated, and adapted, ad infinitum, through an effective control and evaluation system.

A major study can be an overwhelming task. Therefore, we recommend dividing it into six phases, based on the classic work of Morgan (1963): consideration, analysis, synthesis, choosing among alternatives, control, and evaluation. See Figure 2.7. We briefly present a number of questions and issues that should be addressed in each phase.

2.11.1 Consideration

This first phase focuses on conceptualizing the problem and covers a number of strategic and organizational issues, as well as some detailed modeling concerns. The questions and issues include:

◼ What are the important operations objectives for this firm? (Recall the four objectives—cost, quality, delivery, and flexibility—discussed in Chapter 1.) In other words, one should perform at least a brief audit of the operations strategy of the firm.
◼ Who has the overall responsibility for inventory management and production planning and scheduling? Identify the capabilities and limitations of this person as they pertain to the potential goals of the study.

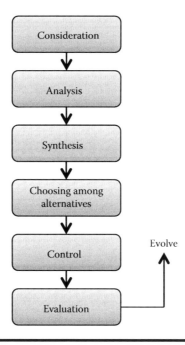

Figure 2.7   Phases of a consulting study.

■ What is the organizational structure? In particular, what is the relationship between operations and marketing?

■ What are annual sales of the organization?

■ What is the current average aggregate inventory level in dollars? This and the previous question help to identify the magnitude of savings possible.

■ What are the current inventory management and production planning and scheduling procedures? How are decisions made, and by whom? Are any models used?

■ What are the computer resources and skills available?

■ What is the SKU? Is it a completed item, or a subassembly that will be finished to order? Are there other alternatives to holding the finished product inventory?

■ Is the inventory necessary? Perhaps, it is possible to eliminate some inventories that exist only for historical reasons.

■ What modeling strategy (from Section 2.7) seems to be appropriate? One should keep in mind the comments in Section 2.8.

2.11.2 Analysis

The second phase is one of the two modeling stages, and focuses on data collection and a detailed understanding of the uncontrollable and controllable variables. Uncontrollable variables are those, such as the number of items and the level of seasonality, which the consultant or the firm cannot easily change. Controllable variables are those quantities that can be manipulated to achieve the goals. We provide quite a long list that can be used to trigger information gathering. (A reexamination of Table 2.4 may also be helpful.) Questions and issues for uncontrollable variables include:

- How many items are to be studied? If there are thousands of items, it is wise to select, say, every second or fifth item in the SKU list and analyze them first. A study of a small, representative sample can indicate the magnitude of potential savings. It is important, however, that both A and B items are chosen. In addition, it is desirable, if practical, to develop the entire DBV curve. This is easy if current data are available in electronic form. If not, it may not be worth the effort.
- Are the items independent of one another? Dependence may be due to multiple items on a customer order, complementary or substitute items, and shared production equipment.
- What does the supply chain look like? Specifically, can we consider this stocking point in isolation of upstream and downstream locations, or is a multiechelon model necessary due to significant interrelationships with other production or distribution facilities?
- Are customer transaction sizes in single units or batches? If transactions are not necessarily unit sized, are they still discrete or can they take on a continuum of values (e.g., the number of gallons or liters of gasoline demanded by a vehicle in a fill-up)? As we shall see in Chapters 6 and 7, unit-sized transactions are the easiest to model.
- Is demand deterministic or variable? If it is variable, is there a known probability distribution that adequately describes it?
- Do customers arrive at a constant or variable rate? Again, if it is variable, can it be described by a known probability distribution? These latter three bullet points often are considered together by looking at the total demand over some time period, such as a day, week, or month.
- Is the average demand seasonal, or is it somewhat constant over the year?
- What historical data are available? Forecasting models are the most accurate when there is sufficient historical data to calibrate the model.
- How many customers are served? If the number is small, demands may not be independent across time. For example, a large customer order today would reduce the probability of a customer order tomorrow.
- How many, and how powerful, are the firm's competitors?
- What is the typical customer promise time? If customers do not demand immediate delivery, it may be possible to reduce inventories by holding stocks of components, and then assembling to order.
- How is customer service measured? What are the service-level targets?
- How many suppliers serve the firm? Do they offer quantity discounts? Is supply seasonal? Are all the units received of adequate quality to be used?
- What is the replenishment lead time (and its variability) for each item studied?
- Is there a concern about obsolescence or spoilage?
- What is the number, location, and capacity of each storage and manufacturing facility?
- What are the costs that we identified in Section 2.6.1 (A, v, r, and so on)?

For production operations, the following questions are also important:

- How many components go into each end item?
- Through how many processing stages do the products go?
- What is the capacity of the production equipment? How many shifts, hours per week, and weeks per year? Are there any complete shutdowns during the year?
- What are the setup or changeover times for each part? Is there a sequence dependency in these times? What are the processing rates for each part?

- How reliable is the production equipment? Are breakdowns a problem? Does the firm regularly schedule preventive maintenance?
- What is the layout of the factory? In other words, it should be positioned on the product-process matrix.
- What are the yields of the equipment? (Yield is the proportion of a batch that is of acceptable quality.)

Now we examine controllable variables:

- If there is more than one stocking or production location, where should each item be stored or produced?
- How frequently should inventory levels be reviewed?
- When should each item be reordered or produced?
- How much should be ordered or produced? In what sequence, or in what groupings?
- How much productive capacity should be established? Should we recommend hiring, firing, or overtime?
- Should the product price be adjusted?
- Should the service targets be changed?
- What transportation modes, and other factors should be changed?

2.11.3 Synthesis

The second modeling stage, which is the third phase of the process, attempts to bring together the vast amount of information gathered in the previous phase. The consultant now should establish relationships among objectives, uncontrollable variables, and the means of control. For example, how is the expected cost per unit time related to the demand rate, ordering cost, order quantity, and so on? A mathematical objective function stated in terms of controllable variables is the most common result of this phase, although it is not always so. This synthesis phase is a primary focus of the remainder of the book, and therefore we will not devote more space to it here.

2.11.4 Choosing among Alternatives

This phase is the second primary focus of this book. Once we have an objective function, or established relationships among objectives, uncontrollable, and controllable variables, it is necessary to determine reasonable settings of the controllable variables by manipulating the model. This can be done with calculus, search methods, and so on. If the decision is repetitive, it is desirable to develop mathematical decision rules that provide the settings of the controllable variables in terms of the uncontrollable variables. For example, the economic order quantity, to be discussed in Chapter 4, determines the order quantity as a function of A, v, r, and the annual demand rate. These decision rules should also allow for other factors that are not explicitly considered in the model. (See Section 1.5 of Chapter 1.) Moreover, the model should accommodate sensitivity analysis so that managers can get a feel for the result of changes in the input data. If the decision is not repetitive, there is no need to develop a framework for repeated use. A "one-off" decision is adequate.

2.11.5 Control

The control phase is primarily focused on implementation of the decision rules, or more generally, the choices among alternatives. Here, the consultant must ensure that the system functions properly on a tactical or short-range basis. One aspect will be discussed in more detail in Section 2.11.8. Other questions and issues include:

- Training of the staff who will use the new system.
- How often should the values of the control parameters (e.g., the order quantity) be recomputed?
- How will the uncontrollable variables be monitored?
- How will the firm keep track of the inventory levels? In Section 2.11.9, we discuss this issue in more depth.
- How will exceptions be handled? For example, when and how should managers override the decision rules?
- What provision will be made for monitoring sales against forecasts? There are two reasons for overseeing the forecasting system—(a) to take action to prevent serious shortages or overages of stock; and (b) possibly to change the forecast model.
- What actions should be taken when stockouts occur or are imminent? Does the firm have provision for expediting, alerting sales people, finding alternative sources of supply, and so on?
- In the production environment, how will actual production be monitored against scheduled production?

2.11.6 Evaluation

The final phase is evaluation, or measuring how well the control system functions. If the consultant gathered data on actual costs and customer service prior to implementation, which is extremely wise, he or she can now compare them with the actual values under the new system. Likewise, to test the validity of any models, it is useful to compare predicted with actual costs and service. Breaking down the components of the total cost can yield an insight into the sources of any discrepancies.

2.11.7 General Comments

Even a carefully executed study can run into major difficulties that are not the responsibility of the consultant. It is possible, for example, that an external forecasting system provides forecasts that are worse than anticipated. These may reflect badly on the new inventory or production system. The consultant should attempt to identify the reasons for unusually bad, or unusually good, performance. If the model or data are wrong, adjustments can be made. On the other hand, there may be resistance from people at the firm that led to poor performance.

It should be clear that implementation is a long, drawn-out process whose importance and duration are both often underestimated. *Gradual* implementation, accompanied by extensive education, is essential. Where possible, the so-called *pilot approach* should be first utilized. Specifically, the new system should be first implemented on a trial basis on a limited class of items (e.g., on only one of the several product lines manufactured by the firm). As noted above, provision should be made for manual overrides of the output of computerized decision rules. In this regard, Bishop (1974) describes a successful implementation where the apparent key factor that led to the adoption was a design option within the new decision system that allowed manual overrides of computer decisions by the manager whenever he disagreed with them. "Early in the operation of the system, approximately 80% of the items were overridden by the use of this feature . . . a year later this proportion had declined to less than 10%." According to Bishop, the inventory control manager, his staff, and his boss used the new decision system simply because they felt that they were in charge, rather than at the mercy, of the innovation.

2.11.8 Transient Effects

Each time the parameters of a decision model are adjusted, a series of transient effects result that must be carefully monitored and controlled. Consider the recommendation of an analyst that affects the reallocation of inventory tied up in safety stocks in such a manner that *total inventory investment is not changed in the long run.* The reorder points for some SKUs are raised, for others the reorder points are lowered, and for the balance there is no change. If the management implemented all the new reorder points at the *same* time, some of the newly calculated reorder points, raised from the previous levels, would immediately trigger orders. If there were many changes, a sudden wave of orders could overload a company's manufacturing plant or order department and also result, after a lead time, in an increase in on-hand dollar inventory investment.

The obvious solution is, of course, not to make all the changes at the same time. Some form of phased implementation strategy that lengthens the period of transition and thereby dampens the disruptions in the system must be chosen. To determine an acceptable transition strategy, the following need to be balanced:

1. The cost of carrying extra inventories during the transition period
2. The expense of a higher than normal ordering rate
3. The opportunity cost of delaying benefits from a new production/inventory decision system that is not implemented all at once
4. Any benefits resulting from the better than normal service level resulting from extra inventory on hand during the transition period

In practice, the changeover to new decision rules is often carried out at the time when an SKU is reordered. This works well if different SKUs are ordered at "random" times. Alternatively, revised decision rules are implemented in phases, starting with A items.

In either case, it is best to try to plan a large transition carefully; in particular, one should closely monitor the resulting stockout rate, the net increase in orders placed, and the rise in total dollar inventory. It is unlikely that the costs of transition can ever be determined and balanced analytically. Therefore, a simulation of transient effects that result from alternative implementation strategies should be undertaken whenever the scope of the problem warrants such an expense.

2.11.9 Physical Stock Counts

Physical stock counts, taken annually in many organizations, are intended to satisfy the auditors that account records reasonably represent the value of inventory investment. But the accuracy of stock records is also an important aspect of production/inventory decisions. Most of the decision systems described in this book would be seriously crippled by inaccurate records. See Swann (1989), DeHoratius (2006), and DeHoratius and Raman (2008).

As a check on records, some physical counting of SKUs actually in stock has always been deemed necessary. But physical counts have proven to be expensive and time consuming.[*] The authors recall the words of a general manager during a tour of a steel-manufacturing operation: "I am deliberately letting the inside inventory drop down to facilitate the annual stock counting." Performing a physical stock count has in the past in many companies involved could involve shutting down all operations, and thereby losing valuable productive time. Because of this large opportunity cost, many companies have adopted different procedures that achieve the same purpose. Given a limited budget to be spent on corroborating inventory records with the physical inventory, a system that more effectively rations the available clerical resources is known as *cycle counting*.

In cycle counting, as the name might suggest, a physical inventory of each particular item is taken once during each of its replenishment cycles. There are several versions of cycle counting, but probably the most effective one involves counting the remaining physical stock of an SKU just as a replenishment arrives. This has two key advantages. First, the stock is counted at its lowest level. Second, a clerk is already at the stocking position delivering the replenishment, so that a special trip is not needed for counting purposes. Clearly, one drawback is that, if the records are faulty on the high side, we may already have gone into an unexpected stockout position. Note that cycle counting automatically ensures that low-usage SKUs are counted less often than high-usage items because the frequency of reordering increases with sales volume (at least under a decision system based on economic order quantities). Most auditors agree with physically counting low-value items less often than the more expensive ones.

Paperwork, relating to an SKU being counted, can be outstanding at any point in time. When outstanding paperwork exists, a disagreement between counted and recorded inventory may not necessarily mean that the written records are in error. Cycle counts are for this reason in some companies confined to the first thing in the morning, or the last thing in the day, or on the off shift, when outstanding paperwork should be at a minimum. In any case, some form of cutoff rule such as the following is required:

1. If a discrepancy is less than some low percent of the reorder point level, adjust your records to the lower of the two figures and report any loss to accounting.
2. If the discrepancy is greater than some low percent, wait for a few days and request a recount. One hopes by this time that outstanding paperwork will have been processed. If the discrepancy is the same on both counts, we have probably found an error and should adjust the records accordingly. If the second count results in a different discrepancy, the problem requires investigation.

There are some practical disadvantages and problems with cycle counting. Cutoff rules are not easy to follow while normal everyday activity is going on. Checking up on the paperwork in the system, so that stock levels may be properly corroborated, requires considerable perseverance and diplomacy. The responsibility for cycle counting is often assigned to stockroom employees. As a result, too often, cycle counting gets relegated to a lower priority, and may even be discontinued; it may even be discontinued when business activity picks up and stockroom personnel find it difficult to carry out their regular activities as well as stock counting. One solution in such cases is to assign a permanent team to carry out cycle counting all year round. In an interesting study,

---

[*] Counting need not be done in a literal sense. An effective, accurate way of estimating the number of units of certain types of products is the use of weigh scales.

Millet (1994) found that inaccuracies in inventory records were often due to inaccurate physical counts, and could be eliminated by providing incentives for those responsible for counting. See also Hovencamp (1995) and Ziegler (1996).

Cycle counting rations effort in proportion to annual dollar usage. It is important to recognize that not all causes of stock discrepancies are a function of dollar usage. The authors once encountered a situation illustrating this point. Items that could be resold on the black market tended to be out of stock more often. Although it could not be proven, it was possible that some of the stolen parts were being resold back to the company through supposedly legitimate wholesalers. Therefore, the accessibility, physical size, and resalability of items, as well as their dollar usage, had to be considered in establishing procedures for stock counts.

Stratified sampling procedures can also be useful for reducing the required sample sizes, particularly for audit purposes. Buck and Sadowski (1983) suggest this approach.

2.12 Summary

In this chapter, we have provided a number of frameworks for the manager and consultant that are helpful in reducing the complexity of the task of managing inventories and production. In the next and last chapter in this part of the book, we turn to the topic of demand forecasting in that an estimate of future demand is clearly needed for all decisions in the production/inventory area.

Problems

2.1 Briefly describe the inventory management problem(s) faced by a typical small neighborhood restaurant. Recall Problem 1.5. Point out the differences in the inventory problem(s) faced by a particular store in a more specialized chain such as McDonald's or KFC.

2.2 Comment on the following statement: "Optimization models, because they make unrealistic assumptions and are difficult for the user to comprehend, are useless in inventory management or production planning and scheduling."

2.3 Discuss the fundamental differences in the approach required to analyze the following two types of inventory control problems:
    a. A one-opportunity type of purchase that promises large savings
    b. A situation where the options remain the same from one buying opportunity to the next

2.4 It has been argued that the fixed cost of a setup A is not really fixed but can vary throughout the year when the production workload is seasonal. Discuss, and include the implications in terms of the impact on replenishment lot sizes.

2.5 Assuming that you have no data on the total annual dollar movement ($\sum Dv$), which of the following SKUs (and under what circumstances) should be classified as A, B, or C items?
    a. Spare parts for a discontinued piece of manufacturing equipment
    b. Bolts and washers
    c. Subassemblies
    d. Imported SKUs for resale
    e. Motor oil
    f. Perishable food stuffs
    g. A Right Handed PS-37R-01
    h. Widget invented by the company owner's nephew
    i. Platinum bushings

If, in addition, you were supplied with figures on the annual dollar movement (*Dv*) for each SKU, how could such figures affect your classification above? What other information would you like to have? Finally, describe and classify one example of A, B, and C from your experience.

2.6 An inventory, mostly made up of machined parts, consisted of 6000 SKUs valued (at full cost) by the accounting department at $420,000. The company had recently built a new warehouse at a cost of $185,000 that was financed through a 12% mortgage. The building was to be depreciated on a 25-year basis. The company's credit rating was sound, and a bank loan of $50,000 was under negotiation with the bank. The main operating costs per year in the new warehouse were estimated to be as follows:

Municipal and other taxes	$5,732.67
Insurance on building and contents	3,200.00
Heating/air conditioning	12,500.00
Electricity and water	3,200.00
Labor (supervisor, three clerks, and a half-time janitor)[a]	69,000.00
Pilferage	5,000.00
Obsolescence	5,000.00
Total	$103,632.67

[a] Includes 60% manufacturing overhead.

It was estimated that on average, each SKU involved one dollar of labor cost per dollar of material cost.

a. Recommend a value for the carrying cost *r* in $/$/year.
b. Suppose that your recommended value for *r* is accepted by the management and the accounting department. You proceed to calculate the economic order quantities for all 6000 items and discover that the new warehouse is too small to physically accommodate all the inventory that is indicated by your calculations. What action would you take now?
c. Alternatively suppose that your recommended value for *r* is accepted by the management and the accounting department, but that the total dollar investment (based on economic order quantities) that you estimate will be needed is approximately $800,000. The top management is unwilling to have a total investment greater than $500,000. What action would you take under such a circumstance?

2.7 Use Figures 2.8 through 2.10 for this problem. Assume that the raw materials and purchased parts each represent 30% of the cost of goods sold and that value added (labor plus factory overhead) constitutes the remaining 40%. Also assume that each production cycle involves the production of 1000 units of finished goods, each valued (at cost) at $10/unit. In each case depicted in the figures, ascertain the appropriate *overall* value of $\bar{I}v$ (i.e., the average inventory level, expressed in dollars). Also describe, in terms of the production and demand processes, how each of the patterns could result.

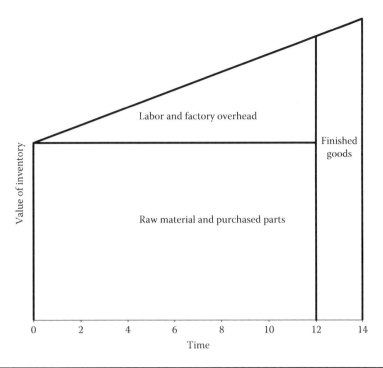

Figure 2.8 Case a for inventory valuation Problem 2.7.

2.8   What parameters that affect the different functional categories of inventories could be changed? For example, cycle stocks are determined in part by the setup cost, and it can be changed. Give other examples, and note how the parameters could change.

2.9   Have a class representative (or representatives) select a random sample of 30 items stocked in your university bookstore. For each item, obtain an estimate of Dv (from records or from the manager of the store). Prepare a DBV table and a graph. See Table 2.1 and Figure 2.1.

2.10 The type of operation characterized by a large volume of identical or similar products flowing through identical stages of production is typical in which of the following industries?
   a. Aerospace
   b. Automobiles
   c. Machine tools
   For such a situation, which type of production planning and scheduling system is appropriate? Explain your reasoning.

2.11 An important problem in the operation of a retail store is the allocation of limited shelf space among a wide variety of products.
   a. Discuss how demand is influenced by the allocation. Can individual items be treated separately in this regard?
   b. How are inventory-related costs of an item affected by the space allocated?

2.12 Consignment stock is stock held by a retailer but owned by the supplier until it is sold by the retailer. For the same SKU, discuss how you would model the associated inventory carrying costs of the retailer (i) without consignment and (ii) with consignment.

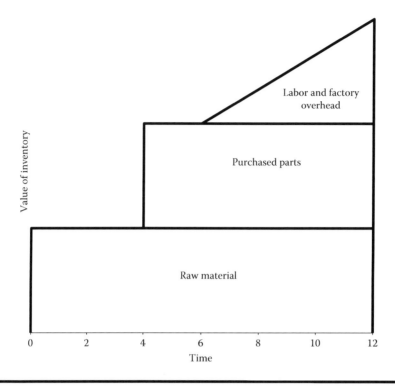

Figure 2.9 Case b for inventory valuation Problem 2.7.

2.13 a. Construct a spreadsheet that permits the development of a DBV table and a graph. Input data should be for each of n items (where, for any particular application, n would be specified by the user):
  i. Item identification
  ii. D_i, the annual usage of item i, in units/year
  iii. v_i, the unit variable cost of item i, in \$/unit
  b. Illustrate the use of your program for the following example involving 20 items:

Item, i	ID	D_i	v_i
1	A	80	422.33
2	B	514	54.07
3	C	19	0.65
4	D	2,442	16.11
5	E	6,289	4.61
6	F	128	0.63
7	G	1,541	2.96
8	H	4	22.05

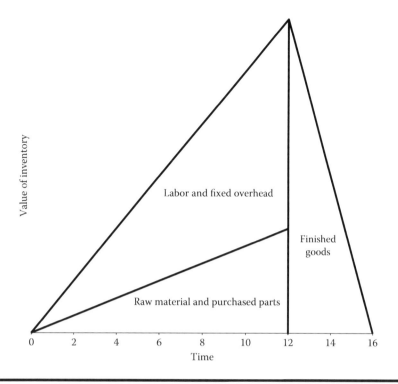

Figure 2.10   Case c for inventory valuation Problem 2.7.

9	I	25	5.01
10	J	2,232	2.48
11	K	2	4.78
12	L	1	38.03
13	M	6	9.01
14	N	12	25.89
15	O	101	59.50
16	P	715	20.78
17	Q	1	2.93
18	R	35	19.52
19	S	1	28.88
20	T	4	29.86

What percentage of the total dollar usage is contributed by the top 20% of the items and by the bottom 50% of the items and

2.14   Repeat Problem 2.13 for the following data:

SKU ID	Annual Usage in Units	Unit Variable Cost ($/Unit)
A	172	27.73
B	8	110.40
C	3	60.60
D	4	19.80
E	2	134.34
F	15	160.50
G	4	49.48
H	3	8.46
I	4	40.82
J	1	34.40
K	212	23.76
L	4	53.02
M	12	71.20
N	2	67.40
O	2	37.70
P	1	28.80
Q	4	78.40
R	12	33.20
S	3	72.00
T	18	45.00
U	50	20.87
V	19	24.40
W	1	48.30
X	10	33.84
Y	27	210.00
Z	33	73.44
1A	48	55.00
1B	5	30.00

SKU ID	Annual Usage in Units	Unit Variable Cost ($/Unit)
1C	210	5.12
1D	27	7.07
1E	10	37.05
1F	12	86.50
1G	48	14.66
1H	117	49.92
1I	12	47.50
1J	94	31.24
1K	4	84.03
1L	7	65.00
1M	2	51.68
1N	4	56.00
1O	12	49.50
1P	2	59.60
1Q	100	28.20
1R	2	29.89
1S	5	86.50
1T	60	57.98
1U	8	58.45

2.15 Select 30 items at random from an inventory population in a local organization. These could include the campus bookstore or supplies in a department office. For each item, ascertain the value of Dv and develop a DBV table and graph. What percent of the total dollar usage is contributed by the top 20% of the items (ranked by dollar usage)?

2.16 One of the major cost categories in this chapter is called "carrying costs." See Section 2.6.1. Until recently people did not appreciate that some quality-related costs could be appropriately included in this category. Indicate why annual quality-related costs might change with the production lot size.

2.17 For a local organization, identify the major cost factors included in Section 2.6.1, including the fixed cost of ordering or setup, A, the carrying charge, r, and the unit variable cost, v, for just one item. Describe the components of each. For A and r, indicate briefly why you decided to include each component. What components did you ignore in your analysis?

Appendix 2A: The Lognormal Distribution

2A.1 Probability Density Function and Cumulative Distribution Function

The density function of a lognormal variable x is given by[*]

$$f_x(x_0) = \frac{1}{bx_0\sqrt{2\pi}} exp\left[-\frac{(\ln x_0 - a)^2}{2b^2}\right] \tag{2A.1}$$

where a and b are the two parameters of the distribution.

We can see that the range of the lognormal is from 0 to ∞. It has a single peak in this range. A sketch of a typical lognormal distribution is shown in Figure 2A.1. The name is derived from the fact that the logarithm of x has a normal distribution (with mean a and standard deviation b).

The cumulative distribution function

$$p_{x\le}(x_0) = \int_{-\infty}^{x_0} f_x(z)dz$$

can be shown to be given by

$$p_{x\le}(x_0) = 1 - p_{u\ge}\left(\frac{\ln x_0 - a}{b}\right) \tag{2A.2}$$

where $p_{u\ge}$ is the probability that a unit normal variable takes on a value of u_0 or larger, a function tabulated in Appendix II.

2A.2 Moments

One can show (See Aitchison and Brown 1957) that the expected value of the jth power (for any real j) of x is given by[†]

$$E(x^j) = m^j exp[b^2 j(j-1)/2] \tag{2A.3}$$

where

$$m = exp[a + b^2/2] \tag{2A.4}$$

is the mean or expected value of x. The variance of x defined by

$$\sigma_x^2 = E(x^2) - [E(x)]^2$$

can be shown, through the straightforward use of Equations 2A.3 and 2A.4, to be

$$\sigma_x^2 = [E(x)]^2[exp(b^2) - 1]$$

---

[*] $f_x(x_0)dx_0$ represents the probability that x takes on a value between x_0 and $x_0 + dx_0$.
[†] In general, $E(x^j) = \int_{-\infty}^{\infty} x_0^j f_x(x_0)dx_0$.

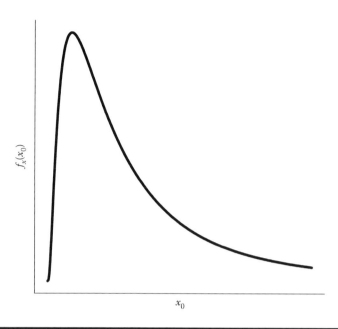

Figure 2A.1 The lognormal probability density function.

Hence, the standard deviation σ_x is given by

$$\sigma_x = E(x)\sqrt{exp(b^2) - 1}$$

and the coefficient of variation by

$$\sigma_x/E(x) = \sqrt{exp(b^2) - 1}$$

the latter depending only on the value of b.

If, as in a DBV of a population of n items, x is actually a discrete variable that takes on the values x_i ($i = 1, 2, \ldots, n$) and the x_i's can be represented by a lognormal distribution with parameters m and b, then

$$E(x^j) = \frac{\sum_{i=1}^{n} x_i^j}{n} \tag{2A.5}$$

and from Equation 2A.3 it follows that

$$\sum_{i=1}^{n} x_i^j = nm^j exp[b^2 j(j-1)/2] \tag{2A.6}$$

Note that the parameter m can be estimated using the sample mean, that is,

$$m = \frac{\sum_{i=1}^{n} x_i}{n} \tag{2A.7}$$

2A.3 Testing the Fit of a Lognormal

One can easily test the fit of a lognormal distribution to a set of data by taking natural logs and then using a normal probability plot routine. Such routines are available in most statistical software packages. See Section 3.5.5 for further discussion. Herron (1976) also provides an alternative approach.

References

Ackoff, R. 1981. The art and science of mess management. *Interfaces 11*(1), 20–26.

Aitchison, J. and J. A. C. Brown. 1957. *The Lognormal Distribution*. Cambridge, England: Cambridge University Press.

Barancsi, É. and A. Chikán. 1990. *Inventory Models*, Volume 16. Dordrecht, the Netherlands: Kluwer Academic Publishers.

Bishop, J. L. 1974. Experience with a successful system for forecasting and inventory control. *Operations Research 22*(6), 1224–1231.

Brown, R. G. 1967. *Decision Rules for Inventory Management*. New York: Holt, Rinehart and Winston.

Brown, R. G. 1982. *Advanced Service Parts Inventory Control*. Norwich, VT: Materials Management Systems, Inc.

Buck, J. R. and R. P. Sadowski. 1983. Optimum stratified cycle counting. *IIE Transactions 15*(2), 119–126.

Chang, Y. S. and P. Niland. 1967. A model for measuring stock depletion costs. *Operations Research 15*(3), 427–447.

Cohen, M. A. and R. Ernst. 1988. Multi-item classification and generic inventory stock control policies. *Production and Inventory Management Journal 29*(3), 6–8.

Cooper, R. and R. S. Kaplan. 1987. How cost accounting systematically distorts product costs. In W. J. J. Bruns and R. S. Kaplan (Eds.), *Accounting and Management: Field Study Perspectives*, pp. 204–228. Cambridge, MA: Harvard Business School Press.

DeHoratius, N. 2006. Inventory record inaccuracy in retail supply chains. In J. J. Cochran, L. A. Cox, P. Keskinocak, J. P. Kharoufeh, and J. C. Smith (Eds.), *Wiley Encyclopedia of Operations Research and Management Science*, pp. 1–14. Hoboken, NJ: John Wiley & Sons, Inc.

DeHoratius, N. and A. Raman. 2008. Inventory record inaccuracy: An empirical analysis. *Management Science 54*(4), 627–641.

Edgeworth, F. Y. 1888. The mathematical theory of banking. *Journal of the Royal Statistical Society 51*, 113–127.

Fishman, G. S. 1978. *Principles of Discrete Event Simulation*. New York: Wiley-Interscience.

Flores, B. E. and D. C. Whybark. 1987. Implementing multiple criteria ABC analysis. *Journal of Operations Management 7*(1–2), 79–85.

Gregory, G., S. A. Klesniks, and J. A. Piper. 1983. Batch production decisions and the small firm. *Journal of the Operational Research Society 34*(6), 469–477.

Hayes, R. and S. Wheelwright. 1979a. Link manufacturing process and product life cycles. *Harvard Business Review 57*(1), 133–140.

Hayes, R. and S. Wheelwright. 1979b. The dynamics of process-product life cycles. *Harvard Business Review 57*(2), 127–136.

Herron, D. P. 1976. Industrial engineering applications of ABC curves. *AIIE Transactions 8*(2), 210–218.

Hillier, F. S. and G. J. Lieberman. 1990. *Introduction to Mathematical Programming*. New York: McGraw-Hill.

Hovencamp, D. P. 1995. Where is it? *APICS* (July), 50–54.

Kaplan, A. J. 1988. Bayesian approach to inventory control of new parts. *IIE Transactions 20*(2), 151–156.

Karmarkar, U. and R. Pitbladdo. 1994. Product-line selection, production decisions and allocation of common fixed costs. *International Journal of Production Economics 34*, 17–33.

Kelton, W. D. and A. M. Law. 2000. *Simulation Modeling and Analysis*. New York: McGraw Hill Boston.

Kotler, P. 1967. *Marketing Management: Analysis, Planning, and Control*. Englewood Cliffs, NJ: Prentice-Hall.

Krupp, J. A. G. 1994. Are ABC codes an obsolete technology? *APICS—The Performance Advantage* (April), 34–35.

Lenard, J. D. and B. Roy. 1995. Multi-item inventory control: A multicriteria view. *European Journal of Operational Research 87*, 685–692.

Little, J. 1970. Models and managers: The concept of a decision calculus. *Management Science 16*(8), 466–485.

Lockett, G. 1981. The management of stocks—Some case histories. *OMEGA 9*(6), 595–604.

Millet, I. 1994. A novena to Saint Anthony, or how to find inventory by not looking. *Interfaces 24*(2), 69–75.

Morgan, J. 1963. Questions for solving the inventory problem. *Harvard Business Review 41*(1), 95–110.

Naddor, E. 1978. Sensitivity to distributions in inventory systems. *Management Science 24*(16), 1769–1772.

Nahmias, S. 1981. Approximation techniques for several stochastic inventory models. *Computers and Operations Research 8*(3), 141–158.

Oral, M., M. Salvador, A. Reisman, and B. Dean. 1972. On the evaluation of shortage costs for inventory control of finished goods. *Management Science 18*(6), B344–B351.

Porteus, E. L. 1990. The impact of inspection delay on process and inspection lot sizing. *Management Science 36*(8), 999–1007.

Powell, S. G. 1995a. Six key modeling heuristics. *Interfaces 25*(4), 114–125.

Powell, S. G. 1995b. Teaching the art of modeling to MBA students. *Interfaces 25*(3), 88–94.

Powell, S. G. and R. J. Batt. 2011. *Modeling for Insight: A Master Class for Business Analysts.* Hoboken, NJ: John Wiley & Sons.

Prasad, S. 1994. Classification of inventory models and systems. *International Journal of Production Economics 34*, 209–222.

Reeves, C. R. 1993. Improving the efficiency of tabu search for machine sequencing problems. *Journal of the Operational Research Society 44*, 375–382.

Reichheld, F. R. and W. E. Sasser Jr. 1990. Zero defections: Quality comes to services. *Harvard Business Review 68*(5), 105–111.

Reynolds, M. P. 1994. Spare parts inventory management. *APICS—The Performance Advantage 4*(4), 42–46.

Rosen, L. S. 1974. *Topics in Managerial Accounting* (2nd ed.). Ryerson, Toronto: McGraw-Hill.

Safizadeh, M. H., L. P. Ritzman, D. Sharma, and C. Wood. 1996. An empirical analysis of the product-process matrix. *Management Science 42*(11), 1576–1591.

Schmenner, R. W. 1993. *Production/Operations Management: From the Inside Out* (5th ed.). New York: Macmillan.

Schonberger, R. J. 1983. Applications of single-card and dual-card kanban. *Interfaces 13*(4), 56–67.

Shank, J. K. and V. Govindarajan. 1988. The perils of cost allocation based on production volumes. *Accounting Horizons 2*, 71–79.

Silver, E. A. 1981. Operations research in inventory management: A review and critique. *Operations Research 29*(4), 628–645.

Silver, E. A. 1996. A concern regarding the revised product-process matrix. *International Journal of Production Research 34*(11), 3285.

Simon, H. A. 1957. *Models of Man, Social and Rational.* New York: John Wiley & Sons.

Singhal, V. R. 1988. Inventories, risk, and the value of the firm. *Journal of Manufacturing and Operations Management 1*, 4–43.

Singhal, V. R. and A. S. Raturi. 1990. The effect of inventory decisions and parameters on the opportunity cost of capital. *Journal of Operations Management 9*(3), 406–420.

Singhal, V. R., A. S. Raturi, and J. Bryant. 1994. On incorporating business risk into continuous review inventory models. *European Journal of Operational Research 75*, 136–150.

Spencer, M. S. and J. F. Cox. 1995. An analysis of the product-process matrix and repetitive manufacturing. *International Journal of Production Research 33*(5), 1275–1294.

Swann, D. 1989. Silver bullets and basics. *Production and Inventory Management Review* (January), 34–35.

Taylor, S., S. Seward, and S. Bolander. 1981. Why the process industries are different. *Production and Inventory Management Journal 22*(4), 9–24.

White, D. 1969. Problems involving infrequent but significant contingencies. *Operational Research Quarterly 20*(1), 45–57.

Zanakis, S., J. Evans, and A. Vazacopoulos. 1989. Heuristic methods and applications: A categorized survey. *European Journal of Operational Research 43*, 88–110.

Ziegler, D. 1996. Selecting software for cycle counting. *APICS November*, 44–47.

Zipkin, P. 1995. Processing networks with planned inventories: Tandem queues with feedback. *European Journal of Operational Research 80*, 344–349.

Chapter 3

Forecasting Models and Techniques

Inventory and production-planning decisions nearly always involve the allocation of resources in the presence of demand uncertainty. For inventory decisions, financial resources must be deployed to procure goods in the anticipation of a future sale of those goods. Similarly, for production planning, capacity must be committed to certain products in anticipation of their future sale. Effective decision making in inventory management and production planning requires accurate descriptions of possible demands in future periods. As noted earlier in Chapter 1, variability can be a key driver of inventory expense. A study at a large high-tech firm revealed that 40% of the inventories in their system were cycle and pipeline stocks, whereas 60% were due to variability. Of the inventories driven by variability, 2% were held due to supplier performance variability, 2% were held due to manufacturing variability, and 96% were held due to demand uncertainty. A similar study at a large chemical producer revealed that 45% of finished goods inventory were held due to variability, again with the majority of that inventory held due to demand variability. These two examples highlight that while the relative impact of demand uncertainty on the inventory will vary across industrial settings, demand uncertainty will almost always be a significant factor for inventory management.

Forecasts can be based on a combination of an extrapolation of what has been observed in the past (what we call statistical forecasting) and informed judgments about future events. Informed judgments can include the knowledge of firm orders from external customers, preplanned shipments between locations within a single organization, or preplanned usage of service parts in preventive maintenance. They also include marketing judgments such as the effects of promotions, competitor reactions, general economic conditions, and so on. For items or groups of related items with accurately maintained historical demand data, we can apply techniques known as *time-series* forecasting models. With a time-series forecasting model, we use the demand history for an item (or group of items) to construct a forecast for the future demand of that item (or group). Such time-series methods are commonly used in practice and, when the models are appropriately selected and calibrated, they can be quite effective in generating high-quality forecasts. The majority of this chapter is dedicated to such time-series methods. In particular, we focus on generating forecasts for a single item using time-series methods.

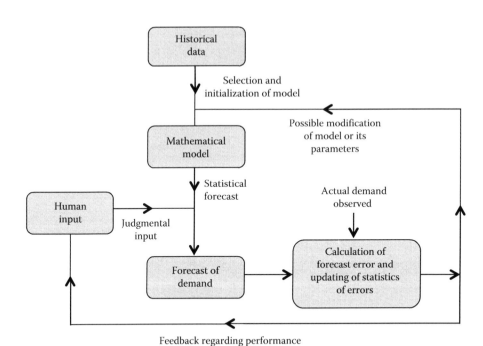

Figure 3.1 A suggested forecasting framework.

The overall framework of a suggested forecasting system is shown in Figure 3.1. The human judgment we have mentioned is a crucial ingredient. Note also that, as the actual demand in a period is observed, it is compared with the earlier forecast so that we can measure the associated error in the forecast. It is important to monitor these errors for at least three reasons. First, the quantity of safety (or buffer) stock needed to provide adequate customer service will depend (as we will see in Chapter 6) on the sizes of the forecast errors. Second, the statistical forecast is based on an assumed underlying mathematical model with specific values for its parameters. The sizes and directions (plus or minus) of the forecast errors should suggest possible changes in the values of the parameters of the model or even in the form of the model itself (e.g., introducing a seasonal component, when it was not already present). Finally, the errors can provide a monitor and feedback on the performance of the subjective input component of the forecasts.

Forecasting is of course a broad topic, and forecasts are used in many contexts from weather predictions, predictions of asset prices, financial or budgeting forecasts, etc. Given the scope of this book, we focus on forecasts of product demand used to support production and inventory resource allocation decisions although we note that many of the forecasting techniques discussed can be applied to other settings. It is often said that the one thing we know about forecasts is that they will be in error. This tends to be true if one views the forecast as a single number— often called a point estimate—but to support an inventory or production decision, we need more than just a single number. We need a richer description of the set of possible demand scenarios. Specifically, we define a forecast as a *description* of future demand scenarios used to *support a resource allocation decision*. Throughout this chapter, we first focus on measuring the performance of the point-estimate forecast. Later, we will discuss how we can use measures of this point-estimate forecasting process to help complete our description.

No matter how forecasts are generated, we are interested in evaluating the quality of the forecasts for two reasons. First, the historical performance of a forecasting process helps us create the description of future demand to support resource allocation decisions. Second, by tracking the performance of a forecasting process over time, we may be able to identify ways to improve that process. This feedback loop is noted in Figure 3.1.

We present some definitions and notation for use throughout this and the following chapters:

- Demand observed in period t is represented by x_t.
- $\hat{x}$ represents a point forecast for a future period. This value will be indexed by two relevant time periods: the period in which the forecast is made and the period the forecast is for. A forecast made at the end of time period t for a future period $t + \tau$ will then be written as $\hat{x}_{t,t+\tau}$. One would typically refer to this as the "lag-τ" forecast since it is a forecast for a time period τ periods in the future. Quite often, we will focus on the "lag-1" forecast—the forecast for one period ahead: $\hat{x}_{t-1,t}$, or $\hat{x}_{t,t+1}$.
- Let $e_{t-\tau,t} = x_t - \hat{x}_{t-\tau,t}$ represent the forecast error associated with the forecast made in period $t - \tau$ for period t. Defined in this way, a negative value indicates that the point forecast was *higher* than the observed demand. As mentioned in the previous point, we will often focus on the "lag-1" forecast, and for notational convenience, we may write the lag-1 error $e_{t-1,t}$ as simply e_t.

This chapter is organized as follows. Section 3.1 discusses the plausible components of mathematical models of a demand time series. Then, in Section 3.2, we present the three steps involved in the use of a mathematical model—namely (1) the selection of a general form of the model, (2) the choice of values for the parameters within the model, and, finally, (3) the use of the model for forecasting purposes. Although the emphasis in this chapter, with regard to statistical forecasting, is on individual-item, short-range forecasting (needed for inventory management and production scheduling), we do briefly present material appropriate to medium-range aggregate forecasting (useful for production planning) in Section 3.3. This is followed, in Section 3.4, by a detailed discussion of several different procedures for short-range forecasting, with particular emphasis on what is called as exponential smoothing. Section 3.5 is concerned with various measures of forecast errors, building on our basic forecast error definitions given above. These measures are needed for establishing safety stocks and also for monitoring the forecasting performance. Section 3.6 addresses how to deal with unusual kinds of demand. Then, in Section 3.7, we have considerable discussion concerning how to encourage and use the human judgment input to forecasting. Issues concerning special classes of items are raised in Section 3.8. Finally, in Section 3.9, we return to tactical and strategic issues, specifically to the choice among different forecasting procedures.

3.1 The Components of Time-Series Analysis

Any time series can be thought of as being composed of five components: level (a), trend (b), seasonal variations (F), cyclical movements (C), and irregular random fluctuations (ϵ). *Level* captures the scale of a time series. See Figure 3.2. If only a level was present, the series would be constant with time. *Trend* identifies the rate of growth or decline of a series over time. See Figure 3.3 that illustrates a positive linear trend. *Seasonal variations* can be of two kinds: (1) those resulting from natural forces, and (2) those arising from human decisions or customs. For example, in the northern United States and in Canada, the intensity of many types of economic activity depends on

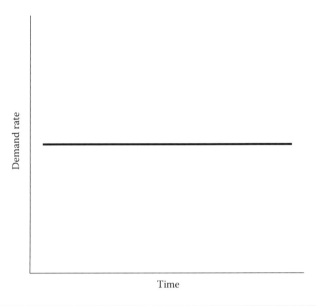

Figure 3.2   Level demand.

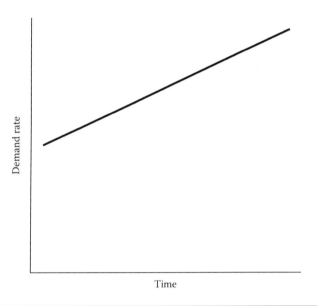

Figure 3.3   Linear trend.

prevailing weather conditions. On the other hand, department store sales increase before the start of the school year, the timing of which is a human decision. Figure 3.4 illustrates an annual seasonal pattern. *Cyclical variations* or alternations between expansion and contraction of economic activity are the result of business cycles. *Irregular fluctuations* in time-series analysis are the residue that remain after the effects of the other four components are identified and removed from the time series. Such fluctuations are the result of unpredictable events such as unusual weather conditions

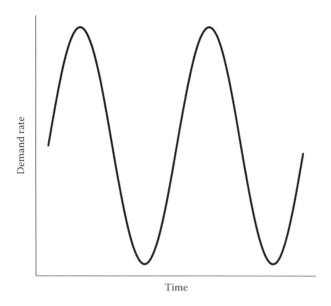

Figure 3.4   Illustrative seasonal pattern.

and unexpected labor strife. They represent our inability to effectively model what else may be affecting the time series we are studying.

Using these concepts, we can formulate a multiplicative model of a time series:

$$\text{Demand} = (\text{Trend})(\text{Seasonal})(\text{Cyclic})(\text{Irregular}) \qquad (3.1)$$

$$\text{Demand} = (b)(F)(C)(\epsilon)$$

or an *additive* model:

$$\text{Demand in period } t = (\text{Level}) + (\text{Trend}) + (\text{Seasonal}) + (\text{Cyclic}) + (\text{Irregular}) \qquad (3.2)$$

$$x_t = a + bt + F_t + C_t + \epsilon_t$$

or a mixed model (partly additive, partly multiplicative). There are statistical procedures for isolating each component. We comment further on this topic later in the chapter, but our discussion will be restricted to models that, in general, do not incorporate cyclical effects because we concentrate on short- to medium-term forecasts.

3.2 The Three Steps Involved in Statistically Forecasting a Time Series

It is conceptually useful to think in terms of three distinct steps involved in statistically forecasting a time series of demands for an individual SKU (or the aggregated demand for a group of related items). In this section, we provide an overview of these steps, and then in the later sections we discuss each in greater detail. The steps are

Step 1 Select an appropriate underlying model of the demand pattern through time.
Step 2 Select the values for the parameters inherent in the model.
Step 3 Use the model (selected in Step 1) and the parameter values (chosen in Step 2) to forecast the future demands.

The choice of the general type of underlying model to use for a particular item (or aggregated group of items) depends very much on cost considerations. The more sophisticated and costly models are really only appropriate for (1) medium- or long-term forecasting of an aggregated time series, and (2) perhaps for short-term forecasts of A items. Trend and particularly seasonal components should be introduced with care. It is often the case with B and C items that there is insufficient reliable historical data to adequately estimate the seasonal components. An analysis of historical data (often woefully unavailable in companies embarking on the introduction of formalized forecasting procedures) should suggest the general types of models (e.g., seasonal or not, additive or multiplicative, etc.) that would appear to be appropriate for a given time series. In addition, managerial knowledge of a particular market is invaluable in suggesting the presence of trends or seasonal patterns.

Once a general form of the model is tentatively selected, it is necessary to estimate the values of any parameters of that model. Estimation can be based on some form of statistical fit to historical data—for example, a least-squares criterion will be discussed in Section 3.3.1. Alternatively, particularly when there is limited or no historical data available, the estimates of parameters are made on essentially a subjective basis—for example, saying that a new item in a family of comparable items will have a seasonal pattern similar to that of well-established members of the family.

In the event that historical data exist, the time series generated by the tentatively selected model (and its estimated parameter values) can be compared with the historical pattern. If the fit is judged to be inadequate, then, it is necessary to iterate back through Steps 1 and 2.

Once we are satisfied with the assumed underlying model, then, in Step 3, it is used, with the estimated values of its parameters, to forecast future demands. This procedure, of course, implicitly assumes that the model and its parameter values will not change appreciably over the period being forecasted. If such changes can be predicted, then, appropriate subjective adjustments in parameter values and in the nature of the model itself are called for.

As we described earlier (see Figure 3.1), as forecasts are made and errors are observed, managers can adjust the estimates of the parameter values and judge the suitability of the model.

3.3 Some Aggregate Medium-Range Forecasting Methods

In this section, we provide a *brief* overview of some of the common models and procedures useful for medium- and longer-term forecasting of an aggregate time series. Such forecasts are an important input to the aggregate production-planning procedures to be discussed in Chapter 14. Our intention is simply to discuss the general nature of the available methodology, and to provide references where further details can be found. Moreover, there are several longer-range forecasting techniques that we will not even mention. An excellent starting point for more information on either medium- or long-term forecasting is the article by Makridakis (2008).

In Section 3.1, we showed both a multiplicative and an additive model (Equations 3.1 and 3.2), each giving demand as a function of level, trend, seasonal, and cyclical components. In actual fact, it may be necessary to first apply a transformation to the demand data in order to achieve a reasonable fit with one of these models. The added complexity of such a transformation is warranted in medium-range forecasting where we are dealing with relatively few (aggregated) time series and forecasts. A class of transformations that has proved to be useful in practice (see Granger and Newbold 2014) is

$$T(x_t) = \ln x_t \tag{3.3}$$

So, $T(x_t)$, as opposed to x_t, would be modeled and forecasted. Then, for any given forecast of $T(x_t)$ the corresponding x_t would be obtained by inverting the transformation, that is, solving Equation 3.3 for x, in terms of T, or,

$$x_t = e^T \tag{3.4}$$

As an example, if $x_t = 20$, $T(x_t)$ would be $\ln(20) = 2.996$. We would use 2.996 in forecasting activities. Then, if a forecast of $T(x_{t+1}) = 3.5$, the actual forecasted demand, x_{t+1}, would be $e^{3.5} = 33.115$.

3.3.1 Regression Procedures

For illustrative purposes, we show only the simple, but common, case of an underlying model that assumes that demand is a linear function of time:

$$x_t = a + bt \tag{3.5}$$

A least-squares criterion is used to estimate values, denoted by $\hat{a}$ and $\hat{b}$, of the parameters a and b. The resulting modeled value of x_t, denoted $\hat{x}_t$, is then given by

$$\hat{x}_t = \hat{a} + \hat{b}t + \epsilon_t \tag{3.6}$$

If there are n actual historical observations x_t (for $t = 1, 2, \ldots, n$), then, the least-squares criterion involves the selection of $\hat{a}$ and $\hat{b}$ to minimize the sum of the squares (S) of the n differences between x_t and $\hat{x}_t$. That is,

$$S = \sum_{t=1}^{n} \left(x_t - \hat{a} - \hat{b}t \right)^2 \tag{3.7}$$

Through the use of calculus we can show that

$$\hat{b} = \frac{\sum_{t=1}^{n} tx_t - \left(\frac{n+1}{2}\right) \sum_{t=1}^{n} x_t}{n(n^2 - 1)/12} \tag{3.8}$$

$$\hat{a} = \sum_{t=1}^{n} x_t/n - \hat{b}(n+1)/2 \tag{3.9}$$

Fitting such a model to the aggregate demand data for the item shown in Table 3.1, we obtain

$$\hat{x}_t = 2961 + 48.68t \tag{3.10}$$

Once the parameters a and b are estimated, the model of Equation 3.6 can be used to forecast (extrapolate) the demand in a future period. For example, the use of Equation 3.10 gives an estimate of demand for the item in October 2015 (period 22) as being 4,032 packages. There are also

Table 3.1 Demand Data for a Sample Item

Year	Month	Period, t	x_t (Packages)
2013	January	1	3,025
	February	2	3,047
	March	3	3,079
	April	4	3,136
	May	5	3,268
	June	6	3,242
	July	7	3,285
	August	8	3,334
	September	9	3,407
	October	10	3,410
	November	11	3,499
	December	12	3,598
2014	January	13	3,596
	February	14	3,721
	March	15	3,745
	April	16	3,650
	May	17	3,746
	June	18	3,775
	July	19	3,906
	August	20	3,973

statistical procedures for estimating the adequacy of the straight-line fit and for developing a confidence interval around the forecast (extrapolated) value for any future period. Details are available in several sources such as Berger (2013).

Our discussion to this point has assumed that the independent variable used in the regression is the time period t. More generally, multiple regression includes two or more exogenous variables. Moreover, in autoregression, the demand in period t is postulated as a function of the demand itself in earlier periods. As an illustration of the latter, we might have

$$x_t = a_1 + a_2 x_{t-1} + \epsilon_t$$

which postulates that the demand in period t is a linear function of the demand in period $t - 1$ plus an error term. A generalization to a system of autoregressive equations in several variables to be forecast, as well as some exogenous variables, leads one into the field of econometric forecasting.

This includes the use of the so-called leading indicators. A variable y is said to be a leading indicator of another variable x if changes in x are anticipated by changes in y. Multiple regression is sometimes referred to as forecasting by association because the variable of interest is forecasted indirectly by its association with other variables. See, for example, Maddala and Lahiri (1992).

3.3.1.1 The Box–Jenkins Approach

Box and Jenkins (see Box et al. 2011) have had a profound effect on the field of statistical forecasting. They suggest a broad class of underlying statistical models of demand patterns, as well as a procedure for selecting an appropriate member of the class based on the historical data available. Conceptually the models are more complex than either the simple regression models (outlined in Section 3.3.1) or the exponential smoothing procedures (to be covered in Section 3.4). However, this complexity (and its associated additional costs) affords an opportunity for more accurate forecasting, which is certainly of interest at the medium-term, aggregate level. In this section, we briefly introduce some aspects of the Box–Jenkins approach.

The so-called autoregressive-moving average (or ARMA) class of models represents the demand in the current period, x_t, by a weighted sum of past demands and unpredictable random components. In mathematical terms:

$$x_t = \phi_1 x_{t-1} + \phi_2 x_{t-2} + \cdots + \phi_p x_{t-p} + \epsilon_t + \theta_1 \epsilon_{t-1} + \theta_2 \epsilon_{t-2} + \cdots + \theta_q \epsilon_{t-q} \quad (3.11)$$

where the ϵ's are the so-called white noise, namely, independent, normally distributed variables with mean 0 and a constant variance, and the ϕ's, θ's, p, and q are constants.

Typically, Equation 3.11 can be simplified. Anderson (1976, p. 45) reports that many stationary time series are adequately represented by ARMA models with $p+q = 2$, that is, at most three terms on the right side of Equation 3.11. As an illustration, the following shows the case of $p = 1$, $q = 1$:

$$x_t = \phi_1 x_{t-1} + \epsilon_t + \theta_1 \epsilon_{t-1} \quad (3.12)$$

Equation 3.11 in fact implies that the x_t series is stationary with time (i.e., its expected value and standard deviation do not change with time). At least two circumstances can invalidate this assumption. First, there may be a trend (linear or higher order) in the underlying process. Second, there may be seasonality. The Box–Jenkins approach can handle both of these circumstances by introducing additional terms into Equation 3.11.

In Section 3.2, we discussed three steps involved in statistically forecasting a time series: selection of a model, estimation of the parameters of the model, and use of the model for forecasting purposes. The Box–Jenkins approach involves considerable elaboration on the first two of these steps, because there is such a rich class of models from which to choose. See Box et al. (2011) for further details.

3.4 Individual-Item, Short-Term Forecasting: Models and Procedures

In this major section of the chapter, we concentrate on procedures for estimating and using the model parameters. These procedures have been found to be reasonably accurate in terms of short-term forecasts for individual SKUs. In addition, they are relatively simple and inexpensive to use, in contrast with the methods of the previous section, and thus are practical to use for the high volume of individual-item, short-term forecasts.

We restrict attention to procedures that are appropriate for at least one of the following assumed underlying models of x_t, the demand in period t:

$$x_t = a + \epsilon_t \text{ (Level model)} \tag{3.13}$$

$$x_t = a + bt + \epsilon_t \text{ (Trend model)} \tag{3.14}$$

$$x_t = (a + bt)F_t + \epsilon_t \text{ (Multiplicative trend-seasonal model)} \tag{3.15}$$

where, as earlier,

a = a level

b = a (linear) trend

F_t = a seasonal coefficient (index) appropriate for period t and the ϵ_t's are independent random variables with mean 0 and constant variance σ^2

Of course, we never know the exact values of the parameters in Equations 3.13 through 3.15. To make matters worse, the true values are likely to change with the passage of time. For example, the arrival of a new competitor in the market is likely to reduce the level or trend by reducing the current market share. Consequently, the procedures we discuss will be concerned with estimating the current values of the parameters of the specific model under consideration. There are two distinct aspects to estimation. First, there is the question of obtaining initial estimates prior to using the model for forecasting purposes. Second, there is the updating of the most recent estimates as we observe the demand (and hence the forecast error) in the latest period. For each procedure, we discuss these two different dimensions of estimation.

3.4.1 The Simple Moving Average

3.4.1.1 Underlying Demand Model

The simple moving average method is appropriate when demand is modeled as in Equation 3.13:

$$x_t = a + \epsilon_t \tag{3.16}$$

that is, a level with random noise. As mentioned above, the parameter a is not really known and is subject to random changes from time to time.

3.4.1.2 Updating Procedure

The simple N-period moving average, as of the end of period t, is given by

$$\bar{x}_{t,N} = (x_t + x_{t-1} + x_{t-2} + \cdots + x_{t-N+1})/N \tag{3.17}$$

where the x's are actual, observed demands in the corresponding periods. The estimate of a as of the end of period t is then

$$\hat{a}_t = \bar{x}_{t,N} \tag{3.18}$$

We can show that this estimate of a results from minimizing the sum of squares of errors over the preceding N periods (see Problem 3.3). A slightly simpler version (for updating purposes) of the

N-period moving average is easily developed from Equation 3.17 as

$$\bar{x}_{t,N} = \bar{x}_{t-1,N} + (x_t - x_{t-N})/N \qquad (3.19)$$

Note from Equation 3.17 that the moving average is simply the mean of the N most recent observations. Provided that the model of Equation 3.16 holds, we know from basic probability that $\hat{x}_{t,N}$ or $\hat{a}_t$ is distributed with mean a and standard deviation $\sigma_\epsilon/\sqrt{N}$. The larger the N is, the more precise will be the estimate, provided Equation 3.16 continues to apply over the entire set of periods. However, if there is a change in the parameter a, a small value of N is preferable because it will give more weight to recent data, and thus will pick up the change more quickly. Typical N values that are used range from 3 to as high as 12.

3.4.1.3 Initialization

In terms of initialization, an N-period moving average should not really be used until there are N periods of history available. In actual fact, if early forecasts are essential, one can use a smaller number of periods in the initial stages.

3.4.1.4 Forecasting

Because of the (level) nature of the model of Equation 3.16, it is appropriate to use our estimate $\hat{a}_t$ as a forecast at the end of period t for any future period:

$$\hat{x}_{t,t+\tau} = \hat{a}_t \qquad (3.20)$$

where $\hat{x}_{t,t+\tau}$ is the forecast, made at the end of period t, of demand in period $t + \tau$ (for $\tau = 1, 2, 3, \ldots$). This procedure can be developed on a spreadsheet quite easily, as is illustrated next. See also Problem 3.14.

3.4.1.5 Illustration

For a product, PSF-008, columns 1–3 of Table 3.2 show the demand data during 2013. The 5-month moving averages, beginning as of the end of May 2013, are shown in column 4 and the associated 1-month ahead forecasts in column 6. We comment further on the results of Table 3.2 in the next section when we discuss the topic of simple exponential smoothing.

3.4.1.6 General Comments

The moving average approach requires the storage of N periods of historical data that must be stored for each SKU. In addition, the weight given to each of these N data points is equal while all older data points are disregarded. Intuitively, we may want to weight more recent observations more heavily, but it seems inappropriate to completely discount the earlier data. It is possible to modify the moving average approach with different weights. The exponential smoothing approach in the next section does precisely this in a compact way that is easy to implement.

Table 3.2 Moving Average and Simple Exponential Smoothing Forecasts for Product PSF-008, Using $N = 5$ and $\alpha = 1/3$ (All Data in 100 Square Meters)

Year (1)	Month t (2)	Demand x_t (3)	5-Month Moving Average $\bar{x}_{t,5}$ (4)	Exponentially Smoothed Average (5)	Forecasts	
					$\hat{x}_{t,t+1} = \bar{x}_{t,5}$ (6)	$\hat{x}_{t,t+1} = \hat{a}_t$ (7)
2013	1	52				
	2	48				
	3	36		(50.0)		
	4	49		(49.7)		
	5	65	50.0	(54.8)	50.0	54.8
	6	54	50.4	54.5	50.4	54.5
	7	60	52.8	56.3	52.8	56.3
	8	48	55.2	53.6	55.2	53.6
	9	51	55.6	52.7	55.6	52.7
	10	62	55.0	55.8	55.0	55.8
	11	66	57.4	59.2	57.4	59.2
	12	62	57.8	60.1	57.8	60.1
2014	13					

3.4.2 Simple Exponential Smoothing

Simple exponential smoothing is probably the most widely used statistical method for short-term forecasting. As we shall see, it is intuitively appealing and very simple to employ. An excellent reference on this topic is Gardner (2006).

3.4.2.1 Underlying Demand Model

As in the case of moving averages, the basic underlying demand pattern assumed is that of Equation 3.13, which, for convenience, we repeat here:

$$x_t = a + \epsilon_t \tag{3.21}$$

However, later, we will mention that the procedure is actually appropriate for a somewhat wider range of demand models.

3.4.2.2 Updating Procedure

We wish to estimate the parameter a in Equation 3.21. Recall that in Section 3.3.1, we used a least-squares regression on historical data to estimate parameters of the underlying models. However, we pointed out that regular least squares gives equal weight to all historical data considered. Simple exponential smoothing can be derived by selecting $\hat{a}_t$, the estimate of a at the end of period t, in order to minimize the following sum of discounted squares of residuals:

$$S' = \sum_{j=0}^{\infty} d^{j+1}(x_{t-j} - \hat{a}_t)^2 \tag{3.22}$$

where
$\quad d = $ a discount factor $(0 < d < 1)$
$\quad x_{t-j} = $ the actual demand in period $t - j$

Note that, because $d < 1$, the weighting given to historical data decreases geometrically as we go back in time. Exponential smoothing would really be more appropriately named geometric smoothing (an exponential curve is a continuous approximation to geometrically decaying points). The minimization of Equation 3.22 is carried out in the Appendix to this chapter. The resulting estimate of a satisfies the following updating formula:

$$\hat{a}_t = \alpha x_t + (1 - \alpha)\hat{a}_{t-1} \tag{3.23}$$

where $\alpha = 1 - d$ is known as the smoothing constant.

In addition, it can be shown (see, e.g., Johnson and Montgomery 1974) that $\hat{a}_t$ is an unbiased estimate of a; that is, $E\hat{a}_t = a$.

Equation 3.23 can be manipulated as follows:

$$\hat{a}_t = \hat{a}_{t-1} + \alpha(x_t - \hat{a}_{t-1})$$

or

$$\hat{a}_t = \hat{a}_{t-1} + \alpha e_t \tag{3.24}$$

where
$\quad e_t = $ the error in period t between the actual demand x_t and the forecast $\hat{a}_{t-1}$ made at the end of the previous period $t - 1$

There is an intuitive appeal to Equation 3.24. We adjust the forecast for period t in the direction of the previous error, with α throttling the strength of this adjustment. If the forecast $\hat{a}_{t-1}$ from the previous period $(t - 1)$ was too low (the error is positive), it makes sense that we would want to increase the next forecast, $\hat{a}_t$, which we do by adding αe_t to the previous forecast.

Note again that these equations can be developed quite easily on a spreadsheet. (See Problem 3.15.) Equation 3.24 states that the new estimate of the level is the old estimate plus a fraction of the most recent error. Practically speaking, we don't ever really know whether any particular error results from a random fluctuation or whether a significant shift in a has occurred. As a result, the simple exponential smoothing model hedges by assuming that only a fraction of the forecast error should be used to revise the estimate of a. Note from Equation 3.23 that as α tends toward 0, our estimate is unchanged by the latest piece of information, while as α moves toward 1, more and more weight is placed on the demand observed in the last period. In Section 3.4.5, we comment on the selection of an appropriate value of the smoothing constant.

The updating procedure of Equation 3.23 can also be shown to minimize the mean-squared error (MSE) of the one-period-ahead forecast for the following models that are both somewhat more general than that of Equation 3.21:

Case 1 (Details can be found in Harrison 1967)

$$x_t = a_t + \epsilon_t$$

with

$$a_t = a_{t-1} + \delta_t$$

where the δ_t's are random changes with mean 0 and independent of the ϵ's. That is, the level is subject to random changes through time (which is what we have implicitly assumed earlier).

Case 2 The x_t's are generated by a modification of the autoregressive process discussed in Section 3.3.1. Specifically,

$$x_t = x_{t-1} + \epsilon_t = \theta_1 \epsilon_{t-1}$$

See Goodman (1974) for details.

The geometric weighting of historical data in exponential smoothing can also be seen by substituting a corresponding relation for $\hat{a}_{t-1}$ in Equation 3.23, and then repeating for $\hat{a}_{t-2}$, etc. This gives

$$
\begin{aligned}
\hat{a}_t &= \alpha x_t + (1-\alpha)[\alpha x_{t-1} + (1-\alpha)\hat{a}_{t-2}] \\
&= \alpha x_t + \alpha(1-\alpha)x_{t-1} + (1-\alpha)^2[\alpha x_{t-2} + (1-\alpha)\hat{a}_{t-3}] \\
&= \alpha x_t + \alpha(1-\alpha)x_{t-1} + \alpha(1-\alpha)^2 x_{t-2} + \cdots \\
\hat{a}_t &= \sum_{j=0}^{\infty} \alpha(1-\alpha)^j x_{t-j}
\end{aligned}
\tag{3.25}
$$

We see the contrast with moving averages where the last N observations were given equal weight. In addition, the recursion relationship of Equation 3.23 shows that only one value of a and the most recent x_t need to be retained. From the weighting of the terms in Equation 3.25, we can ascertain (see Problem 3.4) that the average age of the data encompassed in $\hat{a}_t$ is $1/\alpha$ periods. In an N-month moving average, the average age of the data used is $(N+1)/2$. Equalizing the average ages gives the following relationship between α and N:

$$\alpha = 2/(N+1) \tag{3.26}$$

3.4.2.3 Initialization

Where significant historical data exist, one simply uses the average demand in the first several periods as the initial estimate of a. The numerical illustration will demonstrate this point. The case where inadequate history exists will be addressed in Section 3.8.1.

3.4.2.4 Forecasting

As mentioned earlier, $\hat{a}_t$ is an unbiased estimate, as of the end of period t, of the parameter a, where the latter, according to Equation 3.21, is the expected demand in any future period. Thus, our forecast, made at the end of period t, for any future period $t + \tau$ is

$$\hat{x}_{t,t+\tau} = \hat{a}_t \qquad (3.27)$$

For deciding when to place a replenishment order and how much to order, we will be interested in the total forecast over several unit time periods (e.g., the replenishment lead time may be 4 months, while the forecast update period is 1 month). If we let $\hat{x}_{t,t,u}$ be the forecast, made at the end of period t, of total demand from time t (end of period t) to time u, then from Equation 3.27 we have

$$\hat{x}_{t,t,u} = (u - t)\hat{a}_t \qquad (3.28)$$

3.4.2.5 Illustration

We use the same example of product PSF-008 as we used earlier to illustrate a 5-period moving average. In addition, for comparison, we use Equation 3.26 to select the smoothing constant that gives the same average age of the data; that is,

$$\alpha = 2/(5 + 1) = 1/3$$

(In actual fact, as will be discussed later, one would normally not use such a high value of α.)

In Table 3.2, the average of the first 5 months' sales (50.0) is taken as the initial value for the exponentially smoothed average and plotted opposite $t = 3$, the midpoint of the 5-month range of the average. Then, the exponentially smoothed average in column 5 for month 4 is calculated by using Equation 3.23:

$$\hat{a}_t = \frac{1}{3}x_4 + \frac{2}{3}\hat{a}_3$$

$$= \frac{1}{3}49 + \frac{2}{3}50 = 49.7$$

The same procedure was repeated for all subsequent months. For months 4 and 5, no exponentially smoothed forecasts were issued to allow the effect of setting initial conditions at $\hat{a}_3 = 50.0$ to be smoothed out. This is commonly referred to as the "run-in period." Rather than using $\hat{a}_3 = 50.0$, we could instead set $\hat{a}_0 = 50.0$, and thus make the run-in time equal to 5 periods.

Using Equation 3.27, "exponential smoothing" forecasts were calculated for each period from period 6 onward. These are shown in column 7 of Table 3.2. Note from columns 6 and 7 that the exponential smoothing forecast outperforms the 5-month moving average forecast for 5 of the 7 months of 2013. This appears to be a result of the fact that by using $\alpha = 1/3$, the simple exponential smoothing forecast is made more responsive to recent sales data variation than the 5-month moving average. Since the time series x_t appears to be undergoing changes, this degree of responsiveness appears to be warranted for the particular example shown.

3.4.2.6 General Comments

In summary, simple exponential smoothing is an easy to understand, effective procedure to use when the underlying demand model is composed of level and random components. Even when the underlying demand process is more complicated, exponential smoothing can be used as part of an updating procedure. For example, Steece and Wood (1979) use a Box–Jenkins model for the aggregate demand time series of an entire class of items, and then use simple exponential smoothing to update the estimate of a particular item's demand as a fraction of the aggregate.

3.4.3 Exponential Smoothing for a Trend Model

The simple smoothing procedure presented in the previous section is based on a model without a trend and therefore is inappropriate when the underlying demand pattern involves a significant trend. A somewhat more complicated smoothing procedure is needed under such circumstances.

3.4.3.1 Underlying Demand Model

The basic underlying model of demand is the pattern of Equation 3.14 that we repeat here.

$$x_t = a + bt + \epsilon_t \tag{3.29}$$

Again, we will make a reference later to somewhat more general models for which the suggested smoothing procedure is also appropriate.

3.4.3.2 Updating Procedure

Holt (2004) suggests a procedure that is a natural extension of simple exponential smoothing

$$\hat{a}_t = \alpha_{HW} x_t + (1 - \alpha_{HW})(\hat{a}_{t-1} + \hat{b}_{t-1}) \tag{3.30}$$

$$\hat{b}_t = \beta_{HW}(\hat{a}_t - \hat{a}_{t-1}) + (1 - \beta_{HW})\hat{b}_{t-1} \tag{3.31}$$

where α_{HW} and β_{HW} are smoothing constants and, for the demand model of Equation 3.29, the difference $\hat{a}_t - \hat{a}_{t-1}$ is an estimate of the actual trend in period t. (We use the subscripts HW to represent Holt-Winters since these smoothing constants will be part of a procedure of Winters, to be discussed in Section 3.4.4.)

Harrison (1967), among others, shows that the Holt procedure minimizes the expected one-period-ahead, mean-square forecast error for the trend model of Equation 3.29—in fact, for a more general model where independent random changes in both a and b are possible for each period, in addition to the random noise (the ϵ's).

We use Holt's updating as part of a popular procedure, to be described in Section 3.4.4, for dealing with an underlying model involving both trend and seasonal components. However, in this section, where there is no seasonality, we recommend instead a single-parameter procedure suggested by Brown (1963), which turns out to be a special case of Holt's method. Besides involving only a single smoothing parameter, Brown's procedure again has the intuitively appealing property of being derived from minimizing the sum of geometrically weighted forecast errors (similar to the basis for the simple exponential smoothing procedure of Equation 3.23). Brown's updating equations are (see Problem 3.6)

$$\hat{a}_t = [1 - (1 - \alpha)^2]x_t + (1 - \alpha)^2(\hat{a}_{t-1} + \hat{b}_{t-1}) \tag{3.32}$$

$$\hat{b}_t = \left[\frac{\alpha^2}{1 - (1 - \alpha)^2}\right](\hat{a}_t - \hat{a}_{t-1}) + \left[1 - \frac{\alpha^2}{1 - (1 - \alpha)^2}\right]\hat{b}_{t-1} \tag{3.33}$$

where α is the single smoothing constant ($\alpha = 1 - d$ again, where d is the discount factor used in the geometrically weighted sum of squares).

Note that Equations 3.32 and 3.33 are the special case of Holt's updating equations (Equations 3.30 and 3.31), where

$$\alpha_{HW} = [1 - (1 - \alpha)^2] \tag{3.34}$$

$$\beta_{HW} = \frac{\alpha^2}{1 - (1 - \alpha)^2} \tag{3.35}$$

In Section 3.4.5, we address the issue of the selection of the α value.

3.4.3.3 Initialization

Typically, the initial values of the level (a) and the trend (b) are ascertained by doing a regular (unweighted) least-squares regression on the historical data available. Suppose that there are n periods of data available. Smoothing will begin with the next period after this preliminary data. Thus, we would like to have the level a computed relative to an origin at the end of the last period of the fitted data. That is, we should number the historical periods $0, -1, -2, -3, \ldots, -(n - 1)$ going back in time. In other words, we wish to select a_0 and b_0 to minimize

$$S = \sum_{t=-n+1}^{0} [x_t - (\hat{a}_0 + \hat{b}_0 t)]^2$$

Through the use of calculus, we obtain

$$\hat{a}_0 = \frac{6}{n(n + 1)} \sum_t tx_t + \frac{2(2n - 1)}{n(n + 1)} \sum_t x_t \tag{3.36}$$

and

$$\hat{b}_0 = \frac{12}{n(n^2 - 1)} \sum_t tx_t + \frac{6}{n(n + 1)} \sum_t x_t \tag{3.37}$$

where all the summations range over the integers $-(n - 1), -(n - 2), \ldots, -2, -1, 0$.

Where this initial regression is a relatively straightforward matter, the question arises as to why we don't simply redo a regression after each additional period of information instead of resorting to the smoothing procedure of Equations 3.32 and 3.33. There are two reasons. First, the regression gives equal weight to all historical information, whereas the smoothing reduces the weight geometrically as we go back in time. Second, and less important, smoothing is less computationally intensive, and therefore is more appropriate on a repetitive basis for short-term forecasting. As we shall see, smoothing can be done by adding a single line to a spreadsheet for each new demand data point.

The case where there is insufficient history for regression initialization will be treated in Section 3.8.1.

3.4.3.4 Forecasting

Because of the underlying model of Equation 3.29, we know that

$$\hat{x}_{t,t+\tau} = \hat{a}_t + \hat{b}\tau \tag{3.38}$$

where
$\hat{x}_{t,t+\tau}$ = the forecast, made at the end of period t, of the demand in period $t + \tau$

Using Equation 3.38, we can rewrite Equations 3.30 and 3.31 in a manner similar to Equation 3.24 (for simple exponential smoothing) to highlight the intuition behind the adjustments to the level and trend estimates $\hat{a}_t, \hat{b}_t$.

$$\hat{a}_t = (\hat{a}_{t-1} + \hat{b}_{t-1}) + \alpha_{HW}[x_t - (\hat{a}_{t-1} + \hat{b}_{t-1})]$$
$$= \hat{x}_{t-1,t} + \alpha_{HW} e_t$$

That is, the new estimate for the level $\hat{a}_t$ is the old estimate for the level *at time t* plus an adjustment based on the error, moderated by the smoothing constant. The trend-updating equation can be rewritten for a similar intuition.

$$\hat{b}_t = \hat{b}_{t-1} + \beta_{HW}[(\hat{a}_t - \hat{a}_{t-1}) - \hat{b}_{t-1}]$$

Here, the "error" used to adjust the new trend estimate is not the forecast error but the difference between the old trend estimate and the recently estimated value for the trend obtained by taking the difference between level estimates in periods $t - 1$ and t.

To make a forecast for *cumulative* demand over a horizon, say from time t (end of period t) to time u (end of period u), is

$$\hat{x}_{t,t,u} = \sum_{\tau=1}^{u-t} (\hat{a}_t + \hat{b}_t\tau) \tag{3.39}$$

When u is not exactly an integer, the last term in the summation should cover an appropriate fraction of the last period.

3.4.3.5 Illustration

Consider another product PSF-016 where it is reasonable to assume that demand is growing, approximately linearly, with time. Table 3.3 shows demand data for 15 consecutive months. Suppose that the first 6 months of data (January 2013 through June 2013) are used to estimate $\hat{a}_0$ and $\hat{b}_0$. The use of Equations 3.36 and 3.37 gives

$$\hat{a}_0 = 27.19 \quad \text{and} \quad \hat{b}_0 = 1.34$$

Suppose we do not use a run-in period and directly use these to forecast demand in July 2013 ($t = 1$), employing Equation 3.38. This gives us $\hat{x}_{0,1} = 28.53$. The resulting error in July 2013

Table 3.3   Illustration of Forecasting with a Trend Model for PSF-016 (All Data in 100 Square Meters)

Year	Month	t	x_t	$e_t = x_t - \hat{x}_{t-1,t}$	$\hat{a}_t$	$\hat{b}_t$	$\hat{x}_{t,t+1} = \hat{a}_t + \hat{b}_t$
2013	January	−5	20				
	February	−4	25				
	March	−3	21				
	April	−2	22				
	May	−1	27				
	June	0	28		27.19	1.34	28.53
	July	1	27	−1.53	27.68	1.17	28.85
	August	2	30	1.15	29.49	1.30	30.79
	September	3	34	3.21	32.57	1.66	34.23
	October	4	25	−9.23	29.10	0.63	29.73
	November	5	25	−4.73	27.10	0.10	27.21
	December	6	26	−1.21	26.54	−0.03	26.51
2014	January	7	36	9.49	31.78	1.03	32.81
	February	8	41	8.19	37.36	1.94	39.29
	March	9	39	−0.29	39.13	1.90	41.03

is $27 - 28.53$ or -1.53. The parameters a and b are updated through the use of Equations 3.32 and 3.33 and a forecast $\hat{x}_{1,2}$ is made at the end of July 2013, and so on. Continuing in this fashion, we obtain the results in the last four columns of Table 3.3. (Note that in this, and all other tables in this chapter, we have used a spreadsheet to compute the values shown. For presentation purposes, we report numbers rounded to the nearest hundredth or so. Because the spreadsheet retains many more digits than those shown, the results computed from the rounded numbers may differ from the results reported in the tables.)

3.4.3.6 A Model with Damped Trend

Sometimes, the data are so noisy, or the trend is so erratic, that a linear trend is not very accurate, especially when forecasting several periods ahead. Gardner and McKenzie (1985) introduce a *damped trend* procedure that works particularly well in these situations. The method follows Equations 3.30 and 3.31 closely, and is quite easy to use. The updating and forecasting formulas are

$$\hat{a}_t = \alpha_{HW} x_t + (1 - \alpha_{HW})(\hat{a}_{t-1} + \phi \hat{b}_{t-1}) \tag{3.40}$$

$$\hat{b}_t = \beta_{HW}(\hat{a}_t - \hat{a}_{t-1}) + (1 - \beta_{HW})\phi \hat{b}_{t-1} \tag{3.41}$$

$$\hat{x}_{t,t+\tau} = \hat{a}_t + \sum_{i=1}^{\tau} \phi^i \hat{b}_t \tag{3.42}$$

where ϕ is a dampening parameter. Note the similarity of these equations with Equations 3.30, 3.31, and 3.38. In addition to fitting a and b, we must also search for the best ϕ using historical data. If $0 < \phi < 1$, the trend is damped and as τ gets large, the forecast approaches an asymptote that is the horizontal line

$$\hat{a}_t + \hat{b}_t \phi / (1 - \phi)$$

If $\phi = 1$, the trend is linear, and therefore is identical with the model presented in this section. If $\phi > 1$, the trend is exponential, which Gardner and McKenzie (1985) note is dangerous in an automated forecasting system. The authors show ways to reduce the number of parameters, analogous to Brown's procedure presented in Equations 3.32 and 3.33. They also test the model and find that it performs better than the simple linear trend models in most cases, particularly as the forecast horizon grows. The related references include Lewandowski (1982), Gardner and McKenzie (1985, 1989), and Gardner (1988, 1990).

3.4.3.7 Numerical Illustration

Using the data from Table 3.3, we can compute the forecasts using the damped trend procedure. Using Equation 3.34, $\alpha_{HW} = 0.556$, because $\alpha = 1/3$, and using Equation 3.35, $\beta_{HW} = 0.2$. Let us assume that a simple search for the best ϕ reveals that the smallest sum of squared errors is obtained when $\phi = 0.4$. Also, to be consistent with the previous example, let $\hat{a}_0 = 27.19$ and $\hat{b}_0 = 1.34$. Then, from Equation 3.42, the forecast for July is $\hat{x}_{0,1} = 27.19 + 0.4(1.34) = 27.73$. The resulting error in July 2013 is $27 - 27.73$ or -0.73. The updated parameters a and b are found from Equations 3.40 and 3.41, and are 27.32 and 0.46, respectively. Then, a forecast $\hat{x}_{1,2}$ is made at the end of July 2013, and so on. Continuing in this fashion, we obtain the results in the last four columns of Table 3.4. The reader can verify that the sum of squared errors for this example is somewhat smaller using the damped trend procedure than the linear trend approach.

Now consider a longer-term forecast. Specifically, let us forecast the demand for October 2013 (period 4) at the end of June 2013 (period 0). Using the linear trend model, and Equation 3.38, we find the forecast to be

$$\hat{x}_{t,t+\tau} = \hat{a}_t + \hat{b}_t \tau = 27.19 + 1.34(4) = 32.56$$

The damped trend model, using Equation 3.42 gives

$$\hat{x}_{t,t+\tau} = 27.19 + \sum_{i=1}^{4} (0.4)^i (1.34) = 28.06$$

Because the actual demand for October is 25, the damped trend model is clearly more accurate for this example. Note, however, that the linear trend model is more accurate when predicting September's demand (in June). As will be discussed in Section 3.4.5, we recommend some form of search to choose an appropriate value of the dampening parameter. If it turns out to be 1, we have simply replicated the linear trend model.

3.4.4 Winters Exponential Smoothing Procedure for a Seasonal Model

In many organizations, there are a number of individual SKUs that exhibit demand patterns with significant seasonality. Examples include lawnmowers, skiing equipment, fertilizer, heating fuel,

Table 3.4   Illustration of Forecasting with a Trend Model Using Damped Trend

Year	Month	t	x_t	$e_t = x_t - \hat{x}_{t-1,t}$	$\hat{a}_t$	$\hat{b}_t$	$\hat{x}_{t,t+1} = \hat{a}_t + \hat{b}_t$
2013	January	−5	20				
	February	−4	25				
	March	−3	21				
	April	−2	22				
	May	−1	27				
	June	0	28		27.19	1.34	27.73
	July	1	27	−0.73	27.32	0.46	27.51
	August	2	30	2.49	28.89	0.46	29.08
	September	3	34	4.92	31.81	0.73	32.10
	October	4	25	−7.10	28.16	−0.50	27.96
	November	5	25	−2.96	26.31	−0.53	26.10
	December	6	26	−0.10	26.05	−0.22	25.96
2014	January	7	36	10.04	31.54	1.03	31.95
	February	8	41	9.05	36.98	1.42	37.54
	March	9	39	1.46	38.35	0.73	38.64

and ice cream. For individual items, the most common forecasting update period when seasonality is present is 1 month. However, for illustration, we use a quarterly (trimonthly) update period. Incidentally, the procedure to be discussed is potentially useful in aggregate, medium-range forecasting, in which case quarterly periods are more meaningful.

3.4.4.1 Underlying Model

The underlying model assumed is that of Equation 3.15

$$x_t = (a + bt)F_t + \epsilon_t \tag{3.43}$$

where
  a = the level
  b = the linear trend
  F_t = a seasonal index (coefficient) appropriate for period t
  ϵ_t = independent random variables with mean 0 and constant variance σ^2

The season is assumed to be of length P periods and the seasonal indices are normalized so that, at any point, the sum of the indices over a full season is exactly equal to P. With an annual season, $P = 12$ for a 1-month forecast interval, and $P = 4$ for a quarterly interval. Seasonal models are also applicable for a weekly "season" (e.g., demand for medical services at the emergency facilities of a hospital), or even a daily "season" (e.g., customer demand for electrical energy as a function of the time of the day). The special case when all the F's equal unity represents a nonseasonal trend model.

3.4.4.2 Updating Procedure

To use the model of Equation 3.43 repetitively for forecasting purposes, we must have estimates of a, b, and the F's. Winters (1960) first suggested the following procedure, which is not optimal in the sense, for example, of minimizing mean-square forecast errors. However, it is intuitively appealing and is a natural extension of the Holt procedure for a trend model (see Equations 3.30 and 3.31). Specifically, the parameters are updated according to the following three equations:

$$\hat{a}_t = \alpha_{HW}(x_t/\hat{F}_{t-P}) + (1 - \alpha_{HW})(\hat{a}_{t-1} + \hat{b}_{t-1}) \tag{3.44}$$

$$\hat{b}_t = \beta_{HW}(\hat{a}_t - \hat{a}_{t-1}) + (1 - \beta_{HW})\hat{b}_{t-1} \tag{3.45}$$

$$\hat{F}_t = \gamma_{HW}(x_t/\hat{a}_t) + (1 - \gamma_{HW})\hat{F}_{t-P} \tag{3.46}$$

where α_{HW}, β_{HW}, and γ_{HW} are the three smoothing constants that all lie between 0 and 1. The selection of their values will be covered in Section 3.4.5.

In Equation 3.44, the term $x_t/\hat{F}_{t-P}$ reflects an estimate of the *deseasonalized* actual demand in period t in that $\hat{F}_{t-P}$ was the estimate of the seasonal index for the most recent (P periods earlier) equivalent period in the seasonal cycle. The rest of the terms in Equations 3.44 and 3.45 are equivalent to those shown earlier in the nonseasonal Holt procedure (see Equations 3.30 and 3.31). In Equation 3.46, the term provides an estimate of the seasonal factor based on the latest demand observation. Thus, Equation 3.46, consistent with all the earlier exponential smoothing equations, reflects a linear combination of the historical estimate and the estimate based on the latest piece of data. At all times, we wish to have the sum of the indices through an entire season added to P. Thus, when a specific index is updated, we renormalize all indices to achieve this equality. A numerical illustration will be shown after a discussion of the initialization and forecasting procedures.

As with simple exponential smoothing and exponential smoothing with a trend, we can rewrite the updating equations to show how the old estimate is updated.

$$\hat{a}_t = (\hat{a}_{t-1} + \hat{b}_{t-1}) + \alpha e_t/F_{t-P}$$

$$\hat{b}_t = \hat{b}_{t-1} + \beta_{HW}[(\hat{a}_t - \hat{a}_{t-1}) - \hat{b}_{t-1}]$$

$$\hat{F}_t = F_{t-P} + \gamma_{HW}(x_t/\hat{a}_t - F_{t-p})$$

The level $\hat{a}$ is updated in a similar way to exponential smoothing with a trend where the only difference is that the error used is adjusted by the seasonal index F. Since the estimates a and b are already seasonally adjusted, the updating equation for the trend estimate $\hat{b}$ is the same as in exponential smoothing with a trend. The new seasonal index is calculated by updating the index from one season ago (period $t - P$) and adjusting based on the difference between the most recent observation of the seasonal effect $x_t/\hat{a}_t$ and the previous seasonal index.

3.4.4.3 Initialization

With both trend and seasonal factors present, the initialization becomes considerably more complicated. One has to properly separate out the trend and seasonal effects in the historical data. Several different initialization procedures have been suggested, but we restrict our attention to the most commonly used method, the so-called *ratio to moving average procedure* (which differs from Winters' original suggestion). This approach can effectively handle changes in the underlying trend during the historical period. In addition, it tends to eliminate cyclical effects (which we are not explicitly considering). Further details are available in Hamburg and Young (1994).

Ideally, from a statistical standpoint, it is desirable to have several seasons worth of data because each specific seasonal period (e.g., the month of June) occurs only once per season (year). This is especially true for slow-moving items, or for items with highly erratic demand, because the noise in the data can obscure the underlying seasonality. However, when using too much history, one runs the risk of the seasonal pattern having changed during the history, so that the early portion is no longer representative of the current, let alone, future conditions. As a compromise, we suggest a minimum of 4 complete seasons, using 5 or 6 (with care) if that much is available. The case where insufficient data exist will be treated in Section 3.8.1.

There are several steps in the procedure. We use the example, shown in Table 3.5 and Figure 3.5, of quarterly sales (in 2011–2014) for product EDM-617.

Step 1 Initial Estimation of Level (Including Trend) at Each Historical Period. The demand in any historical period is assumed to be composed of trend (incorporating the level), seasonal, and random components. To estimate the seasonal indices, we must first remove the trend effect. The trend point (i.e., the level) for any particular period t is estimated by a moving average of a full season (i.e., P periods) centered at period t. A full season is used in order to have the moving average free of seasonal effects. Because the most common seasons have an even number of periods ($P = 4$ or 12), the standard P-period moving average ends up being centered between two periods, not right at the middle of a period as desired. To overcome this problem, we take the average of two consecutive moving averages, as illustrated in Table 3.5. In columns 5 and 6, we show 4-period moving averages that are computed from consecutive time periods. For example, the first entry of column 5, 54.25, is the average of the first four data points (43, 57, 71, and 46), and therefore is centered between quarters 2 and 3 of 2011. The first entry of column 6, 56.00, is the average of the four periods beginning at quarter 2 of 2011 (or, 57, 71, 46, and 50), and therefore is centered between quarters 3 and 4 of 2011. Column 7 is the average of these two averages, and, therefore, is the centered 4-period moving average for quarter 3 of 2011. These centered moving averages are also portrayed in Figure 3.5.

Step 2 Estimates of Seasonal Factors. The estimate of the seasonal factor (column 8 of Table 3.5) for any particular historical period t is obtained by dividing the demand x_t (column 4) by the centered moving average (column 7). Note that, because of the moving average procedure, we cannot obtain estimates for the first half-season (first two quarters of 2011) nor for the last half-season (last two quarters of 2014). Estimates still include the random components. In an effort to dampen the random effect, we average the seasonal factors for similar periods in different years. For example, in Table 3.5, the average of the 3rd quarter factors is (1.288 + 1.365 + 1.286)/3 or 1.313. The averages obtained in this fashion are shown in column 2 of Table 3.6. The total of the averages need not add up to exactly P (4 in our example). Thus,

Table 3.5   Initialization of Seasonal Model for Item EDM-617

Year (1)	Quarter (2)	Period t (3)	Demand x_t (4)	Moving Average 1 (5)	Moving Average 2 (6)	Centered 4-Period Moving Average (7)	Estimate of F_t (8) = (4) − (7)
2011	1	−15	43				
	2	−14	57				
	3	−13	71	54.25	56.00	55.13	1.288
	4	−12	46	56.00	57.00	56.50	0.814
2012	1	−11	50	57.00	60.50	58.75	0.851
	2	−10	61	60.50	60.75	60.63	1.006
	3	−9	85	60.75	63.75	62.25	1.365
	4	−8	47	63.75	67.00	65.38	0.719
2013	1	−7	62	67.00	67.25	67.13	0.924
	2	−6	74	67.25	67.50	67.38	1.098
	3	−5	86	67.50	66.25	66.88	1.286
	4	−4	48	66.25	67.25	66.75	0.719
2014	1	−3	57	67.25	71.25	69.25	0.823
	2	−2	78	71.25	74.50	72.88	1.070
	3	−1	102				
	4	0	61				

we normalize to obtain the estimates of seasonal indices that total to P. The results are shown in column 3 of Table 3.6.

Step 3 Estimating $\hat{a}_0$ and $\hat{b}_0$. We still need estimates of the level and trend terms in Equation 3.43. Using the seasonal indices found in Step 2, we deseasonalize the data as shown in Table 3.7. (Once again, we have rounded the data in the table, but the spreadsheet that computes column 6 retains many more digits.) The results are also shown graphically in Figure 4.8. They are reasonably close to the centered moving averages found in Table 3.5, because, in this example, the seasonal coefficients found in Table 3.5 do not fluctuate very much from year to year. In any event, a regression line is now fit to the data of column 6 in Table 3.7 using the procedure of Equations 3.36 and 3.37 of Section 3.4.3. This gives us estimates $\hat{a}_0$ and $\hat{b}_0$ as of the end of 2014 (the end of period 0). The resulting line is shown in Figure 3.6; the parameter values are

$$\hat{a}_0 = 76.866 \quad \text{and} \quad \hat{b}_0 = 1.683$$

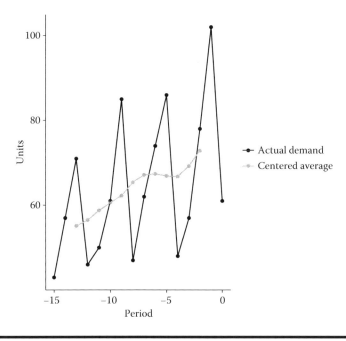

Figure 3.5 Historical data for item EDM-617.

Again, one could use an earlier origin (with $\hat{a}_0$ appropriately adjusted) and an associated run-in period through the rest of the historical data. Also, in the event that the plotted data indicate a shift in trend part way through the history, the regression line should only be fit to the latest data (beyond the shift). (See Williams (1987) for an alternate way of initializing the parameters.)

Table 3.6 Initial Seasonal Indices for Item EDM-617

Quarter i (1)	Average Estimate (2)	Normalized Index F_i (3)
1	0.866	0.869
2	1.058	1.061
3	1.313	1.317
4	0.751	0.753
Total	3.988	4.000

Table 3.7 Calculations to Estimate $\hat{a}_0$ and $\hat{b}_0$ Item EDM-617

Year (1)	Quarter (2)	Period t (3)	Demand x_t (4)	F_t (5)	Estimate of Level (6) = (4)/(5)
2011	1	−15	43	0.869	49.509
	2	−14	57	1.061	53.701
	3	−13	71	1.317	53.908
	4	−12	46	0.753	61.091
2012	1	−11	50	0.869	57.569
	2	−10	61	1.061	57.469
	3	−9	85	1.317	64.538
	4	−8	47	0.753	62.419
2013	1	−7	62	0.869	71.385
	2	−6	74	1.061	69.717
	3	−5	86	1.317	65.297
	4	−4	48	0.753	63.747
2014	1	−3	57	0.869	65.628
	2	−2	78	1.061	73.485
	3	−1	102	1.317	77.445
	4	0	61	0.753	81.012

3.4.4.4 Forecasting

The underlying model includes level, trend, and seasonal factors. As of the end of period t we have

$$\hat{x}_{t,t+\tau} = (\hat{a}_t + \hat{b}_t \tau) \hat{F}_{t+\tau-P} \tag{3.47}$$

where $\hat{x}_{t,t+\tau}$ is the forecast, made at the end of period t, of demand in period $t + \tau$ and $\hat{F}_{t+\tau-P}$ is the most recent estimate of the seasonal index for period $t + \tau$. (This assumes $\tau \leq P$; the last *direct* update of $\hat{F}_{t+\tau-P}$ was made in period $t + \tau - P$. However, when any other seasonal index is updated, renormalization could cause $\hat{F}_{t+\tau-P}$ to change somewhat.)

As earlier, the forecast of cumulative demand from the beginning of period $t + 1$ through to the end of period u is given by

$$\hat{x}_{t,t,u} = \sum_{\tau=1}^{u-t} \hat{x}_{t,t+\tau} \tag{3.48}$$

where again, if u is not an integer, the last term in the summation reflects an appropriate proportion of the last period.

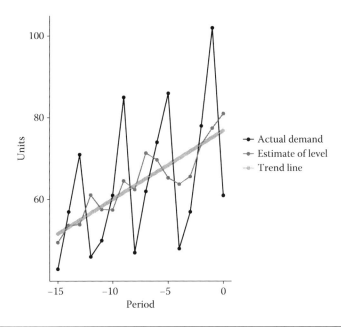

Figure 3.6   Determination of trend line for item EDM-617.

3.4.4.5 Illustration

We again use the item EDM-617. First, consider the forecast $\hat{x}_{0,1}$ of the demand in the first quarter of 2015 made at the end of 2014. The previous first quarter was $t = -3$. From Table 3.6, we thus have $\hat{F}_{-3} = 0.869$. Also $\hat{a}_0 = 76.866$ and $\hat{b}_0 = 1.683$. Thus, from Equation 3.47 we have

$$\hat{x}_{0,1} = (76.866 + 1.683)0.869 = 68.222 \text{ units}$$

Similarly, the forecasts at the end of 2014 for the other three quarters of 2015 are

$$\hat{x}_{0,2} = (76.866 + 1.683(2))1.061 = 85.161 \text{ units}$$
$$\hat{x}_{0,3} = (76.866 + 1.683(3))1.317 = 107.886 \text{ units}$$
$$\hat{x}_{0,4} = (76.866 + 1.683(4))0.753 = 62.946 \text{ units}$$

Suppose that the actual demand in the first quarter of 2015 was 75 units. The level, trend, and seasonality factors can now be updated. Assume that the smoothing constants α_{HW}, β_{HW}, and γ_{HW} have values of 0.2, 0.1, and 0.3, respectively. From Equation 3.44, the updated value of the level is given by

$$\hat{a}_1 = \alpha_{HW}(x_1/\hat{F}_{-3}) + (1 - \alpha_{HW})(\hat{a}_0 + \hat{b}_0)$$
$$= 0.2(75/0.866) + 0.8(76.866 + 1.683) = 80.110$$

Next, from Equation 3.45, the revised estimate of the trend is given by

$$\hat{b}_1 = \beta_{HW}(\hat{a}_1 - \hat{a}_0) + (1 - \beta_{HW})\hat{b}_0$$
$$= 0.1(80.110 - 76.866) + 0.9(1.683) = 1.839$$

Then, by using Equation 3.46, we obtain the updated estimate of the seasonality factor for the first quarter as

$$\hat{F}_1 = \gamma_{HW}(x_1/\hat{a}_1) + (1 - \gamma_{HW})\hat{F}_{-3}$$
$$= 0.3(75/80.110) + 0.7(0.869) = 0.889$$

Because of the change in this seasonality factor, we renormalize the four seasonality factors so that they again add to 4:

$$1.061 + 1.317 + 0.753 + 0.889 = 4.020$$

Therefore,

$$\hat{F}_{-2} = 1.061 \left(\frac{4.00}{4.02}\right) = 1.056$$

$$\hat{F}_{-1} = 1.317 \left(\frac{4.00}{4.02}\right) = 1.310$$

$$\hat{F}_{-0} = 0.753 \left(\frac{4.00}{4.02}\right) = 0.749$$

$$\hat{F}_1 = 0.889 \left(\frac{4.00}{4.02}\right) = \underline{\underline{0.884}}$$

$$Total = 4.000$$

Note that demand for period 1 was forecast (at the end of period 0) as 68.222 units. Actual demand turned out to be 75 units; that is, demand was underforecast by 6.778 units. Because of this forecast error, the level was adjusted upward from 76.866 to 80.110—an increase of 3.244 units, considerably more than was predicted by the original trend value of 1.683 units. Moreover, the estimate of the trend was increased from 1.683 to 1.839, and the estimate of the seasonality factor was raised from 0.869 to 0.884.

The new estimates of the parameters can now be used in Equation 3.47 to provide a revised set of forecasts as of the end of period 1 (the first quarter of 2015). To illustrate, the new forecast of demand in the second quarter of 2015 is

$$\hat{x}_{1,2} = (\hat{a}_1 + \hat{b}_1)\hat{F}_{-2}$$
$$= (80.110 + 1.839)(1.056) = 86.544$$

The forecasts for the remaining two quarters of 2015 are also shown in Table 3.8.

Table 3.8   Effect of Forecast Error in Period 1 on Forecasts of Demand for Later Periods

Calendar Time	Period t	Forecast Calculated at the End of Period 0	Forecast Calculated at the End of Period 1
1st quarter 2015	1	68.222	75.0 (actual)
2nd quarter 2015	2	85.161	86.544
3rd quarter 2015	3	107.886	109.796
4th quarter 2015	4	62.946	64.149

Once the figure for actual sales in quarter 2 of 2015 becomes available, the above procedures must be repeated—the parameters (level, trend, and seasonal factors) revised and a new set of revised forecasts issued.

3.4.4.6 General Comments

Clearly, updating, forecasting, and particularly initialization are considerably more complicated with a seasonal model than was the case with the earlier nonseasonal models. Implementing the equations on a spreadsheet also requires a bit more creativity. (See Problem 3.17.) Thus, the decision to use a seasonal model should be taken with care. In the initialization stage, if the initial F's are all close to unity or fluctuate widely from season to season, then, a seasonal model is probably inappropriate. (See Sections 3.6 and 3.7; Chatfield 1978, 2013; McLeavey et al. 1981; Rhodes 1996.)

There are other, quite different approaches to dealing with seasonality. In particular Brown (1981, Chapter 9), among others, argues for the use of a transcendental model that is a mix of sine waves. This approach has the merit of being more parsimonious (fewer parameters than seasonal indices) and hence tends to be more stable, particularly when the demand level is quite low. However, we feel that this advantage is usually outweighed by the much more intuitively understandable nature of the seasonal indices model.

Finally, there are certain "seasonal" effects whose changing nature can be predicted. One of the best examples is the Easter holiday that for some years occurs in March and some in April. Although the monthly data can be distorted by this effect, the distortion can be anticipated, so that it is necessary for human input to modify the forecasts. Regular promotions are another example; they may occur each summer, but at slightly different times. (Section 9.7.1 addresses some of these issues for style goods.) Another anomalous situation arises where it is crucial (perhaps for production-scheduling purposes) to use monthly data to forecast the demand rate per day. Under these circumstances, one should take into account of the varying numbers of days from month to month. In other circumstances, it is simply adequate to consider the differing numbers of days as being part of the contribution to differing seasonal factors.

3.4.5 Selection of Smoothing Constants

We now address the choice of the smoothing constants for the various smoothing procedures discussed earlier. For the moment, we restrict our attention to a static choice, that is, values that will essentially remain unchanged over time for any particular SKU. In Section 3.5.4, the case where the smoothing constants can vary with time, so-called *adaptive smoothing*, will be treated.

There is a basic trade-off in the choice of the values of the smoothing constants. Small values, which give little weight to the most recent data, hence considerable weight to the history, are appropriate under stable conditions—they effectively smooth out most of the random noise. On the other hand, if the parameters of the underlying demand model are changing appreciably with time, then higher values of the smoothing constants are in order, because they give more weight to recent data and thus more quickly bring the parameter estimates into line with the true changed values. Unfortunately, we are never really sure whether forecast errors are a result of random (transient) noise or a real change in one or more of the parameter values in the underlying demand model. Therefore, a compromise must be made in choosing the values of the smoothing constants. Harrison (1967) quantifies this compromise under special assumptions about the inherent variability of the model parameters. (See also Adshead and Price 1987; Johnston and Boylan 1994.) An additional point related to the same issue is that, all other factors being equal, the smoothing constants should increase with the length of the forecast update period, because the likelihood of changes in the model parameters increases with the passage of time.

In general, we suggest the use of a search experiment to establish reasonable values of the smoothing constants. For each of a number of representative items, the following experiment should be conducted. The available demand history is divided into two sections. The first section is used to initialize the model parameters in the updating procedure selected. Then, with these initial values, the smoothing is carried through the second portion of the data, using a specific combination of values for the smoothing constants. Some appropriate measure of effectiveness (e.g., the mean of the squared forecast errors to be discussed in Section 3.5) is evaluated. This is repeated, on a grid search basis, to find the combination of values of smoothing constants that minimizes the measure of effectiveness across a class of similar items. In actual fact, exact minimization is not required because the measures of effectiveness always tend to be quite insensitive to deviations of the smoothing constants from their best values. Furthermore, we wish to use only a few sets of smoothing constants, each set being employed for a broad class of items.

Moreover, any case where minimization occurs when unusually large values are used for the smoothing constants should be viewed with caution. As an illustration, a large value (>0.3) of α in the simple smoothing procedure should raise the question of the validity of the assumed underlying level model. The high smoothing constant, in this case, suggests that perhaps a trend model is more appropriate. Further discussion on the aforementioned search can be found in Winters (1960), and Gardner and Dannenbring (1980).

Keeping these suggestions in mind, we now present, for each of the smoothing procedures discussed earlier, ranges of reasonable values of the parameters found through our experience and that of several other individuals, as reported in the literature.

3.4.5.1 Simple Exponential Smoothing

Recall from Equation 3.23 that the updating procedure is

$$\hat{a}_t = \alpha x_t + (1 - \alpha)\hat{a}_{t-1}$$

The likely range of α is 0.01–0.30 with a compromise value of 0.10 often being quite reasonable.

3.4.5.2 Smoothing for a Trend Model

Earlier, we suggested the use of Brown's single parameter updating method. For reference purposes, we repeat Equations 3.32 and 3.33.

$$\hat{a}_t = [1 - (1 - \alpha)^2]x_t + (1 - \alpha)^2(\hat{a}_{t-1} + \hat{b}_{t-1}) \tag{3.49}$$

$$\hat{b}_t = \left[\frac{\alpha^2}{1 - (1 - \alpha)^2}\right](\hat{a}_t - \hat{a}_{t-1}) + \left[1 - \frac{\alpha^2}{1 - (1 - \alpha)^2}\right]\hat{b}_{t-1} \tag{3.50}$$

The range of α values prescribed under simple smoothing is also applicable here.

3.4.5.3 Winters Seasonal Smoothing

Repeating Equations 3.44 through 3.46, the updating procedure is

$$\hat{a}_t = \alpha_{HW}(x_t/\hat{F}_{t-P}) + (1 - \alpha_{HW})(\hat{a}_{t-1} + \hat{b}_{t-1}) \tag{3.51}$$
$$\hat{b}_t = \beta_{HW}(\hat{a}_t - \hat{a}_{t-1}) + (1 - \beta_{HW})\hat{b}_{t-1} \tag{3.52}$$
$$\hat{F}_t = \gamma_{HW}(x_t/\hat{a}_t) + (1 - \gamma_{HW})\hat{F}_{t-P} \tag{3.53}$$

As pointed out earlier (in Equations 3.34 and 3.35), Equations 3.49 and 3.50 are a special case of the Holt procedure where

$$\alpha_{HW} = [1 - (1 - \alpha)^2 2] \tag{3.54}$$

and

$$\beta_{HW} = \frac{\alpha^2}{1 - (1 - \alpha)^2} \tag{3.55}$$

This suggests that reasonable guidelines for the choices of α_{HW} and β_{HW} in Equations 3.51 and 3.52 are provided by substituting the above suggested α values into Equations 3.54 and 3.55. The results are shown in Table 3.9 along with the suggested values of γ_{HW}, the latter found through experimentation. Again, we emphasize that, where possible, a search experiment should be conducted with demand data for the particular population of items involved.[*]

Table 3.9 Reasonable Values of Smoothing Constants in the Winters Procedure

	Underlying α Value	α_{HW}	β_{HW}	γ_{HW}
Upper end of range	0.30	0.51	0.176	0.50
Reasonable single value	0.10	0.19	0.053	0.10
Lower end of range	0.01	0.02	0.005	0.05

[*] For stability purposes (see McClain and Thomas 1973), the value of γ_{HW} should be kept well below that of α_{HW}.

3.5 Measuring the Performance of a Forecasting Process

No matter how forecasts are generated, we are interested in evaluating the quality of the forecasts for two reasons. First, the historical performance of a forecasting process helps us create the description of future demand to support resource allocation decisions. Second, by tracking the performance of a forecasting process over time, we may be able to identify ways to improve that process. When we speak of the *accuracy* of a forecast, we generally mean some measure of how far away our point-estimate forecast is from the actual value. A challenge is trying to separate the *dispersion* of forecast errors from the *bias*. Measures of bias tell us how far off the forecasts are from the target *on average*, but they do not tell us how widely or narrowly scattered the forecast errors are.

Significant bias may signal that the parameters of the underlying demand model have been inaccurately estimated or that the model itself is incorrect. Alternatively, it may be human adjustments to forecasts that are causing the bias. We revisit the impact of human judgment in Section 3.7. Where forecasts are used for production planning and inventory management decision making, either of these situations is undesirable; removing bias from the forecast will improve the performance.

A common measure of bias is the average error over some time horizon, say n periods, where periods could be days, weeks, months, etc. As noted above, we represent the forecast error for lag-τ forecasts by $e_{t-\tau,t}$.

$$\bar{e} = \text{Average Error} = \sum_{t=1}^{n} e_{t-\tau,t}/n \qquad (3.56)$$

A closely related measure sometimes used in practice is the *sum* (rather than the average) of forecast errors. This is sometimes referred to as the running sum of forecast errors.[*]

$$C_n = \text{Sum of Errors} = \sum_{t=1}^{n} e_{t-\tau,t} \qquad (3.57)$$

Both $\bar{e}$ and C_n allow positive and negative errors to cancel each other out over time. So, with either of these measures, if their value is close to zero, we say that the forecasting process is *unbiased*; forecasts (point estimates) are not *consistently* too high or *consistently* too low. "Close to zero" is not a terribly precise term obviously; so, we need to be concerned with scaling or interpreting $\bar{e}$ and C_n to try to decide whether our forecast is biased. The challenge is that due to the random component in a demand pattern, it is not easy to detect an underlying bias; the noise tends to camouflage it. We return to this issue in Section 3.9.2 after introducing measures of dispersion and accuracy.

The standard deviation of forecast errors for a forecasting process (forecasting for a unit period), σ_1, gives us a measure of *dispersion* of forecast errors

$$\sigma_1 = \sqrt{\frac{\sum_{t=1}^{n}(e_{t-\tau,t} - \bar{e})^2}{(n-1)}}$$

---

[*] Another measure of bias commonly used in practice is the average of actual values divided by forecasts. If over time this value is close to 100%, we conclude that our forecasting process is unbiased. This kind of measure is more frequently used as a summary performance measure for many forecasts aggregated.

where $\bar{e}$ is the average forecast error defined in Equation 3.56. σ_1 tells how variable forecast errors are but does not speak to bias.

It is critical to note that none of these measures by themselves gives a complete description that could support a resource allocation decision. To see why, suppose we were evaluating a forecasting process that generated forecasts that were always *exactly* 1,000 units too high (e.g., $x_t - \hat{x}_{\tau,t} = -1{,}000$). The average error $\bar{e} = -1{,}000$ and the standard deviation of forecast errors $\sigma_\tau = 0$. This forecasting process has no dispersion or variability but is clearly quite biased. Alternatively, suppose the forecast errors alternated between 1,000 and −1,000. Such a forecasting process is unbiased (average error $\bar{e} = 0$) but has high dispersion. As we will discuss further below, when forecast bias is detected, identifying and eliminating the cause of the bias must be a priority.

3.5.1 Measures of Forecast Accuracy

In this section, we introduce three widely used measures of forecast accuracy, MSE, mean absolute deviation (MAD), and mean absolute percent error (MAPE). As will be demonstrated in Chapter 6, for purposes of establishing the safety stock of an individual item (to provide an appropriate level of customer service), we will need an estimate of the standard deviation (σ_L) of the errors of forecasts of total demand over a period of duration L (the replenishment lead time). In this section, we concentrate on estimating σ_1, the standard deviation of errors of forecasts of demand made for a unit period. (A unit period is a forecast update interval.) In Section 3.5.2, we discuss the conversion of the estimate of σ_1 to an estimate of σ_L. The procedures to be discussed are appropriate for A and B items. C items will be handled by simpler methods to be presented later.

Suppose that we have two types of information for each of n unit time periods, specifically (1) the actual observed demands, $x_1, x_2, \ldots, x_n$ and (2) the one-period-ahead forecasts (by some particular forecasting procedure) $\hat{x}_{0,1}, \hat{x}_{1,2}, \ldots, \hat{x}_{n-1,n}$ where, as earlier, $\hat{x}_{t,t+1}$ is the forecast, made at the end of period t, of demand in period $t + 1$. The measure of variability that is often used (minimized) in fitting of squared errors of a straight line to historical data is the MSE. When forecasts are unbiased, the MSE is directly related to σ_1, as we shall see shortly. Because of these properties, and because it can be computed easily on a spreadsheet, we recommend its use. The MSE for one-period-ahead forecasts is given by

$$\text{MSE} = \frac{1}{n} \sum_{t=1}^{n} (x_t - \hat{x}_{t-1,t})^2 = \sum_{t=1}^{n} (e_{t-1,1})^2 \tag{3.58}$$

A second measure of variability, the MAD, was originally recommended for its computational simplicity. With the widespread availability of computers, however, the MAD is of less practical importance. Nevertheless, it is intuitive, and still frequently used in practice. For the n periods of data, the estimate of the MAD for one-period-ahead forecasts would be

$$\text{MAD} = \frac{1}{n} \sum_{t=1}^{n} |x_t - \hat{x}_{t-1,t}| = \frac{1}{n} \sum_{t=1}^{n} |e_{t-1,t}| \tag{3.59}$$

The term $x_t - \hat{x}_{t-1,t}$ is the error ($e_{t-1,t}$) or the deviation in the forecast for period t. The vertical lines indicate that we take the absolute value of the difference. Finally, the mean of the absolute deviations results by the summation and division by n.

The MAPE is another intuitive measure of variability. It generally is not affected by the magnitude of the demand values, because it is expressed as a percentage. However, it is not appropriate if demand values are very low. For example, a forecast of 1 unit of demand matched with an actual value of 2 units shows an error of 100%. The manager should decide whether a reported error of 100%, or of 1 unit, is more desirable. The MAPE is given by

$$\text{MAPE} = \frac{1}{n}\sum_{t=1}^{n}\left|\frac{x_t - \hat{x}_{t-1,t}}{x_t}\right| = \frac{1}{n}\sum_{t=1}^{n}\left|\frac{e_t}{x_t}\right| \tag{3.60}$$

3.5.1.1 Numerical Illustration

Consider an item with the demands and forecasts as shown for six periods in Table 3.10. The MSE is computed using Equation 3.58.

$$\text{MSE} = \frac{1}{6}\sum_{t=1}^{6}(e_t^2) = 206/6 = 34.33$$

Using Equation 3.59, the estimate of MAD is

$$\text{MAD} = \frac{1}{6}\sum_{t=1}^{6}|e_t| = 32/6 = 5.33$$

Again, as seen in Figure 3.7, the MAD has the appealing graphical interpretation of being the average distance (ignoring direction) between the actual demand in a period and the forecast of that demand.

Table 3.10   Illustration of Computation of MAD, MAPE, and MSE

Period t	1	2	3	4	5	6	Total	Average
Actual demand x_t	100	87	89	87	95	85		
One-period-ahead forecast made at the end of the previous period $\hat{x}_{t-1,t}$	90	92	91	91	90	91		
Error or deviation $e_t = x_t - \hat{x}_{t-1,t}$	10	−5	−2	−4	5	−6		
Squared deviation e_t^2	100	25	4	16	25	36	206	34.33
Absolute deviation $|e_t|$	10	5	2	4	5	6	32	5.33
Percent deviation $|e_t/x_t|$	10.0%	5.7%	2.2%	4.6%	5.3%	7.1%	34.9%	5.82%

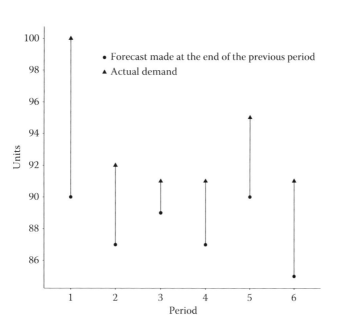

Figure 3.7   Illustration of the MAD.

The MAPE is computed using Equation 3.60

$$\text{MAPE} = \frac{1}{n}\sum_{t=1}^{n}\left|\frac{e_t}{x_t}\right| = 5.82\%$$

3.5.1.2 Conversion to σ_1

To find the standard deviation of forecast errors, for the purpose of setting safety stock levels, we can use a simple relationship between the value of σ_1 and the MSE:

$$\sigma_1 = \sqrt{\text{MSE}} \tag{3.61}$$

This relationship assumes that forecasts are unbiased. As noted above, removing bias from a forecasting process should be a priority. If bias remains, the approach in Equation 3.61 will tend to overestimate σ_1. So, with unbiased forecasts, a reasonable estimate of σ_1, denoted as $\hat{\sigma}_1$, is the square root of the MSE calculated from the sample data. We illustrate how to initialize the MSE below.

The relationship of σ_1 to the MAD is not as simple. However, for the normal distribution and unbiased forecasts, we can show that the following relationship holds between the true MAD (the average absolute deviation from the mean) and the true standard deviation (See the Appendix of this chapter for a derivation.):

$$\sigma = (\sqrt{\pi/2})\text{MAD} = 1.25\ \text{MAD} \approx \sqrt{\text{MSE}}$$

Thus, in cases where the normal distribution is appropriate, one can estimate σ_1 via

$$\hat{\sigma}_1 = 1.25 \text{ MAD}$$

For several other common distributions, the theoretical conversion factor from the MAD to σ is not very different from 1.25. However, the factor can vary enough so that the safety stock obtained may not provide the required level of service. See Jacobs and Wagner (1989). Therefore, we recommend using the MSE.

3.5.1.3 Updating of MSE

In theory, Equation 3.58 could be used to update the estimate of MSE each time an additional period's information became available. Instead, we advocate a simple exponential smoothing form of updating

$$\text{MSE}_t = \omega(x_t - \hat{x}_{t-1,t})^2 + (1 - \omega)\text{MSE}_{t-1} \tag{3.62}$$

where MSE_t is the estimate of MSE at the end of period t and ω is a smoothing constant. (A small value of ω, namely between 0.01 and 0.10, should normally be used.)

There are two reasons for using Equation 3.62 instead of Equation 3.58. First, storage of an indefinitely long history is required for the use of Equation 3.58, whereas only the most recent MSE_{t-1} need be stored for Equation 3.62. Second, in using Equation 3.58, each period of history is given equal weight. In contrast, geometrically decaying weights occur through the use of Equation 3.62. In fact, these arguments are identical to those presented earlier in the chapter when we advocated exponential smoothing, instead of repeated least-squares regression, for updating the forecasts themselves.

3.5.1.4 Initialization of MSE

For each of the forecasting procedures discussed in Section 3.4, we suggested a method for estimating initial values of the parameters of the underlying model of demand, using sufficient available historical data. For such a situation, the initial estimate of MSE is obtained through the use of the following equation that is closely related to Equation 3.58:

$$MSE_0 = \sum_{t=-(n-1)}^{0} (x_t - \tilde{x}_t)/(n - p) \tag{3.63}$$

where the n historical periods are numbered from 0 (the most recent period) back through $-1, -2, \ldots, -(n-1)$, and $\tilde{x}_t$ is the estimate of x_t resulting from the underlying demand model when the estimates of its p parameters are obtained from the historical data. (E.g., if the underlying model was a linear trend, $x_t = a + bt$, then p would be 2 and $\tilde{x}_t$ would be $\hat{a}_0 + \hat{b}_0 t$.)

The denominator in Equation 3.63 is $n - p$, rather than n, to reflect that our estimate of MSE is based on an ex-post fit to the data, where the data have already been used to estimate the p parameters of the model.

The case where insufficient historical information exists will be treated in Section 3.8.1.

Table 3.11   Illustration of Computation of MSE$_0$ for the Example of Table 3.3

	$\hat{a}_0 = 27.19$		$\hat{b}_0 = 1.34$
t	x_t	$\tilde{x}_t = \hat{a}_0 + \hat{b}_0 t$	$(x_t - \tilde{x}_t)^2$
-5	20	20.48	0.23
-4	25	21.82	10.12
-3	21	23.16	4.67
-2	22	24.50	6.27
-1	27	25.85	1.33
0	28	27.19	0.66
		Total	23.28

3.5.1.5 Illustration

In Table 3.3, we showed the initialization of a trend model using six periods of historical information. The estimated values of the two parameters were $\hat{a}_0 = 27.19$ and $\hat{b}_0 = 1.34$. The original demand data and the $\tilde{x}_t$ values are shown in Table 3.11. By using Equation 3.63, we obtain

$$\text{MSE}_0 = \sum_{t=-5}^{0} (x_t - \tilde{x}_t)^2 / (6 - 2)$$
$$= (23.28)/4 = 5.82$$

3.5.2 *Estimating the Standard Deviation of Forecast Errors over a Lead Time*

As mentioned earlier, the replenishment lead time is unlikely to be equal to the forecast update interval. Thus, we need a method of converting $\hat{\sigma}_1$ (our estimate of σ_1) into $\hat{\sigma}_L$ (the corresponding estimate of σ_L). The exact relationship between σ_L and σ_1 depends in a complicated fashion on the specific underlying demand model, the forecast-updating procedure, and the values of the smoothing constants used. See, for example, Brown (1963), Harrison (1967), and Johnston and Harrison (1986). One of the reasons for the complexity is that the smoothing procedure introduces a degree of dependence between the forecast errors in separate periods. Fortunately, however, we have found that for most inventory systems, the following model satisfactorily captures, empirically, the required relationship:

$$\hat{\sigma}_L = L^c \hat{\sigma}_1 \tag{3.64}$$

where

$\hat{\sigma}_L$ = estimate of the standard deviation of forecast errors over a lead time of duration L basic (forecast update) periods (L need not be an integer)

$\hat{\sigma}_1$ = estimate of the standard deviation of forecast errors over one basic (forecast update) period

c = coefficient that must be estimated empirically

Table 3.12   Forecasts Simulated to Estimate *c*

Length of L	Forecasts		Number of Forecasts for Each SKU
1	$\hat{x}_{t,t+1}$	$t = 0, 1, 2, \ldots, 35$	36
2	$\sum_{\tau=1}^{2} \hat{x}_{t,t+\tau}$	$t = 0, 2, 4, \ldots, 34$	18
3	$\sum_{\tau=1}^{3} \hat{x}_{t,t+\tau}$	$t = 0, 3, 6, \ldots, 33$	12
$\vdots$	$\vdots$	$\vdots$	$\vdots$
6	$\sum_{\tau=1}^{6} \hat{x}_{t,t+\tau}$	$t = 0, 6, 12, \ldots, 30$	6

To check the reasonableness of Equation 3.64 for a particular firm, we proceed as follows. We select, say, 10 representative SKUs. We should pick, say, 6 from the A category and 4 from the B category, for which a number of years of historical monthly demand data are available. Assume that we have 7 years of data. Using the first 4 years of data, we initialize the Winters exponential smoothing model by the procedures described in Section 3.4.4. That is, for each of the 10 representative SKUs, we obtain the data needed to initialize the Winters forecast equation (Equation 3.47):

$$\hat{x}_{t,t+\tau} = (\hat{a}_t + \hat{b}_t \tau)\hat{F}_{t+\tau-P} \tag{3.65}$$

Then, for the fifth, sixth, and seventh years of the demand data, Equation 3.65, along with the Winters updating procedures described in Equations 3.44 through 3.46, is used to issue the forecasts listed in Table 3.12. In all cases, the model parameters are updated monthly. For each forecast in Table 3.12, the forecast figure is compared to the actual demand that resulted over the immediate period of duration L, and the associated forecast error is computed by using

$$\text{Forecast error} = e_t(L) = \sum_{\tau=1}^{L} \hat{x}_{t,t+\tau} - \sum_{\tau=1}^{L} x_{t+\tau} \tag{3.66}$$

For each value of L, the sample standard deviation of forecast errors is computed and used as an estimate of $\hat{\sigma}_L$

$$\hat{\sigma}_L = s_L = \left\{ \frac{1}{n-1} \sum_t [e_t(L) - \bar{e}(L)]^2 \right\}^{1/2}$$

where
$\bar{e}(L) = \sum_t e_t(L)/n$ is the average error for the L under consideration
n = number of lead times of length L used (see the right-hand column of Table 3.12)

As a result of all the above calculations, there are 10 values (one for each of the SKUs) of $\hat{\sigma}_L$ for each of $L = 1, 2, 3, \ldots, 6$. For each SKU, the 5 values of $\hat{\sigma}_L/\hat{\sigma}_1$ are computed for $L = 2, 3, 4, 5, 6$. The 50 values of $\log(\hat{\sigma}_L/\hat{\sigma}_1)$ are then plotted against $\log L$, as shown in Figure 3.8. Equation 3.64 can be written as

$$\hat{\sigma}_L/\hat{\sigma}_1 = L^c \text{ or } \log(\hat{\sigma}_L/\hat{\sigma}_1) = c \log L$$

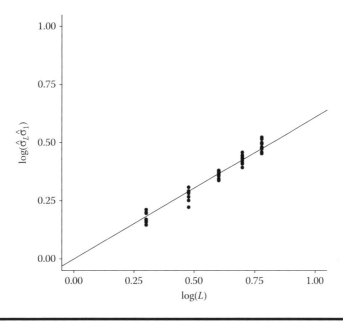

Figure 3.8 **Relative standard deviations of forecast errors versus lead time.**

Therefore, the slope of the line through the origin that gives a reasonable fit to the 50 data points is an estimate of c. The line with slope $c = 0.6$, plotted in Figure 3.8, gives a reasonable fit for the data shown. So, for these data, we choose to use this approximation, which gives

$$\log(\hat{\sigma}_L/\hat{\sigma}_1) = 0.60 \log L$$
$$\hat{\sigma}_L = L^{0.60}\hat{\sigma}_1 = \sqrt{L}\hat{\sigma}_1 \qquad (3.67)$$

The result of Equation 3.67 could also be obtained by assuming that in an L period, the forecast errors in consecutive periods are independent and each has standard deviation σ_1. In actual fact, both of these assumptions are violated to some extent. Nevertheless, the empirical behavior of Equation 3.67 is often quite reasonable. However, we hasten to recommend that an analysis, such as that leading to Figure 3.8, be carried out for any system under consideration. If a straight line through the origin is a reasonable fit, then, its slope, which need not be 0.6, gives the estimate of c in Equation 3.64.

3.5.3 Monitoring Bias

As noted above, we use the word *bias* to indicate that, on average, the forecasts are substantially above or below the actual demands. A bias signals some fundamental problem in the forecasting process. This could be the result of manual intervention in the forecasts or, if we are simply evaluating forecasts coming from a statistical model, bias would indicate that the parameters of the underlying demand model have been inaccurately estimated or that the model itself is incorrect. Where forecasts are used for production planning and inventory management decision making, either of these situations is undesirable. Because of the random component in a demand pattern, it is not easy to detect an underlying bias; the noise tends to camouflage it. In general, the forecast

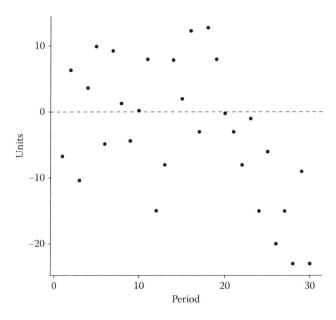

Figure 3.9   An illustration of bias in forecast errors.

error, e_t, should fluctuate around zero. A simple graph of forecast errors can indicate whether they are showing a trend on the high or low side. Figure 3.9 illustrates a case in which there is no bias until about month 20, and then the errors show a negative trend. In this section, we discuss three more rigorous measures of bias; then, in Section 3.5.4, we briefly present the subsequent corrective actions.

3.5.3.1 Cumulative Sum of Forecast Errors

Harrison and Davies (1964) suggest the use of cumulative sum techniques to monitor the bias in a forecasting procedure. Specifically, the cumulative sum of forecast errors is computed recursively using

$$C_t = C_{t-1} + e_t \qquad (3.68)$$

where
   C_t = cumulative sum of forecast errors as of the end of period t, see Equation 3.57
   $e_t = x_t - \hat{x}_{t-1,t}$ = one-period-ahead forecast errors in period t

It is advisable to normalize with respect to the MSE; that is, use

$$U_t = C_t/\sqrt{\mathrm{MSE}_t} \qquad (3.69)$$

where U_t is the cumulative sum tracking signal, and MSE_t is the MSE at the end of period t.

   U_t can take on any value, which is positive or negative. Nevertheless, if the forecasting procedure is unbiased, U_t should fluctuate around 0. An interesting property of the cumulative sum C_t is that on a graphical plot, if a straight line fit to it is reasonable over several periods, then, the slope of the straight line gives us an estimate of the average bias of a single-period forecast. More advanced versions of the cumulative sum technique are based on graphical techniques that detect a certain amount of bias within a specific number of periods. Further details are available in Gardner (2006) and Chatfield (2013) among others.

3.5.3.2 Smoothed Error Tracking Signal

Trigg (1964) suggests the use of a *smoothed error tracking signal* employing the MAD, although it can be adjusted for the $\sqrt{\text{MSE}}$. This tracking signal has some nice intuitive properties, which will be discussed below. However, it also has a potential drawback in that there can be a significant lag before an actual change in the model (or its parameter values) is identified. Moreover, it is not easy to identify after the fact as to when exactly the change took place. The cumulative sum is more effective in both senses. The smoothed error tracking signal is defined by

$$T_t = z_t / MAD_t \tag{3.70}$$

where
T_t = value of the smoothed error tracking signal at the end of period t
MAD_t = mean absolute deviation of one-period-ahead forecast errors as of the end of period t (updated in a like manner to the MSE in Equation 3.62)
z_t = smoothed forecast error at the end of period t

We update z in a fashion analogous to MSE. Note that the signs of the errors are retained

$$z_t = \omega(x_t - \hat{x}_{t-1,t}) + (1 - \omega)z_{t-1} \tag{3.71}$$

where ω is a smoothing constant.

For an unbiased forecasting procedure, the smoothed forecast error z_t should fluctuate about zero. Consider two consecutive forecast errors, one of which is positive, and the other is negative. We can see from Equation 3.71 that in the computation of z_t the effects of these two errors will tend to cancel one another. In contrast, in the computation of MAD_t, where the absolute value of e_t is taken, no canceling occurs. Continuing this line of reasoning, we can write that

$$-MAD_t \leq z_t \leq MAD_t$$

The limiting equalities are achieved if all errors are positive or if all errors are negative. From Equation 3.70, the equivalent limits on the tracking signal T_t are

$$-1 \leq T_t \leq 1$$

Moreover, T_t will be close to one of the limits if most of the errors are of the same sign, which is an indication that the forecasting system is biased (most of the forecasts are too high or most of the forecasts are too low).

3.5.3.3 Illustration

We illustrate both the cumulative sum and the smoothed tracking signal with the same example. Consider the item pictured in Figure 3.9 and let $t = 22$. At the end of period 21, suppose that we have the following values:

$$\hat{x}_{21,22} = 100 \text{ units of demand in period } t$$
$$\text{MAD}_{21} = 6.25$$
$$z_{21} = 1.71$$

and

$$\text{MSE}_{21} = 57.95$$
$$C_{21} = 19.30$$

Therefore, $T_{21} = 1.71/6.25 = 0.27$ and $U_{21} = 19.30/\sqrt{57.95} = 2.54$. Now, it happens that demand in period 22 is 92 units, so the forecast error is -8. New values are computed, assuming that we use $\omega = 0.1$. For the cumulative sum, we update MSE, C, and U as follows:

$$\text{MSE}_{22} = 0.1(92 - 100)^2 + 0.9(57.95)$$
$$= 0.1(64) + 0.9(57.95) = 58.55 \text{ units}$$
$$C_{22} = 19.30 - 8 = 11.30$$
$$U_{22} = 11.30/\sqrt{58.55} = 1.48$$

Note that the MSE has increased due to the larger-than-usual error. The cumulative sum has decreased by 8, and the tracking signal has decreased, because the error was negative. The cumulative sums are also illustrated in Figure 3.10. (We have connected the points for clarity.) Note how they turn sharply negative due to the negative bias in the forecast errors after period 20.

Using a smoothing approach similar to that in Equations 3.62 and 3.71 is

$$\text{MAD}_{22} = 0.1|92 - 100| + 0.9(6.25)$$
$$= 0.1(8) + 0.9(6.25) = 6.43 \text{ units}$$

Also Equation 3.71 gives

$$z_{22} = 0.1(92 - 100) + 0.9(1.71)$$
$$= 0.1(-8) + 0.9(1.71) = 0.74 \text{ units}$$

So $T_{22} = 0.74/6.43 = 0.12$. Note again that the relatively large negative forecast error in period 22 has increased MAD slightly but has caused z to become closer to 0 because z_{21} was positive.

3.5.3.4 Serial Correlation (or Autocorrelation) of Forecast Errors

The serial correlation of a time series measures the degree of dependence between successive observations in the series. With an appropriate forecasting model, there should be negligible correlation between separate forecast errors. If significant positive (negative) correlation exists, then, this indicates that forecast errors of the same (opposite) sign tend to follow one another. This, in turn,

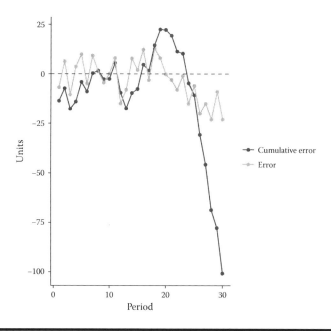

Figure 3.10   An illustration of cumulative sums with bias in the forecast errors.

signals the likelihood of errors in the specification of the demand model or its parameter values. Further details on the computation and properties of serial correlation can be found in several books on probability and statistics (see, e.g., Cox and Miller 1977).

3.5.4 Corrective Actions in Statistical Forecasting

As discussed above, there are at least three different measures of potential bias caused by incorrect specification of the underlying demand model or inaccurate estimates of the values of its parameters. We say "potential" because large absolute values of the tracking signals and substantial serial correlation can, from time to time, be simply generated from the random component in the demand pattern. In this regard, McKenzie (1978) develops statistical properties of the smoothed error tracking signal. The choice of threshold (or critical) values of the signals or serial correlation is completely analogous to the choice of the boundaries between acceptance and rejection in statistical hypothesis testing. There are two types of potential errors. First, we can take a corrective action when nothing has really changed. Second, we can do nothing when, in fact, a fundamental change has taken place. The choice of the threshold (or critical) values of the bias measures involves a, usually implicit, trade-off between these two types of errors.

There are two general types of corrective actions that can be used. The first, which tends to be utilized in a more automated mode, is to increase the values of the smoothing constants in the hope that the form of the underlying model is still correct and all that is needed is to more quickly smooth the parameter estimates closer to their true, but unknown, values. The second type of corrective action is more substantive. It typically involves human intervention to make significant changes in the parameter estimates or in the form of the demand model itself. We discuss each in turn.

3.5.4.1 Changing the Smoothing Constants: Adaptive Forecasting

The general idea of adaptive smoothing is that the smoothing constants are increased (i.e., smoothing becomes faster) when the tracking signal, T or U, gets too far away from 0, while smoothing is decreased as T or U returns closer to 0. We discuss one of the better known adaptive-forecasting techniques, and then reference others.

Eilon and Elmaleh (1970) develop an approach to adaptive smoothing that consists of setting a smoothing constant to a fixed value, and then making forecasts for a specified number of periods (called the evaluation interval). At the end of the evaluation interval, forecast errors are computed using the demand time series that actually occurred. Forecast errors are also computed for a range of permitted values of the smoothing constant (0.1–0.9) to see what size forecast errors would have resulted if one of these values had been used over the evaluation interval. The permitted value of the constant that would have minimized the forecast error variance over the historical evaluation interval is then used until the next evaluation interval. The procedure is repeated periodically. Alternative procedures and enhancements can be found in Chow (1965), Trigg and Leach (1967), Roberts and Reed (1969), Montgomery (1970), Whybark (1973), Flowers (1980), Gardner (1983), Fliedner et al. (1986), and Williams (1987).

Unfortunately, although all these procedures have an intuitive appeal, substantial research findings (see, e.g., Chatfield 1978; Ekern 1981; Flowers 1980; Gardner and Dannenbring 1980) suggest that adaptive methods are not necessarily better than regular, nonadaptive smoothing. In particular, there are dangers of instabilities being introduced by the attempt at adaptation. Consequently, we suggest that careful testing be undertaken with actual time series from the organization under study before any automatic adaptive procedure is adopted.

3.5.4.2 Human Intervention

Where sufficient bias exists or where smoothing constants remain at high levels, it may be appropriate to call for human intervention. Sufficient bias could be defined by

$$|T_t| > f$$

where f is a threshold fraction, or by

$$|C_t| > k\sqrt{\text{MSE}}$$

where k is some threshold multiple.

The choice of f or k implicitly involves the trade-off between the two types of errors mentioned above—namely (1) not intervening when such an action is necessary, and (2) intervening when it is unnecessary. Lower thresholds would be used for A items than for B items, because the first type of error is relatively more expensive for A items; for B items, we would like to keep interventions on an exception basis. In actual practice, a reasonable way to establish f or k is by trial and error in order to provide a reasonable intervention workload in an aggregate sense. In other words, if f or k is too low, managers will be overwhelmed by having to intervene so frequently. Plausible initial values are $f = 0.4$ and $k = 4$.

Significant assistance can be provided to the intervener through a display of several periods of recent actual demands and forecasts made by the most current version of the model. The individual should attempt to ascertain the reason for the recent poor performance. There may be factors, external to the model, that have changed recently, such as

1. A temporary effect—for example, a sales promotion.
2. A more permanent effect—for example, a new competitive product or a new market penetration.

A temporary effect could lead to human override of forecasts for a few periods. On the other hand, a more permanent effect could be handled by manually adjusting parameter values or even by changing the nature of the demand model. If the visual display indicates that the change occurred for several periods in the past, it may be appropriate to reinitialize the model using only data since the suspected time of the change. See Hogarth and Makridakis (1981), Ashton and Ashton (1985), Lawrence et al. (1986), and Heath and Jackson (1994). Some of these papers provide forecasting models that capture both historical demand and human expertise. In Section 3.7, we provide some guidelines for incorporating human judgment.

3.5.5 Probability Distributions of Forecast Errors

As we will see in Chapter 6, the selection of safety stocks to provide adequate customer service will depend not just on the standard deviation of forecast errors over the replenishment lead time but, instead, on the whole probability distribution of these errors. Of particular concern will be the probabilities of large positive values of the errors (demand minus forecasts) because such errors can lead to serious stockout situations.

Through most of this book, we recommend the use of a normal distribution of forecast errors for three reasons. First, empirically the normal distribution usually provides a better fit to data than most other suggested distributions. Second, particularly if the lead time (or lead time plus review interval) is long, forecast errors in many periods are added together, so that we would expect a normal distribution through the Central Limit Theorem. Finally, the normal distribution leads to analytically tractable results. Nevertheless, particularly for expensive, slow-moving items, it may be appropriate to use an alternative distribution such as the exponential, Gamma, Poisson, or negative binomial. In any event, before making a choice, one should at least perform a relatively simple test of the fit of the normal distribution to the observed forecast errors for each of a number of representative items. There is a special normal probability paper such that the cumulative distribution of a normal distribution will appear as a straight line on it. Thus, one can plot the empirical cumulative distribution and observe how closely it is fit by a straight line. In addition, many statistical software packages are available for personal computers. These packages usually contain procedures, including the goodness-of-fit (or chi-square) test and the Kolmogorov–Smirnov test, for testing the fit of data to a hypothesized distribution. See, for example, Marx and Larsen (2006).

3.6 Handling Anomalous Demand

In general, one should have a healthy suspicion of data that are purported to represent actual customer demand period by period. First, in most organizations, sales or even just shipments to customers are recorded as opposed to demand. If not properly interpreted, a series of periods in which a stockout situation existed would imply no demand in each of those periods. Furthermore, if backordering is permitted, the periods immediately following the stockout may show sales or shipments well above the actual demands in those periods. For approaches to forecasting when stockouts distort the demand data, see Wecker (1978), Bell (1981), Taylor and Thomas (1982), Lovejoy (1990), Agrawal and Smith (1996), Nahmias (1994), Lariviere and Porteus (1999), and Lau and Lau (1996a).

Furthermore, recording and transmission errors can distort the demand picture. Certain types of demand also may not be normally handled on a routine basis by the production/stocking point under consideration and thus should not be incorporated in the forecast updating. For example, unusually large customer orders may be shunted back to the higher level of supply. Orders larger than a certain size may be directly shipped from a central warehouse rather than by the usual mode of transshipment through a regional branch warehouse. Finally, there may be certain temporary changes in the demand pattern (e.g., effects of promotions and price changes) that should be filtered out prior to updating.

A useful aid in monitoring anomalous demand is to have an exception signal trigger human investigation when the "demand" stated for a particular period differs from the forecast by more than $k\sqrt{\text{MSE}}$ where a k value of 3–4 is not unreasonable. (Note that this signal is for a single demand value, whereas the thresholds in Section 3.5.4 were for a tracking signal and a cumulative sum.)

3.7 Incorporation of Human Judgment

An overall framework for forecasting was shown in Figure 3.1. In it, the input of human judgment played a key role. Such intervention is quite common in practice. For example, both Sanders and Manrodt (2003a) and Fildes and Goodwin (2007) report results from surveys of forecasters indicating that judgment plays a substantial role in sales forecasting in practice. Lawrence et al. (2006) provide an excellent discussion of research on judgmental input to forecasts, and Armstrong (2001b) is an excellent, in-depth resource for several key concepts in sales forecasting, including concepts for effectively incorporating human judgment.

In Sections 3.5.4 and 3.6, we have already seen the evidence of manual intervention in an essentially reactive fashion, that is, responding to signals of forecast bias or unusual demand levels. Now, we concentrate on the proactive use of human judgment. However, clearly there is considerable overlap in that a reactive response will often initiate a proactive input.

Obviously managerial judgment as an input is essential in medium-range, aggregate forecasting. The decisions concerning work force levels, operating hours, seasonal inventory levels, and so forth, are of major importance to an organization and thus justify the devotion of a considerable amount of management time. Moreover, there are relatively few time series to be forecasted because of the aggregation. One approach that has proved to be helpful is the Delphi method (see Basu and Schroeder (1977) for an illustration). Most of the remaining discussion will concentrate on shorter-range forecasting, but many of the ideas are also applicable in the medium range.

3.7.1 Factors Where Judgment Input Is Needed

There are a number of factors, normally not included in a statistical forecasting model, where judgment input is clearly needed. These factors can be categorized along two different dimensions: first, factors that are external to the organization versus those that are internally influenced; second, factors that have a temporary effect versus a more or less permanent effect. Our discussion will concentrate on the first dimension.

Factors external to the organization include

1. The general economic situation (e.g., the inflation rate, the cost of borrowing capital, and the unemployment rate).

2. Government regulations such as subsidies, import duties and/or quotas, pollution restrictions, safety standards, etc.
3. Competitor actions.
4. Potential labor stoppages at the customer level (users of our products).
5. Consumer preferences (particularly with regard to style goods).

The important factors internal to the organization include

1. Price changes.
2. Promotions.
3. Advertising.
4. Engineering changes that, for example, will improve the reliability of a product, thus reducing the demand for service parts.
5. Introduction of substitute products. A classic example was the effect on the sales of slide rules that resulted from the introduction of hand calculators.
6. Opening a new distribution outlet for an existing product.
7. The pipeline-filling effect associated with the introduction of a new product. A temporary surge of demand will occur simply to initially place appropriate levels of the inventory at the various stocking points.

Related to internal factors is the need to ensure that the total (in dollars) of statistical forecasts of individual items agrees reasonably well with the aggregate financial plan of the organization.

Some success has been achieved in introducing some of the above judgmental factors into a form of the statistical forecasting model known as Bayesian forecasting. The general approach is discussed by Harrison and Stevens (1971), Harrison and Stevens (1976), Johnston (1980), Fildes (1983), Azoury (1985), Kaplan (1988), Bradford and Sugrue (1991), Karmarkar (1994), and Eppen and Iyer (1997).

Fildes and Goodwin (2007) discuss four case studies and go on to suggest practices for maximizing the value of judgmental input. Their recommendations, with which we wholeheartedly agree, include limiting adjustments to statistical forecasts, requiring the forecaster to document their reasons and assumptions when an adjustment is made, and tracking the impact of behavioral adjustments. Fildes et al. (2009) provide an example of the insights from tracking behavioral adjustments in practice. Their work shows that small adjustments to forecasts tend to hurt forecasting performance while large adjustments improve the performance. This is primarily due to the fact that large adjustments are made only when a manager has significant and relevant information.

3.7.2 Guidelines for the Input and Monitoring of Judgment

It is advisable to think in terms of a small committee being responsible for the judgmental input. Ideally, the membership should include an individual with a broad perspective (e.g., from strategic planning), as well as representatives from marketing, production/inventory management, and engineering/design. Meetings should be held on the order of once per month.

This committee is likely to be concerned with groups or families of items, perhaps by geographic region. The starting point in each case should be the objective (statistical) forecast, aggregated across the appropriate group of items. In addition, recent demand history and the associated forecasts should be available. The committee should be able to easily adjust the aggregate forecast. Then, a computer routine should simply prorate a family adjustment to the individual-item forecasts according to the recent fractions of family demand represented by the individual items.

It is often helpful to ensure that the committee develop and record their forecasts individually before talking as a group. Otherwise, the more influential members of the committee might sway the others. The committee can then take the average and variance of the independent forecasts. Further discussion on this topic can be found in Fisher et al. (1994) and Fisher and Ittner (1999). Gaur et al. (2007) show how judgmental forecasts can be effectively used to develop estimates for the mean and variance of demand.

It is important to provide feedback on the performance of the judgmental input. This is accomplished by retaining both the original statistical forecast and the revised (through judgment) forecast. When the actual demand materializes, one can then show the statistical forecast error and the error of the judgmental forecast. In fact, a report can be developed listing the difference of the two errors (likely normalized by dividing by $\sqrt{\text{MSE}}$) for each family of items forecasted. The listing can be ranked by relative performance of the two methods. The intention should not be to externally evaluate the human input. Rather, the objective should be to provide a mechanism for the committee members themselves to learn from their earlier successes and failures.

The above discussion focuses on trying to maximize the value of judgment in improving forecast accuracy. In many instances, there are other motives that may lead to judgmental influence on the forecast. Galbraith and Merrill (1996) describe a survey of executives that examines why forecasts are changed. They find that some executives change the forecast to influence the firm's stock price, while others are attempting to influence internal resource allocations. In some cases, of course, they are simply trying to improve the forecast with better information. This is an interesting article with good warnings for users and developers of forecasts. Oliva and Watson (2011) discuss a case study for a high-tech manufacturer where the organization structure and forecasting process led to poor forecasting performance and poor resource allocation decisions based on the forecasts. Oliva and Watson (2009) further document how such poor alignment can be addressed by creating a common information technology (IT) system that contains all relevant information such as sales, forecasts, and assumptions as well as adopting what they call a consensus-forecasting process where a series of meetings are held with relevant stakeholders in the organization to come to an agreement on the forecast. In their case, and in most cases, the relevant stakeholders include representatives from operations planning, marketing, and finance. Such a process is commonly a part of what is known as the Sales and Operations Planning (S&OP) process. The S&OP process is discussed further in Section 12.2.

3.8 Dealing with Special Classes of Individual Items

The procedures discussed to this stage in the chapter should be of use for a major portion of the items handled by a typical organization. However, there are special classes of items where caution must be exercised. The intention of this section is to provide some insights concerning each of the several special classes. Additional information about forecasting style goods items will be covered in Chapter 9.

3.8.1 Items with Limited History

Here, we finally address the issue of initialization of the exponential smoothing procedures of Section 3.4, when there is insufficient historical data to use the methods suggested earlier for obtaining estimates of the parameters a, b, the Fs, and MSE. There are two reasons for an item having insufficient historical data. First, the item may be a relatively new product (i.e., early in its product life cycle), as discussed in Section 2.5.1. Second, when a new control system is introduced in an organization, there may simply not be enough historical records even though the item has had sales for a considerable period of time.

For an item with insufficient history, a knowledgeable individual (likely a member of the judgmental input committee) should estimate

1. Total demand in the first year
2. Total demand in the second year
3. Seasonal pattern (if appropriate)

The first two give an estimate of the level and trend. The third most likely would be provided by saying that the item's seasonal behavior should be the same as an existing individual item or group of items.

It is important to provide a graph of the implied 2-year demand pattern so that adjustments can be made, if necessary. Higher-than-normal smoothing should be used in the first several periods because of the relative inaccuracy of the initial estimates of the parameters.

The MSE (or σ) of an item with insufficient history is estimated using an empirical relationship between σ and the demand level, a, found to hold for more established items in the same inventory population. In particular, a relationship that has been observed to give a reasonable fit for many organizations is of the form

$$\sigma_1 = c_1 a^{c_2} \tag{3.72}$$

or equivalently

$$\log \sigma_1 = \log c_1 + c_2 \log a \tag{3.73}$$

where c_1 and c_2 are regression coefficients typically with $0.5 < c_2 < 1$. The relationship of Equation 3.73 will plot as a straight line on log–log graph paper.

Sometimes, Equation 3.72 is modified to estimate the standard deviation on a dollar basis

$$\sigma_1 v = c_1 (av)^{c_2} \tag{3.74}$$

where v is the unit variable cost of an item.

This relationship can be used on an on-going basis as needed to estimate the standard deviation of any C items because these items may not be statistically forecasted, and therefore their errors would not be recorded. Relationships other than Equation 3.72 have been suggested in the literature. See, for example, Hausman and Kirby (1970), Stevens (1974), Shore (1995, 1998), and Kurawarwala and Matsuo (1996). The papers by Shore provide a simple procedure for approximating the distribution when data are extremely limited.

3.8.1.1 Illustration

Consider a new item with an estimated demand of 300 units per year. The person responsible for forecasting this item used several hundred other items similar to this one to find c_1 and c_2 by regression, with the following result:

$$\sigma_1 = 1.04(D/12)^{0.65}$$

where D is the annual demand. Therefore, the initial estimate of σ_1 is

$$\sigma_1 = 1.04(300/12)^{0.65} = 8.43 \text{ units}$$

and the initial estimate of MSE is

$$\text{MSE} = (8.43)^2 = 71.02$$

3.8.2 Intermittent and Erratic Demand

The exponential smoothing forecast methods described earlier have been found to be ineffective where transactions (not necessarily unit sized) occur on a somewhat infrequent basis. To illustrate this point, let us consider a particularly simple situation where only transactions of a single size j occur exactly every n periods of time. Suppose that simple exponential smoothing (level only) is used and the updating is done every unit period. Note that immediately after a demand transaction occurs, the forecast will exceed the average demand per period, because of the weight given to the last demand of size j. Because there is zero demand for the next $n - 1$ periods, the forecast decreases thereafter, dropping below the average demand by the end of the cycle. In other words, for a deterministic demand pattern, exponential smoothing produces a sawtooth-type forecast.

This illustration is a symptom of an undesirable performance of exponential smoothing under the more general situation of infrequent transactions, not necessarily all of the same size. For such a situation, it is preferable to forecast two separate components of the demand process—namely, the time between consecutive transactions and the magnitude of individual transactions. That is, we let

$$x_t = y_t z_t \tag{3.75}$$

where

$$y_t = \begin{cases} 1 \text{ if a transaction occurs} \\ 0 \text{ otherwise} \end{cases}$$

and z_t is the size of the transaction.

If we define a quantity n as the number of periods between transactions and if demands in separate periods are considered to be independent, then, the occurrence or nonoccurrence of a transaction in a period can be considered as a Bernoulli process with the probability of occurrence being $1/n$. That is

$$\Pr\{y_t = 1\} = 1/n$$

and

$$\Pr\{y_t = 0\} = 1 - 1/n$$

Within this framework and assuming that transaction sizes are normally distributed, Croston (1972) recommends the following reasonable updating procedure. If $x_t = 0$ (i.e., no demand occurs),

1. Transaction size estimates are not updated, that is, $\hat{z}_t = \hat{z}_{t-1}$
2. $\hat{n}_t = \hat{n}_{t-1}$

If $x_t > 0$ (i.e., a transaction occurs),

1. $\hat{z}_t = \alpha x_t + (1 - \alpha)\hat{z}_{t-1}$
2. $\hat{n}_t = \alpha n_t + (1 - \alpha)\hat{n}_{t-1}$

where

    n_t = number of periods since the last transaction

    $\hat{n}_t$ = estimated value of n at the end of period t

    $\hat{z}_t$ = estimate, at the end of period t, of the average transaction size

Croston argues that forecasts for replenishment purposes will usually be needed immediately after a transaction. He shows that the forecast at that time, namely,

$$\hat{z}_t / \hat{n}_t$$

is preferable to that obtained by simple exponential smoothing for two reasons:

1. It is unbiased, whereas the simple smoothing forecast is not.
2. It has a lower variance than does the simple smoothing forecast.

However, Croston warns that the infrequent updating (only when a transaction occurs) introduces a marked lag in responding to actual changes in the underlying parameters. Therefore, he rightfully stresses the importance of control signals to identify deviations.

A key quantity to use in establishing the safety stock for an intermittently demanded item is MSE(z), the mean square error of the sizes of *nonzero*-sized transactions. This quantity is only updated each time a transaction occurs. The updating equation is of the form

$$\text{New MSE}(z) = \omega(x_t - \hat{z}_{t-1})^2 + (1 - \omega)\text{Old MSE}(z) \qquad x_t > 0 \qquad (3.76)$$

In establishing the safety stock, one must also take into account of the fact that there is a nonzero chance of no transaction occurring in a period. Willemain et al. (1994) show that Croston's method is superior to exponential smoothing, and also note that the lognormal distribution is preferable to the normal for transaction sizes. Further discussion on Croston's method can be found in Hyndman et al. (2008). Johnston (1980) suggests an alternate method for directly estimating σ_1 for an intermittent demand item. See also Wright (1986), Johnston (1993), Johnston and Boylan (1994, 1996), Anderson (1994), Walton (1994), and Syntetos and Boylan (2005).

An item is said to have an erratic demand pattern if the variability is large relative to the mean. Brown (1977) operationalizes this by saying that demand is erratic when $\hat{\sigma}_1 > \hat{a}$ where $\hat{\sigma}$ is obtained by a fit to history, as in Section 3.5.1. In such a case, he suggests restricting the underlying model to the case of only a level and a random component. See also Muir (1980), Watson (1987), and Willemain (2004).

3.8.3 Replacement or Service Parts

In the case of replacement or service parts, we have more information than is usually available in a demand-forecasting situation. In particular, the requirements for replacement parts depend on the failure rate of operating equipment, which, in turn, depends on the time stream of the introduction of new (or repaired) pieces of equipment as well as the distribution of the service life of the equipment. Although it is relatively easy to state the above relationship, it is quite another thing to operationalize it. In particular, the mathematics become extremely involved. Also, it is neither easy to maintain the required records nor to estimate the service-life distribution. Keilson and Kubat (1984) develop tractable results for the special case where new equipment is introduced according to a time-varying Poisson distribution and failures also occur according to a Poisson process. A closely related problem is the forecasting of the returns of rented items or containers (e.g., soft drink bottles, beer kegs, etc.). See Goh and Varaprasad (1986), Kelle and Silver (1989), Fleischmann et al. (1997), Guide and Wassenhove (2001), and the references in Section 11.5.

3.8.4 Terminal Demand

Here, we are dealing with exactly the opposite situation from that in Section 3.8.1. Now, we are nearing the end of the life cycle of the item. It has been empirically observed (see, e.g., Brown et al. 1981; Fortuin 1980) that in the terminal phase of demand, the pattern tends to decrease geometrically with time. That is, the demand in period $t + 1$ is a constant fraction (f) of the demand in period t. Mathematically, we then have

$$x_{t+1} = f x_t$$

or, more generally,

$$x_t = f^t x_0 \qquad t = 0, 1, 2, \ldots \qquad (3.77)$$

In this last equation, the time origin has been placed at the beginning of, or somewhere during, the terminal phase of the pattern.

Taking the logarithms of Equation 3.77 reveals that

$$\log x_t = t \log f + \log x_0 \qquad (3.78)$$

Thus, a plot of x_t versus t on a semilogarithmic graph should produce a straight line with slope $\log f$. Such a line can be fit by standard regression techniques. Brown (1981) develops an expression for the variance of an extrapolated forecast into the future as a function of the residual variance of the historical fit, the number of historical data points used in the fit, and how far into the future the projection is made.

If the origin is placed at the current period, and the current level is x_0, then, an estimate of the all-time future requirements using Equation 3.77 is given by

$$ATR = \sum_{\tau=1}^{\infty} \hat{x}_{0,\tau}$$

$$= \sum_{\tau=1}^{\infty} f^\tau x_0 = \frac{f}{1-f} x_0 \qquad (3.79)$$

3.8.4.1 Illustration

Consider a particular spare part for a late-model car. Suppose that the demand has begun dropping off geometrically with a factor $f = 0.81$ and the demand level in the current year is 115 units. From Equation 3.79, the all-time future requirements are estimated to be

$$ATR = \frac{0.81}{0.19}(115) \approx 490 \text{ units}$$

For other information on this topic, including a discussion of the sizing of replenishment batches, see Hill et al. (1989), and Section 8.5.

3.9 Assessing Forecasting Procedures: Tactics and Strategy

In this section, we first address a tactical question—namely, measuring and comparing the statistical accuracy of different forecasting procedures. Then, we return to the issues of a more strategic nature.

3.9.1 *Statistical Accuracy of Forecasts*

The first point to note is that there are several possible measures of forecast accuracy—for example, the MSE, the MAD, the sample variance, the sample bias, any of these normalized by the mean demand level, and so forth. No single measure is universally best. In fact, forecast accuracy is only a surrogate for overall production/inventory system performance (encompassing the uses of the forecasts). The surrogate is usually justified in terms of keeping the evaluation procedure of a manageable size. We return to this point in Section 3.9.2.

Besides the choice of several possible statistical measures of forecast accuracy, there is the added problem of not even being able to make valid statistical claims about the relative sizes of the same measure achieved by two different forecasting procedures. The reason for this is that the two series of forecast errors are correlated to an unknown extent. (See Peterson 1969.)

To complicate matters, it can be shown that a linear combination of forecasts from two or more procedures can outperform all of those individual procedures. Some research has even shown that attempts to find optimal weights when combining different forecasts are not worth the effort. A simple average provides very good results. (See Kang 1986.) In theory, the best combination of procedures depends on the value of the measure of variability (e.g., the MSE) for each individual procedure, as well as the correlation between the results of the procedures. However, it is difficult to accurately estimate these quantities. Hence, if combinations of forecasts are to be considered, we recommend trying combinations of only two at a time, and starting with equal weights. References on combinations of forecasts include Bates and Granger (1969), Dickinson (1975), Lewis (1978), Makridakis and Winkler (1983), Armstrong (1984, 2001a), Ashton and Ashton (1985), Flores and Whybark (1986), Georgoff and Murdick (1986), Lawrence et al. (1986), Gupta and Wilton (1987), Bunn (1988), Sanders and Ritzman (1989), Schmittlein et al. (1990), and Timmermann (2006).

In developing the relative measures of accuracy, it is crucial to use an ex-ante, as opposed to an ex-post, testing procedure. In an ex-post approach, the forecast model and updating procedure are tested on the same data from which the model form or smoothing constant values were selected. In contrast, the ex-ante method involves testing with an entirely different set of data. This is usually accomplished by using a first part of the historical information for fitting purposes, and the remaining portion for testing. Performance on ex-post tests can be a very poor indicator of ex-ante performance, and it is the latter that is of practical interest. See Fildes and Howell (1979), and Makridakis (1990) for further discussion and references.

Numerous tests that have measured the statistical accuracy of different forecasting procedures have been reported in the literature. However, of primary interest is the comprehensive study undertaken by Makridakis et al. (1982), subsequently followed by Makridakis et al. (1993) and Makridakis and Hibon (2000). It involved some 1,001 time series including both micro- and macrolevel data as well as monthly, quarterly, and annual forecast update intervals. All the common statistical forecasting procedures (from very simple to quite complex) were tested, and a number of measures of statistical accuracy were obtained for each combination of procedure and time series. Some of the salient findings were the following:

1. Significant differences in forecast accuracy occurred as a function of the micro-/macroclassification. On microdata (the case in individual-item forecasting), the relatively simple methods did much better than the statistically sophisticated methods. The latter were better, however, for macrodata (the case for medium or even longer-term forecasting).
2. For horizons of 1–6 periods ahead, simple exponential smoothing, Holt and Brown trend procedures, and the Winters method performed well. For horizons of 7 or more periods, a more complicated procedure, developed by Lewandowski (1979), showed the best performance.
3. Deseasonalization of the data by the ratio to moving average method (which we suggested in Section 3.4.4) did as well as much more sophisticated deseasonalization procedures.
4. Adaptive methods did not outperform nonadaptive methods.
5. A simple average of six of the simpler statistical methods did somewhat better than any of the individual methods, and also outperformed a combination of the methods that in theory was supposed to give the best results.
6. In general, sophisticated methods did not produce more accurate results than did the simpler methods.

In summary, the results of this major international study lend strong support to the forecasting procedures that we have advocated in this chapter. See also Fildes (1989) and Armstrong (2001b).

3.9.2 Some Issues of a More Strategic Nature

As Fildes (1979) points out, two questions dominate any assessment of a forecasting procedure. First, are the results statistically satisfactory (the topic covered in Section 3.9.1)? Second, will the procedure, once developed, be used and perform in a cost-effective fashion? In Fildes's words, "it is the latter that is often of more concern to the practicing forecaster." This view is shared by the authors of the Study Guide for the Certification Program of the American Production and Inventory Control Society (see Blagg et al. 1980):

> Forecasting is not done for its own sake; it is meaningful only as it relates to and supports decision making within the production system. Hence, managerial considerations regarding cost, effectiveness, and appropriateness of the forecasting function must be a part of the domain of forecasting.

Earlier, we mentioned that it has often been advocated that the statistical accuracy of a forecasting method should be used as a surrogate for the portion of system costs associated with forecast errors. Unfortunately, studies have shown that traditional error measures do not necessarily relate to overall system costs. Thus, where possible, a choice between forecasting procedures should be based partly on a comparative simulation test of system costs using historical data or a trial run on new data.

Closely related to this point is the desirable property of robustness (Fildes 1979). The choice of a particular demand model and updating procedure should again be at least partially based on an insensitivity to the data quality, missing observations, outliers, and so forth. We hope that the suggestions made in Section 3.6 would benefit any selected procedure from a standpoint of robustness.

A study by Adam and Ebert (1976) (see also Hogarth and Makridakis 1981) compares objective (statistical model) forecasts with judgmental forecasts. Interestingly enough, the objective forecasts, on the average, outperformed the subjective forecasts. However, Bunn and Wright (1991) note that expert, informed judgmental forecasts have higher quality than many researchers have previously asserted. Therefore, it is essential to solicit human input to the forecasting process. To have meaningful subjective input, the manager must be able to understand the underlying assumptions and general idea of the statistical model. This is an important reason for why we have placed so much emphasis on the relatively simple exponential smoothing procedures. Moreover, in Chapters 10 and 15, where there will be a more explicit dependence of demand among different SKUs (multi-echelon inventory systems and multistage production processes), we will emphasize a different type of forecasting that involves the explosion of known future usage.

Related to the usefulness of a forecasting system, Makridakis and Wheelwright (1979) list a number of potential organizational blocks to improved forecasting. Of specific interest here are (1) bias and (2) credibility and communication. By bias, we mean that in many organizations, there is incentive for a forecast to not represent the most likely outcome but, instead, to serve as a self-fulfilling goal. For example, sales personnel may deliberately provide low forecasts if they are rewarded for exceeding the target levels based, at least in part, on the forecasted values. By credibility, we mean that the forecasts may lack relevance, or appear to lack relevance, which often is a result of a lack of proper communication between the forecast analyst and the decision maker. Additional information is provided by Sanders and Manrodt (1994, 2003b) who find that the lack of relevant data and low organizational support are the primary reasons that formal methods are not used as extensively in U.S. corporations.

In conclusion, we recall the experience of one of the authors who was once asked to mediate between competing forecasting groups at a large international corporation. Each group was obviously very sophisticated in mathematical time-series analysis and very clever in coaxing the most obscure of patterns out of historical data through spectral analysis. However, none of the nine highly trained mathematician forecasters involved had ever really closely examined a retail outlet or wholesaler through which the products that they forecasted were being sold. To most of them, products were characterized by amplitudes and phase angles.

Top managers for their part were somewhat intimidated by all the apparent sophistication (as was the author initially). Little collaboration existed between the users and the producers of forecasts. Both management and analysts ignored the fact that one really cannot foretell the future with great accuracy. Tactics without any overall strategy prevailed. On closer examination, it became clear that the management was not taking the mathematicians' forecasts seriously because they did not understand how they were being generated. The problem that brought matters to a head was the entrance of an aggressive competitor into the market who claimed through massive advertising that the company's basic product concept was technologically obsolete. It took for top management several weeks to react because they were unable to determine the extent of the impact on their sales until wholesalers and retailers started to refuse previously planned shipments.

Eventually the situation was resolved as follows. Instead of trying to foretell the future, top management agreed to set up better procedures that would allow the company to quickly react to forecast errors as soon as their magnitude was determined. This involved a basic change in management philosophy. No longer were forecast errors a surprise event. They were fully anticipated, and everyone in the decision hierarchy prepared contingency plans in response. All agreed that it was at least equally as important to develop techniques for responding to errors, as it was to try to measure and predict them. In the process, six of the high-powered time-series analysts were laid off, resulting in a substantial reduction of overall costs associated with forecasting as well as in a better forecasting strategy.

Problems

3.1 Suppose that the following were sales of beer in the Kitchener-Waterloo area by month for the 3 years shown:

	Jan.	Feb.	Mar.	Apr.	May	June	July	Aug.	Sept.	Oct.	Nov.	Dec.
2012	73	69	68	64	65	98	114	122	74	56	72	153
2013	80	81	83	69	91	140	152	170	97	122	78	177
2014	95	66	81	100	93	116	195	194	101	197	80	189

  a. Plot the data roughly to scale.
  b. Is there a trend present?
  c. Is there any seasonality? Discuss.
  d. Are there any anomalies? Discuss.
  e. What is your forecast of demand for January 2015? For June 2015?

3.2 Southern Consolidated manufactures gas turbines. Some of the parts are bought from outside, while the important ones are made in the company's plant. CZ-43 is an important part that wears out in approximately 2 years and has to be replaced. This means CZ-43 has to be manufactured not only to meet demand at the assembly line but also to satisfy its demand as a spare part. The gas turbine was first sold in 2002 and the demand for CZ-43 for 2002–2015 is given below. Forecast the total demand for CZ-43 for 2016.

Year	Spare Part	Assembly	Demand
2002	3	87	90
2003	6	48	54
2004	14	43	57
2005	30	31	61
2006	41	49	90
2007	46	31	77
2008	42	37	79
2009	60	42	102
2010	58	47	105
2011	67	32	99
2012	70	25	95
2013	58	29	87
2014	57	30	87
2015	85	23	108

3.3 Under the assumption of a level model, that is,

$$x_t = a + \epsilon_t$$

show that the choice of a to minimize

$$S = \sum_{j=t-N+1}^{t} (x_j - a_t)^2$$

where x_j is the actual demand in period j, is given by

$$\hat{a}_t = (x_t + x_{t-1} + \cdots + x_{t-N+1})/N$$

3.4 Prove that the average age of the data used in the estimate of Equation 3.25 is given by $1/\alpha$.

3.5 The number of violent crimes in each of the last 6 weeks in the Hill Road Precinct has been:

Week j	1	2	3	4	5	6
Incidents x_j	83	106	95	91	110	108

a. Use a 2-week moving average to forecast the incidents in each of weeks 7 and 10. Also compute the MSE based on the incidents and forecasts for periods 1–6. To initialize, assume $x_{-1} = x_0 = 100$.

b. Detective Bick Melker suggests the use of simple (single) exponential smoothing with $\alpha = 0.1$. Use such a procedure to develop forecasts of incidents in weeks 7 and 10 for Hill Road. Initialize with $\hat{a}_0 = 100$. Also compute the MSE and MAD.

c. Without redoing all the computations, what would be the forecast of incidents in week 7 using exponential smoothing if $\hat{a}_0$ were 90 instead of 100?

3.6 Brown's updating equations, developed from discounted least squares, for the case of a trend model are actually given by

$$\hat{a}_t = x_t + (1 - \alpha)^2 (\hat{a}_{t-1} + \hat{b}_{t-1} - x_t)$$

and

$$\hat{b}_t = \hat{b}_{t-1} - \alpha^2 (\hat{a}_{t-1} + \hat{b}_{t-1} - x_t)$$

Show that these two equations are equivalent to Equations 3.32 and 3.33.

3.7 The demand for a particular model of tablet at a particular outlet of Great Buy has had the following pattern during the past 15 weeks:

Week t	1	2	3	4	5	6	7	8	9	10	11	12	13	14	15
Demand x_t	34	40	34	36	40	42	46	50	55	49	49	53	63	69	65

A systems analyst at Great Buy believes that demand should be trending upward; thus, she wishes to evaluate the use of exponential smoothing with a trend model.

a. Use the first 5 weeks of data in Equations 3.36 and 3.37 to obtain estimates of $\hat{a}$ and $\hat{b}$ as of the end of week 5.

b. Using an α value of 0.15, employ exponential smoothing to update $\hat{a}$ and $\hat{b}$ and show the forecasts for weeks $6, 7, \ldots, 16$ made at the start of each of these periods. Also, what is the forecast, made at the end of week 15, of the demand in week 20?

3.8 Union Gas reports the following (normalized) figures for gas consumption by households in a metropolitan area of Ontario:

	Jan.	Feb.	Mar.	Apr.	May	June	July	Aug.	Sept.	Oct.	Nov.	Dec.
2012	196	196	173	105	75	39	13	20	37	73	108	191
2013	227	217	197	110	88	51	27	23	41	90	107	172
2014	216	218	193	120	99	59	33	37	59	95	128	201
2015	228											

a. Using a simple (single) exponential smoothing model (with $\alpha = 0.10$), determine the forecast for consumption in (1) February 2015 and (2) June 2016.

b. Do the same using the Winters seasonal smoothing model with $\alpha_{HW} = 0.20$, $\beta_{HW} = 0.05$, and $\gamma_{HW} = 0.10$. Use the first 2 years of data for initialization purposes.

c. Plot the demand data and the two sets of forecasts roughly to scale. Does there appear to be a point of abrupt change in the demand pattern? Discuss your results briefly.

3.9 For purposes of calculating the safety stock of a particular item, we are interested in having an estimate of the MAD or MSE of forecast errors. No historical forecasts are available, but historical monthly demand data for 3 years are available. An analyst on your staff has proposed using the MAD with the following procedure for estimating it:

Step 1 Plot the 3 years of data.
Step 2 Fit the best straight line to it (either by eye or through a statistical fit).
Step 3 For each month find the absolute deviation

$$|\epsilon_t| = |x_t - \tilde{x}_t|$$

where x_t is the actual demand in month t and $\tilde{x}_t$ is the value read off the straight line (found in Step 2) at month t.
Step 4 MAD = average value of $|\epsilon_t|$ over the 36 months.

a. Discuss why the MAD found by the above procedure might be lower than that actually achievable by statistical forecasting.

b. Discuss why the MAD found by the above procedure might be substantially higher than that actually achievable by a forecasting procedure. Hint: The use of diagrams may be helpful.

c. Why might the MSE be a better choice?

3.10 Demand for a new breakfast cereal at a medium-sized grocery store is given in the table below:

Month	Demand	Month	Demand	Month	Demand
1	165	9	357	17	585
2	201	10	410	18	702
3	239	11	603	19	775
4	216	12	449	20	652
5	251	13	479	21	586
6	269	14	569	22	760
7	299	15	623	23	788
8	332	16	591	24	817

   a. Using a simple (single) exponential smoothing model (with $\alpha = 0.15$), determine the forecast for consumption in (1) month 25 and (2) month 30.

   b. Do the same using the trend model with $\alpha_{HW} = 0.20$ and $\beta_{HW} = 0.05$. Use the first year of data for initialization purposes.

   c. Plot the demand data and the two sets of forecasts roughly to scale. Discuss your results briefly.

3.11 In current times, give an illustration of an item fitting into each of the following special classes not using the specific examples in the book (use your everyday experiences at work or elsewhere):

   a. Limited history

   b. Intermittent demand

   c. Erratic demand

   d. Replacement parts

   e. Terminal demand

3.12 Suppose that the demand rate for an item could be modeled as a continuous linearly decreasing function of time

$$x_t = a - bt \quad a > 0, \quad b > 0$$

where $t = 0$ is the present time. Compute an estimate of the all-time future requirements.

3.13 Suppose you have been hired by a company to install a forecasting system for about 5,000 B items. You suggest the use of exponential smoothing and the manager to whom you report is not happy about this. He says "Exponential smoothing is nothing but a massaging of historical data. It takes no account of our pricing strategy (e.g., special sales at certain times of the year), our competitor's actions, etc."

   a. How would you respond to this criticism?

   b. If the manager insisted on having pricing effects built into the forecasting model, briefly outline how you would proceed to do this, including what data you would attempt to obtain.

3.14 Develop a spreadsheet that computes the simple 5-period moving average and uses it for forecasting. Include the computation of the forecast error.

3.15 Develop a spreadsheet that performs the simple level exponential smoothing operation and uses it for forecasting. Include the computation of the forecast error.

3.16 Develop a spreadsheet that performs exponential smoothing with trend and uses it for forecasting. Include the computation of the forecast error.

3.17 Develop a spreadsheet that performs exponential smoothing with seasonality and uses it for forecasting. Include the computation of the forecast error.

3.18 For a local business, find several reasons why demand in the next few periods might be very different from demand in the recent past. How should the forecast be modified?

3.19 For a local business, find several factors external to the firm that might influence the forecast of demand. How should the forecast be modified?

3.20 A summer student working at a small local firm has been assigned the task of forecasting. The president tells her that for some products, there is not enough relevant historical demand to use any of the traditional forecasting techniques described in this chapter. What are the possible reasons the president could give? What should the summer student do? If the student wanted to use historical demand of a related item to forecast one of the items that does not have enough relevant history, what factors should she consider?

3.21 One of the major advantages of the new computer-aided manufacturing technology is that it permits a shorter design and manufacturing lead time, that is, a faster market response. Describe the impact of the faster market response on (i) forecasting requirements and (ii) the required inventory levels.

3.22 A common way of testing forecasting procedures consists of monitoring how a particular procedure responds to a step function in demand or a one period transient change in demand level. (Needless to say, the step and transient functions are only two of the many functions that can be used.)

Time	Step Function	Transient Change
1	100	100
2	100	100
3	150	150
4	150	100
5	150	100
6	150	100
7	150	100
8	150	100
9	150	100
10	150	100

a. Use both a simple and a trend exponential smoothing model (trying both $\alpha = 0.1$ and $\alpha = 0.3$) to track both of the above demand patterns. For initial conditions, assume at the start of time 1 (end of period 0) that $\hat{a} = 100, \hat{b} = 0$.

b. Plot the corresponding forecast errors under each procedure.

3.23 Demand for a given product has been tracked for 5 years, and the results are shown below:
   a. Use the first 4 years of demand to establish a simple (single) exponential smoothing model (with $\alpha = 0.10$). At the end of month 48, what is your forecast for month 49? Month 60?
   b. Employ exponential smoothing with trend (with $\alpha_{HW} = 0.15$ and $\beta_{HW} = 0.05$) to update $\hat{a}$ and $\hat{b}$ and show the forecasts for weeks $49, 50, \ldots, 60$ made at the start of each of these periods.
   c. Plot the demand data and the forecasts roughly to scale. Does there appear to be a point of abrupt change in the demand pattern? Discuss your results briefly.

Month	Demand	Month	Demand	Month	Demand
1	239	21	261	41	165
2	186	22	223	42	119
3	164	23	368	43	176
4	179	24	231	44	177
5	229	25	219	45	200
6	219	26	142	46	173
7	259	27	179	47	260
8	208	28	188	48	192
9	201	29	237	49	136
10	186	30	126	50	184
11	166	31	237	51	239
12	196	32	234	52	219
13	191	33	300	53	174
14	229	34	174	54	208
15	169	35	187	55	241
16	254	36	154	56	220
17	155	37	211	57	194
18	158	38	129	58	197
19	210	39	207	59	218
20	143	40	175	60	252

3.24 A company sells a group of items having a seasonal demand pattern. It is reasonable to assume that all items in the group have the same seasonal pattern. The forecasting procedure currently in use is as follows: From historical data on the whole group, reasonably accurate estimates are available for the fraction of annual demand that occurs in each of the 12 months. Let us call these $F_1, F_2, \ldots, F_{12}$ and let

$$C_k = \sum_{j=1}^{k} F_j$$

be the cumulative fraction of annual demand historically experienced through month k, $k = 1, 2, \ldots, 12$.

Consider a particular item i and suppose that its actual demand in month j is D_{ij}. At the end of month k, the annual demand projection for item i, denoted by P_{ik}, is given by

$$P_{ik} = \frac{\sum_{j=1}^{k} D_{ij}}{C_k}$$

and the forecast for a single future month m in the same year is

$$P_{ikm} = p_{ik} F_m$$

a. Illustrate the above forecasting method with annual rate projections and 1-month-ahead forecasts made at the end of each of Jan., Feb., ..., Dec. and Jan. for an item having the following monthly demands in a particular 13-month period:

	Jan	Feb										Dec	Jan
Month, j	1	2	3	4	5	6	7	8	9	10	11	12	13
D_j	20	20	81	87	106	215	147	125	84	128	122	28	77

The historically determined $F_i's$ for the associated group of items are

Month, j	1	2	3	4	5	6	7	8	9	10	11	12	13
F_j		0.03	0.04	0.06	0.07	0.09	0.16	0.14	0.12	0.11	0.09	0.06	0.03

b. The inventory manager claims that the method is very poor in the early months of a year, but gets better and better as one approaches the end of the year. Does this make sense? If so, how would you suggest modifying the method so as to improve matters early in the year? (There is no need to do a lot of numerical work here.)

3.25 Monthly demand at a local auto parts store for automobile batteries is shown in the table below:

Month	Demand	Month	Demand	Month	Demand
1	34	9	38	17	58
2	44	10	44	18	54
3	42	11	36	19	43
4	30	12	46	20	48
5	46	13	42	21	40
6	44	14	30	22	46
7	56	15	52	23	52
8	50	16	48	24	54

a. Using a simple (single) exponential smoothing model (with $\alpha = 0.15$), determine the forecast for consumption in (1) month 25 and (2) month 30.
b. Do the same using the Winters seasonal smoothing model with $\alpha_{HW} = 0.20$, $\beta_{HW} = 0.05$, and $\gamma_{HW} = 0.10$. Use the first year of data for initialization purposes.
c. Plot the demand data and the two sets of forecasts roughly to scale. Discuss your results briefly.

Appendix 3A: Derivations

3A.1 Derivation of Simple Exponential Smoothing

From Equation 3.22, we know that we wish to select $\hat{a}_t$ in order to minimize

$$S' = \sum_{j=0}^{\infty} d^{j+1}(x_{t-j} - \hat{a}_t)^2$$

By setting $dS'/d\hat{a}_t = 0$, we obtain

$$-2d \sum_{j=0}^{\infty} d^{j}(x_{t-j} - \hat{a}_t) = 0$$

or

$$\hat{a}_t \sum_{j=0}^{\infty} d^{j} = \sum_{j=0}^{\infty} d^{j}x_{t-j} \tag{3A.1}$$

Now

$$\sum_{j=0}^{\infty} d^{j} = \frac{1}{1-d}$$

Therefore, Equation 3A.1 becomes

$$\hat{a}_t = (1 - d) \sum_{j=0}^{\infty} d^j x_{t-j} \tag{3A.2}$$

$$\hat{a}_t = (1 - d) \left(x_t + \sum_{j=1}^{\infty} d^j x_{t-j} \right)$$

By substituting $r = j - 1$ in the summation, we obtain

$$\hat{a}_t = (1 - d)x_t + (1 - d)d \sum_{j=0}^{\infty} d^r x_{t-1-r} \tag{3A.3}$$

From Equation 3A.2, we see that

$$\hat{a}_{t-1} = (1 - d) \sum_{r=0}^{\infty} d^r x_{t-1-r}$$

By substituting into Equation 3A.3

$$\hat{a}_t = (1 - d)x_t + d\hat{a}_{t-1}$$
$$= \alpha x_t + (1 - \alpha)\hat{a}_{t-1}$$

where $\alpha = 1 - d$.

3A.2 Relationship between MAD and Standard Deviation for Unbiased Forecasts

Let MAD represent mean absolute deviation of an unbiased forecast and let σ represent the standard deviation of forecast errors (also for an unbiased forecast). When forecast errors are normally distributed, the following relationship holds:

$$\sqrt{\frac{\pi}{2}} \text{MAD} = \sigma$$

To derive this relationship, note that the unbiased forecast would be mean demand, $\hat{x}$. So, mean absolute deviation can be expressed as

$$\text{MAD} = \int_{-\infty}^{\hat{x}} (\hat{x} - x_0) f_x(x_0) dx_0 + \int_{\hat{x}}^{\infty} (x_0 - \hat{x}) f_x(x_0) dx_0$$

where $f_0(x_0)$ is the probability density function of the normal distribution

$$f_x(x_0) = \frac{1}{\sigma \sqrt{2\pi}} e^{-\frac{(x_0 - \hat{x})^2}{2\sigma^2}}$$

Note the following:

$$f'_x(x_0) = \frac{1}{\sigma\sqrt{2\pi}}e^{-\frac{(x_0-\hat{x})^2}{2\sigma^2}}\left(\frac{-(x_0-\hat{x})}{\sigma^2}\right) = \left(\frac{\hat{x}-x_0}{\sigma^2}\right)f_x(x_0)$$

or equivalently,

$$\sigma^2 f'_x(x_0) = (\hat{x} - x_0)f(x_0)$$

which implies

$$\int (\hat{x} - x_0)f_x(x_0)dx_0 = \int \sigma^2 f'_x(x_0)dx_0 = \sigma^2 f_x(x_0)$$

Using this, we can rewrite MAD as

$$\mathrm{MAD} = \int_{-\infty}^{\hat{x}} \sigma^2 f'_x(x_0)dx_0 - \int_{\hat{x}}^{\infty} \sigma^2 f'_x(x_0)dx_0$$

$$= \sigma^2 \int_{-\infty}^{\hat{x}} f'_x(x_0)dx_0 - \sigma^2 \int_{\hat{x}}^{\infty} f'_x(x_0)dx_0$$

$$= \sigma^2 \left[(f_x(\hat{x}) - 0) - (0 - f_x(\hat{x})) \right]$$

$$= \sigma^2 \frac{2}{\sigma\sqrt{2\pi}}$$

$$\mathrm{MAD} = \sigma\sqrt{\frac{2}{\pi}}$$

Rearranging, this gives the relationship

$$\mathrm{MAD}\sqrt{\frac{\pi}{2}} = \sigma \approx 1.25 MAD$$

References

Adam, E. E. and R. J. Ebert. 1976. A comparison of human and statistical forecasting. *AIIE 8*, 120–127.

Adshead, N. S. and D. H. R. Price. 1987. Demand forecasting and cost performance in a model of a real manufacturing unit. *International Journal of Production Research 25*(9), 1251–1265.

Agrawal, N. and S. A. Smith. 1996. Estimating negative binomial demand for retail inventory management with unobservable lost sales. *Naval Research Logistics 43*(6), 839–861.

Anderson, J. R. 1994. Simpler exponentially weighted moving averages with irregular updating periods. *Journal of the Operational Research Society 45*(4), 486–487.

Anderson, O. D. 1976. *Times Series Analysis and Forecasting: The Box–Jenkins Approach*. London: Butterworths.

Armstrong, J. S. 1984. Forecasting by extrapolation: Conclusions from 25 years of research. *Interfaces 14*(6), 52–66.

Armstrong, J. S. 2001a. Combining forecasts. In J. S. Armstrong (Ed.), *Principles of Forecasting*, pp. 417–439. New York: Springer.

Armstrong, J. S. 2001b. *Principles of Forecasting: A Handbook for Researchers and Practitioners*, Volume 30. New York: Springer Science & Business Media.

Ashton, A. H. and R. H. Ashton. 1985. Aggregating subjective forecasts: Some empirical results. *Management Science 31*(12), 1499–1508.

Azoury, K. S. 1985. Bayes solution to dynamic inventory models under unknown demand distribution. *Management Science 31*(9), 1150–1160.

Basu, S. and R. C. Schroeder. 1977. Incorporating judgments in sales forecasts: Application of the Delphi method at American Hoist & Derrick. *Interfaces 7*(3), 18–27.

Bates, J. and C. Granger. 1969. The combination of forecasts. *Operational Research Quarterly 20*(4), 451–468.

Bell, P. C. 1981. Adaptive forecasting with many stockouts. *Journal of the Operational Research Society 32*(10), 865–873.

Berger, J. O. 2013. *Statistical Decision Theory and Bayesian Analysis*. New York: Springer Science & Business Media.

Blagg, B., G. Brandenburg, R. Conti, D. Fogarty, T. Hoffman, R. Martins, J. Muir, P. Rosa, and T. Vollmann. 1980. APICS Certification Program Study Guide and Review Course Outline.

Box, G. E., G. M. Jenkins, and G. C. Reinsel. 2011. *Time Series Analysis: Forecasting and Control*, Volume 734. Hoboken, NJ: John Wiley & Sons.

Bradford, J. W. and P. K. Sugrue. 1991. Inventory rotation policies for slow moving items. *Naval Research Logistics 38*, 87–105.

Brown, G. G., A. Geoffrion, and G. Bradley. 1981. Production and sales planning with limited shared tooling at the key operation. *Management Science 27*(3), 247–259.

Brown, R. G. 1963. *Smoothing, Forecasting and Prediction*. Englewood Cliffs, NJ: Prentice-Hall.

Brown, R. G. 1977. *Materials Management Systems*. New York: Wiley-Interscience.

Brown, R. G. 1981. *Confidence in All-Time Supply Forecasts*. Norwich, VT: Materials Management Systems, Inc.

Bunn, D. and G. Wright. 1991. Interaction of judgemental and statistical forecasting methods: Issues and analysis. *Management Science 37*(5), 501–518.

Bunn, D. W. 1988. Combining forecasts. *European Journal of Operational Research 33*, 223–229.

Chatfield, C. 1978. The Holt-Winters forecasting procedure. *Applied Statistics 27*(3), 264–279.

Chatfield, C. 2013. *The Analysis of Time Series: An Introduction*. London: CRC Press.

Chow, W. M. 1965. Adaptive control of the exponential smoothing constant. *Journal of Industrial Engineering 16*(5), 314–317.

Cox, D. R. and H. D. Miller. 1977. *The Theory of Stochastic Processes*, Volume 134. Boca Raton, FL: CRC Press.

Dickinson, J. P. 1975. Some comments on the combination of forecasts. *Operational Research Quarterly 24*(2), 253–260.

Eilon, S. and J. Elmaleh. 1970. Adaption limits in inventory control. *Management Science 16*(18), B533–B548.

Ekern, S. 1981. Adaptive exponential smoothing revisited. *Journal of the Operational Research Society 32*(9), 775–782.

Eppen, G. D. and A. V. Iyer. 1997. Improved fashion buying using Bayesian updates. *Operations Research 45*(6), 805–819.

Fildes, R. 1979. Quantitative forecasting—The state of the art: Extrapolative models. *Journal of the Operational Research Society 30*(8), 691–710.

Fildes, R. 1983. An evaluation of Bayesian forecasting. *Journal of Forecasting 2*, 137–150.

Fildes, R. 1989. Evaluation of aggregate and individual forecast method selection rules. *Management Science 35*(9), 1056–1065.

Fildes, R. and P. Goodwin. 2007. Against your better judgment? How organizations can improve their use of management judgment in forecasting. *Interfaces 37*(6), 570–576.

Fildes, R., P. Goodwin, M. Lawrence, and K. Nikolopoulos. 2009. Effective forecasting and judgmental adjustments: An empirical evaluation and strategies for improvement in supply-chain planning. *International Journal of Forecasting 25*(1), 3–23.

Fildes, R. and S. Howell. 1979. On selecting a forecasting model. In S. Makridakis and S. C. Wheelwright (Eds.), *TIMS Studies in the Management Sciences*, Volume 12, pp. 297–312. Amsterdam: North Holland.

Fisher, M. and C. D. Ittner. 1999. The impact of product variety on automobile assembly operations: Empirical evidence and simulation analysis. *Management Science 45*(6), 771–786.

Fisher, M. L., J. H. Hammond, W. R. Obermeyer, and A. Raman. 1994. Making supply meet demand in an uncertain world. *Harvard Business Review May–June*, 83–93.

Fleischmann, M., J. M. Bloemhof-Ruwaard, R. Dekker, E. Van der Laan, J. A. Van Nunen, and L. N. Van Wassenhove. 1997. Quantitative models for reverse logistics: A review. *European Journal of Operational Research 103*(1), 1–17.

Fliedner, E., B. Flores, and V. Mabert. 1986. Evaluating adaptive smoothing models: Some guidelines for implementation. *International Journal of Production Research 24*(4), 955–970.

Flores, B. and D. C. Whybark. 1986. A comparison of focus forecasting with averaging and exponential smoothing. *Production and Inventory Management Journal 27*(3), 96–103.

Flowers, A. D. 1980. A simulation study of smoothing constant limits for an adaptive forecasting system. *Journal of Operations Management 1*(2), 85–94.

Fortuin, L. 1980. The all-time requirement of spare parts for service after sales—Theoretical analysis and practical results. *International Journal of Operations and Production Management 1*(1), 59–70.

Galbraith, C. S. and G. B. Merrill. 1996. The politics of forecasting: Managing the truth. *California Management Review 38*(2), 29–43.

Gardner, E. S. 1983. Approximate decision rules for continuous review inventory systems. *Naval Research Logistics Quarterly 30*, 59–68.

Gardner, E. S. 1990. Evaluating forecast performance in an inventory control system. *Management Science 36*(4), 490–499.

Gardner, E. S. 2006. Exponential smoothing: The state of the art—Part II. *International Journal of Forecasting 22*(4), 637–666.

Gardner, E. S. and D. Dannenbring. 1980. Forecasting with exponential smoothing: Some guidelines for model selection. *Decision Sciences 11*(2), 370–383.

Gardner, E. S. and E. McKenzie. 1985. Forecasting trends in time series. *Management Science 31*(10), 1237–1246.

Gardner, E. S. and E. McKenzie. 1989. Seasonal exponential smoothing with damped trends. *Management Science 35*(3), 372–376.

Gardner, E. S. J. 1988. A simple method of computing prediction intervals for time series forecasts. *Management Science 34*(4), 541–546.

Gaur, V., S. Kesavan, A. Raman, and M. L. Fisher. 2007. Estimating demand uncertainty using judgmental forecasts. *Manufacturing and Service Operations Management 9*(4), 480–491.

Georgoff, D. and R. Murdick. 1986. Manager's guide to forecasting. *Harvard Business Review 62*(1), 110–120.

Goh, T. N. and N. Varaprasad. 1986. A statistical methodology for the analysis of the life-cycle of reusable containers. *IIE Transactions 18*(1), 42–47.

Goodman, D. 1974. A goal programming approach to aggregate planning of production and work force. *Management Science 20*(12), 1569–1575.

Granger, C. W. J. and P. Newbold. 2014. *Forecasting Economic Time Series*. San Diego, CA: Academic Press.

Guide, V. D. R. and L. N. Wassenhove. 2001. Managing product returns for remanufacturing. *Production and Operations Management 10*(2), 142–155.

Gupta, S. and P. Wilton. 1987. Combination of forecasts: An extension. *Management Science 33*(3), 356–372.

Hamburg, M. and P. Young. 1994. *Statistical Analysis for Decision Making*. Belmont, CA: Wadsworth Publishing Company.

Harrison, P. and C. Stevens. 1971. A Bayesian approach to short-term forecasting. *Operational Research Quarterly 22*(4), 341–362.

Harrison, P. and C. Stevens. 1976. Bayesian forecasting. *Journal of the Royal Statistical Society B38*, 205–228.

Harrison, P. J. 1967. Exponential smoothing and short-term sales forecasting. *Management Science 13*(11), 821–842.

Harrison, P. J. and O. Davies 1964. The use of cumulative sum (CUSUM) techniques for the control of product demand. *Operations Research 12*(2), 325–333.

Hausman, W. H. and R. M. Kirby. 1970. Estimating standard deviations for inventory control. *AIIE Transactions 11*(1), 78–81.

Heath, D. and P. Jackson. 1994. Modeling the evolution of demand forecasts with application to safety stock analysis in production/distribution systems. *IIE Transactions 26*(3), 17–30.

Hill, A., V. Giard, and V. Mabert. 1989. A decision support system for determining optimal retention stocks for service parts inventories. *IIE Transactions 21*(3), 221–229.

Hogarth, R. and S. Makridakis. 1981. Forecasting and planning: An evaluation. *Management Science 27*(2), 115–138.

Holt, C. C. 2004. Forecasting seasonals and trends by exponentially weighted moving averages. *International Journal of Forecasting 20*(1), 5–10.

Hyndman, R., A. B. Koehler, J. K. Ord, and R. D. Snyder. 2008. *Forecasting with Exponential Smoothing: The State Space Approach*. Berlin: Springer Science & Business Media.

Jacobs, R. A. and H. M. Wagner. 1989. Lowering inventory systems costs by using regression-derived estimators of demand variability. *Decision Sciences 20*(3), 558–574.

Johnson, L. A. and D. C. Montgomery. 1974. *Operations Research in Production Planning, Scheduling and Inventory Control*, Vol. 6. New York: John Wiley & Sons.

Johnston, F. R. 1980. An interactive stock control system with a strategic management role. *Journal of the Operational Research Society 31*(12), 1069–1084.

Johnston, F. R. 1993. Exponentially weighted moving average (EWMA) with irregular updating periods. *Journal of the Operational Research Society 44*(7), 711–716.

Johnston, F. R. and J. E. Boylan. 1994. How far ahead can an EWMA model be extrapolated. *Journal of the Operational Research Society 45*(6), 710–713.

Johnston, F. R. and J. E. Boylan. 1996. Forecasting for items with intermittent demand. *Journal of the Operational Research Society 47*, 113–121.

Johnston, F. R. and P. J. Harrison. 1986. The variance of lead-time demand. *Journal of the Operational Research Society 37*(3), 303–308.

Kang, H. 1986. Unstable weights in the combination of forecasts. *Management Science 32*(6), 683–695.

Kaplan, A. J. 1988. Bayesian approach to inventory control of new parts. *IIE Transactions 20*(2), 151–156.

Karmarkar, U. 1994. A robust forecasting technique for inventory and leadtime management. *Journal of Operations Management 12*(1), 45–54.

Keilson, J. and P. Kubat. 1984. Parts and service demand distribution generated by primary production growth. *European Journal of Operational Research 17*(2), 257–265.

Kelle, P. and E. A. Silver. 1989. Purchasing policy of new containers considering the random returns of previously issued containers. *IIE Transactions 21*, 349–354.

Kurawarwala, A. A. and H. Matsuo. 1996. Forecasting and inventory management of short life cycle products. *Operations Research 44*(1), 131–150.

Lariviere, M. A. and E. L. Porteus. 1999. Stalking information: Bayesian inventory management with unobserved lost sales. *Management Science 45*(3), 346–363.

Larsen, R. J. and M. L. Marx. 2012. *An Introduction to Mathematical Statistics and Its Applications*. Boston: Prentice Hall.

Lau, H. and A. Lau. 1996. Estimating the demand distributions of single-period items having frequent stockouts. *European Journal of Operational Research 92*, 254–265.

Lawrence, M., P. Goodwin, M. O'Connor, and D. Önkal. 2006. Judgmental forecasting: A review of progress over the last 25 years. *International Journal of Forecasting 22*(3), 493–518.

Lawrence, M. J., R. H. Edmurdson, and M. J. O'Connor. 1986. Accuracy of combining judgemental and statistical forecasts. *Management Science 32*(12), 1521–1532.

Lewandowski, R. 1979. *La Prevision a Court Terme*. Paris: Dunod.

Lewandowski, R. 1982. Sales forecasting by FORSYS. *Journal of Forecasting 1*, 205–214.

Lewis, C. D. 1978. The versatility of the exponential smoother's portfolio of skills. *Production and Inventory Management Journal 19*(2), 53–66.

Lovejoy, W. S. 1990. Myopic policies for some inventory models with uncertain demand distributions. *Management Science 36*(6), 724–738.

Maddala, G. S. and K. Lahiri. 1992. *Introduction to Econometrics*, Volume 2. New York: Macmillan.

Makridakis, S. 1990. Sliding simulation: A new approach to time series forecasting. *Management Science 36*(4), 505–512.

Makridakis, S., A. Andersen, R. Carbone, R. Filder, M. Hibon, R. Lewandowski, J. Newton, E. Parzen, and R. L. Winkler. 1982. The accuracy of extrapolation (time series) methods: Results of a forecasting competition. *Journal of Forecasting 1*, 111–153.

Makridakis, S., C. Chatfield, M. Hibon, M. Lawrence, T. Mills, K. Ord, and L. F. Simmons. 1993. The M2-competition: A real-time judgmentally based forecasting study. *International Journal of Forecasting 9*(1), 5–22.

Makridakis, S. and M. Hibon. 2000. The M3-competition: Results, conclusions and implications. *International Journal of Forecasting 16*(4), 451–476.

Makridakis, S. and S. C. Wheelwright. 1979. Forecasting: Framework and overview. In S. Makridakis and S. C. Wheelwright (Eds.), *TIMS Studies in the Management Sciences*, Volume 12, pp. 1–15. Amsterdam: North Holland.

Makridakis, S., S. C. Wheelwright, and R. J. Hyndman. 2008. *Forecasting Methods and Applications*. Hoboken, NJ: John Wiley & Sons.

Makridakis, S. and R. L. Winkler. 1983. Averages of forecasts: Some empirical results. *Management Science 29*(9), 987–996.

McClain, J. and L. J. Thomas. 1973. Response-variance tradeoffs in adaptive forecasting. *Operations Research 21*(2), 554–568.

McKenzie, E. 1978. The monitoring of exponentially weighted forecasts. *Journal of the Operational Research Society 29*(5), 449–458.

McLeavey, D. W., T. S. Lee, and E. E. Adam. 1981. An empirical evaluation of individual item forecasting models. *Decision Sciences 12*(4), 708–714.

Montgomery, D. C. 1970. Adaptive control of exponential smoothing parameters by evolutionary operation. *AIIE Transactions 2*(3), 268–269.

Muir, J. W. 1980. Forecasting items with irregular demand. In *American Production and Inventory Society Conference Proceedings*, Vol. 23, pp. 143–145, Los Angeles.

Nahmias, S. 1994. Demand estimation in lost sales inventory systems. *Naval Research Logistics 41*, 739–757.

Oliva, R. and N. Watson. 2009. Managing functional biases in organizational forecasts: A case study of consensus forecasting in supply chain planning. *Production and Operations Management 18*(2), 138–151.

Oliva, R. and N. Watson. 2011. Cross-functional alignment in supply chain planning: A case study of sales and operations planning. *Journal of Operations Management 29*(5), 434–448.

Peterson, R. 1969. A note on the determination of optimal forecasting strategy. *Management Science 16*(4), B165–B169.

Rhodes, P. 1996. Forecasting seasonal demand. *APICS—The Performance Advantage April*, 50–54.

Roberts, A. and R. Reed. 1969. The development of a self-adaptive forecasting technique. *AIIE Transactions 1*(4), 314–322.

Sanders, N. and K. Manrodt. 1994. Forecasting practices in US corporations: Survey results. *Interfaces 24*(2), 92–100.

Sanders, N. R. and K. B. Manrodt. 2003a. The efficacy of using judgmental versus quantitative forecasting methods in practice. *OMEGA 31*(6), 511–522.

Sanders, N. R. and K. B. Manrodt. 2003b. Forecasting software in practice: Use, satisfaction, and performance. *Interfaces 33*(5), 90–93.

Sanders, N. R. and L. P. Ritzman. 1989. Some empirical findings on short-term forecasting: Technique complexity and combinations. *Decision Sciences 20*(3), 635–640.

Schmittlein, D. C., J. Kim, and D. G. Morrison. 1990. Combining forecasts: Operational adjustments to theoretically optimal rules. *Management Science 36*(9), 1044–1056.

Shore, H. 1995. Identifying a two-parameter distribution by the first two sample moment (partial and complete). *Journal of Statistical Computation and Simulation 52*(1), 17–32.

Shore, H. 1998. Approximating an unknown distribution when distribution information is extremely limited. *Communications in Statistics—Simulation and Computation 27*(2), 501–523.

Steece, B. and S. Wood. 1979. An ARIMA-based methodology for forecasting in a multi-item environment. *Management Science 25*, 167–187.

Stevens, C. 1974. On the variability of demand for families of items. *Operational Research Quarterly 25*(3), 411–419.

Syntetos, A. A. and J. E. Boylan. 2005. The accuracy of intermittent demand estimates. *International Journal of Forecasting 21*(2), 303–314.

Taylor, P. F. and M. E. Thomas. 1982. Short term forecasting: Horses for courses. *Journal of the Operational Research Society 33*(8), 685–694.

Timmermann, A. 2006. Forecast combinations. *Handbook of Economic Forecasting 1*, 135–196.

Trigg, D. 1964. Monitoring a forecasting system. *Operational Research Quarterly 15*, 271–274.

Trigg, D. and A. Leach. 1967. Exponential smoothing with an adaptive response rate. *Operational Research Quarterly 18*(1), 53–59.

Walton, J. H. D. 1994. Inventory control—A soft approach? *Journal of the Operational Research Society 45*(4), 485–486.

Watson, R. 1987. The effects of demand-forecast fluctuations on customer service and inventory cost when demand is lumpy. *Journal of the Operational Research Society 38*(1), 75–82.

Wecker, W. 1978. Predicting demand from sales data in the presence of stockouts. *Management Science 24*(10), 1043–1054.

Whybark, D. C. 1973. A comparison of adaptive forecasting techniques. *Logistic Transportation Review 9*(1), 13–26.

Willemain, T. R., C. N. Smart, and H. F. Schwarz. 2004. A new approach to forecasting intermittent demand for service parts inventories. *International Journal of Forecasting 20*, 375–387.

Willemain, T. R., C. N. Smart, J. H. Shockor, and P. A. DeSautels. 1994. Forecasting intermittent demand in manufacturing: A comparative evaluation of Croston's method. *International Journal of Forecasting 10*, 529–538.

Williams, T. M. 1987. Adaptive Holt-Winters forecasting. *Journal of the Operational Research Society 38*(6), 553–560.

Winters, P. 1960. Forecasting sales by exponentially weighted moving averages. *Management Science 6*, 324–342.

Wright, D. J. 1986. Forecasting data published at irregular time intervals using an extension of Holt's method. *Management Science 32*(4), 499–510.

REPLENISHMENT SYSTEMS FOR MANAGING INDIVIDUAL ITEM INVENTORIES WITHIN A FIRM

In the first part of this book, we have laid groundwork by specifying the environment, goals, constraints, relevant costs, and forecast inputs of production/inventory decisions. Now, we turn our attention to the detailed logic for short-range decision making in situations of a strictly inventory (i.e., nonproduction) nature. We are concerned primarily with individual items or SKUs. However, as discussed in Chapter 2, aggregate considerations will often still play a crucial role by examining the consequences across a population of inventoried items of using different values of certain policy parameters. Management can select the aggregate operating point on an exchange curve, thus implying values of the parameters that are to be used in the item by item decision rules.

Within this part of the book, we deal with the so-called B items, which, as we have seen earlier, comprise the majority of items in a typical inventory situation. Unlike for C items, which will be handled in Section III, there are usually significant savings to be derived from the use of a reasonably sophisticated control system. On the other hand, the potential savings per item are not as high as those for individual A items. The combination of these two factors suggests the use of a hybrid system that combines the relative advantages of human and computer. The management-by-exception control system that we describe is reasonably sophisticated, and usually automated, but requires human intervention on a relatively infrequent basis. Such automated control systems, when applied in practice, have resulted in significant improvements in the control of B items.

In Chapter 4, we first answer the question of how much to replenish in the context of an approximately level demand pattern. Next, in Chapter 5, we consider the more general situation of a demand pattern, such as that seen with seasonal items, where the mean or forecast demand changes appreciably with time. This more general case considerably complicates the analysis. Therefore, the suggested decision system is based on a reasonable heuristic decision rule.

Chapter 6 is concerned, for the most part, with the situation where the average demand rate stays approximately level with time, but there is a random component present. That is, the mean or forecast demand stays constant or changes slowly with time, but there definitely are appreciable forecast errors. When demand is known only in a probabilistic sense, two additional questions must be answered: "How often should we review the stock status?" and "When should we place a replenishment order?"

All the material in Chapters 4 through 6 ignores the possible benefits of coordinating the replenishments of two or more items. We return to this important topic in Chapters 10 and 11 of Section IV.

Chapter 4

Order Quantities When Demand Is Approximately Level

In this chapter, we address the question of how large a replenishment quantity to use for a single item when conditions are rather stable (any changes in parameters occur slowly over time), and there is relatively little or no uncertainty concerning the level of demand. This simplified situation is a reasonable approximation of reality on certain occasions. More important, however, the results obtained turn out to be major components of the decision systems when parameters change with time, demand is probabilistic or multiple items are managed jointly. In addition, the goal is not to have a completely automated replenishment system, but to provide decision support to a planner. This is essential for several reasons, including (1) giving the planner the ability to incorporate factors not included in the underlying mathematical model, and (2) cultivating a sense of accountability and responsibility on the part of the decision makers.

In Section 4.1, we discuss the rather severe assumptions needed to derive the economic order quantity (EOQ). This is followed in Section 4.2 by the actual derivation. In Section 4.3, we show that the total relevant costs are not seriously affected by somewhat large deviations of the replenishment quantity away from the minimizing value. This robust property of the EOQ is important from an operational standpoint as there are many factors in practice that may make a quantity other than the calculated EOQ convenient. Section 4.4 introduces a tabular aid that groups items into broad categories to make implementation easier, particularly in a noncomputerized system. The important situation of quantity discounts is handled in Section 4.5. This is followed, in Section 4.6, by a discussion of the effects of inflation on the order quantity. Section 4.7 deals with several types of limitations on replenishment quantities ignored in the earlier development of the EOQ. In Section 4.8, we address the production situation in which the entire replenishment batch does not arrive at one time. Other important factors in selecting the best reorder quantity are included in Section 4.9. Finally, the derivation of the basic EOQ includes two cost parameters: (1) the carrying charge r, and (2) the fixed cost per replenishment A. As discussed in Chapter 2, one or both of the values of these parameters may be explicitly evaluated or implicitly specified. They may be implicitly specified by looking at the aggregate consequences of using different values of r (or A/r), and then selecting a suitable aggregate operating point, which implies an r (or A/r) value. Section 4.10 shows how to develop the exchange curves, a crucial step in this implicit approach.

4.1 Assumptions Leading to the Basic EOQ

First, we will state the assumptions. Some of these assumptions are quite strong and may initially appear to be quite limiting but, as we will see, the EOQ forms an important building block in the majority of decision systems that we advocate. In addition, Zheng (1992) has shown that the fractional cost penalty of using the EOQ instead of a more complex, but optimal, order quantity when demand is random is less than 1/8, and is typically much smaller. (Axsäter (1996) has shown that the penalty is less than 11.8%.) Further discussion will be directed to these points later in this chapter and in the subsequent chapters.

1. The demand rate is constant and deterministic.
2. The order quantity need not be an integral number of units, and there are no minimum or maximum restrictions on its size.
3. The unit variable cost does not depend on the replenishment quantity; in particular, there are no discounts in either the unit purchase cost or the unit transportation cost.
4. The cost factors do not change appreciably with time; in particular, inflation is at a low level.
5. The item is treated entirely independently of other items; that is, benefits from joint review or replenishment do not exist or are simply ignored.
6. The replenishment lead time is of zero duration; as we will see in Section 4.9.1, extension to a deterministic, nonzero replenishment lead time creates no problem.
7. No shortages are allowed.
8. The entire order quantity is delivered at the same time.
9. The planning horizon is very long. In other words, we assume that all the parameters will continue at the same values for a long time.

All these assumptions will be relaxed later in this chapter, or elsewhere in the book. Situations where the demand rate changes appreciably with time (i.e., the first assumption is violated) will be treated in Chapters 5 and 8. Problem 4.5 deals with the case in which shortages are allowed. (See also Montgomery et al. (1973) and Moon and Gallego (1994) for a treatment of the case in which there is a mixture of lost sales and backorders. More recently, Bijvank and Vis (2011) review inventory models with lost sales and backordering.) We often refer to "changing the givens," meaning that the static version of the problem may not be appropriate over time. Rather, some of the assumptions, costs, or other data may, and should, change. For example, in this chapter, we generally assume that the setup cost, A, is given. Setup reduction efforts, however, may significantly reduce this value, yielding enormous benefits. Likewise, supplier portals, electronic data interchange (EDI), or other technological infrastructure may reduce the fixed cost of ordering from suppliers, creating new optimal order quantities and significantly lower costs. Sometimes, it is possible to negotiate with suppliers to reduce the minimum order size required for a quantity discount, or it may be possible to negotiate with transportation firms to use smaller trucks to reduce the minimum order size.[*]

---

[*] See Moinzadeh et al. (1997) for a discussion of the impact on traffic of small lot sizes.

The traditional approach to the EOQ has been to optimize the order quantity without questioning the input data. The "changing the givens" approach recognizes the possibility of changing the parameters. In many cases, the most powerful benefit to the firm is through lowering the setup cost, rather than optimizing the order quantity. We argue, however, that the best approach is to initially optimize the order quantity given the parameters as they are, and then devote management (and labor) effort to changing the givens. A fastener manufacturer that supplies U.S. automotive assemblers had setup times on some machines that took from 12 to 20 hours. One of the authors worked as a consultant to this firm and developed (nearly) optimal order quantities. At the same time, however, factory supervisors and operators were working hard to lower the setup times. The use of the new order quantities would lower costs, but not nearly as much as lower setup times would. Unfortunately, the setup reduction process was very slow, and the firm would have sacrificed savings if management had delayed changing order quantities until this process was finished. As setup times decrease, new order quantities should be calculated. Numerous authors have addressed these issues. See, for instance, Porteus (1985, 1986a,b), Silver (1992), Hong et al. (1996), Spence and Porteus (1987), Van Beek and Van Putten (1987), Freeland et al. (1990), Neves (1992), Gallego and Moon (1995), Bourland et al. (1996), Van der Duyn Schouten et al. (1994), and Freimer et al. (2006). We will discuss these issues further in the context of manufacturing in Section IV.

4.2 Derivation of the EOQ

In determining the appropriate order quantity, we use the criterion of minimization of total relevant costs; relevant in the sense that they are truly affected by the choice of the order quantity. We emphasize that in some situations, there may be certain, perhaps intangible, relevant costs that are not included in the model to be discussed. This is one of the reasons we mention the need for a manual override capability. In addition, cost minimization need not be the only appropriate criterion. For example, Buzacott and Zhang (2004) investigate how inventory decisions are affected by the current financial state of the firm. Singhal (1988) also closely examines the connection to the financial state of the firm and argues that inventories are risky, and thus increase the firm's opportunity cost of capital. Other work of interest includes Gerchak and Wang (1994b), which investigates the issue of demand changing as a result of inventory decisions. More recently, Buzacott (2013) reflects further on similar issues.

Returning to the consideration of costs, as discussed in Chapter 2, there are five fundamental categories of costs:

1. Basic production or purchase costs
2. Inventory carrying costs
3. Costs of insufficient capacity in the short run
4. Control system costs
5. Costs of changing work force sizes and production rates

The fifth category is not relevant for the item-by-item control we are considering here. Also, given that we are restricting our attention to a particular type of control system (an order quantity system), the control system costs are not influenced by the exact value of the order quantity. Therefore, they need not be taken into account when selecting the value of the order quantity.

Because of the deterministic nature of the demand, the only cost relevant to the third category would be that which is caused by the decision maker deliberately choosing to run short of the inventory before making a replenishment. For now, we ignore the possibility of allowing planned shortages.

Finally, we are left with only the first two types of costs being relevant to the economic selection of the replenishment quantity.

Before proceeding further, let us introduce some notation:

Q = the replenishment order quantity, in units.

D = the demand rate of the item, in units/unit time.

T = the time between placement (and receipt) of replenishment orders.

A = the fixed cost component (independent of the magnitude of the replenishment quantity) incurred with each replenishment, in dollars. (This cost component has been discussed at length in Chapter 2.)

v = the unit variable cost of the item. This is not the selling price of the item, but rather its value (immediately after the replenishment operation now under consideration) in terms of raw materials and value added through processing and assembly operations. We often think of v as the value of the item on the shelf, including material handling. The dimensions are $/unit.

r = the carrying charge, the cost of having one dollar of the item tied up in the inventory for a unit time interval (normally 1 year); that is, the dimensions are $/$/unit time. (Again, this cost component was discussed at length in Chapter 2.)

$TRC(Q)$ = the total relevant costs per unit time, that is, the sum of those costs per unit time that can be influenced by the order quantity Q. The dimensions are $/unit time.

Because the parameters involved are assumed not to change with time, it is reasonable (and, indeed, mathematically optimal) to think in terms of using the same order quantity, Q, each time that a replenishment is made. Furthermore, because (1) demand is deterministic, (2) the replenishment lead time is zero, and (3) we have chosen not to allow planned shortages, it is clear that each replenishment will be made when the inventory level is exactly at zero. A graph of the inventory level with time is shown in Figure 4.1.

Note that the time between replenishments is given by $T = Q/D$, the time to deplete Q units at a rate of D units per unit time. (D is the usage over a period of time; normally 12 months is used.) Therefore, the number of replenishments per unit time is D/Q. Say a three-ohm resistor has the following data: $D = 2,400$ units per year and $Q = 800$ units. The time between replenishments is $800/2,400 = 0.33$ year, or 4 months. The number of replenishments per year is $2,400/800 = 3$. Associated with each of these is a replenishment cost given by $A + Qv$, where one of our assumptions ensures that the unit variable cost v does not depend on Q. Therefore, the replenishment costs per unit time (C_r) are given by

$$C_r = (A + Qv)D/Q$$

or

$$Cr = AD/Q + Dv \tag{4.1}$$

We can see that the second component in Equation 4.1 is independent of Q and, thus, can have no effect on the determination of the best Q value. (It represents the constant acquisition

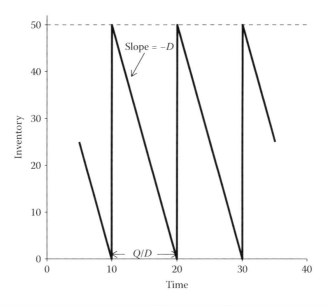

Figure 4.1   Behavior of the inventory level with time.

cost of the item per unit time which cannot be affected by the magnitude of the order quantity.) Therefore, it will be neglected in future discussions.[*]

As discussed in Chapter 2, the common method of determining the costs of carrying the inventory over a unit time period is through the relation

$$C_c = \bar{I} v r$$

where $\bar{I}$ is the average inventory level, in units. The average height of the sawtooth diagram of Figure 4.1 is $Q/2$. Therefore,

$$C_c = Q v r / 2 \qquad (4.2)$$

Combining Equations 4.1 and 4.2 and neglecting the Dv term, as discussed above, we have that the total relevant costs per unit time are given by

$$TRC(Q) = AD/Q + Qvr/2 \qquad (4.3)$$

The two components of Equation 4.3 and their total are plotted for an illustrative numerical example in Figure 4.2. We can see that the replenishment costs per unit time decrease as Q increases (there are fewer replenishments), whereas the carrying costs increase with Q (a larger Q means a larger average inventory). The sum of the two costs is a *u*-shaped function with a minimum that can be found in a number of ways. (Adding the neglected Dv term would simply shift every point on the total cost curve up by Dv; hence, the location of the minimum would not change.) One

---

[*] The Dv component will be of crucial interest in Section 4.5 where, in the case of quantity discounts, v depends on Q.

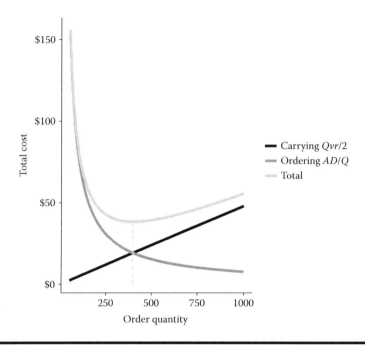

Figure 4.2   Costs as functions of the replenishment quantity.

convenient way to find the minimum is to use the necessary condition that the tangent or slope of the curve is zero at the minimum:

$$\frac{d\,\mathrm{TRC}(Q)}{dQ} = 0$$

that is,

$$\frac{vr}{2} - \frac{AD}{Q^2} = 0$$

$$Q_{\mathrm{opt}} \quad \text{or} \quad \mathrm{EOQ} = \sqrt{\frac{2AD}{vr}} \tag{4.4}$$

A sufficient condition for a minimum, namely, that the second derivative is positive, is also satisfied:

$$\frac{d^2\,TRC(Q)}{dQ^2} = 2AD/Q^3 > 0 \text{ for any } Q > 0$$

Equation 4.4 defines the EOQ also known as the Wilson lot size. This is one of the earliest and most well-known results of inventory theory (see, e.g., Harris (1913), reprinted as Harris (1990), and Erlenkotter (1990)). Note that Equation 4.4 illustrates, among other points, that the preferred order quantity goes up as the square root of demand, rather than being directly proportional to the *D* value. Substituting the square root expression for the EOQ in Equation 4.4 back into Equation 4.3, we find that the two cost components are equal at the EOQ and we obtain the simple and useful result:

$$\mathrm{TRC}(\mathrm{EOQ}) = \sqrt{2ADvr} \tag{4.5}$$

It should be emphasized that the equality of the two cost components ($Qvr/2$ and AD/Q) at the point where their sum is minimized is a very special property of the particular cost functions considered here; it certainly does not hold in general.

The EOQ minimizes the total relevant costs under a given set of circumstances. It may in fact be more appropriate to first find ways of driving down the fixed cost A or the carrying charge r. This would lower the entire total cost curve in Figure 4.2. Also note from Equation 4.4 that a lower A value will reduce the EOQ, and thus the average inventory level.

As noted earlier, for any given order quantity Q, the time between replenishments $T = Q/D$. Using this relationship, we can write the total relevant costs for this model as a function of the time supply T rather than the quantity. Substituting $T = Q/D$ in Equation 4.3:

$$TRC(T) = A/T + DvrT/2 \qquad (4.6)$$

Using Equation 4.4, we find the expression for the time between orders at the EOQ:

$$T_{EOQ} = \frac{EOQ}{D} = \sqrt{\frac{2A}{Dvr}} \qquad (4.7)$$

A frequently used measure of inventory performance is the inventory turnover ratio, defined as the annual demand rate divided by the average inventory level. This ratio gives the number of times the average inventory is sold or "turned" over some given time period. For any given quantity Q, we calculate the inventory turnover ratio as

$$TR = \frac{D}{\bar{I}} = \frac{D}{Q/2} = \frac{2D}{Q} \qquad (4.8)$$

At the EOQ, this inventory turnover ratio is

$$TR_{EOQ} = \sqrt{\frac{2Dvr}{A}} \qquad (4.9)$$

a result that is again seen to depend on the individual item factors D, v, and A. In particular, all other things remaining constant, we see that the turnover ratio increases in proportion to the square root of the demand rate, D.

4.2.1 Numerical Illustration

Consider a 3-ohm resistor used in the assembly of an automated processor of an x-ray film. The demand for this item has been relatively stable over time at a rate of 2,400 units/year. The unit variable cost of the resistor is \$0.40/unit and the fixed cost per replenishment is estimated to be \$3.20. Suppose further that an r value of 0.24 \$/\$/year is appropriate to use. Then Equation 4.4 gives

$$EOQ = \sqrt{\frac{2 \times \$3.20 \times 2{,}400 \text{ units/year}}{\$0.40/\text{unit} \times 0.24\$/\$/\text{year}}}$$
$$= 400 \text{ units}$$

Note that the dimensions appropriately cancel. This would not be the case, for example, if D and r were defined using different time units (e.g., monthly demand with annual carrying charge). Equation 4.5 reveals that the total relevant costs per year for the resistor are

$$\text{TRC(EOQ)} = \sqrt{2 \times \$3.20 \times 2{,}400 \,\text{units/year} \times \$0.40/\text{unit} \times 0.24\$/\$/\text{year}}$$
$$= \$38.40/\text{year}$$

A check on the computations is provided by noting that substitution of $Q = 400$ units in Equation 4.3 also produces a total cost \$38.40/year with each of the two terms equal at \$19.20/year. If this item has an order quantity of 4 months of time supply, or 800 units, we can evaluate the benefit of changing to the EOQ of 400. Substitution of $Q = 800$ into Equation 4.3 reveals that the total relevant costs for the item are \$48/year or some 25% higher than the costs incurred under the use of the EOQ. Our experience indicates that savings of this magnitude are surprisingly common.

We can use other expressions developed in this section to evaluate the time between orders and inventory turnover at the EOQ. The time between orders at the EOQ is given by Equation 4.7:

$$T_{\text{EOQ}} = \sqrt{\frac{2 \times \$3.20}{2{,}400 \,\text{units/year} \times \$0.40/\text{unit} \times 0.24\$/\$/\text{year}}} = 0.167 \,\text{year, or 2 months}$$

The inventory turnover ratio at the EOQ, following Equation 4.9, is

$$TR_{\text{EOQ}} = \sqrt{\frac{2Dvr}{A}} = \sqrt{\frac{2 \times 2{,}400 \,\text{units/year} \times \$0.40/\text{unit} \times 0.24\$/\$/\text{year}}{\$3.20}} = 12 \,\text{turns per year}$$

Note that, at the EOQ, we are ordering a 2-month supply, but we are turning the average inventory 12 times per year or once every month. This is because the 2-month supply ($Q = 400$) is the *maximum* inventory level while the average inventory level (used to calculate turns) is half this quantity.

4.3 Sensitivity Analysis

An important property of the EOQ model is that total relevant costs are rather insensitive to deviations from the optimal lot size given by the EOQ. Referring back to Figure 4.2, note that the total cost curve is quite flat in the neighborhood of the EOQ. This indicates that reasonable-sized deviations from the EOQ will have little impact on the total relevant costs incurred. Here, we explicitly examine the increase in the total relevant cost associated with choosing a lot size other than the EOQ. Suppose we use a quantity Q' that deviates from the EOQ according to the following relation:

$$Q' = (1 + p)\text{EOQ}$$

that is, $100p$ is the percentage deviation of Q' from the EOQ. The percentage cost penalty (PCP) for using Q' instead of the EOQ is given by

$$\text{PCP} = \frac{\text{TRC}(Q') - \text{TRC(EOQ)}}{\text{TRC(EOQ)}} \times 100$$

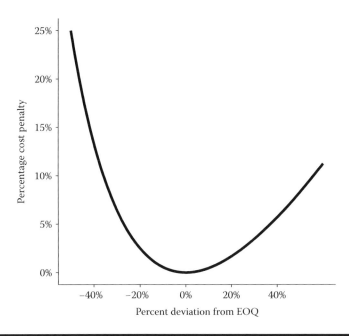

Figure 4.3   Cost penalty for using a $Q' = \text{EOQ}(1 + p)$ value different from the EOQ.

As shown in Section 4A.1 of the Appendix to this chapter,

$$\text{PCP} = \left(\frac{p^2}{2(1 + p)} \right) \times 100 \tag{4.10}$$

This expression is plotted in Figure 4.3. It is clear that even for values of p significantly different from zero, the associated cost penalties are quite small.

To illustrate, consider the 3-ohm resistor that was used as a numerical example in Section 4.2. The EOQ was found to be 400 units. Suppose that, instead, a value of 550 units was used; 550 is 1.375×400; thus, $p = 0.375$. Equation 4.10 or Figure 4.3 reveals that this 37.5% deviation from optimal lot size results in a percentage increase in the total relevant costs of only 5.1%.

The insensitivity of total costs to the exact value of Q used has two important implications. First, the use of an incorrect value of Q can result from inaccurate estimates of one or more of the parameters D, A, v, and r that are used to calculate the EOQ.[*] The conclusion is that it is not worth making accurate estimates of these input parameters if considerable effort is involved; in most cases inexpensive, crude estimates should suffice. Second, certain order quantities may have additional appeal over the EOQ (for nonfinancial reasons or because of physical constraints, such as pallet size or truckload size, which are not included in the basic EOQ model). The relative flatness of the total cost curve near the EOQ indicates that such values can be used provided that they are reasonably close to the EOQ. This seemingly trivial result is one of the important cornerstones of

---

[*] It is worth noting that the cost penalty can actually be zero even if D, A, v, and r are inaccurately estimated, as long as the resulting ratio AD/vr is correct. For example, if both D and v are overestimated by the same percentage, the obtained EOQ will be the same as that resulting from the use of the correct D and v values.

operational inventory control. Without this flexibility, implementable decision systems would be much more difficult to design. Detailed references on sensitivity analysis include Dobson (1988), Juckler (1970), and Lowe and Schwarz (1983).

4.4 Implementation Aids

In spite of the widespread availability of inexpensive computers, our experience indicates that many firms still manually control inventories. Many other firms use simple spreadsheet models, similar to those described throughout this book. We now discuss a table that can facilitate implementation because it groups multiple items into a few broad categories and can be developed easily on a spreadsheet. While it is not too difficult to develop a spreadsheet model to calculate the EOQ values for a group of items, the approach here simplifies implementation. Furthermore, while the focus of this chapter is lot size decisions for a single item, the approach outlined in this section facilitates the coordination of replenishment across multiple items, a topic that we address in Chapter 10.

The basic idea is that one value of the order quantity, typically expressed as a time supply, is used over a range of one or more of the parameters. This makes it easier for an inventory manager to track and control many items. There is an economic trade-off involved, however. The larger the number of ranges, the smaller the percentage cost penalties resulting from not using the exact EOQ, but of course, the difficulty of compiling and using the tables increases with the number of ranges involved. Further relevant discussion can be found in Chakravarty (1981), Crouch and Oglesby (1978), Donaldson (1981), and Maxwell and Singh (1983).

To illustrate, let us consider a table usable for a set of items having the same values of A and r. Suppose we deal with a group of items with $A = \$3.20$ and $r = 0.24$ \$/\$/year. Furthermore, let us assume that management feels that the replenishment quantity of any item should be restricted to one of nine possible time supplies—namely, 1/4, 1/2, 3/4, 1, 2, 3, 4, 5, 6, and 12 months. Table 4.1 shows the months of supply to use as a function of the annual dollar usage (Dv) of the item under consideration. A numerical illustration of the use of the table will be given shortly. First, let us discuss how such a table is constructed.

Table 4.1 Tabular Aid for Use of EOQ (for A = 3.20 and r = 0.24 \$/\$/year)

For Annual Dollar Usage (Dv) in This Range	Use This Number of Months of Supply
$30{,}720 \le Dv$	1/4 ($\approx$ 1 week)
$10{,}240 \le Dv < 30{,}720$	1/2 ($\approx$ 2 weeks)
$5{,}120 \le Dv < 10{,}240$	3/4 ($\approx$ 3 weeks)
$1{,}920 \le Dv < 5{,}120$	1
$640 \le Dv < 1{,}920$	2
$320 \le Dv < 640$	3
$160 \le Dv < 320$	4
$53 \le Dv < 160$	6
$Dv < 53$	12

The use of T months of supply is equivalent to the use of a quantity $Q = DT/12$, provided D is measured in units/year that are the most common units. The use of Equation 4.3 gives

$$\text{TRC(using } T \text{ months)} = \frac{DTvr}{24} + \frac{12A}{T}$$

Now, consider two adjacent allowable values of the months of supply, call them T_1 and T_2. Equating the total relevant costs using T_1 months with that using T_2 months, we obtain the value of Dv at which we are indifferent to using T_1 and T_2:

$$\frac{DT_1vr}{24} + \frac{12A}{T_1} = \frac{DT_2vr}{24} + \frac{12A}{T_2}$$

This reduces to

$$(Dv)_{indifference} = \frac{288A}{T_1 T_2 r}$$

To illustrate, for the given values of A and r in the example, the indifference point for 1 month and 2 months is

$$(Dv)_{indifference} = \frac{288 \times 3.20}{1 \times 2 \times 0.24} = \$1{,}920/\text{year}$$

In a similar fashion, we can develop the rest of Table 4.1. It should be emphasized that more than one value of A (and perhaps r) is likely to exist across a population of items. In such a case, a set of tables would be required, one for each of a number of possible A (and, perhaps, r) values.

4.4.1 Numerical Illustration

Consider a product that is a box of 100 sheets of 8×11 in. stabilization paper. This product has been observed to have a relatively constant demand rate (D) of 200 boxes/year. The unit variable cost (v) is \$16/box. Also assume that it is reasonable to use $A = \$3.20$ and $r = 0.24$ \$/\$/year. We have

$$Dv = \$3{,}200/\text{year}$$

The table indicates that a 1-month supply should be used. Therefore,

$$Q = \frac{D}{12} = 16.7$$

say, 17 boxes; that is, the paper should be replenished in orders of 17 boxes. (The exact EOQ for this item is 18 boxes and the increase in costs for using 17 instead turns out to be only 0.3%.)

4.5 Quantity Discounts

A number of assumptions were made in Section 4.1 in order to derive the basic EOQ. One of the most severe of these was that the unit variable cost v did not depend on the replenishment quantity. In many practical situations, quantity discounts (on the basic purchase price or transportation

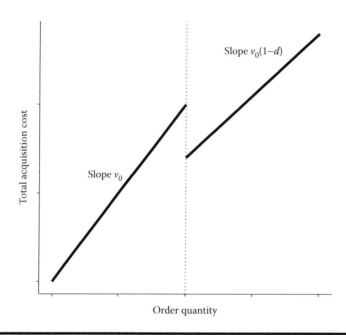

Figure 4.4 **"All-units" quantity discount.**

costs) exist, and taking advantage of these can result in substantial savings. (See e.g., Katz et al. 1994). We must be able to modify the EOQ to cope with these conditions.

Dolan (1987) provides a number of examples of quantity discounts, including pricing for Sealed Air Corporation's protection packaging, and prices for advertising in the *Los Angeles Times*. See also Wilson (1993). These authors show that discount structures are quite varied. We restrict our attention to the most common type of discount structure, namely, that of an "all-units" discount. (One less common situation of an incremental discount structure has been treated by Hax and Candea (1984) and Güder et al. (1994). See Benton and Park (1996) and Problem 4.24 also.)[*]

For the case of a single breakpoint, the unit variable cost in all-units discounts behaves as follows:

$$Dv = \begin{cases} v_0 & 0 \le Q < Q_b \\ v_0(1-d) & Q_b \le Q \end{cases} \qquad (4.11)$$

where v_0 is the basic unit cost without a discount and d is the discount expressed as a decimal, given on all units when the replenishment quantity is equal to or greater than the breakpoint, Q_b. The total acquisition cost as a function of Q is shown in Figure 4.4. Note the discontinuity at the breakpoint.

Now, it is essential to include the Dv component in a cost expression that is to be used for determining the best replenishment quantity. Proceeding exactly as in Section 4.2, but retaining the Dv component, we end up with two expressions for the total relevant costs, that is, for $0 \le Q < Q_b$ (and thus no discount)

$$\text{TRC}(Q) = AD/Q + Qv_0 r/2 + Dv_0 \qquad (4.12)$$

---

[*] Das (1988) presents a general discount structure that combines the features of all-units and incremental discounts.

When $Q_b \leq Q$, the discount d is applicable:

$$\mathrm{TRC}_d(Q) = AD/Q + Qv_0(l - d)r/2 + Dv_0(1 - d) \qquad (4.13)$$

Although Equations 4.12 and 4.13 are valid for nonoverlapping regions of Q, it is useful, in deriving an algorithm for finding the best Q, to compare the two cost expressions at the same value of Q. A term-by-term comparison of the right-hand sides of the equations reveals that Equation 4.13 gives a lower TRC (Q) than does Equation 4.12 for the same Q. Therefore, if the lowest point on the (4.13) curve is a valid one (i.e., at least as large as Q_b), it must be the optimum, since it is the lowest point on the lower curve.

Note that for nonnegative values of the discount d, the EOQ at the standard price v_0 must be lower than the EOQ calculated with the discount, although the difference in these quantities is typically quite small. This means, if the EOQ with the discount is below Q_b, the EOQ without the discount must also be below Q_b. In the usual situation where both EOQs (calculated with and without the discount) are below the breakpoint, the decision of whether to go for the discount hinges on a trade-off between extra carrying costs versus a reduction in the acquisition costs (primarily the reduction in the unit value, but also a benefit from fewer replenishments per unit time). Where the reduction in acquisition costs is larger than the extra carrying costs, the use of Q_b is preferable, as shown in part a of Figure 4.5. Where the extra carrying costs dominate, the best solution is still the EOQ (with no discount), as shown in part b of Figure 4.5. However, there is one other possibility—namely, that Q_b is relatively low, and it may be attractive to go to the local minimum (low point) on the curve of Equation 4.13. This is the EOQ (with discount) and this case is shown in part c of Figure 4.5.

Taking advantage of the above property of Equation 4.13 being the lower curve, the steps of an efficient algorithm for finding the best value of Q are

Step 1 Compute the EOQ when the discount d is applicable, that is,

$$\mathrm{EOQ}(d) = \sqrt{\frac{2AD}{v_0(1 - d)r}}$$

Step 2 Compare EOQ(d) with Q_b. If EOQ(d) $\geq Q_b$, then EOQ(d) is the best order quantity (case c of Figure 4.5). If EOQ(d) $< Q_b$, go to Step 3.

Step 3 Evaluate

$$\mathrm{TRC(EOQ)} = \sqrt{2ADv_0r} + Dv_0$$

and TRC(Q_b), using Equation 4.13.

If TRC(EOQ) < TRC(Q_b), the best order quantity is the EOQ without a discount (case b of Figure 4.5), given by

$$\mathrm{EOQ(no\ discount)} = \sqrt{\frac{2AD}{v_0r}}$$

If TRC(EOQ) > TRC (Q_b), the best order quantity is Q_b (case a of Figure 4.5).

This logic can easily be extended to the case of several breakpoints with increasing discounts. The best order quantity is always at a breakpoint or at a feasible EOQ. Kuzdrall and Britney (1982) have used a somewhat different approach that is useful when there is a large number of different

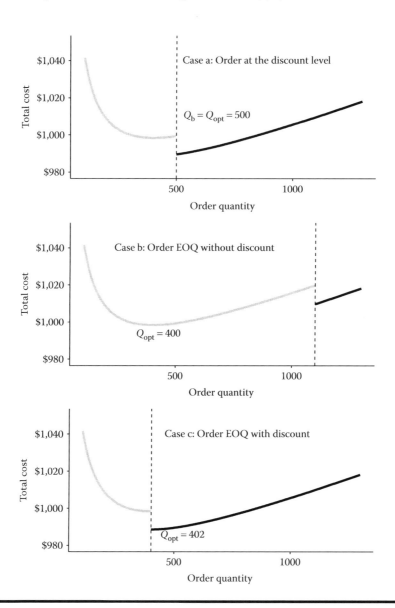

Figure 4.5   Total relevant costs under "all-units" discount.

discounts. Their method estimates the supplier's fixed and variable costs from a given schedule of discounts and breakpoints. See also Goyal (1995) for a simple procedure.

4.5.1 Numerical Illustrations

Consider three components used in the assembly of the x-ray film processor mentioned in Section 4.2. The supplier offers the same discount structure for each of the items, and discounts are based on replenishment sizes of the *individual* items. The relevant characteristics of the items are given below.

Item	D (Units/Year)	v_0 ($/Unit)	A ($)	r ($/$/Year)
A	416	14.20	1.50	0.24
B	104	3.10	1.50	0.24
C	4,160	2.40	1.50	0.24

Because of convenience in manufacturing and shipping, the supplier offers a 2% discount on any replenishment of 100 units or higher of a single item.

The computations are as follows:

4.5.2 Item A (An Illustration of Case a of Figure 4.5)

Step 1 EOQ (discount) = 19 units < 100 units.
Step 2 EOQ (discount) < Q_b; therefore, go to Step 3.
Step 3

$$\text{TRC(EOQ)} = \sqrt{2 \times 1.50 \times 416 \times 14.20 \times 0.24} + 416 \times 14.20$$
$$= \$5,972.42/\text{year}$$
$$\text{TRC}(Q_b) = \text{TRC}(100) = \frac{100 \times 14.20 \times 0.98 \times 0.24}{2} + \frac{1.50 \times 416}{100}$$
$$+ 416 \times 14.20 \times 0.98$$
$$= \$5,962.29/\text{year}$$

TRC(EOQ) > TRC (Q_b). Therefore, the best order quantity to use is Q_b, that is, 100 units.

4.5.3 Item B (An Illustration of Case b of Figure 4.5)

Step 1 EOQ (discount) = 21 units < 100 units.
Step 2 EOQ (discount) < Q_b; therefore, go to Step 3.
Step 3

$$\text{TRC(EOQ)} = \sqrt{2 \times 1.50 \times 104 \times 3.10 \times 0.24} + 104 \times 3.10$$
$$= \$337.64/\text{year}$$
$$\text{TRC}(Q_b) = \text{TRC}(100) = \frac{100 \times 3.10 \times 0.98 \times 0.24}{2} + \frac{1.50 \times 104}{100}$$
$$+ 104 \times 3.10 \times 0.98$$
$$= \$353.97/\text{year}$$

TRC(EOQ) < TRC (Q_b). Therefore, use the EOQ without a discount; that is,

$$\text{EOQ} = \sqrt{\frac{2 \times 1.50 \times 104}{3.10 \times 0.24}} \approx 20 \text{ units}$$

4.5.4 Item C (An Illustration of Case c of Figure 4.5)

Step 1

$$\text{EOQ(discount)} = \sqrt{\frac{2 \times 1.50 \times 4{,}160}{2.40 \times 0.98 \times 0.24}} = 149 \text{ units} > 100 \text{ units}$$

Step 2 EOQ (discount) is greater than Q_b. Therefore, the Q to use is 149 units (perhaps rounded to 150 for convenience).

We also note that it is very easy to compute Equations 4.12 and 4.13 on a spreadsheet. Using an IF statement to determine which formula to apply based on Q, the optimal value can be found quickly. It is also possible to use Solver in an Excel spreadsheet to minimize the total cost. However, because the TRC functions (illustrated in Figure 4.5) in general have more than one local minimum, use several starting values in Solver to avoid finding just a local minimum.

In an all-units discount situation, a possible strategy is to buy Q_b units to achieve the discount; then, dispose of some of the units, at some unit cost (or revenue), to reduce inventory carrying costs. An application of this philosophy is the case where an entire car of a train is reserved, but not completely filled, by a shipment. Sethi (1984) has analyzed this more general situation.

Much work has been done to further our knowledge of quantity discount situations. Dolan (1987) examines motivations for suppliers to offer discounts and provides several in-depth examples. Monahan (1984), Lal and Staelin (1984), Rosenblatt and Lee (1985), Lee (1986), Dada and Srikanth (1987), Kim and Hwang (1988), Min (1991), Parlar and Wang (1994), and Lu (1995) address the quantity discount problem from the supplier's perspective. For various pricing and discount structures, the supplier's own order quantity and the discounts it offers are determined. Das (1984) provides a simple solution for the problem when a premium is charged for bulk purchases. This work is particularly applicable in developing countries. Grant (1993) describes a simple program for quantity discount decisions used at a small JIT manufacturer who was biased against large quantities, and therefore against taking any quantity discounts. Finally, Abad (1988) examines the all-units discount offered to a retailer whose demand is a function of the price the retailer charges. An algorithm for finding the optimal order quantity and price is given. See also Burwell et al. (1990), Weng and Wong (1993), Weng (1995a,b), Lu and Posner (1994), Kim (1995), and Carlson et al. (1996).

Managers, buyers, and purchasing agents, among others, often think in terms of a group of items purchased from a particular vendor, rather than dealing with single items in isolation. Thus, a potentially important discount can be achieved by having a total order, involving two or more items, exceeding some breakpoint level. This topic will be addressed in Chapter 10, where we will be concerned with coordinated replenishment strategies.

It should be mentioned that, in trying to achieve a quantity discount, one must take into account the implications of a large time supply, a topic to be discussed in Section 4.7.

4.6 Accounting for Inflation

One of the assumptions in the derivation of the EOQ was that the inflation rate was at a negligible level. But, in recent times, many countries have been confronted with fluctuating inflation rates that often have been far from negligible.[*] In this section, we investigate the impact of inflation on the choice of replenishment quantities.

There are several options available for modeling the effects of inflation on costs and revenues. We restrict our attention to the case where the fixed replenishment cost A and the unit variable cost v are affected in the same fashion by inflation. A more elaborate model would be necessary if inflation had a different impact on the two types of costs. Moreover, the effects on revenue depend on the choice of pricing policy adopted by the organization (possibly subject to government regulation). In Section 4.6.1, we investigate the case where (selling) price changes are made independent of the replenishment strategy. A special case is where the unit selling price p is assumed to increase continuously at the inflation rate (in the same fashion as A and v). Then, in Section 4.6.2, the result will be shown for the situation where the price is adjusted only once for each replenishment lot. Thus, in this latter case, revenue, as well as costs, now depends on the sizes of the replenishments.

An *exact* analysis in the presence of inflation would be extremely complicated. The reason for this is that costs varying with time, in principle, should lead to the replenishment quantity changing with time. At any point in time, our primary concern is with choosing the value of *only* the very next replenishment. To obtain a tractable analytic result, we assume that all future replenishments will be of the same size as the current replenishment, contrary to what the above discussion would suggest. This assumption is not as serious as it would first appear. In particular, at the time of the next replenishment, we would recompute a *new* value for that replenishment, reflecting any changes in cost (and other) parameters that have taken place in the interim. Second, we employ discounting so that assumptions about future lot sizes should not have an appreciable effect on the current decision. This is an illustration of a heuristic solution procedure, a method, based on analytic and intuitive reasoning, that is not optimal but should result in a reasonable solution. Finally, we do not treat the case where quantity discounts are available. When discounts are available, the best order quantity may be very insensitive to inflation because the breakpoint value often is the best solution over a wide range of parameter values (see Gaither 1981).

4.6.1 Price Established Independent of Ordering Policy

Because the price is established independent of the ordering policy (perhaps pricing is set by market forces external to the company under consideration), we can restrict our attention, as earlier, to the minimization of costs. However, because costs are changing with time, we cannot compare costs over a typical year as in Section 4.2. Instead, we resort to a common tool of investment analysis, namely, the use of the present value (PV) of the stream of future costs.

Suppose that the continuous discount rate is denoted by r; that is, a cost of c at time t has a PV of ce^{-rt} (e.g., Ross et al. 2008). We denote the continuous inflation rate by i. Thus, if A and v are the cost factors at time zero, then, their values at time t are Ae^{it} and ve^{it}, respectively. With a demand rate of D and an order of size Q, we have time between replenishments of Q/D. At each replenishment, we incur costs of $A + Qv$. If the first replenishment is received at time 0, the PV of

---

[*] In some cases, inflation is so high that all cash is quickly used to buy inventories because the increase in the price of the goods exceeds the interest one can earn on invested capital. This section assumes that inflation is at a more reasonable level.

the stream of costs is given by[*]

$$
\begin{aligned}
PV(Q) &= A + Qv + (A + Qv)e^{iQ/D}e^{-rQ/D} + (A + Qv)e^{2iQ/D}e^{-2rQ/D} + \cdots \\
&= (A + Qv)(1 + e^{-(r-i)Q/D} + e^{-2(r-i)Q/D} + \cdots) \\
&= (A + Qv)\frac{1}{1 - e^{-(r-i)Q/D}}
\end{aligned}
\tag{4.14}
$$

To find a minimum, we set $d\text{PV}(Q)/dQ = 0$. As shown in Section 4A.2 of the Appendix of this chapter, the result is

$$
e^{(r-i)Q/D} = 1 + \left(\frac{A}{v} + Q\right)\left(\frac{r-i}{D}\right)
$$

For $(r-i)Q/D \ll 1$, which is a reasonable assumption, this equation can be simplified further by approximating the exponential term ($e^x = 1 + x + x^2/2$ for $x \ll 1$). Again, in the Appendix, we obtain the following result:

$$
Q_{\text{opt}} = \sqrt{\frac{2AD}{v(r-i)}} = \text{EOQ}\sqrt{\frac{1}{1 - i/r}}
\tag{4.15}
$$

that is, the EOQ multiplied by a correction factor. Note that, when there is negligible inflation ($i = 0$), this is nothing more than the EOQ. Hence, our justification for using the symbol r for the discount rate. Buzacott (1975) has developed a somewhat more exact decision rule, for which the result of Equation 4.15 is usually a good approximation. See also Jesse et al. (1983), Kanet and Miles (1985), and Mehra et al. (1991) who include a term for storage costs exclusive of the opportunity cost of capital. Sarker and Pan (1994) determine the optimal order quantity and allowable shortages in a finite replenishment inventory system under inflationary conditions. Also, Rachamadugu (1988) incorporates a discounted total cost function.

In an inflationary environment when price is set independent of the ordering policy, the profit per unit increases with time since the last replenishment; thus, intuitively, the replenishment quantity should be larger than under no inflation. Equation 4.15 indicates this type of behavior.

The cost penalty associated with using the EOQ (that ignores inflation) can be found from

$$
\text{PCP} = \frac{\text{PV}(\text{EOQ}) - \text{PV}(Q_{\text{opt}})}{\text{PV}(Q_{\text{opt}})} \times 100
\tag{4.16}
$$

where $\text{PV}(Q)$ is given by Equation 4.14. The PCP increases as i approaches r but, as shown in this chapter's Appendix, it reaches a limiting value of $100\sqrt{Ar/2Dv}$, which tends to be quite small, although it should be emphasized that this is a percentage of the *total* cost including purchasing costs, not just ordering and holding costs.

---

[*] The last step in this derivation uses the fact that

$$
1 + a + a^2 + \cdots = \frac{1}{1 - a} \text{ for } 0 \leq a < 1
$$

Table 4.2 Effects of Inflation on Choice of Order Quantity for 3-Ohm Resistor

i	0	0.02	0.05	0.1	0.2	0.22	0.24
EOQ$_{opt}$	400	418	450	524	980	1,386	∞
PCP of Using EOQ	0	0.0025	0.021	0.105	0.687	1.26	2.0

4.6.1.1 Numerical Illustration

Let us reconsider the 3-ohm resistor that was used in Section 4.2. Its parameter values were

$D = 2{,}400$ units/year	$v = \$0.40$/unit
$A = \$3.20$	$r = 0.24$ \$/\$/year

The EOQ under no inflation was found to be 400 units. The use of Equation 4.15 produces the behavior shown in Table 4.2 for different values of i. The percentage cost penalties computed from Equation 4.16 are also presented in Table 4.2 and are seen to be extremely small.

4.6.2 Price Set as a Fixed Fractional Markup on Unit Variable Cost

Let us denote the unit selling price established by the company at time t as $p(t)$. Suppose that a replenishment order is paid for at time t at unit variable cost $v(t)$; the next order will be paid for at time $t + T$. Then a fixed fractional markup (f) implies

$$p(t + \tau) = v(t)(1 + f) \quad 0 \leq \tau < T \tag{4.17}$$

where $p(t + \tau)$ is the price at time $t + \tau$. That is, the same unit price is used throughout the entire time period covered by each replenishment.

$v(t)$ depends on the inflation rate as in the previous section; thus, the unit selling price also depends on the inflation rate and on the order quantity Q, the latter in that $T = Q/D$ in Equation 4.17. Consequently, both costs *and revenues* are influenced here by the choice of the order quantity Q. An appropriate criterion is to select Q in order to maximize the PV of revenues minus costs. Buzacott (1975) uses an analysis similar to that presented[*] to obtain

$$Q_{opt} = \sqrt{\frac{2AD}{v(r + fi)}} = \text{EOQ}\frac{1}{\sqrt{1 + fi/r}} \tag{4.18}$$

When the markup for the sale of a *whole* replenishment lot is set as a fixed fraction of the unit variable cost at the *start* of a cycle and costs are increasing because of inflation, intuitively the order quantity should be reduced from the EOQ position in order to be able to keep the selling price more closely in line with a markup f on the current unit variable cost. Equation 4.18 reflects this

---

[*] An added complexity is that the revenue is a continuous function of time that must be discounted.

intuitive behavior. It is seen that, as the product fi increases, Q_{opt} becomes smaller relative to the EOQ.

4.7 Limits on Order Sizes

The basic EOQ that we derived earlier considers only the costs of replenishing and carrying the inventory. All the variables used reflect only financial considerations. There are a number of possible physical constraints, not explicitly included in the model, that might prevent the use of the so-called best solution derived from the model. We look at some of these constraints in Sections 4.7.1 through 4.7.3.[*]

4.7.1 Maximum Time Supply or Capacity Restriction

1. The shelf life (SL) of the commodity may be an important factor. Some items have an SL dictated by perishability, which we investigate more fully in Chapter 9. The SL of other items may be more unusual. For instance, certain computers, generators, and so on have a limited warranty that starts from the date of delivery, and not the date of first use. Previously we derived the EOQ expressed as a time supply as

$$T_{EOQ} = \frac{EOQ}{D} = \sqrt{\frac{2A}{Dvr}}$$

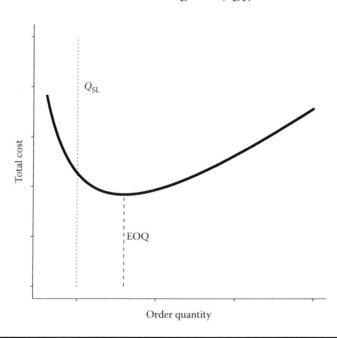

Figure 4.6 Case of an SL constraint.

---

[*] Das (1990) provides an algorithm for computing optimal order sizes for several of these cases. See also Moon and Yun (1993) who applies discounted cash flow analysis to the finite horizon case.

As illustrated graphically in Figure 4.6, if this quantity exceeds the allowable shelf life SL, then, the best feasible order quantity to use (i.e., it produces the lowest feasible point on the cost curve) is the SL quantity itself,

$$Q_{SL} = D(SL)$$

2. Even without an SL limitation, an EOQ that represents a very long time supply may be unrealistic for other reasons. A long time supply takes us well out into the future where demand becomes uncertain and obsolescence can become a significant concern. This is an important constraint on class C items, as we will see in Chapter 8. A good example of obsolescence would be an impending engineering change.
3. There may be a storage capacity limitation on the replenishment size for an item (e.g., the entire replenishment must fit in a dedicated storage bin). Equivalently, in production situations (to be discussed in Chapters 10 through 16), there may be a finite production capacity available for the item or a group of related items. A capacity limitation is mathematically equivalent to an SL restriction.

4.7.2 Minimum Order Quantity

The supplier may specify a minimum allowable order quantity. A similar situation exists in an in-house production operation where there is a lower limit on an order quantity that can be realistically considered. If the EOQ is less than this quantity, then, the best allowable order quantity is the supplier or production minimum.

4.7.3 Discrete Units

The EOQ, as given by Equation 4.4, will likely result in a nonintegral number of units. However, it can be shown mathematically (and is obvious from the form of the total cost curve of Figure 4.2) that the best integer value of Q has to be one of the two integers surrounding the best (noninteger) solution given by Equation 4.4. For simplicity, we recommend simply rounding the result of Equation 4.4, that is, the EOQ, to the nearest integer.[*]

Certain commodities are sold in pack sizes containing more than one unit; for example, many firms sell in pallet-sized loads only, say 48 units (e.g., cases). Therefore, it makes sense to restrict a replenishment quantity to integer multiples of 48 units. In fact, if a new unit, corresponding to 48 of the old unit, is defined, then, we are back to the basic situation of requiring an integer number of (the new) units in a replenishment quantity.

There is another possible restriction on the replenishment quantity that is very similar to that of discrete units. This is the situation where the replenishment must cover an integral number of periods of demand. Again, we simply find the optimal continuous Q expressed as a time supply (see Equation 4.7) and round to the nearest integer value of the time supply.

---

[*] Strictly speaking, one should evaluate the total costs, using Equation 4.3, for each of the two surrounding integers and pick the integer having the lower costs. However, rounding produces a trivial percentage cost penalties except possibly when the EOQ is in the range of only one or two units. In addition, the data that are used in Equation 4.3 often have a significantly larger error than the error introduced by rounding to the suboptimal integer.

4.8 Finite Replenishment Rate: The Economic Production Quantity

One of the assumptions inherent in the derivation of the EOQ was that the whole replenishment quantity arrives at the same time. If, instead, we assume that it becomes available at a rate of m per unit time (the production rate of the machinery used to produce the item), then, the sawtoothed diagram of Figure 4.1 is modified to that of Figure 4.7. All that changes from the earlier derivation is the average inventory level that is now $Q(1 - D/m)/2$. The total relevant costs are given by

$$\text{TRC}(Q) = \frac{AD}{Q} + \frac{Q(1 - D/m)vr}{2}$$

and the best Q value is now the finite replenishment economic order quantity (FREOQ), also known as the economic production quantity (EPQ) or the economic run quantity.

$$\text{FREOQ} = \sqrt{\frac{2AD}{vr(1 - D/m)}}$$

$$= \text{EOQ} \underbrace{\frac{1}{\sqrt{1 - D/m}}}_{\text{correction factor}} \qquad (4.19)$$

As shown in Equation 4.19, the EOQ is multiplied by a correction factor. The magnitude of this correction factor for various values of D/m is shown in Figure 4.8. Also shown are the cost penalties caused by ignoring the correction and simply using the basic EOQ. It is clear that D/m

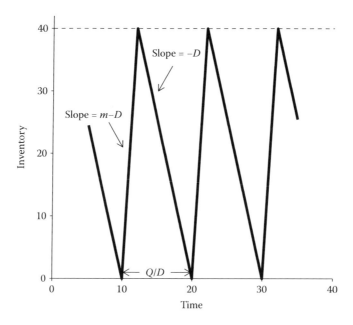

Figure 4.7   Case of a finite replenishment rate.

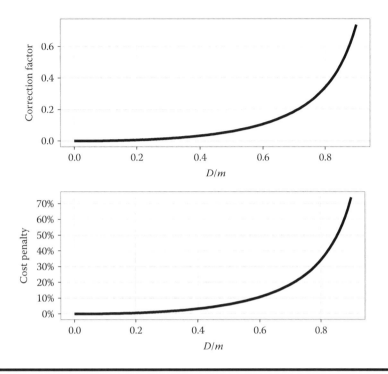

Figure 4.8 Use of EOQ when there is a finite replenishment rate.

has to be fairly large before the cost penalty becomes significant; for example, even when D/m is 0.5 (i.e., the demand rate is one-half of the production rate), the cost penalty for neglecting the finite nature of the production rate is only 6.1%.

Note that as D/m gets very small (equivalent to instantaneous production of the entire lot), the FREOQ reduces to the EOQ as it should. Moreover, as D/m tends to unity, Equation 4.19 suggests that we produce an indefinitely large lot. This makes sense in that demand exactly matches the production rate. For $D/m > 1$, the model breaks down because the capacity of the machine is insufficient to keep up with the demand. However, the model is still useful under these circumstances. To illustrate, suppose that $D = 1.6\,m$. Here, one would dedicate one machine to the product (with no changeover costs) and produce the remaining $0.6\,m$ in finite lot sizes on a second machine, where the latter machine's capacity would likely be shared with other items. An example, encountered by one of the authors in a consulting assignment, involved the same fastener manufacturer mentioned earlier in the chapter. One part had enough demand to more than fully use the capacity of one machine. At the time, however, this part was manufactured on several different machines that also made other parts. Managers were working to rearrange the production schedule to dedicate a machine entirely to the part, while producing any excess demand for that part on other shared machines. Further discussion on the sharing of production equipment among several items will be presented in Chapter 10.[*]

---

[*] See Billington (1987), Kim et al. (1992), and Li and Cheng (1994) for "changing the givens" in the EPQ model.

4.9 Incorporation of Other Factors

In this section, we briefly discuss four further modifications of the basic EOQ that result from the relaxation of one or more of the assumptions required in the earlier derivation. The treatment of another important situation—namely, where the quantity supplied does not necessarily exactly match the quantity ordered—can be found in Silver (1976). Other references on this topic can be found in Chapter 7. For the case in which the amount of stock displayed on the shelf influences the demand, see Paul et al. (1996).

4.9.1 Nonzero Constant Lead Time That Is Known with Certainty

As long as demand remains deterministic, the introduction of a known, constant, nonzero replenishment lead time (L) presents no difficulty. The inventory diagram of Figure 4.1 is unchanged. When the inventory level hits DL, an order is placed. It arrives exactly L time units later, just as the inventory hits zero.[*] If the lead time is longer than the time between replenishments, the order should be placed when the on-hand plus on-order level drops to DL.

The costs are unaltered so that the best order quantity is still given by Equation 4.4. For example, suppose the lead time for the 3-ohm resistor (from Section 4.2) is 3 weeks. The EOQ is 400, and the weekly demand is 48 (based on a 50-week year). When the inventory on hand falls to $48 \times 3 = 144$ units, the order should be placed. Figure 4.9 illustrates this concept. (A nonzero

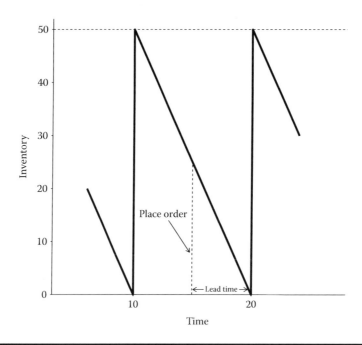

Figure 4.9   Case of a positive lead time.

---

[*] If the lead time is longer than the time between replenishments, the order should be placed when the on-hand plus on-order level drops to DL.

lead time considerably complicates matters when demand is probabilistic; and when lead times themselves can be probabilistic, the analysis also gets even more involved. These cases are discussed in Chapter 6.)

4.9.2 Nonzero Payment Period

We have implicitly assumed, by simply looking at average costs per unit time, that payment is immediate upon the receipt of goods. In practice, one can have a period of t years (t, of course, is a fraction) by the end of which the payment must be received. As in Section 4.6, we now must worry about the timing of expenses, so we again use a PV analysis. We assume that the fixed cost, A, is incurred immediately. The PV of repeatedly using an order quantity of size Q is then

$$PV(Q) = A + Qve^{-rt} + PV(Q)e^{-rQ/D}$$

In a manner closely paralleling Section 4.6, we find

$$Q_{\text{opt}} = \sqrt{\frac{2AD}{ve^{-rt}r}} = \text{EOQ}\sqrt{e^{rt}}$$

This result has a simple intuitive interpretation. The unit variable cost v is reduced to ve^{-rt}, because we retain the associated dollars for a time t. Recognizing that rt is likely to be substantially less than unity, we have

$$Q_{\text{opt}} \approx EOQ \left(1 + \frac{rt}{2}\right)$$

so that the effect of the nonzero payment period is quite small and approximately linear in t. Other treatments of repetitive payment periods are provided by Thompson (1975), Kingsman (1983), and Arcelus and Srinivasan (1993). For the case where the supplier offers a one-time larger payment period ($u > t$), Silver (1999) shows that a simple result holds, namely that the appropriate time supply to use should be $u - t$ years larger than the ongoing best time supply (i.e., Q_{opt}/D). Davis and Gaither (1985) also consider a one-time opportunity to extend the payment.

4.9.3 Different Types of Carrying Charge

In Section 4.2, we assumed that the carrying charge was directly proportional to the average inventory level measured in dollars. We calculated the inventory turnover ratio using similar logic. Modifications to this inventory accounting are possible. Each can be handled with the resulting order quantity expression being somewhat more complex than Equation 4.4.

To illustrate, consider the situation where costs depend on area or volume considerations as well as the value of the inventory. (It is intuitive that it should cost more per unit time to store a dollar of feathers than a dollar of coal.) Suppose that there is a charge of w dollars per unit time per cubic foot of space allocated to an item. Assume that this space must be sufficient to handle the maximum inventory level of the item. One situation where such a change would be appropriate is where items are maintained in separate bins, and a specific bin size (which must house the maximum inventory) is allocated to each item. The best replenishment quantity under these circumstances is

$$\sqrt{\frac{2AD}{2hw + vr}}$$

where h is the volume (in cubic feet) occupied per unit of the item.

4.9.4 Multiple Setup Costs: Freight Discounts

Consider the situation of an item in which the costs of a replenishment depend on the quantity in the following fashion:

Q Range	Cost
$0 < Q \le Q_0$	$A + Qv$
$Q_0 < Q \le 2Q_0$	$2A + Qv$
$2Q_0 < Q \le 3Q_0$	$3A + Qv$

One interpretation is that the item is produced in some type of container (e.g., a vat or barrel) of fixed capacity Q_0 and there is a fixed cost A associated with the use of each container. Such a fixed cost may arise from a need to clean or repair the vat between uses. Another interpretation relates to transportation costs where Q_0 might be the capacity of a shipping container or rail car. Under the above cost structure, Aucamp (1982) has shown that the best solution is either the standard EOQ or one of the two surrounding integer multiples of Q_0. Early research incorporating freight cost into EOQ decisions by Baumol and Vinod (1970) highlights the importance of potentially including freight cost in inventory decisions. This paper investigates the trade-off between speed of delivery and cost in an EOQ-type context. Lee (1986) generalizes Aucamp's work by allowing for a cost of A_j for the jth load. Therefore, there is a possible advantage to increasing the number of loads. Knowles and Pantumsinchai (1988) address a situation in which the vendor offers products only in various container sizes. Larger discounts are offered on larger containers. Tersine and Barman (1994) examine incremental and all-units discounts for both purchase price and freight cost. Russell and Krajewski (1991) address the situation in which firms overdeclare the amount of freight shipped in order to reduce the total freight cost. They include a fixed and incremental cost for both purchase and freight. See also Carter and Ferrin (1995). See Gupta (1994) for a treatment of the case in which the ordering cost increases as the replenishment quantity increases. Finally, Van Eijs (1994) examines the problem in a multi-item context with random demands, a problem we will address in Chapter 10. See also Carter et al. (1995a) and Shinn et al. (1996). Thomas and Tyworth (2006), Mendoza and Ventura (2008), and Toptal and Bingöl (2011) investigate the impact of freight costs, specifically with truckload and less-than-truckload (LTL) costs, on the choice of replenishment lot size.

A common theme in much of this work is that the explicit inclusion of freight costs in the EOQ model can dramatically impact the choice of order quantity due to economies of scale in transportation. As an illustration, let us revisit the example from Section 4.2 of the 3-ohm resistor (stable demand $D = 2,400$ units/year, unit variable cost $v = \$0.40$/unit, fixed cost per replenishment $A = \$3.20$, and holding cost rate $r = 0.24$ \$/\$/year), but now include freight charges. Furthermore, suppose the transportation provider charges a lower rate per unit for larger orders as follows:

Q Range	Freight Cost per Unit
$0 \geq Q < 250$	$0.030
$250 \geq Q < 500$	$0.025
$500 \geq Q < 1,000$	$0.020
$1,000 \geq Q$	$0.015

Now that freight costs vary with Q, those costs should be included in the total relevant costs per unit time. In a manner quite similar to quantity discounts, these freight cost break points cause discontinuities that must be evaluated as shown in Figure 4.10. When these freight costs are included in this example, the cost minimizing lot size is $Q = 500$; it is worth ordering more than the EOQ of 400 in order to obtain the freight savings.

One nuance worth mentioning is that in some freight settings—LTL carriage for example—carriers[*] may charge the shipper for more units than are actually shipped if it results in a lower cost.

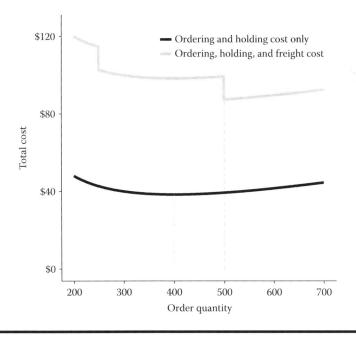

Figure 4.10   Total costs with and without the freight costs included.

[*] Without getting too distracted with transportation terminology, a *carrier* moves goods while a *shipper* hires a carrier to move goods. Coyle et al. (2015) provide an extensive detail on the modes of transportation including the terminology as well as the pricing structure.

In this case, if the quantity shipped is 225 units, the list price in that range is $0.03/unit for a cost per shipment of $6.75 and an annual shipping cost of $72.00; however, with a quantity of 225, it is better for the shipper to be *charged* as if 250 units were being shipped, giving a cost per shipment of (250)($0.025) = $6.25 and an annual shipping cost of ($6.25)(2,400/225) = $66.67. Our experience suggests that when this situation arises, it is frequently worth increasing the shipping quantity to equal the billed quantity, as is the case in this example.

4.9.5 A Special Opportunity to Procure

An important situation, often faced by both producers and merchants, is a one-time opportunity to procure an item at a reduced unit cost. For example, this may be the last opportunity to replenish before a price rise.

Because of the change in one of the parameters (the unit cost), the current order quantity can differ from future order quantities. (All these future quantities will be identical—namely, the EOQ with the new unit cost.) Thus, the simple sawtooth pattern of Figure 4.1 no longer necessarily applies. Hence, we cannot base our analysis on the average inventory level and the number of replenishments in a typical year (or other units of time). Instead, in comparing the two possible choices for the current value of the order quantity Q, strictly speaking, we should compare costs out to a point in time where the two alternatives leave us in an identical inventory state. To illustrate, suppose that the demand rate is 100 units/month and the new EOQ will be 200 units. Furthermore, suppose that two alternatives under consideration for the current order quantity are 200 and 400 units. As shown in Figure 4.11, an appropriate horizon for cost comparison would be 4 months because, for both alternatives, the inventory level would be zero at that point in time. A comparison of costs out to 1.95 months, for example, would be

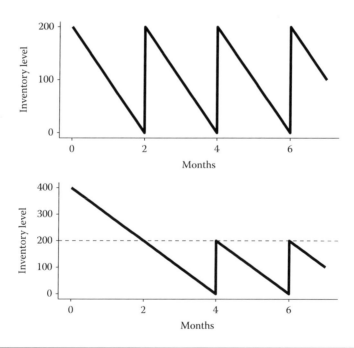

Figure 4.11 Comparison of two alternatives when the order quantity changes with time.

biased in favor of the 200-unit alternative because the imminent (0.05 months later) setup cost under this alternative would not be included. An exact (PV) comparison for all alternatives is rather complex mathematically, particularly when there is a continuum of choices for the order quantity.

Instead, we recommend an approximate approach, suggested by Naddor (1966). Let the decision variable, the current order quantity, be denoted by Q. Let the current unit cost be v_1 and the future unit cost be v_2 ($v_2 > v_1$). The EOQ, after the price rise, is given by Equation 4.4 as

$$\text{EOQ}_2 = \sqrt{\frac{2AD}{v_2 r}} \qquad (4.20)$$

Furthermore, from Equation 4.3, the costs per unit time are then

$$\text{TRC(EOQ}_2) = \sqrt{2ADv_2 r} + Dv_2 \qquad (4.21)$$

If the current order quantity is of size Q, then, it will last for Q/D units of time. The average inventory during this period is $Q/2$. Hence, the total costs out to time Q/D are

$$\begin{aligned} TC(Q) &= A + Qv_1 + \frac{Q}{2}v_1\frac{Q}{D}r \\ &= A + Qv_1 + \frac{Q^2 v_1 r}{2D} \end{aligned} \qquad (4.22)$$

Consider the extreme case where Q was set equal to 0 so that the price increase would occur and we would immediately start ordering the new EOQ_2. Under this strategy, the total costs out to some time T would be, from Equation 4.21,

$$T\sqrt{2ADv_2 r} + DTv_2$$

It is reasonable to select the value of Q to maximize

$$F(Q) = (Q/D)\sqrt{2ADv_2 r} + Qv_2 - TC(Q) \qquad (4.23)$$

the improvement in the total cost out to time Q/D achieved by ordering Q at the old price instead of not ordering anything at the old price. It can be shown that the use of this criterion need not lead to the exact cost-minimizing solution. Again, we are opting for a heuristic procedure for selecting the value of the decision variable (in this case Q). A convenient approach to maximizing $F(Q)$ is to set $dF(Q)/dQ = 0$. From Equations 4.22 and 4.23

$$\begin{aligned} F(Q) &= \frac{Q}{D}\sqrt{2ADv_2 r} + Qv_2 - A - Qv_1 - \frac{Q^2 v_1 r}{2D} \\ \frac{dF(Q)}{dQ} &= \frac{1}{D}\sqrt{2ADv_2 r} + v_2 - v_1 - \frac{Qv_1 r}{D} = 0 \end{aligned} \qquad (4.24)$$

or

$$Q_{\text{opt}} = \frac{\sqrt{2ADv_2 r}}{v_1 r} + \frac{v_2 - v_1}{v_1}\frac{D}{r}$$

that is,

$$Q_{\text{opt}} = \frac{v_2}{v_1} \text{EOQ}_2 + \frac{v_2 - v_1}{v_1} \frac{D}{r} \qquad (4.25)$$

The above analysis implicitly assumed that the initial inventory (at the moment that the special opportunity replenishment was to be made) was 0. If instead the inventory was I_0, then, Lev and Soyster (1979) show that Q_{opt} becomes the result of Equation 4.25 minus I_0; that is, Equation 4.25 gives the appropriate order-up-to-level. Taylor and Bradley (1985) generalize to the case of future announced price increases; that is, the firm has time to adjust several order cycles in advance. See also Tersine (1996).

Other analyses of this problem are presented by Brown (1982) and McClain and Thomas (1980). The latter authors, in particular, use a marginal approach that very quickly gives a decision rule. However, the rule, unlike Equation 4.25, is not easily extendible to the case of multiple items, a case to be discussed in a moment. Miltenburg (1987) has shown that the Brown and Naddor decision rules give very similar results on a broad range of problems. Yanasse (1990) arrives at Equation 4.25 through a slightly different approach.[*]

Lev et al. (1981) extend the analysis to include the possibility of multiple cost or demand parameter changes in two distinct time periods. Ardalan (1994) finds the optimal price and order quantity for a retailer who is offered a temporary price reduction. The retailer's demand increases as its price decreases. Finally, Brill and Chaouch (1995) discuss the EOQ model when a random shock can affect the demand rate. See Goyal et al. (1991) for a review of the literature in this area. See also Aucamp and Kuzdrall (1989), Aull-Hyde (1992, 1996), Tersine and Barman (1995), Arcelus and Srinivasan (1995), Kim (1995), and Moinzadeh (1997).

4.9.5.1 Numerical Illustration

Consider a particular Toronto-based x-ray film dealer supplied by a film manufacturer. Suppose that the manufacturer announces a price increase for a product XMF-082 from \$28.00/box to \$30.00/box. The dealer uses approximately 80 boxes per year and estimates the fixed cost per order to be \$1.50 and the carrying charge as 0.20 \$/\$/year. From Equation 4.20

$$\text{EOQ}_2 = \sqrt{\frac{2 \times 1.50 \times 80}{30.00 \times 0.2}} \approx 6 \text{ boxes}$$

Then, Equation 4.25 gives

$$Q_{\text{opt}} = \frac{30.00}{28.00}(6) + \frac{2.00}{28.00}\frac{80}{0.2}$$
$$= 35.3, \text{say } 35 \text{ boxes}$$

The increase in unit price from \$28.00/box to \$30.00/box (a 7% increase) causes a one-time procurement quantity of 35 boxes instead of 6 boxes (an increase of almost 500%!), a much higher sensitivity than in the basic EOQ itself (see Section 4.3).

---

[*] One inventory software package artificially "ages" the inventory by multiplying all reorder points by a factor (say 1.3) when a price increase is announced. Then, orders are placed for all products that have fallen below these new temporary reorder points. If the cost of the orders is too high, the factor is reduced. We will have more to say on reorder points in Chapters 6 and 7.

Substitution of each of $Q = 6$ and $Q = 35$ into Equation 4.24 gives

$$F(6) = \$12.09$$

and

$$F(35) = \$42.23$$

so that the one-time cost savings of using the order quantity of 35 boxes is (42.23–12.09) or $30.14.

Two additional points are worth making:

1. The one-time purchase is likely to represent a large time supply. Obsolescence and other forms of uncertainty may dictate an upper limit on this time supply. Many high-technology firms face rapid obsolescence—the price of their products falls quickly with time. They are therefore very reluctant to carry much stock.
2. The multi-item case, in which a vendor offers a one-time price break on a range of products, is obviously of interest. In such a situation, there is likely to be a constraint on the total amount that can be spent on the one-time purchase or on the total inventory (because of warehouse capacity). For illustrative purposes, let us consider the case of a constraint on the total amount that can be spent. For notation let

$$D_i = \text{usage rate of item } i, \text{ in units/year}$$
$$A_i = \text{fixed setup cost of item } i, \text{ in dollars}$$
$$v_{1i} = \text{unit cost of item } i \text{ in the special opportunity to buy, in \$/unit}$$
$$v_{2i} = \text{unit cost of item } i \text{ in the future, in \$/unit}$$
$$Q_i = \text{amount, in units, of item } i \text{ to be purchased in the special buy}$$
$$\text{EOQ}_{2i} = \text{economic order quantity of item } i \text{ in units, under the future unit cost}$$
$$n = \text{number of items in the group under consideration (we assume in the sequel that the items are numbered } 1, 2, 3, \ldots, n)$$
$$W = \text{maximum total amount, in dollars, that can be spent on the special buy}$$

Then, the procedure for selecting the $Q'_i s$ is as follows:

Step 1 Determine the unconstrained best $Q'_i s$ by the analog of Equation 4.25, namely,

$$Q_i^* = \frac{v_{2i}}{v_{1i}} \text{EOQ}_{2i} + \frac{v_{2i} - v_{1i}}{v_{1i}} \frac{D_i}{r} \quad i = 1, 2, \ldots, n \tag{4.26}$$

Step 2 Calculate the total dollar value of the unconstrained $Q'_i s$ and compare with the constraining value; thus from Equation 4.26

$$\sum_{i=1}^{n} Q_i^* v_{1i} = \sum_{i=1}^{n} \left[v_{2i} \text{EOQ}_{2,i} + (v_{2i} - v_{1i}) D_i / r \right] \tag{4.27}$$

$\sum_{i=1}^{n} Q_i^* v_{1i} \leq W$, we use the Q_i^*'s of Step 1. If not, we go to Step 3.

Step 3 Compute the constrained best Q_i values. (The derivation is shown in Section 4A.3 of the Appendix of this chapter):

$$Q_i^* = \frac{v_{2i}}{v_{1i}}\left(\text{EOQ}_{2i} + \frac{D_i}{r}\right) - \frac{D_i}{\sum_{j=1}^n D_j v_{1j}}$$

$$\times \left[\sum_{j=1}^n v_{2j}\left(\text{EOQ}_{2j} + \frac{D_j}{r}\right) - W\right] \quad i = 1, 2, \ldots, n \quad (4.28)$$

If $Q_i^* < 0$, set $Q_i^* = 0$ and solve for all other items.

Note in Equation 4.28 that as the total budget decreases, the term $(D_i W / \sum_{j=1}^n D_j v_{1j})$ decreases linearly. Thus, the item with the highest demand rate, D_i, will have its Q_i^* decrease fastest. This point is also evident from the fact that $(d(Q_i^* v_i)/dW) = -(D_i v_{1i}/\sum_{j=1}^n D_j v_{1j})$. For more discussion on this topic, see Chapter 10, which deals with coordinating the purchase or production of multiple items.

4.10 Selection of the Carrying Charge (*r*), the Fixed Cost per Replenishment (*A*), or the Ratio *A/r* Based on Aggregate Considerations: The Exchange Curve

Often, it is difficult to explicitly determine an appropriate value of the carrying charge *r* or the fixed cost per replenishment *A*. An alternate method, which we discussed in Chapter 2, takes into account of the aggregate viewpoint of management. In this and the later sections, we derive the relationships and graphs that we earlier asked you to accept on faith. For a population of inventoried items, the management may impose an aggregate constraint of one of the following forms:

1. The average total inventory cannot exceed a certain dollar value or volume.[*]
2. The total fixed cost (or total number) of replenishments per unit time (e.g., 1 year) must be less than a certain value.[†]
3. The maximum allowed backorder delay may not exceed a certain value.
4. The firm should operate at a point where the trade-off (exchange) between the average inventory and the cost (or number) of replenishments per unit time is at some reasonable prescribed value.

We restrict our attention to the case where the fixed cost per replenishment for item *i*, denoted by A_i, cannot be determined explicitly, but it is reasonable to assume that a common value of *A* holds (at least approximately) for all items in the portion of the inventory population under consideration. We designate the demand rate, unit variable cost, and order quantity of item *i* by

[*] Sometimes, a firm allows its different entities to determine their optimal budgets, and then, it cuts them all by a common factor. Because this "cut across the board" rule is in general not optimal, Rosenblatt and Rothblum (1994) identify the conditions for it to provide globally optimal solutions. See also Pirkul and Aras (1985).
[†] See Aucamp (1990).

D_i, v_i, and Q_i, respectively. We also let n be the number of items in the population. (The case where the A_i's are explicitly predetermined can be treated in a similar fashion.)

As shown in Section A4.4 of the Appendix of this chapter, if we use an EOQ for each item, we obtain the total average cycle stock in dollars,

$$\text{TACS} = \sqrt{\frac{A}{r}} \frac{1}{\sqrt{2}} \sum_{i=1}^{n} \sqrt{D_i v_i} \qquad (4.29)$$

and the total number of replenishments per unit time,

$$N = \sqrt{\frac{r}{A}} \frac{1}{\sqrt{2}} \sum_{i=1}^{n} \sqrt{D_i v_i} \qquad (4.30)$$

Both TACS and N depend on the value of the ratio A/r. Multiplication of Equations 4.29 and 4.30 gives

$$(\text{TACS})(N) = \frac{1}{2} \left(\sum_{i=1}^{n} \sqrt{D_i v_i} \right)^2 \qquad (4.31)$$

which is a hyperbola. Moreover, the division of Equation 4.29 by Equation 4.30 gives

$$\frac{\text{TACS}}{N} = \frac{A}{r} \qquad (4.32)$$

so that any point on the hyperbolic curve implies a value of A/r. Of course, if either A or r is known explicitly, the implicit value of A/r implies a value of the one remaining unknown parameter.

To summarize, when an EOQ strategy[*] is used for each item, the management can select a desired point on the tradeoff curve (with associated aggregate conditions), thus implying an appropriate value of r, A, or A/r. The latter parameters can now be thought of as management control variables. In fact, in the late 1980s, a major U.S. automotive firm used a value of 40% for r. When questioned why the value was so high, the reply was, "We want to decrease inventory." In other words, the management artificially increased r so that computerized inventory control routines would force down inventory levels. It should be noted that the use of r as a policy variable is closely related to a technique known as LIMIT (lot-size inventory management interpolation technique) discussed by Eaton (1964). See also Chakravarty (1986).

4.10.1 Exchange Curve Illustration

A small firm stocks just 20 items, all of which are purchased from several suppliers in Germany. Therefore, it is realistic to think in terms of a fixed setup cost A, which does not vary appreciably

---

[*] When an aggregate constraint is to be satisfied, it is not obvious that the best strategy for an individual item is to use an EOQ-type order quantity. In fact, the more mathematically oriented reader may wish to verify that, for example, minimization of the total number of replenishments per unit time subject to a specified total average stock (or minimization of total average stock subject to a specified total number of replenishments per unit time) does indeed lead to an EOQ-type formula. If a Lagrange multiplier approach is used, the multiplier turns out to be identical with r/A (or A/r). An introduction to the use of Lagrange multipliers in constrained optimization is provided in Appendix I.

from item to item. Based on a detailed study of costs and policy factors, the management decided to set the value of A at $2.10. A spreadsheet was used to generate the exchange curve of Figure 4.12. The current operating procedure (estimated by using the most recent replenishment quantity actually used for each of the items in the division) is also shown in Figure 4.12.

It is clear that improvements are possible through the use of EOQs with a suitable value of A/r. (This is because the current logic does not explicitly take the economic factors into account—some inexpensive items are ordered too frequently while some fast movers are not ordered often enough.) At one extreme, by operating at point P (with an implied A/r of 11.07), the same total average cycle stock is achieved as by the current rule but the total number of replenishments per year is reduced from 1700 to 759 (a 55.4% reduction). Point Q (with an implied A/r of 2.21) represents the opposite extreme where the same total number of replenishments is achieved as by the current rule, but the total average cycle stock is cut from $8,400 to $3,749 (a 55.4% reduction). Decreases, compared with the current rule, are possible in both aggregate measures at any point between P and Q. Suppose the management has agreed that a total average cycle stock of $7,000 is a reasonable aggregate operating condition. The associated A/r value (using Equation 4.29) is 7.69. Since A has been explicitly determined, we have that

$$r = A \div A/r = 2.10 \div 7.69 = 0.273 \$/\$/\text{year}$$

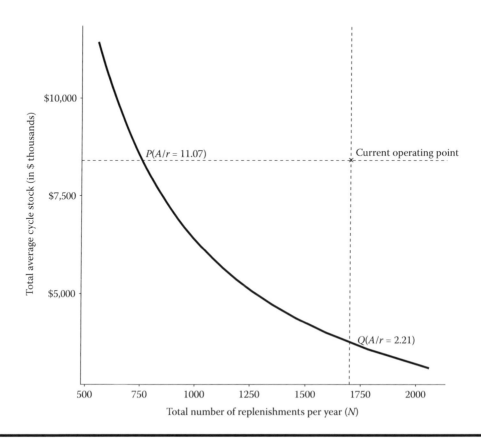

Figure 4.12 Aggregated consequences of different *A/r* values—an exchange curve.

Rather than using a three-decimal value for r, we choose to round to 0.27 $/$/year for convenience in subsequent calculations.

To summarize, exchange curves permit us to show improvements (in terms of aggregate performance measures) over the current operating practices. From a top management perspective, this is far more appealing than the fact that the EOQ minimizes the total costs on an individual item basis (particularly when the individual cost factors are so difficult to estimate). In Section 4A.6 of the Appendix to this chapter, we describe a powerful estimation procedure for exchange curves that is based on the lognormal distribution. (See the Appendix to Chapter 2 as well.)

4.11 Summary

In this chapter, we have specified a procedure for selecting the appropriate size of a replenishment quantity under demand conditions that are essentially static in nature; that is, we assume that the demand rate is at a constant level that does not change with time. Under a rigorous set of additional assumptions, the resulting order quantity is the well-known square root, EOQ.

We argued that the exact values of one or both of the parameters r and A may be very difficult to obtain. In such a case, we showed that a value of r, A, or A/r can be implicitly specified by selecting a desired operating point on an aggregate exchange curve of the total average stock versus the total number or costs of replenishments per year.

Sizable deviations of the order quantity away from the optimum value produce rather small cost errors. This robust feature of the EOQ helps justify its widespread use in practice. Often, the limited savings achievable do not justify the effort required to develop overly sophisticated variations of the EOQ. Nevertheless, as conditions change with time, it becomes necessary to recompute the appropriate value of the order quantity.

Several simple modifications of the basic EOQ have been discussed; these modifications make the results more widely applicable. In the next chapter, we develop a practical procedure for coping with situations where the average or forecast demand rate varies appreciably with time. Three other important extensions will be left to the later chapters, namely,

1. Situations where demand is no longer deterministic (Chapter 6).
2. Situations where savings in replenishment costs are possible through the coordination of two or more items at the same stocking point (Chapter 10).
3. Situations (referred to as multiechelon) where the replenishment quantities of serially related production or storage facilities should not be treated independently (Chapter 11).

In each of these circumstances, we will find that the basic EOQ plays an important role.

Problems

4.1   a. For the item used as an illustration in Section 4.2, suppose that A was changed to $12.80. Now find
    i. EOQ in units
    ii. EOQ in dollars
    iii. EOQ as a months of supply
      Does the effect of the change in A make sense?

b. For the same item, suppose that A was still \$3.20 but r was increased from 0.24/year to 0.30/year. Find the same quantities requested in part (a). Does the effect of the change in r make sense?

4.2 Demand for a product at a depot is at a constant rate of 250 units per week. The product is supplied to the depot from the factory. The factory cost is \$10 per unit, while the total cost of a shipment from the factory to the depot when the shipment has size M is given by

$$\text{Shipment cost} = \$50 + \$2M$$

The inventory carrying cost in \$/\$/week is 0.004. Determine the appropriate shipment size.

4.3 A firm can produce and package their own guitar strings, or buy prepacked rolls of "discount" guitar strings. If they load their own strings, there is a production setup cost of \$20. The finished product is valued at \$1.23 a set, and the production rate is 500 sets/day. "Discount" strings cost the firm \$1.26 per set and the fixed ordering charge is \$3/order. In either case, an inventory carrying charge of 0.24 \$/\$/year would be used by the company. Demand for this item is 10,000 sets per year. From the standpoint of replenishment and carrying costs, what should the company do? What other considerations might affect the decision?

4.4 A manufacturing firm located in Calgary produces an item in a 3-month time supply. An analyst, attempting to introduce a more logical approach to selecting run quantities, has obtained the following estimates of the characteristics of the item:

$D = 4{,}000$ units/year
$A = \$5$
$v = \$4$ per 100 units
$r = 0.25$ \$/\$/year
Note: Assume that the production rate is much larger than D.

a. What is the EOQ of the item?
b. What is the time between consecutive replenishments of the item when the EOQ is used?
c. The production manager insists that the $A = \$5$ figure is only a guess. Therefore, he insists on using his simple 3-month supply rule. Indicate how you would find the range of A values for which the EOQ (based on $A = \$5$) would be preferable (in terms of a lower total of replenishment and carrying costs) to the 3-month supply.

4.5 Suppose that all the assumptions of the EOQ derivation hold except that we now allow backorders (i.e., we deliberately let the stock level run negative before we order; backordered requests are completely filled out of the next replenishment). Now, there are two decision variables: Q and s (the level below zero at which we place an order).

a. Using a sketch and geometrical considerations, find the average on-hand inventory level and the average level of backorders.
b. Suppose that there is a cost $B_2 v$ per unit backordered (independent of how long the unit is backordered) where B_2 is a dimensionless factor. Find the best settings of Q and s as a function of A, D, v, r, and B_2.
c. Repeat the previous part but now with a cost $B_3 v$ per unit backordered per unit time. The dimensions of B_3 are equivalent to those for r.

4.6 The famous Ernie of "Sesame Street" continually faces replenishment decisions concerning his cookie supply. The Cookie Monster devours the cookies at an average rate of 200 per day. The cookies cost $0.03 each. Ernie is getting fed up with having to go to the store once a week. His friend, Bert, has offered to do a study to help Ernie with his problem.

    a. If Ernie is implicitly following an EOQ policy, what can Bert say about the implicit values of the two missing parameters?

    b. Suppose that the store offered a special of 10,000 cookies for $200. Should Ernie take advantage of the offer? Discuss. (Hint: Consult your local TV listing for the timing of and channel selection for "Sesame Street.")

4.7 U. R. Sick Labs manufactures penicillin. Briefly discuss the special considerations required in establishing the run quantity of such an item.

4.8 Table 4.1 was developed based on management specifying permitted values of the order quantity expressed as a time supply. Suppose instead that order quantities (in units) were specified.

    a. Develop an indifference condition between two adjacent permitted values, namely Q_1 and Q_2.

    b. For A = $10 and r = 0.25 $/$/year, develop a table, similar to Table 4.1 (although the variable involving D and v need not necessarily be Dv), for Q values of 1, 10, 50, 100, 200, 500, and 1,000 units.

    c. For an item with D = 500 units/year and v = 0.10 $/unit, what is the exact EOQ? What Q does the table suggest using? What is the cost penalty associated with the use of this approximate Q value?

4.9 A mining company routinely replaces a specific part on a certain type of equipment. The usage rate is 40 per week and there is no significant seasonality. The supplier of the part offers the following all-units discount structure:

Range of Q	Unit Cost
$0 < Q < 300$	$10.00
$300 \leq Q$	9.70

    The fixed cost of a replenishment is estimated to be $25 and a carrying charge of 0.26 $/$/year is used by the company.

    a. What replenishment size should be used?

    b. If the supplier was interested in having the mining company acquire at least 500 units at a time, what is the largest unit price they could charge for an order of 500 units?

4.10 The supplier of a product wants to discourage large quantity purchases. Suppose that all the assumptions of the basic EOQ apply except that a reverse quantity discount is applicable; that is, the unit variable cost is given by

$$v = \begin{cases} v_0 & 0 \leq Q < Q_b \\ v_0(1+d) & Q_b \leq Q \end{cases} \text{ where } d > 0$$

a. Write an expression (or expressions) for the total relevant costs per year as a function of the order quantity Q. Introduce (and define) whatever other symbols you feel are necessary.
b. Using graphical sketches, indicate the possible positions of the best order quantity (as is done in Figure 4.5 for the regular discount case).
c. What is the best order quantity for an item with the following characteristics:
Demand rate = 50,000 units/year
Fixed setup cost per replenishment = $10
$v = \$1.00/\text{unit}$ $d = 0.005$
Selling price = $1.44/unit
Carrying charge = 0.20 $/$/year
$Q_b = 1{,}500$ units

4.11 A particular product is produced on a fabrication line. Using the finite replenishment EOQ model, the suggested run quantity is 1,000 units. On a particular day of production, this item is being run near the end of a shift and the supervisor, instead of stopping at 1,000 units and having to set up for a new item (which would make her performance look bad in terms of the total output), continues to produce the item to the end of the shift, resulting in a total production quantity of 2,000 units. Discuss how you would analyze whether or not she made a decision that was attractive from a company standpoint. Include comments on the possible intangible factors.

4.12 Consider an inventory with only four items with the following characteristics:

Item i	D_i (Units/Year)	v_i ($/Unit)
1	7,200	2.00
2	4,000	0.90
3	500	5.00
4	100	0.81

The inventory manager vehemently argues that there is no way that he can evaluate A and r; however, he is prepared to admit that A/r is reasonably constant across the items. He has been following an ordering policy of using a 4-month time supply of each item. He has been under pressure from the controller to reduce the average inventory level by 25%. Therefore, he is about to adopt a 3-month time supply.
a. Develop an EOQ aggregate exchange curve.
b. What are the values of TACS and N for $A/r = 200$?
c. What A/r value gives the same TACS as the current rule?
d. What A/r value gives the same N as the proposed rule?
e. Use the curve to suggest an option or options open to the manager that is (are) preferable to his proposed action.

4.13 In a small distribution firm, the controller was advocating a min-max system with specific rules for computing the minimum and maximum levels of each item. Suppose that for each item the values of A, v, D, L, and r could be reasonably accurately ascertained.

    a. Indicate how you would estimate the approximate cost savings of selecting the order quantity of an item through EOQ considerations rather than by the controller's method. (Hint: Assume that the order quantity is approximately equal to the difference between the maximum and minimum levels.)

    b. Illustrate for the following item: $A = \$3.20$; $v = \$10$; $D = 900$ units/year; $L = 10$ weeks; and $r = 0.24$ \$/\$/year.

4.14 The Dartmouth Bookstore sells 300 notebooks per term. Each notebook carries a variable cost of \$0.50, while the fixed cost to restock the shelves amounts to \$4.00. The carrying charge of one notebook tied in the inventory is \$0.08 per \$/term.

    a. Calculate the EOQ.

    b. Calculate the total relevant cost associated with the EOQ calculated in the previous part.

    c. If the Dartmouth Bookstore instead chooses to order 300 notebooks instead of the EOQ, calculate the PCP of this order size.

    d. Calculate the best order quantity now considering a 40% discount on orders greater than 300.

    e. Calculate the optimal order quantity if the bookstore decides to incorporate an inflation rate of 2% in the decision.

    f. Assume that a particular notebook occupies 0.10 cubic foot of space, for which the bookstore figures that it costs them \$2 per term per cubic foot. Calculate the best replenishment quantity under these circumstances.

    g. Now assume that the unit cost of the same notebook increases from \$0.50 to \$0.60. Calculate the new EOQ.

4.15 Consider an item whose demand rate is 100 units per year and that costs \$40 per unit. If the fixed cost of ordering is \$50 and the carrying charge is 25%, answer the following:

    a. Find the EOQ.

    b. The purchasing manager is trying to understand the reasons for the high cost of ordering. In this process, she wonders how low A must become for the EOQ to equal 1 unit. What value of A will she find?

4.16 An item is produced on a particular machine. The setup time to change over production from some other item to this one is quite large. The machine is run for a single 10-hour shift each day. Changeovers can be done, if desired, in the off-shift time. When a production run of the item is initiated, an obvious decision is how much to produce. How is this decision similar to the one of ordering an item from the supplier where the supplier only sells the item in a rather large pack-size (or container)?

4.17 Monopack is a manufacturer of a single product. The company has been concerned about the rising inflation rate in the economy for sometime. Currently, Monopack does not incorporate the effects of inflation into its calculation of its EOQ. The company is interested in finding out if inflation will impact its choice of replenishment sizes. Characteristics of the product were found to be

    $D = 3{,}000$ units per year

    $A = \$5.45$

    $v = \$0.60$/unit

    $r = 20\%$

    Note: Assume that the product price is established independent of the ordering policy and that inflation affects A and v in the same manner.

    a. Calculate the EOQ in units.

    b. Calculate the optimum replenishment quantities when the inflation rate is $i = 0.2\%, 4\%, 9\%$, and 15%.

    c. Calculate the PCP associated with using the EOQ (ignoring inflation) for all values of i.

    d. What is the limiting value of the PCP?

4.18 A firm buys a product from a supplier using the EOQ to determine the order quantity. The supplier recently called and asked the buyer to purchase in larger lots so that the supplier's production process can be made more efficient. Specifically, assume that the buyer will buy k times EOQ units, where $k > 1$. The per unit cost must be reduced to make this attractive to the buyer. What is the fractional discount at which the buyer will be indifferent between the old EOQ at the old price and buying k EOQ at the new price?

4.19 A retailer sells about 5,000 units per year of a particular item. The following data has been found for this item:

    $A = \$100$

    $v = \$2$ per unit

    $r = 0.28 \ \$/\$/\text{year}$

    The retailer is now using an order quantity of 500 units. A keen analyst observes that savings could be realized by instead using the EOQ. The retail manager says to forget about mathematics—he has a way of reducing the setup cost to only $25. He argues for doing this and retaining the current order quantity. Who is right? Discuss.

4.20 The Andy Dandy Candy Company has asked you to recommend an EOQ for their Jelly Bean product. They have provided you with the following information:

    a. The only costs involved are the cost of ordering and the cost of holding the inventory that is equal to the average amount of capital times $0.30. Back orders are not allowed, and there are no other costs.

    b. The company always has a sale when a shipment of Jelly Beans arrives, until half of the shipment is gone, at which time the company raises its price.

    This implies that there are two known, continuous, constant rates of demand D_2 and D_1, D_1 being the rate during the sale, with $D_1 > D_2$. As a consultant, you are requested to recommend an EOQ formula for this situation. In your report, you must specifically cover the following points:

    a. A graphical representation of the inventory process. (What is the fraction of time spent at the rate D_1?)

    b. A total cost equation, defining the terms.

    c. A brief explanation of the method of obtaining the optimal order quantity.

    d. A numerical example using $A = \$20$; $vr = \$2$; $D_1 = 30$; and $D_2 = 15$.

    e. How much should be ordered in the above numerical example if $D_1 = D_2 = 30$?

4.21 A supplier offers a manufacturer an item at a 5% discount on orders of 500 units or more, and a 15% discount on orders of 1,000 units or more. From past experience, it appears that demand for this item is 3,075 units/year. If the ordering cost is $50, the basic price (without discount) of the item is $2, and an r value of 0.20 $/$/year is used, how often should the manufacturer order from the supplier? What order quantity should be used?

4.22 A supplier offers the following discount structure on purchases of any single item:

0 < Q < 1,000	$5.00 per unit
1,000 ≤ Q < 2,000	$4.90 per unit
2,000 ≤ Q	$4.75 per unit

The discounts apply to all units. For each of the following items treated separately, what is the appropriate order quantity to use, assuming a common value of $r = 0.30$ $/$/year?

Item	D (Units/Year)	A ($)
1	10,000	25
2	1,000	25
3	4,000	25
4	130,000	25

4.23   A company stocks three items having the following characteristics:

Item i	D_i (Units/Year)	v_i ($/Unit)	Current Q_i(Units)
1	1,000	5.00	250
2	4,000	0.80	1,000
3	200	4.00	50

a. Develop an exchange curve of the total average stock (in $) versus the total number of replenishments per year under an EOQ strategy treating A/r as the policy parameter.
b. Plot the current operating position on the graph.
c. For what values of A/r would the EOQ strategy produce an improvement on both aggregate dimensions?

4.24   Consider a quantity discount problem in which the form of the discount differs from the all-units case that we have discussed in this chapter. Specifically, the supplier offers an incremental discount such that the lower price applies only to the number of units in the particular discount interval, and not to all units ordered.

a. Using the notation introduced in Section 4.5, write a total cost expression as a function of the order quantity, Q, for the case of just two discount intervals (the first is the regular price, and the second is the discount price).
b. Write a total cost expression for any number, i, discount intervals.
c. Define an algorithm for finding the optimal order quantity.
d. For the following case, find the optimal order quantity.

$0 < Q < 100$	\$20.00 per unit
$100 \leq Q$	\$19.50 per unit

Also $r = 0.30$ \$/\$/year; $A = \$24$; and $D = 1,200$ units/year.

4.25 a. Develop a spreadsheet to compute the EOQ and the total relevant cost associated with any order quantity.

b. Repeat part (a) for the case of the all-units quantity discount.

c. Repeat part (a) for the case of the incremental discount described in Problem 4.24.

d. Repeat part (a) for the case of a finite replenishment rate.

e. Repeat part (a) for the EOQ that accounts for an inflation rate.

4.26 The May-Bee Company distributes fire-fighting equipment. Approximately 50% of the company's SKUs are imported and the rest are acquired from domestic suppliers. May-Bee sales have been increasing at a steady rate and future trends appear to be very optimistic. The company is wholly owned by a Mr Brown who relies heavily on his bank for working capital. His current credit limit is set at \$150,000 and he is experiencing some resistance from his banker in getting more cash to increase his investment in inventories to meet an anticipated increase in sales. In the past, Brown has used only simple decision rules for ordering from his suppliers. He orders a 3-week supply of SKUs that cost more than \$10 per unit and a 6-week supply of items that cost less than \$10 per unit. Last year, May-Bee's average inventory was around \$1,000,000. He paid \$19,500 interest on his bank loan, \$80,000 rent for his warehouse, and \$15,000 to insurance companies. Brown believes that it would be profitable to increase his bank loan because he feels that he could probably earn 14–15% on the borrowed money. The following six items represent 30% of May-Bee's annual dollar sales volume and according to Brown are his bread and butter:

SKU Number	Value/Unit v (\$)	Annual Demand (Units)
1	15.78	6,400
2[a]	40.50	2,200
3[a]	18.30	4,500
4	8.40	3,500
5[a]	9.70	6,000
6	25.00	4,800

[a] Items imported from the same supplier.

Brown gets directly involved with the ordering of imported items because for these items he has to go to the bank to get a letter of credit and on the average spends about 60 minutes per order for doing so. An order for imported items had in the past included from 1 to 10 SKUs. The table below gives a summary of the costs involved in placing an order. These estimates were supplied by the persons involved, who guessed at the amount of time involved in each activity.

Hourly Person Involved	Rate ($)	Time per Order (Minutes)	
		Domestic[a]	Imported
Typist	3.00	20	24
Inventory clerk	3.20	20	20
Receiver	4.50	30	30
Bookkeeper	4.20	20	20
Mr Brown	20.00	20	60

[a] Domestic orders usually included 1–5 SKUs.

Apart from the above, there was a charge of $10 per order that represented the bank charge for the letter of credit for imported items whereas the bank charges were only $0.25 per order for items acquired from domestic suppliers.

a. The above cost description was compiled by the bookkeeper. In your opinion are there any costs that have been left out? Comment.
b. Compute May-Bee's carrying cost (r).
c. Compute the cost of placing an order for a single domestic and for a single imported SKU.
d. Compute the EOQ for all six items.
e. Compute the total cycle stock for the six items under the present ordering rules.
f. Compute the total relevant costs for the six items using EOQs.
g. Compute the difference in the total relevant costs under 4.26e and 4.26f. Estimate the change in the total relevant costs for all the 100 items.
h. Compute the total number of replenishments for the sample of six items under the present system and under the EOQ policy.
i. Draw an exchange curve of the total relevant costs versus N for the imported items. From the graph, calculate the increase or decrease in the total relevant costs if the combined total number of orders for the three imported items is changed to 36 per year.
j. Compute the total relevant costs (ordering costs plus holding costs) under the present system and under the EOQ policy. What are the estimated savings (percent) if the EOQ policy is applied throughout the whole population?

Appendix 4A: Derivations

4A.1 Penalty for Using an Erroneous Value of the Replenishment Quantity

The percentage cost penalty is

$$PCP = \frac{TRC(Q') - TRC(EOQ)}{TRC(EOQ)} \times 100 \qquad (4A.1)$$

where

$$Q' = (1 + p)\text{EOQ} = (1 + p)\sqrt{\frac{2AD}{vr}}$$

Substituting this Q' expression into the expression for TRC from Equation 4.3, we obtain

$$\text{TRC}(Q') = (1 + p)\sqrt{\frac{ADvr}{2}} + \frac{1}{1 + p}\sqrt{\frac{ADvr}{2}}$$

$$= \sqrt{2ADvr}\frac{1}{2}\left(1 + p + \frac{1}{1 + p}\right) \tag{4A.2}$$

Also from Equation 4.5

$$\text{TRC}(\text{EOQ}) = \sqrt{2ADvr} \tag{4A.3}$$

Substituting the results of Equations 4A.2 and 4A.3 into Equation 4A.1 gives

$$\text{PCP} = \left[\frac{1}{2}\left(1 + p + \frac{1}{1 + p}\right) - 1\right] \times 100$$

or

$$\text{PCP} = \left(\frac{1}{1 + p} - 1 + p\right) = \frac{p^2}{2(1 + p)} \times 100$$

4A.2 Order Quantity under Inflation

From Equation 4.14:

$$PV(Q) = (A + Qv)\frac{1}{1 - e^{-(r-i)Q/D}}$$

$$\frac{dPV(Q)}{dQ} = \frac{\left(1 - e^{-(r-i)Q/D}\right)v - (A + Qv)\left(\frac{r-i}{D}\right)e^{-(r-i)Q/D}}{\left[1 - e^{-(r-i)Q/D}\right]^2} = 0$$

The numerator must be zero, which (dividing through by v) implies

$$1 - e^{-(r-i)Q/D} - \left(\frac{A}{v} + Q\right)\left(\frac{r-i}{D}\right)e^{-(r-i)Q/D} = 0$$

at the optimal order quantity Q. This simplifies to

$$e^{(r-i)Q/D} = 1 + \left(\frac{A}{v} + Q\right)\left(\frac{r-i}{D}\right)$$

Using the second-order Taylor's approximation for e^x at $x = 0$, $e^x \approx 1 + x + x^2/2$, gives

$$1 + \frac{(r-i)Q}{D} + \frac{(r-i)^2Q^2}{2D^2} \approx 1 + \left(\frac{A}{v}\right)\left(\frac{r-i}{D}\right) + \frac{Q(r-i)}{D}$$

Solving for the optimal quantity (given the approximation to e^x) gives

$$Q_{opt} = \sqrt{\frac{2AD}{v(r-i)}}$$

From Equations 4.14 and 4.16, the PCP for using the EOQ instead of the Q_{opt} is

$$PCP = \left[\frac{A + EOQv}{A + Q_{opt}v} \frac{1 - e^{-(r-i)Q_{opt}/D}}{1 - e^{-(r-i)EOQ/D}} - 1 \right] \times 100$$

Substituting from Equation 4.15 gives

$$PCP = \left[\frac{A + \sqrt{\frac{2ADv}{r}}}{A + \sqrt{\frac{2ADv}{(r-i)}}} \left(\frac{1 - e^{-(r-i)\sqrt{2A/(r-i)Dv}}}{1 - e^{-(r-i)\sqrt{2A/rDv}}} \right) - 1 \right] \times 100 \qquad (4A.4)$$

As $x \to 0$, we know that $e^{-x} \cong 1 - x$. Using this result in Equation 4A.4 leads, after some simplification, to

$$PCP = \left[\frac{\sqrt{\frac{Ar}{2Dv}} + 1}{\sqrt{\frac{Ar}{2Dv}} + \sqrt{\frac{r}{r-i}}} \frac{\sqrt{r}}{\sqrt{r-1}} - 1 \right] \times 100$$

$$= \left[\frac{\sqrt{\frac{Ar}{2Dv}} + 1}{\sqrt{\frac{A(r-i)}{2Dv}} + 1} - 1 \right] \times 100$$

As $i \to r$, we obtain

$$PCP = \sqrt{Ar/2Dv} \times 100$$

4A.3 Cost Penalty Incurred by Using EOQ When There Is a Finite Replenishment Rate

Recall from Equation 4.19 that the optimal order quantity for the case when the replenishment rate is finite can be expressed as the EOQ times a correction factor of $1/\sqrt{1 - D/m}$. Figure 4.8 plots this correction factor as well as the PCP incurred when using the EOQ. To make this PCP calculation, we first find the percentage deviation in quantity p as follows:

$$EOQ = (1 + p)FREOQ$$

$$EOQ = (1 + p)EOQ\frac{1}{\sqrt{1 - D/m}}$$

$$\sqrt{1 - D/m} = (1 + p)$$

$$p = \sqrt{1 - D/m} - 1$$

Using this value for p, we can apply the percentage deviation in quantity from the PCP formula in Equation 4.10.

4A.4 Multi-Item Special Opportunity to Procure

As in Section 4.9.5, we would like to pick each Q_i so as to maximize

$$\frac{Q_i}{D_i}\sqrt{2A_iD_iv_{2i}r} + Q_iv_{2i} - A_i - Q_iv_{1i} - \frac{Q_i^2v_{1i}r}{2D_i}$$

However, there is the constraint

$$\sum_{i=1}^{n} Q_iv_{1i} = W \tag{4A.5}$$

which, given that we have reached Step 3, must be binding.

We use a standard constrained optimization procedure, the Lagrange multiplier approach (see Appendix I at the end of the book). We maximize

$$L(Q_1,\ldots,Q_n,M) = \sum_{i=1}^{n}\left[\frac{Q_i}{D_i}\sqrt{2A_iD_iv_{2i}r} + Q_iv_{2i} - A_i - Q_iv_{1i} - \frac{Q_i^2v_{1i}r}{2D_i}\right]$$
$$- M\left(\sum_{i=1}^{n} Q_iv_{1i} - W\right)$$

where M is a Lagrange multiplier. Taking partial derivatives and setting them to zero

$$\frac{\partial L}{\partial M} = 0$$

gives Equation 4A.5, and

$$\frac{\partial L}{\partial Q_i} = 0 \quad (j = 1, 2, \ldots, n)$$

leads to

$$Q_j^* = \frac{v_{2j}}{v_{1j}}\text{EOQ}_{2j} + \frac{v_{2j} - (M+1)v_{1j}}{v_{1j}}\frac{D_j}{r} \quad j = 1, 2, \ldots, n \tag{4A.6}$$

where Q_j^* is the best constrained value of Q_j. Therefore,

$$Q_j^*v_{1j} = v_{2j}\text{EOQ}_{2j} + \left[v_{2j} - (M+1)v_{1j}\right]\frac{D_j}{r} \tag{4A.7}$$

Summing Equation 4A.7 over j and using Equation 4A.5 gives

$$\sum_{j=1}^{n} v_{2j}\text{EOQ}_{2j} + \sum_{j=1}^{n}\left[v_{2j} - (M+1)v_{1j}\right]\frac{D_j}{r} = W \tag{4A.8}$$

This is solved for $M + 1$ and the result is substituted into Equation 4A.6, producing Equation 4.28 of the main text.

4A.5 Exchange Curve Equations When A/r Is Constant across All Items

Suppose that we have n items in the population and we designate item i's characteristics by v_i and D_i. The total average cycle stock in dollars

$$\text{TACS} = \sum_{i=1}^{n} \frac{Q_i v_i}{2}$$

and the total number of replenishments per unit time

$$N = \sum_{i=1}^{n} D_i / Q_i$$

If we use the EOQ for each item, we have, from Equation 4.4

$$Q_i = \sqrt{\frac{2AD_i}{v_i r}}$$

Therefore,

$$\text{TACS} = \sum_{i=1}^{n} \sqrt{\frac{AD_i v_i}{2r}} = \sqrt{\frac{A}{r}} \frac{1}{\sqrt{2}} \sum_{i=1}^{n} \sqrt{D_i v_i}$$

which is Equation 4.29. Also,

$$N = \sum_{i=1}^{n} \sqrt{\frac{D_i v_i r}{2A}} = \sqrt{\frac{r}{A}} \frac{1}{\sqrt{2}} \sum_{i=1}^{n} \sqrt{D_i v_i}$$

which is Equation 4.30.

4A.6 The EOQ Exchange Curve Using the Lognormal Distribution

From the Appendix to Chapter 2, we know that if for a population of n items (numbered $i = 1, 2, \ldots, n$) a variable x has a lognormal distribution with parameters m and b, then

$$\sum_{i=1}^{n} x_i^{j} = nm^{j} exp\left[b^2 j(j-1)/2\right] \qquad (4A.9)$$

Suppose a quantity of interest (e.g., the cycle stock in dollars) for item i, call it y_i, can be expressed in the form

$$y_i = \sum_{h=1}^{H} c_h (D_i v_i)^{j_h} \qquad (4A.10)$$

where the c_h's and j_h's are coefficients and H is a constant, that is, a linear combination of the annual dollar usage raised to various powers. Let

$$Y = \sum_{i=1}^{n} y_i \qquad (4A.11)$$

be the aggregate across the population of the quantity of interest (e.g., the total cycle stock in dollars). Then, because the annual dollar usage is approximately lognormally distributed (with parameters m and b), we have from Equations 4A.9 through 4A.11

$$Y = n \sum_{h=1}^{H} c_h m^{j_h} exp \left[b^2 j_h (j_h - 1)/2 \right] \qquad (4A.12)$$

The cycle stock of item i in dollars is

$$\frac{Q_i v_i}{2} = \sqrt{\frac{A_i D_i v_i}{2r}}$$

$$= \sqrt{\frac{A}{2r}} (D_i v_i)^{1/2}$$

assuming the same A for all the items.

This is a special case of Equation 4A.9 with $H = 1$, $c_1 = \sqrt{A/2r}$, and $j_1 = 1/2$. Thus, from Equation 4A.12 that the total cycle stock in dollars is given by

$$TACS = n\sqrt{A/2r} m^{1/2} exp(-b^2/8)$$

Suppose for a hypothetical firm we have that $n = 849$, $m = 2{,}048$, and $b = 1.56$. Thus

$$TACS = (849)\sqrt{A/2r}(2{,}048)^{1/2} exp[-(1.56)^2/8]$$
$$= 20{,}042\sqrt{A/r} \qquad (4A.13)$$

The number of replenishments per year of item i is

$$D_i/Q_i = D_i v_i/Q_i v_i$$
$$= D_i v_i/\sqrt{2AD_i v_i/r}$$
$$= \sqrt{\frac{r}{2A}} (D_i v_i)^{1/2}$$

As above, we have that the total number of replenishments per year N is given by

$$N = n\sqrt{r/2A} m^{1/2} exp(-b^2/8)$$

and for the hypothetical firm

$$N = 20{,}042\sqrt{r/A} \qquad (4A.14)$$

Multiplication of Equations 4A.13 and 4A.14 gives

$$(\text{TACS})N \approx 401,700,000 \qquad\qquad (4\text{A}.15)$$

which is the lognormal approximation of the exchange curve of TACS versus N. The exact exchange curve is

$$(\text{TACS})N = 406,500,000$$

The latter was developed by a complete item-by-item computation. The lognormal approximation of Equation 4A.15 falls a negligible amount above the exact curve—such a small amount that one would have difficulty in discerning any difference between the curves.

References

Abad, P. L. 1988. Determining optimal selling price and lot size when the supplier offers all-unit quantity discounts. *Decision Sciences 19*, 622–634.

Arcelus, F. J. and G. Srinivasan. 1993. Delay of payments for extraordinary purchases. *Journal of the Operational Research Society 44*, 785–795.

Arcelus, F. J. and G. Srinivasan. 1995. Discount strategies for one-time-only sales. *IIE Transactions 27*, 618–624.

Ardalan, A. 1994. Optimal prices and order quantities when temporary price discounts result in increase in demand. *European Journal of Operational Research 72*, 52–61.

Aucamp, D. C. 1982. Nonlinear freight costs in the EOQ problem. *European Journal of Operational Research 9*, 61–63.

Aucamp, D. C. 1990. Aggregate backorder exchange curves. *IIE Transactions 22*(3), 281–287.

Aucamp, D. C. and P. J. Kuzdrall. 1989. Order quantities with temporary price reductions. *Journal of the Operational Research Society 40*(10), 937–940.

Aull-Hyde, R. 1992. Evaluation of supplier-restricted purchasing options under temporary price discounts. *IIE Transactions 24*(2), 184–186.

Aull-Hyde, R. L. 1996. A backlog inventory model during restricted sale periods. *Journal of the Operational Research Society 47*, 1192–1200.

Axsäter, S. 1996. Using the deterministic EOQ formula in stochastic inventory control. *Management Science 42*(6), 830–834.

Baumol, W. J. and H. D. Vinod. 1970. An inventory theoretic model of freight transport demand. *Management Science 16*(7), 413–421.

Benton, W. C. and S. Park. 1996. A classification of literature on determining the lot size under quantity discounts. *European Journal of Operational Research 92*, 219–238.

Bijvank, M. and I. F. Vis. 2011. Lost-sales inventory theory: A review. *European Journal of Operational Research 215*(1), 1–13.

Billington, P. J. 1987. The classic economic production quantity model with setup cost as a function of capital expenditure. *Decision Sciences 18*(1), 25–42.

Bourland, K., S. Powell, and D. Pyke. 1996. Exploiting timely demand information to reduce inventories. *European Journal of Operational Research 92*(2), 239–253.

Brill, P. H. and B. A. Chaouch. 1995. An EOQ model with random variations in demand. *Management Science 41*(5), 927–936.

Brown, R. G. 1982. *Advanced Service Parts Inventory Control.* Norwich, VT: Materials Management Systems, Inc.

Burwell, T. H., D. S. Dave, K. E. Fitzpatrick, and M. R. Roy. 1990. A note on determining optimal selling price and lot size under all-unit quantity discounts. *Decision Sciences 21*, 471–474.

Buzacott, J. A. 1975. Economic order quantities with inflation. *Operational Research Quarterly 26*(3), 553–558.

Buzacott, J. A. 2013. Then and now—50 years of production research. *International Journal of Production Research 51*(23–24), 6756–6768.

Buzacott, J. A. and R. Q. Zhang. 2004. Inventory management with asset-based financing. *Management Science 50*(9), 1274–1292.

Carlson, M. L., G. J. Miltenburg, and J. J. Rousseau. 1996. Economic order quantity and quantity discounts under date-terms supplier credit: A discounted cash flow approach. *Journal of the Operational Research Society 47*, 384–394.

Carter, J. R. and B. G. Ferrin. 1995. The impact of transportation costs on supply chain management. *Journal of Business Logistics 16*(1), 189–212.

Carter, J. R., B. G. Ferrin, and C. R. Carter. 1995. On extending Russell and Krajewski's algorithm for economic purchase order quantities. *Decision Sciences 26*(6), 819–829.

Chakravarty, A. K. 1981. Multi-item inventory aggregation into groups. *Journal of the Operational Research Society 32*(1), 19–26.

Chakravarty, A. K. 1986. Quantity discounted inventory replenishments with limited storage space. *INFOR 24*(1), 12–25.

Crouch, R. and S. Oglesby. 1978. Optimization of a few lot sizes to cover a range of requirements. *Journal of the Operational Research Society 29*(9), 897–904.

Coyle, J. J., R. A. Novack, and B. Gibson. 2015. *Transportation: A Global Supply Chain Perspective*. Australia: Cengage Learning.

Dada, M. and K. N. Srikanth. 1987. Pricing policies for quantity discounts. *Management Science 33*, 1247–1252.

Das, C. 1984. A graphical approach to price-break analysis-3. *Journal of the Operational Research Society 35*(11), 995–1001.

Das, C. 1988. Some tips for discount buyers. *Production and Inventory Management Journal 29*(2), 23–26.

Das, C. 1990. An algorithm for selecting quantity discounts from realistic schedules. *Journal of the Operational Research Society 41*(2), 165–172.

Davis, R. and N. Gaither. 1985. Optimal ordering policies under conditions of extended payment privileges. *Management Science 31*(4), 499–509.

Dobson, G. 1988. Sensitivity of the EOQ model to parameter estimates. *Operations Research 36*(4), 570–574.

Dolan, R. J. 1987. Quantity discounts: Managerial issues and research opportunities. *Marketing Science 6*(1), 1–22.

Donaldson, W. A. 1981. Grouping of inventory items by review period. *Journal of the Operational Research Society 32*(12), 1075–1076.

Eaton, J. A. 1964. New—The LIMIT technique. *Modern Materials Handling 19*(2), 38–43.

Erlenkotter, D. 1990. Ford Whitman Harris and the economic order quantity model. *Operations Research 38*(6), 937–946.

Freeland, J. R., J. P. Leschke, and E. N. Weiss. 1990. Guidelines for setup reduction programs to achieve zero inventory. *Journal of Operations Management 9*(1), 85–100.

Freimer, M., D. Thomas, and J. Tyworth. 2006. The value of setup cost reduction and process improvement for the economic production quantity model with defects. *European Journal of Operational Research 173*(1), 241–251.

Gaither, N. 1981. The effects of inflation and fuel scarcity upon inventory policies. *Production and Inventory Management Journal 22*(2), 37–48.

Gallego, G. and I. Moon. 1995. Strategic investment to reduce setup times in the economic lot scheduling problem. *Naval Research Logistics 42*(5), 773–790.

Gerchak, Y. and Y. Wang. 1994. Periodic review inventory models with inventory-level-dependent demand. *Naval Research Logistics 41*(1), 99–116.

Goyal, S. K. 1995. A one-vendor multi-buyer integrated inventory model: A comment. *European Journal of Operational Research 82*, 209–210.

Goyal, S. K., G. Srinivasan, and F. J. Arcelus. 1991. One time only incentives and inventory policies. *European Journal of Operational Research 54*, 1–6.

Grant, M. 1993. EOQ and price break analysis in a JIT environment. *Production and Inventory Management Journal 34*(3), 64–69.

Güder, F., J. Zydiak, and S. Chaudhry. 1994. Capacitated multiple item ordering with incremental quantity discounts. *Journal of the Operational Research Society 45*(10), 1197–1205.

Gupta, O. K. 1994. An inventory model with lot-size dependent ordering cost. *Production Planning and Control 5*(6), 585–587.

Harris, F. W. 1913. How many parts to make at once. *Factory, The Magazine of Management 10*, 135–136, 152.

Harris, F. W. 1990. How many parts to make at once. *Operations Research 38*(6), 947–950.

Hax, A. C. and D. Candea. 1984. *Production and Inventory Management*. Englewood Cliffs, NJ: Prentice-Hall.

Hong, J., S. Kim, and J. C. Hayya. 1996. Dynamic setup reduction in production lot sizing with nonconstant deterministic demand. *European Journal of Operational Research 90*, 182–196.

Jesse Jr., R., A. Mitra, and J. Cox. 1983. EOQ formula: Is it valid under inflationary conditions? *Decision Sciences 14*(3), 370–374.

Juckler, F. 1970. *Modeles de Gestion des Stocks et Couts Marginaux*. Louvain, Belgium: Vandeur.

Kanet, J. J. and J. A. Miles. 1985. Economic order quantities and inflation. *International Journal of Production Research 23*(3), 597–608.

Katz, P., A. Sadrian, and P. Tendick. 1994. Telephone companies analyze price quotations with Bellcore's PDSS software. *Interfaces 24*(1), 50–63.

Kim, D. H. 1995. A heuristic for replenishment of deteriorating items with a linear trend in demand. *International Journal of Production Economics 39*, 265–270.

Kim, K. H. and H. Hwang. 1988. An incremental discount pricing schedule with multiple customers and single price break. *European Journal of Operational Research 35*, 71–79.

Kim, S. L., J. C. Hayya, and J.-D. Hong. 1992. Setup reduction in the economic production quantity model. *Decision Sciences 32*(2), 500–508.

Kingsman, B. G. 1983. The effect of payment rules on ordering and stockholding in purchasing. *Journal of the Operational Research Society 34*(11), 1085–1098.

Knowles, T. W. and P. Pantumsinchai. 1988. All-units discounts for standard container sizes. *Decision Sciences 19*, 848–857.

Kuzdrall, P. and R. Britney. 1982. Total setup lot sizing with quantity discounts. *Decision Sciences 13*(1), 101–112.

Lal, R. and R. Staelin. 1984. An approach for developing an optimal discount pricing policy. *Management Science 30*(12), 1524–1539.

Lee, C. Y. 1986. The economic order quantity for freight discount costs. *IIE Transactions 18*(3), 318–320.

Lev, B. and A. L. Soyster. 1979. An inventory model with finite horizon and price changes. *Journal of the Operational Research Society 30*(1), 43–53.

Lev, B., H. Weiss, and A. Soyster. 1981. Optimal ordering policies when anticipating parameter changes in EOQ systems. *Naval Research Logistics Quarterly 28*(2), 267–279.

Li, C. and T. C. E. Cheng. 1994. An economic production quantity model with learning and forgetting considerations. *Production and Operations Management 3*(2), 118–131.

Lowe, T. J. and L. B. Schwarz. 1983. Parameter estimation for the EOQ lot size model: Minimax and expected value choices. *Naval Research Logistics 30*, 367–376.

Lu, L. 1995. A one-vendor multi-buyer integrated inventory model. *European Journal of Operational Research 81*, 312–323.

Lu, L. and M. Posner. 1994. Approximation procedures for the one-warehouse multi-retailer system. *Management Science 40*(10), 1305–1316.

Maxwell, W. L. and H. Singh. 1983. The effect of restricting cycle times in the economic lot scheduling problem. *IIE Transactions 15*(3), 235–241.

McClain, J. O. and L. J. Thomas. 1980. *Operations Management: Production of Goods and Services*. Englewood Cliffs, NJ: Prentice-Hall.

Mehra, S., S. P. Agrawal, and M. Rajagopalan. 1991. Some comments on the validity of EOQ formula under inflationary conditions. *Decision Sciences 22*, 206–212.

Mendoza, A. and J. A. Ventura. 2008. Incorporating quantity discounts to the EOQ model with transportation costs. *International Journal of Production Economics 113*(2), 754–765.

Miltenburg, G. J. 1987. Co-ordinated control of a family of discount-related items. *INFOR 25*(2), 97–116.

Min, K. J. 1991. Inventory and quantity discount pricing policies under maximization. *Operations Research Letters 11*, 187–193.

Moinzadeh, K. 1997. Replenishment and stocking policies for inventory systems with random deal offerings. *Management Science 43*(3), 334–342.

Moinzadeh, K., T. D. Klastorin, and E. Berk. 1997. The impact of small lot ordering on traffic congestion in a physical distribution system. *IIE Transactions 29*(8), 671–680.

Monahan, J. 1984. A quantity discount pricing model to increase vendor profits. *Management Science 30*(6), 720–726.

Montgomery, D., M. Bazaraa, and A. Keswani. 1973. Inventory models with a mixture of backorders and lost sales. *Naval Research Logistics Quarterly 20*(2), 255–263.

Moon, I. and W. Yun. 1993. An economic order quantity model with a random planning horizon. *The Engineering Economist 39*(1), 77–86.

Moon, I. and G. Gallego. 1994. Distribution free procedures for some inventory models. *Journal of Operational Research Society 45*(6), 651–658.

Naddor, E. 1966. *Inventory Systems.* New York: John Wiley & Sons.

Neves, J. S. 1992. Average setup cost inventory model: Performance and implementation issues. *International Journal of Production Research 30*(3), 455–468.

Parlar, M. and Q. Wang. 1994. Discounting decisions in a supplier–buyer relationship with a linear buyer's demand. *IIE Transactions 26*(2), 34–41.

Paul, K., T. K. Datta, K. S. Chaudhuri, and A. K. Pal. 1996. An inventory model with two-component demand rate and shortages. *Journal of the Operational Research Society 47*, 1029–1036.

Pirkul, H. and O. Aras. 1985. Capacitated multiple item ordering problem with quantity discounts. *IIE Transactions 17*(3), 206–211.

Porteus, E. L. 1985. Investing in reducing setups in the EOQ model. *Management Science 31*(8), 998–1010.

Porteus, E. L. 1986a. Investing in new parameter values in the discounted EOQ model. *Naval Research Logistics Quarterly 33*(1), 39–48.

Porteus, E. L. 1986b. Optimal lot sizing, process quality improvement and setup cost reduction. *Operations Research 34*(1), 137–144.

Rachamadugu, R. 1988. Error bounds for EOQ. *Naval Research Logistics 35*, 419–425.

Rosenblatt, M. J. and H. L. Lee. 1985. Improving profitability with quantity discounts under fixed demand. *IIE Transactions 17*(4), 388–395.

Rosenblatt, M. J. and U. G. Rothblum. 1994. The optimality of the "cut across the board" rule applied to an inventory model. *IIE Transactions 26*(2), 102–108.

Ross, S. A., R. Westerfield, and B. D. Jordan. 2008. *Fundamentals of Corporate Finance.* New York: Tata McGraw-Hill Education.

Russell, R. M. and L. J. Krajewski. 1991. Optimal purchase and transportation cost lot sizing for a single item. *Decision Sciences 22*(4), 940–952.

Sarker, B. and H. Pan. 1994. Effects of inflation and the time value of money on order quantity and allowable shortage. *International Journal of Production Economics 34*, 65–72.

Sethi, S. P. 1984. A quantity discount model with disposals. *International Journal of Production Research 22*(1), 31–39.

Shinn, S. W., H. Hwang, and S. S. Park. 1996. Joint price and lot size determination under conditions of permissible delay in payments and quantity discounts for freight cost. *European Journal of Operational Research 91*, 528–542.

Silver, E. A. 1976. Establishing the order quantity when the amount received is uncertain. *INFOR 14*(1), 32–39.

Silver, E. A. 1992. Changing the givens in modelling inventory problems: The example of just-in-time systems. *International Journal of Production Economics 26*, 347–351.

Silver, E. A. 1999. Simple replenishment rules when payment periods (dating deals) are offered. *Production and Inventory Management Journal 40*(4), 1.

Singhal, V. R. 1988. Inventories, risk, and the value of the firm. *Journal of Manufacturing and Operations Management 1*, 4–43.

Spence, A. M. and E. L. Porteus. 1987. Setup reduction and increased effective capacity. *Management Science 33*(10), 1291–1301.

Taylor, S. and C. Bradley. 1985. Optimal ordering strategies for announced price increases. *Operations Research 33*(2), 312–325.

Tersine, R. and S. Barman. 1994. Optimal lot sizes for unit and shipping discount situations. *IIE Transactions 26*(2), 97–101.

Tersine, R. J. 1996. Economic replenishment strategies for announced price increases. *European Journal of Operational Research 92*, 266–280.

Tersine, R. J. and S. Barman. 1995. Economic purchasing strategies for temporary price discounts. *European Journal of Operational Research 80*, 328–343.

Thomas, D. J. and J. E. Tyworth. 2006. Pooling lead-time risk by order splitting: A critical review. *Transportation Research Part E: Logistics and Transportation Review 42*(4), 245–257.

Thompson, H. 1975. Inventory management and capital budgeting: A pedagogical note. *Decision Sciences 6*, 383–398.

Toptal, A. and S. O. Bingöl. 2011. Transportation pricing of a truckload carrier. *European Journal of Operational Research 214*(3), 559–567.

Van Beek, P. and C. Van Putten. 1987. OR contributions to flexibility improvement in production/inventory systems. *European Journal of Operational Research 31*, 52–60.

Van der Duyn Schouten, F. A., M. J. G. van Eijs, and R. Heuts. 1994. The value of supplier information to improve management of a retailer's inventory. *Decision Sciences 25*(1), 1–14.

Van Eijs, M. J. G. 1994. Multi-item inventory systems with joint ordering and transportation decisions. *International Journal of Production Economics 35*, 285–292.

Weng, Z. K. and R. T. Wong. 1993. General models for the supplier's all-unit quantity discount policy. *Naval Research Logistics (NRL) 40*(7), 971–991.

Weng, Z. K. 1995a. Channel coordination and quantity discounts. *Management Science 41*(9), 1509–1522.

Weng, Z. K. 1995b. Modeling quantity discounts under general price-sensitive demand functions: Optimal policies and relationships. *European Journal of Operational Research 86*(2), 300–314.

Wilson, R. 1993. *Nonlinear Pricing*. New York: Oxford University Press.

Yanasse, H. H. 1990. EOQ systems: The case of an increase in purchase cost. *Journal of the Operational Research Society 41*(7), 633–637.

Zheng, Y. S. 1992. On properties of stochastic inventory systems. *Management Science 38*(1), 87–103.

Chapter 5

Lot Sizing for Individual Items with Time-Varying Demand

In Chapter 4, we developed the economic order quantity (EOQ) (and various modifications of it) by essentially assuming a *level, deterministic* demand rate. In this chapter, we relax this assumption and allow the average demand rate to vary with time. In a manner similar to Chapter 4, here, we focus on settings where replenishment quantities are not constrained by capacity. We analyze the implications of finite capacity extensively in Chapters 13 through 15. As compared to the EOQ setting, the time-varying demand situations addressed in this chapter encompass a broader range of practical situations, including:

1. Multiechelon assembly operations where a firm schedule of finished products, exploded back through the various assembly stages, leads to production requirements at these earlier levels. These requirements are relatively deterministic but almost always vary appreciably with time. (This is a topic to which we return in Chapter 15 under the heading of Material Requirements Planning, or Material Resources Planning [MRP].)
2. Production to contract, where the contract requires that certain quantities have to be delivered to the customer on specified dates.
3. Items having a seasonal demand pattern. (Artificial seasonality can be induced by pricing or promotion actions.)
4. Replacement parts for an item that is being phased out of operation. Here, the demand rate drops off with time. (In some cases, particularly toward the end of its life, such an item is better treated by the Class C procedures to be discussed in Chapter 8.)
5. More generally, items with known patterns in demand that are expected to continue.
6. Parts for preventive maintenance where the maintenance schedule is accurately known.

This topic has been termed *lot sizing* in the academic and practitioner literatures.

In Section 5.1, we point out how much more complex the analysis becomes when we allow the demand rate to vary with time. Section 5.2 introduces three different approaches, namely: (1) straightforward use of the EOQ (even though one of the key assumptions on which it is based is violated), (2) an exact optimal procedure, and (3) an approximate heuristic method. In Section 5.3, the assumptions common to all three approaches are laid out, as well as a numerical example to be used for illustrative purposes throughout Sections 5.4 through 5.6. Further details on each of the three approaches are presented in Sections 5.4, 5.5, and 5.6, including a discussion of when to use each method. In Section 5.7, we describe a procedure for handling quantity discounts when demand varies with time. Then, in Section 5.8, we again focus our attention on aggregate considerations by means of the exchange curve concept.

5.1 The Complexity of Time-Varying Demand

When the demand rate varies with time, we can no longer assume that the best strategy is always to use the same replenishment quantity; in fact, this will seldom be the case. (A similar difficulty arose in Chapter 4 when economic factors changed with time, such as the conditions of inflation or a special opportunity to buy.) An exact analysis becomes very complicated because the diagram of the inventory level versus time, even for a constant replenishment quantity, is no longer the simple repeating sawtooth pattern that we saw in Figure 4.1 of Chapter 4. This prevents us from using simple average costs over a typical unit period as was possible in the EOQ derivation. Instead, we now have to use the demand information over a finite period, extending from the present, when determining the appropriate value of the current replenishment quantity. This period is known as the planning *horizon* and its length can have a substantial effect on the total relevant costs of the selected strategy. Moreover, all other factors being equal, we would prefer to have the planning horizon as short as possible; the farther into the future we look for demand information, the less accurate it is likely to be.

It should be kept in mind, however, that in actual applications, a rolling schedule procedure is almost always used. Specifically, although replenishment quantities may be computed over the entire planning horizon, *only the imminent decision is implemented*. Then, at the time of the next decision, new demand information is appended to maintain a constant length of horizon. This topic is discussed in depth in Chapter 15 when we present one of the primary applications of lot sizing, or materials resources planning (MRP).

As indicated by the examples of time-varying demand discussed earlier, the demand can be either continuous with time or can occur only at discrete equispaced points in time. The former represents a stream of small-sized demands where the arrival rate varies with time, whereas the latter corresponds to something such as weekly shipments where a large demand quantity is satisfied at one time. The decision system that we propose is not affected by whether the demand is continuous or discrete with respect to time; all that will be needed is the total demand in each basic period. A common case is one in which the demand rate stays constant throughout a period, only changing from one period to another. An illustration is shown in Figure 5.1. As discussed in Chapter 3, demand forecasts are usually developed in this form, for example, a rate in January, another rate in February, and so on.

Another element of the problem that is important in selecting the appropriate replenishment quantities is whether replenishments must be scheduled at specified discrete points in time (e.g., replenishments can only be scheduled at intervals that are integer multiples of a week), or whether they can be scheduled at any point in continuous time. Considering a single item in isolation, the second case is probably more appropriate if the demand pattern is continuous with time. However, to ease scheduling coordination problems in multi-item situations, it usually makes sense to limit the opportunity times for the replenishment of each item. Furthermore, if the demand pattern is such that all the requirements for each period must be available at the beginning of that period, it is appropriate to restrict the replenishment opportunities to the beginnings

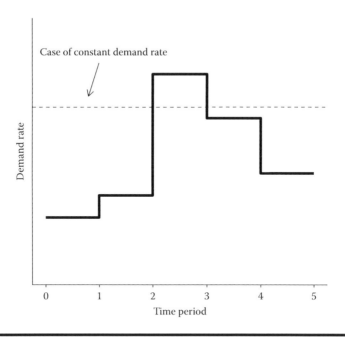

Figure 5.1   Demand pattern when the rate stays constant through each period.

of periods. This latter assumption will be made in developing the logic of the decision system that we propose. (The case where replenishments can take place at any time has been treated by Silver and Meal 1969.)

Still another factor that can materially influence the logic in selecting the replenishment quantities is the duration of the demand pattern. A pattern with a clearly specified end resulting from a production contract is very different from the demand situation where no well-defined end is in sight. In the former case, it is important to plan to reach the end of the pattern with a low (if not zero) level of the inventory. In contrast, for the continuing demand situation, if we use a finite planning horizon, there is no need to force the solution to include a low inventory level at the end of the horizon. The remaining inventory can be used in the interval beyond the horizon, where demand will continue (albeit, at a possibly different rate).

5.2 The Choice of Approaches

There are essentially three approaches to dealing with the case of a deterministic, time-varying demand pattern:

1. *Use of the basic EOQ.* This is a very simple approach. Use a fixed EOQ, based on the average demand rate out to the horizon, anytime a replenishment is required. As would be expected, this approach makes sense when the variability of the demand pattern is low; that is, the constant demand rate assumption of the fixed EOQ is not significantly violated.
2. *Use of the exact best solution to a particular mathematical model of the situation.* As we will see, under a specific set of assumptions, this approach, known as the Wagner–Whitin algorithm,

minimizes the total of certain costs. (For reasons that will become evident later in the chapter, we purposely avoid the use of the words "total relevant costs" here.)

3. *Use of an approximate or heuristic method.* The idea here is to use an approach that captures the essence of the time-varying complexity but at the same time remains relatively simple for the practitioner to understand, and does not require lengthy computations.

Many MRP systems offer a variety of approaches, although the most common is the EOQ, or some other fixed order quantity. These systems do not typically recommend an approach. Rather, the user must specify which technique to use. The discussion in this chapter, as well as in Chapter 15, should help inform these decisions.

5.3 General Assumptions and a Numerical Example

5.3.1 The Assumptions

We now specify a number of assumptions that will be made (at least implicitly) in all three of these approaches. Certain of these assumptions can be relaxed by somewhat minor modifications of the decision methodology; such modifications will be demonstrated later.

1. The demand rate is given in the form of $D(j)$ to be satisfied in period j ($j = 1, 2, \ldots, N$), where the planning horizon is at the end of period N. Of course, the demand rate may vary from one period to the next, but it is assumed as known. The case of probabilistic, time-varying demand will be addressed in Chapter 7.
2. The entire requirements of each period must be available at the beginning of that period. Therefore, a replenishment arriving partway through a period cannot be used to satisfy that period's requirements; it is cheaper, in terms of the reduced carrying costs, to delay its arrival until the start of the next period. Thus, replenishments are constrained to arrive at the beginnings of periods.
3. The unit variable cost does not depend on the replenishment quantity; in particular, there are no discounts in either the unit purchase cost or the unit transportation cost.
4. The cost factors do not change appreciably with time; in particular, inflation is at a negligibly low level.
5. The item is treated entirely independently of other items; that is, the benefits from joint review or replenishment do not exist or are ignored.
6. The replenishment lead time is known with certainty (a special case being zero duration) so that delivery can be timed to occur right at the beginning of a period.
7. No shortages are allowed.
8. The entire order quantity is delivered at the same time.
9. For simplicity, it is assumed that the carrying cost is only applicable to the inventory that is carried over from one period to the next. It should be emphasized that all the three approaches can easily handle the situation where carrying charges are included on the material during the period in which it is used to satisfy the demand requirements but, for practical purposes, this is an unnecessary complication.

These assumptions, except for 1, 2, and 9, are identical with those required in the derivation of the basic EOQ (as seen in Section 4.1 of Chapter 4). There are problems associated with these assumptions, as discussed in Chapter 4. Nevertheless, the methods described below provide a good starting point for most of the useful techniques.

Because of the assumption of deterministic demand, it is clear, from assumptions 2, 6, and 7, that an appropriate method of selecting replenishment quantities should lead to the arrival of replenishments only at the beginning of periods when the inventory level is exactly zero.

5.3.2 A Numerical Example

A plant in Germany uses the following simple decision rule for ascertaining production run quantities: "Each time a production run is made, a quantity sufficient to satisfy the total demand in the next 3 months is produced." A Canadian subsidiary of the German firm is the only customer of the German plant for a seasonal product PSF-007, a 25-by-30-cm lithographic film. The basic unit of the product is a box of 50 sheets of film. Canadian requirements by month (shifted to take into account of the shipping time) in the upcoming year are given in Table 5.1:[*]

We can see that the demand pattern has two peaks, one in late spring, and the other in autumn.

Table 5.1   Monthly Requirements

Month	Sequential Number	Requirements (Boxes)
January	1	10
February	2	62
March	3	12
April	4	130
May	5	154
June	6	129
July	7	88
August	8	52
September	9	124
October	10	160
November	11	238
December	12	41
		Total = 1,200

---

[*] These requirements are often termed "net requirements" because any inventory on hand and on order is "netted out" before determining the requirements. We shall assume that there is no inventory on hand or on order. Equivalently, we could assume that any initial inventory has been "netted out." For example, an initial inventory level of 50 and January requirements of 60 would produce the same net requirements shown in Table 5.1.

Table 5.2   Results of Using the Company's "Three Month" Rule on the Numerical Example

Month	1	2	3	4	5	6	7	8	9	10	11	12	Total
Starting inventory	0	74	12	0	283	129	0	176	124	0	279	41	–
Replenishment	84	–	–	413	–	–	264	–	–	439	–	–	1,200
Requirements	10	62	12	130	154	129	88	52	124	160	238	41	1,200
Ending inventory	74	12	0	283	129	0	176	124	0	279	41	0	1,118

Total replenishment costs $= 4 \times \$54 =$	$216.00
Total carrying costs $= 1{,}118$ box-months $\times \$20/\text{box} \times 0.02\ \$/\$/\text{month} =$	$447.20
Total replenishment plus carrying costs $=$	$663.20

The German plant estimates the fixed setup cost (A) per replenishment to be approximately $54, and the carrying charge (r) has been set by the management at 0.02 $/$/month. The unit variable cost (v) of the film is $20/box.[*]

The production-planning department of the German plant would like to establish the size of the first production run (that was needed by January 1), and also estimate the timing and sizes of future production quantities. The word "estimate" is used because the Canadian subsidiary could conceivably revise its requirements, particularly those late in the year, but still give adequate time to permit adjustment of the later portions of the production schedule.

The use of the company's "3 month" decision rule leads to the replenishment schedule and the associated costs shown in Table 5.2. There are four replenishments covering the 12 months out to the horizon. The total relevant costs are $663.20. The average month-ending inventory is 1,118/12 or 93.17 boxes. Therefore, the turnover ratio is 1,200/93.17 or 12.9.

5.4 Use of a Fixed EOQ

When the demand rate is approximately constant, we have advocated, in Chapter 4, the use of the basic EOQ. One possible approach to the case of a time-varying rate is to simply ignore the time variability, thus continuing to use the EOQ. (An alternative, closely related, approach is to use a fixed time supply equal to the EOQ time supply, T_{EOQ}. This approach will be discussed further in Section 5.6.2.) To be more precise, the average demand rate ($\bar{D}$) out to the horizon (N periods), or to whatever point our forecast information extends, is evaluated and the EOQ

$$\text{EOQ} = \sqrt{\frac{2A\bar{D}}{vr}}$$

---

[*] See Rachamadugu and Schriber (1995) for an analysis of "changing the givens"—specifically, finding the best run size when setup times decrease because of learning. See also Diaby (1995) and Tzur (1996).

is used anytime a replenishment is needed. Actually $\bar{D}$ can simply be based on a relatively infrequent estimate of the average demand per period and need not necessarily be reevaluated at the time of each replenishment decision. *To account for the discrete opportunities to replenish, at the time of a replenishment the EOQ should be adjusted to exactly satisfy the requirements of an integer number of periods.* A simple way to do this is to keep accumulating periods of requirements until the closest total to the EOQ is found. As discussed in Section 5.1, and as we will see in more detail in Section 5.5, it is desirable to plan to receive a replenishment order when the inventory is low (or zero) as this keeps inventory holding costs down. Alternatively, the quantity can be fixed, ensuring that the factory floor always sees the same production quantity. While this will lead to an increase in inventory holding costs, there may be efficiencies in production, packaging, shipping, or similar activities associated with always producing the same quantity.

To illustrate, for our numerical example,

$$\bar{D} = \frac{\text{Total requirements}}{12 \text{ months}} = 100 \text{ boxes/month}$$

Therefore,

$$\text{EOQ} = \sqrt{\frac{2 \times \$54 \times 100 \text{ boxes/month}}{\$20/\text{box} \times 0.02\$/\$/\text{month}}} \approx 164 \text{ boxes}$$

Consider the selection of the replenishment quantity at the beginning of January. The following table is helpful:

	January	February	March	April
Requirements	10	62	12	130
Cumulative requirements to the end of the month	10	72	84	214

The EOQ of 164 boxes lies between 84 and 214, and 214 is closer to 164 than is 84. Therefore, the first replenishment quantity is 214 boxes, lasting through the end of April. The detailed results of applying the fixed EOQ approach to the numerical example are as shown in Table 5.3. We can see from Tables 5.2 and 5.3 that the fixed EOQ approach, compared to the company's "three month" rule, reduces the total of replenishment and carrying costs from \$663.20 to \$643.20 or about 3%, a rather small saving. The turnover ratio has been increased from 12.9 to $1,200 \times 12/528$ or 27.3, but the replenishment costs have also increased.

5.5 The Wagner–Whitin Method: An "Optimal" Solution under an Additional Assumption

Wagner and Whitin (1958), in a classic article, developed an algorithm that guarantees an optimal (in terms of minimizing the total costs of replenishment and carrying inventory) selection of

Table 5.3   Results of Using the Fixed EOQ Approach on the Numerical Example

Month	1	2	3	4	5	6	7	8	9	10	11	12	Total
Starting inventory	0	204	142	130	0	0	0	52	0	0	0	0	–
Replenishment	214	–	–	–	154	129	140	–	124	160	238	41	1,200
Requirements	10	62	12	130	154	129	88	52	124	160	238	41	1,200
Ending inventory	204	142	130	0	0	0	52	0	0	0	0	0	528

Total replenishment costs = 8 × \$54 = \$432.00
Total carrying costs = 528 box-months × \$20/box × 0.02 \$/\$/month = \$211.20
Total replenishment plus carrying costs = \$643.20

replenishment quantities under the set of assumptions that were listed in Section 5.3.1.[*] They also identified one additional assumption that may be needed—namely, that either the demand pattern terminates at the horizon or else the ending inventory must be prespecified.[†]

5.5.1 The Algorithm

The algorithm is an application of dynamic programming, a mathematical procedure for solving sequential decision problems.[‡]

The computational effort, often prohibitive in dynamic programming formulations, is significantly reduced because of the use of two key properties (derived by Wagner and Whitin) that the optimal solution must satisfy:

Property 1 A replenishment only takes place when the inventory level is zero (previously discussed in Section 5.3).

Property 2 There is an upper limit to how far before a period j we would include its requirements, $D(j)$, in a replenishment quantity. Eventually, the carrying costs become so high that it is less expensive to have a replenishment arrive at the start of period j than to include its requirements in a replenishment from many periods earlier.

Suppose we define $F(t)$ as the total costs of the best replenishment strategy that satisfies the demand requirements in periods $1, 2, \ldots, t$. To illustrate the procedure for finding $F(t)$ and the

---

[*] In this chapter, we focus on the case where demand is deterministic. See for example, Sox et al. (1997), Vargas (2009), and Kang and Lee (2013) for extensions to the stochastic demand setting.

[†] The Appendix to this chapter contains additional details of the Wagner–Whitin model, including an integer-programming formulation as well as the dynamic programming model.

[‡] The problem here is a sequential decision problem because the outcome of the replenishment quantity decision at one point has effects on the possible replenishment actions that can be taken at later decision times; for example, whether or not we should replenish at the beginning of March depends very much on the size of the replenishment quantity at the beginning of February. Vidal (1970) provides an equivalent linear-programming formulation and an associated efficient solution procedure. See also Federgruen and Tzur (1991, 1994, 1995), Heady and Zhu (1994), and Wagelmans et al. (1992). Pochet and Wolsey (2006) discuss the techniques for improving the efficiency for numerous production-planning models.

associated replenishments, we again use the example of Table 5.1. In the following table, we replicate the requirements given in Table 5.1.

Month	Sequential Number	Requirements (Boxes)
January	1	10
February	2	62
March	3	12
April	4	130
May	5	154
June	6	129

Month	Sequential Number	Requirements (Boxes)
July	7	88
August	8	52
September	9	124
October	10	160
November	11	238
December	12	41
		Total = 1,200

$F(1)$ is the total costs of a replenishment of size 10 at the start of January, which is simply the setup cost A or $54. To determine $F(2)$, we have two possible options to consider:

Option 1 Replenish 10 boxes at the start of January and 62 boxes at the start of February.

Option 2 Replenish enough (72 boxes) at the start of January to cover the requirements of both January and February.

Costs of Option 1 = (Costs of best plan from the start to the end of January) + (Costs of a replenishment at the start of February to meet February's requirements)

$$= F(1) + A$$
$$= \$54 + \$54$$
$$= \$108$$

Costs of Option 2 = (Setup cost for January replenishment) + (Carrying costs for February's requirements)

$$= \$54 + 62 \text{ boxes} \times \$0.40/\text{box/month} \times 1 \text{ month}$$

$$= \$54 + \$24.80$$
$$= \$78.80$$

The cost of the second option is less than that of the first. Therefore, the best choice to satisfy the requirements of January and February is Option 2 and $F(2) = \$78.80$.

To satisfy requirements through to the end of March, there are three options, namely, where we position the last replenishment:

Option 1 Cover to the end of February in the best possible fashion and replenish 12 boxes at the start of March.

Option 2 Cover to the end of January in the best possible fashion and replenish 74 boxes at the start of February.

Option 3 Have a single replenishment of 84 boxes at the start of January.

Costs of Option 1 $= F(2) + A$

$$= \$78.80 + \$54$$
$$= \$132.80$$

Costs of Option 2 $= F(1) + A +$ (Carrying costs for March's requirements)

$$= \$54 + \$54 + (12 \times 0.40 \times 1)$$
$$= \$112.80$$

Costs of Option 2 $= A +$ (Carrying costs for February's requirements) $+$ (Carrying costs for March's requirements)

$$= \$54 + (62 \times 0.40 \times 1) + (12 \times 0.40 \times 2)$$
$$= \$88.40$$

Therefore, Option 3, a single replenishment at the start of January, is best in terms of meeting requirements through to the end of March.

We continue forward in this fashion until we complete period N (here $N = 12$). For any specific month t, there are t possible options to evaluate. It is important to note that the method, as just described, requires an ending point where it is known that the inventory level is to be at zero or some other specified value.

To illustrate the effect of property 2 discussed at the beginning of this section, consider the options open to us in terms of meeting requirements through to the end of July (month 7). The discussion in the previous paragraph would indicate that we must evaluate seven options. However, consider the requirements of 88 boxes in the month of July. The carrying costs associated with these requirements, if they were included in a replenishment at the beginning of June, would be

$$88 \text{ boxes} \times \$0.40/\text{box/month} \times 1 \text{ month or } \$35.20$$

Similarly, if these requirements were instead included in a replenishment at the beginning of May, the carrying costs associated with the 88 units would be

$$88 \text{ boxes} \times \$0.40/\text{box/month} \times 2 \text{ months or } \$70.40$$

which is larger than $54, the fixed cost of a replenishment. Therefore, the best solution could never have the requirements of July included in a replenishment at the beginning of May or earlier. It would be less expensive to have a replenishment arrive at the beginning of July. Thus, we need to consider only two options—namely, having the last replenishment at the start of either June or July.

If we were interested in computing the size of only the first replenishment quantity, then, it may not be necessary to go all the way out to the horizon (month N). The use of property 2 shows that, if for a period j, the requirements $D(j)$ are so large that the cost of carrying the inventory needed to meet $D(j)$ (for one period) exceeds the cost of a replenishment,

$$D(j)vr > A$$

or equivalently,

$$D(j) > A/vr$$

then the optimal solution will have a replenishment at the beginning of period j; that is, the inventory must go to zero at the start of period j. Therefore, the earliest j where this happens can be used as a horizon for the calculation of the first replenishment. In the numerical example,

$$A/vr = \frac{\$54}{\$20/\text{boxes} \times 0.02\$/\$/\text{month}} = 135\,\text{box} - \text{months}$$

We can see from the demand pattern that the earliest month with $D(j) > 135$ is May. Therefore, the end of April can be considered as a horizon in computing the replenishment quantity needed at the beginning of January. It should be emphasized that it is possible for a given problem instance that all per-period requirements are below this threshold.

Table 5.4 illustrates a simple spreadsheet used to calculate the optimal solution. Notice that, because of the ease of computation on the spreadsheet, we have included all possible production levels for each month, even though we could have eliminated certain values based on the insights from the previous paragraph. Notice also how the minimum value in the column for month 12 is highlighted. The underlined values represent the number of months of production. For instance, the underline from month 12 to month 11 (or $501.20 to $484.80) indicates that production in month 11 will meet the requirements in months 11 and 12. The next step is to look for the minimum value in month 10 (which again is highlighted), and underline values to the left, if possible. Continuing in this manner, we develop the results shown in Table 5.5.

There are seven replenishments and the total costs amount to $501.20. Comparison with Tables 5.2 and 5.3 shows that the Wagner–Whitin algorithm produces the total costs that are some 24.4% lower than those of the company's three-month rule and some 22.1% lower than those of a fixed EOQ strategy. Furthermore, the turnover ratio has been increased to $1,200 \times 12/308$ or 46.8.

5.5.2 Potential Drawbacks of the Algorithm

As mentioned earlier, the Wagner–Whitin algorithm is guaranteed to provide a set of replenishment quantities that minimize the sum of replenishment plus carrying costs out to a specified horizon. (In fact, a generalized form permits including backordering as well as replenishment costs that depend on the period in which the replenishment occurs.) As we have seen, it is possible to develop

Table 5.4 Spreadsheet Table for the Wagner–Whitin Algorithm on the Numerical Example

| Demand | 10 | 62 | 12 | 130 | 154 | 129 | 88 | 52 | 124 | 160 | 238 | 41 |
Month	1	2	3	4	5	6	7	8	9	10	11	12
1	54.00	78.80	**88.40**	244.40	490.80	748.80	960.00	1,105.60	1,502.40	2,078.40	3,030.40	3,210.80
2		108.00	112.80	216.80	401.60	608.00	784.00	908.80	1,256.00	1,768.00	2,624.80	2,788.80
3			132.80	184.80	308.00	462.80	603.60	707.60	1,005.20	1,453.20	2,214.80	2,362.40
4				**142.40**	204.00	307.20	412.80	496.00	744.00	1,128.00	1,794.40	1,925.60
5					196.40	**248.00**	318.40	380.80	579.20	899.20	1,470.40	1,585.20
6						250.40	285.60	327.20	476.00	732.00	1,208.00	1,306.40
7							302.00	**322.80**	422.00	614.00	994.80	1,076.80
8								339.60	389.20	517.20	802.80	868.40
9									**376.80**	440.80	631.20	680.40
10										**430.80**	526.00	558.80
11											484.80	**501.20**
12												538.80

Table 5.5 Results of Using the Wagner–Whitin Algorithm or the Silver–Meal Heuristic on the Numerical Example

Month	1	2	3	4	5	6	7	8	9	10	11	12	Total
Starting inventory	0	74	12	0	0	129	0	52	0	0	0	41	
Replenishment	84	0	0	130	283	0	140	0	124	160	279	0	1,200
Requirements	10	62	12	130	154	129	88	52	124	160	238	41	1,200
Ending inventory	74	12	0	0	129	0	52	0	0	0	41	0	308

Total replenishment costs $= 7 \times \$54 = \$378.00$

Total carrying costs $= 308$ box-months $\times \$20/\text{box} \times 0.02\ \$/\$/\text{month} = \$123.20$

Total replenishment plus carrying costs $= \$501.20$

spreadsheet solutions to the dynamic programming model; and with linear programming add-ins to spreadsheets, it is possible to use that formulation and find an optimal solution. Modified versions of the algorithm have been successfully implemented. See, for example, Potamianos and Orman (1996). In general, however, the algorithm has received extremely limited acceptance in practice. The primary reasons for this lack of acceptance are:

1. The relatively complex nature of the algorithm that makes it more difficult for the practitioner to understand than other approaches.
2. The possible need for a well-defined ending point for the demand pattern. (This would be artificial for the typical inventoried item where termination of demand is not expected in the near future.) As shown above, such an ending point is not needed when there is at least one period whose requirements exceed A/vr. Moreover, all information out to the end point may be needed for computing even the initial replenishment quantity. In this connection, considerable research effort (see e.g., Eppen et al. 1969; Kunreuther and Morton 1973; Blackburn and Kunreuther 1974; Lundin and Morton 1975) has been devoted to ascertaining the minimum required length of the planning horizon to ensure the selection of the optimal value of the initial replenishment.
3. A related issue is the fact that the algorithm is often used in conjunction with MRP software. Because MRP typically operates on a rolling schedule, the replenishment quantities chosen should not change when new information about future demands becomes available. Unfortunately, the Wagner–Whitin approach does not necessarily have this property (which Baker (1989) defines as the *insulation* property). This is potentially a critical flaw in this approach.[*]
4. The necessary assumption that replenishments can be made only at discrete intervals (namely, at the beginning of each of the periods). This assumption can be relaxed by subdividing the periods; however, the computational requirements of the algorithm go up rapidly with the number of periods considered. (In contrast, the other two methods are easily modified to account for continuous opportunities to replenish.)

---

[*] Sahin et al. (2013) review the literature on production planning with rolling horizons.

When taking into account of the total relevant costs, particularly including system control costs, it is not at all surprising that the algorithm has received such limited acceptability. We do not advocate its use for B items. (Let us not lose sight of the fact that, in this chapter, we are still talking about this class of items as opposed to Class A items. For the latter, considerably more control expense is justified because of the higher potential savings per item.) Instead, it is appropriate to resort to simpler heuristic methods that result in reduced control costs that more than offset any extra replenishment or carrying costs that their use may incur.

5.6 Heuristic Approaches for a Significantly Variable Demand Pattern

As indicated by the results in the previous two sections, the Wagner–Whitin approach does substantially better than a fixed EOQ, at least for the illustrative numerical example. However, as mentioned in Section 5.5.2, the Wagner–Whitin approach has drawbacks from the practitioner's standpoint. Therefore, the natural question to ask is: "Is there a simpler approach that will capture most of the potential savings in the total of replenishment and carrying costs?" A number of researchers (see, e.g., Diegel 1966; DeMatteis 1968; Gorham 1968; Mendoza 1968; Silver and Meal 1973; Donaldson 1977; Groff 1979; Karni 1981; Wemmerlöv 1981; Brosseau 1982; Freeland and Colley 1982; Boe and Yilmaz 1983; Mitra et al. 1983; Bookbinder and Tan 1984; Baker 1989; Coleman and McKnew 1990; Triantaphyllou 1992; Teng 1994; Hariga 1995a; Goyal et al. 1996; Zhou 1996; Schulz 2011) have suggested various decision rules, some of which have been widely used in practice.[*] In this section, we review several heuristics that, as we will find, accomplish exactly what we desire. Moreover, in numerous test examples, some of these heuristics have performed extremely well. Because they are so frequently seen in practice, we also discuss some commonly used heuristics that perform less well.

As mentioned earlier, the fixed EOQ approach should perform suitably when the demand pattern varies very little with time. For this reason, we advocate the use of the somewhat more complicated heuristics only when the pattern is significantly variable. An operational definition of "significantly variable" will be given in Section 5.6.7.

We begin with the Silver–Meal (or least period cost) heuristic (see Silver and Meal 1973), using it to illustrate the basic concepts. Then we briefly describe other heuristics.

5.6.1 The Silver–Meal, or Least Period Cost, Heuristic

The Silver–Meal heuristic selects the replenishment quantity in order to replicate a property that the basic EOQ possesses when the demand rate is constant with time, namely, the *total relevant costs per unit time for the duration of the replenishment quantity are minimized*. If a replenishment arrives at the beginning of the first period and it covers requirements through to the end of the *T*th period, then, the criterion function can be written as follows:

$$\frac{\text{(Setup cost)} + \text{(Total carrying costs to end of period } T)}{T}$$

---

[*] Wee (1995) develops an exact solution to the problem of finding a replenishment policy for an item that faces declining demand, and that deteriorates over time. See Hariga (1994), Ting and Chung (1994), Hariga (1995b), Bose et al. (1995), and Chapter 8 as well. See Hill (1995) for the case of increasing demand, and Hariga (1996) for the case of inflationary conditions.

This is a reasonable criterion and it has the desirable feature of not including, in the present replenishment, a large requirement well in the future (inclusion of such a requirement would make the "costs per unit time" measure too high). It is not difficult to develop numerical examples in which the use of the criterion does not lead to the overall optimal solution, particularly for the case where the demand pattern has a well-defined ending point. Fortunately, this is not a major drawback because our primary concern is with demand patterns that do not have a clearly defined ending point in the near future. (This follows from our basic definition of Class B items.)

Because we are constrained to replenishing at the beginnings of periods, the best strategy must involve replenishment quantities that last for an integer number of periods. Consequently, we can think of the decision variable for a particular replenishment as being the time T that the replenishment will last, with T constrained to integer values. The replenishment quantity Q, associated with a particular value of T, is

$$Q = \sum_{j=1}^{T} D(j) \tag{5.1}$$

provided we set the time origin so that the replenishment is needed and arrives at the beginning of period 1. According to the chosen criterion, we wish to pick the T value that minimizes the total relevant costs per unit time of replenishment and carrying inventory over the time period T.

Let the total relevant costs associated with a replenishment that lasts for T periods be denoted by $\mathrm{TRC}(T)$. These costs are composed of the fixed replenishment cost A and the inventory carrying costs. We wish to select T to minimize the total relevant costs per unit time, $\mathrm{TRCUT}(T)$, where

$$\mathrm{TRCUT}(T) = \frac{\mathrm{TRC}(T)}{T} = \frac{A + \text{carrying costs}}{T} \tag{5.2}$$

If $T = 1$, there are no carrying costs (we only replenish enough to cover the requirements of period 1), that is,

$$\mathrm{TRCUT}(T) = \frac{A}{1} = A$$

If the setup cost A is large, this may be unattractive when compared with including the second period's requirements in the replenishment, that is, using $T = 2$.

With $T = 2$, the carrying costs are $D(2)vr$, which are the cost of carrying the requirements $D(2)$ for one period. Therefore,

$$\mathrm{TRCUT}(2) = \frac{A + D(2)vr}{2}$$

Now, the setup cost is apportioned across two periods, but a carrying cost is incurred. With $T = 3$ we still carry $D(2)$ for one period, but now we also carry $D(3)$ for two periods. Thus,

$$\mathrm{TRCUT}(3) = \frac{A + D(2)vr + 2D(3)vr}{3}$$

In this case, the setup charge is apportioned across three periods, but this may not be attractive because of the added carrying costs.

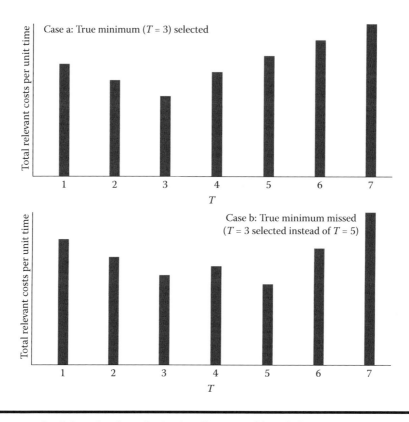

Figure 5.2   Graph of the selection of *T* in the Silver–Meal heuristic.

The basic idea of the heuristic is to evaluate $\mathrm{TRCUT}(T)$ for increasing values[*] of T until, for the first time,

$$\mathrm{TRCUT}(T+1) > \mathrm{TRCUT}(T)$$

that is, the total relevant costs per unit time start increasing. When this happens, the associated T is selected as the number of periods that the replenishment should cover. The corresponding replenishment quantity Q is given by Equation 5.1.

As evidenced by Figure 5.2, this method guarantees only a local minimum in the total relevant costs per unit time, *for the current replenishment*. It is possible that still larger values of T would yield still lower costs per unit time since we stop testing with the first increase in costs per unit time. We could protect against this eventuality by computing the ratio for a few more values of T, but the likelihood of improvement in most real cases is small. Furthermore, let us not lose sight of the fact that "costs per unit time for the first replenishment" is only a surrogate criterion.

---

[*] When some of the D's are zeros, we proceed as follows: Suppose $D(j-1) > 0, D(j) > 0, D(j+1) = 0$, and $D(j+2) > 0$. We evaluate $\mathrm{TRCUT}(j-1)$, and then jump to $\mathrm{TRCUT}(j+1)$, because a replenishment that covers period j will also, trivially, cover the zero requirement in period $j+1$.

It is conceivable that the TRCUT(T) may continue to decrease all the way out to $T = N$. This says that it is appropriate to cover all the requirements out to the horizon with the current replenishment. In such a case, it may be appropriate to cover an even longer period, but the heuristic is unable to provide the answer without forecast information beyond period N.

Except for the unusual situation just described, the performance of the heuristic, in contrast with the Wagner–Whitin algorithm, does not crucially depend on the N value chosen. It tends to use demand data of only the first few periods, an attractive property when one recognizes that the deterministic demand assumption becomes less reasonable as one projects further into the future.

To illustrate the application of the heuristic, let us again use the numerical example of Section 5.3. As previously introduced, the values of A, v, and r and the first part of the requirements pattern are as follows:

$$A = \$54 \quad v = \$20/\text{box} \quad r = 0.02 \ \$/\$/\text{month}$$

Month j	1	2	3	4	5	6...
Requirements D(j)	10	62	12	130	154	129...

The calculations for the first replenishment quantity (assuming that the inventory is zero at the beginning of month 1) are shown in Table 5.6. The heuristic selects a T value of 3 with an associated Q, using Equation 5.1, of

$$Q = D(1) + D(2) + D(3)$$
$$= 10 + 62 + 12$$
$$= 84 \text{ boxes}$$

The table illustrates that the computations are simply multiplications, additions, divisions, and comparisons of numbers.

It turns out for this numerical example (and for a substantial portion of all others tested) that this simple heuristic gives the same solution as the Wagner–Whitin algorithm. Thus, the solution has already been shown in Table 5.5.

Table 5.6 Computations for the First Replenishment Quantity Using the Silver–Meal Heuristic

T	Requirements	Months in Inventory	Carrying Cost	Order Cost	Total Cost	Total Cost ÷ T
1	10	0	$0.00	$54.00	$54.00	$54.00/month
2	62	1	$24.80[a]	$54.00	$78.80	$39.40/month
3	12	2	$34.40[b]	$54.00	$88.40	$29.47/month
4	130	3	$190.40	$54.00	$244.40	$61.10/month

[a] $24.80 = D(2)vr = 62(20)(0.02)$.

[b] $34.40 = D(2)vr + 2D(3)vr = 62(20)(0.02) + 2(12)(20)(0.02)$.

5.6.2 The EOQ Expressed as a Time Supply (POQ)

One approach described earlier (in Section 5.4) was to use a fixed order quantity, based on using the average demand rate in the EOQ equation. Empirically (see, e.g., Brown 1977), where there is significant variability in the demand pattern, better cost performance has been obtained by proceeding slightly differently. The EOQ is expressed as a time supply using, namely,

$$T_{EOQ} = \frac{EOQ}{\bar{D}} = \sqrt{\frac{2A}{\bar{D}vr}} \tag{5.3}$$

rounded to the nearest integer greater than zero. Then, any replenishment of the item is made large enough to cover exactly the requirements of this integer number of periods. Another name for this approach is the periodic order quantity (POQ). This approach has the advantage that arrivals of orders to the factory floor are regular even though the order quantities will be different from order to order. Note that this is in contrast to the fixed order quantity approach that has irregular timing, but constant quantities.

For the numerical example of Section 5.3.2, Equation 5.3 gives

$$T_{EOQ} = \sqrt{\frac{2(54)}{(100)(20)(0.02)}} = 1.64$$

which rounds to 2. The repeated use of a 2-period time supply in the example gives the total costs of $553.60 as compared to $501.20 using the Silver–Meal heuristic.

5.6.3 Lot-for-Lot

Perhaps, the easiest approach is Lot-for-Lot (L4L), which simply orders the exact amount needed for each time period. Thus, inventory holding costs are zero with L4L. Clearly, this approach is not cost effective when fixed replenishment costs are significant. For our example, the total costs are 12 × $54 or $648. We mention L4L here only because it will reappear in the context of MRP where it is commonly used. (See Chapter 15.)

5.6.4 Least Unit Cost

The Least Unit Cost (LUC) heuristic is similar to the Silver–Meal heuristic except that it accumulates requirements until the cost per unit (rather than the cost per period) increases. Calculations for the first order are shown in Table 5.7.

Table 5.8 contains the results for the entire 12-month time. The total cost of $558.80 is 11.5% higher than the optimal solution. Research has shown that this is not unusual performance for the LUC heuristic.

5.6.5 Part-Period Balancing

The basic criterion used here is to select the number of periods covered by the replenishment such that the total carrying costs are made as close as possible to the setup cost, A. (Exact equality is usually not possible because of the discrete nature of the decision variable, T.) To illustrate, for the

Table 5.7 Computations for the First Replenishment Quantity Using the LUC Heuristic

T	Requirements	Months in Inventory	Carrying Cost	Order Cost	Total Cost	Lot Size	TC per Unit
1	10	0	$0.00	$54.00	$54.00	10	$5.40
2	62	1	$24.80	$54.00	$78.80	72	$1.09
3	12	2	$34.40	$54.00	$88.40	84	$1.05
4	130	3	$190.40	$54.00	$244.40	214	$1.14
5	154	4	$436.80	$54.00	$490.80	368	$1.33

numerical example, we have the calculations for the first replenishment quantity (assuming that the inventory is zero at the beginning of month 1):

T	Carrying Costs
1	0
2	$D(2)vr = \$24.80 < \$54$
3	$24.80 + 2D(3)vr = \$34.40 < \$54$
4	$\$34.40 + 3D(4)vr = \$194.40 > \$54$

$34.40 is closer to $54 (the A value) than is $190.40. Therefore, a T value of 3 is selected for the first replenishment. Repeated use of this procedure gives replenishments listed in Table 5.9, with the total costs of $600—again well above the $501.20 found by the Silver–Meal heuristic (Table 5.5).

Table 5.8 Results of Using LUC on the Numerical Example

Month	1	2	3	4	5	6	7	8	9	10	11	12	Total
Starting inventory	0	74	12	0	154	0	88	0	124	0	0	0	
Replenishment	84	0	0	284	0	217	0	176	0	160	238	41	1,200
Requirements	10	62	12	130	154	129	88	52	124	160	238	41	1,200
Ending inventory	74	12	0	154	0	88	0	124	0	0	0	0	452
							Total replenishment costs $= 7 \times \$54 =$						$378.00
			Total carrying costs $= 452$ box-months $\times \$20/box \times 0.02$ \$/\$/month $=$										$180.80
						Total replenishment plus carrying costs $=$							$558.80

Table 5.9   Results of Using PPB on the Numerical Example

Month	1	2	3	4	5	6	7	8	9	10	11	12	Total
Starting inventory	0	74	12	0	154	0	88	0	124	0	238	0	
Replenishment	84	0	0	284	0	217	0	176	0	398	0	41	1,200
Requirements	10	62	12	130	154	129	88	52	124	160	238	41	1,200
Ending inventory	74	12	0	154	0	88	0	124	0	238	0	0	690

Total replenishment costs $= 6 \times \$54 \ = \ \$378.00$

Total carrying costs $= 690$ box-months $\times \$20/\text{box} \times 0.02\ \$/\$/\text{month}\ = \ \$276.00$

Total replenishment plus carrying costs $= \$600.00$

Refinements of the part-period balancing (PPB) method, requiring more computational effort, have been developed (see e.g., DeMatteis 1968; Mendoza 1968). One refinement is called the Look-Ahead/Look-Back technique. The Look-Ahead technique evaluates the cost of moving an order later in time. For instance, because the first order of 84 units cannot be moved to period 2 without incurring a stockout, we first examine the order for 284 units in period 4. If that order were moved to period 5, the 130 units required in period 4 must be produced in period 1 and held for 3 periods—at a cost of 130 units (or parts) × 3 periods = 390 part periods. The 154 units held from period 4 to period 5 are saved—or 154 parts × 1 period = 154 part periods. Therefore, no adjustment is made to that order. We come to a similar conclusion with the order for 217 units in period 6: save 88 units for 1 period (or 88 part periods) at a cost of 129 units held for 2 periods (or 258 part periods).

It is attractive to make an adjustment to the 176 units ordered in period 8. Moving that order to period 9 requires shifting the 52 units needed in period 8 to the order in period 6—at a cost of 52 × 2 = 104 part periods. The cost of holding 124 units from period 8 to period 9 is saved— 124 × 1 = 124 part periods. So, the adjustment would be made; the order in period 6 would now be 217 + 52 = 269 units, and the next order would be 124 units in period 9.

Unfortunately, Wemmerlöv (1983) has shown that the Look-Ahead/Look-Back techniques do not necessarily improve the performance of the PPB heuristic. He also notes that the PPB heuristic generally performs poorly in dynamic environments.

5.6.6 Performance of the Heuristics

Many of these heuristics have been tested against the Wagner–Whitin algorithm, the basic EOQ, and other heuristics on a wide range of examples. Baker (1989) summarizes this literature and reconciles the differences among different papers. Van den Heuvel and Wagelmans (2010) analyze the worst-case behavior of several lot-sizing heuristics. One lesson from all the literature is that most heuristics outperform the fixed EOQ approach. Another lesson, illustrated in Baker's paper, is that the Silver–Meal heuristic incurs an average cost penalty for using the heuristic instead of the "optimal" Wagner–Whitin algorithm of less than 1%; in many cases there is no penalty whatsoever. (See Table 5.10.) Finally, tests in a rolling-horizon environment (see, e.g., Blackburn and Millen 1980) have revealed that frequently the Silver–Meal heuristic actually outperforms the dynamic programming algorithm (because the latter now gives the optimal solution to the wrong problem).

Table 5.10 Performance of Selected Heuristics on 20 Test Problems

Heuristic	Number Optimal	Average Percent Error (%)
POQ	7	10.004
LUC	8	9.303
PPB	16	1.339
Silver–Meal	14	0.943

Source: Adapted from Baker, K. R. 1989. *Journal of Manufacturing and Operations Management* 2(3), 199–221.

5.6.7 *When to Use Heuristics*

These heuristics, although quite simple, are still more involved than the determination of the basic EOQ. We know that the latter is the best replenishment quantity to use when there is no variability in the demand rate. In fact, the variability of the demand pattern should exceed some threshold value before it makes sense to change to a heuristic.

A useful measure of the variability of a demand pattern is the squared coefficient of variation. This statistic, denoted by SCV, is given by

$$\text{SCV} = \frac{\text{Variance of demand per period}}{\text{Square of average demand per period}}$$

which can be computed easily on a spreadsheet. Tests have shown that a threshold value of SCV appears to be in the neighborhood of 0.2, that is, if SCV < 0.2, use a simple EOQ involving $\bar{D}$ as the demand estimate. If SCV ≥ 0.2, use a heuristic.

For the numerical example discussed in this chapter, we get SCV = 0.426, which is greater than the threshold value indicating that a heuristic, rather than the EOQ, should be used. (Recall from Tables 5.3 and 5.5 that the Silver–Meal heuristic produced total relevant costs of 22.1% below those resulting from the use of the EOQ.)

In addition, consider the absolute importance of the item, in terms of the potential savings in replenishment and carrying costs. A useful surrogate for this quantity is the total of these costs per unit time under the use of the EOQ in the EOQ cost model (we know that the EOQ model is, strictly speaking, only an approximation for the case of time-varying demand, but it is simple to use), namely,

$$\sqrt{2A\bar{D}vr}$$

If this quantity was very small, departing from the simple EOQ decision rule would not be justified. However, in this chapter, we are concerned with B items, defined to be those with intermediate values of $\bar{D}v$. Thus, by definition, for such items, the above test quantity will be large enough to ensure that we should not blindly remain with the basic EOQ when the demand rate varies enough.

A word of caution is in order. There are two situations where the use of the heuristic can lead to significant cost penalties:

1. When the demand pattern drops rapidly with time over several periods.
2. When there are a large number of periods having no demand.

Silver and Miltenburg (1984) have suggested modifications of the heuristic to cope with these circumstances. Baker (1989) also notes other heuristics that perform slightly better at little computational cost. When unit production costs vary over time, the heuristic can also perform quite poorly. **?** develop an approximation that performs well in this case.

5.6.8 Sensitivity to Errors in Parameters

In Section 4.3 of Chapter 4, it was illustrated that, for the case of the basic EOQ, the total cost curve is quite shallow in the neighborhood of the best order quantity. Thus, even substantial deviations of the order quantity away from its best value tend to produce small percentage cost penalties. One case of such deviations is errors in the parameter (cost or demand) values. We concluded that the costs were relatively insensitive to errors in the input parameters for the case of the EOQ formulation.

Fortunately, tests have revealed that a similar phenomenon exists for the Silver–Meal heuristic. The results of one such test are shown in Figure 5.3. The basic data used (taken from Kaimann 1969) were

$$A/vr = 150$$

Period	1	2	3	4	5	6	7	8	9	10	11	12
Requirements	10	10	15	20	70	180	250	270	230	40	0	10

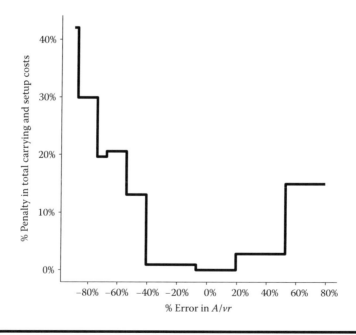

Figure 5.3   Illustration of the insensitivity of heuristic results to errors in cost parameters.

A/vr was deliberately changed from its correct value. For this incorrect value of A/vr, the Silver–Meal heuristic was used to compute the replenishment quantities. The true value of A/vr was then employed to compute the costs associated with this sequence of replenishments. These costs were compared with those of the replenishment pattern resulting from using the correct value of A/vr. This was repeated for a number of different percentage errors in A/vr. To illustrate, in Figure 5.3, we can see that if A/vr is erroneously set at a value of 40% that is too high (i.e., at a value of 210 instead of the correct level of 150), then, the cost penalty is less than 2%. The plot is not a smooth curve (as was the case in Figure 4.3 of Chapter 4) because here the replenishment opportunities are discrete in nature. Nonetheless, in all cases tested, the percentage cost penalties, even for reasonable-sized errors in A/vr, were quite small. Pan (1994) reaches the same conclusion.

5.6.9 Reducing System Nervousness

In a rolling-horizon environment, as new demand information becomes available, the timing and sizes of future replenishments may be altered. Frequent alterations of this type may be unattractive from an operational standpoint. Carlson et al. (1979) and Kropp et al. (1983) have introduced the idea of a penalty cost for changing a previously established schedule; they have shown how to incorporate this idea into the Silver–Meal heuristic and the Wagner–Whitin algorithm. As a by-product, this modification permits the handling of the situation where the setup cost varies from period to period. See also Blackburn et al. (1986), Sridharan and Berry (1990), Sridharan et al. (1987), Kropp and Carlson (1984), DeMatta and Guignard (1995), and Toklu and Wilson (1995). We will have more to say on this topic in Chapter 15.

5.7 Handling of Quantity Discounts

An important extension of the heuristic is to permit the inclusion of quantity discounts. As in Chapter 4, we illustrate for the case of an all-units fractional discount d if $Q \geq Q_b$. The presence of a discount removes the validity of one of the key properties of a solution first mentioned in Section 5.5—namely, that replenishments are received only when the inventory level is 0. Sometimes, it may indeed be appropriate to order exactly Q_b, which need not cover precisely an integer number of periods of requirements.

A revised heuristic, developed by Bregman and Silver (1993), involves a three-step procedure. First, determine the value of T, the replenishment quantity expressed as a time supply, which gives the lowest cost per period ignoring the discount possibility. Denote this lowest cost time supply, T'. If $T' \geq T_b$, where T_b is the time supply covered by a replenishment of size Q_b, then T' is used. The second step is needed if $T' < T_b$ and it involves checking if the use of T_b (i.e., taking advantage of the discount) is attractive. If so, the third step is to check if it is attractive to cover the integer T value above T_b. More formally,

Step 1 Find the first minimum ignoring the discount possibility.
Find the lowest integer T that gives a local minimum of

$$\text{TRCUT}(T) = \begin{cases} \dfrac{A + vr\sum_{j=1}^{T}(j-1)D(j)}{T} & T < T_b \\[3mm] \dfrac{A + v(1-d)r\sum_{j=1}^{T}(j-1)D(j)}{T} & T \geq T_b \end{cases} \tag{5.4}$$

If TRCUT($T + 1$) is less than or equal to TRCUT(T), continue trying $T + 2$, and so on. If there is a $D(T + 1) = 0$, then, immediately compare TRCUT($T + 1$) to TRCUT($T - 1$) because TRCUT($T + 1$) will always be less than TRCUT(T) in this situation. Denote the minimizing T value as T'. If $T' \geq T_b$, then use $T = T'$ with

$$Q = \sum_{j=1}^{T'} D(j) \tag{5.5}$$

If $T' < T_b$, then go to Step 2.

Step 2 Check if the discount is attractive when $T' < T_b$.

Evaluate the modified total relevant costs per unit time (modified to take into account of the differences in the unit costs)

$$\text{MTRCUT}(T') = \frac{A + vr \sum_{j=1}^{T'}(j - 1)D(j) + vd \sum_{j=1}^{T'} D(j)}{T'} \tag{5.6}$$

$$\text{MTRCUT}(T_b) = \frac{A}{[T_b]} + \frac{v(1 - d)r \sum_{j=1}^{[T_b]}(j - 1)D(j) + v(1 - d)r[T_b]\left(Q_b - \sum_{j=1}^{[T_b]} D(j)\right)}{T_b} \tag{5.7}$$

where $[T_b]$ is the integer portion of T_b (e.g., if $T_b = 3.2$, then $[T_b] = 3$). If MTRCUT(T_b) > MTRCUT(T'), then use T'. If MTRCUT(T_b) $\leq$ MTRCUT(T'), then go to Step 3.

The $vd \sum D(j)$ term in Equation 5.6 represents the discount dollar savings passed up in using T'. The A term in Equation 5.7 is divided by $[T_b]$ because a new order will be required $[T_b]$ periods later even though some stock will still be on hand. The vr terms in the numerators of Equations 5.6 and 5.7 represent the relevant carrying costs.

Step 3 Check if it is attractive to cover $[T_b] + 1$ periods.

To determine if the added carrying costs for an additional period are less than the savings from avoiding a fixed replenishment cost for the additional period, compute TRCUT($[T_b] + 1$) using Equation 5.4 for $T > T_b$. If MTRCUT(T_b) < TRCUT($[T_b] + 1$), use T_b with $Q = Q_b$. If MTRCUT(T_b) $\geq$ TRCUT($[T_b] + 1$), use

$$Q = \sum_{j=1}^{[T_b]+1} D(j) \tag{5.8}$$

Bregman and Silver (1993) have shown that this heuristic performs extremely well, and that its computation time is quite small compared to other heuristics.

Other work on this topic includes Callarman and Whybark (1981), Benton (1983, 1985, 1991), La Forge (1985), Chung et al. (1987), Christoph and La Forge (1989), Federgruen and Lee (1990), Bregman (1991), and Hariga (1995c). Hu et al. (2004) investigate heuristics for the incremental quantity discount setting. They propose a modified version of the Silver–Meal heuristic that works quite well, particularly when the horizon is not too short, on a large set of test problems.

5.8 Aggregate Exchange Curves

As has been discussed earlier in the book, certain parameter values, such as r or A/r, may be selected implicitly by the top management; they specify a reasonable operating situation (such as an inventory budget) for all items in the inventory. In Chapter 4, for the case of the basic EOQ, it was relatively easy to develop an exchange curve because of the analytic relationships between (1) the total average cycle stock and A/r (see Equation 4.29), and (2) the total number of replenishments per unit time and A/r (see Equation 4.30).

For the case of time-varying demand, no such simple results exist because of the discrete nature of the decision variable T, the number of periods of requirements to be included in a replenishment. Therefore, in this case, in order to develop an exchange curve, we must proceed as follows.

A representative sample of items is selected from the population of items under consideration. The exact sample size to select depends on the number and characteristics of items in the population that are to be controlled by the replenishment batch size heuristic or algorithm (i.e., have a squared coefficient of variation exceeding 0.2).

A value of A/r is selected. Then, for each item in the sample, the heuristic, say Silver–Meal, is used to determine the replenishment quantities over a reasonable time period such as 12 months. This implies a certain number of replenishments and an average inventory level (in dollars) for the item. Summing these quantities across the items of the sample produces the total number of replenishments and the total average inventory of the sample. These figures must be appropriately scaled up to correspond with the total group of items to be controlled by the heuristic. This must be repeated for several values of A/r to develop the tradeoff curve.

An additional point is worth mentioning here. It is often desirable to have an exchange curve for all the items in the population. In such a case, the results for the Silver–Meal heuristic items must be added to those for the items to be replenished according to the basic EOQ (Section 4.10 of Chapter 4). This produces one composite curve as a function of A/r.

5.9 Summary

In this chapter, we have provided a decision system for coping with the relaxation of another of the key assumptions inherent in the use of the basic EOQ. The system permits the practitioner to deal, in a realistic way, with time-varying demand patterns. The Silver–Meal heuristic, as well as some other approaches, has been found to be robust. A guideline has been provided to aid the decision maker in choosing when to use the simple EOQ approach.

Time-varying demand patterns will play a crucial role in the Material Requirements Planning framework to be discussed in Chapter 15. In addition, an extension of the Silver–Meal heuristic to permit multiple items within a capacitated production context will be discussed in Chapter 10.

In the next chapter, we return to the context of an essentially level demand pattern but, for the first time, incorporate uncertainty in the demand rate. In Chapter 7, a brief mention will be made of the treatment of the combination of time-varying and uncertain demand.

Problems

5.1   A sale of polarizing filters is held twice annually by a photographic products distributor. The demand pattern for a particular size of filter for the past year is as follows:

Jan.	Feb.	Mar.	Apr.	May	June
21	29	24	86	31	38
July	Aug.	Sept.	Oct.	Nov.	Dec.
45	39	31	78	29	32

It is anticipated that demand for the next year will follow this pattern; hence, these figures are being used as the "best estimates" of forthcoming sales. Demand will also continue in future years. The cost of these filters is $8.65, ordering costs are approximately $35, and the carrying cost is 0.24 $/$/year.

Calculate the squared coefficient of variation and select the appropriate order quantities.

5.2 The demand pattern for another type of filter is

Jan.	Feb.	Mar.	Apr.	May	June
18	31	23	95	29	37
July	Aug.	Sept.	Oct.	Nov.	Dec.
50	39	30	88	22	36

These filters cost the company $4.75 each; ordering and carrying costs are as in Problem 5.1. The squared coefficient of variation equals 0.33. Use the Silver–Meal heuristic to determine the sizes and timing of replenishments of stock.

5.3 a. For the following item having zero inventory at the beginning of period 1, develop the magnitude of the *first* replenishment *only* using the LUC method.

Item characteristics:

$$A = \$50 \quad v = \$2/\text{unit} \quad r = 0.05 \ \$/\$/\text{period}$$

Period j	1	2	3	4	5	6	7
D(j)	200	300	500	500	400	400	300

b. A marked difference between this method and that of Silver–Meal exists in the dependence on $D(1)$. Briefly discuss. In particular, suppose $D(1)$ was much larger than any of the other $D(j)$'s. What effect would this have on the best T for
   i. Silver–Meal?
   ii. LUC?

5.4 Consider an item with the following properties:

$A = \$20 \quad v = \$2/\text{unit} \quad r = 0.24 \ \$/\$/\text{year}$

At time 0, the inventory has dropped to zero and a replenishment (with negligible lead time) must be made. The demand pattern for the next 12 months is

Month j	1	2	3	4	5	6
Demand (units) $D(j)$	50	70	100	120	110	100
Month j	7	8	9	10	11	12
Demand (units) $D(j)$	100	80	120	70	60	40

All the requirements of each month must be available at the beginning of the month. Replenishments are restricted to the beginnings of the months. No shortages are allowed. Using each of the following methods, develop the pattern of replenishments to cover the 12 months and the associated total costs of each pattern (do not bother to count the costs of carrying $D(j)$ during its period of consumption, namely, period j). In each case, the size of the last replenishment should be selected to end month 12 with no inventory.
a. Fixed EOQ (rounded to the nearest integer number of months of supply; i.e., each time the EOQ, based on the average demand through the entire 12 months, is adjusted so that it will last for exactly an integer number of months).
b. A fixed time supply (an integer number of periods) based on the EOQ expressed as a time supply, using the average demand rate for the 12 months.
c. On each replenishment, the selection of Q (or, equivalently, the integer T), which minimizes the costs per unit of quantity ordered to cover demand through T.
d. The Silver–Meal heuristic.
e. One replenishment at the start of month 1 to cover all the requirements to the end of month 12.
f. A replenishment at the beginning of every month.
   Hint: For each case, it would be helpful to develop a table with at least the following columns: (1) Month, (2) Replenishment Quantity, (3) Starting Inventory, (4) Demand, and (5) Ending Inventory.
5.5 Consider an item with the following deterministic, time-varying demand pattern:

Week	1	2	3	4	5	6
Demand	50	80	180	80	0	0
Week	7	8	9	10	11	12
Demand	180	150	10	100	180	130

Suppose that the pattern terminates at week 12. Let other relevant characteristics of the item be:

Inventory carrying cost per week (incurred only on units carried over from 1 week to the next) is $0.20/unit.

Fixed cost per replenishment is $30.

Initial inventory and replenishment lead time are both zero.

Perform an analysis similar to that of Problem 5.4 (of course, substituting "week" for "month").

5.6   Consider an item with the following declining demand pattern:

Period	1	2	3	4	5	6	7	8	9	10
Demand	600	420	294	206	145	101	71	50	35	25
Period	11	12	13	14	15	16	17	18	19	20
Demand	17	12	9	6	5	3	2	2	1	1

There is no further demand after period 20.

$$A = \$50 \quad v = \$2.50/\text{unit} \quad r = 0.02 \ \$/\$/\text{period}$$

Perform an analysis similar to that of Problem 5.4.

5.7   A Calgary-based company (Piedmont Pipelines) has been awarded a contract to lay a long stretch of pipeline. The planned schedule for the laying of pipe can be considered to be accurately known for the duration of the project (18 months). Because of weather conditions and the availability of labor, the rate of placement of the pipe is not constant throughout the project. The supplier of the pipe offers a discount structure for purchases. Outline an analysis that would assist Piedmont in deciding on the purchase schedule for the piping.

5.8   Consider a population of items having essentially deterministic but time-varying demand patterns. Suppose that one was interested in developing an exchange curve of total average cycle stock (in dollars) versus the total number of replenishments per year as a function of A/r (the latter being unknown, but constant for all the items). An analyst has suggested using the approach of Section 4.10 with

$$D_i = \text{annual demand for item } i$$

Discuss whether or not this is a reasonable suggestion.

5.9   Suppose you were a supplier for a single customer having a deterministic, but time-varying demand pattern. Interestingly enough, the time-varying nature is actually of benefit to you!

   a.   Making use of the Silver–Meal heuristic, indicate how you would decide on the maximum discount that you would be willing to give the customer in return for him/her maintaining his/her current pattern instead of providing a perfectly level demand pattern. Assume that replenishments can take place only at the beginning of periods.

   b.   Illustrate for the numerical example of Problem 6.5. Suppose that the current selling price is $60/unit.

5.10   Consider a company facing a demand pattern and costs as follows:

Month	Sequential Number	Requirements (Units)
January	1	20
February	2	40
March	3	110
April	4	120
May	5	60
June	6	30
July	7	20
August	8	30
September	9	80
October	10	120
November	11	130
December	12	40
	Total	800

A	$25.00
r (per month)	0.05$/$ (carrying costs are very high in this industry)
v	$4.00

Using a "3-month" decision rule, the replenishment schedule and associated costs are as follows:

Month	1	2	3	4	5	6	7	8	9	10	11	12	Total
Starting inventory	0	150	110	0	90	30	0	110	80	0	170	40	
Replenishment	170	0	0	210	0	0	130	0	0	290	0	0	800
Requirements	20	40	110	120	60	30	20	30	80	120	130	40	800
Ending inventory	150	110	0	90	30	0	110	80	0	170	40	0	780

Total replenishment costs	$100.00
Total carrying costs	$156.00
Total replenishment + carrying	$256.00

a. Construct a replenishment schedule and calculate the associated costs using the fixed EOQ method.
b. Repeat using the Wagner–Whitin algorithm.
c. Repeat using the Silver–Meal heuristic.
d. Repeat using the LUC method.
e. Repeat using the PPB method.
f. Repeat using the POQ method.

5.11 Consider a company facing a demand pattern and costs as follows:

Month	Sequential Number	Requirements (Units)
January	1	200
February	2	300
March	3	180
April	4	70
May	5	50
June	6	20
July	7	10
August	8	30
September	9	90
October	10	120
November	11	150
December	12	190
	Total	1,410

A	$160.00
r (per month)	0.10$/$
v	$5.00

    a. Construct a replenishment schedule and calculate the associated costs using the fixed EOQ method.

    b. Repeat using the Wagner–Whitin algorithm.

    c. Repeat using the Silver–Meal heuristic.

    d. Repeat using the LUC method.

    e. Repeat using the PPB method.

    f. Repeat using the POQ method.

5.12 What are we trying to achieve by using the Silver–Meal or any of the other heuristics in this chapter? Note that we are not asking for the criterion used in the Silver–Meal heuristic.

5.13 Consider an item with the following demand pattern:

Period j	1	2	3	4	5
$D(j)$ (units)	50	80	65	100	30

      The initial inventory at the start of period 1 is 0 and a replenishment, to cover an integer number (T) of period, must be placed. Suppose that the use of the Silver–Meal heuristic leads to the choice of $T = 2$. What can you say about the value of the ratio A/vr where r is on a per-period basis? Hint: Use the normalized total relevant costs per unit time by dividing through by vr.

5.14 Develop a spreadsheet for the Wagner–Whitin algorithm as shown in Table 5.4.

    a. Simply perform the calculations on the spreadsheet, finding the optimal solution manually.

    b. Now use the spreadsheet to find the optimal solution.

    c. Repeat parts 5.14a and 5.14b for the Silver–Meal heuristic.

    d. Repeat parts 5.14a and 5.14b for the LUC method.

    e. Repeat parts 5.14a and 5.14b for the PPB method.

5.15 In this chapter, we consider *stocking* problems in which stock is ordered in anticipation of depletion by demand processes. A different kind of problem arises in transportation and distribution systems such as freight railroads, express and delivery services, freight forwarders, etc., in which the inventory accumulates according to supply processes and is depleted by dispatches. Consider such a *dispatching* problem, in which an inventory of dissimilar items accumulates according to a time-varying process. A periodic review is assumed, whereby at the end of each time period, there is an opportunity to dispatch, that is, to reduce the inventory to zero. A cost trade-off similar to the stocking problem is at issue, that is, there are inventory holding costs, and a significant, fixed dispatching cost.

      We suppose the system starts (at the beginning of period 1) with an initially empty inventory (i.e., a dispatch cleared the system at the end of period 0). In the tth period, $t = 1, 2, \ldots$, we let W_t denote the set of items supplied to the inventory whose first opportunity for dispatch is at the end of period t, and $HC(W_t)$ shall denote the cost of holding items W_t in the inventory for one period. (E.g., if the set of items W_t are kept in the inventory for 3 periods before dispatching, the associated holding costs are $3\,HC(W_t)$). We let A denote the fixed cost per dispatching event. A dispatching policy is desired, that is, a determination in each time period of whether to dispatch or not dispatch in that period.

a. Develop a decision procedure, such as the Silver–Meal heuristic, based on minimizing the relevant costs per unit time between dispatch moments.

b. Illustrate your procedure (for just the first two dispatches) with the following example involving the dispatching of freight trains from a switchyard terminal:

At a terminal, freight cars with a particular destination W are switched to departure tracks where they accumulate until a train to destination W is dispatched. The holding costs per hour vary from car to car according to ownership, age, car type, etc., but we suppose that an information system at the terminal allows the calculation of holding costs for a given group of cars. Let W_t denote the set of cars destined to W switched in hour t, and let $HC(W_t)$ denote the hourly holding costs for W_t. We consider the case where a train was dispatched to W just before hour 1, and the switching results of the next several hours are as follows:

Hour, t	1	2	3	4	5	6	7
Holding cost, $HC(W_t)$	220	85	145	210	310	175	300

The fixed cost of dispatching a train to W is $A = 800$.

5.16 Consider an item where the assumptions of Section 5.3.1 are reasonable (including $L = 0$), but with one important exception, namely that there is a significant chance of the perishability of material kept too long in the inventory. Specifically, consider an order that arrives at time 0 (the beginning of period 1). The units of this order can become unusable at the end of each period that they are kept in stock. In particular, it is reasonable to assume that, for any period, there is a fixed chance p that any good inventory at the start of that period, which is still left over at the end of the period, will not be usable for satisfying the demand in future periods. Note that we are assuming that at the end of a period, one of the two things happens: (1) all of the remaining inventory is still usable, or (2) all of the remaining inventory has become worthless (with 0 salvage value). In the latter case, a new replenishment must be immediately placed (remember $L = 0$) to avoid a shortage in the next period. Assume that, if spoilage occurs right at the end of a period, then carrying costs are still incurred for the (now useless) inventory carried over into the next period but that this inventory is discarded before the end of that next period.

a. What should tend to happen to the best value of T as p increases?

b. Develop a heuristic decision rule (in the spirit of the Silver–Meal heuristic) for choosing an appropriate order quantity at time 0. Use the notation consistent with the text. Make sure that your heuristic gives reasonable results for the limiting cases of $p = 0$ and $p = 1$.

c. Illustrate your rule with the following numerical example (but only choose the first-order quantity, the one at time 0). Also, determine the order quantity that would be found by the Silver–Meal heuristic (i.e., assuming $p = 0$) and comment on any difference from the value obtained using the heuristic with $p > 0$.

$$A = \$50, \quad v = \$10/\text{unit}, \quad r = 0.02\$/\$/\text{period}, \quad p = 0.01$$

j	1	2	3	4	5	6...
$D(j)$ units	200	100	80	50	30	100...

5.17 Consider a company facing a demand pattern and costs as follows:

Month	Sequential Number	Requirements (Units)
January	1	350
February	2	200
March	3	0
April	4	150
May	5	500
June	6	600
July	7	450
August	8	350
September	9	200
October	10	0
November	11	150
December	12	200
	Total	3,150

A	$50.00
r (per month)	0.02$/$/month
v	$65.00

a. Construct a replenishment schedule and calculate the associated costs using the fixed EOQ method.
b. Repeat using the Wagner–Whitin algorithm.
c. Repeat using the Silver–Meal heuristic.
d. Repeat using the LUC method.
e. Repeat using the PPB method.
f. Repeat using the POQ method.

5.18 Suppose that there is a finite replenishment rate m (as discussed in Section 4.8).
a. Modify the Silver–Meal heuristic to take account of this fact.

    Note: For simplicity, assume that m is large enough so that, if a replenishment is initiated at the start of period j, then, the demand for period j, $D(j)$, can be met from that replenishment. In addition, assume that it is now necessary to include in the analysis the carrying costs of $D(j)$ during its period of consumption, j. To do this, assume a uniform demand rate during any single period (without the extra cost included, the finite replenishment rate would have no effect on the decision).

b. For the item of Problem 5.4 suppose that $m = 150$ units/month. Now, use the modified heuristic to find the timing and sizes of the replenishments.

c. For this example, what is the percentage cost penalty over the 12-month horizon associated with ignoring the finite m value, that is, using the unmodified heuristic?

Appendix 5A: Dynamic Programming and Linear Programming Formulations

Define the following notation:

$D(j)$ = requirements in period j.
I_j = ending inventory in period j.
$F(j)$ = total costs of the best replenishment strategy that satisfies the demand requirements in periods $1, \ldots, j$.
Q_j = replenishment quantity in period j.

The problem can be formulated as an integer program, and solved quite easily with a spreadsheet (e.g., the Solver in Excel).

$$\text{Minimize TRC} = \sum_{j=1}^{T} [A\delta(A_j) + I_j vr]$$

subject to

$$I_j = I_{j-1} + Q_j - D(j) \quad j = 1, \ldots, T$$
$$Q_j \geq 0 \quad j = 1, \ldots, T$$
$$I_j \geq 0 \quad j = 1, \ldots, T$$

where $\delta(Q_j) = \begin{cases} 0 & \text{if } Q_j = 0 \\ 1 & \text{if } Q_j > 0 \end{cases}$

The equivalent dynamic programming formulation is

$$F(j) = \min \left\{ \min_{1 \leq t < j} \left[a + \sum_{h=t}^{j-1} \sum_{k=h+1}^{j} rvD(k) + F(t-1) \right], A + F(j-1) \right\}$$

where $F(1) = A$, $F(0) = 0$, and $\sum_{h=t}^{j-1} \sum_{k=h+1}^{j} rvD(k)$ represents the inventory carrying cost of periods $t + 1$ through j.

$A + F(j-1)$ represents the case in which a setup is performed in period j, increased by the best solution through period $j - 1$.

Ferreira and Vidal (1984) provide an equivalent linear-programming model.

Let $c_{ij} = \sum_{k=i}^{j-1} rvI_k$ for all $i < j \leq T$.

Define

$$z_{ij} = Q_{ij}/D(j) \text{ for all } i \leq j$$
$$C_{ij} = c_{ij}D(j) \text{ for all } i \leq j$$
$$Y_i = \delta \left(\sum_{j=1}^{T} Q_{ij} \right) = \delta \left(\sum_{j=1}^{T} z_{ij} \right) \text{ for all } i$$

c_{ij} is the cost of producing in period i for the subsequent use in period j

z_{ij} is the fraction of units to be used in period j that should be produced in period i

Now, relax the requirement that Y_i should be 0 or 1 (and write y_i instead of Y_i). The linear-programming model with decision variables y_i and z_{ij} is

$$\min \left(\sum_{j=1}^{T} \sum_{i=1}^{j} C_{ij}z_{ij} + \sum_{i=1}^{T} Ay_i \right)$$

$$\sum_{i=1}^{j} z_{ij} = 1 \text{ for all } j$$

$$y_i - z_{ij} \geq 0 \text{ for all } i \leq j$$

$$z_{ij}, \; y_i \geq 0 \text{ for all } i,j$$

References

Baker, K. R. 1989. Lot-sizing procedures and a standard data set: A reconciliation of the literature. *Journal of Manufacturing and Operations Management 2*(3), 199–221.

Benton, W. C. 1983. Purchase quantity discount procedures and MRP. *Journal of Purchasing and Materials Management 19*(1), 30–34.

Benton, W. C. 1985. Multiple price breaks and alternative purchase lot-sizing procedures in material requirements planning systems. *International Journal of Production Research 23*(5), 1025–1047.

Benton, W. C. 1991. Safety stock and service levels in periodic review inventory systems. *Journal of the Operational Research Society 42*(12), 1087–1095.

Blackburn, J., D. Kropp, and R. Millen. 1986. Comparison of strategies to dampen nervousness in MRP systems. *Management Science 32*(4), 413–429.

Blackburn, J. and H. Kunreuther. 1974. Planning and forecast horizons for the dynamic lot size model with backlogging. *Management Science 21*(3), 215–255.

Blackburn, J. and R. Millen. 1980. Heuristic lot-sizing performance in a rolling-schedule environment. *Decision Sciences 11*, 691–701.

Boe, W. J. and C. Yilmaz. 1983. The incremental order quantity. *Production and Inventory Management 2nd Quarter, 24*, 94–100.

Bookbinder, J. H. and J. Y. Tan. 1984. Two lot-sizing heuristics for the case of deterministic time-varying demands. *International Journal of Operations and Production Management 5*(4), 30–42.

Bose, S., A. Goswami, and K. S. Chaudhuri. 1995. An EOQ model for deteriorating items with linear time-dependent demand rate and shortages under inflation and time discounting. *Journal of the Operational Research Society 46*, 771–782.

Bregman, R. L. 1991. An experimental comparison of MRP purchase discount methods. *Journal of the Operational Research Society 42*(3), 235–245.

Bregman, R. L. and E. A. Silver. 1993. A modification of the Silver–Meal heuristic to handle MRP purchase discount situations. *Journal of the Operational Research Society 44*(7), 717–723.

Brosseau, L. J. A. 1982. An inventory replenishment policy for the case of a linear decreasing trend in demand. *INFOR 20*(3), 252–257.

Brown, R. G. 1977. *Materials Management Systems.* New York: Wiley-Interscience.

Callarman, T. and D. C. Whybark. 1981. Determining purchase quantities for MRP requirements. *Journal of Purchasing and Materials Management 17*(3), 25–30.

Carlson, R., J. Jucker, and D. Kropp. 1979. Less nervous MRP systems: A dynamic economic lot-sizing approach. *Management Science 25*(8), 754–761.

Christoph, O. B. and R. L. La Forge. 1989. The performance of MRP purchase lot-size procedures under actual multiple purchase discount conditions. *Decision Sciences 20*, 348–358.

Chung, C. S., D. T. Chiang, and C. Y. Lu. 1987. An optimal algorithm for the quantity discount problem. *Journal of Operations Management 7*, 165–177.

Coleman, B. J. and M. A. McKnew. 1990. A technique for order placement and sizing. *Journal of Purchasing and Materials Management 26*(2), 32–40.

DeMatta, R. and M. Guignard. 1995. The performance of rolling production schedules in a process industry. *IIE Transactions 27*, 564–573.

DeMatteis, J. 1968. An economic lot-sizing technique I: The part-period algorithm. *IBM Systems Journal 7*(1), 30–38.

Diaby, M. 1995. Optimal setup time reduction for a single product with dynamic demands. *European Journal of Operational Research 85*, 532–540.

Diegel, A. 1966. A linear approach to the dynamic inventory problem. *Management Science 12*(7), 530–540.

Donaldson, W. A. 1977. Inventory replenishment policy for a linear trend in demand—An analytic solution. *Operational Research Quarterly 28*(3ii), 663–670.

Eppen, G. D., F. J. Gould, and B. P. Pashigian. 1969. Extensions of the planning horizon theorem in the dynamic lot size model. *Management Science 15*(5), 268–277.

Federgruen, A. and C.-Y. Lee. 1990. The dynamic lot size model with quantity discount. *Naval Research Logistics 37*, 707–713.

Federgruen, A. and M. Tzur. 1991. A simple forward algorithm to solve general dynamic lot sizing models with n periods in $0(n \log n)$ or $0(n)$ time. *Management Science 37*(8), 909–925.

Federgruen, A. and M. Tzur. 1994. Minimal forecast horizons and a new planning procedure for the general dynamic lot sizing model: Nervousness revisited. *Operations Research 42*, 456–469.

Federgruen, A. and M. Tzur. 1995. Fast solution and detection of minimal forecast horizons in dynamic programs with a single indicator of the future: Applications to dynamic lot-sizing models. *Management Science 41*(5), 874–893.

Ferreira, J. A. S. and R. V. V. Vidal. 1984. Lot sizing algorithms with applications to engineering and economics. *International Journal of Production Research 22*(4), 575–595.

Freeland, J. R. and J. L. J. Colley. 1982. A simple heuristic method for lot-sizing in a time-phased reorder system. *Production and Inventory Management 23*(1), 15–22.

Gorham, T. 1968. Dynamic order quantities. *Production and Inventory Management Journal 9*(1), 75–79.

Goyal, S. K., M. A. Hariga, and A. Alyan. 1996. The trended inventory lot sizing problem with shortages under a new replenishment policy. *Journal of the Operational Research Society 27*, 1286–1295.

Groff, G. 1979. A lot sizing rule for time-phased component demand. *Production and Inventory Management Journal 20*(1), 47–53.

Hariga, M. 1995a. An EOQ model for deteriorating items with shortages and time-varying demand. *Journal of the Operational Research Society 46*, 398–404.

Hariga, M. 1995b. Comparison of heuristic procedures for the inventory replenishment problem with a linear trend in demand. *Computers and Industrial Engineering 28*(2), 245–258.

Hariga, M. 1995c. Effects of inflation and time-value of money on an inventory model with time-dependent demand rate and shortages. *European Journal of Operational Research 81*, 512–520.

Hariga, M. A. 1994. Economic analysis of dynamic inventory models with non-stationary costs and demand. *International Journal of Production Economics 36*, 255–266.

Hariga, M. A. 1996. Optimal EOQ models for deteriorating items with time varying demand. *Journal of the Operational Research Society 47*, 1228–1246.

Heady, R. B. and H. Zhu. 1994. An improved implementation of the Wagner–Whitin algorithm. *Production and Operations Management 3*(1), 55–63.

Hill, R. M. 1995. Inventory models for increasing demand followed by level demand. *Journal of the Operational Research Society 46*, 1250–1259.

Hu, J., C. Munson, and E. A. Silver. 2004. A modified Silver–Meal heuristic for dynamic lot sizing under incremental quantity discounts. *Journal of the Operational Research Society 55*(6), 671–673.

Kaimann, R. A. 1969. E.O.Q. vs. dynamic programming—Which one to use for inventory ordering. *Production and Inventory Management Journal 10*(4), 66–74.

Kang, H.-Y. and A. H. Lee. 2013. A stochastic lot-sizing model with multi-supplier and quantity discounts. *International Journal of Production Research 51*(1), 245–263.

Karni, R. 1981. Maximum part-period gain (MPG)—A lot sizing procedure for unconstrained and constrained requirements planning systems. *Production and Inventory Management Journal 22*(2), 91–98.

Kropp, D., R. Carlson, and J. Jucker. 1983. Heuristic lot-sizing approaches for dealing with MRP system nervousness. *Decision Sciences 14*(2), 156–169.

Kropp, D. and R. C. Carlson. 1984. A lot-sizing algorithm for reducing nervousness in MRP systems. *Management Science 20*(2), 240–244.

Kunreuther, H. and T. Morton. 1973. Planning horizons for production smoothing with deterministic demands: I. *Management Science 20*(1), 110–125.

La Forge, R. L. 1985. MRP lot sizing with multiple purchase discounts. *Computer and Operations Research 12*(6), 579–587.

Lundin, R. and T. Morton. 1975. Planning horizons for the dynamic lot size model: Zabel vs protective procedures and computational results. *Operations Research 23*(4), 711–734.

Mendoza, A. 1968. An economic lot-sizing technique II: Mathematical analysis of the part-period algorithm. *IBM Systems Journal 7*(1), 39–46.

Mitra, A., J. Cox, J. Blackstone, and R. Jesse. 1983. A re-examination of lot sizing procedures for requirements planning systems: Some modified rules. *International Journal of Production Research 21*, 471–478.

Pan, C.-H. 1994. Sensitivity analysis of dynamic lot-sizing heuristics. *OMEGA 22*(3), 251–261.

Pochet, Y. and L. A. Wolsey. 2006. *Production Planning by Mixed Integer Programming*. New York: Springer.

Potamianos, J. and A. J. Orman. 1996. An interactive dynamic inventory-production control system. *Journal of the Operational Research Society 47*, 1017–1028.

Rachamadugu, R. and T. J. Schriber. 1995. Optimal and heuristic policies for lot sizing with learning in set-ups. *Journal of Operations Management, 13*, 229–245.

Sahin, F., A. Narayanan, and E. P. Robinson. 2013. Rolling horizon planning in supply chains: Review, implications and directions for future research. *International Journal of Production Research 51*(18), 5413–5436.

Schulz, T. 2011. A new Silver–Meal based heuristic for the single-item dynamic lot sizing problem with returns and remanufacturing. *International Journal of Production Research 49*(9), 2519–2533.

Silver, E. and J. Miltenburg. 1984. Two modifications of the Silver–Meal lot sizing heuristic. *INFOR 22*(1), 56–68.

Silver, E. A. and H. Meal. 1969. A simple modification of the EOQ for the case of a varying demand rate. *Production and Inventory Management Journal 10*(4, 4th Qtr), 52–65.

Silver, E. A. and H. C. Meal. 1973. A heuristic for selecting lot size quantities for the case of a deterministic time-varying demand rate and discrete opportunities for replenishment. *Production and Inventory Management Journal 2nd Quarter, 14*, 64–74.

Sox, C., L. J. Thomas, and J. McClain. 1997. Coordinating production and inventory to improve service. *Management Science 43*(9), 1189–1197.

Sridharan, V. and L. Berry. 1990. Freezing the master production schedule under demand uncertainty. *Decision Sciences 21*(1), 97–120.

Sridharan, V., W. L. Berry, and V. Udayabhanu. 1987. Freezing the master production schedule under rolling planning horizons. *Management Science 33*(9), 1137–1149.

Teng, J. 1994. A note on inventory replenishment policy for increasing demand. *Journal of the Operational Research Society 45*(11), 1335–1337.

Ting, P. and K. Chung. 1994. Inventory replenishment policy for deteriorating items with a linear trend in demand considering shortages. *International Journal of Operations and Production Management 14*(8), 102–110.

Toklu, B. and J. M. Wilson. 1995. An analysis of multi-level lot-sizing problems with a bottleneck under a rolling schedule environment. *International Journal of Production Research 33*(7), 1835–1847.

Triantaphyllou, E. 1992. A sensitivity analysis of a (ti, Si) inventory policy with increasing demand. *Operations Research Letters 11*(3), 167–172.

Tzur, M. 1996. Learning in setups: Analysis, minimal forecast horizons, and algorithms. *Management Science 42*(12), 1732–1743.

Van den Heuvel, W. and A. P. Wagelmans. 2010. Worst-case analysis for a general class of online lot-sizing heuristics. *Operations Research 58*(1), 59–67.

Vargas, V. 2009. An optimal solution for the stochastic version of the Wagner–Whitin dynamic lot-size model. *European Journal of Operational Research 198*(2), 447–451.

Vidal, R. V. V. 1970. Operations research in production planning. IMSOR, Technical University of Denmark, Lyngby, Denmark, 95–99.

Wagelmans, A., S. van Hoesel, and A. Kolen. 1992. Economic lot sizing: An 0(n log n) algorithm that runs in linear time in the Wagner–Whitin case. *Operations Research 40*(1), S145–S156.

Wagner, H. and T. M. Whitin. 1958. Dynamic problems in the theory of the firm. *Naval Research Logistics Quarterly 5*, 53–74.

Wee, H.-M. 1995. A deterministic lot-size inventory model for deteriorating items with shortages and a declining market. *Computers and Operations Research 22*(3), 345–356.

Wemmerlöv, U. 1981. The ubiquitous EOQ—Its relations to discrete lot sizing heuristics. *International Journal of Operations and Production Management 1*, 161–179.

Wemmerlöv, U. 1983. The part-period balancing algorithm and its look ahead–look back feature: A theoretical and experimental analysis of a single stage lot-sizing procedure. *Journal of Operations Management 4*(1), 23–39.

Zhou, Y. 1996. Optimal production policy for an item with shortages and increasing time-varying demand. *Journal of the Operational Research Society 47*, 1175–1183.

Chapter 6

Individual Items with Probabilistic Demand

In the preceding two chapters, which dealt with the determination of replenishment quantities, the decision rules resulted from analyses that assumed deterministic demand patterns. In several places, we showed that the relevant costs associated with the selection of order quantities were relatively insensitive to inaccuracies in the estimates of the various factors involved. However, the costs of insufficient capacity in the short run—that is, the costs associated with shortages or with averting them—were not included in the analyses. When demand is no longer assumed to be deterministic, these costs assume a much greater importance. Clearly, the assumption of deterministic demand is inappropriate in many production and distribution situations. Therefore, this chapter is devoted to the theme of how to develop control systems to cope with the more realistic case of probabilistic demand. We initially restrict our attention to the case (comparable to Chapter 4) where the *average* demand remains approximately constant with time. We address the more difficult setting of time-varying probabilistic demand in Chapter 7.

The introduction of uncertainty in the demand pattern significantly complicates the inventory situation from a conceptual standpoint. This, together with the plethora of the possible shortage-costing methods or customer-service measures, has made the understanding and acceptance of decision rules for coping with probabilistic demand far less frequent than is merited by the importance of the problem. It is safe to say (as will be illustrated in Section 6.11.2 of this chapter) that in most organizations, an appropriate reallocation of buffer (or safety) stocks (which are kept to meet the unexpected fluctuations in demand) can lead to a significant improvement in the service provided to customers.

It should be emphasized that in this and several succeeding chapters, we are still dealing with a single-stage problem. A control system different from those proposed here is usually in order for a multiechelon situation. We have more to say on this topic in Chapters 11 and 15.

We begin in Section 6.1 with a careful discussion of some important issues and terminology relevant to the case of probabilistic demand. In Section 6.2, we return to the topic of classification of items into A, B, or C categories. In Section 6.3, the important dichotomy of continuous versus periodic review is discussed. This is followed in Section 6.4 by an explanation of the four most common types of control systems. Section 6.5 is concerned with the many different ways of measuring service or costing shortages that may be relevant to any particular stocking situation. Section 6.6 deals with an illustrative selection of the reorder point for one type of continuous-review system (order-point, order-quantity system). This section builds the conceptual foundation for the remainder of the chapter. In Section 6.7, we present a detailed treatment of the order-point, order-quantity system, showing the decision rules for a wide range of methods of measuring service and costing shortages. Section 6.8 addresses the implied performance that is obtained when a reorder point is chosen by a different service or cost objective. Section 6.9 is more briefly devoted to another commonly used control system—namely, the periodic-review, order-up-to-level system. This is followed, in Section 6.10, by a discussion of how to deal with appreciable variability in the replenishment lead time. Finally, Section 6.11 presents exchange curves displaying the total buffer stock plotted against the aggregate measures of service as a function of a policy variable; the policy variable is the numerical value of a particular service level or shortage cost.

6.1 Some Important Issues and Terminology

6.1.1 Different Definitions of Stock Level

When demand is probabilistic, it is useful to conceptually categorize the inventories as follows:

1. *On-hand (OH) stock* This is stock that is physically on the shelf; it can never be negative. This quantity is relevant in determining whether a particular customer demand is satisfied directly from the shelf.
2. *Net stock* is defined by the following equation:

$$Net\ stock = (On\ hand) - (Backorders) \tag{6.1}$$

This quantity can become negative (namely, if there are backorders). It is used in some mathematical derivations and is also a component of the following important definition.
3. *Inventory position* (sometimes also called the *available* stock)[*] The inventory position is defined by the relation

$$Inventory\ position = (On\ hand) + (On\ order)$$
$$- (Backorders) - (Committed) \tag{6.2}$$

The on-order stock is that stock which has been requisitioned but not yet received by the stocking point under consideration. The "committed" quantity in Equation 6.2 is required if such stock cannot be used for other purposes in the short run. As we will see, the inventory position is a key quantity in deciding when to replenish.
4. *Safety stock (SS)* The SS (or buffer) is defined as the average level of the net stock just before a replenishment arrives. If we planned to just run out, on the average, at the moment when the replenishment arrived, the SS would be zero. A positive SS provides a cushion or buffer against larger-than-average demand during the effective replenishment lead time. Throughout this book, we consistently use the definition provided here, although the appropriate method of calculation, and thus the numerical value of the SS, depends on what happens to

[*] We have chosen to not use the words "available stock" because of the incorrect connotation that such stock is immediately available for satisfying customer demands.

demands when there is a stockout. If no backorders are permitted, net stock (and thus SS) is always nonnegative. With backorders, it is possible that SS, by the definition provided here, could be negative.

6.1.2 Backorders versus Lost Sales

What happens to a customer's order when an item is temporarily out of stock is of obvious importance in inventory control. There are two extreme cases:

1. *Complete backordering.* Any demand, when out of stock, is backordered and filled as soon as an adequate-sized replenishment arrives. This situation corresponds to a captive market, common in government organizations (particularly the military) and at the wholesale–retail link of some distribution systems (e.g., exclusive dealerships).
2. *Complete lost sales.* Any demand, when out of stock, is lost; the customer goes elsewhere to satisfy his or her need. This situation is most common at the retail–consumer link. For example, a person is unlikely to backorder a demand for a loaf of bread. We focus in this chapter on single-item inventory control models, but it is worth noting that when multiple items are available for purchase, a customer facing an out-of-stock situation may choose a substitute product. This would clearly be a lost sale for the originally intended item, but may affect the anticipated demand.

In most practical situations, there is a combination of these two extremes, whereas most inventory models have been developed for one or the other of the extremes. Nevertheless, most of these models serve as reasonable approximations because the decisions they yield tend to be relatively insensitive to the degree of backordering possible in a particular situation. This is primarily a consequence of the common use in practice of high customer-service levels; high service levels imply infrequent *stockout* occasions. When we use the term stockout, we mean a stockout occasion or event. The number of units backordered or lost is a measure of the impact of the stockout.

We now show that the numerical value of the SS depends on the degree to which backorders or lost sales occur. Consider a particular replenishment lead time in which a stockout occurs. Under complete backordering, if demand occurs during the stockout, the net stock will be negative just before the next replenishment arrives. On the other hand, if all demands that occur during the stockout are lost, then, the net stock will remain at the zero level throughout the stockout period. In other words, in a cycle when a stockout occurs, the value of the net stock just before the replenishment arrives depends on whether backorders can occur. Because SS is defined to be the *average* net stock just before a replenishment arrives, its numerical value is influenced by whether backordering is actually possible.

6.1.3 Three Key Issues to Be Resolved by a Control System under Probabilistic Demand

The fundamental purpose of a replenishment control system is to resolve the following three issues or problems:

1. How often the inventory status should be determined.
2. When a replenishment order should be placed.
3. How large the replenishment order should be.

Under the conditions of deterministic demand (discussed in the previous two chapters), the first issue is trivial because knowing the inventory status at any one point allows us to calculate it at all points in time (at least out to a reasonable horizon). Furthermore, under deterministic demand, the second problem is answered by placing an order such that it arrives precisely when the inventory level hits some prescribed value (usually set at zero). Finally, under deterministic demand, the use of the EOQ, one of its variations, or one of the procedures discussed in Chapter 5 provides the solution to the third problem.

Under probabilistic demand, the answers are more difficult to obtain. Regarding the first point, it takes resources (labor, computer time, etc.) to determine the inventory status. On the other hand, the less frequently the status is determined, the longer is the period over which the system must protect against unforeseen variations in demand in order to provide the desired customer service. The answer to the second problem rests on a trade-off between the costs of ordering somewhat early (hence, carrying extra stock) and the costs (implicit or explicit) of providing inadequate customer service. The factors relevant in answering the third problem are similar to those discussed in the derivation of the basic EOQ, except that under some service criteria, specified by the management, there is an interaction; the answer to the question "When to replenish?" may be affected by the replenishment quantity used.

To respond to these three fundamental issues, a manager must determine several things. We pose these as four questions that managers can use to systematically establish inventory policies.

1. How important is the item?
2. Can, or should, the stock status be reviewed continuously or periodically?
3. What form should the inventory policy take?
4. What specific cost or service objectives should be set?

The next four sections address these four questions in turn.

6.2 The Importance of the Item: A, B, and C Classification

Managers must first establish how critical the item under consideration is to the firm. As mentioned in Chapter 2, many firms segment their items into the categories A, B, or C. Recall that A items make up roughly 20% of the total number of items, but represent 80% of the dollar sales volume. B items comprise roughly 30% of the items, but represent 15% of the dollar volume; C items comprise roughly 50% of the items, and represent only 5% of the dollar volume. Firms often include slow-moving and inexpensive items in the A category if the items are critical to the business. It is possible that including many slow-moving or inexpensive items in the A category could shift the typical 80–20 distribution, but our experience is that this is rarely the case.

The importance of the item helps direct the response to the remaining three questions. The remainder of this chapter is devoted to B items. Chapter 7 covers A items, and Chapter 8 addresses C items.

6.3 Continuous versus Periodic Review

The answer to the problem of how often the inventory status should be determined specifies the review interval (*R*), which is the time that elapses between two consecutive moments at which we know the stock level. An extreme case is where there is continuous review; that is, the stock status is always known. In reality, continuous surveillance is usually not required; instead, each transaction (shipment, receipt, demand, etc.) triggers an immediate updating of the status. Consequently, this type of control is often called transactions reporting. Point-of-sale (POS) data collection systems (involving electronic scanners), which permit transactions reporting, are having a profound impact at the retail level. With periodic review, as the name implies, the stock status is determined only every R time units; between the moments of review, there may be considerable uncertainty as to the value of the stock level. Even with POS scanning, inventory decisions are often made periodically (e.g., end of the day). Thus, from the perspective of the inventory reordering decision, the review period, R, is 1 day.

A common example of periodic-review systems is the soda machine in a dormitory. The driver comes regularly, say once every week, to refill the machine. If the machine stocks out of an item between visits, no action is taken. As another example, note that some small firms assign one person to review inventory and make purchase decisions. Often, this person is so busy that he or she can allocate only one afternoon each week to perform these tasks. As these examples show, physical constraints or limits on time often determine the review interval.

We now comment on the advantages and disadvantages of continuous and periodic review. Items may be produced on the same piece of equipment, purchased from the same supplier, or shipped in the same transportation mode. In any of these situations, coordination of replenishments may be attractive. In such a case, periodic review is particularly appealing because all items in a coordinated group can be given the same review interval (e.g., all items purchased from a particular supplier might be scheduled for review every Thursday). Periodic review also allows a reasonable prediction of the level of the workload on the staff involved. In contrast, under continuous review, a replenishment decision can be made at practically any moment in time; hence the load is less predictable. A rhythmic, rather than random, pattern is usually appealing to the staff.

Another disadvantage of continuous review is that it is generally more expensive in terms of reviewing costs and reviewing errors. This is particularly true for fast-moving items where there are many transactions per unit of time. Today, POS data collection systems have dramatically reduced reviewing costs and errors. For extremely slow-moving items, very little reviewing costs are incurred by continuous review because updates are only made when a transaction occurs. On the other hand, we have the anomalous condition that periodic review may be more effective than continuous review in detecting the spoilage (or pilferage) of such slow-moving items; periodic review forces an occasional review of the situation, whereas, in transactions reporting, no automatic review will take place without a transaction occurring.

The major advantage of continuous review is that, to provide the same level of customer service, it requires less SS (hence, lower carrying costs) than does periodic review. This is because the period over which safety protection is required is longer under periodic review (the stock level has the opportunity to drop appreciably between review instants without any reordering action being possible in the interim).

6.4 The Form of the Inventory Policy: Four Types of Control Systems

Once the manager has determined whether the item falls in the A, B, or C category, and he or she has settled the question of continuous versus periodic review, it is time to specify the *form* of the inventory control policy. The form of the inventory policy will begin to resolve the second and third issues: *When* should an order be placed and *what quantity* should be ordered.

There are a number of possible control systems. The physical operation of the four most common ones will be described in the next section, along with a brief discussion of the advantages and disadvantages of each. Our discussion of the advantages and disadvantages will be rather general in that they are dependent on the specific environment in which the systems are to be implemented.

6.4.1 Order-Point, Order-Quantity (s, Q) System

This is a continuous-review system (i.e., $R = 0$). A fixed quantity Q is ordered whenever the inventory position drops to the reorder point s or lower. Note that the inventory position, and not the net stock, is used to trigger an order. The inventory position, because it includes the on-order stock, takes proper account of the material requested but not yet received from the supplier. In contrast, if net stock was used for ordering purposes, we might unnecessarily place another order today even though a large shipment was due in tomorrow. A good example of ordering on the basis of inventory position is the way a person takes medicine to relieve a headache. After taking the medicine, it is not necessary to take more every 5 minutes until the headache goes away. Rather, it is understood that the relief is *on order*.

The (s, Q) system is often called a two-bin system because one physical form of implementation is to have two bins for storage of an item. As long as units remain in the first bin, demand is satisfied from it. The amount in the second bin corresponds to the order point. Hence, when this second bin is opened, a replenishment is triggered. When the replenishment arrives, the second bin is refilled and the remainder is put into the first bin. It should be noted that the physical two-bin system will operate properly only when no more than one replenishment order is outstanding at any point in time. Thus, to use the system, it may be necessary to adjust Q upward so that it is appreciably larger than the average demand during a lead time.

The advantages of the fixed order-quantity (s, Q) system include that it is quite simple, particularly in the two-bin form, for the stock clerk to understand, that errors are less likely to occur, and that the production requirements for the supplier are predictable. The primary disadvantage of an (s, Q) system is that in its unmodified form, it may not be able to effectively cope with the situation where individual transactions are large; in particular, if the transaction that triggers the replenishment in an (s, Q) system is large enough, then, a replenishment of size Q won't even raise the inventory position above the reorder point. (As a numerical illustration, consider a Q value of 10 together with a demand transaction of size 15 occurring when the position is just 1 unit above s.) Of course, in such a situation, one could instead order an integer multiple of Q where the integer was large enough to raise the inventory position above s.

6.4.2 Order-Point, Order-Up-to-Level (s, S) System

This system again assumes continuous review; and, like the (s, Q) system, a replenishment is made whenever the inventory position drops to the order point s or lower. However, in contrast to the (s, Q) system, a variable replenishment quantity is used, ordering enough to raise the inventory position to the order-up-to-level S. If all demand transactions are unit sized, the two systems are identical because the replenishment requisition will always be made when the inventory position is exactly at s; that is, $S = s + Q$. If transactions can be larger than unit size, the replenishment quantity in the (s, S) system becomes variable. The difference between (s, Q) and (s, S) systems is illustrated in Figure 6.1. The (s, S) system is frequently referred to as a min-max system because

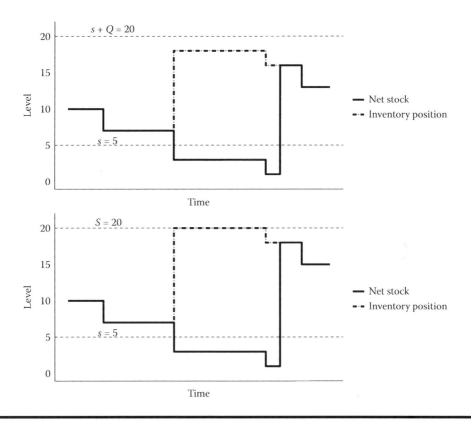

Figure 6.1   Two types of continuous-review systems.

the inventory position, except for a possible momentary drop below the reorder point, is always between a minimum value of *s* and a maximum value of *S*.

The best (s, S) system can be shown to have the total costs of replenishment, carrying inventory, and shortage that are not larger than those of the best (s, Q) system. However, the computational effort to find the best (s, S) pair is substantially greater. Thus, the (s, Q) may be the better choice, except perhaps when dealing with an item where the potential savings are appreciable (i.e., an A item). It is interesting that (s, S) systems are frequently encountered in practice. However, the values of the control parameters are usually set in a rather arbitrary fashion. For B items (and even most A items), mathematical optimality does not make sense; instead, we need a fairly simple way of obtaining reasonable values of *s* and *S*. This will be discussed further in the next chapter, which deals with A items. One disadvantage of the (s, S) system is the variable order quantity. Suppliers could make errors more frequently; and they certainly prefer the predictability of a fixed order quantity, particularly if the predetermined lot size is convenient from a packaging or handling standpoint (e.g., pallet, container, or truckload.)

6.4.3 Periodic-Review, Order-Up-to-Level (R, S) System

This system, also known as a replenishment cycle system, is in common use, particularly in companies without sophisticated computer control. It is also frequently seen when items are ordered from the same supplier, or require resource sharing. The control procedure is that every *R* units of

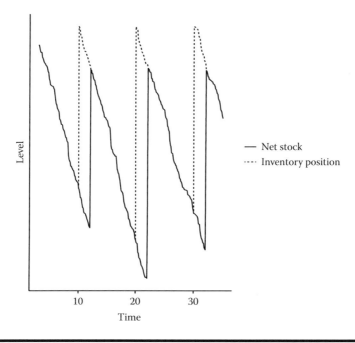

Figure 6.2   The (R, S) system. Orders placed every 10 periods. Lead time of 2 periods.

time (i.e., at each review instant) enough is ordered to raise the inventory position to the level S. A typical behavior of this type of system is shown in Figure 6.2.

Because of the periodic-review property, this system is much preferred to order point systems in terms of coordinating the replenishments of related items. For example, when ordering from overseas, it is often necessary to fill a shipping container to keep shipping costs under control. The coordination afforded by a periodic-review system can provide significant savings. In addition, the (R, S) system offers a regular opportunity (every R units of time) to adjust the order-up-to-level S, a desirable property if the demand pattern is changing with time. The main disadvantages of the (R, S) system are that the replenishment quantities vary and that the carrying costs are higher than in continuous-review systems.

6.4.4 (R, s, S) System

This is a combination of (s, S) and (R, S) systems. The idea is that every R units of time we check the inventory position. If it is at or below the reorder point s, we order enough to raise it to S. If the position is above s, nothing is done until at least the next review instant. The (s, S) system is the special case where $R = 0$, and the (R, S) is the special case where $s = S - 1$. Alternatively, one can think of the (R, s, S) system as a periodic version of the (s, S) system. Also, the (R, S) situation can be viewed as a periodic implementation of (s, S) with $s = S - 1$.

It has been shown (see, e.g., Scarf 1960) that, under quite general assumptions concerning the demand pattern and the cost factors involved, the best (R, s, S) system produces a lower total of replenishment, carrying, and shortage costs than does any other system. However, the computational effort to obtain the best values of the three control parameters is more intense than that

Table 6.1 Rules of Thumb for Selecting the Form of the Inventory Policy

	Continuous Review	*Periodic Review*
A items	(s, S)	(R, s, S)
B items	(s, Q)	(R, S)

for other systems, certainly for class B items. Therefore, for such items, simplified methods often are used to find reasonable values (again, a topic we discuss in the next chapter). This system is also more difficult to understand than some of the previously mentioned systems. (R, s, S) systems are found in practice where R is selected largely for convenience (e.g., 1 day) even when POS equipment permits continuous review of the inventory position.

Table 6.1 gives a simple rule of thumb for choosing the form of the policy. For C items, firms generally use a more manual and simple approach (which can be equivalent to simple (s, Q) or (R, S) systems). Less effort is devoted to inventory management because the savings available are quite small.

6.5 Specific Cost and Service Objectives

When demand (or delivery capability) is probabilistic, there is a definite chance of not being able to satisfy some of the demand on a routine basis directly out of stock. If demand is unusually large, a stockout may occur or emergency actions may be required to avoid a stockout. On the other hand, if demand is lower than anticipated, the replenishment arrives earlier than needed and inventory is carried. Managers have different perspectives on how to balance these two types of risks. There are four possible methods of modeling these management perspectives to arrive at appropriate decision rules. The choice among these should be consistent with the customers' perceptions of what is important. For example, Kumar and Sharman (1992) describe a firm that sets different service objectives for different items based on cost—higher cost items had higher service targets. Their customers held more inventory of lower cost items, allowing the firm to adjust its policies accordingly. It is critical, however, to recognize that a stockout of a $1 item could shut down a customer's operation.

1. *SSs Established through the Use of a Simple-Minded Approach.* These approaches typically assign a common safety factor or a common time supply as the SS of each item. Although more easily understood than most of the other procedures to be discussed, we will find that there is a logical flaw in the use of these methods.
2. *SSs Based on Minimizing Cost.* These approaches involve specifying (explicitly or implicitly) a way of costing a shortage and then minimizing the total cost. For instance, air freight may be used to meet a customer demand, even though it costs more. Holding more inventory reduces the probability that air freight will be required, but increases the inventory holding cost. The cost-minimization approach trades off these costs to find the lowest cost policy. Of the many alternative ways of costing a shortage, several will be illustrated shortly. As discussed in Chapter 2, it is important to include only the relevant costs.
3. *SSs Based on Customer Service.* Recognizing the severe difficulties associated with costing shortages, an alternative approach is to introduce a control parameter known as the service

level. The service level becomes a constraint in establishing the SS of an item; for example, it might be possible to minimize the carrying costs of an item subject to satisfying, routinely from stock, 95% of all demands. Again, there is considerable choice in the selection of a service measure. In fact, often in practice, an inventory manager, when queried, may say that company policy is to provide a certain level of service (e.g., 95%) and yet not be able to articulate exactly what is meant by service. Later, we illustrate several of the more common service measures.

4. *SSs Based on Aggregate Considerations.* The idea of this general approach is to establish the SSs of individual items, using a given budget, to provide the best possible aggregate service across a population of items. Equivalently, the selection of the individual SSs is meant to keep the total investment in stocks as low as possible while meeting a desired aggregate service level.

6.5.1 Choosing the Best Approach

Unfortunately, there are no hard-and-fast rules for selecting the appropriate approach and/or measure of service. Which one to use depends on the competitive environment of the particular company. For instance, if the products are in the mature phase of the product life cycle, competition is often based on cost and delivery performance. The presence of many competitors suggests that the cost of a lost sale is simply the lost contribution margin of that sale. Knowing the explicit cost of a shortage allows the firm to minimize inventory costs (or maximize profit). On the other hand, if the products are relatively new to the market, delivery performance may have significant implications for capturing market share; cost is less important. Managers in these cases tend to set stringent service targets. Other factors that can influence this decision include the amount of substitutability of products (i.e., standardized vs. customized products), whether the products are purchased or manufactured, and how the company actually reacts to an impending shortage—for example, does it expedite? In every case, managers should focus on customer needs and desires. Many firms formulate service objectives and work hard to meet them, only to discover that their service objectives are not consistent with what their customers' desire.

Later, we will see equivalencies between certain service measures and methods of costing shortages (a further reference on this topic is Boylan and Johnston 1994). Moreover, it is quite possible that different types of shortage penalties or service measures are appropriate for different classes of items within the same organization.

Our treatment in this section is not meant to be exhaustive; rather, we wish to present the more common measures used in establishing SSs. Because of the wide variety of options for establishing SSs, we provide a summary guide in Table 6.2. We shall show later in the chapter that some of the criteria listed in Table 6.2 are similar to one another (e.g., P_2 and B_3; B_2 and TBS), or equivalent (e.g., B_1 and expected total stockout occasions per year [ETSOPY]; B_2, expected total value of shortages per year [ETVSPY], and TBS). As we have said elsewhere in this chapter, it is not necessary precisely to prescribe the numerical value of a shortage cost or service level.

Finally, the reader should now be aware that maximizing turnover (i.e., minimizing the level of inventories), in itself, is an inadequate criterion for selecting SSs in that it does not take into account of the impact of shortages.

6.5.2 SSs Established through the Use of a Simple-Minded Approach

We illustrate with two of the most frequently used approaches.

Table 6.2 Summary of Different Methods of Selecting the SSs in Control Systems under Probabilistic Demand

Criterion	Discussed in Section	Sections in Which Decision Rules for (s, Q) System Can Be Found[a]
Equal time supplies	6.5.2	–
Fixed safety factor	6.5.2	6.7.4
Cost (B_1) per stockout occasion	6.5.3	6.7.5
Fractional charge (B_2) per unit short	6.5.3	6.7.6
Fractional charge (B_3) per unit short per unit time	6.5.3	6.7.7
Cost (B_4) per customer line item backordered	6.5.3	6.7.8
Specified probability (P_1) of no stockout per replenishment cycle	6.5.4	6.7.9
Specified fraction (P_2) of demand to be satisfied directly from the shelf (i.e., the fill rate)	6.5.4	6.7.10
Specified ready rate (P_3)	6.5.4	–
Specified average time (TBS) between stockout occasions	6.5.4	6.7.11
Minimization of ETSOPY subject to a specified total safety stock (TSS)	6.5.5	6.7.12
Minimization of expected total value short per year (ETVSPY) subject to a specified TSS	6.5.5	6.7.13

[a] As will be shown in Section 6.9, the decision rules for (R, S) systems are easily obtained from those for (s, Q) systems.

6.5.2.1 Equal Time Supplies

This is a simple, commonly used approach. In fact, a large U.S.-based international consulting firm estimates that 80%–90% of its clients use this approach for setting SS levels. The SSs of a broad group of (if not all) items in an inventory population are set equal to the same time supply; for example, reorder any item when its inventory position minus the forecasted lead time demand drops to a 2-month supply or lower. This approach is seriously in error because it fails to take into account of the difference in the uncertainty of forecasts from item to item. In applying this policy to a number of items, the policy variable is the common number of time periods of supply. In other words, all items in the group have SS set to the same number of periods of supply. One of the authors has worked with a fastener manufacturer that was required by a major U.S. automotive assembler to hold 2 weeks of forecasted demand as SS. This policy worked reasonably well for the assembler, but did not prevent stockouts of items with highly variable demand, and it certainly did not optimize inventory costs.

6.5.2.2 Equal Safety Factors

As we will see later, it is convenient to define the safety stock as the product of two factors as follows:

$$SS = k\sigma_L \tag{6.3}$$

where k is called the safety factor and σ_L, as first defined in Chapter 3, is the standard deviation of the errors of forecasts of total demand over a period of duration L (the replenishment lead time). The equal-safety factors approach uses a common value of k for a broad range of items.

6.5.3 SSs Based on Minimizing Cost

We present four illustrative cases.

6.5.3.1 Specified Fixed Cost (B₁) per Stockout Occasion

Here, it is assumed that the only cost associated with a stockout occasion is a fixed value B_1, independent of the magnitude or duration of the stockout. This cost arises when a firm expedites to avert an impending stockout.

6.5.3.2 Specified Fractional Charge (B₂) per Unit Short

Here, one assumes that a fraction B_2 of unit value is charged per unit short; that is, the cost per unit short of an item i is $B_2 v_i$, where v_i is the unit variable cost of the item. A situation where this type of costing would be appropriate is where units short are made during overtime production (costing a per unit premium).

6.5.3.3 Specified Fractional Charge (B₃) per Unit Short per Unit Time

The assumption here is that there is a charge B_3 per dollar short (equivalently, $B_3 v$ per unit short) per unit time. An example would be where the item under consideration is a spare part and each unit short results in a machine being idled (with the idle time being equal to the duration of the shortage).

6.5.3.4 Specified Charge (B₄) per Customer Line Item Short

In this case, there is a charge B_4 for each customer line item backordered. A customer line item is one of perhaps many lines of a customer's order. Therefore, if a customer orders 10 items, but only nine are in stock, the supplier is short of one customer line item. The assumption here is that there is a charge B_4 per line item short. Many firms impose such penalty payments on their suppliers through a service-level agreement. We revisit the implications of such service-level agreements in Chapter 12.

6.5.4 SSs Based on Customer Service

The following are among the more common measures of service.

6.5.4.1 Specified Probability (P₁) of No Stockout per Replenishment Cycle: Cycle Service Level

Equivalently, P_1 is the fraction of cycles in which a stockout does not occur. A stockout is defined as an occasion when the OH stock drops to the zero level. As we will see later, using a common P_1 across a group of items, is equivalent to using a common safety factor k. P_1 service is often called the *cycle service level*.

6.5.4.2 Specified Fraction (P₂) of Demand to Be Satisfied Routinely from Available Inventory: Fill Rate

The fill rate is the fraction of customer demand that is met routinely; that is, without backorders or lost sales. This form of service has considerable appeal to practitioners (particularly where a significant portion of the replenishment lead time is unalterable; e.g., a branch warehouse where the major part of the lead time is the transit time by rail car). It can be shown under certain conditions that the use of the B_3 shortage-costing measure leads to a decision rule equivalent to that for the P_2 service measure, where the equivalence is given by the relation

$$P_2 = \frac{B_3}{B_3 + r} \tag{6.4}$$

with r, as earlier, being the carrying charge.[*]

6.5.4.3 Specified Fraction of Time (P₃) during Which Net Stock Is Positive: Ready Rate

The ready rate is the fraction of time during which the net stock is positive; that is, there is some stock on the shelf. The ready rate finds common application in the case of equipment used for emergency purposes (e.g., military hardware or bank computer system). Under Poisson demand, this measure is equivalent with the P_2 measure. Solving for the optimal inventory policy in this case is more complex in general than for other service measures, and will not be pursued in this chapter. See Schneider (1981) for a discussion of the ready rate and other service measures. Hopp et al. (1999) employ this service measure, using approximations similar to those developed in this chapter, for a distribution center that stocks 30,000 parts.

6.5.4.4 Specified Average Time (TBS) between Stockout Occasions

Equivalently, one could use the reciprocal of TBS, which represents the desired average number of stockout occasions per year. If each stockout occasion is dealt with by an expediting action, then, a specific TBS value can be selected to result in a tolerable number of expediting actions.

Boylan and Johnston (1994) define other service measures, including a fraction of order lines filled, and illustrate the relationships among them. See also Aardal et al. (1989). A generalization of the fill rate is the fraction of demand that must be met within a specified time (see Problem 6.8 as well as Van der Veen 1981, 1984). Another generalization is the fraction of demand met in each

---

[*] This relationship holds when all unmet demand is backlogged and specific assumptions are made about how demand occurs over time. See Rudi et al. (2009).

period, rather than as a long-run average. See Chen and Krass (1995). Song (1998) investigates the order fill rate, or the probability that an entire customer order, composed of possibly more than one item, is met from the shelf. See also Ernst and Pyke (1992). Rudi et al. (2009) investigate how the method of inventory accounting (continuous-time vs. end-of-period accounting) affects the appropriate service-level choice.

6.5.5 SSs Based on Aggregate Considerations

We present two common methods of establishing SSs based on aggregate considerations. In both cases, rather than just aggregating cost or service effects across individual items, the item-level measure can be weighted by a factor known as the essentiality of the item. For example, if a particular item was deemed twice as important as another from the perspective of inventory availability, then it essentiality would be double that of the other item.

6.5.5.1 Allocation of a Given Total Safety Stock among Items to Minimize the ETSOPY

As will be shown in Section 6A.4 of the Appendix of this chapter, allocating a fixed total safety stock (TSS) among several items to minimize the expected total number of stockout occasions per year leads to a decision rule for selecting the safety factor of each item based on B_1. In fact, the decision rule is identical with that obtained by assuming a value (the same for all items) of B_1, and then selecting the safety factor to keep the total of carrying and stockout costs as low as possible. This allocation interpretation is probably more appealing to management.

6.5.5.2 Allocation of a Given TSS among Items to Minimize the ETVSPY

It can be shown that allocating a fixed TSS among a group of items to minimize the ETVSPY (in dollars) leads to a decision rule for selecting the safety factor that is identical with either

1. The one obtained by assuming a value (the same for all items) of the factor B_2 and then selecting the safety factor for each item to minimize the total of carrying and shortage costs.
2. The one obtained by specifying the same average time (TBS) between stockout occasions for every item in the group.

Again, this aggregate view of allocating a limited resource may be considerably more appealing to management than attempting to explicitly ascertain a B_2 value or the somewhat arbitrary specification of a TBS value.

6.6 Two Examples of Finding the Reorder Point *s* in a Continuous-Review, Order-Point, Order-Quantity (*s*, *Q*) System

We now present a simple illustration of the use of SS to protect against a stockout over a replenishment lead time. Then, we use an empirical distribution to build intuition about the operation of an (*s*, *Q*) system. In Section 6.7, we will present decision rules for a broad range of service measures and methods of costing shortages.

6.6.1 Protection over the Replenishment Lead Time

In a continuous-review system, a replenishment action can be taken immediately after any demand transaction. Once we place an order, a replenishment lead time (L) elapses before the order is available for satisfying customer demands. Therefore, we want to place an order when the inventory is still adequate to protect us over a replenishment lead time. If the order is placed when the inventory position is at exactly the reorder point s, then, a stockout will not occur by the end of the lead time if and only if the total demand during the lead time is less than the reorder point. If the expected demand over the lead time is exactly equal to s, and the lead time demand distribution is symmetric, we would expect to stock out in half of all replenishment cycles. If s is larger than the expected lead time demand, we will stock out less often, but we will carry more inventory. Because SS is defined as the average level of net stock just before a replenishment order arrives, SS will be positive if s is larger than the expected lead time demand. This is illustrated in Figure 6.3, where, for simplicity in presentation, we have assumed that at most one replenishment order is outstanding at any moment. In the figure, the reorder point is 40 units. In the first lead time shown, the total demand is 18 so that the net stock just before the replenishment arrives is $(40 - 18)$ or 22; that is, no stockout occurs. In contrast, in the second lead time, the total demand is 47, resulting in a total backorder of $(47 - 40)$ or 7 units when the replenishment arrives.

Note that above, we assumed that the replenishment action is taken when the stock level is exactly at the reorder point s. Even for the situation of continuous review, this may not necessarily be the case. To illustrate, suppose at a certain moment the stock level for the item in Figure 6.3 was 42 units and a demand transaction of 5 units occurred. This would immediately drive the inventory level to 37 units, which is an undershoot of the reorder point by 3 units. It is evident that, strictly speaking, no stockout occurs if and only if the sum of the undershoot and the total demand in

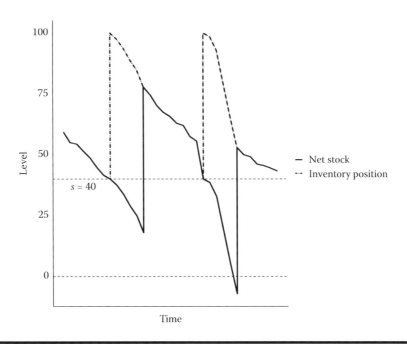

Figure 6.3   The occurrence of a stockout in an (s, Q) system.

the replenishment lead time is less than the reorder point. Except where explicitly noted otherwise, we make the simplifying assumption that the undershoots are small enough to be neglected. We explicitly consider undershoots in the next chapter.

This, and the subsequent, discussion assumes that the manager will not take expediting action when faced with an impending stockout. In other words, the replenishment lead time is fixed and cannot be changed. In many cases, however, it is possible to expedite some or all of the replenishment order. If the cost of expediting is low relative to the cost of carrying the inventory, managers should carry less inventory and expedite more often. Likewise, if the cost of expediting is low relative to the cost of stocking out, managers should provide very high service by expediting when necessary. When expediting is expensive, however, more inventory should be carried, and it may be optimal to provide somewhat lower service. See Moinzadeh and Nahmias (1988) for extensions to the standard (s, Q) policy for this case. Also, see Moinzadeh and Schmidt (1991) for the case in which $Q = 1$. Jain et al. (2010, 2011) investigate the impact of the capability to expedite on continuous- and periodic-review models, respectively, and Veeraraghavan and Scheller-Wolf (2008) develop an easily implementable policy for a periodic review inventory system with two modes of supply.

6.6.2 An Example Using a Discrete Distribution

To develop intuition, we now present a stylized example using a discrete demand distribution. We will formalize the treatment of some of these issues in the next section, where we present a more realistic demand distribution. In answering the four questions introduced in Section 6.1, assume that we have a B item, and that it is possible to review the inventory status continuously. Based on Table 6.1, we choose an (s, Q) policy. The lead time demand distribution is presented in Table 6.3.

Assume that the lead time is 1 week, and that the order quantity is 20 units. We can see from the demand distribution that the expected demand per week (or per lead time) is 2.2 units.[*] Thus, if the firm operates for 50 weeks per year, the annual demand is $50(2.2) = 110$ units.

Let us also assume that the firm has established that the fixed cost per order, A, is \$18, and the cost per unit of inventory held per year is \$10. Furthermore, assume that all unmet demand is lost rather than backordered. This will simplify our example as net stock for this setting must

Table 6.3   Lead Time Demand

Lead Time Demand	Fraction of Occurrences at This Value	Cumulative Fraction at This Value or Lower
0	0.1	0.1
1	0.2	0.3
2	0.3	0.6
3	0.2	0.8
4	0.2	1

---

[*] The expected value of a discrete, nonnegative random variable, X, is $\sum_{x=0}^{\infty} x \Pr(X = x)$, where $\Pr(X = x)$ is the probability that X takes on the value x.

be nonnegative. If we wanted to achieve a cycle service level, P_1, of 90%, it is evident from the lead time demand distribution that a reorder point of 4 is necessary to be certain that no stockout occurs in (at least) 90% of replenishment cycles. A reorder point of 3 would provide a P_1 of only 80%. Thus, if the P_1 requirement were, say, 75%, a reorder point of 3 would be sufficient.

Now, let us consider in more detail a shortage-costing method using the cost per unit short, B_2v where the firm has established that B_2v is $20. The total annual cost is comprised of ordering cost, holding cost, and shortage cost.

6.6.2.1 Ordering Cost

First, let us determine how many orders are placed each year. In a similar fashion to Chapter 4, the number of orders (or replenishment cycles) per year is

$$\text{Annual demand/Order size} = 110/20 = 5.5 \text{ orders per year}$$

Because the cost per order is $18, the annual fixed cost of ordering is $18(5.5) = $99. See Section 6A.6 of the Appendix to the chapter for a more technical discussion on finding the expected costs.

6.6.2.2 Holding Cost

To determine the holding cost, we refer to aspects of the development of the EOQ from Chapter 4. Recall that the inventory level follows the familiar sawtooth of Figure 6.4. The average inventory OH, in that simple case, is $Q/2$. Now, when we add the possibility of SS and probabilistic demand,

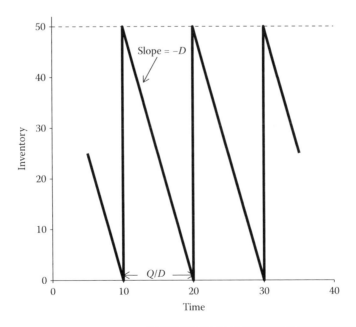

Figure 6.4   Behavior of inventory level with time.

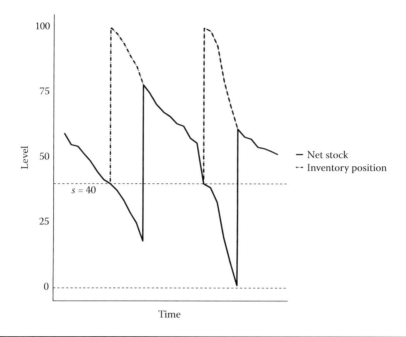

Figure 6.5   Behavior of inventory level with time: probabilistic demand.

the picture looks more like Figure 6.5. Intuitively, imagine the sawtooth of Figure 6.4 floating on a "buffer"—which we call SS. Hence, we can informally suggest that the average inventory OH is $Q/2$ + safety stock. (We shall present a more formal treatment in the next section.)

Now we know that $Q/2$ is $20/2 = 10$. But what is the SS? For the moment, let the reorder point, s, be 2. Then when lead time demand is 0, we will have 2 units left when the replenishment order arrives. When lead time demand is 1 unit, we will have 1 unit left. When lead time demand is 2, 3, or 4, there is no OH inventory when the replenishment arrives. Since no backorders are allowed, OH inventory and net stock are the same. Table 6.4 fills out the details.

Thus, the safety stock, or the expected amount OH (or net stock) just before the replenishment arrives, is 0.4 units $((0.1)(2) + (0.2)(1))$. So the expected annual holding cost is

$$(Q/2 + SS)(\$10 \text{ per unit}) = (10.4)(\$10) = \$104$$

Table 6.4   Safety Stock

Lead Time Demand	Probability	Amount OH Just before Replenishment Arrives
0	0.1	2
1	0.2	1
2	0.3	0
3	0.2	0
4	0.2	0

Note that we could express the expected SS as

$$\sum_{x=0}^{s-1}(s-x)\Pr(X=x) \tag{6.5}$$

if X is the lead time demand.

6.6.2.3 Shortage Cost

To determine the shortage cost, we need to establish how many units we would expect to be short in each replenishment cycle; and we follow an approach similar to that for the holding cost. If s is 2, and lead time demand is 0, there will be no shortages. This occurs with probability 0.1. If demand is 3, say, we will be one unit short. Again, a simple table describes the events (Table 6.5).

Notice that when lead time demand is 2, and the reorder point is 2, there are no shortages. No customers are turned away without having their demand satisfied, even though there is no inventory left when the replenishment order arrives. The expected number of units short in a cycle is therefore 0.6 units. Because there are $110/20 = 5.5$ cycles per year, and the cost per unit short is \$20, the expected annual shortage cost is

(Expected number of units short per cycle) (Number of cycles per year)
(Cost per unit short) $= (0.6)(5.5)(\$20) = \$66$

Note that we could express expected units short as

$$\sum_{x=s+1}^{\infty}(x-s)\Pr(X=x) \tag{6.6}$$

6.6.2.4 Total Cost

The total annual cost is therefore $\$99 + \$104 + \$66 = \$269$. We can now search for the optimal value of s. These equations and search can be easily implemented on a spreadsheet. The results are given in Table 6.6. We can see that the optimal reorder point is 4.

There are several weaknesses of the simple procedure illustrated here:

Table 6.5   Shortages

Lead Time Demand	Probability	Units Short
0	0.1	0
1	0.2	0
2	0.3	0
3	0.2	1
4	0.2	2

Table 6.6   Optimal Reorder Point

Reorder Point	Ordering Cost	Shortage Cost	Inventory Cost	Total Cost
0	$99	$242	$100	$441
1	$99	$143	$101	$343
2	$99	$66	$104	$269
3	$99	$22	$110	$231
4	$99	$0	$118	$217
5	$99	$0	$128	$227

1. Demand information is rarely presented in the manner described in this section. If it is, there is a good chance that the right tail of the distribution (extremely high demand events) will not be adequately represented, unless we have information from a considerable number of historical lead times. Because service levels are often high, it is this right tail that is the most important. Thus, this simple procedure may not be accurate. (We will have more to say about a similar procedure in Section 6.10.)

2. If the replenishment lead time changes, for some reason, the current information about lead time demand would be of limited value. (We will have more to say about changing the givens in Section 6.7.1.)

To help overcome these weaknesses, it is helpful to fit a member of a family of probability distributions to the available data. In the next section, we do this for the so-called normal probability distribution. Nevertheless, the basic ideas presented in this section are the conceptual foundation for the more formal and extensive treatment presented next.

6.7 Decision Rules for Continuous-Review, Order-Point, Order-Quantity (s, Q) Control Systems

As discussed in Section 6.5, there is a wide choice of criteria for establishing SSs. The choice of a criterion is a strategic decision, to be executed on a relatively infrequent basis and involving senior management directly (the exchange curves of Section 6.11 will be helpful in this regard). Once a criterion (and the implied associated decision rule) is chosen, there is then the tactical issue of the selection of a value of the associated policy variable (e.g., choosing that the target value of P_2 should be 98%).

Besides the choice of a suitable criterion, there is also a choice of the probability distribution of lead time demand. As discussed in Section 3.5.5, from a pragmatic standpoint, we recommend, at least for most B items, the use of a normal distribution of lead time demand. In Section 6.7.14, we discuss the use of other distributions. In particular, we will see that the Gamma distribution can be used almost as easily as the Normal to fit demand during lead time and may be a better choice in certain circumstances.

The assumptions, notation, general approach to the selection of the reorder point, and the portion of the derivation of the rules common to all criteria will be presented prior to showing the individual decision rules. Further details of some of the derivations can be found in the Appendix of this chapter.

6.7.1 Common Assumptions and Notation

There are a number of assumptions that hold independent of the method of costing shortages or measuring service. These include:

1. Although demand is probabilistic, the average demand rate changes very little with time. Although this assumption may appear to be somewhat unrealistic, note that the decision rules can be used adaptively (i.e., the parameters are updated with the passage of time). The situation where the demand rate changes over time is addressed further in Section 7.6.

2. A replenishment order of size Q is placed when the inventory position is exactly at the order point s. This assumption is tantamount to assuming that all demand transactions are of unit size or else that the undershoots of the order point are of negligible magnitude compared with the total lead time demand. (In Chapter 7, we will consider a situation where the undershoots are not neglected.)

3. If two or more replenishment orders for the same item are simultaneously outstanding, then, they must be received in the same order in which they were placed; that is, crossing of orders is not permitted. A special case satisfying this assumption is where the replenishment lead time is a constant.

4. Unit shortage costs (explicit or implicit) are so high that a practical operating procedure will always result in the average level of backorders being negligibly small when compared with the average level of the OH stock. In practice, this is not a very limiting assumption since, if the level of backorders far exceeds the average OH stock, we are really managing a make (or buy)-to-order business rather than a make-to-stock business.

5. Forecast errors have a normal distribution with no bias (i.e., the average value of the error is zero) and a known standard deviation σ_L for forecasts over a lead time L. In actual fact, as discussed in Chapter 3, the forecast system only provides us with an estimate $\hat{\sigma}_L$ of σ_L. However, extensive simulation studies by Ehrhardt (1979) have revealed that performance is not seriously degraded by using $\hat{\sigma}_L$ instead of σ_L (see Problem 6.5). (Recall that we use the standard deviation of forecast errors rather than the standard deviation of the true lead time demand distribution.) Appendix II, at the end of the book, is devoted to a discussion of the normal distribution, and a graphical representation of a typical distribution is shown in Figure 6.6.

6. Where a value of Q is needed, it is assumed to have been predetermined. In most situations, the effects of the two decision variables, s and Q, are not independent; that is, the best value of Q depends on the s value, and vice versa. However, as will be shown in Chapter 7, where we will take a closer look at the simultaneous determination of the two control parameters, the assumption of Q being predetermined without the knowledge of s makes very good practical sense, particularly for B items. See also Zheng (1992) and Axsäter (1996).

7. The operating costs of the control system do not depend on the specific value of s selected.

Throughout the book, we have been emphasizing the possibility, and requirement, to "change the givens." In Chapter 4, we noted that there are many ways to reduce the optimal order quantity, such as the use of technological infrastructure (such as EDI) to effectively reduce the ordering cost.

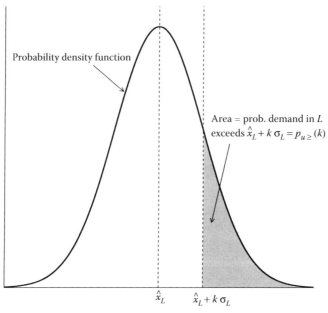

Figure 6.6   Normally distributed forecast errors.

Here, we comment on ways to reduce SS by changing the givens. It will become evident below that some of the important drivers for the amount of SS are the replenishment lead time, the variability of demand, and the service level required. Many firms have focused (rightly) on reducing SS by reducing the replenishment lead time. Choosing a supplier that is closer to your facility and shipping via a faster transportation mode such as air freight are the two ways of reducing this lead time. Improving forecast accuracy and providing customer incentives for specific purchase times and quantities are examples of effective ways to reduce demand variability. Occasionally, it is possible to reduce the required service level by better understanding customer needs and expectations. Once again, our recommendation is to optimize inventory levels given the parameters as they are; and then devote resources to changing the givens. We shall see in Section 6.11.2 that in many cases, it is possible to reduce inventory costs, and improve the service, simultaneously. This is why we suggest initially optimizing inventory levels before applying resources to changing the givens. Then, however, managers should focus attention on reducing inventories by removing the reasons for them.

The common notation includes:

D = demand per year in units/year.

$G_u(k)$ = a special function of the unit normal (mean 0, standard deviation 1) variable.[*] $G_u(k)$ is used in finding the expected shortages per replenishment cycle (ESPRC). See Equation 6.6 for intuition on this function.

---

[*] $G_u(k)$ or $p_{u\geq}(k)$ can be obtained from k, or vice versa, using a table lookup (a table of the unit normal distribution is shown in Appendix II) or a rational approximation presented in Appendix III. Also, as discussed in Appendix III, $p_{u\geq}(k)$ can also be obtained on Excel spreadsheet from Microsoft: given k, $p_{u\geq}(k) = 1 - \text{NORMSDIST}(k)$. Given $p_{u\geq}(k)$, $k = \text{NORMSINV}(1 - p_{u\geq}(k))$.

k = safety factor.

L = replenishment lead time, in years.

$p_{u\geq}(k)$ = probability that a unit normal (mean 0, standard deviation 1) variable takes on a value of k or larger. (A graphical interpretation of $p_{u\geq}(k)$ is shown in Figure 6.6.) Note that $p_{u\geq}(k)$ is often expressed as $1 - \Phi(k)$, where $\Phi(k)$ is the cumulative distribution function (or the left tail) of the unit normal evaluated at k.

Q = prespecified order quantity, in units.

r = inventory carrying charge, in \$/\$/year.

s = order point, in units.

SS = safety stock, in units.

v = unit variable cost, in \$/unit.

$\hat{x}_L$ = forecast (or expected) demand over a replenishment lead time, in units.

σ_L = standard deviation of errors of forecasts over a replenishment lead time, in units.

6.7.2 General Approach to Establishing the Value of s

In Section 6.6.2, we set about directly determining the required reorder point s. Here, it will turn out to be more appropriate to work indirectly using the following relationships:

$$\text{Reorder point, } s = \hat{x}_L + (\text{Safety stock}) \tag{6.7}$$

and

$$\text{Safety stock} = k\sigma_L \tag{6.8}$$

where k is known as the safety factor. The determination of a k value leads directly to a value of s through the use of these two relations.[*]

The general logic used in computing the appropriate value of s (via k) is portrayed in Figure 6.7. Of particular note is the manual override. The user should definitely have the option of adjusting the reorder point to reflect factors that are not included in the model. However, care must be taken to not again adjust for factors for which manual adjustments were already made as part of the forecasting system (see Figure 3.1 of Chapter 3).

Figure 6.7 suggests specifying the lowest allowable value for k. Management must specify this value because the inventory calculations may recommend, in effect, holding no inventory at all. When the shortage cost is lower than the inventory holding cost, for instance, the logical choice would be to backorder Q customers and then fill their demands with an order of size Q. However, managers may resist providing such poor service. (It is not unlikely that, in this case, the shortage cost was incorrectly specified.) Johnson et al. (1995) point out, however, that k can be negative in multilevel distribution systems. Retailers hold buffer stocks, so that their supplier, a central warehouse, is not required to provide high levels of service. In any case, managers may set a limit on the service provided by specifying that k must be greater than, say, 0.

---

[*] The definition of safety stock in Equation 6.8 assumes that unmet demand is backlogged; thus, net stock can be negative. Below, we discuss the adjustments to safety stock calculations when unmet demand is partially or completely lost.

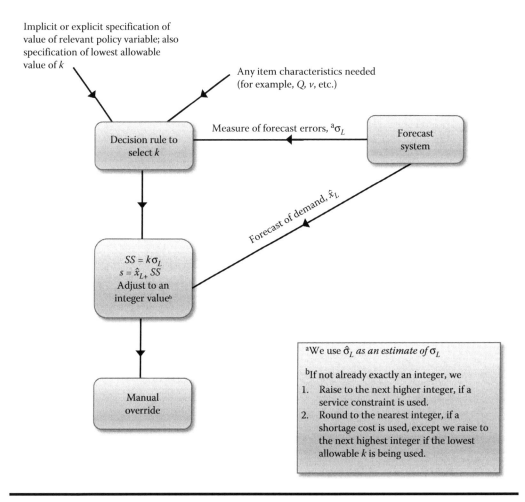

Figure 6.7 General decision logic used in computing the value of s.

6.7.3 Common Derivation

Recall that an (s, Q) system operates in the following manner. Any time that the inventory position drops to s or lower, a replenishment of size Q is placed. Because of the assumption of no crossing of orders, if an order is placed at some time t when the inventory position is at level s, then, all previous orders outstanding at that time must have arrived prior to the moment (call it $t + L$) at which the current order arrives. Furthermore, any orders placed after the current one cannot arrive before $t + L$. In other words, all of the inventory position of s units at time t, and no other stock, must have reached the stocking shelf by time $t + L$. Therefore, the service impact of placing the current order when the inventory position is at the level s is determined by whether the total demand x in the replenishment lead time exceeds s. (If the OH level happens to be very low at time t, a stockout may be incurred early in the replenishment cycle before an earlier outstanding order arrives. However, this event is independent of the current replenishment, the one at time t, and should not be considered in evaluating the consequences of using an order point s for the current order.)

If the demand (x) in the replenishment lead time has a probability density function $f_x(x_0)$ defined such that

$$f_x(x_0)dx_0 = \text{Pr\{total demand in the lead time lies between } x_0 \text{ and } x_0 + dx_0\}}$$

then the above arguments lead to the following three important results:

1. Safety stock is the expected net stock just before the replenishment arrives. If unmet demand is backlogged,

$$SS = \int_0^\infty (s - x_0)f_x(x_0)dx_0 = s - \hat{x}_L \quad (6.9)$$

This has a particularly simple interpretation for the case where no more than one order is ever outstanding. The average OH inventory level just before a replenishment arrives is equal to the inventory level when the replenishment is placed reduced by the average demand during the lead time.

If unmet demand is lost (no backorders), net stock cannot be negative and an adjustment must be made to the SS calculation:

$$E(\text{OH just before a replenishment arrives}) = \int_0^s (s - x_0)f_x(x_0)dx_0$$

When expected unmet demand is small, the difference in the SS calculation from incorporating this adjustment is also small. We discuss in point 3 below how to operationalize this adjustment.

2. Prob {stockout in a replenishment lead time}

$$= \text{Pr}\{x \geq s\}$$

$$= \int_s^\infty f_x(x_0)dx_0 \quad (6.10)$$

= the probability that lead time demand is at least as large as the reorder point (illustrated graphically in Figure 6.6).

3. Expected shortage per replenishment cycle,

$$ESPRC = \int_s^\infty (x_0 - s)f_x(x_0)dx_0 \quad (6.11)$$

Using the expression in Equation 6.11, we can develop an expression for SS for the lost sales case mentioned in point 1 above.

$$
\begin{aligned}
E(\mathrm{OH}) &= \int_0^s (s - x_0) f_x(x_0)\, dx_0 \\
&= \int_0^s (s - x_0) f_x(x_0)\, dx_0 + \int_s^\infty (s - x_0) f_x(x_0)\, dx_0 - \int_s^\infty (s - x_0) f_x(x_0)\, dx_0 \\
&= s - \hat{x}_L + \int_s^\infty (x_0 - s) f_x(x_0)\, dx_0 \\
&= s - \hat{x}_L + \mathrm{ESPRC}
\end{aligned}
\tag{6.12}
$$

From here, we proceed assuming the simpler case of backlogged demand (with SS given by $s - \hat{x}_L$), but note that it is a simple adjustment to the SS expression (the inclusion of ESPRC) to accommodate the lost sales case.

Now, the mean rate of demand is constant with time. Therefore, on average, the OH level drops linearly during a cycle from $(s - \hat{x}_L + Q)$ right after a replenishment arrives to $(s - \hat{x}_L)$ immediately before the next replenishment arrives. Thus

$$
E(\mathrm{OH}) \approx \frac{Q}{2} + (s - \hat{x}_L) = \frac{Q}{2} + k\sigma_L
\tag{6.13}
$$

where, as mentioned earlier, we have chosen to express the safety stock as the multiple of two factors:

$$
\mathrm{SS} = k\sigma_L
\tag{6.14}
$$

A useful graph of Equation 6.13 is shown in Figure 6.8. One other common feature, independent of the service measure or shortage-costing method used, is the expected number of replenishments per year. Each replenishment is of size Q and the mean rate of demand is D, as earlier. Therefore,

$$
\text{Expected number of replenishments per year} = \frac{D}{Q}
\tag{6.15}
$$

To this point, the results hold for any probability distribution of lead time demand (or forecast errors). To proceed further requires specifying the particular distribution. As shown in Appendix II, when forecast errors are assumed to be normally distributed and the SS is expressed as in Equation 6.14, then Equations 6.10 and 6.11 simplify to

$$
\Pr\{\text{stockout in a replenishment lead time}\} = p_{u \geq}(k)
\tag{6.16}
$$

and

$$
\mathrm{ESPRC} = \sigma_L G_u(k)
\tag{6.17}
$$

At this point, the safety factor derivations diverge based on the particular shortage cost or service measure used. Some illustrative details can be found in the Appendix of this chapter. We now turn

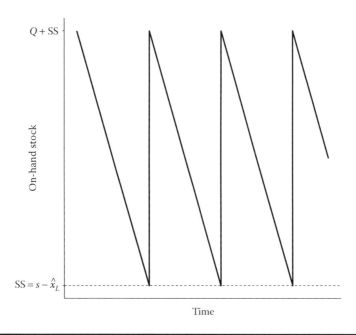

Figure 6.8   Average behavior of OH stock in an (s, Q) system.

to the presentation of the individual decision rules. In each case, we first present the rule. In most cases, this is followed by a numerical illustration of its use, and then by a discussion of the behavior of the rule in terms of how the safety factor k varies from item to item in a population of items. As discussed in Section 6.5, there is a wide choice of criteria for establishing SSs; our treatment is only meant to be illustrative of the many possibilities.

6.7.4 Decision Rule for a Specified Safety Factor (k)

Once a k value is specified, the choice of s immediately follows from Equations 6.7 and 6.8:
   The Rule

Step 1 Safety stock, $SS = k\sigma_L$.
Step 2 Reorder point, $s = \hat{x}_L + SS$, increased to the next higher integer (if not already exactly an integer).

6.7.5 Decision Rule for a Specified Cost (B₁) per Stockout Occasion

Two approaches to the derivation of the following rule are shown in Section 6A.1 of the Appendix of this chapter. Here, we simply present the total cost function and then give the decision rule. The total cost is again comprised of ordering, holding, and shortage costs. Relevant expected annual ordering (or replenishment) and holding costs are, as before,

$$C_r = AD/Q$$

and

$$C_c = \left(\frac{Q}{2} + k\sigma_L\right) vr$$

The expected stockout costs C_s per year are obtained by multiplying three factors together—namely, (1) the expected number of replenishment cycles per year, (2) the probability of a stockout per cycle, and (3) the cost per stockout. Thus, using Equations 6.15 and 6.16, we have

$$C_s = \left(\frac{D}{Q}\right) p_{u\geq}(k)B_1$$

The total cost is the sum of these three terms. Because spreadsheets have the standard normal distribution built in, computing the total cost for a given k is quite simple. Searching for the optimal k is easy as well; but when there are thousands of items, it is useful to have a simple decision rule:

Step 1 Is

$$\frac{DB_1}{\sqrt{2\pi}Qv\sigma_L r} < 1? \tag{6.18}$$

where Q has been predetermined, presumably by one of the methods of Chapter 4, B_1 is expressed in dollars (or any other currency), and all the other variables (with units consistent such that the left-hand side of the equation is dimensionless) are as defined in Section 6.7.1. If yes, then go to Step 2. If no, then continue with[*]

$$k = \sqrt{2\ln\left(\frac{DB_1}{\sqrt{2\pi}Qv\sigma_L r}\right)} \tag{6.20}$$

If Equation 6.20 gives a value of k lower than the minimum allowable value specified by the management, then go to Step 2. Otherwise, proceed directly to Step 3.

Step 2 Set k at its lowest allowable value (specified by management).

Step 3 Reorder point $s = \hat{x}_L + k\sigma_L$, rounded to the nearest integer (except raised to the next highest integer if Step 2 was used).

6.7.5.1 Numerical Illustration

Suppose an item is one of a number for which a B_1 value of $300 has been specified. Other relevant characteristics of the item are

$D = 200$ units/year
$Q = 129$ units

---

[*] An alternative to the use of Equation 6.20 is to find by a table lookup the k value that satisfies

$$f_u(k) = \frac{Qv\sigma_L r}{DB_1} \tag{6.19}$$

where $f_u(k)$ is the probability density function of a unit normal variable evaluated at k. (See Table II.1 in Appendix II.)

$v = \$2/\text{unit}$
$\hat{x}_L = 50$ units
$\sigma_L = 21.0$ units
$r = 0.24$ \$/\$/year
$A = \$20/\text{order}$

Step 1

$$\frac{DB_1}{\sqrt{2\pi}Qv\sigma_L r} = \frac{200\,\text{units/year} \times \$300}{\sqrt{2\pi} \times 129\,\text{units} \times \$2/\text{unit} \times 21/\text{units} \times 0.24/\text{yr}}$$
$$= 18.4 > 1$$

Hence, from Equation 6.20

$$k = \sqrt{2\ln(18.4)} = 2.41$$

Step 3 (Step 2 is bypassed in this example.)

$$s = 50 + 2.41(21) = 101\,\text{units}$$

The high value of B_1 has led to a rather large safety factor.

6.7.5.2 Discussion

Notice in Figure 6.9 the behavior of the carrying, shortage, and total costs as k is varied from 1.3 to 3.3 for the numerical example.* Now suppose that σ_L is reduced from 21 to 7 units due to a closer relationship with customers. The effects of changing this "given" are shown graphically in the same figure. (The shortage costs are the same for both values of σ_L, as are the ordering costs AD/Q.) For k values greater than 1.7, the entire total cost curve for $\sigma_L = 7$ is below the low point of the $\sigma_L = 21$ curve. In other words, reducing σ_L from 21 to 7, without much attempt to optimize over k, is much more beneficial than choosing the best value of k for σ_L fixed at 21.

We can see from Equation 6.20 that k decreases as σ_L or v goes up. Intuitively, the behavior of k with v makes sense under the assumed stockout-costing mechanism. If there is only a fixed cost per stockout occasion that is the same for all the items, then, it makes sense to allocate a greater proportional SS to the less-expensive items where an adequate level of protection is achieved with relatively little investment. Furthermore, as shown in Section 6A.1 of the Appendix of this chapter, it follows from Equation 6.20 that k decreases as Dv increases (this is not obvious from a quick look at the equations because both $\sigma_L v$ and Qv depend on Dv). What this means is that higher safety factors are provided to the slower-moving items.

The appearance of Q in the decision rule for finding the safety factor k indicates that a change in Q will affect the required value of the SS. Such effects of Q were ignored in the derivations of Chapter 4. For B items, this turns out to be a reasonable approximation, as we will see in Chapter 7, where such interactions between Q and k will be more closely examined in the context of control of A items.

* The total costs shown for both $\sigma_L = 7, 21$ include ordering cost AD/Q with $A = \$20$.

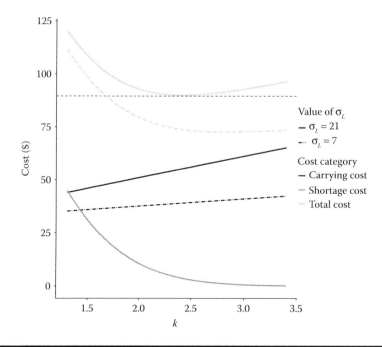

Figure 6.9 Effects of changing σ_L with B_1 costing (dashed lines for $\sigma_L = 7$).

There is no solution to Equation 6.20 when the condition of Equation 6.18 is satisfied. In such a situation, as shown in the derivation in Section 6A.1 of the Appendix, the best solution is the lowest allowable value of k. This situation may arise when B_1 is low relative to vr; that is, when the shortage cost is low relative to the holding cost.

Note that in establishing the reorder point in Step 3, we advocate rounding to the nearest integer rather than always going to the next highest integer. This is because, in contrast with service measures, here, we are not bound by a service constraint.

6.7.6 Decision Rule for a Specified Fractional Charge (B_2) per Unit Short

Derivations of the following decision rule can be found in McClain and Thomas (1980). Again, we provide the cost function and then the decision rule. The total expected annual cost is

$$TC = AD/Q + \left(\frac{Q}{2} + k\sigma_L\right)vr + \frac{B_2 v\sigma_L G_u(k)D}{Q}$$

The shortage cost term is composed of the ESPRC, $\sigma_L G_u(k)$, the number of cycles per year, D/Q, and the cost per unit short, $B_2 v$.

Step 1 Is

$$\frac{Qr}{DB_2} > 1?$$

(6.21)

where Q has been predetermined and the units of the variables are such that the left-hand side of the equation is dimensionless (B_2 itself is dimensionless). If yes, then go to Step 2. If no, then continue with the following. Select k so as to satisfy

$$p_{u \geq}(k) = \frac{Qr}{DB_2} \tag{6.22}$$

If the use of Equation 6.22 gives a k value lower than the minimum allowable safety factor specified by the management, then go to Step 2. Otherwise, move to Step 3.

Step 2 Set k at its lowest allowable value.

Step 3 Reorder point, $s = \hat{x}_L + k\sigma_L$, rounded to the nearest integer (except raised to the next highest integer in the event that Step 2 has been used).

6.7.6.1 Numerical Illustration

A firm wishes to allocate a fixed amount of SS among a number of products to keep the total value of backorders per year as low as possible. Management feels that it is reasonable to use a B_2 value of 0.25; that is, each unit short incurs a cost equal to 25% of its unit value. The replenishment quantity of an item under consideration has been predetermined at 85 units. This value is determined from annual demand of 200, a fixed order cost of \$21.50, a carrying charge of 20%, and a unit cost of \$6. Other quantities of interest include:

$$\hat{x}_L = 50 \text{ units}$$
$$\sigma_L = 10 \text{ units}$$

Step 1

$$\frac{Qr}{DB_2} = \frac{85 \text{ units} \times 0.2/\text{year}}{200 \text{ units/year} \times 0.25} = 0.34 < 1$$

Then, Equation 6.22 gives

$$p_{u \geq}(k) = 0.34$$

From Table II.1 in Appendix II (or from the function NORMSINV(1–0.34) in Excel) $k \approx$ 0.41 (presumably larger than the lowest allowable value, hence we go to Step 3).

Step 3

$$s = 50 + 0.41(10) = 54.1 \rightarrow 54 \text{ units}$$

The expected total cost of this policy is

$$TC = AD/Q + \left(\frac{Q}{2} + k\sigma_L\right)vr + \frac{B_2 v \sigma_L G_u(k) D}{Q} = \$51 + \$56 + \$8 = \$115$$

where $G_u(k) = 0.2270$ from Appendix II. Note that holding and ordering costs are not equal, unlike in Chapter 4. We can see from Figure 6.10 the form of the total cost as a function of k. With $B_2 = 0.25$, the decrease in total cost is small as k is increased from 0 to its optimal value ($k \approx 0.41$). Costs can increase sharply if k is set far too high. Now with $B_2 = 1.0$ shortages are much more expensive, and it is beneficial to hold more SS. The result

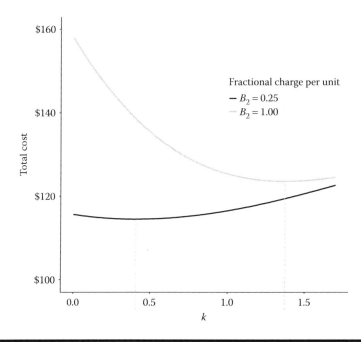

Figure 6.10   Sensitivity of total cost to the safety factor (with $B_2 = 0.25, 1.00$).

is that the decrease in cost as k increases from 0 is much more pronounced. With $B_2 = 1.0$, the optimal k is approximately 1.37.

6.7.6.2 Discussion

As seen in Equation 6.22, the safety factor k increases as B_2/r increases, as one would expect. Moreover, under the use of EOQs, we found that Q/D decreases as Dv increases. Therefore, at least under the use of EOQs, we can see from Equation 6.22 that k increases as Dv increases; that is, under this decision rule, larger safety factors are given to the faster-moving items, all other things being equal.

The $p_{u\geq}(k)$ value represents a probability; hence, there is no solution to Equation 6.22 when the right-hand side of the equation exceeds unity; that is, when Equation 6.21 is satisfied. It can be shown (in a manner paralleling the derivation in Section 6A.1 of the Appendix of this chapter) that, when this happens, the lowest permissible value of the safety factor should be used. Intuitively, the condition is satisfied if the shortage cost rate B_2 is smaller than the holding cost rate, r (for $Q = D$, to be precise). If $r = B_2$, the condition is satisfied if the order quantity is larger than the annual demand.

6.7.7 Decision Rule for a Specified Fractional Charge (B_3) per Unit Short per Unit Time

As mentioned in Section 6.5.4, there is an equivalence between the B_3 costing measure and the P_2 service measure. Thus, we merely present the decision rule here, leaving further comment to the P_2

section (the derivation involves differentiating the total relevant costs with respect to k and using the results in Hadley and Whitin 1963). See also Shore (1986).

Step 1 Select the safety factor k that satisfies

$$G_u(k) = \frac{Q}{\sigma_L}\left(\frac{r}{B_3 + r}\right) \tag{6.23}$$

where Q has been predetermined, presumably by one of the procedures of Chapter 4 (and must be expressed in the same units as σ_L), and B_3 and r are in \$/\$/unit time. Make sure that the k value is at least as large as the lowest allowable (management specified) value (e.g., zero) of the safety factor.

Step 2 Reorder point $s = \hat{x}_L + k\sigma_L$, rounded to the nearest integer (but increased to the next highest integer if the minimum k value is used in Step 1).

6.7.8 Decision Rule for a Specified Charge (B_4) per Customer Line Item Short

Assume that customers order a number of products from a given supplier. The supplier may be interested in minimizing the total cost when there is a charge per line item short. If $\hat{z}$ is the average number of units ordered per customer line item, we can write the following equation for the approximate expected total relevant cost (due to Hausman 1969):

$$\text{ETRC} = AD/Q + \left(\frac{Q}{2} + k\sigma_L\right)vr + B_4\frac{D\sigma_L G_u(k)}{Q\hat{z}} \tag{6.24}$$

Step 1 Select the safety factor k that satisfies

$$p_{u\geq}(k) = \frac{Qrv\hat{z}}{B_4 D} \tag{6.25}$$

where Q has been predetermined, presumably by one of the procedures of Chapter 4 (and must be expressed in the same units as σ_L), and r is in \$/\$/unit time. Make sure that the k value is at least as large as the lowest allowable value (e.g., zero) of the safety factor.

Step 2 Reorder point $s = \hat{x}_L + k\sigma_L$, rounded to the nearest integer (but increased to the next highest integer if the minimum k value is used in Step 1).

6.7.9 Decision Rule for a Specified Probability (P_1) of No Stockout per Replenishment Cycle

Suppose management has specified that the probability of no stockout in a cycle should be no lower than P_1 (conversely, the probability of a stockout should be no higher than $1 - P_1$). Then, we have the following simple decision rule (whose derivation is shown in Section 6A.2 of the Appendix of this chapter).

Step 1 Select the safety factor k to satisfy

$$p_{u\geq}(k) = 1 - P_1 \tag{6.26}$$

where

$$p_{u\geq}(k) = \text{Pr\{Unit normal variable (mean 0, standard deviation 1) takes on a value of } k \text{ or larger\}, a widely tabulated function (see Table II.1 in Appendix II)}$$

Step 2 Safety stock, $SS = k\sigma_L$.

Step 3 Reorder point, $s = \hat{x}_L + SS$, increased to the next higher integer (if not already exactly an integer).

6.7.9.1 Numerical Illustration

Suppose that the forecast system generates the following values:

$$\hat{x}_L = 58.3 \text{ units}$$
$$\sigma_L = 13.1 \text{ units}$$

Management also desires a service level of $P_1 = 0.90$. From Equation 6.26, we know that

$$p_{u\geq}(k) = 0.10$$

From Table II.1 in Appendix II (or from the function $\text{NORMSINV}(1 - p_{u\geq}(k))$ on Excel)

$$k \approx 1.28$$
$$SS = k\sigma_L = 1.28 \times 13.1 \text{ units}$$
$$= 16.8 \text{ units}$$

Reorder point, $s = 58.3 + 16.8 = 75.1$, say 76 units.

6.7.9.2 Discussion of the Decision Rule

From Equation 6.26, we can see that the safety factor k depends only on the value of P_1; in particular, it is independent of any individual-item characteristics such as the order quantity Q. Therefore, all items for which we desire the same service level, P_1, will have identical values of the safety factor k. Thus, we see an equivalence between two criteria, namely, using a specified value of k and a specified value of P_1.

Because of the discrete nature of the reorder point, we are not likely to be able to provide the exact level of service desired since this usually would require a noninteger value of s. Therefore, a noninteger value of s found in Equation 6.26 is rounded up to the next higher integer, with the predicted service level then being slightly higher than required. Such differences are often less pronounced than the error in the input data.

The fact that the safety factor does not depend on any individual-item characteristics may necessitate reexamining the meaning of service here. Recall that service under the measure used in this section is prob no stockout per replenishment cycle. Consider two items, the first being replenished 20 times a year, the other once a year. If both of them are given the same safety factor so that both have a probability of 0.10 of stockout per replenishment cycle, then, we would expect $20 \times (0.10)$ or 2 stockouts per year for the first item and only 1 stockout every 10 years (0.1 per year) for the second item. Therefore, depending on management's definition of service, we, in fact, may not be giving the same service on these two items. Rules based on other service measures will now be presented.

6.7.10 Decision Rule for a Specified Fraction (P₂) of Demand Satisfied Directly from Shelf

The following decision rule is derived in Section 6A.2 of the Appendix to this chapter.

Step 1 Select the safety factor k that satisfies[*]

$$G_u(k) = \frac{Q}{\sigma_L}(1 - P_2) \tag{6.28}$$

where Q has been predetermined, presumably by one of the procedures of Chapter 4 (and must be expressed in the same units as σ_L). The other relevant variables are defined in Section 6.7.1.

Make sure that the k value is at least as large as the lowest allowable value of the safety factor.

Step 2 Reorder point, $s = \hat{x}_L + k\sigma_L$, increased to the next higher integer (if not already exactly an integer).

A brief comment on the order quantity is in order. It is common practice to use the EOQ for the order quantity in an (s, Q) system. Although this practice is often very close to the optimal, Platt et al. (1997) have shown that a simple modification of the order quantity outperforms the EOQ when sequentially choosing the order quantity and reorder point. They recommend an order quantity based on the limit as the ratio $EOQ/\sigma \to \infty$, when using a fill rate as the performance measure:

$$Q = \frac{1}{P_2}\sqrt{\frac{2AD}{rv} + \sigma_L^2}$$

Equation 6.28 underestimates the true fill rate if σ_L is large relative to Q, because Equation 6.28 double counts backorders from a previous cycle that are not met at the start of the next cycle. A more accurate formula for small values of Q/σ_L is

$$G_u(k) - G_u(k + Q/\sigma_L) = \frac{Q}{\sigma_L}(1 - P_2) \tag{6.29}$$

For further discussion of the fill rate, see Silver (1970), Johnson et al. (1995), Zipkin (1986), Platt et al. (1997), and Disney et al. (2015).

---

[*] Equation 6.28 applies for the case of complete backordering. The only difference for the case of complete lost sales is that $1 - P_2$ is replaced by $(1 - P_2)/P_2$:

$$G_u(k) = \frac{Q}{\sigma_L}\left(\frac{1 - P_2}{P_2}\right) \tag{6.27}$$

6.7.10.1 Numerical Illustration

Consider a particular type of developing liquid distributed by a manufacturer of laboratory supplies and equipment. Management has specified that 99% of demand is to be satisfied without backordering. A replenishment quantity of 200 gallons has been predetermined and the forecast system provides us with $\hat{x}_L = 50$ gallons and $\sigma_L = 11.4$ gallons.

Step 1 Equation 6.28 yields

$$G_u(k) = \frac{200}{11.4}(1 - 0.99) = 0.175$$

and Table II.1 in Appendix II gives

$$k = 0.58$$

Step 2

$$s = 50 + 0.58(11.4) = 56.6 \rightarrow 57 \text{ gallons}$$

In this case, Equation 6.29 gives the same result as Equation 6.28. For the case of complete lost sales (rather than backordering), the use of Equation 6.27 would lead us to the same value of the reorder point. Recall that $G_u(k)$ can be obtained using a rational approximation presented in Appendix III at the end of the book. This is especially appropriate for spreadsheet use, and is extremely accurate.

6.7.10.2 Discussion

Intuitively, we would expect that the required SS would increase if (1) Q decreased (more opportunities for stockouts), (2) σ_L increased (higher uncertainty of forecasts), or (3) P_2 increased (better service desired). If any of these changes take place, $G_u(k)$ decreases, but as seen from the table in Appendix II, a decrease in $G_u(k)$ implies an increase in k, that is, exactly the desired behavior. These relationships can be seen in Figure 6.11.

In addition, on average, σ_L tends to increase with D; therefore, the increase in k with increasing σ_L says that, on the average under this decision rule, faster-moving items get higher safety factors than do the slower-moving items.

If the right-hand side of Equation 6.28 is large enough, a negative value of k is required to produce the equality; to give a service level as poor as P_2, one must deliberately plan to be short on the average when a replenishment arrives. In particular, if Q is large, many customers have their demand satisfied before there is even an opportunity for a stockout. When a replenishment order is placed, so many demands have been met from the shelf that a stockout will only serve to lower P_2 to the target. Management may find this intolerable and, instead, set a lower limit on k (e.g., zero). If k is set at zero when Equation 6.28 calls for a negative value of k, the service provided will be better than P_2.

The P_2 value is usually quite close to unity. Therefore, as evidenced by the numerical example, Equations 6.27 and 6.28 normally give very similar values of $G_u(k)$, that is, of k itself. In other words, the value of the safety factor is little influenced by whether the model assumes complete backordering or complete lost sales (or any mix of these two extremes).

Finally, a comparison of Equations 6.23 and 6.28 shows the analogy between the P_2 and B_3 measures mentioned earlier.

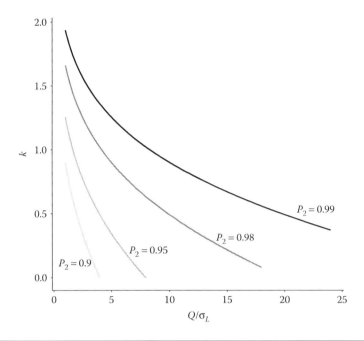

Figure 6.11   Sensitivity of the P_2 service measure.

6.7.11 Decision Rule for a Specified Average Time (TBS) between Stockout Occasions

Step 1 Is

$$\frac{Q}{D(\text{TBS})} > 1? \tag{6.30}$$

where Q has been predetermined (presumably by the methods of Chapter 4) and the units of the variables are such that the left side is dimensionless. If yes, then go to Step 2. If no, then continue with the following. Select the safety factor k to satisfy

$$p_{u\geq}(k) = \frac{Q}{D(\text{TBS})} \tag{6.31}$$

If the resulting k is lower than the minimum allowable value specified by the management, then go to Step 2. Otherwise, move to Step 3.

Step 2 Set k at its lowest permitted value.

Step 3 Reorder point, $s = \hat{x}_L + k\sigma_L$ raised to the next highest integer (if not already exactly an integer).

6.7.11.1 Numerical Illustration

For the item used in Section 6.7.9, suppose that

$\hat{x}_L = 58.3$ units
$\sigma_L = 13.1$ units

$D = 200$ units/year

$Q = 30$ units

and the management specifies a desired TBS value of 2 years. Then, Equation 6.31 gives

$$p_{u\geq}(k) = \frac{300}{200(2)} = 0.075$$

From Table II.1 of Appendix II, we have

$$k = 1.44$$

Then,

$$SS = 1.44x13.1 = 18.9 \text{ units}$$

and,

$$s = 58.3 + 18.9 = 77.2, \text{ say } 78 \text{ units}$$

6.7.11.2 Discussion

A comparison of Equations 6.22 and 6.31 reveals that there is an equivalence between using the B_2 costing method and using the TBS service measure. Specifically, we have, for equivalence, that

$$\text{TBS} = \frac{B_2}{r}$$

From Equation 6.31, we see that increasing TBS leads to a lower value of $p_{u\geq}(k)$, and hence a larger value of the safety factor k, which is certainly an appealing behavior.

6.7.12 Decision Rule for the Allocation of a TSS to Minimize the ETSOPY

As shown in Section 6A.4 of the Appendix of this chapter, allocation of a total safety stock (TSS) among a group of items to minimize the ETSOPY leads to a decision rule identical to that under the B_1 shortage-costing method. Thus, simply use the procedure of Section 6.7.5, adjusting the value of B_1 until the total implied SS is equal to the given quantity to be allocated. (Conversely, when we discuss SS exchange curves later in this chapter, we will see that the selection of an aggregate operating point on an appropriate exchange curve will imply a value of B_1 to be used). The total number of stockout occasions per year is simply the product of the probability of a stockout in a replenishment cycle multiplied by the number of cycles per year:

$$p_{u\geq}(k)\left(\frac{D}{Q}\right)$$

6.7.13 Decision Rule for the Allocation of a TSS to Minimize the ETVSPY

In a fashion similar to that shown in Section 6A.4 of the Appendix of this chapter, it can be shown that there is an equivalence between the decision rules for three different criteria:

1. Allocating a TSS among a group of items to minimize the expected total value of shortages per year (ETVSPY)

2. Use of a B_2 shortage cost
3. Use of the TBS service measure

Thus, for the allocation criterion, one needs to only use the procedure of Section 6.7.6, adjusting the value of B_2 until the total implied SS is equal to the given quantity to be allocated. The total value of shortages per year is the product of the number of shortages per replenishment cycle, multiplied by the value of those shortages, multiplied by the number of cycles per year:

$$\sigma_L G_u(k) v \left(\frac{D}{Q} \right)$$

6.7.14 Nonnormal Lead Time Demand Distributions

As mentioned earlier, for pragmatic reasons, we have focused on the case of normally distributed demands (or forecast errors) in this chapter. The normal distribution often provides a good empirical fit to the observed data: it is convenient from an analytic standpoint, it is widely tabulated and is now built into spreadsheets, and, finally, the impact of using other distributions is usually quite small (see, e.g., Naddor 1978; Fortuin 1980; Lau and Zaki 1982; Tyworth and O'Neill 1997), particularly when recognizing the other inaccuracies present (estimates of the parameters of the distribution, estimates of cost factors, and so on). Tyworth and O'Neill (1997) and Lau and Zaki (1982) show that, while the errors in SS may be high when using the normal, the errors in total cost are quite low. If the ratio $\sigma_L/\hat{x}_L$ is greater than 0.5, it may be desirable to use another distribution for lead time demand, such as the Gamma. (When this ratio equals 0.5, the probability of negative values of demand for a normal distribution is 2.3%; but the real issue in choosing the reorder point is the right tail of the distribution.) Therefore, we often use the following rule of thumb: If the ratio $\sigma_L/\hat{x}_L$ is greater than 0.5, consider a distribution other than the normal. However, so long as this ratio is less than 0.5, the normal is probably an adequate approximation. Finally, if daily demand is not normal, the normal will generally be a good approximation for lead time demand if the lead time is at least several days in duration. (If the lead time itself follows some probability distribution, consult Section 6.10.) As discussed in Chapter 3, it is advisable to at least test the normal distribution using data from a sample of items.

We now comment specifically on three other distributions—namely, the Gamma, the Laplace, and the Poisson. As discussed by Burgin (1975), the Gamma distribution has considerable intuitive appeal for representing the distribution of lead time demand. In particular, if the demand distribution is skewed to the right, or if the ratio $\sigma_L/\hat{x}_L$ is greater than 0.5, use of the Gamma should be considered. The distribution is not as tractable as the normal; thus, considerable effort has been devoted to developing approximations and tables (see Burgin and Norman 1976; Das 1976; Taylor and Oke 1976, 1979; Johnston 1980; Van der Veen 1981; Tyworth et al. 1996). We derive an expression for expected units short per cycle in the Appendix to this chapter, and we present a simple spreadsheet approach for finding reorder points using the Gamma distribution in Appendix III.

The Laplace or pseudoexponential distribution was first proposed by Presutti and Trepp (1970). It is analytically very simple to use and may be an appropriate choice for slow-moving items (see Ng and Lam 1998).

The Poisson distribution is also a candidate for slower-moving items. In fact, in Chapter 7, we will illustrate its use for expensive, low-usage items. Many other distributions of lead time demand have been proposed in the literature. We now simply provide a listing of several of them accompanied by suitable references:

1. Exponential: Brown (1977)
2. Geometric: Carlson (1982)
3. Logistic: Fortuin (1980), Van Beek (1978)
4. Negative binomial: Ehrhardt (1979), Agrawal and Smith (1996)
5. Pearson: Kottas and Lau (1980)
6. Tukey's Lambda: Silver (1977)
7. Split-normal (for asymmetric lead time demand): Lefrancois (1989)

In addition, there has been a significant amount of research on approaches that can model a number of distributions. Zipkin (1988) describes the use of phase-type distributions that can approximate many other distributions. Strijbosch and Heuts (1992) provide nonparametric approximations. See also Sahin and Sinha (1987), Shore (1995), and Kumaran and Achary (1996). A powerful method of fitting many distributions using mixtures of Erlang distributions is presented by Tijms (1986). Much work has been done on distribution-free approaches. The original work by Scarf (1958) proposed finding the worst-possible distribution for each decision variable and then finding the optimal inventory policy for that distribution. Thus, it is a conservative approach. It requires only that the mean and variance of the lead time demand are known. Bulinskaya (1990), Gallego (1992), Gallego and Moon (1993), and Moon and Choi (1994, 1995) have advanced the research, and have shown that, in some cases, applying a distributional form incorrectly can lead to large errors. The distribution-free approach could generate significant savings. If the distribution appears to be very different from a known form, one should consider the distribution-free formulas, or those described in the next paragraph. Silver and Rahnama (1986) find the cost penalty for using the mean and standard deviation from a statistical sample rather than the true values. Silver and Rahnama (1987) show that the safety factor, k, should be biased upward when using a statistical sample.

Finally, we briefly present a procedure that appears to be quite useful for a number of different distributions. Lau and Lau (1993) observe that the optimal order quantities in (s, Q) systems are very similar regardless of the distribution of demand. Likewise, the cumulative distribution function of the lead time demand distributions evaluated at the reorder point s, or $F_x(s) = \int_0^s f_x(x_0)dx_0$, is quite close. Building on these insights, they develop a simple approach for the case of B_2 shortage costing, relatively few stockout occasions, and $Q > s$. They employ the uniform distribution, which is easy to evaluate, for use in finding the order quantity. Then, they use the actual lead time demand distribution for finding the reorder point.

Let $g_2 = 2DB_2/r$ and find

$$Q = \sqrt{\frac{\text{EOQ}^2(g_2)}{g_2(g_2 - 4\sigma_L\sqrt{3})}}$$

and

$$F_x(s) = 1 - 2\sqrt{\frac{\text{EOQ}^2}{g_2(g_2 - 4\sigma_L\sqrt{3})}}$$

Finally, use the actual lead time demand distribution to find s from $F_x(s)$. See Problem 6.26 for an example. Lau and Lau demonstrate that the cost error of using this approach rather than the exact approach is nearly always less than 1%. See also Shore (1995) for a similar approach.

6.8 Implied Costs and Performance Measures

When choosing a cost minimization or a service objective approach, it is important to realize that the obtained inventory policy necessarily implies values for alternate cost or service measures. These should be reported to the management if there is a possibility that the alternate measures are of interest. An example will be helpful.

Assume that the fill rate, $P_2 = 0.98$, is chosen as the service objective, and that the item in question has the following characteristics:

$D = 4000$ units/year
$A = \$20.25$
$r = 0.30 \ \$/\$/\text{year}$
$v = \$6/\text{unit}$
$L = 1$ week
$\hat{x}_L = 80$ units
$\sigma_L = 20$ units

We can calculate the EOQ to be 300 units, and using Equation 6.28 we find that $G_u(k) = 0.30$. Therefore, $k = 0.22$ and $s = 84.4$, or rounded to $s = 85$. Now, it is possible that managers are also interested in the probability of a stockout in a replenishment cycle. For this (s, Q) policy of (85, 300), we know that $k = (85 - 80)/20 = 0.25$. Note that the actual k is greater than 0.22 because we rounded s up to 85. Thus, $p_{u \geq}(k) = 0.4013$ (from Table II.1 in the Appendix II). So $P_1 = 1 - 0.4013 = 0.5987$. In other words, a fill rate of 98% yields a cycle service level of only 59.87%. In this case, the relatively large order quantity implies that many customers have their demands met from the shelf before a lead time ever begins. Thus, frequent stockouts at the end of the replenishment cycle do not significantly lower the fill rate. They do, on the other hand, imply a relatively low cycle service level. See Snyder (1980). This may be acceptable to management; but it may also be problematic, and the inventory policy should be adjusted.

A similar approach can be used to determine the implied values for costs, such as B_2. In the preceding example, the fact that $p_{u \geq}(k) = 0.4013$ implies, using Equation 6.22, that $B_2 = 0.056$—or the cost of a unit short is 5.6% of the unit value. Again, this value may be consistent with management's perspective, or it may seem to be completely in error. If the latter is true, it would be wise to consider other inventory policies that are more consistent.

In general, we can see that for any input performance objective (B_1, B_2, P_1, and so on), we will obtain a value for k, which will be used to find the reorder point, s. The value of k can then be used to find the implied values of any of the other input performance objectives.

6.9 Decision Rules for Periodic-Review, Order-Up-to-Level (R, S) Control Systems

Recall that in an (R, S) system every R units of time, a replenishment order is placed of sufficient magnitude to raise the inventory position to the order-up-to-level S. Fortunately, for such systems, there will be no need to repeat all the detail that was necessary for (s, Q) systems in the previous section, because there is a rather remarkable, simple analogy between (R, S) and (s, Q) systems. Specifically, the (R, S) situation is exactly equivalent to the (s, Q) situation if one makes the following transformations:

(s, Q)	(R, S)
s	S
Q	DR
L	$R + L$

It therefore follows that the decision rule for determining the S value in an (R, S) system for a particular selection of shortage cost or service measure is obtained from the corresponding rule for determining s in an (s, Q) system by simply making the above three substitutions. Prior to proving this equivalence, we will discuss the choice of R as well as the key time period involved in the selection of S.

6.9.1 The Review Interval (R)

In an (R, S) control system, a replenishment order is placed every R units of time; and when computing the value of S, we assume that a value of R has been predetermined. The determination of R is equivalent to the determination of an EOQ expressed as a time supply, except for two minor variations. First, the cost of reviewing the inventory status must be included as part of the fixed setup cost A. Second, it is clear that it would be senseless to attempt to implement certain review intervals, for example, 2.36 days; that is, R is obviously restricted to a reasonably small number of feasible discrete values. In fact, as mentioned in Section 6.3, the value of R is often dictated by external factors, such as the frequency of truck deliveries.

6.9.2 The Order-Up-to-Level (S)

The key time period over which protection is required is now of duration $R + L$, instead of just a replenishment lead time L. This is illustrated in Figure 6.12 and Table 6.7 with an example where $S = 50$ units and where two consecutive orders (called X and Y) are placed at times t_0 and $t_0 + R$, respectively, and arrive at $t_0 + L$ and $t_0 + R + L$, respectively. In selecting the order-up-to-level at time t_0 we must recognize that, once order X has been placed, no other later orders (in particular Y) can be received until time $t_0 + R + L$. Therefore, the order-up-to level at time t_0 must be sufficient to cover demand through a period of duration $R + L$. A stockout could occur in the early portion of the period (prior to $t_0 + L$) but that would be a consequence of the setting of the order-up-to-level on the order preceding order X. What is of interest to us is that a stockout will occur toward the end of the period (after time $t_0 + L$) if the total demand in an interval of length $R + L$ exceeds the order-up-to-level S. Another way of putting this is that any stockouts up to time $t_0 + L$ will not be influenced by our ordering decision at time t_0; however, this decision certainly influences the likelihood of a stockout at time $t_0 + R + L$. Of course, this does not mean that we would necessarily take no action if at time t_0 a stockout appeared likely by time $t_0 + L$. Probably, one would expedite any already-outstanding order or try to get the current order (that placed at time t_0) delivered as quickly as possible. Note, however, that this still does not influence the size of the current order.

We should note that is often the case, in our experience, that an impending stockout generates an overreaction on the part of the person responsible for purchasing. The current order size is inflated, causing excess inventory when the order arrives. If the impending stockout is due to a real

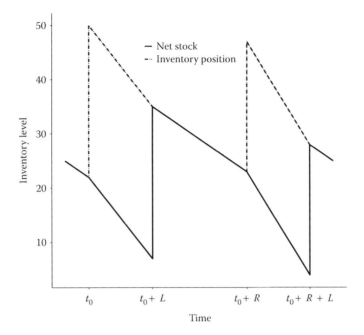

Figure 6.12   The time period of protection in an (R, S) system.

Table 6.7   Illustration of Why $(R + L)$ Is the Crucial Time Interval in an (R, S) System (Illustrated for $S = 50$)

Event	Time or Interval[a]	Demand	$(OH - BO)$[b] Net Stock	On Order	Inventory Position
Place order for 30	$t_0 - \varepsilon$	–	20	–	20
	$t_0 - \varepsilon$ to $t_0 + \varepsilon$		20	30	50[(a)]
Order of 30 arrives	$t_0 + \varepsilon$ to $t_0 + L - \varepsilon$	15[(b)]	5	30	35
	$t_0 + L - \varepsilon$ to $t_0 + L + \varepsilon$	–	35	–	35
Place order for 27	$t_0 + L + \varepsilon$ to $t_0 + R - \varepsilon$	12[(c)]	23	–	23
	$t_0 + R - \varepsilon$ to $t_0 + R + \varepsilon$	–	23	27	50
Order of 27 arrives	$t_0 + R + \varepsilon$ to $t_0 + R + L - \varepsilon$	19[(d)]	4[(e)]	27	31
	$t_0 + R + L - \varepsilon$ to $t_0 + R + L + \varepsilon$	–	31	–	31

$$(e) = (a) - \underbrace{[(b) + (c) + (d)]}_{\text{Total demand in } t_0 \text{ to } t_0 + R + L}$$

[a] ε is a shorthand notation for a very small interval of time. Thus $t_0 - \varepsilon$ represents a moment just prior to time t_0, while $t_0 + \varepsilon$ represents a moment just after time t_0.

[b]   (On hand) – (Backorders).

increase in demand, the value of S should be recomputed. Otherwise, the manager should resist the temptation to inflate inventories.

As mentioned above, a stockout will occur at the end of the current cycle (i.e., at time $t_0 + R + L$) if the total demand in an interval of duration $R + L$ exceeds S. A little reflection shows a very close analogy with the determination of s in the continuous-review system. In fact, the only difference is the duration of the time period of concern, namely, $R + L$ instead of just L.

6.9.3 Common Assumptions and Notation

The assumptions are very similar to those listed in Section 6.7.1, with suitable replacements of $R + L$ for L, DR for Q, and S for s. We note only two additional assumptions:

1. There is a negligible chance of no demand between reviews; consequently, a replenishment order is placed at every review.
2. The value of R is assumed to be predetermined. In most situations, the effects of the two decision variables, R and S, are not independent, that is, the best value of R depends on the S value, and vice versa. However, as will be shown in Chapter 7, it is quite reasonable for practical purposes when dealing with B items to assume that R has been predetermined without knowledge of the S value.

Unlike in the (s, Q) situation, the assumption of unit-sized demand transactions is not needed here.

We mention only notation that has not been used extensively thus far in this chapter:

$R =$ prespecified review interval, expressed in years
$S =$ order-up-to-level, in units
$\hat{x}_{R+L} =$ forecast (or expected) demand over a review interval plus a replenishment lead time, in units
$\sigma_{R+L} =$ standard deviation of errors of forecasts over a review interval plus a replenishment lead time, in units

6.9.4 Common Derivation

We abbreviate this section because it closely parallels the development of (s, Q) systems. Because of Assumption 1, we have

$$(\text{Number of reviews per year}) = 1/R$$

and

$$(\text{Number of replenishment orders placed per year}) = 1/R \qquad (6.32)$$

We now present the relevant equations for SS and stockouts. (Reference to Figure 6.12 will be helpful.) Recall that two orders, X and Y, are considered. If the demand (x) in $R + L$ has a probability density function $f_x(x_0)$ defined such that $f_x(x_0)dx_0 = \Pr\{\text{Total demand in } R + L \text{ lies}$

between x_0 and $x_0 + dx_0$} then the reasoning of Section 6.9.2 leads to the following three important results:

1. Safety stock $= E$ (Net stock just before order Y arrives)

$$= \int_0^\infty (S - x_0) f_x(x_0) dx_0$$

that is,

$$SS = S - \hat{x}_{R+L} \tag{6.33}$$

2. Prob {Stockout in a replenishment cycle}

$$\Pr\{x \geq S\} = \int_S^\infty f_x(x_0) dx_0 \tag{6.34}$$

the probability that the total demand during a review interval plus lead time is at least as large as the order-up-to-level.

3. Expected shortage per replenishment cycle, ESPRC

$$= \int_S^\infty (x_0 - S) f_x(x_0) dx_0 \tag{6.35}$$

On the average, the on-hand level is

$$E(OH) \approx S - \hat{x}_{R+L} + DR/2 \tag{6.36}$$

It is convenient to again set

$$SS = k\sigma_{R+L} \tag{6.37}$$

If forecast errors are normally distributed, the results of Appendix II give that Equations 6.34 and 6.35 reduce to

$$\Pr\{\text{Stockout in a replenishment cycle}\} = p_{u\geq}(k) \tag{6.38}$$

and

$$ESPRC = \sigma_{R+L} G_u(k) \tag{6.39}$$

A comparison of pairs of equations follows:

Compare	With
6.33	6.9
6.39	6.13
6.37	6.14
6.32	6.15
6.38	6.16
6.39	6.17

The comparison reveals the validity of our earlier assertion that the (R, S) situation is exactly equivalent to the (s, Q) situation if one makes the transformations listed at the beginning of this section.

6.10 Variability in the Replenishment Lead Time Itself

In all the decision systems discussed in this chapter, the key variable in setting a reorder point or order-up-to-level is the total demand in an interval of length $R + L$ (which reduces to just L for the case of continuous review). So far, our decisions have been based on a known replenishment lead time L with the only uncertain quantity being the demand rate during L or $R + L$. If L itself is not known with certainty, it is apparent that increased SS is required to protect against this additional uncertainty.[*]

It should be noted that where the pattern of variability is known, for example, seasonally varying lead times, there is no additional problem because the lead time at any given calendar time is known and the SS (and reorder point) can be adjusted accordingly. Where lead times are increasing in a known fashion, such as in conditions of reduced availability of raw materials, again, SSs (and reorder points) should be appropriately adjusted. We discuss these issues in more detail in Chapter 7.

Every reasonable effort should be made to eliminate variability in the lead time, and one can use the models developed in this section to quantify the value of reduction in lead time variability. A key component of the lead time in which uncertainty exists is the shipping time from the supplier to the stocking point under consideration. Choices among transportation modes can affect the variability, as well as the average duration, of the lead time (see Herron 1983, and Problem 6.10). Ordering process and information technology improvements can also reduce the average lead time and lead time variability. Blackburn (2012) and De Treville et al. (2014) provide recent discussions of the potential value of reductions in lead time and lead time variability.

---

[*] When inventory OH just after the arrival of a replenishment order may not be positive, de Kok (1990) has shown that the transformation of Equation 6.29 is not straightforward. In fact, the more accurate equation for P_2 is

$$P_2 = 1 - \frac{\sigma_{R+L} G_u(k) - \sigma_L G_u\left(\frac{\sigma_{R+L}}{\sigma_L} k + \frac{\hat{x}_R}{\sigma_L}\right)}{\hat{x}_R} \tag{6.40}$$

Lead time improvements, of course, require close cooperation with suppliers. In return for firm commitments well ahead of time, a reasonable supplier should be prepared to promise a more dependable lead time. The projection of firm commitments well into the future is not an easy task. We return to this point in Chapter 15 in the context of Material Requirements Planning, which offers a natural means of making such projections. Firms also have been increasingly employing strategic alliances—long-term relationships with close partners—to facilitate the required cooperation.

Where there is still some residual variability in the lead time that cannot be eliminated, two basic courses of action are possible:

1. Measure the distribution of the lead time (or $R + L$), and the distribution of demand per period, separately. Then combine the two to obtain the distribution of total demand over the lead time.
2. Try to ascertain the distribution of total demand over the full lead time by measuring or estimating actual demand over that period.

In each case, there are at least two ways to use the information. First, assume (or fit) a specific functional form—such as the Normal—to the distribution of total demand over lead time. Or, second, use the exact distribution to set an inventory policy.

6.10.1 Approach 1: Use of the Total Demand over the Full Lead Time

We begin by discussing a simple, but effective, method that is useful when it is possible to measure the total demand over the lead time. This procedure, due to Lordahl and Bookbinder (1994), uses these measurements directly and is based on the distribution-free properties of order statistics. The objective is to find the reorder point, s, when the parameters, and form, of the lead time demand distribution are unknown and P_1 is the desired performance measure. The procedure requires only that the user records, over time, the actual demand during lead time. These steps describe the procedure:

1. Rank order the observed lead time demand from the least to the greatest. So $x_{(1)} \le x_{(2)} \le \cdots \le x_{(n)}$ if there are n observations of lead time demand.
2. Let $(n + 1)P_1 = y + w$, where $0 \le w < 1$, and y is an integer.
3. If $(n + 1)P_1 > n$, set $s = x_{(n)}$. Otherwise, set $s = (1 - w)x_{(y)} + wx_{(y+1)}$, a weighted average of two of the observations. In effect, the reorder point is estimated from two of the larger values in the sample.

6.10.1.1 Numerical Illustration

A new entrepreneurial firm is trying to set reorder points for several items that are important to its business. Because lead times are roughly 2–3 weeks, and order sizes are not too large, the manager has gathered a number of observations of lead time demand. These are 15, 18, 12, 22, 28, 25, 22, 23, 11, and 19. The manager has set $P_1 = 0.85$.

Step 1 Rank order the data. 11, 12, 15, 18, 19, 22, 22, 23, 25, 28.
Step 2 $n = 10$
$P_1 = 0.85$

$(n+1)P_1 = (11)(0.85) = 9.35.$

Thus $y + w = 9.35.$

So $y = 9$ and $w = 0.35.$

Step 3 $(n+1)P_1 < n$, so

$s = (1 - w)x_{(y)} + wx_{(y+1)}$

$x_{(y)}$ = the 9th observation = 25

$x_{(y+1)}$ = the 10th observation = 28

$s = (1 - 0.35)25 + 0.35(28) = 26.05$ or 26.

Lordahl and Bookbinder (1994) show that this procedure is better than using the normal distribution in many cases. In particular, if the firm must achieve the targeted service level, this approach works well.

Karmarkar (1994) provides a similar methodology in which the general shape of the demand distribution is ignored because it is irrelevant to the SS decision. Rather, attention is focused on the right tail of the distribution. His method uses a mixture of two distributions, with the right tail modeled by a translated exponential. The method works quite well in simulation experiments.

The second way to apply Approach 1 is to measure the mean and variance of the total demand over the lead time, and apply a specific functional form. Then, if the total demand appears to fit a normal distribution, the methods from this chapter can be used to find the appropriate reorder point.

6.10.2 Approach 2: Use of the Distribution of Demand Rate per Unit Time Combined with the Lead Time Distribution

The second approach assumes that the lead time (L) and the demand (D) in each unit time period are independent random variables, and requires measurements or estimates of each. The assumption of independence is probably a reasonable approximation to reality despite the fact that, in some cases, high demand is likely to be associated with long lead times (i.e., positive correlation), because of the heavy workload placed on the supplier.[*] Likewise, low demand can be associated with long lead times (i.e., negative correlation) because the supplier has to wait longer to accumulate sufficient orders for his or her desired run size. If L and D are assumed to be independent random variables, then

$$E(x) = E(L)E(D) \tag{6.41}$$

and

$$\sigma_x = \sqrt{E(L)\mathrm{var}(D) + [E(D)]^2\mathrm{var}(L)} \tag{6.42}$$

where x, with mean $E(x)$ and standard deviation σ_x, is the total demand in a replenishment lead time, in units; L, with mean $E(L)$ and variance var (L), is the length of a lead time (L is the number of unit time periods, i.e., just a dimensionless number); and D with mean $E(D)$ and variance var (D) is the demand, in units, in a unit time period. Derivations for Equations 6.41 and 6.42 are in Section 6A.5 of the Appendix.

Where variability in L exists, the $E(x)$ and σ_x quantities found from Equations 6.41 and 6.42 should be used in place of $\hat{x}_L$ and σ_L in all the decision rules of this chapter.

---

[*] Wang et al. (2010) analyze the case where demand per period and lead times are correlated.

6.10.2.1 Numerical Illustration

We first illustrate the use of this approach when the combined distribution is (or is assumed to be) normally distributed. Let us consider two situations:

Situation 1: No Uncertainty in the Lead Time Itself (i.e., var (L) = 0 in Equation 6.42)
Suppose the unit time period is 1 week with $E(L) = 4$ and $E(D) = 100$ units and var(D) = 300 (units)^2. Then Equations 6.41 and 6.42 give

$$E(x) = 400 \text{ units}$$

and

$$\sigma_x = 34.64 \text{ units}$$

Then, to provide a desired probability (P_1) of no stockout per replenishment cycle equal to 0.95, we require a reorder point of $s = E(x) + k\sigma_x = 400 + 1.64(34.64) = 456.81$, say, 457 units.

Situation 2: Uncertainty in the Lead Time
Suppose that $E(L)$, $E(D)$, and var(D) are as above, so that $E(x)$, the expected total demand over a lead time, is still 400 units. However, now we assume that L itself is random, in particular

$$\text{var}(L) = 1.44$$

(The standard deviation of L is $\sqrt{\text{var}(L)}$, or 1.2.) Then, Equation 6.42 gives

$$\sigma_x = \sqrt{(4)(300) + (100)^2(1.44)} = 124.90 \text{ units}$$

Now we find that the reorder point must be 605 units. In other words, the uncertainty in L has increased the reorder point by 148 units. Knowing the values for v and r would permit us to calculate the additional carrying costs implied by the extra 148 units. These extra costs would provide a bound on how much it would be worth to reduce or eliminate the variability in L.

6.10.3 Nonnormal Distributions

The expressions for mean and standard deviation of demand during lead time in Equations 6.41 and 6.42 do not necessarily assume a Normal distribution. So, these values could be used in conjunction with another distribution, such as the Gamma distribution (e.g., Tyworth et al. 1996; Moors and Strijibosch 2002; Leven and Segerstedt 2004). Such an approach may be justified when the demand during lead time is positively skewed or the ratio of $\sqrt{\text{var}(L)}/E(L)$ is large[*]. Use of the Gamma distribution is a good alternative in these cases. In Appendix III, we present the details of how one can implement the use of the Gamma distribution in a spreadsheet.

When combining the demand and lead time distribution using the approach in the previous section, it is possible to draw on some compelling research. Eppen and Martin (1988) show by an example that if the lead time takes on just a few values, say 2 or 4 weeks with equal probability, the use of the normal distribution provides very poor and unusual results. In fact, the reorder point in their example actually increases when the lead time decreases. (See Problem 6.27.)

---

[*] Note that if this ratio is much greater than 0.4, there is a significant chance of negative demand values.

Other research on the problem of variable lead time includes Burgin (1972), McFadden (1972), Magson (1979), Ray (1981), Mayer (1984), Bagchi and Hayya (1984), Bagchi et al. (1984, 1986), Zipkin (1988), Bookbinder and Lordahl (1989), Tyworth (1992), Chen and Zheng (1992), Federgruen and Zheng (1992), Wang and Rao (1992), Strijbosch and Heuts (1994), Heuts and de Klein (1995), and Keaton (1995). Most of this research assumes that the lead time and demand distributions are given (even if not known). Wu and Tsai (2001) and Chu et al. (2004) investigate the more complicated but more flexible approach of using mixtures of distributions for modeling lead times.

It is often possible, however, to "change the givens," and some work on this problem has been done. Das (1975) shows that the total relevant costs are more sensitive to changes in the expected lead time if $\sqrt{\text{var}(L)}/E(L) > 1$. Otherwise, they are more sensitive to changes in the variance of lead time. This study has implications for where improvement resources should be allocated—to reduce the average lead time, or to reduce the variability of lead time. See also Nasri et al. (1990), Paknejad et al. (1992), Song (1994a,b), Lee et al. (2000), and Ray et al. (2004).

6.11 Exchange Curves Involving SSs for (s, Q) Systems

As was discussed in Chapters 1 and 2, management is primarily concerned with aggregate measures of effectiveness. In Chapter 4, exchange curves were developed that showed the aggregate consequences, in terms of the total average cycle stock versus the total number (or costs) of replenishments per year, when a particular policy (the EOQ) is used across a population of items. The curve was traced out by varying the value of the policy variable A/r. In a similar fashion, exchange curves can be developed showing the aggregate consequences, in terms of TSS (in dollars) versus aggregate service, when a particular decision rule for establishing SSs (i.e., reorder points) is used across a population of items.

Exchange curves are also useful as a means of determining the preferred level of service for a given investment in the inventory. Many inventory formulas assume that the firm has an infinite amount of working capital available. Obviously, this is never true. If the firm can achieve its service objectives without being constrained by dollar investment in the inventory, it is in an enviable position. Unfortunately, many firms face constraints on the amount of money available for investment in inventories, and must trade off inventory investment with service.

We begin, in Section 6.11.1, our discussion of exchange curves with an example of the inventory/service trade-off for a single item. Then, we show, in Section 6.11.2, a simple three-item example that illustrates the aggregate benefits that are usually possible in moving away from the commonly used approach of basing SSs (or reorder points) on equal time supplies. We also describe the general method of developing exchange curves.

As in Chapter 4, there is an input policy variable for each type of decision rule. For example, the management may decide to compute reorder points based on a common P_2 objective—that is, each item's reorder point is determined using the formulas from Section 6.7.10. Management's selection of an aggregate operating point, based on the exchange curve developed, implies a value of the policy variable—say, $P_2 = 0.97$. The use of this implicit value in the corresponding individual-item decision rule then leads to the desired aggregate consequences. An outline of the derivation of exchange curves is presented in Section 6.11.3. We restrict attention to (s, Q) systems but, because of the analogy developed in Section 6.9, the same type of approach can be used for (R, S) systems. In each case, note that by developing the formulas on a spreadsheet, it is straightforward to use a data table or similar functionality to quickly build exchange curves.

Management's ability to meaningfully utilize the SS versus service exchange curve hinges on their capacity to think of inventories as being split into the two components of cycle stocks and SSs and to be able to make separate tradeoffs of cycle stocks versus replenishments and SSs versus aggregate service. Based on our experience, these conditions are not always met. Therefore, in Section 6.11.4, we discuss the so-called composite exchange curves that show a three-way exchange of the total stock (cycle plus safety, in dollars) versus the number of replenishments per year versus aggregate service.

6.11.1 Single Item Exchange Curve: Inventory versus Service

When a manager is faced with a constraint on the amount of money available to invest in the inventory, it is necessary to determine the best customer service possible for the given investment. Often, however, the constraint is not rigid—it is possible to divert funds from other activities if it is determined that the gains in customer service are worth the loss in those other activities. An exchange curve, even for a single item, is extremely valuable in such situations. Figure 6.13 presents one example, using a fill rate (P_2) service objective and dollar investment in SS. Notice that the relationship is not linear; in fact, there are diminishing returns. In other words, to attain perfect customer service, the cost is exceedingly high. Also, notice that because of a relatively large order quantity, the actual SS required for low service levels is very low. The authors have used curves like

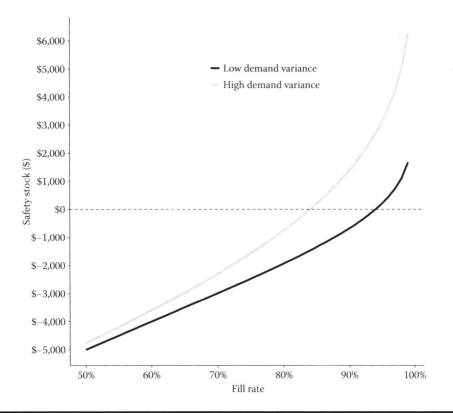

Figure 6.13   A single item exchange curve.

Figure 6.13 to show clients how much money is required to achieve their desired level of customer service. Alternatively, the manager can choose a service level based on the capital available.

It is critical, however, when using exchange curves in this way, to consider the possibility of changing the givens. What would happen, for instance, if the lead time were reduced? Or, what would be the effect of a closer relationship to the customers so that demand variability were reduced? Figure 6.13 illustrates the result of reducing demand variability by half for the particular item considered. The savings in SS are significant, especially for high service levels, implying that managers should consider investing resources to reduce demand variability.

6.11.2 An Illustration of the Impact of Moving away from Setting Reorder Points as Equal Time Supplies

Single item exchange curves can be useful; but, by far, the more common use of exchange curves is with multiple items. We now briefly discuss the general approach to developing exchange curves, an approach that is easy to use even with spreadsheets. As noted in Section 6.8, when an input objective is specified for an item, implied measures of performance are obtained. To develop exchange curves, one simply specifies an input objective (such as B_1, B_2, P_1, or P_2). The computations required to find the reorder point for each item involve finding a k value for each item. This k value then determines the implied value of the other performance measures, such as B_1, B_2, P_1, or P_2. Managers may also be interested in TSS, the expected total number of stockout occasions per year (ETSOPY), or the expected total value of shortages per year (ETVSPY). See Table 6.8. These can then be graphed on an exchange curve.

It is not uncommon for organizations to use the following type of rule for setting reorder points: reorder when the inventory position has dropped to a 2-month time supply. Without going into detail, we wish to show the impact of moving away from this type of decision rule. To illustrate, let us consider three typical items produced and stocked by MIDAS International (a parent company, serving Canadian as well as other subsidiaries and direct customers throughout the world). Relevant characteristics of the items are given in Table 6.9. The order quantities, assumed as prespecified, are as shown. The current reorder points are each based on a 2-month time supply (i.e., $D/6$).

Assuming normally distributed forecast errors, it can be shown (by the procedure to be discussed in Section 6.11.3) that, under the use of the current reorder points, the SSs, the expected

Table 6.8 Exchange Curve Methodology

Input		Outputs
		B_1
		P_2
		P_1
B_1 or B_2 or P_1 or P_2 or ...	$\longrightarrow k \longrightarrow$	P_2
		TSS
		ETSOPY
		ETVSPY

Table 6.9   Three-Item Example

Item Identification	Demand Rate D (Units/Year)	Unit Value v ($/Unit)	Lead Time L (Months)	σ_L (Units)	Current Order Quantity Q (Units)	Current Reorder Point s (Units)
PSP-001	12,000	20	1.5	300	2,000	2,000
PSP-002	6,000	10	1.5	350	1,500	1,000
PSP-003	4,800	12	1.5	200	1,200	800

stockout occasions per year, and the expected values short per year are as listed in columns (2), (3), and (4) of Table 6.10. If the TSS of $14,900 is instead allocated among the items to give equal probabilities of stockout per replenishment cycle (i.e., equal P_1 values), the resulting expected stockout occasions per year and the expected values short per year are as in columns (6) and (7) of the table. The reallocation has reduced both aggregate measures of disservice substantially without increasing inventory investment. The reason for this is that the current equal-time supply strategy does not take into account of the fact that the forecasts for PSP-002 and PSP-003 are considerably more uncertain relative to their average demand during the lead time than PSP-001. Note the shift of SS away from PSP-001 to the other items under the revised strategy. The marked service improvement for PSP-002 and PSP-003 outweighs the reduction in service of PSP-001 caused by this reallocation of SS. The type of service improvement indicated here is typical of what can be achieved in most actual applications of logical decision rules for computing reorder points.

For either of the two service objectives shown in Table 6.10, we can achieve even greater improvement by determining reorder points using a B_1 or B_2 rule. As discussed in Section 6.7.12,

Table 6.10   Numerical Example of Shifting away from Reorder Points as Equal Time Supplies

	Equal Time Supply Reorder Points			Reorder Points Based on Equal Probabilities of Stockout per Cycle		
Item No.	Safety Stock[a]	Expected Stockout Occasions per Year	Expected Total Value Short per Year	Safety Stock	Expected Stockout Occasions per Year	Expected Total Value Short per Year
(1)	(2)	(3)	(4)	(5)	(6)	(7)
1	$10,000	0.287	$714	$7,513	0.632	$1,813
2	2,500	0.950	1,952	4,382	0.421	705
3	2,400	0.635	800	3,005	0.421	484
Totals	$14,900	1.871/year	$3,466/year	$14,900	1.474/year	$3,002/year

[a]   Under the given strategy safety stock (in units) $= s - \hat{x}_L = s - DL$. Then, safety stock (in $) $= (s - DL)v$.

Table 6.11   Numerical Example of Setting Reorder Points with B_1 and B_2 Rules

Item No.	Reorder Points Based on a B_1 Rule			Reorder Points Based on a B_2 Rule		
	Safety Stock	Expected Stockout Occasions per Year	Expected Total Value Short per Year	Safety Stock	Expected Stockout Occasions per Year	Expected Total Value Short per Year
(1)	(2)	(3)	(4)	(5)	(6)	(7)
1	$6,852	0.760	$2,272	$8,210	0.514	$1,415
2	4,387	0.420	703	3,969	0.514	898
3	3,660	0.254	266	2,721	0.514	616
Totals	$14,900	1.435/year	$3,240/year	$14,900	1.541/year	$2,929/year

a B_1 rule will minimize the expected total stockouts per year for a given inventory budget. As indicated in Section 6.7.13, a B_2 will minimize ETVSPY for a given inventory budget. Keeping the same order quantities as above (see Table 6.9), we can calculate the SS allocations that minimize these service objectives for this numerical example. These allocations are shown in Table 6.11.

It is not a coincidence in Table 6.11 that the expected stockout occasions per year are the same for all three items when using a B_2 rule. Rearrangement of Equation 6.22 gives

$$\text{ESOPY} = \left(\frac{D}{Q}\right) p_{u\geq}(k) = \frac{r}{B_2}$$

So, if items are subject to the same holding cost rate r and the same fractional charge per unit short B_2, the expected stockout occasions per year will be the same.

We can compare the performance of these different approaches for setting reorder points across a wide range and generate a multi-item exchange curve as shown in Figures 6.14 and 6.15. As we would expect, allocating SS according to a B_1 rule generates the best (lowest SS levels for a given level of service) exchange curve when the performance measure is the expected stockout occasions per year. In addition, this figure indicates how poorly a simplistic rule such as equal time supply can perform. Figure 6.15 shows an analogous result where an allocation based on a B_2 rule performs best when we seek to minimize the expected total value short.

6.11.3 Derivation of the SS Exchange Curves

We now provide a more formal treatment of exchange curve derivation. In all cases, for a given value of the relevant policy variable, the corresponding point on each exchange curve is found in the following manner. First, the value of the safety factor k_i is ascertained for each item by use of the appropriate decision rule. Then, any number of output performance measures are computed

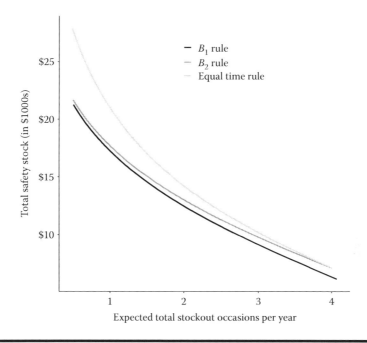

Figure 6.14   Exchange curves of TSS versus the expected total stockout occasions.

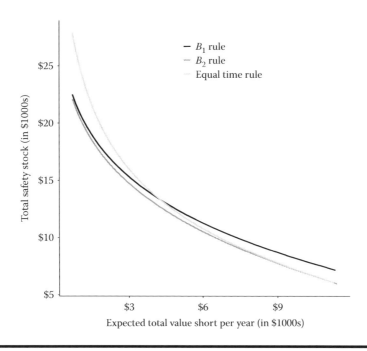

Figure 6.15   Exchange curves of TSS versus the expected total value short.

for each item i:

$$\text{Safety stock (in dollars) of item } i = SS_i v_i = k_i \sigma_{L_i} v_i \qquad (6.43)$$

$$\text{Expected stockout occasions per year} = \frac{D_i}{Q_i} p_{u\geq}(k_i) \text{ for item } i \qquad (6.44)$$

$$\text{Expected value short per year} = \frac{D_i}{Q_i} \sigma_{L_i} v_i G_u(k_i) \text{ for item } i^* \qquad (6.45)$$

$$\text{Expected fill rate} = 1 - \frac{\sigma_{L_i} G_u(k_i)}{Q_i} \text{ for item } i \qquad (6.46)$$

$$\text{Expected cycle service level} = 1 - p_{u\geq}(k_i) \text{ for item } i \qquad (6.47)$$

These equations are, of course, based on an assumption of normally distributed errors. However, one could develop comparable results for any of the other distributions discussed in Section 6.7.14.

Equation 6.44 follows from multiplying the expected number of replenishment cycles per year (Equation 6.15) by the probability of a stockout per cycle (Equation 6.16). Similarly Equation 6.45 is a result of the product of the expected number of replenishment cycles per year (Equation 6.15) and the expected shortage per replenishment cycle, expressed in dollars (v_i times Equation 6.17). Finally, each of the these quantities is summed across all items. In the last two cases, it is common to take a demand-weighted summation:

$$\text{Total safety stock (in dollars)} = \sum_i k_i \sigma_{L_i} v_i \qquad (6.48)$$

$$\text{Expected total stockout occasions per year (ETSOPY)} = \sum_i \frac{D_i}{Q_i} p_{u\geq}(k_i) \qquad (6.49)$$

$$\text{Expected total value short per year (ETVSPY)} = \sum_i \frac{D_i}{Q_i} \sigma_{L_i} v_i G_u(k_i) \qquad (6.50)$$

$$\text{Demand-weighted fill rate} = \frac{\sum_i D_i \left(1 - \frac{\sigma_{L_i} G_u(k_i)}{Q_i}\right)}{\sum_i D_i} \qquad (6.51)$$

$$\text{Demand-weighted cycle service level} = \frac{\sum_i D_i \left(1 - p_{u\geq}(k_i)\right)}{\sum_i D_i} \qquad (6.52)$$

The results of Equations 6.48 and 6.49 give a point on the first exchange curve (e.g., Figure 6.14) and those of Equations 6.48 and 6.50 give a point on the second curve (e.g., Figure 6.15) for the particular SS decision rule. Repeating the whole process for several values of the policy

---

* Strictly speaking, the expected value short per year for item i is

$$= \frac{D_i}{Q_i} \sigma_{L_i} v_i \left[G_u(k_i) - G_u\left(k_i + \frac{Q_i}{\sigma_{L_i}}\right) \right]$$

The approximation of Equation 6.45 is quite accurate as long as $Q_i/\sigma_{L_i} > 1$, because then the additional term becomes negligible. When $Q_i/\sigma_{L_i} < 1$, it is advisable to use the more accurate expression. See Equation 6.29.

variable generates a set of points for each of the two curves. Similar curves can be generated for demand-weighted fill rate and demand-weighted cycle service level.

In some cases, it is not necessary to go through all the calculations of Equations 6.48 through 6.52. To illustrate, for the case of a specified fraction (P_2) of demand to be satisfied directly off the shelf for all items,

$$\text{Expected total value short per year} = (1 - P_2) \sum_i D_i v_i$$

provided, of course, that no individual k_i has to be adjusted up to a minimum allowable value (thus providing a better level of service than P_2 for that particular item).

6.11.4 Composite Exchange Curves

For any decision rule for selecting SSs, we can combine the EOQ exchange curve showing the total average cycle stock versus the number of replenishments per year (recall Figure 4.12) with any of the SS exchange curves (TSS vs. aggregate service level). The result is a set of composite exchange curves showing a three-way trade-off of the total average stock versus the number of replenishments per year versus the aggregate service level. A separate line is required for each combination of decision rule and aggregate service measure. For the three-item numerical example from the previous section, we have shown an illustrative set of curves (for three levels of only one aggregate service measure) in Figure 6.16. They represent the use of the P_2 decision rule and an EOQ ordering strategy.

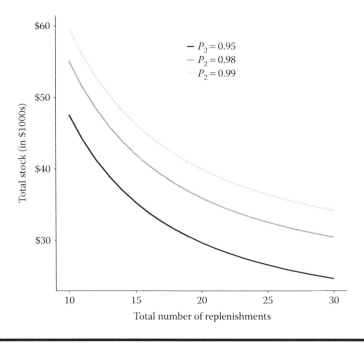

Figure 6.16 Composite exchange curves with a P_2 rule.

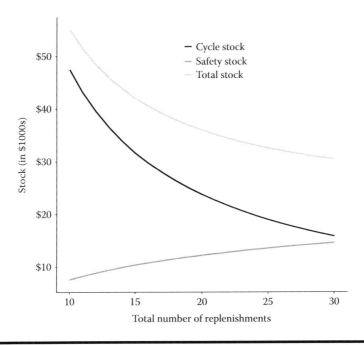

Figure 6.17 **Breakdown of cycle stock and SS for an exchange curve with a P_2 rule.**

Each curve of Figure 6.16 decreases as the number of replenishments N increases for a given aggregate service level (the value of P_2). This is a consequence of the reduction in cycle stock that results from more frequent replenishment dominating the increase in SS needed to maintain the desired P_2 level.

Figure 6.17 shows the behavior of cycle stock and SS separately for $P_2 = 0.98$. It is clear in this figure that more frequent replenishments reduce cycle stock and increase SS. It is quite possible that in some settings, the increase in SS exceeds the reduction in cycle stock, and the aggregate curves may eventually increase as the number of replenishments increases.

Selection of a desired aggregate operating point on a composite exchange curve implies:

1. A value of A/r to be used in the EOQ decision rule (this is not shown directly on the graphs; instead, we show the total number of replenishments per year N. A/r can be found from N using Equation 4.30 of Chapter 4).
2. A value of the parameter to be used in the particular SS decision rule under consideration (e.g., P_1, TBS, B_1, etc.).

For extensions of these concepts see Gardner and Dannenbring (1979) and Gardner (1983).

6.12 Summary

In this chapter, we have introduced the reader to the fundamentals of inventory control under probabilistic demand. For a variety of shortage costs and service measures, we have shown the details of two common types of decisions systems:

1. (s, Q) systems
2. (R, S) systems

The material presented included the actual decision rules, illustrative examples, and a discussion of each of the rules. Again, aggregate exchange curve concepts turned out to be an important aspect of the analysis. In Chapter 7, we will address two other, somewhat more complicated control systems—namely, (s, S) and (R, s, S)—in the context of A items.

It should now be apparent that, even for the case of a single item considered in isolation at a single stocking point, it is not an easy task to cope with uncertainty. It is little wonder that *usable* theory for more complex probabilistic situations (e.g., coordinated control of related items at a single location, multi-echelon control of a single item, etc.) is not extensive. We will have more to say on these issues in Chapters 10, 11, 15, and 16.

Problems

6.1 It is reasonable to assume that demand over its replenishment lead time of an item, ISF-086, is normally distributed with $\hat{x}_L = 20$ units and $\sigma_L = 4.2$ units.
   a. For a desired service level of $P_1 = 0.90$, compute the required safety factor, SS, and reorder point.
   b. Repeat for $P_1 = 0.94, 0.98$, and 0.999.
   c. Prepare a rough sketch of SS versus P_1. Is the curve linear? Discuss.

6.2 The demand for a certain product has shown to be relatively stable; its sales are not subject to any seasonal influence, and demand is not expected to change in the future. In fact, during the last 10 lead times, the following demand was observed:

64	51	48	32	93	21	47	57	41	46

   Assuming that this demand is a good representative of total demands in future lead times and a reorder point of 51 is used, compute:
   a. The various possible shortage sizes and their probabilities of occurrence (HINT: Assume that these are the only values that the demand can take, each with a 0.1 probability of occurring).
   b. The expected (or average) shortage.
   c. What reorder point would you use if the average shortage was to be no larger than 5 units? What is the associated probability of a stockout?

6.3 Consider an item with $A = \$25; Dv = \$4,000/\text{year}; \sigma_L v = \$100; B_1 = \$30$; and $r = 0.10 \$/\$/\text{year}$.
   a. Find the following:
      i. EOQ, in dollars
      ii. k, using the B_1 criterion
      iii. SS, in dollars
      iv. Annual cost of carrying SS
      v. Total average stock, in dollars
      vi. Expected number of stockout occasions per year
      vii. Expected stockout costs per year

b. The cost equation used to develop the rule for the B_1 criterion is

$$\text{ETRC}(k) = k\sigma_L vr + \frac{D}{Q}B_1 p_{u \geq}(k) \qquad (6.53)$$

As seen in part 6.3a, the two components of Equation 6.53 are not equal at the optimal k value. (This is in contrast to what we found for the EOQ analysis in Chapter 4.) Why are the two components not equal? What quantities are equal at the optimal k value?

c. By looking at the basic equations for the items requested in (a), discuss how each would be affected (i.e., whether it would be increased or decreased) by an increase in the r value.

6.4 Consider an item with $A = \$25$; $D = 4,000$ units/year; $v = \$8$/unit; $L = 1$ month; $\sigma_L = 100$ units; TBS $= 6$ months; and $r = 0.16$ \$/\$/year.
   a. Compute the reorder quantity of the (s, Q) policy using the EOQ.
   b. Compute the safety factor, SS, and reorder point.
   c. Graph TBS versus ETVSPY (the expected total value short per year).

6.5 Consider an item that has an average demand rate that does not change with time. Suppose that demands in consecutive weeks can be considered as independent, normally distributed variables. Observations of total demand in each of 15 weeks are as follows:

89	102	107	146	155	64	78	122
78	119	76	80	60	115	86.	

a. Estimate the mean and standard deviation of demand in a 1-week period and use these values to establish the reorder point of a continuously reviewed item with $L = 1$ week. A B_2 value of 0.3 is to be used, and we have $Q = 1,000$ units, $v = \$10$/unit, $r = 0.24$ \$/\$/year, and $D = 5,200$ units/year.

b. In actual fact, the above-listed demands were randomly generated from a normal distribution with a mean of 100 units and a standard deviation of 30 units. With these true values, what is the appropriate value of the reorder point? What is the percentage cost penalty associated with using the result of part 6.5a instead?

6.6 Canadian Wheel Ltd. establishes SSs to provide a fraction of demand satisfied directly from stock at a level of 0.94. For a basic item with an essentially level average demand, the annual demand is 1,000 units and an order quantity of 200 units is used. The supplier of the item ensures a constant lead time of 4 weeks to Canadian Wheel. The current purchase price of the item from the supplier is \$0.80/unit. Receiving and handling costs add \$0.20/unit. The supplier offers to reduce the lead time to a new constant level of 1 week, but in doing so, she will increase the selling price to Canadian by \$0.05/unit. Canadian management is faced with the decision of whether or not to accept the supplier's offer.
   a. Qualitatively discuss the economic trade-off in the decision. (Note: Canadian would not increase the selling price of the item nor would they change the order quantity.)
   b. Quantitatively assist management in the decision. (Assume that $\sigma_t = \sigma_1 t^{1/2}$ and $\sigma_4 = 100$ units where σ_4 is the standard deviation of forecast errors over the current lead

time of 4 weeks. Also assume that forecast errors are normally distributed and that the annual carrying charge is 20%.)

6.7   The 4N Company reacts to shortages at a particular branch warehouse in the following fashion: Any shortage is effectively eliminated by bringing material in from the plant by air freight at a cost of $2.50/kg. Assume that the average demand rate of an item is essentially constant with time and that forecast errors are normally distributed. Assume that an (s, Q) system is to be used and that Q, if needed, is prespecified. To answer the following questions, introduce whatever symbols are necessary:

a.   What are the expected shortage costs per cycle?

b.   What are the expected shortage costs per year?

c.   What are the expected total relevant costs per year associated with the safety factor k?

d.   Develop an equation that the best k must satisfy.

e.   If we had two items with all characteristics identical except that item 1 weighed more (on a unit basis) than item 2, how do you intuitively think that the two safety factors would compare? Verify this using the result of part (d).

6.8   The Norwich Company has a policy, accepted by all its customers, of being allowed up to 1 week to satisfy each demand. Norwich wishes to use a P_2 service measure with a specified P_2 value. Consider an item controlled by an (s, Q) system with Q prespecified. Suppose that the lead time L was equal to the grace period.

a.   Under these circumstances, what s value is needed to give $P_2 = 1.00$?

b.   More generally, consider an item with a desired $P_2 = 0.99$ and a Q value of 500 units. Other (possibly) relevant characteristics of the item are

$D = 1{,}000$ units/year (the average is relatively level throughout the year)

$v = \$5$/unit

$r = 0.20$ \$/\$/year

$L = 5$ weeks

$\sigma_1 = 10$ units (for a 1-week period); $\sigma_t = \sigma_1 t^{1/2}$

How would you modify the logic for computing the s value to take into account of the 1-week "grace" period? Illustrate with this item.

c.   For the item of part (b), how would you estimate the annual cost savings of having the grace period?

6.9   A retail electronic-supply outlet orders once every 2 weeks. For a particular item with unit value (v) of $3, it is felt that there is a cost of $1 for each unit demanded when out of stock. Other characteristics of the item are

$L = 1$ week

$r = 30\%$ per year

a.   Consider a period of the year where $\hat{x}_1 = 30$ units and $MSE_1 = 6$ units (where the unit time period is 1 week). What order-up-to-level S should be used? (Assume $\sigma_t = \sigma_1 t^{1/2}$.)

b.   Suppose the average demand rate changed slowly throughout the year. Suggest a reasonable way for the retailer to cope with this situation.

6.10   A particular company with production facilities located in Montreal has a large distribution warehouse in Vancouver. Two types of transportation are under consideration for shipping material from the plant to the remote warehouse, namely,

Option A: Send all material by railroad freight cars.

Option B: Send all material by air freight.

If only one of the two pure options can be used, *outline* an analysis that you would suggest to help the management choose between the two options. Indicate the types of data that you would require. *Intuitively* would it ever make sense to use both options, that is, to use rail freight for part of the shipments and air freight for the rest? Briefly explain why or why not.

6.11 You have signed a consignment contract to supply a chemical compound to a major customer. This compound costs you $32.50/gallon and generates a contribution of $12.50/gallon. Delivery costs you $175 in shipping and handling. Demand for this year is estimated at 2,000 gallons, standard deviation is 400 gallons. The contract says that the compound is inventoried at your client's facility: he will draw the material from that stock as needed and pay you at such time. It also says that if the client finds that inventory empty when needed, you are subject to a fee of $0.325/gallon/day short. Inventory carrying cost is 0.20 $/$/year.

a. Determine the optimal (s, Q) policy, assuming a stochastic lead time, $L = 30$ days, standard deviation of lead time = 6.

b. Your customer has realized that controlling the stockout charge is too burdensome. Two alternative charges have been proposed:
   i. A flat charge of $1,000 per stockout occasion.
   ii. A $65 charge per gallon short per cycle.
   Under each alternative, what would your policy be? Are the proposed changes acceptable?

6.12 A fast-food restaurant has just launched a new hamburger recipe. All of its products are prepared in small batches, targeting a fill rate $P_1 = 0.7$ at lunch time. The manager observed the demand while the first few batches of the new sandwich were prepared. They were:

3	7	2	0	10	5
4	8	1	6	11	8

a. For the next few days (until better information is gathered) what reorder point should be used?

b. What additional information would you need in order to decide the size of the next batch, now that you have the reorder point?

6.13 Consider an (r, S) policy where the management adopts a stockout charge B_2 per unit short. Is such a policy sensitive to the individual item's parameters?

6.14 a. A company manager has decided that the inventory control policy "should have zero stockouts." Is this a realistic policy to pursue? In particular, discuss the implications in terms of various types of costs.

b. On occasion, this company has filled an important customer order by supplying a competitor's product at below the retail price. What form of shortage costing would it be appropriate to apply under such circumstances? Does this type of strategy make sense?

6.15 Safety Circuits Inc. stocks and sells small electronics products using an (R, S) type of control system. A manager reviews the stock monthly. The supplier delivers at their location following a lead time of 10 days. Inventory carrying charge is 0.12 $/$/year. The manager has observed that demand for the XRL line is as follows:

Item i	Demand D_i (Packs/Year)	v_i ($/Pack)	$\sigma_{i,1}$ (Packs)
XRL-1	1,200	10	35
XRL-2	350	35	50
XRL-3	700	17	40

where $\sigma_{i,1}$ is the standard deviation of yearly demand for item i. Suppose that a TSS of $1,200 is to be allocated among the three items. Consider the following service measures:
1. Same k for the 3 items
2. Same P_1 for the 3 items
3. Same TBS for the 3 items
4. Same B_1 for the 3 items
5. Same B_2 for the 3 items

For each, determine
a. How the $1,200 is allocated among the three items?
b. ETSOPY
c. ETVSPY (expected total value short per year)

Note: Assume that forecast errors are normally distributed and the same r value is applicable to all three items. Negative safety factors are not permitted.

6.16 Safety Circuits Inc. is considering investing in a scanner such that they can move from an (R, S) to an (s, Q) inventory policy (see Problem 6.15). It is expected that with a continuous- review policy, the firm could reduce inventory management costs significantly: holding and shortage costs could be reduced. Management believes that for each unit short, there is an implicit charge $B_2 = 0.15$.
   a. Determine the safety factor, SS, and reorder point for each of the three products, assuming that the same lot sizes are adopted.
   b. Determine the total relevant costs for the XRL product line under an (s, Q) policy and $B_2 = 0.15$.
   c. Determine the total relevant costs for the XRL product line under an (R, S) policy and $B_2 = 0.15$.
   d. Assume that the scanner costs $30,000, and its operating costs are immaterial. The XRL product line is a good representative of the products carried by SFI, and that it comprises 1% of the total average stock for the firm. Calculate the time before the investment breaks even (disregard cost of money). Would you recommend the change? Discuss.

6.17 Waterloo Warehouse Ltd. (WWL) acts as a distributor for a product manufactured by Norwich Company. WWL uses an (s, Q) type of control system. Demands on Waterloo can be assumed to be unit sized. Norwich has a particular way of delivering the quantity Q: They deliver $Q/2$ at the end of the lead time of 1 month and the other $Q/2$ at the end of 2 months. Assume that this business has the following characteristics:

$Q = 100$ units
$s = 40$ units
$\sigma_1 = 10$ units ($\sigma_1 =$ the standard deviation of forecast errors over 1 month)
$D = 360$ units/year

$v = \$1.00/\text{unit}$

$r = 0.25 \$/\$/\text{year}$

a. What is the numerical value of the SS? Explain your answer.

b. Suppose that a replenishment is triggered at $t = 0$, and no order is outstanding at this time. Describe in words what conditions must hold in order for there to be no stockout up to time $t = 2$ months.

c. Suppose that Norwich offered WWL the option to receive the whole quantity Q after a lead time of 1 month, but at a higher unit price. Outline the steps of an analysis to decide whether or not the option should be accepted.

6.18 True Colors Co. purchases a number of chemical additives for its line of colors and dyes. After a recent classification of stock, the inventory manager has to decide the purchase criteria for the B items. Consider the pigment Red-105: demand forecast for the next 12 months of 200 gallons, with a standard deviation of 50 gallons. They cost \$25/gallon at the supplier's plant, in addition to \$30 per order for shipping and handling; expected lead time is 45 days. It is believed that each day costs the firm \$5 per gallon short. Inventory carrying charge is 0.18 \$/\$/year.

a. What (s, Q) policy should be adopted for this product. What is the implicit TBS?

b. Looking through some of the recent orders, the manager observed that, in fact, 90% of the shipments arrived in 30 days, the remainder in 75 days (the later arrivals due to unpredictable losses at the supplier). Should the inventory policy be adjusted to account for this lead time uncertainty? Discuss.

6.19 Consider a group of n items that are each being independently controlled using an (R, S) type of system (periodic-review, order-up-to-level). Let R_i be the prespecified value (in months) of R for item i. Suppose that we wish to allocate a total given SS in dollars (call it W) among the n items to minimize the dollars of expected backorders per unit time summed across all n items.

a. Develop the decision rule for selecting k_i, the safety factor for item i. Assume that forecast errors are normally distributed.

b. Illustrate for the following two-item case ($r = 0.2 \$/\$/\text{year}$; $W = \$1,180$; and $L = 1$ month):

Item i	v_i ($/Units)	A_i ($)	Demand D_i (Units/Year)	R_i (Months)	σ_{R_i} (Units)	σ_{L_i} (Units)	σ_{R+i+L_i} (Units)
1	1	2.00	3,000	1.0	200	200	300
2	2	5.00	15,000	0.5	500	300	700

6.20 Consider the following product line:

Item i	Demand D_i (Units/Year)	v_i ($/Units)	$\sigma_{i,1}$ (Units)	A ($)
1	100	250.00	45	100.00
2	200	30.00	140	100.00
3	300	250.00	210	150.00
4	60	210.00	35	150.00

Inventory carrying charges are 0.12 $/$/year, and the lead time is 3 months. Management feels that a fractional charge of 0.3 for each item short should be considered. Assume that lead time demand is normally distributed.

a. Determine the optimal (s, Q) policy, identifying SS, reorder point, and reorder quantity for each item using the EOQ.

b. After more careful study, management realized that lead time demand is better described by the Gamma distribution. Adjust the (s, Q) policy in the previous question to account for the actual distribution.

c. Data from the previous 10 cycles give the actual lead time demand for item 3 as follows:

0	32	308	77	60	105	6	81	208	8

What would have been the actual shortage per year under each of the suggested policies?

6.21 Consider three items with the following characteristics:

Item i	Demand D_i (Units/Year)	v_i ($/Units)	Q_i (Units)	σ_{L_i} (Units)
1	1,000	1.00	200	100
2	500	0.80	200	100
3	100	2.00	100	30

Suppose that a TSS of $120 is to be allocated among the three items. Consider the following service measures (shortage-costing methods):

1. Same P_1 for the 3 items
2. Same P_2 for the 3 items
3. Same B_1 for the 3 items
4. Same B_2 for the 3 items
5. Same TBS for the 3 items
6. Minimization of total expected stockout occasions per year
7. Minimization of total expected value short occasions per year
8. Equal time supply SSs

For each, determine

a. How the $120 is allocated among the three items
b. ETSOPY
c. ETVSPY

Note: Assume that forecast errors are normally distributed, that the same r value is applicable to all three items, and that negative safety factors are not permitted.

6.22 Botafogo Circuits imports some electronic components and redistributes them in the Brazilian market. This includes some difficulties, such as delivery uncertainty. Using regular air freight, shipping and handling costs $250 per item ordered, but the delivery may take anywhere from 10 to 30 days. Because of this uncertainty, the manager contacted a number of carriers looking for more reliable alternatives. Two distinct bids stood up:

a. An express importer would deliver any order within 15 days, at a cost of $350 per item ordered.

b. A regular shipper would pick orders every 15th of the month and deliver them on the 15th of the next month, for a cost of $200 per order plus $0.50 per unit, for any number of items ordered.

Demand for these items can be approximated by a Normal distribution. The following table shows the mean and standard deviation of the yearly demand, as well as the unit costs for some of these items:

Item i	Demand D_i (Packs/Year)	v_i ($/Pack)	$\sigma_{i,1}$ (Packs)
XRL-10	1,200	40.00	860
XRL-12	1,250	35.00	500
XRL-15	1,800	22.00	120
XRL-20	700	17.00	400

In its internal accounts, the manager considers a fractional charge of 0.50 for each unit short, and an inventory carrying charge of 0.30 $/$/year. Describe the inventory policy that should be used with each of the transportation alternatives. Which alternative should the firm use?

6.23 Consider an item under (s, Q) control. Basic item information is as follows: $D = 40,000$ units/year; $A = \$20$; $r = 0.25$ $/$/year; and $v = \$1.60$/unit. All demand when out of stock is backordered. The EOQ is used to establish the Q value. A service level of 0.95 (demand satisfied without backorder) is desired. The item's demand is somewhat difficult to predict and two forecasting procedures are possible.

System	Cost to Operate per Year ($/Year)	σ_L (Units)
A (complex)	200	1,000
B (simple)	35	2,300

Which forecasting system should be used? Discuss. *Note:* Forecast errors can be assumed to be normally distributed with zero bias for both models.

6.24 Consider two items with the following characteristics:

Item	Unit Value ($/Unit)	Demand D (Units/Year)	x_L (Units)	σ_L (Units)
1	10.00	300	100	10
2	1.00	300	100	35

Suppose that the inventory controller has set the SS of each of these items as a 1-month time supply.

a. What are the SSs in units? In dollars?
b. What is the P_1 value associated with each item?
c. Reallocate the same TSS (in dollars) so that the two items have the same value of P_1.
d. What reduction in TSS is possible if both items have $P_1 = 0.95$?

6.25 Upper Valley Warehouse adopts an (s, Q) inventory policy where all items meet a safety factor $k = 1.2$. Some of the B items enjoy a deterministic lead time of 1 month from their suppliers. They are the following products:

Item i	Demand D_i (Units/Year)	v_i ($/Units)	$\sigma_{i,1}$ (Units)	A ($)
1	1,000	50.00	300	100.00
2	2,000	30.00	700	100.00
3	1,000	100.00	700	100.00
4	800	80.00	500	120.00
5	1,000	35.00	400	120.00
6	3,000	150.00	800	120.00
7	600	210.00	200	150.00

Inventory carrying cost is 0.12 $/$/year. Reorder quantity is calculated by the EOQ.
a. Determine
    i. Implicit TBS for each product
    ii. ETSOPY
    iii. ETVSPY (expected total value short per year)
    iv. SS in dollars
b. The firm has decided to allocate $50,000 into the SS of these products. What should be the individual safety factor, SS, and the reorder point under each of the following policies:
    i. Same TBS for the 7 items
    ii. Minimize ETSOPY
    iii. Minimize ETVSPY
6.26 Consider an item with the following characteristics:
    $D = 1,000$ units/year
    $L = 4$ weeks
    $A = \$20$
    $v = \$5$ per unit
    $r = 0.20$ $/$/year
    $\hat{x}_L = 80$ units
    $\sigma_L = 8$ units
    $B_2 = 0.8$
a. Use the approach of Lau and Lau in Section 6.7.14 to find the optimal order quantity. Then find $F_x(s)$. Finally, assume that demand is normal, and find k and s.
b. Use the approach of Section 6.7.6 to find the optimal s and Q.
6.27 Consider the item discussed in Approach 2 in Section 6.10.2. Recall that with $\text{var}(L) = 0$, the reorder point is 457 units if P_1 is 0.95. Now suppose the firm actually reduces the lead time by working with its supplier. Unfortunately, the reduced lead time is not always achieved. In particular, the lead time is 2 weeks half of the time, but remains at 4 weeks half

the time. So

$$L = 4 \text{ with probability } 0.5$$
$$\text{and} \quad L = 2 \text{ with probability } 0.5$$

Thus, $E(L) = 3$, and $\text{var}(L) = 0.5(2-3)^2 + 0.5(4-3)^2 = 1$.

a. Use Equations 6.41 and 6.42 to find the total lead time demand, and then find the reorder point assuming that the total demand is normally distributed.

b. Eppen and Martin (1988) provide an alternative solution that relies on the actual distributions. Let k_1 be the safety factor when the lead time is 2 weeks; and let k_2 be the safety factor when the lead time is 4 weeks. Find an expression for the reorder point as a function of k_1 and another expression for the reorder point as a function of k_2. Now find the optimal reorder point by noting that the weighted average service level is 95%.

6.28 Tower Bookstore orders books from one particular supplier every Tuesday. Once the order is placed, it takes 3 days for delivery, and an additional four days to get the books on the shelf. This season there is a book—*The One Minute Parent*—that is selling especially well. The purchasing manager has gathered some data that she anticipates will be representative of demand for the next several months. It turns out that weekly demand has been 75 books on average, with a standard deviation of 5. The carrying charge is 18% per dollar invested in books, and the manager has determined that the charge per unit short is 30% of the sales price. Each copy costs Tower $5; and they apply a 50% markup to books of this type. Find the correct ordering policy.

6.29 A materials manager for Missoula Manufacturing has provided the following data for one of their widely used components:

$Q^* = 50$	$D = 500$	$rv = \$10$	$A = \$25$

Lead time demand for the component is normal with a mean of 25 and a variance of 25. What is the expected quantity of backordered items per year if the manager uses an optimal (s, Q) policy?:

a. The desired service level (the probability of no stockout during the lead time) is 90%.

b. The desired service level is 95%.

Appendix 6A: Some Illustrative Derivations and Approximations

6A.1 B_1 Shortage Costing

We show two different approaches: the first using the total relevant costs; the second based on a marginal analysis. The latter is simpler but does not permit sensitivity analysis (in that a total cost expression is not developed).

6A.1.1 Total Cost Approach

We recall from Section 6.7.5 the expression for the expected total relevant costs per year as a function of the control parameter k. Let us denote this function by ETRC (k). There are three relevant components of costs: (1) replenishment (C_r), (2) carrying (C_c), and (3) stockout (C_s). So,

$$\text{ETRC}(k) = C_r + C_c + C_s$$

$$= AD/Q + (Q/2 + k\sigma_L)vr + \frac{DB_1}{Q}p_{u\geq}(k)$$

We wish to select the k value that minimizes ETRC (k). A convenient approach is to set

$$\frac{d\text{ETCR}(k)}{dk} = 0$$

that is,

$$\sigma_L vr + \frac{DB_1}{Q}\frac{dp_{u\geq}(k)}{dk} = 0 \tag{6A.1}$$

But

$$\frac{dp_{u\geq}(k)}{dk} = -f_u(k)$$

(The derivative of the cumulative distribution function is the density function as shown in Appendix II.) Therefore, Equation 6A.1 gives

$$f_u(k) = \frac{Qv\sigma_L r}{DB_1} \tag{6A.2}$$

Now the density function of the unit normal is given by

$$f_u(k) = \frac{1}{\sqrt{2\pi}}exp(-k^2/2)$$

Therefore, Equation 6A.2 gives

$$\frac{1}{\sqrt{2\pi}}exp(-k^2/2) = \frac{Qv\sigma_L r}{DB_1} \tag{6A.3}$$

which solves for

$$k = \sqrt{2\ln\frac{DB_1}{\sqrt{2\pi}Qv\sigma_L r}} \tag{6A.4}$$

However, it is necessary to proceed with caution. Setting the first derivative to zero does not guarantee a minimum. In fact, Equation 6A.4 will have no solution if the expression inside the square root is negative. This will occur if the argument of the logarithm is less than unity. A case where this happens is shown in the lower part of Figure 6A.1. In such a situation, the model says that the lower the k is, the lower the costs are. The model assumes a linear savings in carrying costs with decreasing k. Obviously, this is not true indefinitely. The practical resolution is to set k at its lowest allowable value.

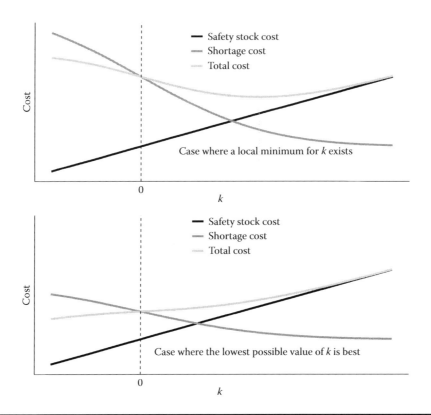

Figure 6A.1   Behavior of the expected total relevant costs for the case of a fixed cost per stockout.

6A.1.2 Marginal Analysis

Let us consider the effects of increasing s by a small amount, say Δs. The extra carrying costs per year are given by

$$\text{Marginal cost/year} = \Delta svr \tag{6A.5}$$

Savings result when, in a lead time, the total demand lies exactly between s and $s + \Delta s$, so that the extra Δs has avoided a stockout. On the average, there are D/Q replenishment occasions per year. Thus,

$$\text{Expected marginal savings/year} = \frac{D}{Q} B_1 \Pr\{\text{Demand lies in interval } s \text{ to } s + \Delta s\}$$

$$= \frac{D}{Q} B_1 f_x(s) \Delta s \tag{6A.6}$$

Now consider a particular value of s. If the marginal cost associated with adding Δs to it exceeds the expected marginal savings, we would not want to add the Δs. In fact, we might want to go below the particular s value. On the other hand, if the expected marginal savings exceeded the marginal cost, we would definitely want to increase s by Δs. However, we might wish to continue increasing s even higher. This line of argument leads us to wanting to operate at the s value, where the marginal cost of adding Δs exactly equals the expected marginal savings. Thus, from Equations 6A.5 and 6A.6, we wish to select s so that

$$\Delta svr = \frac{D}{Q} B_1 f_x(s) \Delta s$$

or

$$f_x(s) = \frac{Qvr}{DB_1} \tag{6A.7}$$

This is a general result that holds for any probability distribution of lead time demand. For the normal distribution, we have that

$$f_x(s) = \frac{1}{\sqrt{2\pi}\sigma_L} exp\left[-(s - \hat{x}_L)^2 / 2\sigma_L^2\right] \tag{6A.8}$$

Moreover, from Equations 6.7 and 6.8,

$$s = \hat{x}_L + k\sigma_L \tag{6A.9}$$

Substitution of Equations 6A.8 and 6A.9 into Equation 6A.7 gives the same result as Equation 6A.3.

6A.1.3 Behavior of k as a Function of Dv

Equation 6A.4 shows that k increases as $D/Qv\sigma_L$ or $Dv/Qv\sigma_L v$ increases. Now, empirically it has been found that, on the average,

$$\sigma_L v \approx c_1 (Dv)^{c_2}$$

where c_1 and c_2 are constants that depend on the particular company involved but, in all known cases, c_2 lies between 0.5 and 1. Also assuming an EOQ form for Qv, we have

$$Qv = c_3 \sqrt{Dv}$$

Therefore, we have that k increases as

$$\frac{Dv}{c_3 \sqrt{Dv} c_1 (Dv)^{c_2}} \quad \text{or} \quad \frac{(Dv)^{1/2 - c_2}}{c_1 c_3}$$

increases. But, because $c_2 > \frac{1}{2}$, the exponent of Dv in the above quantity is negative. Therefore, k goes down as Dv increases.

6A.2 P_1 Service Measure

Suppose the reorder point is expressed as the sum of the forecast demand plus SS, as shown in Equation 6.7. Furthermore, suppose, as in Equation 6.8, that the SS is expressed as the product of two factors

1. The safety factor k—the control parameter
2. σ_L, the standard deviation of forecast errors over a lead time

Then from Equation 6.16, we have

$$\Pr\{\text{stockout in a lead time}\} = p_{u\geq}(k) \qquad (6A.10)$$

If the desired service level is P_1, then we must have

$$\Pr\{\text{stockout in a lead time}\} = 1 - P_1 \qquad (6A.11)$$

It follows from Equations 6A.10 and 6A.11 that we must select k to satisfy

$$p_{u\geq}(k) = 1 - P_1$$

which is Equation 6.26.

Had we not chosen the indirect route of using Equations 6.7 and 6.8, we would have ended up with a more cumbersome result than Equation 6.26, namely,

$$p_{u\geq}[(s - \hat{x}_L)/\sigma_L] = 1 - P_1$$

6A.3 P_2 Service Measure

6A.3.1 Complete Backordering

Because each replenishment is of size Q, we can argue as follows:

$$\text{Fraction backordered} = \frac{\text{Expected shortage per replenishment cycle, ESPRC}}{Q}$$

Fraction of demand satisfied directly from shelf = $1 - $ (Fraction backordered).
Therefore, we want

$$P_2 = 1 - \frac{\text{ESPRC}}{Q} \qquad (6A.12)$$

Substituting from Equation 6.17, we have

$$P_2 = 1 - \frac{\sigma_L G_u(k)}{Q}$$

which reduces to

$$G_u(k) = \frac{Q}{\sigma_L}(1 - P_2)$$

which is Equation 6.28.

6A.3.2 Complete Lost Sales

If demands when out of stock are lost instead of backordered, then, the expected demand per cycle is no longer just Q, but rather is increased by the expected shortage per replenishment cycle, ESPRC. Therefore, we now have

$$\text{Fraction of demand satisfied directly from shelf} = 1 - \frac{\text{ESPRC}}{Q + \text{ESPRC}}$$

The derivation then carries through exactly as above, resulting in Equation 6.27.

6A.4 Allocation of a TSS to Minimize the ETSOPY

The problem is to select the safety factors, the k_i's, to minimize

$$\sum_{i=1}^{n} \frac{D_i}{Q_i} p_{u\geq}(k_i)$$

subject to

$$\sum_{i=1}^{n} k_i \sigma_{L_i} v_i = Y$$

where n is the number of items in the population and Y is the TSS expressed in dollars. One method of the solution is to use a Lagrange multiplier M (see Appendix I), that is, we select the k_i's to minimize

$$L(k_i\text{'s}, M) = \sum_{i=1}^{n} \frac{D_i}{Q_i} p_{u\geq}(k_i) - M\left(Y - \sum_{i=1}^{n} k_i \sigma_{L_i} v_i\right)$$

This is accomplished by setting all the partial derivatives to zero. Now

$$\frac{\partial L}{\partial k_i} = -\frac{D_i}{Q_i} f_u(k_i) + M \sigma_{L_i} v_i = 0$$

or

$$f_u(k_i) = M \frac{Q_i v_i \sigma_{L_i}}{D_i}$$

But, this has the exact same form as Equation 6A.2 with $M \equiv r/B_1$. Therefore, selection of a particular M value implies a value of B_1. Also, of course, a given M value leads to a specific value of Y.

The economic interpretation of the Lagrange multiplier M is that it represents the marginal benefit, in terms of reduced total expected stockouts, per unit increase in the total dollar budget available for SSs. r is the cost to carry one dollar in SS for a year. B_1 is the cost associated with a stockout occasion. From a marginal viewpoint, we wish to operate where

$$\text{(Cost to carry last dollar of safety stock)} =$$
$$\text{(expected benefit of last dollar of safety stock)}$$

that is,

$$r = (\text{Reduction in total expected stockouts per year}) \times (\text{Cost per stockout})$$
$$= MB_1$$

thus, $M = r/B_1$, as we found above.

6A.5 Standard Deviation of Demand during Lead Time

Demand in any period t is assumed to be a random variable D_t with mean $E(D)$ and standard deviation $\sqrt{\text{var}(D)}$. In this analysis, demands in different periods are assumed to be independent and identically distributed and that the random variable for lead time L takes on only integer values. As in Equations 6.41 and 6.42, x has mean $E(x)$ and standard deviation σ_x, and lead time L has mean $E(L)$ and variance $\text{var}(L)$.

Demand during the lead time x is a sum of random variables D_t where the number of terms in the sum is also random (L).

$$x = \sum_{i=1}^{L} D_i$$

We can find both the mean and standard deviation of X by conditioning on the value of the lead time L. First, for the mean,

$$E[x] = E_L[E_D[x|L]]$$
$$= E_L[(L)E(D)]$$
$$= E(D)E(L)$$

Now, for the standard deviation, note that for any random variable Z, we can find the variance of Z by conditioning on another random variable (e.g., Ross 1983) say Y, as follows:

$$\text{var}[Z] = E_Y[\text{var}_Z(Z|Y=y)] + \text{var}_Y[E_Z(Z|Y=y)]$$

So, again conditioning on the value of the lead time L

$$\sigma_x^2 = E_L[\text{var}_D(x|L)] + \text{var}_L[E_D(x|L)]$$
$$= E_L[(L)\text{var}(D)] + \text{var}_L[E(D)(L)]$$
$$= (E(L))\text{var}(D) + [E(D)]^2\text{var}(L)$$
$$\sigma_x = \sqrt{E(L)\text{var}(D) + [E(D)]^2\text{var}(L)}$$

6A.6 Two Approaches to Determining Expected Costs per Unit Time

Throughout the book, the basic approach we use in establishing the expected total relevant costs per unit time will follow these steps:

1. Determine the average inventory level ($\bar{I}$), in units.
2. Carrying costs per year $= \bar{I}vr$.
3. Determine the average number of replenishments per year (N).
4. Replenishment costs per year $= NA$.
5. Determine the expected shortage costs per replenishment cycle, say W.
6. Expected shortage costs per year $= NW$.
7. Total expected costs per year $=$ sum of the results of steps 2, 4, and 6.

If the inventory process has a natural regenerative point where the same state repeats periodically, then, a different approach, using a key result of renewal theory, is possible. Each time that the regenerative point is reached, a cycle of the process is said to begin. The time between consecutive regenerative points is known as the length of the cycle, which we denote by T_c.

There is a set of costs (carrying, replenishment, and shortage) associated with each cycle. Let us denote the total of these costs by C_c. This quantity is almost certainly a random variable because its value depends on the demand process that is usually random in nature.

A fundamental result of renewal theory (see, e.g., Ross 1983), which holds under very general conditions, is

$$\text{Expected total relevant costs per unit time} = \frac{E(C_c)}{E(T_c)}$$

The key point is that in a complicated process, it may be easier to evaluate each of $E(C_c)$ and $E(T_c)$ instead of the left-hand side of this equation directly. See Silver (1976) for an illustration of this approach.

6A.7 Expected Units Short per Cycle with the Gamma Distribution

Suppose demand follows a Gamma distribution described by shape parameter α and scale parameter β, implying a mean of $\alpha\beta$ and standard deviation of $\sqrt{\alpha}\beta$. Assuming a reorder point of s, we write expected units short per replenishment cycle as

$$\text{ESPRC} = \int_s^\infty (x - s)f(x; \alpha, \beta)dx$$

where $f(x; \alpha, \beta)$ represents the probability density function of the Gamma distribution with parameters α and β

$$f(x; \alpha, \beta) = \frac{x^{\alpha-1}e^{-x/\beta}}{\beta^\alpha \Gamma(\alpha)}$$

Γ is the Gamma function that has the property $x\Gamma(x) = \Gamma(x + 1)$.

Note that

$$xf(x; \alpha, \beta) = x \frac{x^{\alpha-1} e^{-x/\beta}}{\beta^{\alpha} \Gamma(\alpha)}$$

$$= \alpha\beta \frac{x^{\alpha} e^{-x/\beta}}{\beta \beta^{\alpha} \alpha \Gamma(\alpha)}$$

$$= \alpha\beta \frac{x^{\alpha} e^{-x/\beta}}{\beta^{\alpha+1} \Gamma(\alpha+1)}$$

$$= \alpha\beta f(x; \alpha+1, \beta)$$

Using this fact,

$$\text{ESPRC} = \int_{s}^{\infty} (x-s) f(x; \alpha, \beta) dx$$

$$= \int_{s}^{\infty} xf(x; \alpha, \beta) dx - s[1 - F(s; \alpha, \beta)]$$

$$= \alpha\beta[1 - F(s; \alpha+1, \beta)] - s[1 - F(s; \alpha, \beta)]$$

References

Aardal, K., Ö. Jonsson, and H. Jönsson. 1989. Optimal inventory policies with service-level constraints. *Journal of the Operational Research Society* 40(1), 65–73.

Agrawal, N. and S. A. Smith. 1996. Estimating negative binomial demand for retail inventory management with unobservable lost sales. *Naval Research Logistics* 43(6), 839–861.

Axsäter, S. 1996. Using the deterministic EOQ formula in stochastic inventory control. *Management Science* 42(6), 830–834.

Bagchi, U., J. Hayya, and J. K. Ord. 1984. Modeling demand during lead time. *Decision Sciences* 15, 157–176.

Bagchi, U. and J. C. Hayya. 1984. Demand during lead time for normal unit demand and Erlang lead time. *Journal of the Operational Research Society* 35(2), 131–135.

Bagchi, U., J. C. Hayya, and C.-H. Chu. 1986. The effect of lead-time variability: The case of independent demand. *Journal of Operations Management* 6(2), 159–177.

Blackburn, J. 2012. Valuing time in supply chains: Establishing limits of time-based competition. *Journal of Operations Management* 30(5), 396–405.

Bookbinder, J. and A. Lordahl. 1989. Estimation of inventory re-order levels using the bootstrap statistical procedure. *IIE Transactions* 21(4), 302–312.

Boylan, J. E. and F. R. Johnston. 1994. Relationships between service level measures for inventory systems. *Journal of the Operational Research Society* 45(7), 838–844.

Brown, R. G. 1977. *Materials Management Systems*. New York: Wiley-Interscience.

Bulinskaya, E. V. 1990. Inventory control in case of unknown demand distribution. *Engineering Costs and Production Economics* 19, 301–306.

Burgin, T. 1972. Inventory control with normal demand and gamma lead times. *Operational Research Quarterly* 23(1), 73–80.

Burgin, T. A. 1975. The Gamma distribution and inventory control. *Operational Research Quarterly* 26(3i), 507–525.

Burgin, T. A. and J. M. Norman. 1976. A table for determining the probability of a stockout and potential lost sales for a gamma distributed demand. *Operational Research Quarterly 27*(3i), 621–631.

Carlson, P. 1982. An alternate model for lead-time demand: Continuous-review inventory systems. *Decision Sciences 13*(1), 120–128.

Chen, F. Y. and D. Krass. 2001. Inventory models with minimal service level constraints. *European Journal of Operational Research 134*(1), 120–140.

Chen, F. and Y. Zheng. 1992. Waiting time distribution in (T,S) inventory systems. *Operations Research Letters 12*, 145–151.

Chu, P., K.-L. Yang, S.-K. Liang, and T. Niu. 2004. Note on inventory model with a mixture of back orders and lost sales. *European Journal of Operational Research 159*(2), 470–475.

Das, C. 1975. Effect of lead time on inventory: A study analysis. *Operational Research Quarterly 26*(2i), 273–282.

Das, C. 1976. Approximate solution to the (Q, r) inventory model for gamma lead time demand. *Management Science 22*(9), 1043–1047.

de Kok, A. G. 1990. Hierarchical production planning for consumer goods. *European Journal of Operational Research 45*, 55–69.

De Treville, S., I. Bicer, V. Chavez-Demoulin, V. Hagspiel, N. Schürhoff, C. Tasserit, and S. Wager. 2014. Valuing lead time. *Journal of Operations Management 32*(6), 337–346.

Disney, S. M., G. J. Gaalman, C. P. T. Hedenstierna, and T. Hosoda. 2015. Fill rate in a periodic review order-up-to policy under auto-correlated normally distributed, possibly negative, demand. *International Journal of Production Economics 170*, 501–512.

Ehrhardt, R. 1979. The power approximation for computing (s, S) inventory policies. *Management Science 25*(8), 777–786.

Eppen, G. D. and R. K. Martin. 1988. Determining safety stock in the presence of stochastic lead time and demand. *Management Science 34*(11), 1380–1390.

Ernst, R. and D. Pyke. 1992. Component part stocking policies. *Naval Research Logistics 39*, 509–529.

Federgruen, A. and Y.-S. Zheng. 1992. An efficient algorithm for computing an optimal (r, Q) policy in continuous review stochastic inventory systems. *Operations Research 40*(4), 808–813.

Fortuin, L. 1980. The all-time requirement of spare parts for service after sales—Theoretical analysis and practical results. *International Journal of Operations and Production Management 1*(1), 59–70.

Gallego, G. 1992. A minmax distribution free procedure for the (Q, R) inventory model. *Operations Research Letters 11*(1), 55–60.

Gallego, G. and I. Moon. 1993. The distribution free newsboy problem: Review and extensions. *Journal of the Operational Research Society 44*(8), 825–834.

Gardner, E. and D. Dannenbring. 1979. Using optimal policy surfaces to analyze aggregate inventory tradeoffs. *Management Science 25*(8), 709–720.

Gardner, E. S. 1983. Approximate decision rules for continuous review inventory systems. *Naval Research Logistics Quarterly 30*, 59–68.

Hadley, G. and T. M. Whitin. 1963. *Analysis of Inventory Systems*. Englewood Cliffs, NJ: Prentice-Hall, Inc.

Hausman, W. 1969. Minimizing customer line items backordered in inventory control. *Management Science 15*(12), 628–634.

Herron, D. P. 1983. Improving productivity in logistics operations. In R. L. Schultz (Ed.), *Applications of Management Science*, pp. 49–85. Greenwich, CT: JAI Press.

Heuts, R. and J. de Klein. 1995. An (s, q) inventory model with stochastic and interrelated lead times. *Naval Research Logistics 42*(5), 839–859.

Hopp, W. J., R. Q. Zhang, and M. L. Spearman. 1999. An easily implementable hierarchical heuristic for a two-echelon spare parts distribution system. *IIE Transactions 31*(10), 977–988.

Jain, A., H. Groenevelt, and N. Rudi. 2010. Continuous review inventory model with dynamic choice of two freight modes with fixed costs. *Manufacturing and Service Operations Management 12*(1), 120–139.

Jain, A., H. Groenevelt, and N. Rudi. 2011. Periodic review inventory management with contingent use of two freight modes with fixed costs. *Naval Research Logistics 58*(4), 400–409.

Johnson, M. E., H. Lee, T. Davis, and R. Hall. 1995. Expressions for item fill rates in periodic inventory systems. *Naval Research Logistics 42*(1), 57–80.

Johnston, F. R. 1980. An interactive stock control system with a strategic management role. *Journal of the Operational Research Society 31*(12), 1069–1084.

Karmarkar, U. 1994. A robust forecasting technique for inventory and lead time management. *Journal of Operations Management 12*(1), 45–54.

Keaton, M. 1995. Using the Gamma distribution to model demand when lead time is random. *Journal of Business Logistics 16*(1), 107–131.

Kottas, J. and H. S. Lau. 1980. The use of versatile distribution families in some stochastic inventory calculations. *Journal of the Operational Research Society 31*(5), 393–403.

Kumar, A. and G. Sharman. 1992. We love your product, but where is it? *Sloan Management Review 33*(2), 93–99.

Kumaran, M. and K. K. Achary. 1996. On approximating lead time demand distributions using the generalised l-type distribution. *Journal of the Operational Research Society 47*, 395–404.

Lau, A. H.-L. and H.-S. Lau. 1993. A simple cost minimization procedure for the (Q, R) inventory model: Development and evaluation. *IIE Transactions 25*(2), 45–53.

Lau, H.-S. and A. Zaki. 1982. The sensitivity of inventory decisions to the shape of lead time-demand distribution. *IIE Transactions 14*(4), 265–271.

Lee, H. L., K. C. So, and C. S. Tang. 2000. The value of information sharing in a two-level supply chain. *Management Science 46*(5), 626–643.

Lefrancois, P. 1989. Allowing for asymmetry in forecast errors: Results from a Monte-Carlo study. *International Journal of Forecasting 5*, 99–110.

Leven, E. and A. Segerstedt. 2004. Inventory control with a modified Croston procedure and Erlang distribution. *International Journal of Production Economics 90*(3), 361–367.

Lordahl, A. E. and J. H. Bookbinder. 1994. Order statistic calculation, costs, and service in an (s, Q) inventory system. *Naval Research Logistics 41*(1), 81–97.

Magson, D. 1979. Stock control when the lead time cannot be considered constant. *Journal of the Operational Research Society 30*, 317–322.

Mayer, R. M. 1984. Optimum safety stocks when demands and lead times vary. *Journal of Purchasing and Materials Management 20*(1), 27–32.

McClain, J. O. and L. J. Thomas. 1980. *Operations Management: Production of Goods and Services*. Englewood Cliffs, NJ: Prentice-Hall.

McFadden, F. R. 1972. On lead time demand distributions. *Decision Sciences 3*, 106–126.

Moinzadeh, K. and S. Nahmias. 1988. A continuous review model for an inventory system with two supply modes. *Management Science 34*(6), 761–773.

Moinzadeh, K. and C. P. Schmidt. 1991. An $(S - 1, S)$ inventory system with emergency orders. *Operations Research 39*(2), 308–321.

Moon, I. and S. Choi. 1994. The distribution free continuous review inventory system with a service level constraint. *Computers and Industrial Engineering 27*(1–4), 209–212.

Moon, I. and S. Choi. 1995. The distribution free newsboy problem with balking. *Journal of the Operational Research Society 46*, 537–542.

Moors, J. J. A. and L. W. G. Strijbosch. 2002. Exact fill rates for (R, s, S) inventory control with gamma distributed demand. *Journal of the Operational Research Society 53*, 1268–1274.

Naddor, E. 1978. Sensitivity to distributions in inventory systems. *Management Science 24*(16), 1769–1772.

Nasri, F., J. F. Affisco, and M. J. Paknejad. 1990. Setup cost reduction in an inventory model with finite-range stochastic lead times. *International Journal of Production Research 28*(1), 199–212.

Ng, K. Y. K. and M. N. Lam. 1998. Standardisation of substitutable electrical items. *Journal of the Operational Research Society 49*(9), 992–997.

Paknejad, M. J., F. Nasri, and J. Affisco. 1992. Lead-time variability reduction in stochastic inventory models. *European Journal of Operational Research 62*, 311–322.

Platt, D. E., L. W. Robinson, and R. B. Freund. 1997. Tractable (Q, R) heuristic models for constrained service levels. *Management Science 43*(7), 951–965.

Presutti, V. and R. C. Trepp. 1970. More Ado about EOQ. *Naval Research Logistics Quarterly 17*(2), 243–251.

Ray, D. L. 1981. Assessing U.K. manufacturing industry's inventory management performance. *Focus on PDM (Journal of the Institute of Physical Distribution Management) 27*, 5–11.

Ray, S., Y. Gerchak, and E. M. Jewkes. 2004. The effectiveness of investment in lead time reduction for a make-to-stock product. *IIE Transactions 36*(4), 333–344.

Ross, S. M. 1983. *Stochastic Processes*. New York: John Wiley & Sons.

Rudi, N., H. Groenevelt, and T. R. Randall. 2009. End-of-period vs. continuous accounting of inventory-related costs. *Operations Research 57*(6), 1360–1366.

Sahin, I. and D. Sinha. 1987. Renewal approximation to optimal order quantity for a class of continuous-review inventory models. *Naval Research Logistics 34*, 655–667.

Scarf, H. 1958. A min-max solution of an inventory problem. In K. Arrow, S. Karlin, and H. Scarf (Eds.), *Studies in The Mathematical Theory of Inventory and Production*, pp. 201–209. California: Stanford University Press.

Scarf, H. 1960. The optimality of (S, s) policies in the dynamic inventory problem. In K. J. Arrow, S. Karlin, and P. Suppes (Eds.), *Mathematical Methods in the Social Sciences*, pp. Chapter 13. Stanford: Stanford University Press.

Schneider, H. 1981. Effect of service-levels on order-points or order-levels in inventory models. *International Journal of Production Research 19*(6), 615–631.

Shore, H. 1986. General approximate solutions for some common inventory models. *Journal of the Operational Research Society 37*(6), 619–629.

Shore, H. 1995. Fitting a distribution by the first two moments (partial and complete). *Computational Statistics and Data Analysis 19*, 563–577.

Silver, E. 1970. Some ideas related to the inventory control of items having erratic demand patterns. *Canadian Operations Research Society Journal 8*(2), 87–100.

Silver, E. A. 1976. Establishing the order quantity when the amount received is uncertain. *INFOR 14*(1), 32–39.

Silver, E. A. 1977. A safety factor approximation based on Tukey's Lambda distribution. *Operational Research Quarterly 28*(3ii), 743–746.

Silver, E. A. and M. R. Rahnama. 1986. The cost effects of statistical sampling in selecting the reorder point in a common inventory model. *Journal of the Operational Research Society 37*(7), 705–713.

Silver, E. A. and M. R. Rahnama. 1987. Biased selection of the inventory reorder point when demand parameters are statistically estimated. *Engineering Costs and Production Economics 12*, 283–292.

Snyder, R. D. 1980. The safety stock syndrome. *Journal of the Operational Research Society 31*(9), 833–837.

Song, J. 1994a. The effect of leadtime uncertainty in a simple stochastic inventory model. *Management Science 40*(5), 603–613.

Song, J. 1994b. Understanding the lead-time effects in stochastic inventory systems with discounted costs. *Operations Research Letters 15*, 85–93.

Song, J.-S. 1998. On the order fill rate in a multi-item, base-stock inventory system. *Operations Research 46*(6), 831–845.

Strijbosch, L. W. G. and R. Heuts. 1992. Modelling (s, Q) inventory systems: Parametric versus non-parametric approximations for the lead time demand distribution. *European Journal of Operational Research 63*, 86–101.

Strijbosch, L. W. G. and R. Heuts. 1994. Investigating several alternatives for estimating the lead time demand distribution in a continuous review inventory model. *Kwantitatieve Methoden 15*(46), 57–75.

Taylor, P. B. and K. H. Oke. 1976. Tables for stock control—Problems of formulation and computation. *Operational Research Quarterly 27*(3ii), 747–758.

Taylor, P. B. and K. H. Oke. 1979. Explicit formulae for computing stock-control levels. *Journal of the Operational Research Society 30*(12), 1109–1118.

Tijms, H. C. 1986. *Stochastic Modelling and Analysis: A Computational Approach*. New York: John Wiley & Sons.

Tyworth, J. E. 1992. Modeling transportation–inventory trade-offs in a stochastic setting. *Journal of Business Logistics 13*(2), 97–124.

Tyworth, J. E., Y. Guo, and R. Ganeshan. 1996. Inventory control under gamma demand and random lead time. *Journal of Business Logistics 17*(1), 291–304.

Tyworth, J. E. and L. O'Neill. 1997. Robustness of the normal approximation of lead-time demand in a distribution setting. *Naval Research Logistics 44*(2), 165–186.

Van Beek, P. 1978. An application of the logistic density on a stochastic continuous review stock control model. *Zeitschrift für Operations Research Band 22*, B165–B173.

Van der Veen, B. 1981. Safety stocks—An example of theory and practice in OR. *European Journal of Operational Research 6*(4), 367–371.

Van der Veen, B. 1984. Safety stocks and the paradox of the empty period. *European Journal of Operational Research 16*(1), 19–33.

Veeraraghavan, S. and A. Scheller-Wolf. 2008. Now or later: A simple policy for effective dual sourcing in capacitated systems. *Operations Research 56*(4), 850–864.

Wang, M. C. and S. S. Rao. 1992. Estimating reorder points and other management science applications by bootstrap procedure. *European Journal of Operational Research 56*, 332–342.

Wang, P., W. Zinn, and K. L. Croxton. 2010. Sizing inventory when lead time and demand are correlated. *Production and Operations Management 19*(4), 480–484.

Wu, J.-W. and H.-Y. Tsai. 2001. Mixture inventory model with back orders and lost sales for variable lead time demand with the mixtures of normal distribution. *International Journal of Systems Science 32*(2), 259–268.

Zheng, Y. S. 1992. On properties of stochastic inventory systems. *Management Science 38*(1), 87–103.

Zipkin, P. 1986. Inventory service level measures: Convexity and approximations. *Management Science 32*(8), 975–981.

Zipkin, P. 1988. The use of phase-type distributions in inventory-control models. *Naval Research Logistics 35*, 247–257.

SPECIAL CLASSES OF ITEMS

In Section II, we were concerned with routine control of B items, that is, the bulk of the items in a typical population of stock-keeping units. Now, we turn our attention to special items, those at the ends of the dollar-volume (Dv) spectrum, namely, the A and C groups as well as those items not having a well-defined continuing demand pattern. The most important items, the A group, are handled in Chapter 7. Chapter 8 is concerned with slow-moving and low-dollar-volume items, the C group. Then, in Chapter 9, we deal with style goods and perishable items, that is, SKUs that can be maintained in inventory for only relatively short periods of time. The well-known newsvendor problem is covered in this chapter. The presentation of this portion of the book continues in the same vein as Section II; again, we are concerned primarily with managing individual-item inventories, although we do discuss several cases of coordinating across multiple items.

Chapter 7

Managing the Most Important Inventories

In this chapter, we devote our attention to the relatively small number of important items typically classified in the A group. For such items, additional effort may be warranted in both selecting control parameters and managing inventory. While not all important items are necessarily high-volume items, we restrict our attention in this chapter to items that are moderate to high volume. Chapter 8 addresses the management of slow-moving items of all levels of importance. Section 7.1 presents a brief review of the nature of A items. This is followed, in Section 7.2, by some general guidelines for the control of such items. Section 7.3 deals with faster-moving items and a more elaborate set of decision rules, which, in an (s, Q) system, take account of the service (shortage)-related effects in selecting the value of the order quantity Q.[*] Next, in Sections 7.4 and 7.5, we present control logic for two somewhat more complicated systems whose use may be justified for A items. In Section 7.6, we address the issue of demand that is not stationary over time. Finally, in Section 7.7, we briefly discuss the research that analyzes the case of multiple suppliers.

7.1 Nature of Class A Items

Recall from Chapter 2 that class A items are defined to be those at the upper end of the spectrum of importance. That is, the total costs of replenishment, carrying stock, and shortages associated with such an item are high enough to justify a more sophisticated control system than those proposed for the less important B items in Chapters 4 through 6. This usually means a high annual dollar usage (a high value of Dv). By *usage* we mean retail customer demand as well as usage, say, of spare parts in production. In addition, one or more other factors may dictate that an item be placed in the A category. For example, an item, although its sales are relatively low, may be essential to round out the product line. To illustrate, a company may carry a high-end item for its prestige value. Poor customer service on such an item may have adverse effects on customers who buy large amounts

---

[*] Equivalently, R in an (R, S) system.

of other products in the line. In essence, the important issue is a trade-off between control system costs (such as the costs of collecting required data, performing the computations, providing action reports, etc.) and the other three categories of costs mentioned above. (For simplicity, let us call this latter group "other" costs.) By definition, these "other" costs for an A item will be appreciably higher than for a B item. On the other hand, *given that the same control system is used* for an A and a B item, the control costs will be essentially the same for the two items. A numerical example is in order.

Consider two types of control systems: one simple, one complex. Suppose we have an A and a B item with estimated "other" costs using the simple system of $900/year and $70/year, respectively. It is estimated that the use of the complex control system will reduce these "other" costs by 10%. However, the control costs of this complex system are estimated to be $40/year/item higher than those incurred in using the simple system. The following table illustrates that the complex system should be used for the A, but not for the B item.

	Using Complex System		
Item	Increase in Control Costs	Reduction in Other Costs	Net Increase in Costs
A	40	$0.1(900) = 90$	−50
B	40	$0.1(70) = 7$	+33

As mentioned above, the factor Dv is important in deciding whether or not to put an item in the A category. However, the type of control to use within the A category should definitely depend on the magnitudes of the individual components, D and v. As we see later in this chapter, a high Dv resulting from a low D and a high v value (low unit sales of a high-value item) implies different control from a high Dv resulting from a high D and a low v value (high unit sales of a low-value item). To illustrate, we would certainly control the inventories of the following two items in different ways:

Item 1: Midamatic film processor, valued at $30,000, which sells at a rate of 2 per year.
Item 2: Developing chemical, valued at $2/liter, which sells at a rate of 30,000 liters per year.

Inventory control for items such as number 1 is addressed in Section 8.1, and for items such as number 2 is examined in Sections 7.3 through 7.5.

7.2 Guidelines for Control of A Items

In Section II, which was concerned primarily with B items, we advocated a management by exception approach. That is, most decisions would be made by routine (computerized or manual) rules. This should not be the case with A items. The potentially high payoff warrants frequent managerial attention to the replenishment decisions of individual items. Nonetheless, decision rules, based on mathematical models, do have a place in aiding the manager. Normally, these models cannot incorporate all the important factors. In such cases, the manager should incorporate subjective factors that were omitted from the model. The art of management is very evident in this type of activity.

We now list a number of suggested guidelines for the control of A items. These may be above and beyond what normally can be incorporated in a usable mathematical decision rule:

1. *Inventory records should be maintained on a perpetual (transactions recording) basis, particularly for the more expensive items.* This need not be through the use of a computer system although this will be the most common setting. Even if computer systems are available for all items, our experience is that manual effort is still required to maintain system and data accuracy; thus, this effort may not be warranted for items of lesser importance.

2. *Keep top management informed.* Frequent reports (e.g., monthly) should be prepared for at least a portion of the A items. These should be made available to senior management for careful review.

3. *Estimate and influence demand.* This can be done in three ways:
   a. Provide manual input to forecasts. For example, call your customers to get advance warning of their needs.
   b. Ascertain the predictability of demand. Where the demand is of a special planned nature (e.g., scheduled overhaul of a piece of equipment) or where adequate warning of the need for a replacement part is given (e.g., a machine part fails but can be rapidly and inexpensively repaired to last for sufficient time to requisition and receive a new part from the supplier), there is no need to carry protection stock. In some cases, information regarding the impending need for replacement parts can be gathered by monitoring a system. Such information can be helpful in anticipating the need for replacement parts. Jardine et al. (2006) discuss such condition-based maintenance approaches. When the demand can occur without warning (i.e., a random breakdown), some protective stock may be appropriate. When the unit value of an item is extremely high, it may make sense to use a pool of spare parts *shared among several companies within the same industry.* An example is in the ammonia manufacturing industry where certain spare parts are maintained in a common pool.
   c. Manipulate a given demand pattern. Seasonal or erratic fluctuations can sometimes be reduced by altering price structures, negotiating with customers, smoothing shipments, and so forth.

4. *Estimate and influence supply.* Again, it may not be advisable to passively accept a given pattern of supply. Negotiations (e.g., establishing blanket orders and freeze periods) with suppliers may reduce the average replenishment lead time, its variability, or both. The idea of a freeze period is that the timing or size of a particular order cannot be altered within a freeze period prior to its due date. In the same vein, it is very important to coordinate A item inventory requirements in-house with the production scheduling department.

5. *Use conservative initial provisioning.* For A items that have a very high *v* value and a relatively low *D* value, the initial provisioning decision becomes particularly crucial. For such items, erroneous initial overstocking (due to overestimating the usage rate) can be extremely expensive. Thus, it is a good idea to be conservative in initial provisioning.

6. *Review decision parameters frequently.* Generally, frequent review (as often as monthly or bimonthly) of order points and order quantities is advisable for A items.

7. *Determine precise values of control quantities.* In Chapters 4 and 8, we consider the use of tabular aids in establishing order quantities for B and C items. We argued that restricting attention to a limited number of possible time supplies (e.g., 1/2, 1, 2, 3, 6, and 12 months)

results in small cost penalties. This is not the case for A items. The percentage penalties may still be small, but even small percentage penalties represent sizable absolute costs. Therefore, order quantities of A items should be based on the most exact analysis possible.

8. *Confront shortages as opposed to setting service levels.* In Chapter 6, we pointed out that it is often very difficult to estimate the cost of shortages; as an alternative, we suggested specifying a desired level of customer service instead. In most cases, we do not advocate this approach for A items because of the typical management behavior in the short run in coping with potential or actual shortages of such items. The response is not usually to sit back passively and accept shortages of A items (or B items, for that matter). Instead, expediting (both in and out of house), emergency air freight shipments, and other actions are undertaken. That is, the shortages are avoided or eliminated quickly. Such actions, with associated costs that can be reasonably estimated, should be recognized when establishing safety stocks to prevent them. On the other hand, it should be pointed out that, in certain situations, customers are willing to wait a short time for delivery. Because A items are replenished frequently, it may be satisfactory to operate with very little on-hand stock (i.e., low safety stock) and, instead, backorder and satisfy such demand from the next order due in very soon.

7.3 Simultaneous Determination of *s* and *Q* for Fast-Moving Items

For fast-moving A items, we can usually rely on the normal distribution. We discuss the use of alternative distributions in Chapter 8 where we address slow-moving items. We covered sequential determination of *s* and *Q* in Chapter 6, noting there that the value of *Q* can often be constrained, or at least influenced, by minimum order sizes or predetermined lot sizes. For very important items, it may be valuable to simultaneously determine *s* and *Q*. In this section, we provide an example of simultaneous determination for the case of a B_1 shortage cost and normal lead time demand.[*]

First, we present the total cost equation for a general lead time demand distribution and a cost per stockout occasion:

$$\text{ETRC}(s, Q) = AD/Q + \left[Q/2 + \int_0^s (s - x_0) f_x(x_0) dx_0 \right] vr + \frac{DB_1}{Q} \int_s^\infty f_x(x_0) dx_0 \qquad (7.1)$$

where the second term in the brackets is the safety stock, and the final term is the shortage cost. See the Appendix of Chapter 6 and Section 6.7.3 for more details. For the case of normal demand, we have an approximation

$$\text{ETRC}(k, Q) = AD/Q + (Q/2 + k\sigma_L)vr + \frac{DB_1}{Q} p_{u\geq}(k) \qquad (7.2)$$

---

[*] Silver and Wilson (1972) develop similar results for a B_2 cost and for (R, S) systems. For P_2, Yano (1985) treats the case of normal demand, and Moon and Choi (1994) addresses the case where we know only the mean and variance of demand. See also Hill (1992). See Problem 7.13 for the formula for the B_2 case. Finally, see Moon and Gallego (1994) for the $/unit short case where we know only the mean and variance of demand.

where, as earlier,

$$s = \hat{x}_L + k\sigma_L$$

and $p_{u\geq}(k)$ is a function giving the probability that a unit normal variable takes on a value of k or larger. (As noted in Appendix III, $p_{u\geq}(k)$ is also available on many spreadsheets.)

7.3.1 Decision Rules

The derivation is shown Appendix of this chapter. All of the underlying assumptions discussed in Section 6.7.1 of Chapter 6 still apply, except that a value of Q is no longer prespecified. The two equations in the two unknowns Q and k (the latter being the safety factor) are

$$Q = \text{EOQ} \sqrt{\left[1 + \frac{B_1}{A} p_{u\geq}(k) \right]} \tag{7.3}$$

and

$$k = \sqrt{2 \ln \left[\frac{1}{2\sqrt{2\pi}} \left(\frac{B_1}{A} \right) \left(\frac{\sigma_L}{Q} \right) \left(\frac{\text{EOQ}}{\sigma_L} \right)^2 \right]} \tag{7.4}$$

where
      EOQ = the basic economic order quantity
      A = fixed cost component incurred with each replenishment
      D = demand rate
      v = unit variable cost of the item
      r = carrying charge
      σ_L = standard deviation of lead time demand
      $p_{u\geq}(k)$ = function of the unit normal variable

Note that Equation 7.4 is identical to the formula given by Equation 6.20 in Chapter 6:

$$k = \sqrt{2 \ln \left[\frac{DB_1}{\sqrt{2\pi} Q v \sigma_L r} \right]} \tag{7.5}$$

A suggested iterative solution procedure is to initially set Q = EOQ; then solve for a corresponding k value in Equation 7.4, and then use this value in Equation 7.3 to find a new Q value, and so forth. In certain cases, Equation 7.4 will have the square root of a negative number. In this case, k is set at its lowest allowable value, as previously discussed in Section 6.7.5, and then Q is obtained using this k value in Equation 7.3. Because of the convex nature of the functions involved, convergence to the true simultaneous solution pair (Q and k) is ensured. For further details concerning convexity, when the solution converges, and so on, refer to Chapter 4 of Hadley and Whitin (1963) as well as to Problem 7.3.

Using functions and equations from Appendix III, it is faster to set up columns for an iteration table. (See the table in the example below.) It is possible, however, simply to use two cells, one for k and one for Q. These are defined in terms of each other—k depends on Q, and Q depends on k. Spreadsheet programs will give a "circular references" warning, but it can be safely ignored. The spreadsheet will perform the number of iterations specified and converge on a solution. In all of our tests, this happened almost instantaneously. See Problem 7.9.

From Equation 7.3, we can see that the simultaneous Q is always larger than the EOQ. This makes sense in view of Equation 7.2. The EOQ is the Q value that minimizes the sum of the first two cost components. However, at that value, the last term in Equation 7.2 is still decreasing as Q increases. Thus, it pays to go somewhat above the EOQ.

Once we have ascertained that the simultaneous Q is larger than the EOQ, it follows from Equation 7.4 that the simultaneous k is smaller than or equal to the sequential k. (They may be equal because of the boundary condition of a lowest allowable k.) Again, this is intuitively appealing—the larger Q value makes stockout opportunities less frequent; hence we can afford to carry less safety stock.

7.3.1.1 Numerical Illustration

For illustrative purposes, suppose that a firm was considering the use of a B_1 shortage cost rule for certain items with a B_1 value of \$32 per shortage occasion. One of these items has the following characteristics:

$D = 700$ containers/year
$v = \$12$/container
$\hat{x}_L = 100$ containers
$\sigma_L = 30$ containers

Using the values of $A = \$3.20$ and $r = 0.24$/year, we have that

$$\text{EOQ} = \sqrt{\frac{2 \times 3.20 \times 700}{12 \times 0.24}} \approx 39.4$$

With the iterative procedure, we obtain

Iteration Number	1	2	3	4	5	6	7	8	9
Q	39.4	53.3	59.1	61.8	63.0	63.5	63.8	64.0	64.0
k	1.39	1.15	1.06	1.02	1.00	0.99	0.98	0.98	0.98

The order quantity (Q) is
$$Q = 64 \text{ containers}$$
and the reorder point (s) is

$$s = \hat{x}_L + k\sigma_L = 100 + 0.98(30) \approx 130 \text{ containers}$$

The total cost from Equation 7.2 is

$$(3.20)(700/64) + \left[\frac{64}{2} + (0.98)(30)\right](12)(0.24) + \frac{700(32)}{64}(0.1635) = \$269.07$$

Robinson (1993) shows that under very general conditions the total relevant cost for the simultaneous solution case is $(Q + k\sigma_L)rv$. In this case, it is $[64 + (0.98)(30)](0.24)(12) = \$269$.

7.3.2 Cost Penalties

The percentage penalty in costs (excluding those of the control system) resulting from using the simpler sequential procedure in lieu of the sophisticated simultaneous approach is given by

$$\text{PCP} = \frac{[\text{ETRC (sequential parameter values)} - \text{ETRC (simultaneous parameter values)}]}{\text{ETRC (simultaneous parameter values)}} \quad (7.6)$$

where PCP is the percentage cost penalty and the ETRC $(\cdot)$ values are expected total relevant costs (excluding system control costs) per unit time using the parameter values in the parentheses. Figure 7.1 shows PCP as a function of two dimensionless parameters in Equation 7.4, EOQ/σ_L and B_1/A (note again the appearance of the EOQ). The curves have been terminated to the left where the simultaneous solution first gives $k = 0$.

7.3.2.1 Numerical Illustration of the Cost Penalty

For the item shown above

$$B_1/A = 10$$

and

$$\text{EOQ}/\sigma_L = 1.3$$

Figure 7.1 shows the percentage cost penalty for the previous example is

$$\text{PCP} \approx 4.21\%$$

Now the sequential parameter values are quite easy to obtain. By the method of Section 6.7.5 $Q = \text{EOQ} = 39$ and k is selected to satisfy Equation 7.3:

$$k = \sqrt{2 \ln \left[\frac{1}{2\sqrt{2\pi}} \left(\frac{B_1}{A} \right) \left(\frac{\text{EOQ}}{\sigma_L} \right) \right]}$$

$$= \sqrt{2 \ln \left[\frac{1}{2\sqrt{2\pi}} (10)(1.3) \right]}$$

$$= 1.39$$

Using Equation 7.2, we have

$$\text{ETRC}(1.39, 30) = \$280.41$$

The absolute savings is $\$280.41 - \$269.07 = \$11.33$/year. A savings of $\$11.33$ per year for an item that has an annual usage of $\$8,400$ seems very small. If, however, the 4.2% savings applies to many other A items, the total saved could be impressive. Therefore, it is worthwhile checking some other items for similar savings.

7.3.3 Further Comments

We have seen that the simultaneous approach requires an iterative solution of two complicated equations in two unknowns. (The situation is even more complicated for (R, S) systems. See

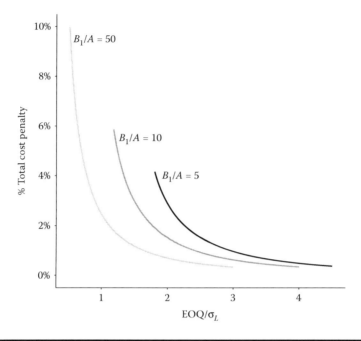

Figure 7.1   Percent cost penalty associated with using the sequential approach for an (s, Q) system for the case of a fixed cost per stockout occasion, B_1.

Donaldson (1984) for a treatment of this case.) This amount of complexity should not be taken lightly when considering the possible use of a simultaneous solution. Even though the solution is very fast once the formulas are built into a spreadsheet, as noted above, we find that the complexity of the development of the formulas, and of the formulas themselves, generates resistance among some managers. A somewhat simpler approach is presented by Platt et al. (1997), who developed an accurate approximation for both s and Q when a fill rate is used. Their approximation can be built on a spreadsheet very easily, and is similar in development to the power approximation to be discussed in Section 7.5.1. When *the simultaneous approach is used*, we would suggest the use of a spreadsheet or another computerized routine. However, let us carefully examine the results of Figure 7.1. Notice that the percentage cost penalty, for a fixed value of B_1/A, decreases as EOQ/σ_L increases, that is, as the variability of demand decreases relative to the EOQ. This is to be expected because the sequential approach computes the order quantity ignoring the variability. Furthermore, the percentage penalties are quite low as long as EOQ/σ_L does not become too small.[*] Now, let us look more closely at the factor EOQ/σ_L.

$$\frac{\text{EOQ}}{\sigma_L} = \sqrt{\frac{2AD}{vr\sigma_L^2}} = \sqrt{\frac{2ADv}{r(\sigma_L v)^2}} \tag{7.7}$$

---

[*] Furthermore, Brown (1977) has suggested a pragmatic approach that eliminates the larger penalties. Specifically, if $\text{EOQ} < \sigma_L$, simply use $Q = \sigma_L$ and then find k sequentially.

As discussed in Chapter 3, some investigations have found evidence that, *on average*, $\sigma_L v$ tends to increase with Dv in the following manner:

$$\sigma_L v \approx c_1 (Dv)^{c_2}$$

where c_2 is greater than 0.5. Substituting this in Equation 7.7 yields

$$\frac{\text{EOQ}}{\sigma_L} \approx \sqrt{\frac{2ADv}{rc_1^2 (Dv)^{2c_2}}} \propto (Dv)^{0.5 - c_2}$$

Because c_2 is larger than 0.5, this says that, on the average, EOQ/σ_L decreases as Dv increases; that is, low values of EOQ/σ_L are more likely with A items than with B or C items. From above, this implies that the larger percentage errors in using the sequential approach are more likely to occur with A items. This, together with the fact that the absolute costs associated with A items are high, indicates that the sophisticated simultaneous procedures may be justified for some A items. A corollary is that the absolute savings of a simultaneous approach are not likely to be justified for B items; hence, we suggested the use of the simpler sequential approach in Chapter 6.

In an interesting study, Gallego (1998) shows that using the following formula for the order size in a B_3 cost situation in conjunction with the optimal s yields a cost no more than 6.07% higher than the optimal. Therefore, a sequential approach may be used. His formula for the order quantity is

$$\text{EOQ}(1 + \sigma_L^2 v (r + B_3)/(2AD))^{0.25}$$

One final comment: If Q is small, as is common when pallet-sized quantities are delivered frequently, an order may not restore the inventory position to be greater than s. In such a case, consider an (s, nQ) system in which an order of a multiple of Q insures that the inventory position is greater than s at all times. See Zheng and Chen (1992) for a treatment of these policies.

7.4 Decision Rules for (s, S) Systems

We turn now to the commonly used (s, S) system. Once again we shall introduce a sequential approach followed by a simultaneous approach. Recall from Chapter 6 that an (s, S) system assumes continuous review. Whenever the inventory position drops to the reorder point s or lower, an order is placed to raise the position to the order-up-to level S. When all transactions are of unit size, then every order is of size $(S - s)$ and is placed when the inventory position is exactly at the level s. We can thus think of an order quantity

$$Q = S - s \tag{7.8}$$

and then the assumptions of an (s, Q) system (see Section 6.7.1 of Chapter 6) hold. Bartakke (1981) has reported on the use of such a system under Poisson (unit-sized) demand at Sperry-Univac. It is *non-unit-sized transactions that complicate the analysis*. Furthermore, in contrast to an (R, S) system where each cycle of fixed length R starts with the inventory position at the level S, here, we have a cycle of random length (how long it takes for the inventory position to drop from S to s or lower). Schneider (1979) shows that using an (s, Q) system in place of an (s, S) system can severely reduce the obtained service level.

We now discuss the sequential and simultaneous approaches for selecting s and S under conditions of non-unit-sized transactions. This is applied to one type of shortage-costing measure (cost of B_1 per stockout occasion) and normally distributed forecast errors. Other references on (s, S) systems include Ward (1978); Archibald (1981); Williams (1982); Sahin (1982), and Federgruen and Zipkin (1984). In addition, the related work of Tijms and Groenevelt (1984) will be discussed in Section 7.5.2.

7.4.1 Simple Sequential Determination of s and S

In this first approach, we choose to neglect the undershoots (how far below s the inventory position is located when an order is placed). In addition, S and s are computed in a sequential fashion as follows. The Q expression of Equation 7.8 is set equal to the economic order quantity as in Chapter 4. Then, given this value of Q, we find s by the procedure of Section 6.7.5 of Chapter 6. Finally, from Equation 7.8, the S value is given by

$$S = s + Q \tag{7.9}$$

7.4.1.1 Numerical Illustration

Consider an item XMF-014 that is a 1,000-sheet box of 8-by-10-inch rapid-process x-ray film. For illustrative purposes, suppose that this firm was considering using an (s, S) control system for this item. Relevant given characteristics of the item include

$D = 1{,}400$ boxes/year
$v = \$5.90$/box
$A = \$3.20$
$r = 0.24$ \$/\$/year
$B_1 = \$150$
$\hat{x}_L = 270$ boxes
$\sigma_L = 51.3$ boxes

The use of the above-outlined procedure leads to

$Q = 80$ boxes
$s = 389$ boxes
$S = 469$ boxes

7.4.2 Simultaneous Selection of s and S Using the Undershoot Distribution

In the second approach, we select s and S simultaneously and also attempt to take account of the nonzero undershoots. A stockout occurs if the sum of the undershoot plus the total lead time demand exceeds s. Thus, the variable in which we are interested is

$$x' = z + x$$

where z is the undershoot and x is the total lead time demand.

We know the distribution of x (assumed to be normal). The distribution of z is quite complex, depending in general on the distance $S - s$ and the probability distribution of transaction sizes.

However, when $S - s$ is considerably larger than the average transaction size, we can make use of a result developed by Karlin (1958), namely,

$$p_z(z_0) = \frac{1}{E(t)} \sum_{t_0=z_0+1}^{\infty} p_t(t_0) \tag{7.10}$$

where

$p_z(z_0)$ = probability that the undershoot is of size z_0
$p_t(t_0)$ = probability that a demand transaction is of size t_0
$E(t)$ = average transaction size

Equation 7.10 can be used to compute the mean and variance of the undershoot variable z. The results are[*]

$$E(z) = \frac{1}{2}\left[\frac{E(t^2)}{E(t)} - 1\right]$$

and

$$\text{var}(z) = \frac{1}{12}\left\{\frac{4E(t^3)}{E(t)} - 3\left[\frac{E(t^2)}{E(t)}\right]^2 - 1\right\}$$

The two variables z and x can be assumed to be independent.[†] Therefore

$$E(x') = E(z) + E(x) = \frac{1}{2}\left[\frac{E(t^2)}{E(t)} - 1\right] + \hat{x}_L \tag{7.11}$$

and

$$\text{var}(x') = \text{var}(z) + \text{var}(x) = \frac{1}{12}\left\{\frac{4E(t^3)}{E(t)} - 3\left[\frac{E(t^2)}{E(t)}\right]^2 - 1\right\} + \sigma_L^2 \tag{7.12}$$

For convenience, we assume that x' has a normal distribution[‡] with the mean and variance given above. With this assumption, the method, whose derivation closely parallels that shown Appendix of this chapter, is as follows:

Step 1 Select k and Q to simultaneously satisfy the following two equations:

$$Q = \text{EOQ}\sqrt{1 + \frac{B_1}{A}p_{u\geq}(k)} - E(z), \text{ rounded to an integer} \tag{7.13}$$

---

[*] Baganha et al. (1999) develop a simple algorithm to compute the undershoot. And Baganha et al. (1996) show that for most common demand distributions, the mean and variance given here are quite close to the actual values if $Q > 2E(t)$. See also Hill (1988) who examines the cost of ignoring the undershoot and develops an approximation for it.

[†] Janssen et al. (1998) show that it is better to measure both the undershoot and the lead time demand, and then use this data in setting reorder points.

[‡] x is normally distributed but z is certainly not. Equation 7.10 reveals that $p_z(z_0)$ is monotonically decreasing with z_0. Hence, the assumption of $(z + x)$ having a normal distribution is only an approximation.

and

$$k = \sqrt{2\ln\left[\frac{1}{2\sqrt{2\pi}}\left(\frac{B_1}{A}\right)\frac{(EOQ)^2}{(Q + E(z))\sigma_{x'}}\right]} \quad (7.14)$$

where

$p_{u\geq}(k)$ = probability that a unit normal variable takes on a value greater than or equal to k

$EOQ = \sqrt{2AD/vr}$ and $\sigma_{x'} = \sqrt{\mathrm{var}(x')}$. Note the similarity of Equations 7.4 and 7.14.

Step 2

$$s = E(x') + k\sigma_{x'}, \text{ rounded to the nearest integer}$$

Step 3

$$S = s + Q$$

7.4.2.1 Numerical Illustration

We use the same item, XMF-014. Now, of course, we need a probability distribution of transaction sizes (or, alternatively, the first three moments $E(t)$, $E(t^2)$, and $E(t^3)$). Suppose for illustrative purposes that the distribution is

t_0	1	2	3	6	12	24	36	72
$p_t(t_0)$	0.25	0.05	0.05	0.1	0.25	0.15	0.1	0.05

$$E(t) = \sum_{t_0} t_0 p_t(t_0) = 14.9$$

$$E(t^2) = \sum_{t_0} t_0^2 p_t(t_0) = 515.7$$

$$E(t^3) = \sum_{t_0} t_0^3 p_t(t_0) = 25,857.2$$

From Equations 7.11 and 7.12

$$E(x') = \frac{1}{2}\left(\frac{515.7}{14.9} - 1\right) + 270$$

$$= 16.81 + 270$$

$$= 286.81$$

$$\mathrm{var}(x') = \frac{1}{12}\left\{\frac{4(25,857.2)}{14.9} - 3\left(\frac{515.7}{14.9}\right)^2 - 1\right\} + (51.3)^2$$

$$= 278.90 + 2,631.69$$

$$= 2,910.59$$

Step 1 Equations 7.13 and 7.14 give that k and Q must satisfy

$$Q = 79.5\sqrt{1 + 46.875 p_{u\geq}(k)} - 16.81$$

and

$$k = \sqrt{2 \ln\left[\frac{1}{2\sqrt{2\pi}}(46.875)\frac{(79.5)^2}{(Q+16.81)(53.95)}\right]}$$

The solution is

$$Q = 87 \text{ boxes}$$
$$k = 2.17$$

Step 2

$$s = 286.81 + 2.17\sqrt{2,910.59}$$
$$= 404 \text{ boxes}$$

Step 3

$$S = 404 + 87 = 491 \text{ boxes}$$

7.4.3 Comparison of the Methods

It is interesting to compare the results of the two methods for the numerical illustration of item XMF-014. Archibald (1976) has developed a search procedure that finds the optimal (s, S) pair for an arbitrary discrete distribution of transaction sizes and also gives the exact expected total relevant costs for any (s, S) pair used (see also Archibald and Silver 1978). Utilizing his procedure, we found the results shown in Table 7.1. Observe that the simultaneous method, which takes account of the undershoot, leads to a solution negligibly close to the optimum. Note that, even with the significant departure from unit-sized transactions shown in the numerical illustration, the sequential approach of Chapter 6 still gives a cost penalty of only 3.5%. Based on results of this type, we suggest the use of the simpler procedures of Chapter 6 unless the demand is erratic in nature (defined in Chapter 3 as when $\hat{\sigma}_1$ exceeds the demand level a, where both are measured over a unit time period). In addition, when a savings of 3%–4% represents a significant amount of money, one should use the simultaneous method or one of the more advanced procedures referenced earlier. Nevertheless, it is often worthwhile to check a small sample of items to see if the potential savings are greater.

Table 7.1   Illustrative Comparison of the Two Methods of Finding Values of *s* and *S*, Item XMF-014

Method	Description	s	S	ETRC ($/Year)	Percent above Best Value
1	Sequential, ignoring undershoot	389	469	350.82	3.5
2	Simultaneous and using undershoot	404	491	339.01	0.0
	Optimal solution	406	493	339.00	–

7.5 Decision Rules for (R, s, S) Systems

As mentioned in Chapter 6, it has been shown that, under quite general conditions, the system that minimizes the total of review, replenishment, carrying, and shortage costs will be a member of the (R, s, S) family. However, the determination of the exact best values of the three parameters is extremely difficult. The problem results partly from the fact that undershoots of the reorder point s are present even if all transactions are unit-sized. This is because we only observe the inventory level every R units of time. Because of the complexity of exact analyses, we again advocate and illustrate the use of heuristic approaches. Some are simple enough that they could also be used for the control of B items. Hence, in Section 7.5.2, we demonstrate the use of a service constraint that would normally not be employed for A items. It is important to note, however, that Zheng and Federgruen (1991) have developed a fast algorithm for finding the optimal s and S for a given R and for discrete demand distributions. This algorithm is best developed using a programming language, or macros with a spreadsheet, rather than a simple spreadsheet. For applications in companies that can afford the programming time, this algorithm would be an excellent choice.

7.5.1 Decision Rule for a Specified Fractional Charge (B₃) per Unit Short at the End of Each Period

We now present a heuristic method, known as the revised power approximation. The original power approximation was suggested by Ehrhardt (1979) and the revision was proposed by Ehrhardt and Mosier (1984). An interesting alternative approach has been developed by Naddor (1975).

The revised power approximation determines values of the two control parameters $Q = S - s$ and s. We first show the approximation and then briefly discuss its nature and its derivation.

7.5.1.1 Revised Power Approximation Decision Rules

Step 1 Compute

$$Q_p = 1.30 \hat{x}_R^{0.494} \left(\frac{A}{vr} \right)^{0.506} \left(1 + \frac{\sigma_{R+L}^2}{\hat{x}_R^2} \right)^{0.116} \tag{7.15}$$

and

$$s_p = 0.973 \hat{x}_{R+L} + \sigma_{R+L} \left(\frac{0.183}{z} + 1.063 - 2.192z \right) \tag{7.16}$$

where

$$z = \sqrt{\frac{Q_p r}{\sigma_{R+L} B_3}} \tag{7.17}$$

$$\hat{x}_R = DR$$
$$\hat{x}_{R+L} = D(R + L)$$

with B_3 in \$/\$ short at the end of a review interval; r in \$/\$/review interval; D in units/year; and R and L in years.

Step 2 If $Q_p/\hat{x}_R > 1.5$, then let*

$$s = s_p \tag{7.18}$$

$$S = s_p + Q_p \tag{7.19}$$

Otherwise, go to Step 3.
Step 3 Compute

$$S_0 = \hat{x}_{R+L} + k\sigma_{R+L} \tag{7.20}$$

where k satisfies

$$p_{u\geq}(k) = \frac{r}{B_3 + r} \tag{7.21}$$

Then

$$s = \text{minimum}\{s_p, S_0\} \tag{7.22}$$

$$S = \text{minimum}\{s_p + Q_p, S_0\} \tag{7.23}$$

7.5.1.2 Numerical Illustration

Consider a particular item having $R = 1/2$ month and $L = 1$ month. Other relevant characteristics are $D = 1,200$ units/year, $A = \$25$, $v = \$2$/unit, $r = 0.24$ \$/\$/year, $B_3 = 0.2$ \$/\$ short/period, and $\sigma_{R+L} = 60$ units.

Step 1

$$\hat{x}_R = RD = \frac{1}{24}(1,200) = 50 \text{ units}$$

$$\hat{x}_{R+L} = (R + L)D = 150 \text{ units}$$

$$r = \frac{0.24}{24} = 0.01\$/\$/\text{review interval}$$

Then, from Equation 7.15, we have

$$Q_p = 1.30(50)^{0.494}[25/(2.00)(0.01)]^{0.506}[1 + (60)^2/(50)^2]^{0.116}$$
$$\approx 367 \text{ units}$$

Next, Equation 7.17 gives

$$z = \sqrt{\frac{(367)(0.01)}{(60)(0.20)}} = 0.553$$

Then, the use of Equation 7.16 leads to

$$s_p = 0.973(150) + 60\left[\frac{0.183}{0.553} + 1.063 - 2.192(0.553)\right]$$
$$\approx 157 \text{ units}$$

---

* If demands are integer valued, then s and S are rounded to the nearest integer. There may also be a management-specified minimum value of s.

Step 2

$$Q_p/\hat{x}_R = 367/50 > 1.5$$

Therefore, from Equations 7.18 and 7.19, we have

$$s = 157 \text{ units}$$
$$S = 157 + 367 = 524 \text{ units}$$

7.5.1.3 Discussion

The rules were derived in the following fashion. Roberts (1962) developed analytic forms for Q and s that hold for large values of A and B_3. Somewhat more general forms of the same nature were assumed containing a number of parameter values; for example,

$$Q_p = a\hat{x}_R^{1-b} \left(\frac{A}{vr}\right)^b \left(1 + \frac{\sigma_{R+L}^2}{\hat{x}_R^2}\right)^c \tag{7.24}$$

where a, b, and c are parameters.

Optimal values of $Q = S - s$ and s were found (by a laborious procedure not practical for routine operational use) for a wide range of representative problems, and the values of the parameters (e.g., a, b, and c in Equation 7.24) were determined by a regression fit to these optimal results. In the case of Q_p, this led to Equation 7.15. Equations 7.16 and 7.17 were found in much the same way.

Step 3 is necessary because, when $Q_p/\hat{x}_R$ is small enough, Roberts' limiting policy is no longer appropriate; in effect, we are reduced to an order every review interval, that is, an (R, S) system. Equation 7.21 can be derived in the same fashion as Equation 6.23 of Section 6.7.7 of Chapter 6, except using the more accurate representation for the expected on-hand stock footnoted in Section 6.7.3.

Note that as $\sigma_{R+L} \to 0$, Equation 7.15 gives a result close to the basic economic order quantity. Extensive tests by Ehrhardt (1979) and Ehrhardt and Mosier (1984) have shown that the approximation performs extremely well in most circumstances.

There has been much research on (R, s, S) policies, including that of Zheng and Federgruen (1991). Porteus (1985) tests many methods for finding near-optimal (s, S) values with B_3 costing, including an older version of the power approximation. He also examines whether using the B_2 method described in Section 6.7.6 works well in the B_3 case. (Not surprisingly, it does not.) The paper provides an algorithm that works very well on a wide range of parameter values. See also Ehrhardt et al. (1981) for the case of autocorrelated demand, which might be observed by a warehouse that supplies retailers who themselves order using (s, S) policies. For a similar approximation for random lead times, see Ehrhardt (1984). See also Naddor (1978).

7.5.2 Decision Rule for a Specified Fraction (P_2) of Demand Satisfied Directly from Shelf

For the case of a fill rate constraint, we present a heuristic procedure (due to Tijms and Groenevelt 1984), which, incidentally, can be adapted for use within the (s, S) context of Section 7.4, as well as for the P_1 and TBS service measures. Schneider (1978, 1981) used a somewhat different approach to derive the same results. We show the decision rule, followed by a numerical illustration, and then provide a brief discussion concerning the rule. The underlying assumptions are basically the same as those presented in Section 6.9.3 for (R, S) systems except, of course, now an order is placed at a review instant only if the inventory position is at or below s. We show illustrative details only for the case of normally distributed demand. This is appropriate to use as long as $(CV)_{R+L} \leq 0.5$ where $(CV)_{R+L} = \sigma_{R+L}/\hat{x}_{R+L}$ is the coefficient of variation of demand over $R + L$. When CV exceeds 0.5, a gamma distribution provides better results (see Tijms and Groenevelt 1984; and Section 6.7.14 of Chapter 6) because, with such a high value of CV, a normal distribution would lead to a significant probability of negative demands.

7.5.2.1 Decision Rule

Select s to satisfy

$$\sigma_{R+L}^2 J_u\left(\frac{s - \hat{x}_{R+L}}{\sigma_{R+L}}\right) - \sigma_L^2 J_u\left(\frac{s - \hat{x}_L}{\sigma_L}\right) = 2(1 - P_2)\hat{x}_R\left[S - s + \frac{\sigma_R^2 + \hat{x}_R^2}{2\hat{x}_R}\right] \quad (7.25)$$

where

$S - s =$ is assumed predetermined (e.g., by the EOQ)
$\hat{x}_t =$ expected demand in a period of duration t
$\sigma_t =$ standard deviation of errors of forecasts of total demand over a period of duration t
$J_u(k) = \int_k^\infty (u_0 - k)^2 f_u(u_0) du_0 =$ another special function of the unit normal distribution

Note that, as shown by Hadley and Whitin (1963),

$$J_u(k) = (1 + k^2)p_{u\geq}(k) - kf_u(k) \quad (7.26)$$

so that $J_u(k)$ can be evaluated easily. We have developed a column in Table II.1 of Appendix II for $J_u(k)$ using the rational approximation described in Appendix III.

Equation 7.25 requires, in general, a trial-and-error-type solution. However, in the event that the desired service level is high enough ($P_2 \geq 0.9$), the demand pattern is relatively smooth and R is not too small compared with L; then the second term on the left side of Equation 7.25 can be ignored. Thus, if we set

$$s = \hat{x}_{R+L} + k\sigma_{R+L} \quad (7.27)$$

the decision rule reduces to selecting k so as to satisfy

$$\sigma_{R+L}^2 J_u(k) = 2(1 - P_2)\hat{x}_R\left[S - s + \frac{\sigma_R^2 + \hat{x}_R^2}{2\hat{x}_R}\right] \quad (7.28)$$

If a table of $J_u(k)$ versus k is available, then finding the appropriate k poses no difficulty. Alternatively, one can use rational functions to find k from $J_u(k)$ as suggested by Schneider (1981) and presented in Appendix III. These are easily implemented on a spreadsheet. Using Appendix III, finding $J_u(k)$ from k is also easy on a spreadsheet.

7.5.2.2 Numerical Illustration

We illustrate a situation where the simpler Equation 7.28 can be used. Suppose a firm is controlling a particular item on a periodic basis with $R = 2$ weeks and the lead time $L = 1$ week. Assume that $S - s$ has been preestablished as 100 units. The forecast update period is 2 weeks and current estimates are $\hat{x}_1 = 50$ units and $\sigma_1 = 15$ units, where the subscript, 1, indicates one forecast interval rather than 1 week. Suppose that $P_2 = 0.95$. First, we evaluate

$$\hat{x}_R = \hat{x}_1 = 50 \text{ units}$$
$$\hat{x}_{R+L} = \hat{x}_{3/2} = 75 \text{ units}$$
$$\sigma_R = \sigma_1 = 15 \text{ units}$$
$$\sigma_{R+L} = \sigma_{3/2} = \sqrt{3/2}\sigma_1 = 18.4 \text{ units}$$
$$(CV)_{R+L} = \frac{\sigma_{R+L}}{\hat{x}_{R+L}} = \frac{18.4}{75} = 0.25 < 0.5$$

Thus, an assumption of normally distributed demand is reasonable. From Equation 7.28, we have

$$(18.4)^2 J_u(k) = 2(1 - 0.95)50 \left[100 + \frac{(15)^2 + (50)^2}{2(50)} \right]$$

or

$$J_u(k) = 1.885$$

Using Table II.1 or Appendix III, we find that

$$k \approx -0.98$$

Thus, from Equation 7.27, the reorder point is given by

$$s = 75 - 0.98(18.4)$$
$$\approx 57 \text{ units}$$

Finally,

$$S = s + (S - s) = 57 + 100 = 157 \text{ units}$$

Incidentally, the use of the more complicated Equation 7.25 produces the same results in this example.

7.5.2.3 Discussion

To derive the decision rule, use, as in Chapter 6,

$$1 - P_2 = \frac{\text{Expected shortage per replenishment cycle}}{\text{Expected replenishment size}} \tag{7.29}$$

As in Section 7.4.2, the key random variable is the total of the undershoot z (of the reorder point) and the demand x in a lead time. The denominator of Equation 7.29 leads to the term in square brackets in Equation 7.25. The numerator in Equation 7.29 is developed based on

$$\int_S^\infty (x' - s) f_{x'}(x_0') dx_0'$$

where $x' = z + x$. The result, after lengthy calculus, turns out to be the left side of Equation 7.25, divided by $2\hat{x}_R$.

For other research in this area, see Chen and Krass (1995) who discuss a "marginal service level" which is the minimum fraction of demand met from the shelf in *any period*. Thus, it is analogous to the fill rate (P_2) but it applies in any period rather than as a long-term average. They also describe analogies for the P_1 case. They show that (s, S) policies are optimal and provide a fast algorithm for finding the optimal values of s and S. Schneider and Ringuest (1990) present a power approximation for a *ready rate* service level. Also, for non-normal distributions, see Section 6.7.14 of Chapter 6. See also Schneider and Rinks (1989).

7.6 Coping with Nonstationary Demand

As discussed in Chapter 5, the average demand rate may vary appreciably with time. This should have an impact on the choice of control parameters such as s and Q for a continuous review system (or analogous parameters for periodic review systems). In some cases, nonstationary demand can be handled directly by the forecasting process from which we get the point forecast $\hat{x}_{t,t+L}$ and standard deviation of forecast errors $\sigma_{t,t+L}$ for demand over the next L periods.[*] Harrison (1967) presents expressions for the standard deviation of forecast errors over the lead time for both simple exponential smoothing and exponential smoothing with trend. It is worth emphasizing here that the degree of nonstationarity in the demand can have a significant impact on the necessary safety stock. As an example, consider a setting where demand is being forecasted with a simple exponential smoothing model. Furthermore, let σ represent the standard deviation of the one-period ahead forecast error and let α be the ideal smoothing constant for the forecasting process.[†] For a given safety factor k, the level of safety stock is

$$k\sigma_{t,t+L} = k\sigma\sqrt{L}\sqrt{1 + \alpha(L-1) + \frac{\alpha^2(L-1)(2L-1)}{6}} \tag{7.30}$$

Note that if demand is stationary (e.g., the average level does not change over time), the ideal forecasting model would be to not smooth forecasts ($\alpha = 0$), and the level of safety stock becomes $k\sigma\sqrt{R+L}$; that is, the required level of safety stock is proportional to the square root of the protection interval $R + L$. When the underlying level of demand varies over time, the ideal smoothing constant α is greater than zero, and more safety stock is needed for the same protection level. This relationship, shown in Figure 7.2, is important to remember since often forecast errors are reported for a fixed time lag, such as one-period ahead, rather than over the lead time. For exponential smoothing with trend (Harrison 1967), where σ is the one-step ahead standard deviation of forecast errors and the ideal smoothing constants are α and β, the required level of safety stock is

$$k\sigma_{t,t+L} = k\sigma\sqrt{\sum_{j=0}^{L-1}\left[1 + j\alpha + j(j+1)\beta/2\right]} \tag{7.31}$$

---

[*] The ideas presented here can also be applied to periodic review systems where $R + L$ would replace L as the required protection period.

[†] Graves (1999) analyzes a base-stock model for this setting.

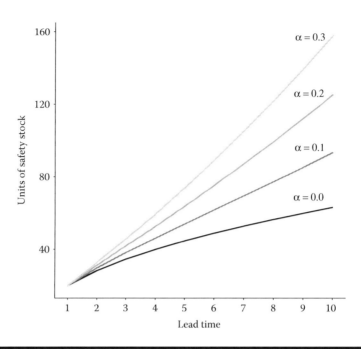

Figure 7.2 **Necessary safety stock as a function of the protection period $R + L$ and the ideal; smoothing constant α.**

The above expressions can be used when demand is best forecasted with simple exponential smoothing or exponential smoothing with trend and one-period ahead forecast errors are measured and reported. An exact analysis of general time-varying and probabilistic demand is far too complicated for routine use in practice. Therefore, again we adopt heuristic approaches. If the average demand rate changes with time, then the values of the control parameters should also change with time. From a pragmatic standpoint, frequent changes in R, the review interval, for a periodic review system are awkward to implement. Thus, changes in R, if made at all, should be very infrequent. One appealing approach to adjusting s and Q (or s and S) with time is to compute these values in a particular period t using demand (forecast) information over the immediately following interval of duration L or $R + L$ as appropriate, but still using an underlying steady-state model. This is equivalent, in the terminology of Chapter 5, to using a rolling horizon of length $R + L$. Silver (1978) uses much the same idea but allows the demand rate to vary within the current horizon. This is accomplished by using safety stock considerations to decide when to place an order, followed by the deterministic Silver–Meal heuristic to select the size of the then current replenishment. Askin (1981) also addresses this problem. In a similar vein, Vargas (1994) develops a stochastic version of the Wagner–Whitin model described in Chapter 5. See also Vargas and Metters (1996).

One paper of interest is by Bollapragada and Morton (1993), who develop a myopic heuristic that is very fast and very accurate. Their heuristic involves precomputing the (s, S) values for various values of mean demand, possibly using methods of this chapter. Then they approximate the nonstationary problem by a stationary one—essentially, averaging the nonstationary demand over an estimate of the optimal time between orders (obtained from the stationary problem) and reading the corresponding (s, S) values from the precomputed table. Tests show that the error of using the heuristic averages only 1.7%. This heuristic is easy to implement, but should be avoided if demand is expected to decline rapidly as a product becomes obsolete. Chapters 8 and 9 address this latter case.

Additional related work on this problem includes Eilon and Elmaleh (1970), Lasserre et al. (1985), LeFrancois and Cherkez (1987), Popovic (1987), Bookbinder and Tan (1988), Zipkin (1989), Wang and Rao (1992), Zinn et al. (1992), Charnes et al. (1995), Marmorstein and Zinn (1993), Cheng and Sethi (1999), Metters (1997), Ernst and Powell (1995), Morton and Pentico (1995), Chen and Krass (1995), and Anupindi et al. (1996).

A complexity to be faced under general, time-varying probabilistic demand, even in some of the simple procedures mentioned above, is the estimation of $\sigma_{t,t+R+L}$, the standard deviation of forecast errors for a forecast, made in period t, of demand over the next $R + L$ periods. In the exponential smoothing models discussed above, even though the level (and perhaps trend) of the demand are varying over time, the standard deviation of forecast errors is not. In general, and unlike in earlier discussions, we cannot make the assumption that $\sigma_{t,t+R+L} = \sigma_{R+L}$. Where the average demand rate changes with time, it is possible and perhaps likely that σ will also vary. A pragmatic approach to the estimation $\sigma_{t,t+R+L}$ is to use an aggregate relationship, first mentioned in conjunction with estimating σ's of new items in Section 3.8.1 of Chapter 3, of the form

$$\sigma_{t,t+R+L} = c_1 \hat{x}_{t,t+R+L}^{c_2} \tag{7.32}$$

where $\hat{x}_{t,t+R+L}$ is the forecast, made in period t, of total demand over the next $R + L$ periods, and c_1 and c_2 are constants estimated for a population of items (as discussed in Chapter 4). Another approach, taken by Bollapragada and Morton (1993), is to assume a constant ratio of the standard deviation to the mean (i.e., $c_2 = 1$).

7.7 Comments on Multiple Sources of Supply and Expediting

Many firms are trying to reduce their number of suppliers because of the cost of managing multiple relationships. It takes time and effort to maintain good relationships with many different vendors. The buying firm must verify the quality of delivered parts, and often is required to work with the vendor to reduce the cost of those parts. Perhaps the most important factor in an industry that competes on time-to-market is the new product development cost. If the firm has multiple suppliers for a given part, the purchasing and engineering staffs must work with many vendors to insure that parts for the new products are of high quality and are ready on time. Nevertheless, from an inventory management perspective, costs can decrease if there are more suppliers. Most research in inventory theory disregards factors such as quality, time to market, and the cost of managing relationships. Unfortunately, in our discussions with managers, even at large and respected automotive firms, the inventory implications of sourcing decisions are often ignored. In this section, we comment briefly on research that should be a useful addition to decisions that are usually made on more qualitative grounds.

If multiple suppliers are used, the inventory equations developed in this chapter and in Chapter 6 must be modified. In general, the dual-sourcing models discussed here seek to balance the use of the two supply modes to get as much of the cost advantage from the slow source while keeping safety stock levels close to what could be achieved based on only the fast supply mode. There are two related challenges with dual-sourcing models. First, the resulting control policy should not be so complicated that it cannot be implemented in practice. Second, the analysis for dual-sourcing systems is typically much more complicated, so one can expect greater effort to be required to identify good values for control parameters. Moinzadeh and Nahmias (1988) developed an approximate model for an inventory system with two possible suppliers, one having a shorter lead time. They defined a procedure for finding approximately optimal values of s and Q for both suppliers. (See also Moinzadeh and Schmidt 1991.) Kelle and Silver (1990) discussed order splitting among two or more vendors so that the average and variance of replenishment lead time are reduced. Lau and Zhao (1993) investigated a similar problem and discovered that the major advantage of order splitting is the reduction of cycle inventory cost. This result contrasts with earlier work that focused on reduction in safety stock. Guo and Ganeshan (1995) provided simple decision rules to find the optimal number of suppliers for particular lead time distributions. See also Sculli and Wu (1981), Ramasesh et al. (1991), Lau and Zhao (1994), Chiang and Benton (1994), Ernst et al. (2007), Fong and Gempesaw (1996), Parlar and Perry (1996), Zhang (1997), Chiang and Chiang (1996), and Feng and Shi (2012).

Veeraraghavan and Scheller-Wolf (2008) developed an approach for a relatively simple "dual index" policy for a periodic review system where orders for the two supply modes are determined based on two target inventory levels. At each ordering instance, two orders are placed: (1) an order with the fast lead time supply source to bring the total of on-hand and on-order inventory that will arrive within the fast lead time up to a specified target inventory level, and (2) an order with the slow lead time supplier to bring on-hand plus on-order inventory up to another (higher) target inventory level. While some computational effort is involved in finding the two control parameters, the authors demonstrate on a broad set of numerical experiments that the best dual-index policy is quite close to optimal. Their model can also accommodate capacity constraints on the supply sources.

Motivated by a setting where supply from China and Mexico was used to meet U.S. demand, Allon and Van Mieghem (2010) developed a "base-surge" approach to managing two supply sources where a constant "base" volume is sourced from the long lead time source and the fast lead time source is used to react to variations in demand ("surge"). For stationary demand, the authors developed an expression for the near-optimal long-term volume to be allocated to the slow source. This prescription may serve as a reasonable estimate even when demand is nonstationary. See also Boute and Van Mieghem (2014), Jain et al. (2010, 2011), and Sheopuri et al. (2010).

A related problem is when to place an emergency order if a replenishment order is late or cannot meet a surge in demand. White and White (1992) investigated the problem of setting the inventory policy for this case, and found that the cost of emergency ordering may be more than offset by reductions in inventory carrying costs. Dhakar et al. (1994) investigated an $(S - 1, S)$ inventory system for high-cost repairable items that has three possible modes of replenishment: normal repair orders, emergency repair orders, and expediting outstanding orders. They use a simulation procedure to find S and an expediting threshold. Kalpakam and Sapna (1995) addressed a lost sales inventory system in which orders are placed at two levels $s(> 0)$ and zero, with ordering quantities $Q > s$ and s, respectively. Finally, in an interesting paper, Song and Zipkin (1996) discussed a system in which replenishment lead times evolve over time as the system changes. They found that a longer lead time does not necessarily imply more inventory. A related work, Ben-Daya and Raouf (1994), addressed an inventory system in which lead time is a decision variable, which can be reduced at a cost. There is a significant amount of research on the inventory problem when the quantity supplied does not necessarily exactly match the quantity ordered. Costa and Silver (1996) have dealt with the situation where supply is uncertain, but different levels of supply can be reserved ahead of time. See also Lee and Moinzadeh (1987, 1989), Gerchak et al. (1988), Yano and Lee (1995), and Wang and Gerchak (1996).

7.8 Summary

In this chapter, we have dealt with the most important class of items, namely, A items. Procedures, more elaborate than those in earlier chapters, have been suggested. Where analytic models cannot incorporate all important factors and still be solvable by deductive reasoning, it may very well be worthwhile in the case of A items to resort to simulation methods that are much broader in applicability but more expensive to use. (See Section 2.7 of Chapter 2.)

In the next chapter, we turn to the opposite end of the spectrum, namely, C items. However, in later chapters, where we deal with coordination among items, A and B items will again play a prominent role.

Problems

7.1 Consider an (s, Q) system of control for a single item and with normally distributed forecast errors.
   a. Find an expression that k must satisfy given a Q value if we wish to have an expected number of stockout occasions per year equal to N.
   b. Derive the two simultaneous equations that Q and k must satisfy if we wish to minimize the total expected costs of replenishment and carrying inventory subject to the expected number of stockout occasions per year being N.
   c. For the following example, develop Q, k, and the associated total relevant costs per year for two strategies:
      i. $Q = $ EOQ and corresponding k
      ii. Simultaneous best (Q, k) pair
         Item characteristics:

$$A = \$5 \qquad\qquad v = \$2/\text{unit}$$
$$r = 0.16\$/\$/\text{year} \qquad D = 1{,}000 \text{ units/year}$$
$$\sigma_L = 80 \text{ units} \qquad N = 0.5/\text{year}$$

7.2 Ceiling Drug is concerned about the inventory control of an important item. Currently, they are using an (s, Q) system where Q is the EOQ and the safety factor k is selected based on a $B_2 v$ shortage-costing method. Relevant parameter values are estimated to be

$$D = 2{,}500 \text{ units/year} \quad B_2 = 0.6$$
$$v = \$10/\text{unit} \qquad\quad \hat{x}_L = 500 \text{ units}$$
$$A = \$5 \qquad\qquad\quad \sigma_L = 100 \text{ units}$$
$$r = 0.25\$/\$/\text{year}$$

   a. What are the Q and s values currently in use?
   b. Determine the simultaneous best values of Q and s.
   c. What is the percent penalty (in the total of replenishment, carrying, and shortage costs) of using the simpler sequential approach?

7.3 Consider a situation where the lowest allowable k value is 0. The B_1 shortage-costing method is appropriate.

a. On a plot of k versus Q/σ_L sketch each of Equations 7.3 and 7.4. (Hint: For each, evaluate $dk/d(Q/\sigma_L)$. Also look at the behavior of each for limiting values of k and Q/σ_L).

b. Attempt to ascertain under precisely what conditions there will be no simultaneous solution of the two equations.

c. Verify, using Equation 7.2, that under such circumstances the cost-minimizing solution is to set $k = 0$ and

$$Q/\sigma_L = EOQ/\sigma_L\sqrt{1 + 0.5B_1/A}$$

7.4 In relation to an (s, Q) inventory control system, an operations analyst has made the following observation:

Suppose we determine Q from the EOQ; then find the safety factor from the "fraction demand satisfied routinely from shelf" criterion. Suppose that through explicit cost considerations we estimate r to be a particular value r_1. Now if we deliberately set r at a lower value r_2, when determining Q and s, we can reduce actual total relevant costs.

a. Could the analyst be telling the truth? Attempt to develop an analytic proof.

b. Illustrate with an item having the following characteristics:

$$Dv = \$4000/\text{year} \quad \sigma_L v = \$500$$
$$A = \$100 \quad P_2 = 0.975$$
$$r_1 = 0.2\$/\$/\text{year}$$

7.5 Company X has an A item which it controls by an (s, S) system. The item has only a few customers. According to the (s, S) policy, if a large order drives the inventory position to or below s, enough is ordered to raise the position to S. Discuss the implicit cost trade-off that the company is making in still raising the position to S even after a large demand has just occurred.

7.6 a. Consider an item for which an (R, s, S) system is used. The review interval is 1 week. Other characteristics include

$$L = 2 \text{ weeks} \qquad r = 0.26\$/\$/\text{year}$$
$$D = 800 \text{ units/year} \qquad B_3 = 0.30\$/\$ \text{ short/week}$$
$$A = \$20 \qquad \sigma_{R+L} = 14.2 \text{ units}$$
$$v = \$1.50/\text{unit}$$

Using the revised power approximation, find the appropriate values of s and S. (Assume that there are exactly 52 weeks per year.)

b. Repeat for another item with the same characteristics except $v = \$150$ /unit.

7.7 An oil distribution depot supplies customers by road tanker. The depot itself is supplied by a unit train with 24 cars, each with a capacity of 75 tons. At the depot, three different products are stored in eight 1,000-ton tanks. At present, three tanks are used for product A, three tanks for product B, and two tanks for product C. Analysis of demand records shows that the average weekly demand for product A is 600 tons with a standard deviation 400 tons, average weekly demand for product B is 500 tons with a standard deviation 300 tons, and average weekly demand for product C is 400 tons with a standard deviation 200 tons. Any seasonal effects can be ignored. The stock level of each product is reported by the depot to the refinery every Monday morning. The distribution office then has to decide whether to replenish the depot the following Friday and, if so, how many cars of the train should be filled with each product. The cost of operating the train is not reduced if any cars are empty and is approximately $3,000 per trip. If a product runs out at the depot, then the road tankers have to be diverted from that depot to another depot at considerably higher cost.

a. How does the limited capacity tankage at the depot affect the applicability of standard inventory control models?

b. Are there any other features of the problem that make standard inventory control models difficult to apply?

c. *Without doing any detailed numerical calculations*, indicate the type of inventory control guidelines that you could propose for the distribution office.

d. *Briefly* indicate what other (if any) types of data you would need. Also, what type of solution procedure (for finding precise values of control parameters within the general control guidelines of part c) do you think would be needed?

7.8 A particular inventory control manager places orders for very expensive items. A report filed by an outside consultant included the following quotes. Comment on each.

a. "He tried to order only enough items to meet actual sales for the 1-month period, one lead time period ahead." Comment in particular the choice of a 1-month period.

b. "To meet unexpected demand he usually ordered 10%–20% more than the number of units he was able to identify through purchase orders or commitments."

7.9 Develop a spreadsheet approach for Equations 7.3 and 7.4. First develop a table for iteration. Then use fewer cells by using circular references.

7.10 Repeat Problem 7.2 using $B_1 = 100 in place of $B_2 = 0.6$.

7.11 Consider the item in Problem 7.2 and assume that the sequential approach is used. Suppose that, through improved forecasting, σ_L can be reduced to 70 units. Up to what extra amount per year should the company be willing to pay to achieve this improved forecasting?

7.12 Air Alberta has been using an (s, Q) control system for a fast-moving spare part for one of its types of aircraft. The properties of this part are

$$D = 1,920 \text{ units/year} \quad A = $15$$
$$v = $20.00/\text{unit} \quad r = 0.2$/$/\text{year}$$
$$\hat{x}_L = 150 \text{ units} \quad \sigma_L = 60 \text{ units}$$

Whenever a shortage is impending, an expediting action is taken, avoiding the shortage. The cost of such an expedite is roughly $200, independent of the magnitude of the impending shortage. The company uses an EOQ as the order quantity.

a. Determine the EOQ and s.

b. What are the expected total relevant costs (replenishing, carrying, and expediting) per year?

c. A management science graduate, working in Air Alberta's operations research department, says that increasing Q to 5% above the EOQ will reduce the expected total relevant costs per year? Can he be right? Discuss.

7.13   The two equations to solve simultaneously for the case of B_2 shortage cost for the (s, Q) system are

$$Q = \sqrt{\frac{2AD + 2DB_2 v \sigma_L G_u(k)}{rv}}$$

$$\text{and } p_{u\geq}(k) = \frac{Qr}{DB_2}$$

For the following item, compute the optimal s and Q.

$L = 2$ weeks $r = 0.26\$/\$/\text{year}$
$D = 800$ units/year $B_2 = 0.30$
$A = \$20$ $\sigma_L = 14.2$ units
$v = \$4.50/\text{unit}$

7.14   An arrangement, sometimes made between a supplier and a customer, is what is called a "standing" order. Under this arrangement, the supplier ships a fixed quantity Q_f at regular intervals (R time units apart) throughout the year. Because of the customer commitment for regular shipments, the supplier offers an attractive unit cost, v_f. However, because demand is variable, special auxiliary orders must be placed from time to time and/or one or more of the standard shipments may have to be postponed for R units of time. Suppose a particular steel products distributor, Salem Inc., is considering a standing order arrangement for a specific product with a supplier, Jericho Steel. The demand rate for this product averages 12,000 units/year. For a standing order arrangement of 1,000 units each month, Jericho offers a unit price (including shipment cost) of $2.00/unit. Without a standing order arrangement, the unit price (including shipment cost) is $2.20/unit. The Salem inventory manager estimates that the fixed cost associated with any standing order is $25, whereas the fixed cost of any other order is $35. Salem uses a carrying charge of 0.26 $/$/year. If a standing order arrangement is adopted, auxiliary orders can only be placed so as to arrive with standing orders (the lead time for such an auxiliary order is 1 month; in fact, this is the lead time of any nonstandard order). The decision to postpone a standard order also must be made 1 month before its scheduled arrival. Assume that demands in non-overlapping periods are independent, normally distributed variables with a standard deviation of 250 units for a 1 month period. Salem wishes to ensure a maximum probability of 0.03 of a stockout on any replenishment.
   a.   If the standing order policy was adopted, indicate how you would assist Salem in deciding on
      i.   When to place an auxiliary order
      ii. When to postpone a standard order
   b.   Indicate how you might use a model to ascertain whether Salem should even adopt the standing order arrangement.

Appendix 7A: Simultaneous Solutions for Two Control Parameters

7A.1 (s, Q) *System with B_1 Cost Penalty*

We let

$$s = \hat{x}_L + k\sigma_L$$

From Equation 7.2, the expected total relevant costs per unit time under use of a safety factor k and an order quantity Q are given by

$$ETRC(k, Q) = AD/Q + (Q/2 + k\sigma_L)vr + \frac{DB_1}{Q}p_{u\geq}(k)$$

To reduce the number of separate parameters that have to be considered, we normalize by multiplying both sides by the constant term $2/vr\sigma_L$. We refer to the result as the normalized total relevant costs, $NTRC(k, Q)$. Thus,

$$NTRC(k, Q) = \frac{2AD}{vrQ\sigma_L} + \frac{Q}{\sigma_L} + 2k + \frac{2AD}{vrQ\sigma_L}\frac{B_1}{A}p_{u\geq}(k)$$

Recall that

$$EOQ = \sqrt{\frac{2AD}{vr}}$$

Therefore,

$$NTRC(k, Q) = \left(\frac{EOQ}{\sigma_L}\right)^2 \frac{\sigma_L}{Q}\left[1 + \frac{B_1}{A}p_{u\geq}(k)\right] + \frac{Q}{\sigma_L} + 2k$$

A necessary condition (unless we are at a boundary) for the minimization of a function of two variables is that the partial derivative with respect to each variable be set to zero.

$$\frac{\partial NTRC(k, Q)}{\partial Q} = -\left(\frac{EOQ}{\sigma_L}\right)^2 \frac{\sigma_L}{Q^2}\left[1 + \frac{B_1}{A}p_{u\geq}(k)\right] + \frac{1}{\sigma_L} = 0$$

or

$$\frac{Q}{\sigma_L} = \frac{EOQ}{\sigma_L}\sqrt{\left[1 + \frac{B_1}{A}p_{u\geq}(k)\right]}$$

$$\frac{\partial NTRC(k, Q)}{\partial k} = 2 - \left(\frac{EOQ}{\sigma_L}\right)^2 \frac{\sigma_L}{Q}\frac{B_1}{A}f_u(k) = 0$$

or

$$f_u(k) = 2\frac{A}{B_1}\frac{Q}{\sigma_L}\left(\frac{\sigma_L}{EOQ}\right)^2$$

that is,

$$\frac{1}{\sqrt{2\pi}}exp(-k^2/2) = 2\frac{A}{B_1}\frac{Q}{\sigma_L}\left(\frac{\sigma_L}{EOQ}\right)^2$$

This simplifies to Equation 7.4.

Note: We have used the fact, proved in Appendix II, that

$$\frac{dp_{u\geq}(k)}{dk} = -f_u(k)$$

References

Allon, G. and J. A. Van Mieghem. 2010. Global dual sourcing: Tailored base-surge allocation to near-and offshore production. *Management Science 56*(1), 110–124.

Anupindi, R., T. E. Morton, and D. Pentico. 1996. The nonstationary stochastic lead-time inventory problem: Near-myopic bounds, heuristics, and testing. *Management Science 42*(1), 124–129.

Archibald, B. 1976. *Continuous Review (s, S) Policies for Discrete, Compound Poisson, Demand Processes*. PhD thesis. University of Waterloo. Waterloo, Ontario, Canada.

Archibald, B. and E. A. Silver. 1978. (s, S) Policies under continuous review and discrete compound Poisson demands. *Management Science 24*(9), 899–909.

Archibald, R. 1981. Continuous review (s, S) policies with lost sales. *Management Science 27*(10), 1171–1177.

Askin, R. G. 1981. A procedure for production lot sizing with probabilistic dynamic demand. *AIIE Transactions 13*(2), 132–137.

Baganha, M. P., D. F. Pyke, and G. Ferrer. 1996. The undershoot of the reorder point: Tests of an approximation. *International Journal of Production Economics 45*, 311–320.

Baganha, M. P., D. F. Pyke, and G. Ferrer. 1999. The residual life of the renewal process: A simple algorithm. *Naval Research Logistics 46*, 435–443.

Bartakke, M. N. 1981. A method of spare parts inventory planning. *OMEGA 9*(1), 51–58.

Ben-Daya, M. and A. Raouf. 1994. Inventory models involving lead time as a decision variable. *Journal of the Operational Research Society 45*(5), 579–582.

Bollapragada, S. and T. Morton. 1993. *The Non-Stationary (s,S) Inventory Problem: Near-Myopic Heuristics, Computational Testing*. PhD thesis. Carnegie-Mellon. Pittsburgh, Pennsylvania, United States.

Bookbinder, J. H. and J. Y. Tan. 1988. Strategies for the probabilistic lot-sizing problem with service-level constraints. *Management Science 34*(9), 1096–1108.

Boute, R. N. and J. A. Van Mieghem. 2014. Global dual sourcing and order smoothing: The impact of capacity and lead times. *Management Science. 61*(9), 2080–2099.

Brown, R. G. 1977. *Materials Management Systems*. New York: Wiley-Interscience.

Charnes, J. M., H. Marmorstein, and W. Zinn. 1995. Safety stock determination with serially correlated demand in a periodic-review inventory system. *Journal of the Operational Research Society 46*(8), 1006–1013.

Chen, F. Y. and D. Krass. 2001. Inventory models with minimal service level constraints. *European Journal of Operational Research 134*(1), 120–140.

Cheng, F. and S. P. Sethi. 1999. Optimality of state-dependent (s, S) policies in inventory models with Markov-modulated demand and lost sales. *Production and Operations Management 8*(2), 183–192.

Chiang, C. and W. Benton. 1994. Sole sourcing versus dual sourcing under stochastic demands and lead times. *Naval Research Logistics 41*(5), 609–624.

Chiang, C. and W. Chiang. 1996. Reducing inventory costs by order splitting in the sole sourcing environment. *Journal of the Operational Research Society 47*, 446–456.

Costa, D. and E. A. Silver. 1996. Exact and approximate algorithms for the multi-period procurement problem where dedicated supplier capacity can be reserved. *Operations-Research-Spektrum 18*(4), 197–207.

Dhakar, T. S., C. P. Schmidt, and D. M. Miller. 1994. Base stock level determination for high cost low demand critical repairable spares. *Computers and Operations Research 21*(4), 411–420.

Donaldson, W. A. 1984. An equation for the optimal value of *T*, the inventory replenishment review period when demand is normal. *Journal of the Operational Research Society 35*(2), 137–139.

Ehrhardt, R. 1979. The power approximation for computing (s, S) inventory policies. *Management Science* 25(8), 777–786.

Ehrhardt, R. 1984. (s, S) Policies for a dynamic inventory model with stochastic lead times. *Operations Research* 32(1), 121–132.

Ehrhardt, R. and C. Mosier. 1984. Revision of the power approximation for computing (s, S) policies. *Management Science* 30(5), 618–622.

Ehrhardt, R. A., C. R. Schultz, and H. M. Wagner. 1981. (s,S) Policies for a wholesale inventory system. In L. B. Schwarz (Ed.), *Multi-Level Production/Inventory Control Systems: Theory and Practice*, Volume 16, pp. 145–161. Amsterdam: North-Holland.

Eilon, S. and J. Elmaleh. 1970. Adaption limits in inventory control. *Management Science* 16(18), B533–B548.

Ernst, R., B. Kamrad, and K. Ord. 2007. Delivery performance in vendor selection decisions. *European Journal of Operational Research* 176(1), 534–541.

Ernst, R. and S. G. Powell. 1995. Optimal inventory policies under service-sensitive demand. *European Journal of Operational Research* 87, 316–327.

Federgruen, A. and P. Zipkin. 1984. An efficient algorithm for computing optimal (s, S) policies. *Operations Research* 32(6), 1268–1285.

Feng, Q. and R. Shi. 2012. Sourcing from multiple suppliers for price-dependent demands. *Production and Operations Management* 21(3), 547–563.

Fong, D. K. H. and V. M. Gempesaw. 1996. Evaluating re-order levels and probabilities of stockout during a lead time for some stock-control models. *Journal of the Operational Research Society* 47, 321–328.

Gallego, G. 1998. New bounds and heuristics for (Q, r) policies. *Management Science* 44(2), 219–233.

Gerchak, Y., R. Vickson, and M. Parlar. 1988. Periodic review production models with variable yield and uncertain demand. *IIE Transactions* 20(2), 144–156.

Graves, S. C. 1999. A single-item inventory model for a nonstationary demand process. *Manufacturing & Service Operations Management* 1(1), 50–61.

Guo, Y. and R. Ganeshan. 1995. Are more suppliers better? *Journal of the Operational Research Society* 46, 892–895.

Hadley, G. and T. M. Whitin. 1963. *Analysis of Inventory Systems*. Englewood Cliffs, NJ: Prentice-Hall, Inc.

Harrison, P. J. 1967. Exponential smoothing and short-term sales forecasting. *Management Science* 13(11), 821–842.

Hill, R. M. 1988. The effect of transfer-pack size on order quantity. *Journal of the Operational Research Society* 39(3), 305–309.

Hill, R. M. 1992. Numerical analysis of a continuous-review lost-sales inventory model where two orders may be outstanding. *European Journal of Operational Research* 62, 11–26.

Jain, A., H. Groenevelt, and N. Rudi. 2010. Continuous review inventory model with dynamic choice of two freight modes with fixed costs. *Manufacturing & Service Operations Management* 12(1), 120–139.

Jain, A., H. Groenevelt, and N. Rudi. 2011. Periodic review inventory management with contingent use of two freight modes with fixed costs. *Naval Research Logistics* 58(4), 400–409.

Janssen, F., R. Heuts, and A. de Kok. 1998. On the (R, s, Q) inventory model when demand is modelled as a compound Bernoulli process. *European Journal of Operational Research* 104(3), 423–436.

Jardine, A. K., D. Lin, and D. Banjevic. 2006. A review on machinery diagnostics and prognostics implementing condition-based maintenance. *Mechanical Systems and Signal Processing* 20(7), 1483–1510.

Kalpakam, S. and K. P. Sapna. 1995. A two reorder level inventory system with renewal demands. *European Journal of Operational Research* 84, 402–415.

Karlin, S. 1958. The application of renewal theory to the study of inventory policies. In K. Arrow, S. Karlin, and H. Scarf (Eds.), *Studies in the Mathematical Theory of Inventory and Production*, Chapter 15. Stanford, CA: Stanford University Press.

Kelle, P. and E. A. Silver. 1990. Safety stock reduction by order splitting. *Naval Research Logistics* 37, 725–743.

Lasserre, J. B., C. Bes, and F. Roubellat. 1985. The stochastic discrete dynamic lot size problem: An open-loop solution. *Operations Research* 33(3), 684–689.

Lau, H. and L. Zhao. 1994. Dual sourcing cost-optimization with unrestricted lead-time distributions and order-split proportions. *IIE Transactions* 26(5), 66–75.

Lau, H.-S. and L.-G. Zhao. 1993. Optimal ordering policies with two suppliers when lead times and demands are all stochastic. *European Journal of Operational Research 68*, 120–133.

Lee, H. L. and K. Moinzadeh. 1987. Two-parameter approximations for multi-echelon repairable inventory models with batch ordering policy. *IIE Transactions 19*(2), 140–149.

Lee, H. L. and K. Moinzadeh. 1989. A repairable item inventory system with diagnostic and repair service. *European Journal of Operational Research 40*, 210–221.

LeFrancois, P. and C. Cherkez. 1987. Adaptive limits for pc-based inventory control. *International Journal of Production Research 25*(9), 1325–1337.

Marmorstein, H. and W. Zinn. 1993. A conditional effect of autocorrelated demand on safety stock determination. *European Journal of Operational Research 68*(1), 139–142.

Metters, R. 1997. Production planning with stochastic seasonal demand and capacitated production. *IIE Transactions 29*(11), 1017–1029.

Moinzadeh, K. and S. Nahmias. 1988. A continuous review model for an inventory system with two supply modes. *Management Science 34*(6), 761–773.

Moinzadeh, K. and C. P. Schmidt. 1991. An $(S - 1, S)$ inventory system with emergency orders. *Operations Research 39*(2), 308–321.

Moon, I. and S. Choi. 1994. The distribution free continuous review inventory system with a service level constraint. *Computers and Industrial Engineering 27*(1–4), 209–212.

Moon, I. and G. Gallego. 1994. Distribution free procedures for some inventory models. *Journal of Operational Research Society 45*(6), 651–658.

Morton, T. E. and D. W. Pentico. 1995. The finite horizon nonstationary stochastic inventory problem: Near myopic bounds, heuristics, testing. *Management Science 41*(2), 334–343.

Naddor, E. 1975. Optimal heuristic decisions for the s, S inventory policy. *Management Science 21*(9), 1071–1072.

Naddor, E. 1978. Sensitivity to distributions in inventory systems. *Management Science 24*(16), 1769–1772.

Parlar, M. and D. Perry. 1996. Inventory models of future supply uncertainty with single and multiple suppliers. *Naval Research Logistics 43*, 191–210.

Platt, D. E., L. W. Robinson, and R. B. Freund. 1997. Tractable (Q, R) heuristic models for constrained service levels. *Management Science 43*(7), 951–965.

Popovic, J. B. 1987. Decision making on stock levels in cases of uncertain demand rate. *European Journal of Operational Research 32*, 276–290.

Porteus, E. L. 1985. Investing in reducing setups in the EOQ model. *Management Science 31*(8), 998–1010.

Ramasesh, R. V., J. K. Ord, J. C. Hayya, and A. Pan. 1991. Sole versus dual sourcing in stochastic lead-time (s, Q) inventory models. *Management Science 37*(4), 428–443.

Roberts, D. 1962. Approximations to optimal policies in a dynamic inventory model. In K. Arrow, S. Karlin, and H. Scarf (Eds.), *Studies in Applied Probability and Management Science*, pp. 207–229. Stanford, CA: Stanford University Press.

Robinson, L. 1993. The cost of following the optimal inventory policy. *IIE Transactions 25*(5), 105–108.

Sahin, I. 1982. On the objective function behavior in (s, S) inventory models. *Operations Research 30*(4), 709–724.

Schneider, H. 1978. Methods for determining the re-order point of an (s, S) ordering policy when a service level is specified. *Journal of the Operational Research Society 29*(12), 1181–1193.

Schneider, H. 1979. The service level in inventory control systems. *Engineering and Process Economics 4*, 341–348.

Schneider, H. 1981. Effect of service-levels on order-points or order-levels in inventory models. *International Journal of Production Research 19*(6), 615–631.

Schneider, H. and J. L. Ringuest. 1990. Power approximation for computing (s, S) policies using service level. *Management Science 36*(7), 822–834.

Schneider, H. and D. Rinks. 1989. Optimal policy surfaces for a multi-item inventory problem. *European Journal of Operational Research 39*, 180–191.

Sculli, D. and S. Y. Wu. 1981. Stock control with two suppliers and normal lead times. *Journal of the Operational Research Society 32*(11), 1003–1009.

Sheopuri, A., G. Janakiraman, and S. Seshadri. 2010. New policies for the stochastic inventory control problem with two supply sources. *Operations Research* 58(3), 734–745.

Silver, E. A. 1978. Inventory control under a probabilistic time-varying, demand pattern. *AIIE Transactions* 10(4), 371–379.

Silver, E. A. and T. G. Wilson. 1972. Cost penalties of simplified procedures for selecting reorder points and order quantities. In *Fifteenth Annual International Conference of the American Production and Inventory Control Society*, Toronto, Canada, pp. 219–234.

Song, J. and P. H. Zipkin. 1996. Evaluation of base-stock policies in multiechelon inventory systems with state-dependent demands. Part II: State-dependent depot policies. *Naval Research Logistics* 43, 381–396.

Tijms, H. and H. Groenevelt. 1984. Simple approximations for the reorder point in periodic and continuous review (s, S) inventory systems with service level constraints. *European Journal of Operational Research* 17, 175–190.

Vargas, V. A. 1994. *The Stochastic Version of the Wagner-Whitin Production Lot-Size Model*. PhD thesis. Emory University. Atlanta, Georgia.

Vargas, V. A. and R. Metters. 1996. Adapting lot-sizing techniques to stochastic demand through production scheduling policy. *IIE Transactions* 28, 141–148.

Veeraraghavan, S. and A. Scheller-Wolf. 2008. Now or later: A simple policy for effective dual sourcing in capacitated systems. *Operations Research* 56(4), 850–864.

Wang, M. C. and S. S. Rao. 1992. Estimating reorder points and other management science applications by bootstrap procedure. *European Journal of Operational Research* 56, 332–342.

Wang, Y. and Y. Gerchak. 1996. Periodic review production models with variable capacity, random yield, and uncertain demand. *Management Science* 42(1), 130–137.

Ward, J. B. 1978. Determining reorder points when demand is lumpy. *Management Science* 24(6), 623–632.

White, C. R. and B. R. White. 1992. Emergency inventory ordering assuming Poisson demands. *Production Planning and Control* 3(2), 168–174.

Williams, B. W. 1982. Vehicle scheduling: Proximity priority searching. *Journal of the Operational Research Society* 33(10), 961–966.

Yano, C. A. 1985. New algorithms for (Q, r) systems with complete backordering using a fill-rate criterion. *Naval Research Logistics Quarterly* 32(4), 675–688.

Yano, C. A. and H. L. Lee. 1995. Lot sizing with random yields: A review. *Operations Research* 43(2), 311–334.

Zhang, A. X. 1997. Demand fulfillment rates in an assemble to-order system with multiple products and dependent demands. *Production and Operations Management* 6(3), 309–324.

Zheng, Y. and F. Chen. 1992. Inventory policies with quantized ordering. *Naval Research Logistics* 39, 285–305.

Zheng, Y. S. and A. Federgruen. 1991. Finding optimal (s, S) policies is about as simple as evaluating a single policy. *Operations Research* 39(4), 654–665.

Zinn, W., H. Marmorstein, and J. Charnes. 1992. The effect of autocorrelated demand on customer service. *Journal of Business Logistics* 13(1), 173–192.

Zipkin, P. 1989. Critical number policies for inventory models with periodic data. *Management Science* 35(1), 71–80.

Chapter 8

Managing Slow-Moving and Low-Value (Class C) Inventories

Slow-moving and low-value items often fall in the C category or so-called "cats and dogs," although as noted in Chapter 7, slow-moving items may be important enough to be included in the A or B category. Collectively, these items usually represent an appreciable percentage of the total number of distinct SKUs, but a small fraction of the total dollar investment in the inventory of a typical company. Most of these items taken singly are relatively unimportant, but, because of their large numbers, appropriate simple control procedures must be used. In this chapter, we discuss such methods.

In Section 8.1, decision rules are presented for important (e.g., high-value or critical) slow-moving items using (s, Q) systems. The special case of intermittent demand for high-unit-value items is discussed in Section 8.2. Section 8.3 spells out in greater detail the nature of C items and the primary objectives in designing methods for controlling them. In Section 8.4, we propose control procedures for low-value items having relatively steady demand. Specifically, we consider, in order, the type of inventory records to use, the selection of the reorder quantity (or reorder interval), and the choice of the reorder point (or order-up-to level). Next, in Section 8.5, we treat the case of items with significantly declining usage patterns, that is, approaching the termination of usage. Section 8.6 is concerned with the important issue of removing totally (or relatively) inactive items from the organization's inventory. Finally, in Section 8.7, we address the related question of whether demands for a particular item should be purchased (made) to order or met from stock.

8.1 Order-Point, Order-Quantity (s, Q) Systems for Slow-Moving A Items

In Chapter 6, for the case of B items, we argued for the almost exclusive use of the normal distribution to represent the distribution of forecast errors over a lead time. For A items, where the potential benefits of using a more accurate representation are higher, we suggest, based on extensive tests (see Tyworth and O'Neill 1997), still using the normal distribution if the average (or forecast) demand in a lead time is high enough (at least 10 units). As mentioned in Chapter 6, we advocate the use of explicit (or implicit) costing of shortages. Therefore, decision rules such as those in Sections 6.7.5 through 6.7.8 of Chapter 6 should be used. If the average,[*] $\hat{x}_L$, is below 10 units, then a discrete distribution, such as the Poisson, is more appropriate. (See the Appendix of this chapter for a discussion of some properties of the distribution and a derivation of expected units short per cycle. See Appendix III for details on how to find expected units short per cycle using common spreadsheet functions.)

The Poisson distribution has but a single parameter, namely, the average demand (in this case, $\hat{x}_L$). Once $\hat{x}_L$ is specified, a value of the standard deviation of forecast errors, σ_L, follows from the Poisson relation

$$\sigma_L = \sqrt{\hat{x}_L} \tag{8.1}$$

Therefore, the Poisson is appropriate to use only when the observed σ_L (for the item under consideration) is quite close to $\sqrt{\hat{x}_L}$. An operational definition of "quite close" is within 10% of $\sqrt{\hat{x}_L}$. When this is not the case, somewhat more complicated discrete distributions such as the negative binomial (see Taylor 1961; Porteus 1985; Aggarwal and Moinzadeh 1994) or a compound Poisson (see, e.g., Adelson 1966; Feeney and Sherbrooke 1966; Mitchell et al. 1983) should be considered.

The discrete nature of the Poisson distribution (in contrast, the normal distribution deals with a continuous variable) is both a blessing and a curse. For slow-moving items, it is important to be able to deal with discrete units. On the other hand, discrete mathematics create problems in implementation, as we will soon see. In most of our discussions in Chapter 6, there were two control parameters per item:

s and Q in an (s, Q) system

R and S in an (R, S) system

Our approach was first to determine one of the parameters—namely, Q or R; then (sequentially) to find the best value of the second parameter, s or S, conditional on the value of Q or R, respectively. The derivation of the value of Q or R ignored the effect that this parameter has on the service level or shortage costs per unit time. To illustrate, consider the (s, Q) control system with a fixed cost B_1 per stockout occasion. From Section 6A.1 of the Appendix of Chapter 6 and Section 6.7.5, the total relevant costs per year, ETRC, are

$$AD/Q + (Q/2)vr + (SS)vr + \frac{DB_1}{Q} \times (\text{Probability of a stockout in a cycle}) \tag{8.2}$$

where SS is the safety stock.

The approach, advocated in Chapter 6, was first to pick the Q value that minimizes the first two terms, ignoring the last two terms; then to choose the best value of safety stock (or $k\sigma_L$ in the case of normal demand). But, clearly, the value of Q influences the shortage costs (the last term). The larger the Q is, the smaller these costs become. The exact approach would be to simultaneously determine

---

[*] In an (R, S) system $\hat{x}_L$, the average demand in a lead time would be replaced by $\hat{x}_{R+L}$, the average demand in a review interval plus a lead time.

SS and Q to minimize the ETRC expression.[*] As we saw in Chapter 7, this exact approach is more involved from both a conceptual and a computational standpoint. Furthermore, the order quantity is no longer the simple, well-understood economic order quantity, and hence is likely to be more difficult to implement. Fortunately, as we saw, the percentage penalty in the total of replenishment, carrying, and shortage costs using the simpler sequential approach tends to be quite small.[†] Thus, in most situations, we were justified in advocating its use for B items. However, even small percentage savings may be attractive for A items. In the following sections, we present first the sequential approach, and then comment on, or present in detail, the simultaneous approach.

We now present rules for two different shortage-costing measures for slow-moving items. In Sections 8.1.1 and 8.1.2, we choose Q and s sequentially. In Section 8.1.3, we present simultaneous selection of s and Q.

8.1.1 B_2 Cost Measure for Very-Slow-Moving, Expensive Items ($Q = 1$)

As mentioned above, the discrete nature of the Poisson distribution causes problems in obtaining an operational decision rule. For one important situation, namely, where the replenishment quantity $Q = 1$, an analysis leads to a decision rule quite easy to implement.[‡] First, let us see under what conditions $Q = 1$ makes sense, for any distribution, including the Poisson. From Equation 4.3 of Chapter 4, the total relevant costs, if an order quantity of size Q is used, are

$$\text{TRC}(Q) = Qvr/2 + AD/Q$$

We are indifferent between $Q = 1$ and $Q = 2$ where

$$\text{TRC}(1) = \text{TRC}(2)$$

that is,

$$vr/2 + AD = vr + AD/2$$

or

$$D = vr/A \qquad (8.3)$$

For any lower value of D, we prefer the use of $Q = 1$. Again note that this rule ignores shortage costs and thus may lead to some errors. (See Problem 8.25.) To illustrate the rule, suppose that $A/r \approx 13.11$. Therefore, we have

$$\text{Use } Q = 1 \text{ if}$$
$$D < 0.0763v$$

[*] Natarajan and Goyal (1994) discuss many methods of setting safety stock and illustrate the interdependence of s and Q.

[†] As noted in Chapter 4, Zheng (1992) shows that the fractional cost penalty of using the EOQ is always less than 1/8, and is usually much less. See also Axsäter (1996).

[‡] When $Q = 1$, an (s, Q) policy is identical to an $(S - 1, S)$ policy because an order is placed to replace units sold every time a demand occurs. See Moinzadeh (1989), Schultz (1990), and Moinzadeh and Schmidt (1991) for additional work on this problem. The latter paper uses information about pipeline orders, which is available in many cases due to supply chain integration and other technological developments.

8.1.1.1 Assumptions behind the Derivation of the Decision Rule

The assumptions include:

1. Continuous-review, order-point, order-quantity system with $Q = 1$. The remaining decision variable is the order point, s, or the order-up-to level, $S = s + 1$.
2. Poisson demand. (The derivation in Section 8A.2 of the Appendix of this chapter does not depend on Poisson demand, however, so that similar decision rules apply for any discrete demand distribution.)
3. The replenishment lead time is a constant L time periods. (The results still hold if L has a probability distribution as long as we use its mean value, $E(L)$. However, the derivation is considerably more complicated in this more general case.)
4. There is complete backordering of demands when out of stock. Karush (1957) and Mitchell (1962) have investigated the case of complete lost sales; the results differ very little from the complete backordering situation. (See Rabinowitz et al. 1995 and Ouyang et al. 1996 for the case of partial backorders.)
5. There is a fixed cost, $B_2 v$, per unit backordered (i.e., a fixed fraction, B_2, of the unit value is charged for each unit backordered).

8.1.1.2 Decision Rule

As shown in Section 8A.2 of the Appendix of this chapter, we are indifferent between reorder points s and $s + 1$ when

$$\frac{p_{p_0}(s + 1 | \hat{x}_L)}{p_{p_0 \leq}(s | \hat{x}_L)} = \frac{r}{DB_2} \tag{8.4}$$

where[*]

$p_{p_0}(s + 1 | \hat{x}_L)$ = probability that a Poisson variable with mean $\hat{x}_L$ takes on the value $s + 1$

$p_{p_0 \leq}(s | \hat{x}_L)$ = probability that a Poisson variable with mean $\hat{x}_L$ takes on a value less than or equal to s

and, as earlier,

r = carrying charge, in \$/\$/unit time

D = demand rate, in units/unit time

B_2 = fraction of unit value charged per unit short

$\hat{x}_L$ = average (forecast) demand in a replenishment lead time

Equation 8.4 is developed by equating the total relevant cost function evaluated at s with the same function evaluated at $s + 1$, and manipulating the resulting equation. (The left-hand side of this equation is sometimes called the "reversed hazard rate," the hazard rate being the ratio of the probability density function to the complement of the cumulative density function.) Equation 8.4 can be implemented relatively quickly on a spreadsheet.

---

[*] $p_{p_0}(x_0 | \hat{x}_L) = \frac{(\hat{x}_L)^{x_0} exp(-\hat{x}_L)}{x_0!} x_0 = 0, 1, 2, \ldots$

8.1.1.3 Numerical Illustration

Consider the case of a high-value item with

$v = \$350/\text{unit}$
$D = 25 \text{ units/year}$
$r = 0.24 \ \$/\$/\text{year}$
$L = 3.5 \text{ weeks} = 3.5/52 \text{ year}$
$A = \$3.20$

and B_2 is estimated to be $1/5$ (i.e., $350/5$ or $\$70$ per unit backordered).

First, we can see from the sentence after Equation 8.3 that $Q = 1$. Now

$$\frac{r}{DB_2} = \frac{0.24}{25(1/5)} = 0.048/\text{unit}$$

and

$$\hat{x}_L = DL = 25\left(\frac{3.5}{52}\right) \approx 1.7 \text{units}$$

The use of Equation 8.4 shows that the best order point is 4. That is, we order one-for-one each time a demand drops the inventory position to 4. Another way of saying this is that the inventory position should always be maintained at the level of 5.

8.1.1.4 Sensitivity to Value of B_2

A precise value of B_2 is not easy to ascertain. Fortunately, there is a range of B_2 values for which each s value holds. To illustrate, consider the item in the above numerical illustration. Assume that D, r, and $\hat{x}_L$ are all known at the indicated values. Repeated use of Equation 8.4 identifies the following regions where each s value shown is best for the given value of $\hat{x}_L$.

s Value	Range of r/DB_2	Corresponding Range of B_2
3	0.064–0.163	0.059–0.150
4	0.021–0.064	0.150–0.456
5	0.006–0.021	0.456–1.636

These results are striking. Note the wide ranges of B_2 for which the same value of s is used. For example, as long as B_2 is anywhere between 0.150 and 0.456, the best s value is 4. Again, we see a situation where the result is insensitive to the value of one of the parameters.

8.1.1.5 More General Case of Q > 1 and a B_2 Cost Structure

This is still the case in which a fraction of the unit cost is charged per unit short and Q and s are determined sequentially. The solution discussed above is only appropriate for $Q = 1$. When the use of Equation 8.4 indicates that the best replenishment quantity is 2 or larger, we can no longer obtain as convenient a solution, at least for the case of B_2 (a fraction of the unit value charged per unit short). An exact analysis (see Problem 8.25) leads to indifference between s and $s + 1$ where the following condition holds:

$$\frac{\sum_{j=1}^{Q} p_{p_0}(s + j|\hat{x}_L)}{p_{p_0 \leq}(s|\hat{x}_L)} = \frac{Qr}{DB_2} \tag{8.5}$$

where

$p_{p_0}(x_0|\hat{x}_L)$ is the probability that a Poisson variable with mean $\hat{x}_L$ takes on the value x_0

$p_{p_0 \leq}(x_0|\hat{x}_L)$ is the probability that the same variable takes on a value less than or equal to x_0

This equation is more challenging to develop on a spreadsheet. The use of another computerized routine is more appropriate.

8.1.2 Case of $Q \geq 1$ and a B_1 Cost Structure

In some cases, it may be more appropriate to assume a fixed cost B_1 per (impending) stockout occasion as opposed to a cost per unit short. For such a case, we have developed indifference curves useful for selecting the appropriate reorder point. As shown in Section 8A.2 of the Appendix of this chapter, we are indifferent between s and $s + 1$ as reorder points when

$$\frac{p_{p_0}(s + 1|\hat{x}_L)}{p_{p_0 \leq}(s|\hat{x}_L)} = \frac{Qvr}{DB_1} \tag{8.6}$$

where all the variables have been defined above. The indifference curves of Figure 8.1 are developed by repeated calculation of Equation 8.6: for a given value of s and $\hat{x}_L$, Equation 8.6 is solved for Qvr/DB_1. For a given value of s, this is repeated for a number of values of $\hat{x}_L$ to produce a curve representing indifference between s and $s + 1$ as a function of $\hat{x}_L$ and Qvr/DB_1. These curves define regions on a graph of Qvr/DB_1 versus $\hat{x}_L$ where we prefer one s value to others.

For a given item, values of Q, v, r, D, and B_1 are used to evaluate the quantity Qvr/DB_1. The point in Figure 8.1 corresponding to this quantity and the appropriate $\hat{x}_L$ indicates the best value of the order point, s, to use.

8.1.3 Simultaneous Determination of s and Q for Slow-Moving Items

As noted above, the sequential approach of first finding Q and then finding s can lead to higher costs. For the case of Poisson demand and various shortage costs, there are efficient methods for finding s and Q simultaneously. In a groundbreaking study, Federgruen and Zheng (1992) provide a very fast algorithm for both B_2 and B_3 shortage costs. (See also Gallego 1996.) An alternative is to build the total cost equation on a spreadsheet and search (perhaps using a data table) for the optimal values. Here, we simply present two total cost equations and refer to the relevant literature for the algorithm. It will be clear from the cost equations that a spreadsheet will not be as effective as a routine written with a macro in a spreadsheet or in another computer programming language.

For the case of B_3 shortage costs, we have

$$\text{ETRC}(s, Q) = AD/Q + \frac{1}{Q} \sum_{y=s+1}^{s+Q} \left\{ rv \sum_{i=0}^{y} (y - i) p_{p_0}(i|\hat{x}_L) + B_3 v \sum_{i=y+1}^{\infty} (i - y) p_{p_0}(i|\hat{x}_L) \right\} \tag{8.7}$$

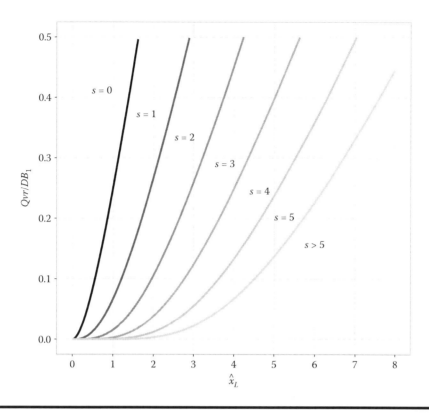

Figure 8.1 **Indifference curves for the reorder point under the B_1 cost measure.**

For the case of B_2 shortage costs, we have

$$\text{ETRC}(s, Q) = AD/Q + \frac{1}{Q} \sum_{y=s+1}^{s+Q} \left\{ rv \sum_{i=0}^{y}(y-i)p_{p_0}(i|\hat{x}_L) + B_2 Dv \sum_{i=y}^{\infty} p_{p_0}(i|\hat{x}_L) \right\} \quad (8.8)$$

See Federgruen and Zheng (1992) for details. We have found that in spite of the speed of the algorithm, and the fast computers available, many managers resist these more complex equations and algorithms. Thus, we suggest that a test be done on a small sample of items to see if the simultaneous solution is worth the added effort.

8.2 Controlling the Inventories of Intermittent Demand Items

In Section 3.8.1, we argued for a special type of forecasting for this kind of item that specifically separates out as two variables (1) the time (n) between demand transactions and (2) the sizes (z) of the transactions. In particular, we advocated estimating $\hat{n}$ (the average number of periods between transactions), $\hat{z}$ (the average size of a transaction), and MSE(z). Now, in selecting values of control variables, such as s, we must take account of the intermittent nature of the demands; specifically, there may now be a significant chance of no demand at all during a lead time. Problem 8.29 deals with the case of an (s, Q) system. When intermittent demand is known well ahead of time, it is more appropriate to use a decision rule based on a deterministic, time-varying demand pattern (see Chapter 5). Ekanayake and Norman (1977) discuss this type of issue in a case study context. Scala et al. (2013) describe using a base-stock model with intermittent demand, and Teunter et al. (2011) provide an adjustment for obsolescence in an intermittent demand setting. Bachman et al. (2016) present an approach for managing highly variable items that has been effectively applied for the U.S. Defense Logistics Agency. See also Croston (1972) and Shenstone and Hyndman (2005). Related works include Silver (1970), Wilcox (1970), Cawdery (1976), Ward (1978), Williams (1982, 1984), Alscher et al. (1986), Etienne (1987), Schultz (1987), Watson (1987), Yilmaz (1992), Hollier et al. (1995), Johnston and Boylan (1996), Syntetos and Boylan (2001), and Snyder et al. (2012).

8.3 Nature of C Items

The categorization of items for control purposes was first discussed in Chapter 2. As a reminder, we repeat some of the important points here.

The primary factor that indicates that an item should be placed in the C category is a low dollar usage (Dv value). The exact cut-off value between the B and C categories should be selected so that somewhere on the order of 30%–50% of all the items are classified in the C category. A low Dv value alone is not sufficient to dictate that an item should be placed in the C category, however. The real requirement is that the annual total of replenishment, carrying, and shortage costs be quite low under any reasonable control strategy. Normally, Dv is a useful surrogate for this total. However, a low-dollar-usage item may have potentially severe shortage penalties associated with it. Some typical examples come to mind:

1. A slow-moving product that rounds out a service line provided to an important customer. Shortages on this item can cause severe reductions in the usage of several faster-moving items.
2. A product that is the "pride and joy" of the president because he or she was instrumental in its development. In this case, there is a high *implicit* cost associated with a shortage.
3. An inexpensive product that has a high *explicit* shortage cost. The authors vividly remember the case of an electronics manufacturer whose limited manufacturing area was clogged with five expensive pieces of custom-made equipment, each the size of two large refrigerators, because the company had once again run out of the plastic control knobs worth 7 cents each.

For a true C item, the low total of replenishment, carrying, and shortage costs implies that, regardless of the type of control system used, we cannot achieve a sizable *absolute* savings in these costs. Therefore, the guiding principle should be to use simple procedures that keep the control costs per SKU quite low. That is, we wish to keep the labor and paperwork per item to a minimum. Equipment that permits electronic capture of data at the point of sale clearly reduces control costs. However, in our opinion, some organizations are overusing this capability. If the actual cost of using the electronic system (including maintenance of the software and updating the inventory records) is quite low, such a system could be a useful tool for C items.[*] Finally, we should be careful in defining C items because an item's classification can change over time. The classification should be reviewed periodically so that important items are not relegated to C item status.

---

[*] In spite of the widespread use of automated data collection systems and even the availability of spreadsheets for computing inventory policies, a remarkable number of firms still operate relatively manually.

8.4 Control of C Items Having Steady Demand

In this section, we deal with items having a relatively low importance and a demand rate that is not changing appreciably with time. Moreover, we assume that it makes sense to actually stock the C items under consideration. In Section 8.7, we address the more fundamental question of stocking an item versus making or buying it to each customer order.

8.4.1 Inventory Records

As a consequence of the primary objective mentioned above, in most cases, it may be most appropriate not to maintain any inventory record of a C item. It may be more desirable simply to rely on an administrative mechanism for reordering, such as placing an order when the last box in a bin is opened. If an inventory record is maintained, it should certainly not require that each transaction be recorded (except for the possible use of electronic point-of-sale data capture). Instead, a rather long review interval, such as 6 months, should be considered. Of course, for demand estimation and order control purposes, a record should be kept of when orders were placed and received.

Stated another way, we are saying that there are two choices for selecting a review interval for a C item:

1. Periodic review with a relatively long interval.
2. Continuous review but with a mechanism for triggering orders that requires neither a physical stock count nor the manual updating (in a ledger) of the stock status. Examples are (i) a two-bin system of control, and (ii) electronic capture of usage data.

8.4.2 Selecting the Reorder Quantity (or Reorder Interval)

Stated bluntly, the frills of the economic order quantity, Silver–Meal heuristic, and so on are completely unwarranted for Class C items. Instead, a simple offshoot of the basic EOQ developed in Chapter 4 should be used. Specifically, one of at most a few possible time supplies should be assigned to each Class C item. The time supply to use should be based as much on shelf life and obsolescence as on setup cost, unit cost, or carrying charge. It is usually reasonable to use a single value of the ratio A/r for all C items. (One convenient way of estimating this value is through the exchange curves of Chapter 4.) To illustrate, consider the following situation. The value of r is found to be 0.24\$/\$/year and a reasonable average value of A is \$3.20. Furthermore, for C items, management has decided that only three possible time supplies—6, 12, and 18 months—are worth considering. Using the approach suggested in Section 4.4 of Chapter 4, we develop the particularly simple decision rule shown in Table 8.1. D would not be estimated through a forecasting system but rather through knowledge of the starting and ending inventories over a convenient time period. For example, the procedure could be as follows in a two-bin (s, Q) system. Consider two consecutive orders A and B. Let A be *received* at time t_A and B be *placed* at time t_B. Let I_A be the inventory level just *before* A is received (a typically low level, hence easy to count). Then we have

$$\text{Demand rate} \approx \frac{I_A + Q - s}{t_B - t_A} \qquad (8.9)$$

Of course, if pilferage losses are significant, we would have an overly high estimate of the demand rate. Needless to say, rather than taking account of these losses, the preferable approach is to reduce the losses themselves.

Table 8.1 Suggested Reorder Time Supplies for a Sample Firm's C Items with Steady Demand

For Annual Dollar Movement (Dv) in This Range ($/Year)	Use This Number of Months of Supply
$53 \leq Dv$	6
$18 \leq Dv < 53$	12
$Dv < 18$	18

8.4.3 Selecting the Reorder Point (or Order-Up-to Level)

Any of the criteria discussed in Chapter 6 could be chosen for selecting the safety factor. (And with spreadsheets, the safety factors can be found easily.) However, we advocate the use of a specific criterion: select the safety factor to provide a specified expected time, TBS, between stockout occasions. We have found this criterion to be particularly appealing to management in most C item contexts. It appears to be a method that managers are comfortable with for expressing their risk aversion. Thinking in terms of an average time between stockouts is apparently more straightforward than dealing with probabilities or fractions. Quite often, many C items are involved in a single customer order. Therefore, to assure a reasonable chance of satisfying the complete customer order, a very high level of service must be used for each C item. Thus, large values of TBS (e.g., 5–100 years) are not unreasonable when we recognize the small added expense of carrying a high safety stock. The decision rule for the TBS criterion (see Section 6.7.11 of Chapter 6) is to select the safety factor k to satisfy[*]

$$p_{u \geq}(k) = \frac{Q}{D(\text{TBS})} \tag{8.10}$$

where

$p_{u \geq}(k) = \Pr\{\text{Unit normal variable (mean 0, standard deviation 1) takes on a value of } k \text{ or larger}\}$

is a widely tabulated function (see Table II.1 in Appendix II) and is a built-in function in many spreadsheets. (See Appendix III.) Then we set the reorder point[†]

$$s = \hat{x}_L + k\sigma_L \tag{8.11}$$

where[‡]

$$\hat{x}_L = DL \tag{8.12}$$

---

[*] In Equation 8.10, a negative k results if $Q/D(TBS) > 0.5$. This is very unlikely to occur if TBS is large, as suggested earlier. In any event, management should specify a minimum allowable value of k.

[†] If a periodic review, order-up-to-level (R, S) system is to be used, then throughout this section, the decision rules should be modified as follows (see Section 6.9):

  1. Replace L by $R + L$
  2. Replace Q/D by R
  3. Replace s by S

[‡] Note that $\hat{x}_L$ is not an output of the forecast system (as mentioned earlier, we are advocating not forecasting C items). D is estimated as discussed in the preceding subsection.

is the expected demand (in units) in the replenishment lead time of length L (years), and σ_L is the estimate of the standard deviation (in units) of forecast errors over L. (We return shortly to a discussion of how σ_L is estimated.)

The appropriate question at this point is: "Who needs all this aggravation, especially for C items?" Equation 8.10 holds for general values of Q/D and TBS. In the context of any company, only a few values of each of these two parameters would be specified. For these few combinations, the equations could be used once and for all to develop a simple table. We illustrate this in Table 8.2, when demand is normally distributed for a sample firm, using the earlier selected Q/D values of 6, 12, and 18 months, and also management-specified values of 10, 20, and 50 years for TBS.

Once k is selected, the reorder point follows from Equation 8.11. However, for this purpose, an estimate of σ_L is required. One point is certain: σ_L should not be developed from observing forecast errors of the specific C item under consideration. Instead, we suggest the use of the Poisson distribution for developing an estimate of σ_L because it typically applies to slow-moving items. This approach provides quick estimates of σ_L even though it may not be exact.

For the Poisson distribution (see Equation 8A.1), we have the simple relation, which again will only approximate the true standard deviation, and therefore safety stock,[*] if demand is not Poisson

$$\sigma_L = \sqrt{\hat{x}_L} = \sqrt{DL} \text{ units} \tag{8.13}$$

Finally, we note that it may be a good idea to keep a record of stockouts to see if the specified TBS is being met.

8.4.3.1 Numerical Illustration

Consider an item that is a slow-moving with $D = 48$ units/year, $L = 2$ months, an order quantity time supply of 12 months, and a specified TBS of 20 years.

$$\hat{x}_L = DL = 48 \text{ units/year} \times 1/6 \text{ year} = 8 \text{ units}$$

From Table 8.2, we read that the safety factor k should be set at 1.64. Suppose we make the Poisson assumption. Then Equation 8.13 gives

$$\sigma_L \approx \sqrt{8} = 2.83 \text{ units}$$

From Equation 8.11 we have that the reorder point is

$$s = 8 + 1.64(2.83) = 12.6, \text{ say 13 units}$$

8.4.4 Two-Bin System Revisited

In Section 6.4 of Chapter 6, we described a particularly simple physical form of an (s, Q) system— the two-bin system. To review, in a two-bin system, the on-hand inventory is segregated into two distinct sections (generically called bins). The capacity of one of the bins (called the reserve bin) is set equal to the reorder point. Demands are satisfied from the other bin until its stock is depleted.

---

[*] Alternatively, the estimating technique in Section 3.8.1 of Chapter 3 may be used.

Table 8.2   Table to Select Safety Factor, *k*, for a Sample Firm

	Order Quantity Expressed as a Time Supply in Months (12Q/D)		
TBS Value (Years)	*6*	*12*	*18*
10	1.64	1.23	1.04
20	1.96	1.64	1.44
50	2.33	2.05	1.88

Opening the reserve bin is a signal to place a replenishment order. When the order arrives, the reserve bin is refilled and sealed. The remainder of the order is placed in the other bin.

The authors have seen some rather clever forms of the bins concept. Examples include: (i) the use of "baggies" in the case of small electronic components, and (ii) color coding the end of an appropriate bar in steel stacked in a distribution warehouse.

To facilitate ordering, a bin reserve tag should be attached to the reserve bin so that when this bin is opened, the clerk will have the tag to remind him/her to report that the bin was opened. In fact, in many cases, the tag itself can be used as an order form. At the very least, dates and order quantities can be recorded on it (permitting an estimate of *D* when it is needed).

Obviously, for satisfactory performance of this system, it is imperative that a tag be promptly submitted to purchasing (or production scheduling) whenever a reserve bin is opened. Supervisors, assemblers, sales personnel, and others must be motivated to follow this procedure while they meet their own objectives in the rush of daily activities. One approach to maintaining this level of discipline is known as *plan for every part (PFEP)*, where the movement of each part is recorded in one database. See Harris (2004) for more detail on PFEP.

8.4.5 Simple Form of the (R, S) System

Smith (1984) reports on the implementation of a simple form of the (R, S) system, known as the semiautomatic stock control system, that is particularly effective for control of C (as well as B) items in a retail environment. Here is how it operates for any particular item, illustrated for a weekly review interval and a lead time less than R.

Periodically (likely less often than every review period, for example, each quarter-year), management specifies a desired value of S and at *only* that time the on-hand stock is counted and an order is placed to raise the inventory position to S. That is, the on-hand plus the order quantity must equal S. Then, at each regular review instant (here each week), the computer simply orders enough to replace sales since the last review. Therefore, the only record that must be maintained is the total sales in the current review interval. Problem 8.3 deals with how a desired increase or decrease in S would be implemented.

8.4.5.1 Numerical Illustration

Consider an item where the desired value of S is 35 units and a physical stock count reveals that the on-hand inventory is 26 units (with nothing on order). Then an order for 9 units would be placed. Suppose that the sales prior to the next review (ordering time) were 12 units. As a result, the next order would simply be for 12 units.

8.4.6 Grouping of Items

In some cases, it may be advantageous to group C items for control purposes. In particular, if a group of items (including some from the A and B categories) are provided by the same supplier, or produced in-house on the same equipment, then coordinated control (to be discussed in Chapter 10) may very well be in order to reduce replenishment costs. That is, when one item in the group needs reordering, several others should likely be included in the order. This is particularly appealing for a C item. By including it in an order initiated by another item, we avoid incurring a full setup cost (A) for its replenishment.

Coordination does not rule out the use of a two-bin system of control for the individual C item. Instead, reserve tags of opened reserve bins are held centrally between designated periodic times at which that group is ordered. Of course, the reserve stock must be appropriately scaled up to provide protection over a length of time equal to the sum of the periodic ordering interval plus the replenishment lead time.

8.5 Control of Items with Declining Demand Patterns

In this section, we deal with items nearing the end of their usage lifetimes (the end of the product life cycle that was discussed in Section 2.5 of Chapter 2). These ideas, however, are applicable beyond just C items. In particular, high-unit-value service parts (i.e., B and A items) eventually enter a period of declining requirements when the parent population is phased out.[*]

8.5.1 Establishing the Timing and Sizes of Replenishments under Deterministic Demand

Brosseau (1982) studied the special, but important, case where the (instantaneous) demand rate is dropping linearly with time. The demand pattern therefore is

$$x_t = a - bt \tag{8.14}$$

where t is time, and a and b are constants. For example, let $a = 10$, $b = 2$, and let t be measured in weeks. Then the demand rate at the end of week 1 is projected to be $10 - 2(1) = 8$, and will be zero at the end of week 5. The total remaining requirements are $a^2/2b$ (or 25 in the example). The optimal strategy turns out to be a function of only a single parameter, M, which is given by

$$M = \frac{Ab^2}{vra^3} \tag{8.15}$$

Note that M gets larger as a/b decreases—that is, as we get closer to the end of the demand pattern.

Brosseau ascertains as a function of M (1) how many replenishments to make and (2) the timing of the replenishments (as fractions of the horizon a/b). These results are shown in the form of a simple table. More importantly, however, he shows that a very simple procedure, based again on the EOQ, produces negligible cost penalties. Specifically, use the basic EOQ, based on the

---

[*] The case of increasing demand was addressed in Chapters 5 and 7.

current demand rate and expressed as a time supply, as long as $M < 0.075$, and use a single last replenishment equal to the total remaining requirements $(a^2/2b)$ when $M \geq 0.075$. The cutoff value of 0.075 turns out to be approximately the value of M where we are indifferent between 1 and 2 replenishments *in the optimal solution*. (See Problem 8.12.)

Smith (1977) and Brown (1977) found similar results for other types of declining demand patterns (including a continuous analog of the geometric pattern described in Section 3.8.4 of Chapter 3)—namely, that use of the EOQ, until it implies an order quantity almost as large as the total remaining requirements, produces excellent results.

We operationalize these results by suggesting the use of the EOQ as long as the total remaining requirements are at least 1.3 EOQ. If not, simply use one last replenishment to cover the remaining requirements. It can be shown that the 1.3 factor is equivalent to Brosseau's cutoff M value for the case of linearly decreasing demand.

For further research on this topic, see Chakravarty and Martin (1989), who addressed the case of exponentially decreasing demand and price reductions. Wee (1995) developed an exact solution to the problem of finding a replenishment policy for an item that faces declining demand, and that deteriorates over time. See also Kicks and Donaldson (1980), Hariga (1994), Ting and Chung (1994), Hariga (1995), Bose et al. (1995), and Chapter 9.

8.5.2 Sizing of the Final Replenishment under Probabilistic Demand

When demand is declining, but probabilistic in nature, the choice of the size of one last replenishment involves a trade-off between (1) the costs of insufficient inventory (reordering, expediting, shortages, etc.) if the total remaining demand exceeds the inventory position after the order is placed, and (2) the costs of overacquisition if the total remaining demand is less than the inventory position. Again, the first category of costs is typically very difficult to establish. So management may resort to the use of a desired service level, which, in turn, is likely to be largely dictated by industry norms (e.g., meeting 95% of all requests for spare parts during a 10-year period after the last sale of a new unit). We present a decision rule based on satisfying a certain fraction of the remaining demand. The derivation is outlined in Section 8A.3 of the Appendix of this chapter. Fortuin (1980) treated other service measures. See also Cattani and Souza (2003) and Pourakbar et al. (2012, 2014).

8.5.2.1 Decision Rule

Select the order-up-to-level[*] S where

$$S = \hat{y} + k\sigma_y \tag{8.16}$$

with $\hat{y}$ being the forecasted total remaining demand,[†] and σ_y the standard deviation of the total remaining demand. (One way to estimate σ_y would be to use the procedure of Brown (1981),

---

[*] $S =$ (current inventory position) + (order quantity).
[†] For the deterministic linear model of Equation 8.14,

$$\hat{y} = \frac{a^2}{2b} \tag{8.17}$$

For the geometrically decaying pattern of Equation 3.77 of Chapter 3, namely

$$x_t = f^t x_0 = f^t a$$

referenced in Section 3.8.2 of Chapter 3.) The safety factor k satisfies

$$G_u(k) = \frac{\hat{y}(1 - P_2)}{\sigma_y} \tag{8.19}$$

where $G_u(k)$ is again the special function of the unit normal distribution tabulated in Table II.1 of Appendix II, and given for spreadsheets in Appendix III.

8.5.2.2 Numerical Illustration

An automotive manufacturer is concerned with the size of the last replenishment of a spare part for a particular make of automobile that was discontinued 3 years ago. A fit to recent historical monthly usage indicates that a geometrically decaying pattern is appropriate with

$$\hat{y} = 83.1 \text{ units}$$
$$\sigma_y = 16.8 \text{ units}$$

Suppose that it is management's decision to select an S value for the last replenishments such that management expects 95% of the remaining demand for the spare part to be met by S. This does not mean that the other 5% would be lost; instead, cost-incurring actions, such as parts conversion, may be taken to satisfy the remaining demand, or the customer (or service department at the dealer) would seek a third-party/aftermarket part.

By using Equation 8.19 , we obtain

$$G_u(k) = \frac{83.1}{16.8}(1 - 0.95) = 0.2473$$

Then, from Table II.1 of Appendix II,

$$k \approx 0.35$$

Hence, from Equation 8.16

$$S = 83.1 + 0.35(16.8)$$
$$\approx 89 \text{ units}$$

8.6 Reducing Excess Inventories

When studying the inventories of many companies, it is common to find a significant percentage of the stocked items that have had absolutely no sales (or internal usage) in the last one or more

---

we have, from Equation 3.79 of Chapter 3,

$$\hat{y} = \frac{fa}{1 - f} \tag{8.18}$$

years. In some industries, the percentage of stocked items that have had no usage in the previous 52 weeks can be as high as 47 %. We refer to these SKUs as dead items.[*]

Of course, any remaining stock of a dead item is, by definition, excess stock. In addition, there are often many slow-moving items for which stock levels are excessive. The reasons for excess inventories can be grouped into two categories. First are errors associated with replenishments, that is, having purchased or produced too much of an item. These include production overruns, unjustified quantity purchases, errors in the transmission of an order request (e.g., 1,000 units instead of a desired 100 units), and inaccurate inventory records (e.g., location errors and "hoarding" of parts). The second class of reasons relates to overestimating the demand rate. Included here are inaccurate forecasts, deliberate changes in sales/marketing efforts, technological obsolescence (such as through an engineering change), and customer cancellations.

Whatever the causes of excess stocks, it is important to be able to (1) identify the associated items and (2) decide on what remedial action to take. Ideally, it would be desirable to be able to anticipate the decrease in usage of an item and take appropriate action to avoid being caught with a large surplus stock. It should be noted that this general problem area is increasing in importance as the rate of technological change increases, causing the life cycle of the typical product to shorten.

8.6.1 Review of the Distribution by Value

Again, we make use of a simple, but powerful, tool of analysis, the distribution by value (DBV) list. Recall from an earlier discussion (Table 2.1 in Chapter 2) if that the DBV is a listing of items in decreasing order of annual dollar usage (Dv value). At the bottom of the table listing will be all of the items that had no sales (or internal usage) over the period from which D was estimated (normally the most recent full year). Furthermore, moving up the table, we immediately encounter the items that have had very low usage.

To establish the DBV, values of D_i and v_i are required for each item i. If, at the same time, a reading on the inventory level I_i of item i can be obtained, then a very useful additional table can be developed as follows.

For each item, the expected time at which the current stock level will be depleted, also called the *coverage* or *runout time*, is computed by the formula

$$CO = \frac{12I}{D} \tag{8.20}$$

where
$\quad CO =$ coverage, in months
$\quad\quad I =$ on-hand inventory, in units
$\quad\quad D =$ expected usage rate, in units/year

The items are shown in Table 8.3 in decreasing order of coverage. Also shown is the inventory (in dollars) of each item and the cumulative percentage of the total inventory as one proceeds down the table. In Table 8.3 (involving a population of 200 items with a total inventory valued at $7,863), we can see that 4.2% of the total inventory is tied up in stock of zero-movers and

---

[*] Occasionally, items are kept in inventory to be used as parts for other products. These items never sell as is, but they are vital to the overall service performance. Clearly, these items should be stocked.

Table 8.3   Items Listed by Coverage

Rank Order	Cumulative Percent of Items	Item Identification	I (Units)	D (Units/Year)	v ($/Unit)	CO (Months)	I_v ($)	Cumulative Inventory — In Dollars	Cumulative Inventory — As Percent of Total
1	0.5	–	150	0	1.80	Infinite	270	270	3.4[a]
2	1.0	–	60	0	0.50	Infinite	30	300	3.8
3	1.5	–	10	0	1.10	Infinite	11	311	4.0
4	2.0	–	5	0	3.20	Infinite	16	327	4.2
5	2.5	–	53	2	1.00	318	53	380	4.8
6	3.0	–	40	3	0.90	160	36	416	5.3
7	3.5	–	64	12	2.00	64	128	544	6.9
8	4.0	–	180	37	0.35	58.4	63	607	7.7
⋮									
								7,861	100.0
199	99.5	–	2	1,000	1.00	0.024	2.00	7,863	100.0
200	100.0	–	0	463	0.20	0	0	7,863	100.0

[a] $3.4 = (270/7{,}863) \times 100.$

some 6.9% is included in items having a coverage of 5 years (60 months) or more. Such information would be of use to management in deciding how serious the overstock situation is and what remedial action to take.[*]

Individual item usage values (D_i's) are needed to construct the distribution by value table. Where these are not readily available, a practical alternative is to simply tour the storage facilities, applying a so-called dust test: any stock that looks overly dusty (or its equivalent) is a candidate for identification as a dead or very-slow-moving item.

8.6.2 Rule for the Disposal Decision

By disposing of some stock, we now achieve benefits of salvage revenue (possibly negative) and reduced inventory carrying costs. On the other hand, we will have to replenish at an earlier time than if no stock disposal was made. With regard to inventory carrying costs, it should be recognized that often the most costly aspect of excess inventory is that it is taking up significant space in a storage area of limited capacity. In addition, there is often a nuisance cost associated with stock that has physically deteriorated. Examples include the presence of rats and vermin, excessive rusting, and so forth. The authors recall the case of manufactured, large, metal piping that was stored in an open yard. One particular item, left standing for a number of years, had slowly sunk into the mud with each spring thaw until it stabilized in a semisubmerged state. High-cost excavation equipment was required to remove the dead stock in order to free up the space it was occupying. In this subsection, we present two approaches to this disposal decision, one that assumes deterministic demand, and one that assumes probabilistic demand.

8.6.2.1 Deterministic Demand

The derivation, presented in Section 8A.4 of the Appendix of this chapter, is based on a deterministic level demand pattern. Alternative decision rules could be developed (1) for a declining, but still deterministic, demand pattern, or (2) for probabilistic demand; the latter case closely parallels the selection of S for the terminal replenishment (see Section 8.5.2).

Let W be the amount of stock to dispose. Then W is given by

$$W = I - \text{EOQ} - \frac{D(v - g)}{vr} \tag{8.21}$$

where I is the current inventory level of the item,

$$\text{EOQ} = \sqrt{\frac{2AD}{vr}}$$

as earlier, and g is the unit salvage value (g can be negative if there is a net cost per unit associated with disposing of stock). Note that when $v = g$, that is, the salvage value is equal to the full unit variable cost, Equation 8.21 says that $I - W$ (which is the amount to leave in stock) is equal to the EOQ. That is, when we get our full money back, we dispose of stock so that on-hand inventory will just equal the EOQ. (See also Problem 8.10.)

---

[*] We worked with one firm in which nearly 1,000 of the 3,600 items in the firm's inventory sold so infrequently that management was aggressively working to eliminate them altogether.

8.6.2.2 Numerical Illustration

The manager of the hardware division of a retail outlet has discovered that the stock level of a particular item is 1,200 units, not 200 units as he had expected. Thus, he feels that some of the stock should be disposed of by means of a sale at 2/3 of unit cost (i.e., $g = (2/3)v$). He estimates the following values for other characteristics of the item:

$D = 400$ units/year
$A = \$20$
$v = \$3.00$/unit
$r = 0.25$ \$/\$/year

Equation 8.21 implies that the disposal amount should be

$$W = 1,200 - \sqrt{\frac{2(20)(400)}{(3.00)(0.25)}} - \frac{400(3.00 - 2.00)}{(3.00)(0.25)} = 521 \text{ units}$$

8.6.2.3 Probabilistic Demand

The second approach due to Rosenfield (1989) focuses on the number of units to be saved (not disposed of) and initially assumes that demands arrive according to a Poisson process and no further replenishments are to be made. If there is one unit demanded per arrival, a simple formula finds the optimal number of units to be saved. Find the largest integer less than

$$\log((V_c + a/i)/(V_u + a/i))/\log(D/D + i) \tag{8.22}$$

where
 V_c = the salvage value at the present time as a percentage of the current value
 V_u = the ultimate salvage value as a percentage of the current value
 i = discount rate
 a = carrying charge excluding capital holding costs, as a percentage of the current value
 D = average demand per year.

8.6.2.4 Numerical Illustration

Consider an item with the following characteristics:

$$V_c = 0.5, \ V_u = 0.7, \ i = 0.12, \ a = 0.09, \ D = 20$$

Find the optimal number of units to save in inventory, if there is always one unit demanded per customer arrival, and demands arrive according to a Poisson process. The use of Equation 8.22 gives

$$\log((V_c + a/i)/(V_u + a/i))/\log(D/D + i)$$
$$= \log((0.5 + 0.09/0.12)/(0.7 + 0.09/0.12))/\log(20/20 + 0.12) = 24.81$$

So the firm should save 24 units. If current inventory is 50 units, they should dispose of 26 units.

Rosenfield also develops formulas for several more complex cases, including items that are perishable. For example, if the batch size distribution is geometric with parameter q, a simple formula is given:

Find the largest integer less than

$$\frac{\log((V_c + a/i)/(V_u + a/i))}{\log(qD/(D+i) + (1-q))}$$

For a geometric distribution, the probability of a batch size of y is $q(1-q)^{y-1}$, and the mean batch size is $1/q$. The variance of the batch size is $(1-q)/q^2$.

8.6.3 Options for Disposing of Excess Stock

Possible courses of action for disposing of excess stock include the following:

1. *Use for other purposes.* For example, in the case of components or raw materials, consider their use, possibly requiring some rework, in different final products.
2. *Shipment of the material to another location.* Where an item is stocked at more than one location, it may have negligible demand at one location but still experience a significant demand elsewhere. The transfer of stock not only frees up space at the "first" location, but it also helps avoid the costly setup (or acquisition) costs associated with satisfying the likely infrequent demands at the other location(s).
3. *Use of stock for promotional purposes.* Excess stock may be converted to a revenue generator by providing free samples of it when customers purchase other products. For example, an appliance store with an overstock of an inexpensive table lamp might offer it free as a bonus to each customer buying a television set. The extreme case of this action is donating the stock to some charitable cause.
4. *Mark-downs or special sales.* The unit price is reduced sufficiently to generate appreciable demand for the product. An example would be the reduced price of wrapping paper or decorations immediately following the Christmas season.
5. *Returns to suppliers at a unit value likely lower than the initial acquisition cost.* A good example is that of a college bookstore returning extra copies of a book to the publishing company's distributor.
6. *Auctions.* These should not be overlooked as a possible means of salvaging some value from excess stock.
7. *Disposal for scrap value.* The best course of action may be to sell the material directly to a scrap dealer, a common transaction when there is a significant metallic component of the item.

Internal uses of the inventory (numbers 1 and 2) often provide better return than trying to move the inventory on the external market. Whatever course of action is selected, the excess stock will not disappear overnight. Therefore, it is important to lay out an approximate timetable for disposal, and then to ensure that there is appropriate follow-up. Also note that whenever inventory is disposed of and generates revenue that is less than the acquisition cost, there is an effect on the firm's income statement. Operating profit may decrease because of the loss taken on the inventory.

Thus, income taxes may decrease, and cash flow may improve. These effects should be examined closely. (See Smith 1987.)

8.7 Stocking versus Not Stocking an Item

Many items with extremely low sales may be minor variations of faster-moving products. In such a case, it can be attractive to eliminate the slow-moving special versions. Of course, decisions of this type, as well as those where there is no substitute product, go beyond the area of production planning and inventory management. Marketing considerations, including customer relations, are obviously relevant. For example, the slow mover may be carried as a service item for an important customer. Furthermore, the appropriate course of action may not be to discontinue selling the item, but rather to make or buy it to order as opposed to stocking it, an issue addressed in the following material.

Given that we have decided to satisfy customer demands for a given SKU, the question we now face is: *Should we make a special purchase from the supplier (or production run) to satisfy each customer-demand transaction or should we purchase (or produce) to stock?* It should be emphasized that this question is not restricted to C items. As we will see, a low demand rate tends to favor nonstocking, but there are a number of other factors that also influence the decision.

Earlier in the book, we have seen that, given that an item is stocked, its total relevant costs per year are quite insensitive to the precise settings of the control variables (e.g., the reorder point and the order quantity). Interestingly enough, more substantial savings can often be achieved through the appropriate answer to the question of whether or not the item should even be stocked.

8.7.1 Relevant Factors

For simplicity, our discussion is in terms of a purchased item, but the concepts are readily adaptable to in-house production. There are a number of factors that can influence the decision to stock or not stock the item. These include:

1. The system cost (file maintenance, forecasting, etc.) per unit time of stocking an item.
2. The unit variable cost of the item both when it is bought for stock and when it is purchased to meet each demand transaction. A more favorable price may be achieved by the regular larger buys associated with stocking. In addition, a premium per unit may be necessary if the nonstocking purchases are made from a competitor.
3. The cost of a temporary backorder associated with each demand when the item is not stocked.
4. The fixed setup cost associated with a replenishment in each context. An account should be taken of possible coordination with other items, because setup costs may be reduced.
5. The carrying charge (including the effects of obsolescence) which, together with the unit variable cost, determines the cost of carrying each unit of inventory per unit time.
6. The frequency and magnitude of demand transactions.
7. The replenishment lead time.

Often, it is best to avoid stocking an item at a given location if there is an internal supplier (a central warehouse, for example) that can hold inventory for several locations. Of course, it is important to examine the replenishment lead time and ascertain that customers are willing to wait the additional transportation time. If not, the cost of a temporary shortage should be included in the analysis. In what follows, we assume that the replenishment lead time is quite short.

8.7.2 Simple Decision Rule

A general model to handle all of the above factors would be of limited value to the typical practitioner because of its complexity. Therefore, in this subsection, we present a simple rule (based on work by Popp 1965), valid under specific assumptions. Even when these assumptions are violated, the rule should still be a useful guideline. In any event, in the next subsection, some extensions will be discussed.[*]

8.7.2.1 Assumptions

1. The unit variable cost is the same under stocking and nonstocking.
2. The fixed setup cost is the same under stocking and nonstocking.
3. In deriving the decision rule to decide on whether or not to stock the item, we allow the order quantity to be a noninteger. Of course, if the item was actually stocked and demands were in integer units, we would use an integer value for the order quantity.
4. The replenishment lead time is negligible; consequently, there is no backordering cost.

Based on comparing the cost of not stocking with the cost of stocking, we establish the following decision rule[†]:

8.7.2.2 The Decision Rule

Do not stock the item if either of the following two conditions holds (otherwise stock the item):

$$c_s > A/E(i) \tag{8.23}$$

or

$$E(t)vr > \frac{E(i)}{2A}\left[\frac{A}{E(i)} - c_s\right]^2 \tag{8.24}$$

where
c_s = system cost, in dollars per unit time, of having the item stocked
A = fixed setup cost, in dollars, associated with a replenishment
$E(i)$ = expected (or average) interval (or time) between demand transactions
$E(t)$ = expected (or average) size of a demand transaction in units
v = unit variable cost of the item, in \$/unit
r = carrying charge, in \$/\$/unit time

If the decision is to stock the item, then the economic order quantity should be used.

$$\text{EOQ} = \sqrt{\frac{2AD}{vr}} \tag{8.25}$$

---

[*] Valadares Tavares and Tadeu Almeida (1983) have presented an even simpler decision rule under somewhat more restrictive assumptions (Poisson demand and a choice solely between stocking 1 unit versus no stocking).
[†] The derivation of this rule is given in Section 8A.5 of the Appendix of this chapter.

where

$$D = E(t)/E(i) \qquad (8.26)$$

is the demand rate, in units per unit time.

The behavior of the rule as a function of the factors involved makes intuitive sense. As the setup cost A goes up, we tend to stock the item. Likewise, as the expected time between transactions decreases, we prefer to stock the item. As v or r increases, we tend not to stock because it becomes too expensive to carry the item in inventory. As c_s increases, we are less likely to prefer stocking the item. Finally, the larger the expected transaction size $E(t)$, the less likely we are to stock the item because the individual special orders satisfy a relatively large demand.

8.7.2.3 Numerical Illustration

A firm has been purchasing a specialty item PDF-088 from a competitor each time that a customer demand has been encountered. The manager of purchasing feels that it would be attractive to make less frequent, larger purchases. Relevant parameter values have been estimated as follows:

$$v = \$4.70/\text{roll}$$
$$E(t) = 1.4 \text{ rolls}$$
$$E(i) = 10 \text{ weeks or } 10/52 \text{ year}$$
$$A = \$2.50$$
$$r = 0.24 \ \$/\$/\text{year}$$
$$c_s = 0.20 \ \$/\text{year}$$

We have

$$E(t)vr = (1.4)(4.70)(0.24) = 1.58 \ \$/\text{year}$$
$$A/E(i) = 2.50/10/52 = 13.00 \ \$/\text{year}$$

The use of Equations 8.22 and 8.23 reveals that the item should indeed be purchased for stock, with the best order quantity being 6 rolls.

8.7.3 Some Extensions

It is rather straightforward (see Problem 8.6) to generalize the decision rule of the previous subsection to allow the unit variable cost and the fixed setup cost to each depend on whether or not the item is purchased for stock. In place of Equations 8.23 and 8.24, we can develop the following:

Do not stock the item if either of the following conditions holds (otherwise stock the item):

$$c_s > \frac{A_{ns}}{E(i)} + \frac{E(t)}{E(i)}(v_{ns} - v_s) \qquad (8.27)$$

or

$$E(t)v_s r > \frac{E(i)}{2A_s}\left[\frac{A_{ns}}{E(i)} + \frac{E(t)}{E(i)}(v_{ns} - v_s) - c_s\right]^2 \qquad (8.28)$$

where

v_s = unit variable cost, in \$/unit, if the item is stocked

v_{ns} = unit variable cost, in \$/unit, if the item is not stocked

A_s = fixed setup cost, in dollars, if the item is stocked

A_{ns} = fixed setup cost, in dollars, if the item is not stocked (the cost of backordering and extra paperwork would be included in A_{ns})

and all the other variables are as defined after Equations 8.22 and 8.23.

Croston (1974) developed a decision rule for the case of a periodic review order-up-to-level (R, S) system. His model assumes a negligible replenishment lead time, at most one demand transaction in each review interval, and a normal distribution of transaction sizes. Shorrock (1978) developed a tabular aid for the stock/no stock decision.

It should be emphasized that the two quantities, $E(t)$ and $E(i)$, needed in Equations 8.23 and 8.24 or Equations 8.27 and 8.28 would be estimated from rather sparse data because the items we are considering are slow-moving. The statistical fluctuations of such limited data could cause an item, whose underlying parameters had really not changed, to pass the stocking test 1 year and fail it in the next. Johnson (1962) proposed the useful idea of two threshold values, one to discontinue stocking an item, the other to institute stocking. The intermediate area helps prevent an item from flipping back and forth between stocking and nonstocking.

8.8 Summary

In this chapter, we have discussed procedures for controlling the inventories of the large number of items with low annual dollar usage as well as important items that are slow-moving. For low-value items, simplicity of control was stressed because of the low annual total of replenishment, carrying, and shortage costs associated with such items. In other words, if the total of replenishment, carrying, and shortage costs for an item is just a few dollars per year, we must use a control system for that item that costs only pennies per year.

A final point worth mentioning is that with C items, in particular, the best method of control may be ascertained from the warehouse or production supervisor who knows, for example, that a lot size of 10 units fits conveniently into a storage bin or into the usual production schedule. Using a decision rule that provides a different quantity, with a trivial annual difference in costs, is hardly worth the risk of alienating the supervisor.

Problems

8.1 Consider an item with the following characteristics:

$v = \$0.40/\text{unit}$

$D = 50$ units/year

$L = 3$ months

In establishing a value of σ, it is reasonable to assume that demand for the item is adequately approximated by a Poisson process. For an (s, Q) control system

a. Find the appropriate Q value using Table 8.1.

b. Find the s value assuming a desired TBS of 20 years and using Table 8.2.

8.2 Suppose for the item in Problem 8.1 we instead decided to use an (R, S) system of control. What should be the values of R and S?

8.3 Suppose that in a retail outlet an item was being controlled by the semiautomatic stock control system described in Section 8.4.5 and that an S value of 35 units had been in use.

   a. Owing to changing conditions, management now feels that it is appropriate to increase S to 40 units. Exactly what, if anything, has to be done to ensure that the system uses $S = 40$ from now on?

   b. Repeat part (a) for a decrease of S from 35 to 28 units.

8.4 The Fly-by-Night airlines has been phasing out one of the older types of aircraft that has been a mainstay of its illustrious fleet. Consequently, it has been observed that the associated usage of a glue used for maintenance of this type of aircraft has been "dropping off" linearly with time. The current level and trend of usage are estimated to be 110 liters/year and minus 45 liters/year/year. Other characteristics of the glue include $A = \$25$, $v = \$4.50$/liter, and $r = 0.3$ \$/\$/year.

   a. If it is now time to replenish (i.e., there is no inventory left), what order quantity should be used?

   b. If the usage rate continues to drop off at the same linear rate, what should be the size of the next replenishment?

8.5 Dr. Hunsen Bunnydew of Puppet Laboratories, Inc. had asked his assistant Test Tube to order in some beakers. Poor Test Tube made a mistake in placing the order, and the lab now has a 20-year supply! Hunsen has informed Test Tube that he must dispose of an appropriate number of beakers. Test Tube's friend, Zongo, has arranged for another medical laboratory to purchase any quantity that Puppet offers, but at \$0.50/beaker, which is 1/3 of what Test Tube paid for each of them. Using the following information, come to Test Tube's aid by suggesting an appropriate amount to sell off.

   $A = \$15$

   $r = 0.2$ \$/\$/year

   $D = 500$ beakers/year

8.6 Develop Equations 8.27 and 8.28.

8.7 For the specialty item discussed in Section 8.7.2, how large would c_s have to be to make it unattractive to stock the item?

8.8 A C item is often purchased from the same supplier as a B or A item. Suppose that such is the case and, based on an EOQ analysis for the B (or A) item involved, an order is placed for the B (or A) item every 2 months. Assume that demand for the C item is essentially constant with time at a rate of 18 units/year. Assume that the item has a unit value of \$3/unit. The *additional* fixed cost of including the C item in an order of the B (or A) item is \$1.20. The carrying charge is 0.24\$/\$/year. It is reasonable to restrict attention to ordering the C item in a quantity that will last for 2 months, 4 months, 6 months, and so on (i.e., an integer multiple of the time supply of the other item). Which of these time supplies is preferred?

8.9 Suppose you have been hired as an inventory control consultant by a client having thousands of inventoried items. The client feels that there is merit in each of the (R, S) and two-bin (s, Q) systems of control and is willing to have some C items under each of the two types of control. Using whatever symbols you feel are appropriate, develop as simple a decision rule as possible for deciding on which type of control to use for each specific item.

8.10 How are the disposal decision of Section 8.6.2 and the special opportunity to buy decision of Section 4.9.5 related? In particular, compare Equations 8.21 and 4.25.

8.11 For the item used in the numerical illustration of Rosenfield's approach in Section 8.6.2:
   a. What happens if the current salvage value is 0.8 instead of 0.5? Explain the result.
   b. What happens if the current salvage value is 0.7 instead of 0.5? Explain the result.
   c. Now suppose that the batch size is geometric with parameter $q = 0.4$ and the current salvage value is 0.5. How many units should be saved?

8.12 For the context studied by Brosseau, namely, a deterministic, linearly decreasing demand pattern, $x_t = a - bt$, use a dynamic programming formulation to show that the optimal strategy does indeed only depend upon the single parameter $M = Ab^2/vra^3$.

8.13 Suppose a company is using the following rule for controlling the stock of C items. Whenever the inventory position drops to 2 months or lower, order a 6 month supply. For each of the following items, what is the implied value of TBS? Assume Poisson demand.
   a. Item 372: $D = 60$ units/year, $L = 1.5$ months
   b. Item 373: $D = 60$ units/year, $L = 3$ months
   c. Item 374: $D = 30$ units/year, $L = 1.5$ months

8.14 Under deterministic level demand and an EOQ strategy, how does the coverage of an item vary throughout a cycle?

8.15 Consider an item with a linearly decreasing demand pattern. Suppose that the current level is 1,000 units/year and the trend is -400/year.
   a. What are the remaining all-time requirements of the item?
   b. Suppose the fixed cost per replenishment is $50, the unit variable cost is $0.80/unit, and the carrying charge is 0.20 $/$/year. Use Brosseau's method to establish the timing and sizes of all remaining replenishments. (Assume the current inventory level is 0.)

8.16 Consider a company with $A = \$4.00$, $r = 0.30\$/\$/$year, and cost of stocking $c_s = \$0.50$/year. There are two items (X and Y) having the same annual demand rate, $D = 10$ units/year and unit value $v = \$6.00$/unit. However, the transactions for item X tend to be larger than those for item Y, namely, the average transaction size for X is 5 units while that for Y is only 2 units.
   a. Show, using Section 8.7, that X should not be stocked and Y should be stocked.
   b. Why are the decisions different for the two items?

8.17 A company, attempting to remove slow movers from its line of stocked items, proposes the following types of rules for deciding on whether or not to stock an item. If an item is currently being stocked and no demands have been observed in the last T years, then the item should not be stocked any longer. On the other hand, if the item is not currently being stocked, we should begin stocking it if a total demand of at least m units is observed in a 1-year period. The quantities T and m are control variables. There is a fixed cost c_1 associated with activating the stocking of an item (preparation of a new form, etc.). Suppose that it is reasonable to assume that demand for an item is Poisson distributed with rate D (where D is not really known with certainty).
   a. For a given value of D, what is the probability that a currently stocked item will be removed from the files precisely T years after a particular demand transaction? Develop a graph of T values, as a function of D, that ensure a probability of 0.90.
   b. Consider an item that has been removed from routine stocking at a time t. For a given value of D, express the probability of activating the item in the very next year (following t) as a function of m.

   c. Without developing explicit decision rules, how do you believe that each of T and m will change as c_1 increases?

   d. In reality, D is changing with time. How might one attempt to recognize this in selecting values of T and m?

8.18 Under the assumption that lead time demand has a Poisson distribution, we have that

$$s = DL + k\sqrt{DL}$$

   For fixed values of Q/D and TBS, indicate how you would develop a graphical aid giving s as a function of DL. Illustrate for $Q/D = 0.5$ year and TBS $= 10$ years. What is the s value for $DL = 3.7$ units?

8.19 For an item that has demand rate $D = 36$ units per year, $L = 1$ month, and an order quantity of 12 units:

   a. Find the reorder point if TBS is specified as 10 years.

   b. How does the result change if TBS $= 2$ years?

   c. How does the result change if TBS $= 20$ years?

8.20 If the inventory of an item is currently at 2,000 units because of a clerk who bought too much, how much should be disposed of if the following data apply to the item?

   $g = 0.5v$
   $D = 600$ units/year
   $A = \$25$
   $v = \$4$/unit
   $r = 0.2$ \$/\$/year

8.21 For the item described below, use Rosenfield's approach in Section 8.6.2 to find the number of units to be saved.

$$V_c = 0.2, \; V_u = 0.5, \; i = 0.14, \; a = 0.07, \; D = 30$$

   Assume that there is always one unit demanded per customer arrival, and that demands arrive according to a Poisson process.

8.22 In an (s, Q) system, it can be shown that the inventory position has a uniform distribution with probability $1/Q$ at each of the integers $s + 1, s + 2, \ldots, s + Q - 1, s + Q$.

   a. Using this result, show, for Poisson demand and the B_2 shortage cost measure, that

$$\text{ETRC}(s) = vr \sum_{x_0=0}^{s} (s - x_0) p_{p_0}(x_0|\hat{x}_L) + B_2 vD \sum_{j=1}^{Q} \frac{1}{Q} \sum_{x_0=s+j}^{\infty} p_{p_0}(x_0|\hat{x}_L)$$

   b. Show that indifference between s and $s + 1$ exists when

$$\frac{\sum_{j=1}^{Q} p_{p_0}(s + j|\hat{x}_L)}{p_{p_0 \le}(s|\hat{x}_L)} = \frac{Qr}{DB_2}$$

(Note: $p_{p_0}(x_0|\hat{x}_L)$ is the p.m.f. of a Poisson variable with mean $\hat{x}_L$.)

8.23 Build a spreadsheet to quickly solve for the indifference point using Equation 8.4.

8.24 An item in a warehouse seems to sell quite slowly—only 20 units per year. However, it is an expensive item, valued at $750. Assume that the variance of the lead time demand is roughly equal to the mean, so a Poisson distribution adequately describes the lead time demand. The following data may be relevant:

$$A = \$7 \quad r = 0.22\$/\$/\text{year} \quad B_2 = 2$$
$$L = 1 \text{ week (assume a 50 week year)}$$

   a. What (s, Q) policy should be used?
   b. Now assume that $A = \$16$. What is the new (s, Q) policy?

8.25 Gregg Olson stocks an important but slow-moving item in his variety store. The relevant data are the following. Demand is 8 per year on average, and the variance of lead time demand is roughly equal to the mean of lead time demand.

   $v = \$400/\text{unit}$
   $A = \$5$
   $r = 0.25/\text{year}$
   $B_2 = 0.6$
   $L = 2 \text{ weeks} = 1/26 \text{ year}$
   a. What (s, Q) policy should Gregg use?
   b. If the methods of Chapter 6 are used, what (s, Q) policy would Gregg use?

8.26 Consider an (s, Q) system with Poisson demand, complete backordering, and $Q = 1$.
   a. Develop an expression for the expected number of units on backorder at any random point in time.
   b. Using a cost of $B_2 v$ per unit backordered per unit time, develop the equation for indifference between s and $s + 1$.

8.27 Because of the industrious nature of the faculty members involved, the Department of Management Sciences at a well-known university consumes a large amount of $81/2 \times 11$ lined paper. One of the secretaries, with the aid of a graduate student, has developed the following procedure: Once a week (at 5 p.m. on Friday), she reviews the stock level. If there are 20 or fewer pads, she orders 100 additional pads, which always arrive late on the next Tuesday. If there are more than 20 pads on the shelf, she does nothing until at least the next Friday. Assume that the faculty's needs follow a Poisson distribution with the following rates (in pads/day):

Day	Sun	Mon	Tue	Wed	Thu	Fri	Sat
Mean	0.5	2	2	2	2	2	0.5

If a professor needs a pad of paper and none are available, he rushes over to the university bookstore (or to an outlet in the local community) and buys the pad.
   a. Indicate how one might arrive at the cost per unit short.
   b. Make a crude estimate of each of (i) the safety stock, (ii) the probability of stockout per replenishment cycle, and (iii) the expected number of pads short per year.

c. Discuss how one could more accurately estimate the quantities in part 8.27b.

8.28 A maker of venetian blinds builds a wide variety of products. Demand tends to be erratic and is difficult to forecast. Each order of a particular blind requires a set of components with some of the components (e.g., cords) being common to many finished items. The company's policy is to satisfy any customer order within 1 week. The actual time dedicated to the satisfaction of an order, including some design work and retrieving the required set of components from inventory, varies between 1 and 3 days. Inventory records are not updated on a perpetual basis but rather the inventories of items are physically reviewed periodically (with R varying from 2 weeks to 5 weeks, depending upon the item). An order-up-to level is used for each component. Make some realistic suggestions for improving the inventory management situation in this company.

8.29 The Acme Company stocks an item where the demand (x) during the lead time has the following probability distribution:

$$\Pr(x = 0) = 0.2$$

is normal with mean 40 units and standard deviation 8 units. In other words, nonzero demands are normally distributed.

a. If an (s, Q) system of control is used with $Q = 100$ units and the desired fraction of demand satisfied from shelf is 0.98, what value of s should be used?

b. What is the associated value of the safety stock?

Appendix 8A: Poisson Distribution and Some Derivations

8A.1 Poisson Distribution

An extensive reference is the book by Haight (1967).

8A.1.1 Probability Mass Function

A Poisson variable is an example of a discrete random variable. Its probability mass function (p.m.f.) is given by

$$p_x(x_0) = \Pr\{x = x_0\} = \frac{a^{x_0} e^{-a}}{x_0!} \quad x_0 = 0, 1, 2, \ldots \tag{8A.1}$$

where a is the single parameter of the distribution. The graph rises to a peak, then drops off except for very low values of a where the peak is at $x_0 = 0$.

8A.1.2 Moments

One can show (see, e.g., Rohatgi 1976) that

1. $E(x) = \sum_{x_0=0}^{\infty} x_0 p_x(x_0) = a$
   that is, the parameter a is the mean of the distribution.
2. $\sigma_x^2 = \sum_{x_0=0}^{\infty} [x_0 - E(x)]^2 p_x(x_0) = a$

Hence, the standard deviation σ_x is given by

$$\sigma_x = \sqrt{a}$$

8A.1.3 Expression for Expected Units Short per Cycle with Poisson Demand

For a Poisson variable x with mean a, the probability mass function is defined above in Equation 8A.1. Define the cumulative distribution function as

$$P_x(s) = \sum_{x_0=0}^{s} p_x(x_0)$$

Expected units short during a replenishment cycle can be calculated as follows:

$$
\begin{aligned}
\text{ESPRC} &= \sum_{x_0=s+1}^{\infty} (x_0 - s) p_x(x_0) = \sum_{x_0=s+1}^{\infty} (x_0 - s) \frac{a^{x_0} e^{-a}}{x_0!} \\
&= \sum_{x_0=s+1}^{\infty} x_0 \frac{a^{x_0} e^{-a}}{x_0!} - \sum_{x_0=s+1}^{\infty} s \frac{a^{x_0} e^{-a}}{x_0!} \\
&= \sum_{x_0=s+1}^{\infty} \frac{(x_0)(a) a^{x_0-1} e^{-a}}{(x_0)(x_0-1)!} - s[1 - P_x(s)] \\
&= a \sum_{x_0=s+1}^{\infty} \frac{a^{x_0-1} e^{-a}}{(x_0-1)!} - s[1 - P_x(s)]
\end{aligned}
$$

By changing the starting point for the index for the summation and adjusting the inner term, the first term of the previous expression can be rewritten as

$$a \sum_{x_0=s+1}^{\infty} \frac{a^{x_0-1} e^{-a}}{(x_0-1)!} = a \sum_{x_0=s}^{\infty} \frac{a^{x_0} e^{-a}}{(x_0)!} = a[1 - P_x(s-1)]$$

leading to the following expression for ESPRC using only the reorder point s, the parameter a, and the cumulative distribution function.

$$\text{ESPRC} = a[1 - P_x(s-1)] - s[1 - P_x(s)] \tag{8A.2}$$

As shown in Appendix III, we can use standard spreadsheet functions to calculate the ESPRC.

8A.2 Indifference Curves for Poisson Demand

8A.2.1 Case of $Q = 1$ and a B_2 Penalty

Suppose an order-up-to-level S is used, or equivalently, an order for one unit is placed when the inventory position drops to the level $S - 1$. In this system, the inventory position is effectively always at the level S. All outstanding orders at a time t must arrive by time $t + L$ and no order placed after t can arrive by $t + L$. Therefore, the net (on-hand minus backorders) stock at time $t + L$ must be equal to the inventory position at time t minus any demand in t to $t + L$; that is,

$$(\text{Net stock at time } t + L) = S - (\text{demand in } L)$$

or

$$p_{NS}(n_0) = \Pr\{x = S - n_0\} \tag{8A.3}$$

where

$p_{NS}(n_0)$ = probability that the net stock at a random point in time takes on the value n_0

x = total demand in the replenishment lead time

Furthermore, with Poisson demand, the probability that a particular demand requires backordering is equal to the probability that the net stock is zero or less; that is,

$$\Pr\{\text{a demand is not satisfied}\} = p_{NS \leq}(0)$$

Using Equation 8A.3, we have

$$\Pr\{\text{a demand is not satisfied}\} = p_{x \geq}(S)$$

The expected on-hand inventory ($\bar{I}$) at the end of the lead time is the expected *positive* net stock at that time,

$$\bar{I} = \sum_{n_0=0}^{S} n_0 p_{NS}(n_0)$$

$$= \sum_{n_0=0}^{S} n_0 p_x(S - n_0)$$

where

$$p_x(x_0) = \text{probability that total time demand is } x_0$$

Substituting, $j = S - n_0$, we have

$$\bar{I} = \sum_{j=0}^{S} (S - j) p_x(j)$$

The expected shortage costs per unit time (C_s) are

$$
\begin{aligned}
C_s &= (\text{cost per shortage}) \times (\text{expected demand per unit time}) \\
&\quad \times \Pr\{\text{a demand is not satisfied}\} \\
&= B_2 v D p_{x \geq}(S)
\end{aligned}
$$

Expected total relevant costs per unit time, as a function of S, are

$$\text{ETRC}(S) = \bar{I}vr + C_s$$

$$= vr \sum_{j=0}^{S}(S-j)p_x(j) + B_2 vD p_{x\geq}(S)$$

Notice the similarity with Equation 8.8 which is for $Q \geq 1$. Notice also that the ordering cost term in Equation 8.8 is not included here, because it drops out in the indifference curve solution. These indifference curves are obtained by equating $\text{ETRC}(S)$ to $\text{ETRC}(S+1)$ for a given value of S. In general, this gives, after simplification,

$$\frac{p_x(S)}{p_{x\leq}(S-1)} = \frac{r}{DB_2} \tag{8A.4}$$

However, because $Q = 1$, we have that

$$s = S - 1$$

Therefore, Equation 8A.4 can be written as

$$\frac{p_x(s+1)}{p_{x\leq}(s)} = \frac{r}{DB_2}$$

Equation 8.4 follows by recognizing that the lead time demand (x) has a Poisson distribution with mean $\hat{x}_L$.

8A.2.2 Case of a B_1 Cost Penalty

We assume that the cost B_1 is incurred only if the demand in the lead time exceeds the reorder point s. The expected total relevant costs per unit time associated with using a reorder point s are

$$\text{ETRC}(s) = vr \sum_{x_0=0}^{s}(s-x_0)p_x(x_0) + \frac{D}{Q}B_1 p_{x>}(s)$$

Again, the fixed cost of ordering is independent of s. Equating $\text{ETRC}(s) = \text{ETRC}(s+1)$ for indifference between s and $s+1$ leads to

$$\frac{p_x(s+1)}{p_{x\leq}(s)} = \frac{Qvr}{DB_1}$$

Again, Equation 8.6 follows by recognizing that the lead time demand (x) has a Poisson distribution with mean $\hat{x}_L$.

8A.3 Decision Rule for the Size of the Last Replenishment

We are interested in the total remaining demand y for the item, where y is a random variable. We approximate its mean value $E(y)$ by the total demand $\hat{y}$ implied by the deterministic representation of the declining demand rate as a function of time (see Equations 8.17 and 8.18). In addition, we make the plausible assumption that y has a normal distribution (although the derivation could be modified to handle any other desired distribution) with standard deviation σ_y. Then, closely paralleling the P_2 service derivation in Section 6A.3 of the Appendix of Chapter 6, we have that the expected shortage over the lifetime

$$ES = \sigma_y G_u(k)$$

and we want

$$\frac{ES}{\hat{y}} = 1 - P_2$$

so that

$$G_u(k) = \frac{\hat{y}(1 - P_2)}{\sigma_y}$$

8A.4 Decision Rule for Disposal of Stock

Assuming a deterministic demand pattern (rate D), the situation is conceptually identical with that of a special opportunity to procure at a reduced unit cost (see Section 4.9.5 of Chapter 4). Let I be the current inventory level and W be the disposal amount. Then, as in Section 4.9.5, we look at costs out to time I/D—that is, when the current inventory would be depleted if no disposal took place. Under the option of no disposal, the costs would be

$$C_{ND} = \frac{I^2 vr}{2D}$$

If we dispose of W, then the remaining stock will run out at time $(I - W)/D$. Assuming an EOQ strategy after that, the relevant costs of the disposal decision out to time I/D are approximately

$$C_D = -gW + \left(\frac{I - W}{D}\right)\left(\frac{I - W}{2}\right)vr + \frac{W}{D}(\sqrt{2ADvr} + Dv)$$

As in Section 4.9.5, we select W to maximize $C_{ND} - C_D$. Using calculus, this leads to the decision rule of Equation 8.21.

8A.5 Decision Rule for Stocking versus Not Stocking

The demand rate per unit time D is given by

$$D = E(t)/E(i) \tag{8A.5}$$

Ignoring the minor effects of the non-unit-sized transactions, we have the same setting as in the derivation of the economic order quantity (Section 4.2 of Chapter 4). Therefore, from Equation 4.4 of Chapter 4 and Equation 8.25, the best order quantity, if the item is stocked, is

$$\text{EOQ} = \sqrt{\frac{2AE(t)}{vrE(i)}}$$

Moreover, from Equation 4.5 of Chapter 4 and including the c_s system cost, the best total relevant costs per unit time, if the item is stocked, are

$$\text{TRC}_s(\text{EOQ}) = \sqrt{2AE(t)vr/E(i)} + c_s \qquad (8\text{A}.6)$$

If the item is not stocked, there is a cost A associated with each transaction which occurs, on the average, every $E(i)$ units of time. Therefore, the total relevant costs per unit time, if the item is not stocked, are

$$\text{TRC}_{ns} = A/E(i) \qquad (8\text{A}.7)$$

Clearly, if c_s itself is larger than $A/E(i)$, then $\text{TRC}_{ns} < \text{TRC}_s$. This gives the condition of Equation 8.23. More generally, requiring

$$\text{TRC}_{ns} = \text{TRC}_s$$

and using Equations 8A.6 and 8A.7 leads to

$$A/E(i) < \sqrt{2AE(t)vr/E(i)} + c_s$$

or

$$2AE(t)vr/E(i) > \left[\frac{A}{E(i)} - c_s \right]^2$$

that is,

$$E(t)vr > \frac{E(i)}{2A} \left[\frac{A}{E(i)} - c_s \right]^2$$

References

Adelson, R. 1966. The dynamic behavior of linear forecasting and scheduling rules. *Operational Research Quarterly 17*(4), 447–462.

Aggarwal, P. and K. Moinzadeh. 1994. Order expedition in multi-echelon production/distribution systems. *IIE Transactions 26*(2), 86–96.

Alscher, J., M. Kühn, and C. Schneeweiss. 1986. On the validity of reorder point inventory models for regular and sporadic demand. *Engineering Costs and Production Economics 10*(1), 43–55.

Axsäter, S. 1996. Using the deterministic EOQ formula in stochastic inventory control. *Management Science 42*(6), 830–834.

Bachman, T. C., P. J. Williams, K. M. Cheman, J. Curtis, and R. Carroll. 2016. PNG: Effective inventory control for items with highly variable demand. *Interfaces 46*(1), 18–32.

Bose, S., A. Goswami, and K. S. Chaudhuri. 1995. An EOQ model for deteriorating items with linear time-dependent demand rate and shortages under inflation and time discounting. *Journal of the Operational Research Society 46*, 771–782.

Brosseau, L. J. A. 1982. An inventory replenishment policy for the case of a linear decreasing trend in demand. *INFOR 20*(3), 252–257.

Brown, R. G. 1977. *Materials Management Systems*. New York: Wiley-Interscience.

Brown, R. G. 1981. *Confidence in All-Time Supply Forecasts*. Materials Management Systems, Inc., Norwich, Vermont.

Cattani, K. D. and G. C. Souza. 2003. Good buy? Delaying end-of-life purchases. *European Journal of Operational Research 146*(1), 216–228.

Cawdery, M. N. 1976. The lead time and inventory control problem. *Operational Research Quarterly 27*(4ii), 971–973.

Chakravarty, A. K. and G. E. Martin. 1989. Discount pricing policies for inventories subject to declining demand. *Naval Research Logistics 36*, 89–102.

Croston, J. D. 1972. Forecasting and stock control for intermittent demands. *Operational Research Quarterly 23*(3), 289–303.

Croston, J. D. 1974. Stock levels for slow-moving items. *Operational Research Quarterly 25*(1), 123–130.

Ekanayake, P. and J. Norman. 1977. Aspects of an inventory control study. *European Journal of Operational Research 1*, 225–229.

Etienne, E. C. 1987. A simple and robust model for computing the service level impact of lot sizes in continuous and lumpy demand contexts. *Journal of Operations Management 7*(1), 1–9.

Federgruen, A. and Y.-S. Zheng. 1992. An efficient algorithm for computing an optimal (r, Q) policy in continuous review stochastic inventory systems. *Operations Research 40*(4), 808–813.

Feeney, G. J. and C. C. Sherbrooke. 1966. The $(S-1, S)$ inventory policy under compound Poisson demand. *Management Science 12*(5), 391–411.

Fortuin, L. 1980. The all-time requirement of spare parts for service after sales—Theoretical analysis and practical results. *International Journal of Operations & Production Management 1*(1), 59–70.

Gallego, G. 1996. Lecture Notes, Production Management.

Haight, F. 1967. *Handbook of the Poisson Distribution*. New York: Wiley.

Hariga, M. 1994. Two new heuristic procedures for the joint replenishment problem. *Journal of the Operational Research Society 45*(4), 463–471.

Hariga, M. 1995. Effects of inflation and time-value of money on an inventory model with time-dependent demand rate and shortages. *European Journal of Operational Research 81*, 512–520.

Harris, C. 2004. The plan for every part (PFEP). http://www.lean.org.

Hollier, R. H., K. L. Mak, and C. L. Lam. 1995. Continuous review (s, S) policies for inventory systems incorporating a cutoff transaction size. *International Journal of Production Research 88*(10), 2855–2865.

Johnson, J. 1962. On stock selection at spare parts stores sections. *Naval Research Logistics Quarterly 9*(1), 49–59.

Johnston, F. R. and J. E. Boylan. 1996. Forecasting for items with intermittent demand. *Journal of the Operational Research Society 47*, 113–121.

Karush, W. 1957. A queueing model for an inventory problem. *Operations Research 5*(5), 693–703.

Kicks, P. and W. A. Donaldson. 1980. Irregular demand: Assessing a rough and ready lot size formula. *Journal of the Operational Research Society 31*(8), 725–732.

Mitchell, C. R., R. A. Rappold, and W. B. Faulkner. 1983. An analysis of air force EOQ data with an application to reorder point calculation. *Management Science 29*(4), 440–446.

Mitchell, G. 1962. Problems of controlling slow-moving engineering spares. *Operational Research Quarterly 13*, 23–39.

Moinzadeh, K. 1989. Operating characteristics of the $(S-1, S)$ inventory system with partial backorders and constant resupply times. *Management Science 35*(4), 472–477.

Moinzadeh, K. and C. P. Schmidt. 1991. An $(S-1, S)$ inventory system with emergency orders. *Operations Research 39*(2), 308–321.

Natarajan, R. and S. K. Goyal. 1994. Safety stocks in JIT environments. *International Journal of Operations & Production Management 14*(10), 64–71.

Ouyang, L., N. Yeh, and K. Wu. 1996. Mixture inventory model with backorders and lost sales for variable lead time. *Journal of the Operational Research Society 47*, 829–832.

Popp, W. 1965. Simple and combined inventory policies, production to stock or to order. *Management Science 11*(9), 868–873.

Porteus, E. L. 1985. Investing in reducing setups in the EOQ model. *Management Science 31*(8), 998–1010.

Pourakbar, M., J. Frenk, and R. Dekker. 2012. End-of-life inventory decisions for consumer electronics service parts. *Production and Operations Management 21*(5), 889–906.

Pourakbar, M., E. Van der Laan, and R. Dekker. 2014. End-of-life inventory problem with phaseout returns. *Production and Operations Management 23*(9), 1561–1576.

Rabinowitz, G., A. Mehrez, C. Chu, and B. E. Patuwo. 1995. A partial backorder control for continuous review (r, Q) inventory system with Poisson demand and constant lead time. *Computers and Operations Research 22*(7), 689–700.

Rohatgi, V. K. 1976. *An Introduction to Probability Theory and Mathematical Statistics*. New York: John Wiley & Sons.

Rosenfield, D. 1989. Disposal of excess inventory. *Operations Research 37*(3), 404–409.

Scala, N. M., J. Rajgopal, and K. L. Needy. 2013. A base stock inventory management system for intermittent spare parts. *Military Operations Research 18*(3), 63–77.

Schultz, C. R. 1987. Forecasting and inventory control for sporadic demand under periodic review. *Journal of the Operational Research Society 38*(5), 453–458.

Schultz, C. R. 1990. On the optimality of the $(S - 1, S)$ policy. *Naval Research Logistics 37*, 715–723.

Shenstone, L. and R. J. Hyndman. 2005. Stochastic models underlying Croston's method for intermittent demand forecasting. *Journal of Forecasting 24*(6), 389–402.

Shorrock, B. 1978. Some key problems in controlling component stocks. *Journal of the Operational Research Society 29*(7), 683–689.

Silver, E. 1970. Some ideas related to the inventory control of items having erratic demand patterns. *Canadian Operations Research Society Journal 8*(2), 87–100.

Smith, B. T. 1984. *Focus Forecasting: Computer Techniques for Inventory Control*. Essex Junction, VT: Oliver Wight Limited Publications, Inc.

Smith, G. 1987. The inventory doctor: Obsolete inventory—Symptoms, treatment, & cure. *Production and Inventory Management Review December 7*, 32–35.

Smith, P. 1977. Optimal production policies with decreasing demand. *European Journal of Operational Research 1*, 355–367.

Snyder, R. D., J. K. Ord, and A. Beaumont. 2012. Forecasting the intermittent demand for slow-moving inventories: A modelling approach. *International Journal of Forecasting 28*(2), 485–496.

Syntetos, A. A. and J. E. Boylan. 2001. On the bias of intermittent demand estimates. *International Journal of Production Economics 71*(1), 457–466.

Taylor, C. 1961. The application of the negative binomial distribution to stock control problems. *Operational Research Quarterly 12*, 81–88.

Teunter, R. H., A. A. Syntetos, and M. Z. Babai. 2011. Intermittent demand: Linking forecasting to inventory obsolescence. *European Journal of Operational Research 214*(3), 606–615.

Ting, P. and K. Chung. 1994. Inventory replenishment policy for deteriorating items with a linear trend in demand considering shortages. *International Journal of Operations & Production Management 14*(8), 102–110.

Tyworth, J. E. and L. O'Neill. 1997. Robustness of the normal approximation of lead-time demand in a distribution setting. *Naval Research Logistics 44*(2), 165–186.

Valadares Tavares, L. and L. Tadeu Almeida. 1983. A binary decision model for the stock control of very slow moving items. *Journal of the Operational Research Society 34*(3), 249–252.

Ward, J. B. 1978. Determining reorder points when demand is lumpy. *Management Science 24*(6), 623–632.

Watson, R. 1987. The effects of demand-forecast fluctuations on customer service and inventory cost when demand is lumpy. *Journal of the Operational Research Society 38*(1), 75–82.

Wee, H.-M. 1995. A deterministic lot-size inventory model for deteriorating items with shortages and a declining market. *Computers and Operations Research 22*(3), 345–356.

Wilcox, J. 1970. How to forecast "lumpy" items. *Production and Inventory Management Journal 11*(1, 1st Qtr), 51–54.

Williams, B. W. 1982. Vehicle scheduling: Proximity priority searching. *Journal of the Operational Research Society 33*(10), 961–966.

Williams, T. M. 1984. Special products and uncertainty in production/inventory systems. *European Journal of Operational Research 15*, 46–54.

Yilmaz, C. 1992. Incremental order quantity for the case of very lumpy demand. *International Journal of Production Economics 26*, 367–371.

Zheng, Y. S. 1992. On properties of stochastic inventory systems. *Management Science 38*(1), 87–103.

Chapter 9

Style Goods and Perishable Items

The developments in Chapters 4 through 8 were all based on the assumption of a demand pattern for an item continuing well into the future (possibly on a time-varying basis)—that is, "a going concern" with the opportunity to store inventory from one selling period to the next. In this chapter, we remove the ability to indefinitely store inventory, due to perishability or obsolescence. Decision situations where this type of framework is relevant include

1. The newsvendor: How many copies of a particular issue of a newspaper to stock?
2. The garment manufacturer (or retailer): What quantity of a particular style good should be produced (or bought for sale) prior to the short selling season?
3. The Christmas tree vendor: How many trees should be purchased to put on sale?
4. The cafeteria manager: How many hot meals of a particular type should be prepared prior to the arrival of customers?
5. The supermarket manager: How much meat or fresh produce should be purchased for a particular day of the week?
6. The administrator of a regional blood bank: How many donations of blood should be sought and how should they be distributed among hospitals?
7. The supplies manager in a remote region (e.g., medical and welfare work in northern Canada): What quantity of supplies should be brought in by boat prior to the long winter freeze-up?
8. The farmer: What quantity of a particular crop should be planted in a specific season?
9. The toy manufacturer: A particular product shows significant potential sales as a "fad" item. How many units should be produced on the one major production run to be made?

These are notoriously difficult decisions. Every Christmas, for instance, news reports focus on the inability to find certain extremely popular toys. At the same time, retailers are left with piles of toys that turned out to be less popular than expected. Many firms have been able to slash lead times by prepositioning raw materials ready for assembly, using air freight to expedite delivery, communicating orders electronically, and so on. Therefore, they can estimate demand closer to the actual selling season. Unfortunately, they are still faced with considerable uncertainty about demand.

In the simplest situation, inventory cannot be carried forward at all from one period to the next. In this context, a single period is relevant for each replenishment decision, thus significantly simplifying the analysis (in part because we generally do not calculate a reorder point). This is in contrast with the earlier, more general, case where inventory is storable from one period to the next. In that context, the effects of a replenishment action can last for several periods and thus can interact with later replenishment actions. If we refer to the lifetime of an item as the length of time until it becomes obsolete or perishes, then the analysis becomes much more complicated as the lifetime increases.

In Section 9.1, we spell out in more detail the characteristics (and associated complexity) of the general style goods problem. The models presented in Sections 9.2 through 9.5 deal with progressively more realistic, but more complicated, versions of the problem. In Section 9.2, we are concerned with the case of a single product with but one replenishment (procurement or production) opportunity. This is followed in Section 9.3 by the important generalization to more than one product, thus permitting the aggregate viewpoint first emphasized in Chapter 2. Again, an exchange curve is a key by-product of the analysis. In Section 9.4, we examine the case when differentiation of the product can be delayed, a technique known as product postponement. Next, in Section 9.5, we provide some brief comments concerning the situation where there are several production periods with capacity restrictions prior to the selling season. Section 9.6 extends the discussion to the case where obsolescence is several periods in the future. Therefore, there are several opportunities to reorder before demand completely disappears. Section 9.7 deals with some other issues relevant to the control of style goods. The chapter is completed by a discussion in Section 9.8 concerning items subject to perishability.

9.1 Style Goods Problem

The main features of the style goods problem are:

1. There is a relatively short (typically no longer than 3 or 4 months) selling season with a well-defined beginning and end.
2. Buyers (at the stocking point) or producers have to commit themselves to a large extent, in terms of how much of each SKU to order or produce, prior to the start of the selling season.
3. There may be one or more opportunities for replenishment after the initial order is placed. Such replenishment actions may be taken prior to the selling season (if the forecast of demand has risen appreciably) or during the early part of the selling season itself (if actual demand to date indicates that the original forecast was considerably low). Occasionally, the season extends several periods in the future; we will discuss such a situation in Section 9.6.
4. Forecasts prior to the season include considerable uncertainty stemming from the long period of inactivity (no sales) between seasons. During this inactive period, the economic conditions or style considerations may have changed appreciably. Consequently, forecast revisions tend to be heavily influenced by firm orders or sales to date in the current season, far more than in the case of reasonably stable items (as discussed in Chapter 3).
5. When the total demand in the season exceeds the stock made available, there are associated underage costs. These may simply be lost profit if the sales are foregone. On the other hand, they may represent the added costs of expediting (e.g., a special order by air freight) or acquiring the material from a competitor at a high unit value.

6. When the total demand in the season turns out to be less than the stock made available, overage costs result. The unit salvage value of leftover inventory at the end of a season is likely to be quite low (well below the unit acquisition cost). It is quite expensive to carry stock over to the next season; moreover, there is no guarantee that such leftover stock will even sell at that time. Markdowns (i.e., reduced selling prices) or transfers (between locations) may be used to avoid leftover stock. (Incidentally, the style goods situation can be thought of as a special case of perishability where there is not physical deterioration but, instead, a marked deterioration in the economic value of the goods as of a particular point in time.)

7. Style goods products are often substitutable. Depending on how finely one defines a *style*, a stockout in one SKU does not necessarily mean a lost sale, because the demand may be satisfied by another product.

8. Sales of style goods are usually influenced by promotional activities, and space allocation in the store, among other things.

   In the simplest form of the style goods problem, wherein we ignore the last two points made above as well as markdowns and transfers, the remaining decision variables are the timing and magnitude of the replenishment quantities. To illustrate the complexity of even this simplified version, we need only refer to the early work of Wadsworth (1959). He considered the case of a style good sold in a fall season with raw material acquisition required in the spring, summer production of unlimited capacity, and a limited amount of fall production (after the total season demand is known exactly) being possible but at a higher unit cost than the summer production. The complex nature of Wadsworth's solution to this problem indicates that an exact analysis of a more realistic style goods problem is out of the question. For this reason, as elsewhere in this book, for all but the simplest problems we advocate the use of heuristic methods designed to generate reasonable solutions.

9.2 Simplest Case: Unconstrained, Single-Item, Newsvendor Problem

9.2.1 Determination of the Order Quantity by Marginal Analysis

Let us consider the situation faced by the owner of a newsstand. Each day he has to decide on the number of copies of a particular paper to stock. There is an underage cost, c_u, associated with each demand that cannot be met, and an overage cost, c_o, associated with each copy that is not sold. Suppose the underage cost was exactly equal to the overage cost. Then it seems intuitively reasonable that he would want to select the number (Q) of copies to purchase such that there was a 50% chance of the total demand being below Q and a 50% chance of it being above Q. But what about the case where $c_u = 2c_o$ or $c_u = 3c_o$? It turns out that his decision rule should be to select the Q^* value that satisfies

$$p_{u<}(Q^*) = \frac{c_u}{c_u + c_o} \tag{9.1}$$

where $p_{x<}(x_0)$ is the probability that the total demand x is less than the value x_0.

Notice that when $c_u = c_o$ Equation 9.1 gives

$$p_{u<}(Q^*) = 0.5$$

which is the intuitive result discussed above.

Table 9.1 Marginal Analysis of the News Vendor Problem

Specific Cost Element	Probability That the Specific Cost Element Is Incurred (or Avoided) by the Acquisition of the Qth Unit	Expected Value of the Specific Cost Element Associated with the Qth Unit (Product of Previous Two Columns)
Overage, c_o	Pr{Demand is less than Q} = $p_{x<}(Q)$	$c_o p_{x<}(Q)$
Underage, c_u	Pr{Demand is greater than or equal to Q} = $1 - p_{x<}(Q)$	$c_u[1 - p_{x<}(Q)]$

When $c_u = 2c_o$, we have from Equation 9.1

$$p_{u<}(Q^*) = \frac{2c_o}{2c_o + c_o} = \frac{2}{3}$$

that is, he should select the order quantity such that the probability of the total demand being smaller than the order quantity is two-thirds.

Now let us reflect on how Equation 9.1 is developed. The simplest derivation is a marginal analysis, as first used in Section 6A.1 of the Appendix of Chapter 6. The argument proceeds as follows.

Consider the Qth unit purchased. It will be sold if, and only if, the total demand equals or exceeds Q; otherwise, an overage cost will be incurred for this Qth unit. However, if the demand equals or exceeds Q, we have avoided an underage cost by having the Qth unit available. In other words, we have the analysis shown in Table 9.1.

Consider a particular Qth unit. If the expected overage cost associated with acquiring it exceeded the expected saving in underage costs, we would not want to acquire that unit. In fact, we might not want to acquire the $(Q - 1)$st, the $(Q - 2)$nd, and so forth. On the other hand, if the expected saving in underage costs exceeded the expected overage cost associated with acquiring the Qth unit, we would want to acquire it. However, we might also wish to acquire the $(Q + 1)$st, the $(Q + 2)$nd, etc. The last unit (Q^*) we would want to acquire is that one where the expected overage cost incurred exactly equaled the expected underage cost saved; that is, Q^* must satisfy

$$c_o p_{x<}(Q^*) = c_u[1 - p_{x<}(Q^*)]$$

or

$$p_{x<}(Q^*) = \frac{c_u}{c_u + c_o}$$

which is Equation 9.1.

The solution of Equation 9.1 is shown graphically in Figure 9.1 where we have plotted the cumulative distribution of demand, $p_{x<}(x_0)$. Locate the value of $c_u/(c_u+c_o)$ on the vertical axis, come across horizontally to the curve, then move down vertically to read the corresponding x_0, which is the Q^* value. There is always a solution because the fraction lies between zero and unity, which is the range covered by the cumulative distribution.

We have just seen that the marginal cost analysis approach leads to the decision rule of Equation 9.1 in a rather straightforward fashion. However, it does not give us an expression for evaluating the expected costs when any particular Q value is used. The alternative approach of minimizing total cost does, but the analysis is more complicated.

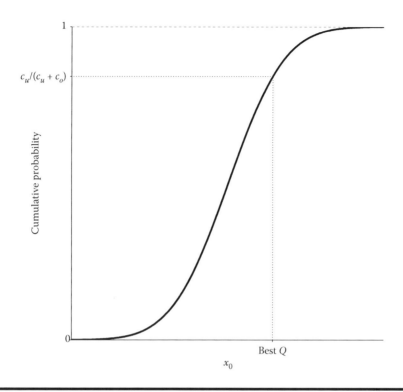

Figure 9.1 Graphical solution of the newsvendor problem.

9.2.2 An Equivalent Result Obtained through Profit Maximization

The above discussion has centered on cost minimization. Instead, let us now approach the problem from a profit maximization standpoint. Let

$v =$ acquisition cost, in dollars/unit

$p =$ revenue per sale (i.e., the selling price), in dollars/unit

B(or $B_2 v$) = penalty (beyond the lost profit) for not satisfying demand, in dollars/unit (B_2, as in Chapter 6, is the fraction of unit value that is charged per unit of demand unsatisfied)

$g =$ salvage value, in dollars/unit

As earlier, we let

$$Q = \text{ quantity to be stocked, in units}$$

and $p_{x<}(x_0)$ represent the cumulative distribution of total demand, the probability that total demand x takes on a value less than x_0.

Then, as shown in Section 9A.1 of the Appendix of this chapter, the decision rule is to select the Q^* that satisfies

$$p_{x<}(Q^*) = \frac{p - v + B}{p - g + B} \tag{9.2}$$

This result is identical with that of Equation 9.1 when we realize that the cost of underage is given by

$$c_u = p - v + B$$

and the cost of overage is

$$c_u = v - g$$

Because the cumulative distribution function is nondecreasing as its argument increases, we can make the following intuitively appealing deductions from Equation 9.2.

1. If the sales price, cost per unit short, or salvage value increases, the best Q increases.
2. If the acquisition cost per unit increases, the best Q decreases.

As shown in Section 9A.1 of the Appendix of this chapter, the expected profit is

$$E[P(Q)] = (p - g)\hat{x} - (v - g)Q - (p - g + B)ES \qquad (9.3)$$

where $\hat{x}$ is expected demand and ES is the expected number of units short, or expected shortages. If we let EO represent expected units left over, we can also represent the expected profit as

$$E[P(Q)] = (p - v)\hat{x} - (v - g)EO - (p - v + B)ES$$
$$= (p - v)\hat{x} - c_o EO - c_u ES \qquad (9.4)$$

This rearrangement shows the expected profit as a "perfect information" expected profit term minus an overstocking loss and an understocking loss. The first term of Equation 9.4 is the expected profit we would receive if we knew what demand would be and chose the quantity to exactly meet that demand.

9.2.3 Case of Normally Distributed Demand

The result of Equation 9.1 or 9.2 is applicable for any distribution of total demand. Again, it is worthwhile, as in Chapter 6, to look at the special case of normally distributed demand. First, such a distribution tends to be applicable in a significant number of situations. Second, as we will see, sensitivity analysis (effects of incorrectly specifying the order quantity Q) can be performed rather easily when demand is normal, in contrast with the case of a general probability distribution.

Suppose the demand is normally distributed with a mean of $\hat{x}$ and a standard deviation of σ_x and we define

$$k = \frac{Q - \hat{x}}{\sigma_x} \qquad (9.5)$$

This is analogous to what we did in Chapter 6. Then, as shown in Section 9A.1 of the Appendix of this chapter, the decision rule of Equation 9.2 transforms to

$$\text{Select } k \text{ such that } p_{u \geq}(k) = \frac{v - g}{p - g + B} \qquad (9.6)$$

where $p_{u \geq}(k)$ is the probability that a unit normal variable takes on a value of k or larger; this is a widely tabulated function (see, e.g., Table II.1 of Appendice II) first introduced in Chapter 6. Then, from Equation 9.5, set

$$Q = \hat{x} + k\sigma_x \qquad (9.7)$$

Furthermore, in this case, as shown in the Appendix of this chapter, we obtain a relatively simple expression for expected profit as a function of Q, namely,

$$E[P(Q)] = (p - g)\hat{x} - (v - g)Q - (p - g + B)\sigma_x G_u \left(\frac{Q - \hat{x}}{\sigma_x} \right) \qquad (9.8)$$

where $G_u(u_0)$ is a tabulated function of the unit normal variable (see Table II.1 of Appendice II) first introduced in Chapter 6. It can also be shown that (see Section A.1 of the Appendix of this chapter):

$$\frac{E[P(Q^*)]}{\sigma_x} = \frac{(p - v)\hat{x}}{\sigma_x} - (p - g + B)f_u(k^*) \qquad (9.9)$$

9.2.3.1 Numerical Illustration

Sales of ski equipment and apparel depend heavily on the snow conditions. A small discount retailer (called The Ski Bum) of ski equipment near some of the large ski areas of New Hampshire and Vermont faces the difficult task of ordering a particular line of ski gloves in the month of May well before the ski season begins. The selling price for these gloves is $50.30 per pair, while the cost to the retailer is only $35.10. The retailer can usually buy additional gloves from competitors at their retail price of $60. Gloves left over at the end of the short selling season are sold at a special discount price of $25 per pair. The owner and manager of the store assumes that demand is approximately normally distributed with a mean of 900 gloves and a standard deviation of 122 gloves.[*] Using our notation

$p = \$50.30/\text{pair}$
$v = \$35.10/\text{pair}$
$g = \$25.00/\text{pair}$
$B = (\$60.00 - 50.30)$ or $\$9.70/\text{pair}$

From Equation 9.6, we want

$$p_{u\geq}(k) = \frac{35.10 - 25}{50.30 - 25 + 9.70} = 0.288$$

---

[*] One way of arriving at an estimate of the standard deviation would be to specify the probability that total demand exceeded a certain value (other than the mean); for example, prob $(x \geq 1{,}100) = 0.05$ leads to the standard deviation of 122 pairs. For x normally distributed,

$$\Pr(x \geq 1{,}100) = p_{u\geq} \left(\frac{1{,}100 - \hat{x}}{\sigma_x} \right) = p_{u\geq} \left(\frac{200}{\sigma_x} \right)$$

But from the table in Appendix II

$$p_{u\geq}(1.645) = 0.05$$

Therefore,

$$\frac{200}{\sigma_x} = 1.645$$
$$\sigma_x = 122$$

Table 9.2   Sensitivity of Expected Profit to Q Value Selected in Example

Q	k	G(k)	Percent Deviation from Best Q (Namely, 968)	E[P(Q)]	Percent of Best Possible Profit
700	−1.64	1.6611	27.7	8,607	71.0
800	−0.82	0.9360	17.4	10,690	87.5
900	0.00	0.3989	7.0	11,980	98.0
925	0.20	0.3069	4.4	12,126	99.2
930	0.25	0.2863	3.9	12,150	99.4
968	0.56	0.1799	0.0	12,220	100.0
1,000	0.82	0.1160	3.3	12,170	99.6
1,030	1.07	0.0728	6.4	12,050	98.6
1,100	1.64	0.0211	13.6	11,570	94.7
1,200	2.46	0.0023	24.0	10,640	87.1

From the table in Appendix II, we have

$$k = 0.56$$

Therefore, Equation 9.7 gives

$$Q = 900 + 0.56(122) = 968.3, \text{ say 968 pairs}$$

The use of Equation 9.8 gives the expected profit for various values of Q; the results are shown in Table 9.2. It is interesting to note that by following the temptation of ordering the expected demand—namely, $Q = 900$—we would be led to a 2.0% reduction in expected profit.

Lau (1980) has shown that, for any demand distribution, to maximize the probability of achieving a target profit, T, one should choose Q^* such that $Q^* = T/(p - v)$. For the example above, to maximize the probability of achieving a profit of $10,690, set $Q^* = 703$, whereas to maximize the probability of achieving a profit of $12,220, set $Q^* = 804$. Note that these order quantities are significantly different from those that provide the same expected profit (from Table 9.2). The optimal probability that profit achieves (or exceeds) the optimal expected profit target is of course $1 - p_{x<}(Q^*)$. See also Sankarasubramanian and Kumaraswamy (1983).

9.2.4 Case of a Fixed Charge to Place the Order

Where there is a high enough fixed setup cost A to place any size order, the best strategy may be to not order anything at all. The decision would be based on comparing the cost of not ordering at all with the expected profit (ignoring the fixed cost as we did earlier) under use of the Q^* value:

$$\text{If } E[P(Q^*)] > A - B\hat{x}, \text{ order } Q^*$$
$$\text{If } E[P(Q^*)] < A - B\hat{x}, \text{ order nothing}$$

If there is inventory on hand at the start of the period, the policy of ordering Q^* should be seen as ordering up to Q^* instead. Observe then that there is a threshold value implied in the decision rule. If A is large enough, do not order. Otherwise, order up to Q^*. Said another way, if beginning inventory is relatively high, the penalty cost associated with not ordering will be relatively low, so do not order. However, if inventory is very low, order up to Q^*. This policy is a single-period version of the (s, S) policy introduced in Chapter 7. See Porteus (1990). As discussed above, we do not advocate explicitly determining s. Rather, for any given initial inventory level, we compare the expected profit of not ordering with the expected profit of using the best order-up-to-level, given that we order.

9.2.5 Case of Discrete Demand

Strictly speaking, when we are dealing with discrete (integer) units of demand and purchase quantity, it is unlikely that there is an integer Q value that exactly satisfies Equation 9.1 or 9.2. To illustrate, suppose we have an item where

$$\frac{c_u}{c_u + c_o} = 0.68$$

and the probability distribution of demand is as shown in the second row of the following table (the third row is derived from the second row):

Number of Units, x_0	1	2	3	4	5	Total
$p_x(x_0) = \Pr(x = x_0)$	0.2	0.3	0.2	0.2	0.1	1
$p_{x<}(x_0) = \Pr(x < x_0)$	0	0.2	0.5	0.7	0.9	–

We can see from the third row that there is no Q that would give $p_{x<}(Q) = 0.68$; the nearest values are $p_{x<}(3) = 0.5$ and $p_{x<}(4) = 0.7$. As shown in Section 9A.1 of the Appendix of this chapter, when discrete units are used, the best value of Q is the smallest Q value that satisfies

$$p_{x\leq}(Q) \geq \frac{c_u}{c_u + c_o} = \frac{p - v + B}{p - g + B} \tag{9.10}$$

9.2.5.1 Numerical Illustration

Consider the case of a Mennonite farmer who raises cattle and sells the beef at the weekly farmer's market in Kitchener-Waterloo, Ontario. He is particularly concerned about a specific expensive cut of meat. He sells it in a uniform size of 5 kg at a price of \$30. He wants to know how many 5-kg units to take to market. Demand is not known with certainty; all that he has available are frequency counts of the total 5-kg units demanded in each of the last 20 market sessions. These are chronologically as follows: 1, 3, 3, 2, 2, 5, 1, 2, 4, 4, 2, 3, 4, 1, 5, 2, 2, 3, 1, and 4. Frequency counts of how many 1's, how many 2's, etc. occurred in the 20 weeks are shown in the second row of Table 9.3. The third row is a set of estimates of the probabilities of the various sized demands based on the assumption that the historical occurrences are representative of the probabilities. For

Table 9.3   Demand Distribution for Farmer's Meat Example

1. Demand in a session x_0	1	2	3	4	5	Total
2. Frequency count $N(x_0)$	4	6	4	4	2	20
3. $p_x(x_0) \approx \frac{N(x_0)}{20}$	0.2	0.3	0.2	0.2	0.1	1.0
4. $p_{x \leq}(x_0)$	0.2	0.5	0.7	0.9	1.0	–

example, in 6 out of 20 weeks, the total demand has been 2 units; therefore, the probability of a total demand of 2 units is approximately 6/20 or 0.3.

The farmer stores the processed beef in a freezer on his farm. He estimates the value of a 5-kg unit to be $19 after he removes it from the freezer and gets it to the market. Once the beef is brought to market, he chooses not to take it home and refreeze it. Therefore, any leftover beef at the market he sells to a local butcher at a discount price of $15 per 5-kg unit. Finally, other than the foregone profit, he feels that there is no additional cost (e.g., loss of goodwill) associated with not satisfying a demand.

According to our notation

$p = \$30/\text{unit}$
$v = \$19/\text{unit}$
$B = 0$
$g = \$15/\text{unit}$

Therefore,

$$\frac{p - v + B}{p - g + B} = \frac{30 - 19 + 0}{30 - 15 + 0} = 0.733$$

The use of the fourth row of Table 9.3, together with the decision rule of Equation 9.10, shows that he should bring four 5-kg units to market. It is interesting to note that the average or expected demand is

$$\hat{x} = \sum_{x_0} x_0 p_x(x_0) = (1)(0.2) + 2(0.3) + 3(0.2) + 4(0.2) + 5(0.1)$$
$$= 2.7 \text{ units}$$

Therefore, because of the economics, the best strategy is to take $(4 - 2.7)$ or 1.3 units above the average demand.

For other non-normal demand distributions, see Scarf (1958), Gallego and Moon (1993), and Moon and Choi (1995). In particular, if the demand distribution is not known, but managers have some sense of the mean and variance of demand, Scarf (1958) and Gallego and Moon (1993) show that using the following formula maximizes the expected profit against the worst possible distribution of demand:

$$Q^* = \hat{x} + \frac{\sigma_x}{2} \left(\sqrt{\frac{p/v - 1}{1 - g/v}} - \sqrt{\frac{1 - g/v}{p/v - 1}} \right) \qquad (9.11)$$

For the numerical example given for the normal distribution, this formula yields an order quantity of 925, versus 968 when the distribution is known to be normal. We can see from Table 9.2 that the cost penalty of using 925 is quite small.

9.3 Single-Period, Constrained, Multi-Item Situation

Now we consider the situation of more than one type of SKU to be stocked for a single period's demand. It should be emphasized that the model to be developed here is based on the assumption of independent demands for the various products. However, there is a space or budget constraint on the group of items. Examples include the following:

1. Several different newspapers sharing a limited space or budget in a corner newsstand.
2. A buyer for a style goods department of a retail outlet who has a budget limitation for a group of items.
3. The provisioning of supplies or spare parts on a spacecraft, submarine, or the like, prior to a long mission without the possibility of resupply.
4. The repair kit taken by a maintenance crew on a routine visit to an operating facility.

We illustrate the solution procedure for the case of a restriction on the total dollar value of the units stocked. The procedure is quite similar if the constraint is on space, capacity, or some other resource. We discuss how one would adapt the approach for some other constrained resource. First, some notation is required. Let

$$n = \text{number of different items (SKUs) involved (numbered } 1, 2, 3, \ldots, n)$$
$$_i p_{x<}(x_0) = \text{Pr\{total demand for item } i \text{ is less than } x_0\}$$
$$v_i = \text{acquisition cost of item } i, \text{ in \$/unit}$$
$$p_i = \text{selling price of item } i, \text{ in \$/unit}$$
$$B_i = \text{penalty (beyond the lost profit) for not satisfying demand of item } i, \text{ in \$/unit}$$
$$g_i = \text{salvage value of item } i, \text{ in \$/unit}$$
$$W = \text{budget available for allocation among the stocks of the } n \text{ items, in dollars}$$

We are assuming the budget is constrained, and that a manager wishes to use up the total budget to maximize the total expected profit of the n items. If this is not necessarily the case, each item i's unconstrained optimal Q_i^* should be ascertained independently using Equation 9.2. If the total $\sum Q_i^* v_i < W$, then the Q_i^*s should be used. If not, the constrained approach must be used to find the Q_is.

As shown in Section 9A.2 of the Appendix of this chapter, a convenient way to solve this problem is through the use of a Lagrange multiplier. This approach is useful for multi-item problems that are connected through some common resource constraint. In this case, it is relatively easy to solve each single-item problem using Equation 9.2, but of course this leads to a solution that exceeds the budget constraint. The Lagrange multiplier approach works by charging a penalty cost M per unit of resource consumed beyond the budget. In this case, the unit of resource is a dollar of the budget.

The resulting decision procedure is[*]

Step 1 Select an initial positive value of the multiplier, M.
Step 2 Determine each Q_i ($i = 1, 2, 3, \ldots, n$) to satisfy

$$i p_{x<}(Q_i) = \frac{p_i - (M+1)v_i + B_i}{p_i - g_i + B_i} \text{ and } Q_i \geq 0 \qquad (9.12)$$

$$\sum_{i=1}^{n} Q_i v_i$$

with W.
If $\sum Q_i v_i \approx W$, stop. The optimal solution has been obtained.
If $\sum Q_i v_i < W$, return to Step 2 with a smaller value of M.
If $\sum Q_i v_i > W$, return to Step 2 with a larger value of M.

The multiplier has an interesting economic interpretation; it is the value (in terms of increased total expected profit) of adding one more dollar to the available budget, W. We revisit this interpretation in the context of the numerical illustration below.

It is important to recognize that as different values of M are used, we are actually tracing out an exchange curve of total expected profit versus total allowed budget (investment in stock). This is illustrated in Figure 9.2, which is an exchange curve for the numerical illustration to be presented momentarily.

Other research on the multiproduct problem includes Nahmias and Schmidt (1984), Lau and Lau (1988, 1994, 1996b), Li et al. (1990, 1991), and Jönsson and Silver (1996). Erlebacher (2000) develops simple expressions for certain special cases such as when all items follow a uniform distribution or all items follow the same demand distribution. Lau and Lau (1996b) show that if demand can never be zero, or has a very long left tail, it may be impossible to solve Equation 9.11 for certain values of M. Therefore, common search methods, such as the one given above, may not work. They provide a more complex method that works for any distribution.

The above formulation assumes that the individual-item stocking decisions have independent effects on the objective function to be maximized or minimized. Smith et al. (1980) have addressed the following more complicated situation, often called the "repair kit problem." Consider the kit of spare parts stocked by a repair crew visiting an off-site location. A particular repair job may require more than one spare part. An appropriate measure of effectiveness is now the fraction of jobs completed without stockout, as opposed to the fraction of needed individual spare parts that are provided. In addition, Mamer and Smith (1982) extend the analysis to the situation where part demands are no longer independent. See also Graves (1982), Hausman (1982), Scudder and March (1984), Mamer and Smith (1985), Mamer and Shogan (1987), and Moon et al. (2000).

---

[*] If the constraint is, instead, of a production nature such that no more than W units can be produced, then the decision procedure is modified in two ways:

1. Equation 9.12 is changed to

$$i p_{x<}(Q_i) = \frac{p_i - v_i - M + B_i}{p_i - g_i + B_i}$$

2. In Step 3, we compare $\sum_{i=1}^{n} Q_i$ with W.

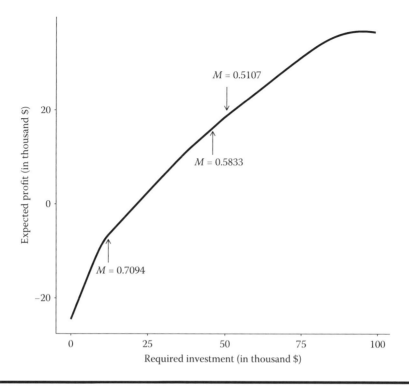

Figure 9.2   Exchange curve for the multi-item, single-period situation.

9.3.1 Numerical Illustration

Suppose The Ski Bum was faced with decisions on four items rather than just the one illustrated in Section 9.2.3. The manager is willing to accept that in each case total demand is normally distributed. For the case of normally distributed demand, Equation 9.12 becomes

$$p_{u \geq}(k_i) = \frac{(M + 1)v_i - g_i}{p_i - g_i + B_i} \tag{9.13}$$

where, as earlier, $p_{u \geq}(k_i)$ is the probability that a unit normal variable takes on a value of k_i or larger (tabulated in Appendix II) and

$$Q_i = \max(0, \hat{x}_i + k_i \sigma_i) \tag{9.14}$$

where

$\hat{x}_i$ = forecast demand for item i

σ_i = standard deviation of the demand for item i (here we have simplified the notation somewhat to avoid a subsubscript)

The relevant parameter values are estimated to be as shown in the top six rows of Table 9.4. The manager has a budget of $70,000 to allocate among these four items. She would like to know how to do this. She is also quite interested in the profit impact if the budget is changed somewhat.

Table 9.4 Numerical Illustration of Multi-Item, Single-Period Problem

	Item 1	Item 2	Item 3	Item 4	Total
p_i $/unit	50.30	40.00	32.00	6.10	–
v_i $/unit	35.10	15.00	28.00	4.80	–
g_i $/unit	25.00	12.50	15.10	2.00	–
B_i $/unit	9.70	0	10.30	1.50	–
$\hat{x}_i$ units	900	800	1,200	2,300	–
σ_i units	122	200	170	200	–
Using $M = 0$ (i.e., no budget restriction):					
Q_i from Equations 9.12 and 9.13	968	1,067	1,211	2,300	–
$E[P(Q_i)]$ from Equation 9.8	12,222	19,100	2,959	2,543	$36,824
$Q_i v_i$ $	33,978	16,006	33,907	11,040	$94,930
Using $M = 0.4$:					
Q_i	840	900	995	2,099	–
$E[P(Q_i)]$	11,325	18,660	1,613	2,334	$33,932
$Q_i v_i$	29,470	13,495	27,861	10,074	$80,900
Using $M = 0.6$:					
Q_i	750	841	0[a]	0[a]	–
$E[P(Q_i)]$	9,723	18,224	−12,360	−3,450	$12,137
$Q_i v_i$	26,331	12,630	0	0	$38,951
Using $M = 0.5093$[b]:					
Q_i	798	867	694	1,995	–
$E[P(Q_i)]$	10,652	18,438	−2,439	2,104	$28,755
$Q_i v_i$	27,996	13,006	19,430	9,575	$70,007

[a] The right-hand side of Equation 9.12 exceeds unity, implying that k_i (or equivalently Q_i) should be set at its lowest possible value. The lowest value for Q_i is zero from Equation 9.14.

[b] By trial and error, it was found that this value of M was required to produce a total used budget of approximately 70,000.

The calculations for several values of the multiplier M are shown in the lower part of Table 9.4 and the results are shown graphically in Figure 9.2. If there is no budget constraint, all items would be stocked, a budget of $94,930 would be needed, and the expected profit would be $36,824. We can see that the expected total profit is $28,755 when the proposed budget of approximately $70,000 is used. However, increasing the budget could lead to a substantial improvement in the total profit, a useful piece of information for the owner. The far right part of the curve shows that the total expected profit would start to decrease if the budget was too high and the manager was required to invest the whole amount in stocks of the four items (a negative value of M applies in this region). Three significant values of M are indicated in Figure 9.2. These values correspond to points at which the critical fractile (Equation 9.13) for one of the products equals zero, indicating it is unattractive to produce and stock that product. As shown in Table 9.4, a value of $M \approx 0.5093$ generated a set of order quantities that consumed the proposed budget of \$70,000. As stated in the previous section, this value of M is the value in terms of expected profit of adding one more dollar to the budget; another dollar of budget would generate approximately \$0.51 in expected profit. Another important set of managerial insights that can be gleaned from this problem formulation relates to comparisons across the items competing for the same budget. Notice that when the budget is quite constrained some products are not sufficiently attractive to order at all. For example, at $M = 0.6$ (leading to a budget of \$38,951), items 3 and 4 are not ordered at all. See Problem 9.26.

9.4 Postponed Product Differentiation

Because of the enormous variety and stiff delivery time requirements characterizing many style goods, such as fashion apparel, firms have been developing creative ways to reduce costs and improve delivery performance. One firm that has generated much interest is Benetton. By changing the order of production—their revised system switched the order from dye then knit, to knit then dye—Benetton was able to gain dramatic operational benefits. By switching the order of operations, Benetton was able to produce "plain" sweaters and then later dye them, only committing them to their final product form when more accurate demand information was available. This approach is known as delayed (or postponed) product differentiation or *postponement*.

Lee and Tang (1998) develop formal models to show when operations reversal is desirable. They show that when the major source of demand uncertainty is due to the option mix, but the total demand for all options is fairly stable, then the sequence of operations should have one or more of the following properties:

1. The first operation has the longer lead time if the value added of the two operations are roughly equal.
2. The value added of the first operation is less than the second operation, if the lead times are the same.
3. The first operation takes longer to perform, but the value added per unit of lead time for the first operation is smaller than or equal to the value added per unit of lead time for the second stage.

They provide several other properties, and they note that if total demand is highly variable, then the operation order suggested by these properties should be exactly reversed. See also Swaminathan and Tayur (1998), Fisher et al. (1999), and Randall and Ulrich (2001) for other examples of using the ideas of postponed product differentiation to manage multiple end items.

To adopt a postponement strategy, a firm typically manufactures a semifinished or base product that can later be finished into a final product. The Benetton setting describes such an example where the base product is the un-dyed sweater. This may require a redesign of the product and/or the process. Lee et al. (1993) discuss such a redesign effort. Ramdas (2003) provides a broader discussion about managing product variety, including a discussion of manufacturing process capabilities that are essential in adopting postponement strategies.

There are two key factors that drive the potential operational improvement obtained by postponing product differentiation. First, by delaying some of the manufacturing value-add, the per-unit overstocking cost is reduced. Second, by delaying the commitment of the product into its final form, a firm can improve performance, either reducing safety stock or improving the level of service. We explore each of these two factors next, providing a numerical illustration in each case of the performance improvement obtained.

9.4.1 Value of Delayed Financial Commitment

We use the subscript 0 to represent the base product (e.g., the un-dyed sweater in the Benetton setting discussed above). Let t represent the additional transformation cost to convert a base product into a finished product. This means the cost to acquire (or produce) a base product and then transform that base product is $v_0 + t$. As before, let g denote the salvage value for the finished product. Let g_0 denote the salvage value for the base product. Here, we assume that the base product can be transformed into the finished item at the time demand is realized. This means that a firm will stock out when supply of the base product is insufficient and will incur the transformation cost only for products that are sold. In Section 9.5, we discuss the substantially more complicated situation when the transformation decision must be made before final demand occurs but after the firm has updated their demand forecast.

With these assumptions, the cost of understocking is $c_u = p - v_0 - t + B$ while the cost of overstocking is $v_0 - g_0$. As in Section 9.2.2, let EO, ES represent expected leftover units and expected units short, respectively. Letting Q_0 represent the quantity of base product acquired, the expected profit is

$$E[P(Q_0)] = (p - v_0 - t)\hat{x} - (v_0 - g_0)EO - (p - v_0 - t + B)ES \qquad (9.15)$$

Following Equation 9.1, we get that the best quantity satisfies:

$$p_{u<}(Q_0^*) = \frac{c_u}{c_u + c_o} = \frac{p - v_0 - t + B}{p - g_0 - t + B} \qquad (9.16)$$

Whether or not there is a benefit to the delayed financial commitment in this single-item setting depends on the relationship between the costs in the original system and this new system. To see why there is potential for savings, let us first simplify to the case where the total unit cost is the same in the two systems, $v = v_0 + t$, and the two salvage values are the same $g_0 = g$. Note that these relationships may not hold in general, but this simplification allows us to understand the source of the potential savings. Under these assumptions, the only difference between the two systems is that the cost of overstocking is lower in the postponed system (v_0 instead of $v_0 + t$) since the cost t is not incurred unless the demand is realized.

9.4.1.1 Numerical Illustration

Suppose The Ski Bum has historically purchased and sold a particular ski/binding package where the bindings are mounted (attached to the skis) by the distributor. The manager assumes that demand is approximately normally distributed with a mean of 100 and a standard deviation of 20.

This particular ski package sells for $500, costs $300, and can be sold at the end of the season for $100. The manager estimates that failing to satisfy a demand will result in a future lost sale of a comparable ski package that would result in a similar margin contribution. So, the manager uses a penalty cost equal to the current product margin of $200. In our notation:

$p = \$500/\text{pair}$
$v = \$300/\text{pair}$
$g = \$100/\text{pair}$
$B = \$200/\text{pair}$

This is a standard newsvendor problem (no delayed product differentiation). Using the same approach as in Section 9.2.3, we obtain the results summarized in Table 9.5.

Now, suppose The Ski Bum management can purchase the skis and bindings separately for a total of $275. It is estimated that it will cost approximately $25 to mount the bindings in the store at the time of customer demand. Management also estimates that they can sell the skis and bindings at the end of the season and recoup the same $100 salvage value. In our notation:

$p = \$500/\text{pair}$
$v_0 = \$275/\text{pair}$
$t = \$25/\text{pair}$
$g_0 = \$100/\text{pair}$
$B = \$200/\text{pair}$

Following Equation 9.16,

$$p_{u<}(Q^*) = \frac{p - v_0 - t + B}{p - g_0 - t + B} = \frac{500 - 275 - 25 + 200}{500 - 100 - 25 + 200} = \frac{400}{575} \tag{9.17}$$

Since demand is assumed to be normally distributed, we can use Appendix II or a spreadsheet formula to get $k = 0.512$. The results are summarized in Table 9.5 for easy comparison to the nonpostponed case.

From Table 9.5, we see that adopting this delayed differentiation results in a slightly higher stocking quantity and slightly higher profit. In this example, the cost of understocking is the same in both systems but the cost of overstocking is lower in the postponed system since the $t = \$25$ transformation cost is not incurred for items purchased but not sold. Since the cost consequences of overstocking are less severe, the firm is willing to purchase more. This also results in an increase in profit of a little more than 2%. Certainly no manager would turn down a 2% increase in profit, but as we will see in the next section, the potential benefits of postponement really kick in when multiple items are served from the same base product.

In some settings, the total unit and transformation cost is higher when postponement is adopted. It can be verified for this example that if the transformation cost is $t = 28.53$, the profit in the two systems is the same. That is, The Ski Bum could tolerate slightly higher total product costs under postponement and still obtain the same expected profit.

9.4.2 Value of Flexibility

We now consider the case where there are multiple end items created from the same base product. First, we address the special case where the cost of understocking $p_i - v_0 - t_i + B_i$ (where t_i is the

Table 9.5   Numerical Illustration of Single-Item, Single-Period Problem with and without Postponed Product Differentiation

	Original System	Postponed System
p \$/unit	500	500
v \$/unit	300	–
v_0 \$/unit	–	275
t \$/unit	–	25
g \$/unit	100	100
B \$/unit	200	200
$\hat{x}_i$ units	100	100
σ_i units	20	20
k	0.431	0.512
Q or Q_0	109	110
$E[P(Q)]$ or $E[P(Q_0)]$	\$15,637	\$15,976

transformation cost for item i) is the same for all products. Making this simplifying assumption allows us to use the standard newsvendor formulation to calculate the optimal base product quantity. We subsequently relax this limiting assumption below and discuss how to address the more general multiple-item setting.

To see why the relative cost of understocking across items is critical to the analysis, consider the situation where a firm discovers they are in short supply after observing demand. They must make a choice of how to prioritize demand for different items. At this point in the decision timeline, the firm knows they are in short supply and thus will not incur overstocking costs. This means they will seek to minimize the understocking cost by first serving the demand for the item with the highest understocking cost, then the next one, and so on. If we assume that all understocking costs are the same, we can formulate the expected profit, and determine the ideal base product quantity by calculating only the *total* expected units short rather than the expected units short for each item.

Let $\hat{x}_i$, σ_i represent mean and standard deviation of demand for item $i = 1, \ldots, n$, and let $\hat{x}_0, \sigma_0$ represent the mean and standard deviation of total demand across all items. We obtain the mean for total demand by adding the individual means, $\hat{x}_0 = \sum_{i=1}^{n} \hat{x}$. Here we will assume that demands for these items are mutually independent, meaning we obtain the variance for total demand by adding the variance of the individual demands, $\sigma_0^2 = \sum_{i=1}^{n} \sigma_i^2$. It may not be the case in practice that the demands are independent.

The cost of overstocking is the same as in the previous section $c_o = v_0 - g_0$. Under our assumption that the cost of understocking is the same for any stockout, $c_u = p_i - v_0 - t_i + B_i$, we can formulate the expected profit as

$$E[P(Q_0)] = \sum_{i=1}^{n} (p_i - v_0 - t_i)\hat{x}_i - c_o EO - c_u ES \qquad (9.18)$$

where EO and ES are expected units left over and expected units short based on Q_0 and the total demand described by $\hat{x}_0, \sigma_0$. Following Equation 9.1, we get that the best quantity satisfies

$$p_{u<}(Q_0^*) = \frac{c_u}{c_u + c_o} = \frac{p - v_0 - t_i + B_i}{p - g_0 - t_i + B_i} \tag{9.19}$$

We now consider the case where costs of understocking are different across items. For this setting, we need to know the expected units short *for each item* rather than just the total expected units short. Letting ES_i represent expected units short for item i, the expected profit function is

$$E[P(Q_0)] = \sum_{i=1}^{n}(p_i - v_0 - t_i)\hat{x}_i - (v_0 - g_0)EO - \sum_{i=1}^{n}(p_i - v_0 - t_i + B_i)ES_i \tag{9.20}$$

Our experience is that for most settings where this newsvendor model is appropriate, demand is occurring over a moderately long selling season and retailers tend to serve demand first-come first-serve. We first address this setting and discuss below the case where the retailer may be able to prioritize the use of the base product for more critical end-item demand.

If the retailer follows a policy of meeting end-item demand as it occurs, we can simplify the expected profit given in Equation 9.20 by using an *average* cost of understocking. We recommend using the demand-weighted average cost of understocking. That is, if item 1 has twice the average demand as item 2, we would give twice the weight to the cost of understocking for item 1. Formally, let $\bar{c}_u$ be the weighted-average cost of understocking, obtained as follows:

$$\bar{c}_u = \frac{\sum_{i=1}^{n}\hat{x}_i(p_i - v_0 - t_i + B_i)}{\sum_{i=1}^{n}\hat{x}_i} \tag{9.21}$$

Now we can use the same, single-item newsvendor approach as before, with $\bar{c}_u$ as the cost of understocking and $(v_0 - g_0)$ as the cost of overstocking.

If a firm has the ability to prioritize use of the base product, it should choose to meet more demand for items with higher understocking costs at the expense of items with lower-understocking costs. To illustrate this effect, we consider the setting where the retailer sees all demands at once. The exact approach requires some effort in determining the expected units short for each item, ES_i. We comment on this approach further below, but first we suggest an approximate approach based on the observation that the item with the lowest understocking cost that will have the most missed sales since demand for that item will always be prioritized last. Let c_u^{min} represent the lowest understocking cost among the items. If we assume that *all* missed sales incurred are for this item, we can modify Equation 9.18 to be

$$E[P(Q_0)] = \sum_{i=1}^{n}(p_i - v_0 - t_i)\hat{x}_i - c_o EO - c_u^{min} ES \tag{9.22}$$

and the optimal solution can then be found using the same newsvendor approach we have already seen several times with overstocking and understocking costs c_o, c_u^{min}. This simplified approach risks overstating the expected profit slightly as we assumed that all missed sales incur the lowest-possible understocking cost.

It is possible to find values for expected units short *for each item*. To illustrate how, suppose we have only two items and that item 1 has the higher understocking cost (meaning demand for item 1 will be met first). As noted above, we can calculate the total *ES* based on Q_0 and $\hat{x}_0, \sigma_0$. Since demand for item 1 will always be met first if possible, we can calculate ES_1 by using the same Q_0 and $\hat{x}_1, \sigma_1$. If we know the total expected units short and the expected units short for item 1, expected units short for item 2 must be the difference between the total and item 1 missed sales, $ES_2 = ES - ES_1$.

9.4.2.1 Numerical Illustration

One of the popular snowboards sold by The Ski Bum comes with either a glossy or matte finish. The type of finish is determined at the time the topsheet is applied to the board. Historically, the retailer has purchased and sold matte and glossy snowboards separately but is now considering switching to a postponement system where the desired finish can be applied in the shop at the time of the customer order.

The snowboard with matte finish sells for $400 while the glossy-finish board sells for $450. The unit costs for either board are $200 if purchased from the manufacturer with the finish already applied. It costs $190 to buy an unfinished board and $20 to complete the finishing in the shop. Any type of board can be sold at the end of season for $50. The manager estimates that failing to satisfy a demand will result in future lost business of $300. In our notation (let subscripts g, m represent glossy and matte):

$$p_g, p_m = \$450, \$400$$
$$v_g, v_m = \$190$$
$$t_g, t_m = \$20$$
$$g_g = \$50$$
$$B_g, B_m = \$300$$

The manager estimates that demand for glossy-finish boards is normally distributed with mean 90 and standard deviation 25 while the demand for matte boards is normally distributed with mean 80 and standard deviation 20. Demand for the boards occurs throughout the winter selling season, so The Ski Bum meets demands as they come and cannot prioritize one over the other. This means the approach of using the weighted average understocking cost is appropriate. With postponement, the cost of understocking for each of the two boards is obtained using Equation 9.21

$$\bar{c}_u = \frac{\hat{x}_g(p_g - v_g + B_g)}{(\hat{x}_g + \hat{x}_m)} + \frac{\hat{x}_m(p_m - v_m + B_m)}{(\hat{x}_g + \hat{x}_m)}$$
$$= \left(\frac{90}{90 + 80}\right)(\$560) + \left(\frac{80}{90 + 80}\right)(\$510) = \$536.47$$

The cost of overstocking an unfinished board is the cost of the board minus the salvage value, $v_0 - g_0 = \$190 - \$50 = \$140$. To obtain the ideal stocking quantity of unfinished boards, we need the mean and standard deviation of *total* demand. The average demand is the sum of the individual-item averages, $\hat{x}_0 = \hat{x}_g + \hat{x}_m = 170$. As noted above, we assume that item demands are independent, so we get the standard deviation by taking the square root of the sum of the variances, $\sigma_0 = \sqrt{\sigma_g^2 + \sigma_m^2} = 32$. The ideal quantity is then obtained with this mean and standard deviation and the critical fractile obtained using the costs of overstocking and<?pag ?> understocking.

Table 9.6 Numerical Illustration of Single-Item, Single-Period Problem with and without Postponed Product Differentiation

	Original System — Glossy	*Original System — Matte*	*Postponed System*
p \$/unit	450	400	450/400
v \$/unit	200	200	–
v_0 \$/unit	–	–	190
t \$/unit	–	–	20
g \$/unit	50	50	50
B \$/unit	300	300	300
$\hat{x}_i$ units	90	80	170
σ_i units	25	20	32
k	0.792	0.736	0.817
Q or Q_0	110	95	196
$E[P(Q)]$ or $E[P(Q_0)]$	\$17,397	\$12,045	\$32,312
ES	3.05	2.69	3.73
EO	22.84	17.41	29.89

$$p_{u<}(Q^*) = \frac{\bar{c}_u}{\bar{c}_u + c_o} = \frac{536.47}{536.47 + 140} = 0.79 \tag{9.23}$$

Since demand is assumed to be normally distributed, we can use Appendix II or a spreadsheet formula to get $k = 0.817$ and an ideal order quantity for the base product of 196. It is straightforward to follow the approaches outlined above to determine quantities and expected profits for the case where the two snowboards are managed separately, without postponement. Results are summarized in Table 9.6 for easy comparison to the nonpostponed case. From Table 9.6, we see the flexibility benefit made possible by delaying product differentiation. Even though the total product and transformation cost is higher, the postponed system generates higher expected profit, \$32,312 instead of \$29,442; a gain of almost 10%. Examining the expected units short and expected units left over (ES, EO), it is clear that the increase in profit comes from reducing the mismatch between supply and demand. Total expected missed sales drops from 5.74 to 3.73, and total units left over drops from 40.25 to 29.89.

The preceding analysis assumed that The Ski Bum served each demand as it arrived and could not prioritize demand for one item over the other. Suppose instead that The Ski Bum had to order snowboards well in advance of the ski season, in the summer perhaps, but all customers would pre-order in the fall. In such a case, The Ski Bum could then determine which demands to meet. It should be clear that they would prefer to sell a glossy board since it has a higher cost of understocking. Using the approach outlined in the previous section, the only change that needs to be made is that the lower cost of understocking of \$510 replaces the weighted-average cost of understocking of \$536.47. This results in a value of k of 0.776, a slightly lower-order quantity of 195 and a slightly higher profit of \$32,453. Of course, we would expect the profit to increase given that missed sales are now only for the matte boards and thus cost only \$510 per missed sale.

9.5 More than One Period in Which to Prepare for the Selling Season

Particularly in a production context (e.g., style goods in the garment industry) there may be an extended length of time (divided, for convenience, into several time periods) in which replenishment commitments are made before the actual selling season begins. There are likely to be production constraints on the total amounts that can be acquired in each time period. Furthermore, forecasts of total demand are almost certain to change during these preseason periods, perhaps as a consequence of firm customer orders being received. We would like to be able to modify the decision rule of the previous section to cope with these new complexities effectively.

Hausman and Peterson (1972) have shown that an "optimal" solution to the problem is out of the question. However, they have proposed two heuristic decision rules that appear to perform quite well, at least on the numerical examples tested. The heuristics are closely related to the single-period procedure of Section 9.3, but allowance is made for production capability in all remaining periods. Related references include Crowston et al. (1973), Bitran et al. (1986), Matsuo (1990), Bradford and Sugrue (1990), and Fisher and Raman (1996). The latter paper allows for updating of forecasts based on sales during the early part of the season and develops a heuristic solution. Experience with a fashion apparel manufacturer is reported. See also Abillama and Singh (1996) who model a situation in which the capacity to produce within a season is limited. Therefore, production is started prior to the season, and any shortages of realized demand are produced if capacity is available in season. They find optimal order quantities for single and multiple products, and describe a number of properties of the optimal solution.

9.6 Multiperiod Newsvendor Problem

In many cases, obsolescence occurs after several time periods, implying that more than one order may be placed as the demand declines or simply evaporates. Section 8.5 of Chapter 8 discussed the case of gradually declining demands and the sizing of the final replenishment. We now introduce literature that addresses the problem of a product that faces sudden obsolescence at some unknown future time.

For multiperiod production/inventory problems, Ignall and Veinott (1969) have developed conditions under which a relatively simple policy, known as a myopic policy, is optimal for the sequence of periods. A myopic policy is one that selects the production quantities for the current period to minimize expected costs in the current period alone. Obviously, a myopic policy is much simpler to compute than is a more general multiperiod policy. However, the multiperiod style goods problem discussed above does not satisfy the conditions for which a myopic policy is optimal.

Joglekar and Lee (1993) provide an exact formulation of the inventory costs for items with deterministic demand that may face sudden obsolescence at an unknown future time. Jain and Silver (1994) generalize the problem somewhat by incorporating the potential for a longer lifetime. At the end of each period, the remaining inventory either becomes worthless or is usable for at least one more period. They use variations of the lot sizing heuristics introduced in Chapter 5 to solve the problem of how much to order each period. See also David and Mehrez (1995) who introduce a problem in which all items become useless at once, but at a random time in the future. See also Rosenfield (1989).

Song and Zipkin (1990) develop a flexible model that can analyze sudden obsolescence as well as gradual declining demand when demand is probabilistic and is distributed as a Poisson random variable. They show that in many cases, an inventory hump occurs. When goods are ordered prior to the decline in demand, inventory reaches a level that is too high for the lower demand rate. It takes some time to work off the excess inventory. Therefore, simple myopic policies can produce significant errors, as can simple adjustments to the inventory holding cost rate. (The latter approach is commonly used. The carrying charge, r, is inflated to account for the possibility of obsolescence.) Song and Zipkin show that inventory policies that explicitly account for obsolescence perform much better than these heuristic methods. Unfortunately, these policies are not easy to compute for most demand distributions. If disposal of stock is possible, the problem becomes somewhat more manageable. Masters (1991), on the other hand, shows that the common approach of inflating r is indeed optimal when demand is deterministic and the time to obsolescence is exponentially distributed. See also Cobbaert and Van Oudheusden (1996).

Related work on multiperiod problems includes Liao and Yang (1994) and Kurawarwala and Matsuo (1996).

9.7 Other Issues Relevant to the Control of Style Goods

In this section, we provide the highlights of other research that has been done related to the control of style goods. For more detailed analyses, see the referenced literature at the end of this chapter.

9.7.1 Updating of Forecasts

This topic has been treated in Chapter 3, but we now reexamine those aspects of particular relevance to the style goods problem. In a multiperiod problem, opportunities exist to update forecasts. Several alternative methods exist, including

1. *Exploitation of the properties of the forecasts made by decision makers.* Hausman (1969) found that under certain circumstances, ratios of successive forecasts turn out to be independent variables with a specific form of probability distribution. Fisher et al. (1994b) discovered that members of the buying committee of a fashion skiwear company tended to listen to a few dominant members in developing estimates of the next season's demand. When the committee members independently forecasted demand, the average of the forecasts tended to be much more accurate. See also Chambers and Eglese (1986), Fisher et al. (1994a), and Fisher and Raman (1996).

2. *Taking advantage of the observation that sales at the retail level tend to be proportional to inventory displayed.* Wolfe (1968) used this approach to revise forecasts based on demands observed early in the selling period. See also Baker and Urban (1988), Bar-Lev et al. (1994), Kurawarwala and Matsuo (1996), Gerchak and Wang (1994), Urban (1995), Padmanabhan and Vrat (1995), and Giri et al. (1996).

3. *Simple extrapolation methods using a particular mathematical form for the cumulative sales as a function of time.* Hertz and Schaffir (1960) assumed a normal-shaped cumulative curve. Gilding and Lock (1980), in the context of a furniture manufacturer, dealt with the case

of an artificial seasonality induced by sales promotions. They found that a Gompertz curve provides a reasonable fit to cumulative sales:

$$Y_{ult} = \ln Y_t + ae^{-bt} \tag{9.24}$$

where

Y_{ult} = represents the ultimate (total) sales of the item

Y_t = cumulative sales as of time t

and a and b are constants. See also Franses (1994).

4. *Bayesian procedures.* In the Bayesian approach, one or more parameters of the probability distribution of demand are assumed unknown. Prior knowledge is encoded in the form of probability distributions over the possible values of these parameters. As demands are observed in the early part of the season, these probability distributions are appropriately modified to take account of the additional information. References include Chang and Fyffe (1971), Murray and Silver (1966), Riter (1967), Harpaz et al. (1982), Azoury (1985), Lovejoy (1990), Lariviere and Porteus (1999), and Eppen and Iyer (1997b). Also, Fisher et al. (1994b) found that in the firm they studied, using information from only the first 20% of the season's sales data significantly improved forecast accuracy.

5. *Use of information about patterns of past product demand in conjunction with estimates of the total life cycle sales of the current product.* Managers can estimate the duration of the product life cycle for a short cycle product based on the experiences with previous versions of the product. Then estimates about total sales, the timing of peak sales (such as in December), and other seasonal variations, can be added to develop specific forecasts of demand. Multiperiod variations of the newsvendor inventory model are then used to control inventory. See Kurawarwala and Matsuo (1998, 1996), and Lau and Lau (1996a). See also Fisher et al. (1994b), who advocated dividing items into those which are more predictable and those which are less predictable. Thus, productive capacity can be devoted to predictable items first. The smaller amounts of capacity available just before, or during, the season can be reserved for less predictable items.

9.7.2 Reorders and Markdowns

Wolfe (1968), using the assumption about sales being proportional to the inventory level, developed methods of determining each of the following:

1. The expected time T_F to sell a fraction F of the initial inventory if a fraction f has been sold by time $t(f < F)$. A style can be tentatively identified as a fast mover (with associated possible reorder) or slow mover (with associated possible markdown) depending on the value of T_F for a particular selected F value.

2. The associated order-up-to-level if a reorder is to be placed at a specific time.

3. The timing of a markdown as a function of the fraction of initial inventory sold to date, the current price, the markdown price, the salvage value of leftover material, and the assumed known ratio of the sales rate after the markdown to that before it.

Feng and Gallego (1995) studied the problem assuming that there is a fixed number of units available for sale in a season. They determined when to lower or raise the price as a function of the time until the end of the season, if there is but one chance to change the price. Based on the number of items left, they provide a threshold time when the price should be changed. Khouja (1995) shows that multiple markdowns provide higher expected profit than a single markdown. See also Carlson (1982a,b), Lau and Lau (1988); Arcelus and Rowcroft (1992), and Khouja (1996).

9.7.3 Reserving Capacity Ahead of Time

It is common in the apparel industry for a firm to reserve some portion of the capacity of one or several suppliers. One fashion-forward apparel firm we have worked with limits its activities to designing garments and arranging to have them manufactured. All manufacturing and logistics is performed by outside contractors. To ensure that enough production capacity is available to meet peak-season demand, this firm agrees to buy a certain number of, say, sport coats from a supplier over the year. Then, specific styles and colors are determined as the season approaches. Jain and Silver (1995) examine this problem for an item with uncertainty in its demand and uncertainty in the capacity of the supplier. Dedicated capacity can be ensured by paying a premium to the supplier. The two decision variables are the replenishment quantity to request and the amount of capacity to reserve. It turns out to be very easy to select the best value of the replenishment quantity. On the other hand, they are only able to characterize the general behavior of the expected profit as a function of the level of dedicated capacity. See also Ciarallo et al. (1994), So and Song (1998), and Costa and Silver (1996). In particular, the latter authors examine a multiperiod version of the Jain and Silver (1995) paper. Bassok and Anupindi (1997) examined the case in which a commitment is made to the supplier to buy a certain minimum quantity over a multiple period horizon. Anupindi and Bassok (1995) studied the case in which the buyer provides a forecast but the supplier may allow some flexibility to deviate from the forecast. Eppen and Iyer (1997a) studied a similar system in which the buyer orders a given number of items, but some are held back by the vendor. If the buyer does not purchase these, a penalty is paid. If they are ordered, they can be purchased at the original price. See also Tsay (1999) and Eppen and Iyer (1997b). Liao and Yang (1994) analyzed a system in which more than one replenishment is triggered at once. Cachon (1999) studied a case in which multiple buyers are forced to place their orders according to their supplier's production schedule so that the variability of demand experienced by the supplier is reduced. Nevertheless, total supply chain costs increase, and the buyers end up incurring most of the higher cost.

9.7.4 Inventory Policies for Common Components

In a manufacturing firm that can assemble finished products to customer order, it is often desirable to hold safety stock of components, rather than of end items. If the assembly time is short relative to the lead time for purchasing or making components, and the value added at the assembly stage is high, holding component inventory makes sense. In addition, if there is significant commonality among components, total safety stocks may be reduced by holding common components and then assembling them to order. Component commonality refers to the situation in which certain components are used to make more than one finished product.

A number of researchers have investigated the question of how much stock to hold when there is component commonality. This literature has usually viewed the problem in a newsvendor context because the long component lead time yields a single decision of how much component inventory to order. A second-stage decision in the problem involves which finished products to assemble once demand has been realized.

Dogramaci (1979) shows that component commonality can lower costs due to a reduction in the forecast error. Baker (1985) shows how the typical relationship between the safety factor and finished product service level is altered when the safety factor is chosen for common components. Choosing a safety factor of components to satisfy, say, 90% of demand will not give a 90% service level of finished products because the common components can be allocated to several products. See also Bagchi and Gutierrez (1992).

Gerchak and Henig (1986) show that (1) the inventory of product-specific components always increases when other components are combined into common parts and the objective is to maximize profit; and (2) the inventory of the common components could increase, decrease, or stay the same. They also show that the multiperiod problem satisfies the conditions for a myopic optimal solution, so that qualitative insights from the newsvendor-type solutions hold in the more general case. Gerchak et al. (1988) extend the analysis to the case in which there is a service-level requirement, and show that property (1) does not necessarily hold. See also Baker et al. (1986). When there is a capacity or storage constraint, Gerchak and Henig (1989) show that property (1) again may not hold. See also Jönsson and Silver (1989a,b), Jönsson et al. (1993).

Eynan and Rosenblatt (1996) relax a prevalent assumption that the common component has the same cost as the product-specific components it replaces. They show that commonality may not be desirable. See also Eynan and Rosenblatt (1997). Vakharia et al. (1996) illustrate how commonality may increase the disruption on the shop floor by increasing the variability of the load at certain work centers.

Eynan and Rosenblatt (1995), in an interesting paper, compare an assemble-in-advance strategy with an assemble-to-order strategy. If products are made in advance, additional holding costs may be incurred, whereas if they are made to order, processing costs are higher. The authors give simple newsvendor-type solutions for the optimal inventory levels. They also provide conditions under which it is optimal to choose each strategy, or a combination of the two. See also Lee and Tang (1997) and Moon and Choi (1997).

Srinivasan et al. (1992) develop approximations and an algorithm for a large-scale commonality problem at IBM. Other references on the commonality problem include Collier (1982), McClain et al. (1984), Guerrero (1985), Baker (1988), Aneja and Singh (1990), Tsubone et al. (1994), Agrawal and Cohen (2001), Balakrishnan et al. (1996), Eynan (1996), and Zhang (1997).

9.7.5 Other Research

We now comment briefly on other research pertaining to the general problem area of this chapter. Gerchak and Mossman (1992) show that the optimal order quantity in the newsvendor problem may actually increase when demand variability decreases. The order quantity increases as demand variability increases if and only if

$$p_{x \leq}(\hat{x}) < \frac{c_u}{c_u + c_o}$$

which is often true. They show that, under quite general conditions, as uncertainty increases, the order quantity always moves away from mean demand, and the optimal cost increases.

Additional research in this area includes Eeckhoudt et al. (1995) and Kim et al. (1995).

9.8 Inventory Control of Perishable Items

In the earlier parts of the book, we have not explicitly dealt with perishability, other than possibly through a rather crude adjustment of the inventory carrying charge. Perishability refers to the physical deterioration of units of a product. Demand continues for further units of the item. In contrast, when obsolescence occurs, there is negligible further demand for the SKU involved.

Perishable items can be divided into two categories, fixed or random, depending on the lifetime of a unit of the item. For items with a fixed lifetime, the utility of each unit stays essentially constant for a fixed period of time, then drops appreciably (perhaps decreed by law). An example is blood stored in blood banks. For items where the lifetime is a random variable, the utility can decrease throughout the lifetime in a fashion that may or may not depend on the age of the unit involved (examples include fresh produce, certain types of volatile chemicals, drugs, etc.). See Abdel-Malek and Ziegler (1988).

For the fixed lifetime case, Nahmias (1982) provides an extensive and excellent review of the research on this topic. Here, we briefly comment on the research since that time. Schmidt and Nahmias (1985) consider $(S - 1, S)$ inventory policies for fixed lifetime items with Poisson demands, and positive replenishment lead times. Ravichandran (1995) analyzes the same system using (s, S) policies, and shows that it is extremely difficult to analyze. Chiu (1995) allows for a positive replenishment lead time and develops approximations for (s, Q) policies. See also Cohen et al. (1983) and Kalpakam and Sapna (1996).

Most research that analyzes the random lifetime problem assumes that inventory on hand deteriorates at a constant rate. In other words, a constant fraction of inventory on hand becomes useless each unit of time. Raafat (1991) reviews some of the literature in this area. Chung and Ting (1994) investigate a system in which demand decreases linearly with time, and where a constant fraction of items deteriorates each period. They develop a good heuristic for the amount of inventory to order. (See also Kim 1995.) Hariga and Benkherouf (1994) expand the problem to include time-varying demand. They employ heuristics like those presented in Chapter 5. Benkherouf (1995) presents an optimal procedure for the case in which demand rates decrease over a known time horizon and items deteriorate at a constant rate. See also Benkherouf and Mahmoud (1996) and Balkhi and Benkherouf (1996). Aggarwal and Jaggi (1995) extend the analysis to allow for delayed payment to the supplier. Kalpakam and Sapna (1994) analyze an (s, S) system with Poisson demands and lifetimes that follow an exponential distribution. Lead times are positive and probabilistic as well. They describe properties of the long-run expected cost. When demand is probabilistic, as in this latter case, the analysis becomes extremely complicated. Rosenfield (1989) examines the probabilistic case when each item can perish at a random time. He determines the optimal amount of inventory to dispose of when demands arrive according to a Poisson process and batch sizes are one unit. See also Padmanabhan and Vrat (1995), Abad (1996), and Chapter 8 of this book.

In an interesting paper, Vaughn (1994) blends the fixed lifetime and random lifetime literatures. In his model, neither the vendor nor the customer can observe whether the item has deteriorated. Therefore, the vendor assigns a "sell by" or "use by" date to each item. The useful life of the item is a random variable, but the shelf life is deterministic and is given by the chosen "sell by" date. This paper examines the interaction between the optimal order up to level S and the optimal "sell by" date.

For most of this research, inventory is depleted according to a first-in-first-out (FIFO) policy; that is, the oldest inventory is used first. FIFO is an optimal issuing policy over a wide range of assumptions, particularly where the issuing organization has complete control over the issuing actions. However, where the customer makes the selection (e.g., in retail food distribution where the expiration date is shown on each unit), a last-in-first-out (LIFO) policy is likely to be observed. See Pierskalla and Roach (1972) and Albright (1976).

9.9 Summary

In this chapter, we have presented some decision logic for coping with items subject to obsolescence or perishability. In the style goods version of the problem, the logic exploited the property of not being able (for economic reasons) to carry stock over from one selling season to the next. With this chapter, we have completed our coverage of individual-item inventory management.

Problems

9.1 Alexander Norman owns several retail fur stores in a large North American city. In the spring of each year, he must decide on the number of each type of fur coat to order from his manufacturing supplier for the upcoming winter season. For a particular muskrat line, his cost per coat is $150 and the retail selling price is $210. He estimates an average sales of 100 coats but with considerable uncertainty, which he is willing to express as a uniform distribution between 75 and 125. Any coat not sold at the end of the winter can be disposed of at cost ($150) to a discount house. However, Norman feels that on any such coat he has lost money because of the capital tied up in the inventory for the whole season. He estimates a loss of $0.15 for every dollar tied up in a coat that must be sold off at the end of the season.

  a. How many coats should he order?

  b. One of the factors contributing to the uncertainty in sales is the unknown level of retail luxury tax on coats that will be established by the government early in the fall. Norman has connections in the government and manages to learn the tax level prior to his buying decision. This changes his probability distribution to a normal form with a mean of 110 and a standard deviation of 15. Now what is the best order quantity? How much was the inside information concerning the tax worth to him?

9.2 Consider the situation where tooling is set up to make a major piece of equipment and spare parts can be made relatively inexpensively. After this one production run of the product and its spares, the dyes, etc. will be discarded. Discuss the relationship between the problem of how many spares to run and the single-period problem.

9.3 Neighborhood Hardware Ltd. acts as a central buying agent and distributor for a large number of retail hardware outlets in Canada. The product line is divided into six major categories, with a different buyer being responsible for each single category. One category is miscellaneous equipment for outdoor work around the home. The buyer for this group, Mr. Harry Lock, seeks assistance from a recently hired analyst, J. D. Smith, in the computer division of the company. In particular, he is concerned with the acquisition of a particular type of small snowblower that must be ordered several months before the winter. Smith, after considerable discussions with Lock, has the latter's agreement with the following data:

Unit acquisition cost is $60.00/unit

Selling price is $100.00/unit

Any units unsold at the end of the winter will be marked down to $51/unit, ensuring a complete clearance and thus avoiding the prohibitive expense of storage until the next season. The probability distribution of regular demand is estimated to be

Hundreds of Units, j_o	3	4	5	6	7	8
$P_j(j_o)$	0.1	0.1	0.4	0.2	0.1	0.1

a. What is the expected demand?
b. What is the standard deviation of demand?
c. To maximize expected profit, how many units should Smith (using a discrete demand model) tell Lock to acquire?
d. What is the expected profit under the strategy of (c).
e. Suppose Smith instead decides to fit a normal distribution, having the same mean and standard deviation, to the above discrete distribution. With this normal model, what is the recommended order quantity, rounded to the nearest hundred units?
f. If the discrete distribution is the true one, what cost penalty is incurred by the use of the somewhat simpler normal model?

9.4 Suppose the supplier in Problem 9.3 offers a discount such that the unit cost is $55 if an order of at least 750 units is placed? Should Neighborhood take advantage of this offer?

9.5 In reality, Mr. Lock of Neighborhood Hardware has to order three different snowblowers from the same supplier. He tells the analyst, Smith, that he will go along with his normal model (whatever that is!) but, in no event, will he allocate more than $70,000 for the acquisition of snowblowers. In addition to item SB-1, described in Problem 9.3, characteristics of the other two items are:

Item	Unit Acquisition Cost ($/Unit)	Selling Price ($/Unit)	Clearance Price ($/Unit)	Demand Mean (Units)	Demand Standard Deviation (Units)
SB-2	80	110	70	300	50
SB-3	130	200	120	200	40

a. What should Smith suggest as the order quantities of the three items?
b. What would be the approximate change in expected profit if Lock agreed to a $5,000 increase in the budget allocated for snowblowers?

   Note: Assume in this problem that there are no quantity discounts available.

9.6 The *Calgary Herald* has a policy of not permitting returns from hotels and stores on unsold newspapers. To compensate for this fact, they have lowered the value of *v* somewhat.

  a. From the standpoint of a particular store, how would you evaluate the effect of a change from v_1, g_1 to $v_2, 0$ where $g_1 > 0$ and $v_2 < v_1$?
  b. Discuss the rationale for the *Herald*'s policy.

9.7 Professor Leo Libin, an expert in inventory theory, feels that his wife is considering the preparation of far too much food for an upcoming family get-together. He feels that the newsvendor formulation is relevant to this situation.
  a. What kinds of data (objective or subjective) should he collect?
  b. His wife says that certain types of leftovers can be put in the freezer. Briefly indicate what impact this has on the analysis.

9.8 Consider a company with N warehouses and a certain amount (W units) of stock of a particular SKU to allocate among the warehouses. The next allocation will not be made until one period from now. Let

$_if_x(x_0) = $ probability density of demand in a period for warehouse $i(i = 1, 2, \ldots, N)$

  Develop a rule for allocating the W units of stock under each of the following costing assumptions (treated separately):
  a. A cost H_i incurred at warehouse i if any shortage occurs.
  b. A cost B_i is charged per unit short at warehouse i.

9.9 Consider a perishable item with a unit acquisition cost v, a fixed ordering cost A, a known demand rate D, and a carrying charge r. The perishability is reflected in the following behavior of the selling price p as a function of the age t of a unit:

$$p(t) = p - bt \quad 0 < t < SL$$

where
  $b = $ positive constant
  $SL = $ shelf life
  Any units of age SL are disposed of at a unit value of g, which is less than both v and $p - b(SL)$. Set up a decision rule for finding the profit maximizing value of the order quantity Q.

9.10 Consider a sporting goods company that carries ski equipment in stores across Canada. Because of unusual weather conditions 1 year, near the end of the ski season, the company finds itself with a surplus of ski equipment in the West and with not enough on hand in the East to meet the demand. At a meeting of marketing managers a person from the East requests that the surplus goods in the West be rush-shipped to the East. On the other hand, a Western manager suggests a discount sale there to get rid of the stock. In an effort to resolve the conflict, the President asks the Management Sciences Group to analyze the problem and report back with a recommended decision within 1 week. Briefly outline an analysis you would undertake, including the types of data that would be required. Also, if possible, suggest another alternative (i.e., differing from the two suggested by the marketing managers).

9.11 Although its derivation in the appendix is for the case of continuous units, Equation 9.3 also holds for the case of discrete demand and units where $ES = \sum_{x_0 > Q}(x_0 - Q)p_x(x_0)$.

a. For an item with the following data, compute the expected profit from Equation 9.3 for Q equal to each possible (nonzero probability) value of x and thus find the best Q value. Verify that the use of Equation 9.10 gives the same result.

Number of Units, x_0	1	2	3	Total
$p_x(x_0) = \text{Pr}(x = x_0)$	0.25	0.4	0.35	1
$p_x < (x_0) = \text{Pr}(x < x_0)$	0	0.25	0.65	1

$p = \$10$
$v = \$5$
$g = \$1$
$B = \$1$

b. Prove analytically that Equation 9.3 holds, in general, for discrete demand units.

9.12 Mr. Jones, a recent MBA graduate, the only entrepreneur in his graduating class, will establish a distinctive newspaper concession on the corner of two busy streets near his alma mater. After conducting in-depth interviews, he has decided to stock three daily newspapers: the *Hong Kong Chronicle*, the *New York Times*, and the *Barrie Bugle*. Having successfully avoided all courses requiring arithmetic during his academic career, he has contacted you to advise him on the kind of ordering policy that he should follow.

a. You first make the assumption that the daily demand for each of the three papers is normally distributed. The mean number of papers sold during the first 4 days of the week appears to be 50% greater than on Fridays and Saturdays. Mr. Jones has a small kiosk (acquired through a student loan from the bank) that can stock only 500 papers each day. In keeping with analytic modeling tradition, you assume, for mathematical convenience, that the cost of being short of any of the three papers on any particular day is $0.06 and the cost of not selling a paper on any particular day is $0.02. Recommend inventory stocking rules for Mr. Jones.

b. Illustrate with the following numerical example for a typical Tuesday:

Paper	Mean Daily Demand	Variance Daily Demand
Chronicle	400	10,000
Times	200	8,100
Bugle	300	10,000

c. After collecting your exorbitant consulting fee, you confess to your bartender, explaining the shortcomings of the numerical example and ordering policy above. Criticize your answers to (a) and (b) in light of reality.

d. How would you ascertain whether it was worthwhile for Jones to go deeper into debt to buy a larger kiosk (capable of holding 1,000 copies)?

9.13 A local vendor of newspapers feels that dissatisfaction of customers leads to future lost sales. In fact, she feels that the average demand for a particular newspaper is related to the service level as follows:

$$\hat{x} = 1{,}000 + 500P$$

where $\hat{x}$ is the average demand in copies/period and P is the fraction of demand that is satisfied. Demands are normally distributed and the standard deviation (per period) is equal to 200, independent of the service level. Other (possibly) relevant factors are:

$$\text{Fixed cost per day to get any papers in} = \$2$$
$$\text{Cost per paper (for vendor)} = \$0.07$$
$$\text{Selling price per paper} = \$0.15$$
$$\text{Salvage value per paper} = \$0.02$$

a. Ignoring the effect of disservice on demand, what is the best number of papers to acquire per period?
b. Taking account of the effect, now what is the best number to acquire?
c. What average profit is the vendor losing if she proceeds as in (a) instead of as in (b)?

9.14 The Food Services Department of a large university is concerned with the decision making related to the purchase of bread for use in the cafeterias of the university. The daily demand can be assumed to be uniformly distributed between 20 and 60 loaves. There are 20 slices in a loaf of bread and the selling price is $0.04 per slice. Any demand for bread when there is none on hand is lost but there is no associated loss of goodwill, etc. Any bread left over at the end of a day is used as ingredients for cooking (e.g., in bread pudding, stuffing, etc.). The value of such use is estimated to be $0.05 per loaf. The supplier, Easton Bakeries, in an effort to induce high consumption offers a discount structure as follows:

$$\text{Cost per loaf for first 50 loaves} = \$0.30$$
$$\text{Cost per any loaf over 50 loaves} = \$0.20$$
$$\text{The fixed cost of a procurement from the bakery} = \$5.00$$

a. Show the expected daily profit as a function of the number of loaves of bread ordered per day.
b. What quantity of bread should be purchased per day?
c. Outline how you would determine the daily loss in profit (compared with the optimal strategy) that would result if the university simply ordered the expected demand?
d. What are some complexities that would have to be included to make the problem formulation more realistic?

9.15 The hardware store in Hanover, New Hampshire orders lawn mowers each January. The cost per mower is $300, and they sell for $425. Any mowers left over at the end of the selling season are sold during a Labor Day Sale for $250 each. Assume that the following data applies:

Demand	0	1	2	3	4	5
Probability	0.10	0.15	0.30	0.20	0.15	0.10

How many mowers should the store order?

9.16 A retail store sells a seasonal product for $10 per unit. The cost to the store is $6 per unit. An end-of-season sale is required for all excess units that are sold at half price. Assume that demand for the units is normal with mean of 500 and standard deviation of 100.

a. What is the recommended order quantity?

b. What is the probability that at least some customers will ask to purchase the product after the outlet is sold out, assuming you use the order quantity recommended in (a)?

c. What is the effect of reducing the standard deviation of demand to 75 (through improved forecasting or through the use of early demand information from some customers)?

9.17 An exclusive local men's fashion store is considering the number and styles of ties to purchase for the upcoming spring and summer season. He has informed you that ties are generally coordinated with other items (suits and shirts). He has also told you that a tie can often make the difference between a sale and a no-sale on a suit. Finally, he has informed you that ties he generally carries range from classic ties to stylish ties to seasonal colored ties. He is unsure of the factors that he should be considering in deciding what stock of ties to carry. Make some suggestions about these factors.

9.18 A small discount retailer of snorkeling equipment must decide how many snorkeling masks to order for this summer season. The selling price for these masks is $20, while the cost to the retailer is only $12. The retailer can usually buy additional masks from competitors at their retail price of $25. Masks left over at the end of the short selling season are sold at a special discount price of $10 per pair. The manager of the store assumes that demand is approximately normally distributed with a mean of 600 masks and a standard deviation of 200 masks.

a. How many should she order?

b. What is the effect on the order quantity if she could reduce the standard deviation to 100?

9.19 In the text, we discussed the basic newsvendor problem where there is but a single ordering opportunity per selling period (day). In fact, we assumed that the order had to be placed prior to the actual selling period (day). Let us now consider the following modified version of the problem. After all the demand is known for a particular day, the newsvendor has an additional opportunity to order any copies that he still needs above and beyond what he ordered prior to the day. These new copies arrive quickly enough to satisfy customers, thus avoiding lost sales and any loss of good will. Let

v_1 = unit acquisition cost for copies ordered prior to the day in question (i.e., by the usual assumed method), $/copy

v_2 = unit acquisition cost for the emergency additional order, $/copy

Assume $g < v_1 < v_2 < p$ where

g is the unit salvage value, $/copy and

p is the selling price, $/copy

Let Q_1 and Q_2 be the order quantities before and after the demand is known. Let x be the demand in a day.

Note: Part (b) is a marginal approach; parts (c) through (e) are total cost approaches toward obtaining the same decision rule.

a. How is Q_2 determined given Q_1 and x?

b. Use a marginal cost approach to develop a decision rule that the best Q_1 must satisfy. Assume that x and Q_1 are continuous variables and that the probability density function of x is denoted by $f_x(x_0)$.

c. Write an expression or expressions for the profit as a function of Q_1 and the specific value of x observed.

d. Develop an expression for the expected profit.

e. Using the result of (d), develop a decision rule that the best value of Q_1 must satisfy.

f. Indicate how the following complexity would alter the development in (a) and (b) only: now there is a probability $\pi(0 < \pi < 1)$ that the emergency order will be filled (before we assumed that $\pi = 1$).

9.20 One of the assumptions in the basic newsvendor problem is that there is a constant underage cost c_u per unit short, independent of the size of the shortage. Suppose instead that there was a cost c_1 if any shortage occurred and an additional cost c_2 if the shortage exceeded some size Y (and no cost per unit short).

a. Why would a marginal analysis now be extremely difficult, if not impossible?

b. Indicate how you would develop a decision rule for determining the best order quantity Q. (Assume that there is an overage cost of c_o per unit leftover). For simplicity, treat Q and demand as continuous variables.

9.21 A manufacturer of summer style goods has two opportunities to make replenishments. The first opportunity, at a unit cost v_1, is during the previous winter when very little in the way of firm orders have yet been received. The second opportunity, at unit cost v_2 (where $v_2 > v_1$), is in the early spring after a (random) number x of firm orders have been received. The ultimate total number of units demanded is denoted by y, and x and y have a joint probability density function $f_{x,y}(x_0, y_0)$. The selling price per unit is p and any units left can be scrapped at a unit value of g. Let Q_1 be the quantity produced in the winter and Q_2 be the quantity produced in the spring.

a. Why would x and y have a joint distribution as opposed to being independent?

b. Develop an expression for the best value of Q_2 given values of Q_1 and x.

c. Outline how you would go about obtaining the best Q_1 value.

9.22 Jack G. Gold, the manager of the bookstore at Lawson College, must decide on how many copies of a particular text to order in connection with an elective course to be offered by Professor Patrick Dunphy of the college's Department of Philosophy. Gold decides to seek the assistance of Professor William F. Jennings of the Department of Industrial Engineering who has some familiarity with inventory models. Jennings ascertains from Gold that historically there has been considerable variance between the number of books that Dunphy says he needs for a course and the actual number purchased by the students. The particular text under consideration is supplied by a publisher that allows up to 20% returns from the bookstore, with the latter only incurring the shipping costs (i.e., the wholesale purchase price is completely refunded). To illustrate, up to 20 copies could be handled in this fashion if the original order was for 100 copies. No refund is possible on any amount beyond the 20%. Gold indicates that any copies that could not be returned would be marked down to

50% of his cost (the wholesale price), which would clear them out. Gold points out that his permitted markup on all books is 15% of his wholesale price. Suppose that, through analysis of historical records and discussions with Gold and Dunphy, Jennings is able to develop a reasonable probability distribution of total demand for the text.

a. Outline how Jennings might develop a decision rule for Gold to use in deciding on how many copies of Dunphy's text to order.

b. Mention some complexities ignored in your analysis.

9.23 Consider a perishable commodity that is outdated in exactly three basic periods. Suppose that an ordering decision is made at the beginning of each period and the replenishment arrives instantaneously. An order-up-to-level (R, S) system is used. The perishability of the item means that any units procured at the start of period t and still on-hand at the end of period $t + 2$ must be discarded. Any demand not satisfied is lost (i.e., there is no backordering). Simulate the behavior of such a system for 20 periods assuming $S = 30$ units and the demand in the 20 consecutive periods are 11, 9, 3, 4, 14, 11, 7, 9, 11, 10, 12, 15, 8, 11, 8, 8, 14, 15, 5, 11. In particular, what is the average of each of the following?

a. Inventory carried forward to the next period

b. Number of units discarded per period

c. Lost sales per period

Assume that at the start of period 1 (before an order is placed) there is no older stock, that is, we place an order for 30 units.

9.24 The demand for a particular spare part appears to be decreasing rapidly. Any demand occurs only in the month of December each year. Because of warranty considerations, the company has guaranteed the availability of spare parts through to (and including) the 2017 year. The inventory of the item in mid-November 2016 is 10 units. The demands in December 2016 and December 2017 are assumed to be independent random variables x and y with probability mass functions as follows:

x_0	0	10	20	30
$p_x(x_0)$	0.2	0.4	0.3	0.1

y_0	0	10	20
$p_y(y_0)$	0.4	0.5	0.1

There is a setup cost of $50 associated with producing any units of the item. The unit variable cost of a produced piece is $20/unit and it costs $4.50 to carry each unit in inventory from December 2016 to December 2017. In any month of December when the total demand exceeds the stock available, additional spare parts are purchased from a competitor at a cost in dollars of $60 + 35z$, where z is the number of units purchased (i.e., the size of the shortage). Any unit left over at the end of December 2017 will have a salvage value of $5/unit. Production each year must take place before the December demand is known.

a. Suppose at the beginning of December 2017 there were no units in stock. Determine the quantity that should be produced at that time and the associated expected costs.

    b. What are two complicating factors that make a standard newsvendor analysis inappropriate for determining the best quantity to produce prior to the December 2016 season (i.e., in late November, 2016).

    c. Briefly outline (but do not attempt to carry out any of the numerical details of) a method for determining the best quantity to produce prior to the December 2016 season.

9.25 a. Build a spreadsheet to determine the order quantity for the basic newsvendor problem for a given discrete demand distribution.

    b. Repeat for the normal distribution.

9.26 Consider the multi-item newsvendor setting described in Section 9.3. Equation 9.12 defines the optimal quantity for a given item i and a particular budget M.

    a. In order for an item i to be stocked, the right-hand side of Equation 9.12 must be positive. Write an inequality with the right-hand side of Equation 9.12 greater than or equal to zero and rearrange this inequality to get a condition where the budget M is isolated on one side of the inequality.

    b. How can you use this inequality to determine which items will be stocked for a given budget level M?

    c. What is the economic interpretation of this inequality?

Appendix 9A: Derivations

9A.1 Basic Newsvendor Results

9A.1.1 Profit Maximization

Each unit purchased costs v, each unit sold produces a revenue of p, each unit disposed as salvage gives a revenue of g, and there is an additional cost B associated with each unit of demand not satisfied. If a quantity Q is stocked and a demand x_0 occurs, the profit is

$$P(Q, x_0) = \begin{cases} -Qv + px_0 + g(Q - x_0) & \text{if } x_0 \leq Q \\ -Qv + pQ - B(x_0 - Q) & \text{if } x_0 \geq Q \end{cases} \tag{9A.1}$$

The expected value of the profit, as a function of Q, is given by

$$E[P(Q)] = \int_0^\infty P(Q, x_0) f_x(x_0) dx_0$$

By substituting from Equation 9A.1, we obtain

$$E[P(Q)] = \int_0^Q [-Qv + px_0 + g(Q - x_0)] f_x(x_0) dx_0 + \int_Q^\infty [-Qv + pQ - B(x_0 - Q)] f_x(x_0) dx_0$$

$$= -Qv + gQ \int_0^Q f_x(x_0) dx_0 + (p - g) \int_0^Q x_0 f_x(x_0) dx_0 + pQ \int_Q^\infty f_x(x_0) dx_0$$

$$- B \int_Q^\infty (x_0 - Q) f_x(x_0) dx_0$$

$$= -Qv + gQ \int_0^Q f_x(x_0) dx_0 + (p-g) \left[\int_0^\infty x_0 f_x(x_0) dx_0 - \int_Q^\infty (x_0 - Q) f_x(x_0) dx_0 \right.$$

$$\left. - Q \int_Q^\infty f_x(x_0) dx_0 \right] + pQ \int_Q^\infty f_x(x_0) dx_0 - B \int_Q^\infty (x_0 - Q) f_x(x_0) dx_0$$

$$= -Qv + gQ \int_0^Q f_x(x_0) dx_0 + (p-g) \left[\hat{x} - ES - Q \int_Q^\infty f_x(x_0) dx_0 \right]$$

$$+ pQ \int_Q^\infty f_x(x_0) dx_0 - B(ES)$$

$$= (p-g)\hat{x} - (v-g)Q - (p-g+B)ES \qquad (9A.2)$$

where *ES* is the expected number of units short. This is Equation 9.3.[*]

We can also represent Equation 9A.2 as perfect information profit minus demand–supply mismatch costs. To do this, we use the following relationship between expected units short *ES* and expected units left over which we denote *EO*.

$$ES - EO = \int_Q^\infty (x_0 - Q) f_x(x_0) dx_0 - \int_0^Q (Q - x_0) f_x(x_0) dx_0$$

$$= \int_Q^\infty (x_0 - Q) f_x(x_0) dx_0 + \int_0^Q (x_0 - Q) f_x(x_0) dx_0 = \hat{x} - Q$$

By this relationship, $Q = \hat{x} - ES + EO$, so Equation 9A.2 becomes

$$E[P(Q)] = (p-g)\hat{x} - (v-g)\left(\hat{x} - ES + EO\right) - (p-g+B)ES$$

$$= (p-v)\hat{x} - (v-g)EO - (p-v+B)ES \qquad (9A.3)$$

Equation 9A.3 is then the profit under perfect information (the per unit margin, $p - v$, times the expected demand) minus the per unit overage cost (unit cost minus salvage value) times expected units left over, and underage cost (per unit margin $p - v$ plus the penalty cost B) times expected units short.

---

[*] Lau (1997) provides a similar development for expected costs and gives formulas for uniform or exponential demands. As shown in Appendix II for the normal distribution, $ES = \sigma_x G_u \left(Q - \hat{x}/\sigma_x \right)$, which gives Equation 9.8.

To find the maximizing value of Q, we set

$$\frac{dE[P(Q)]}{dQ} = 0 \tag{9A.4}$$

By applying Leibnitz's rule,[*] we get $\frac{dES}{dQ} = p_{x<}(Q) - 1$, so that

$$\frac{dE[P(Q)]}{dQ} = 0 = -(v - g) - (p - g + B)[p_{x<}(Q) - 1]$$

or

$$p_{x<}(Q) = \frac{p - v + B}{p - g + B} \tag{9A.5}$$

which is Equation 9.2.

The expected profit at the optimal value of Q is found by substituting the right-hand side of Equation 9A.4 into the expected profit function:

$$E[P(Q^*)] = (p - g)\hat{x} - (v - g)Q^* - (p - g + B)ES$$

$$= (p - g)\hat{x} - (v - g)Q^* - (p - g + B) \int_{Q^*}^{\infty} (x_0 - Q^*)f_x(x_0)dx_0$$

$$= (p - g)\hat{x} - (v - g)Q^* - (p - g + B) \int_{Q^*}^{\infty} x_0 f_x(x_0)dx_0 + (p - g + B)Q^* \int_{Q^*}^{\infty} f_x(x_0)dx_0$$

$$= (p - g)\hat{x} - (v - g)Q^* - (p - g + B) \int_{Q^*}^{\infty} x_0 f_x(x_0)dx_0$$

$$+ (p - g + B)Q^*[1 - p_{x<}(Q^*)]$$

By using Equation 9A.4, we get

$$E[P(Q^*)] = (p - g)\hat{x} - (v - g)Q^* - (p - g + B) \int_{Q^*}^{\infty} x_0 f_x(x_0)dx_0 + (v - g)Q^*$$

$$= (p - g)\hat{x} - (p - g + B) \int_{Q^*}^{\infty} x_0 f_x(x_0)dx_0$$

---

[*] Leibnitz's rule states that $\frac{d}{dy}\int_{a_1(y)}^{a_2(y)} h(x,y)dx = \int_{a_1(y)}^{a_2(y)}[\partial h(x,y)/\partial y]dx + h(a_2(y),y)\frac{da_2(y)}{dy} - h(a_1(y),y)\frac{da_1(y)}{dy}$.

By a result in Appendix II, we obtain

$$E[P(Q^*)] = -B\hat{x} + (p - g + B) \int_0^{Q^*} x_0 f_x(x_0) dx_0$$

It is easily verified that the second derivative is negative, thus ensuring that we have found a profit maximizing value of Q.

9A.1.2 Normally Distributed Demand

If we set

$$k = \frac{Q - x}{\sigma_x} \tag{9A.6}$$

then, as shown in Appendix II, Equation 9A.5 simplifies to

$$p_{x<}(k) = \frac{p - v + B}{p - g + B} \tag{9A.7}$$

where

$$p_{u<}(k) = \Pr\{\text{a unit normal variable takes on a value less than } k\}$$

Using the fact that

$$p_{u\geq}(k) = 1 - p_{u<}(k) \tag{9A.8}$$

Equation 9A.7 can be rewritten as

$$p_{u\geq}(k) = \frac{v - g}{p - g + B}$$

Now we show the derivation of Equation 9.9. Equation 9.8 is

$$E[P(Q)] = (p - g)\hat{x} - (v - g)Q - (p - g + B)\sigma_x G_u\left(\frac{Q - \hat{x}}{\sigma_x}\right)$$

Using the expression for $G_u(k) = [f_u(k) - kp_{u\geq}(k)]$ from Appendix II and the expression from Equation 9A.8, we obtain

$$\begin{aligned}
E[P(Q)] &= (p - g)\hat{x} - (v - g)(\hat{x} + k\sigma_x) - (p - g + B)\sigma_x[f_u(k) - kp_{u\geq}(k)] \\
&= (p - g)\hat{x} - (v - g)\hat{x} - (v - g)k\sigma_x - (p - g + B)\sigma_x f_u(k) + (p - g + B)k\sigma_x p_{u\geq}(k) \\
&= (p - v)\hat{x} - (v - g)k\sigma_x - (p - g + B)\sigma_x f_u(k) + (p - g + B)k\sigma_x\left(\frac{v - g}{p - g + B}\right) \\
&= (p - v)\hat{x} - (v - g)k\sigma_x - (p - g + B)\sigma_x f_u(k) + (v - g)k\sigma_x \\
&= (p - v)\hat{x} - (p - g + B)\sigma_x f_u(k)
\end{aligned}$$

9A.1.3 Discrete Case

The expected total cost in the discrete case is

$$\text{ETC}(Q) = \sum_{x_0=0}^{Q-1} c_0(Q - x_0)p_x(x_0) + \sum_{x_0=Q}^{\infty} c_u(x_0 - Q)p_x(x_0) \tag{9A.9}$$

where

$$p_x(x_0) = \Pr\{x = x_0\}$$

Let

$$\Delta\text{ETC}(Q) = \text{ETC}(Q+1) - \text{ETC}(Q) \tag{9A.10}$$

Then, $\Delta\text{ETC}(Q)$ is the change in expected total cost when we switch from Q to $Q+1$. For a convex cost function (the case here), the best Q (or one of the best Q values) will be the lowest Q where $\Delta\text{ETC}(Q)$ is greater than zero. Therefore, we select the smallest Q for which

$$\Delta\text{ETC}(Q) > 0 \tag{9A.11}$$

Substituting from Equations 9A.9 and 9A.10 leads, after considerable simplification, to Equation 9.10.

9A.2 Constrained Multi-Item Situation

The problem is to select the Q_i's $(i = 1, 2, \ldots, n)$ to maximize

$$\sum_{i=1}^{n} E[P(Q_i)]$$

subject to

$$\sum_{i=1}^{n} Q_i v_i = W \tag{9A.12}$$

where, from Equation 9A.2

$$E[P(Q_i)] = -Q_i v_i + \int_0^{Q_i} [p_i x_0 + g_i(Q_i - x_0)]_i f_x(x_0) dx_0$$

$$+ \int_{Q_i}^{\infty} [p_i Q_i - B_i(x_0 - Q_i)]_i f_x(x_0) dx_0 \tag{9A.13}$$

Equation 9A.13 and

$$_i f_x(x_0) dx_0 = \Pr\{\text{Total demand for product } i \text{ lies between } x_0 \text{ and } x_0 + dx_0\}$$

The Lagrangian approach (see Appendix I) is to select the multiplier M and Q_i's to maximize

$$L(Q_i\text{'s}, M) = \sum_{i=1}^{n} E[P(Q_i)] - M\left[\sum_{i=1}^{n} Q_i v_i - W\right] \tag{9A.14}$$

This is accomplished by setting partial derivatives to zero. Setting the partial derivative with respect to M equal to zero simply produces Equation 9A.12, implying that an optimal solution must exactly consume the budget W.

Substituting Equation 9A.13 into Equation 9A.14, we have

$$L(Q_i\text{'s}, M) = \sum_{i=1}^{n}(-Q_i v_i + I_1 + I_2) - M\left[\sum_{i=1}^{n} Q_i v_i - W\right] \tag{9A.15}$$

where I_1 and I_2 are the two integrals in Equation 9A.13. The key point is that I_1 and I_2 do not involve the v_i's.

Now Equation 9A.15 can be rewritten as

$$L(Q_i\text{'s}, M) = \sum_{i=1}^{n}[-Q_i(M + 1)v_i + I_1 + I_2] + MW$$

The expression in square brackets is the same as $E[P(Q_i)]$, except that v_i is replaced by $(M + 1)v_i$. Therefore, when we set

$$\frac{\partial L}{\partial Q_i} = 0$$

we obtain the same result as the independent $Q*$ for item i except that v_i is replaced by $(M + 1)v_i$; that is, select Q_i to satisfy

$$_ip_{x<}(Q_i) = \frac{p_i - (M + 1)v_i + B_i}{p_i - g_i + B_i} \tag{9A.16}$$

A given value of M implies a value for each Q_i which in turn implies a value of

$$\sum_{i=1}^{n} Q_i v_i$$

Equation 9A.16 shows that the higher the value of M, the lower must be $_ip_{x<}(Q_i)$, that is, the lower must be Q_i.

References

Abad, P. L. 1996. Optimal pricing and lot-sizing under conditions of perishability and partial backordering. *Management Science* 42(8), 1093–1104.

Abdel-Malek, L. L. and H. Ziegler. 1988. Age dependent perishability in two-echelon serial inventory systems. *Computers and Operations Research* 15(3), 227–238.

Abillama, W. R. and M. R. Singh. 1996. *Optimal Order Quantities of Style Products.* PhD thesis, The Tuck School of Business Administration, Dartmouth College.

Aggarwal, S. P. and C. K. Jaggi. 1995. Ordering policies of deteriorating items under permissible delay in payments. *Journal of the Operational Research Society 46*, 658–662.

Agrawal, N. and M. Cohen. 2001. Optimal material control in an assembly system with component commonality. *Naval Research Logistics 48*(5), 409–429.

Albright, S. C. 1976. A Bayesian Approach to a Generalized Secretary Problem. Working Paper. Graduate School of Business, Indiana University.

Aneja, Y. P. and N. Singh. 1990. Scheduling production of common components at a single facility. *IIE Transactions 22*(3), 234–237.

Anupindi, R. and Y. Bassok. 1995. Analysis of supply contracts with total minimum commitment and flexibility. In *2nd International Symposium in Logistics*, University of Nottingham, UK.

Arcelus, F. J. and J. E. Rowcroft. 1992. All-units quantity-freight discounts with disposals. *European Journal of Operational Research 57*, 77–88.

Azoury, K. S. 1985. Bayes solution to dynamic inventory models under unknown demand distribution. *Management Science 31*(9), 1150–1160.

Bagchi, U. and G. Gutierrez. 1992. Effect of increasing component commonality on service level and holding cost. *Naval Research Logistics 39*, 815–832.

Baker, K. R. 1985. Safety stocks and component commonality. *Journal of Operations Management 6*(1), 13–22.

Baker, K. R. 1988. Scheduling the production of components at a common facility. *IIE Transactions 20*(1), 32–35.

Baker, K. R., M. J. Magazine, and H. L. W. Nuttle. 1986. The effect of commonality on safety stock in a simple inventory model. *Management Science 32*(8), 982–988.

Baker, R. C. and T. L. Urban. 1988. Single-period inventory dependent demand models. *OMEGA 16*(6), 605–607.

Balakrishnan, A., R. L. Francis, and S. J. Grotzinger. 1996. Bottleneck resource allocation in manufacturing. *Management Science 42*(11), 1611–1625.

Balkhi, Z. T. and L. Benkherouf. 1996. A production lot size inventory model for deteriorating items and arbitrary production and demand rates. *European Journal of Operational Research 92*, 302–309.

Bar-Lev, S., M. Parlar, and D. Perry. 1994. On the EOQ model with inventory-level-dependent demand rate and random yield. *Operations Research Letters 16*, 167–176.

Bassok, Y. and R. Anupindi. 1997. Analysis of supply contracts with total minimum commitment. *IIE Transactions 29*(5), 373–381.

Benkherouf, L. 1995. On an inventory model with deteriorating items and decreasing time-varying demand and shortages. *European Journal of Operational Research 86*, 293–299.

Benkherouf, L. and M. G. Mahmoud. 1996. On an inventory model with deteriorating items and increasing time-varying demand and shortages. *Journal of the Operational Research Society 47*, 188–200.

Bitran, G., E. Haas, and H. Matsuo. 1986. Production planning of style goods with high setup costs and forecast revisions. *Operations Research 34*(2), 226–236.

Bradford, J. and P. Sugrue. 1990. A Bayesian approach to the two-period style goods inventory problem with single replenishment and heterogeneous Poisson demands. *Journal of the Operational Research Society 41*(3), 211–218.

Cachon, G. P. 1999. Managing supply chain demand variability with scheduled ordering policies. *Management Science 45*(6), 843–856.

Carlson, P. C. 1982a. A system for the control of retail fashion inventories. In *Spring National Conference, Institute of Industrial Engineers*, New Orleans, pp. 505–510.

Carlson, P. C. 1982b. Fashion retailing: Orders, reorders and markdowns. In *Fall National Conference, Institute of Industrial Engineers*, Cincinnati, pp. 315–321.

Chambers, M. and R. Eglese. 1986. Use of preview exercises to forecast demand for new lines in mail order. *Journal of the Operational Research Society 37*(3), 267–273.

Chang, S. H. and D. E. Fyffe. 1971. Estimation of forecast errors for seasonal style-goods sales. *Management Science 18*(2), B89–B96.

Chiu, H. N. 1995. An approximation to the continuous review inventory model with perishable items and lead times. *European Journal of Operational Research 87*, 93–108.

Chung, K. and P. Ting. 1994. On replenishment schedule for deteriorating items with time-proportional demand. *Production Planning & Control 5*(4), 392–396.

Ciarallo, F., R. Akella, and T. Morton. 1994. A periodic review, production planning model with uncertain capacity and uncertain demand—Optimality of extended myopic policies. *Management Science 40*(3), 320–332.

Cobbaert, K. and D. Van Oudheusden. 1996. Inventory models for fast moving spare parts subject to "sudden death" obsolescence. *International Journal of Production Economics 44*, 239–248.

Cohen, M. A., W. P. Pierskalla, and R. J. Sassetti. 1983. The impact of adenine and inventory utilization decisions on blood inventory management. *Transfusion 23*(1), 54–58.

Collier, D. 1982. Aggregate safety stock levels and component part commonality. *Management Science 28*(11), 1296–1303.

Costa, D. and E. A. Silver. 1996. Exact and approximate algorithms for the multi-period procurement problem where dedicated supplier capacity can be reserved. *OR Spektrum 18*(4), 197–207.

Crowston, W. B., M. Wagner, and J. F. Williams. 1973. Economic lot size determination in multi-stage assembly systems. *Management Science 19*(5), 517–527.

David, I. and A. Mehrez. 1995. An inventory model with exogenous failures. *Operations Research 43*(3), 902–903.

Dogramaci, A. 1979. Design of common components considering implications of inventory costs and forecasting. *AIIE Transactions 11*(2), 129–135.

Eeckhoudt, L., C. Gollier, and H. Schlesinger. 1995. The risk-averse (and prudent) newsboy. *Management Science 41*(5), 786–794.

Eppen, G. D. and A. V. Iyer. 1997a. Backup agreements in fashion buying—The value of upstream flexibility. *Management Science 43*(11), 1469–1484.

Eppen, G. D. and A. V. Iyer. 1997b. Improved fashion buying using Bayesian updates. *Operations Research 45*(6), 805–819.

Erlebacher, S. J. 2000. Optimal and heuristic solutions for the multi-item newsvendor problem with a single capacity constraint. *Production and Operations Management 9*(3), 303–318.

Eynan, A. 1996. The impact of demands' correlation on the effectiveness of component commonality. *International Journal of Production Research 34*(6), 1581–1602.

Eynan, A. and M. J. Rosenblatt. 1995. Assemble to order and assemble in advance in a single-period stochastic environment. *Naval Research Logistics 42*, 861–872.

Eynan, A. and M. J. Rosenblatt. 1996. Component commonality effects on inventory costs. *IIE Transactions 28*, 93–104.

Eynan, A. and M. J. Rosenblatt. 1997. An analysis of purchasing costs as the number of products components is reduced. *Production and Operations Management 6*(4), 388–397.

Feng, Y. and G. Gallego. 1995. Optimal starting times for end-of season sales and optimal stopping times for promotional fares. *Management Science 41*(8), 1371–1391.

Fisher, M. and A. Raman. 1996. Reducing the cost of demand uncertainty through accurate response to early sales. *Operations Research 44*(1), 87–99.

Fisher, M., K. Ramdas, and K. Ulrich. 1999. Component sharing in the management of product variety: A study of automotive braking systems. *Management Science 45*(3), 297–315.

Fisher, M. L., J. H. Hammond, W. R. Obermeyer, and A. Raman. 1994a. Accurate response: The key to profiting from QR. *Bobbin February 35*, 48–62.

Fisher, M. L., J. H. Hammond, W. R. Obermeyer, and A. Raman. 1994b. Making supply meet demand in an uncertain world. *Harvard Business Review May–June 72*(3), 83–93.

Franses, P. H. 1994. Fitting a Gompertz curve. *Journal of the Operational Research Society 45*(1), 109–113.

Gallego, G. and I. Moon. 1993. The distribution free newsboy problem: Review and extensions. *Journal of the Operational Research Society 44*(8), 825–834.

Gerchak, Y. and M. Henig. 1986. An inventory model with component commonality. *Operations Research Letters 5*(3), 157–160.

Gerchak, Y. and M. Henig. 1989. Component commonality in assemble-to-order systems: Models and properties. *Naval Research Logistics 36*, 61–68.

Gerchak, Y., M. J. Magazine, and B. Gamble. 1988. Component commonality with service level requirements. *Management Science 34*(6), 753–760.

Gerchak, Y. and D. Mossman. 1992. On the effect of demand randomness on inventories and costs. *Operations Research 40*(4), 804–807.

Gerchak, Y. and Y. Wang. 1994. Periodic review inventory models with inventory-level-dependent demand. *Naval Research Logistics 41*(1), 99–116.

Gilding, D. B. and C. Lock. 1980. Determination of stock for sale promotions. *Journal of the Operational Research Society 31*(4), 311–318.

Giri, B. C., S. Pal, A. Goswami, and K. S. Chaudhuri. 1996. An inventory model for deteriorating items with stock-dependent demand rate. *European Journal of Operational Research 95*, 604–610.

Graves, S. 1982. Using Lagrangian techniques to solve hierarchical production planning problems. *Management Science 28*(3), 260–275.

Guerrero, H. H. 1985. The effect of various production strategies on product structures with commonality. *Journal of Operations Management 5*(4), 395–410.

Hariga, M. and L. Benkherouf. 1994. Optimal and heuristic inventory replenishment models for deteriorating items with exponential time-varying demand. *European Journal of Operational Research 79*, 123–137.

Harpaz, G., W. Y. Lee, and R. L. Winkler. 1982. Optimal output decisions of a competitive firm. *Management Science 28*, 589–602.

Hausman, W. 1969. Minimizing customer line items backordered in inventory control. *Management Science 15*(12), 628–634.

Hausman, W. H. 1982. On optimal repair kits under a job completion criterion. *Management Science 28*(11), 1350–1351.

Hausman, W. H. and R. Peterson. 1972. Multiproduct production scheduling for style goods with limited capacity, forecast revisions and terminal delivery. *Management Science 18*(7), 370–383.

Hertz, D. and K. Schaffir. 1960. A forecasting method for management of seasonal style-goods inventories. *Operations Research 8*(1), 45–52.

Ignall, E. and A. Veinott. 1969. Optimality of myopic inventory policies for several substitute products. *Management Science 15*(5), 284–304.

Jain, K. and E. A. Silver. 1994. Lot sizing for a product subject to obsolescence or perishability. *European Journal of Operational Research 75*, 287–295.

Jain, K. and E. A. Silver. 1995. The single-period procurement problem where dedicated supplier capacity can be reserved. *Naval Research Logistics 42*(6), 915–934.

Joglekar, P. and P. Lee. 1993. An exact formulation of inventory costs and optimal lot size in face of sudden obsolescence. *Operations Research Letters 14*, 283–290.

Jönsson, H., K. Jornsten, and E. A. Silver. 1993. Application of the scenario aggregation approach to a two-stage, stochastic, common component, inventory problem with a budget constraint. *European Journal of Operational Research 68*(2), 196–211.

Jönsson, H. and E. A. Silver. 1989a. Common component inventory problems with a budget constraint: Heuristics and upper bounds. *Engineering Costs and Production Economics 18*, 71–81.

Jönsson, H. and E. A. Silver. 1989b. Optimal and heuristic solutions for a simple common component inventory problem. *Engineering Costs and Production Economics 16*, 257–267.

Jönsson, H. and E. A. Silver. 1996. Some insights regarding selecting sets of scenarios in combinatorial stochastic problems. *International Journal of Production Economics 45*(1), 463–472.

Kalpakam, S. and K. P. Sapna. 1994. Continuous review (s, S) inventory system with random lifetimes and positive leadtimes. *Operations Research Letters 16*, 115–119.

Kalpakam, S. and K. P. Sapna. 1996. A lost sales $(S-1, S)$ perishable inventory system with renewal demand. *Naval Research Logistics 43*, 129–142.

Khouja, M. 1995. The newsboy problem under progressive multiple discounts. *European Journal of Operational Research 84*, 458–466.

Khouja, M. 1996. The newsboy problem with multiple discounts offered by suppliers and retailers. *Decision Sciences 27*(3), 589–599.

Kim, D., D. Morrison, and C. Tang. 1995. A tactical model for airing new seasonal products. *European Journal of Operational Research 84*, 240–264.

Kim, D. H. 1995. A heuristic for replenishment of deteriorating items with a linear trend in demand. *International Journal of Production Economics 39*, 265–270.

Kurawarwala, A. A. and H. Matsuo. 1996. Forecasting and inventory management of short life cycle products. *Operations Research 44*(1), 131–150.

Kurawarwala, A. A. and H. Matsuo. 1998. Product growth models for medium-term forecasting of short life cycle products. *Technological Forecasting and Social Change 57*(3), 169–196.

Lariviere, M. A. and E. L. Porteus. 1999. Stalking information: Bayesian inventory management with unobserved lost sales. *Management Science 45*(3), 346–363.

Lau, A. H.-L. and H.-S. Lau. 1988. The newsboy problem with price-dependent demand distribution. *IIE Transactions 20*(2), 168–175.

Lau, H. 1980. The newsboy problem under alternative optimization objectives. *Journal of the Operational Research Society 31*, 525–535.

Lau, H. 1997. Simple formulas for the expected costs in the newsboy problem: An educational note. *European Journal of Operational Research 100*, 557–561.

Lau, H. and A. Lau. 1994. The multi-product multi-constraint newsboy problem: Applications, formulation and solution. *Journal of Operations Management 13*(2), 153–162.

Lau, H. and A. Lau. 1996a. Estimating the demand distributions of single-period items having frequent stockouts. *European Journal of Operational Research 92*, 254–265.

Lau, H.-S. and A. H.-L. Lau. 1996b. The newsstand problem: A capacitated multiple-product single-period inventory problem. *European Journal of Operational Research 94*(1), 29–42.

Lee, H., C. Billington, and B. Carter. 1993. Hewlett-Packard gains control of inventory and service through design for localization. *Interfaces 23*(4), 1–11.

Lee, H. L. and C. S. Tang. 1997. Modelling the costs and benefits of delayed product differentiation. *Management Science 43*(1), 40–53.

Lee, H. L. and C. S. Tang. 1998. Variability reduction through operations reversal. *Management Science 44*(2), 162–172.

Li, J., H. Lau, and A. H. Lau. 1990. Some analytical results for a two-product newsboy problem. *Decision Sciences 21*, 710–726.

Li, J., H. S. Lau, and A. H. L. Lau. 1991. A two-product newsboy problem with satisficing objective and independent exponential demands. *IIE Transactions 23*(1), 29–39.

Liao, C. and W. Yang. 1994. An inventory system with multiple replenishment scheduling. *Operations Research Letters 15*, 213–222.

Lovejoy, W. S. 1990. Myopic policies for some inventory models with uncertain demand distributions. *Management Science 36*(6), 724–738.

Mamer, J. and A. Shogan. 1987. A constrained capital budgeting problem with application to repair kit selection. *Management Science 33*, 800–806.

Mamer, J. and S. A. Smith. 1982. Optimizing field repair kits based on job completion rate. *Management Science 26*(11), 1328–1333.

Mamer, J. W. and S. A. Smith. 1985. Job completion based inventory systems: Optimal policies for repair kits and spare machines. *Management Science 31*(6), 703–718.

Masters, J. M. 1991. A note on the effect of sudden obsolescence on the optimal lot size. *Decision Sciences 22*(5), 1180–1186.

Matsuo, H. 1990. A stochastic sequencing problem for style goods with forecast revisions and hierarchical structure. *Management Science 36*(3), 332–347.

McClain, J., W. Maxwell, J. Muckstadt, L. J. Thomas, and E. Weiss. 1984. Comment on "aggregate safety stock levels and component part commonality". *Management Science 30*(6), 772–773.

Moon, I. and S. Choi. 1995. The distribution free newsboy problem with balking. *Journal of the Operational Research Society 46*, 537–542.

Moon, I. and S. Choi. 1997. The distribution free procedures for make-to-order, make-to-stock, and composite policies. *International Journal of Production Economics 47*(1), 21–28.

Moon, I., E. A. Silver et al. 2000. The multi-item newsvendor problem with a budget constraint and fixed ordering costs. *Journal of the Operational Research Society 51*(5), 602–608.

Murray, G. R. and E. A. Silver. 1966. A Bayesian analysis of the style goods inventory problem. *Management Science 12*(11), 785–797.

Nahmias, S. 1982. Perishable inventory theory: A review. *Operations Research 30*(4), 680–708.

Nahmias, S. and C. P. Schmidt. 1984. An efficient heuristic for the multi-item newsboy problem with a single constraint. *Naval Research Logistics 31*, 463–474.

Padmanabhan, G. and P. Vrat. 1995. EOQ models for perishable items under stock dependent selling rate. *European Journal of Operational Research 86*, 281–292.

Pierskalla, W. P. and C. Roach. 1972. Optimal issuing policies for perishable inventory. *Management Science 18*(11), 603–614.

Porteus, E. L. 1990. The impact of inspection delay on process and inspection lot sizing. *Management Science 36*(8), 999–1007.

Raafat, F. 1991. Survey of literature on continuously deteriorating inventory models. *Journal of the Operational Research Society 42*(1), 27–37.

Ramdas, K. 2003. Managing product variety; an integrative review and research directions. *Production and Operations Management 12*(1), 79–101.

Randall, T. and K. Ulrich. 2001. Product variety, supply chain structure, and firm performance: Analysis of the U.S. bicycle industry. *Management Science 47*(12), 1588–1604.

Ravichandran, N. 1995. Stochastic analysis of a continuous review perishable inventory system with positive lead time and Poisson demand. *European Journal of Operational Research 84*, 444–457.

Riter, C. 1967. The merchandising decision under uncertainty. *Journal of Marketing 31*, 44–47.

Rosenfield, D. 1989. Disposal of excess inventory. *Operations Research 37*(3), 404–409.

Sankarasubramanian, E. and S. Kumaraswamy. 1983. Optimal ordering quantity for pre-determined level of profit. *Management Science 29*, 512–514.

Scarf, H. 1958. A min-max solution of an inventory problem. In K. Arrow, S. Karlin, and H. Scarf (Eds.), *Studies in the Mathematical Theory of Inventory and Production*, pp. 201–209. California: Stanford University Press.

Schmidt, C. P. and S. Nahmias. 1985. Optimal policy for a two-stage assembly system under random demand. *Operations Research 33*(5), 1130–1145.

Scudder, G. and S. T. March. 1984. On optimizing field repair kits based on job completion rate. *Management Science 30*(8), 1025–1028.

Smith, S., J. Chambers, and E. Shlifer. 1980. Optimal inventories based on job completion rate for repairs requiring multiple items. *Management Science 26*(8), 849–852.

So, K. C. and J.-S. Song. 1998. Price, delivery time guarantees and capacity selection. *European Journal of Operational Research 111*(1), 28–49.

Song, J.-S. and P. H. Zipkin. 1990. Managing inventory with the prospect of obsolescence. *Operations Research 44*(1), 215–222.

Srinivasan, R., R. Jayaraman, R. Roundy, and S. Tayur. 1992. *Procurement of common components in a stochastic environment*. IBM Thomas J. Watson Research Division: RC-18580,12.

Swaminathan, J. and S. R. Tayur. 1998. Managing broader product lines through delayed differentiation using vanilla boxes. *Management Science 44*(12), S161–S172.

Tsay, A. A. 1999. The quantity flexibility contract and supplier-customer incentives. *Management Science 45*(10), 1339–1358.

Tsubone, H., H. Matsuura, and S. Satoh. 1994. Component part commonality and process flexibility effects on manufacturing performance. *International Journal of Production Research 32*(10), 2479–2493.

Urban, T. L. 1995. Inventory models with the demand rate dependent on stock and shortage levels. *International Journal of Production Economics 40*, 21–28.

Vakharia, A. J., D. A. Parmenter, and S. M. Sanchez. 1996. The operating impact of parts commonality. *Journal of Operations Management 14*(1), 3–18.

Vaughn, T. S. 1994. A model of the perishable inventory system with reference to consumer-realized product expiration. *Journal of the Operational Research Society* 45(5), 519–528.

Wadsworth, G. P. 1959. Probability. In *Notes on Operations Research*, pp. 26–29. Cambridge, MA: Technology Press.

Wolfe, H. B. 1968. A model for control of style merchandise. *Industrial Management Review* 9(2), 69–82.

Zhang, A. X. 1997. Demand fulfillment rates in an assemble-to-order system with multiple products and dependent demands. *Production and Operations Management* 6(3), 309–324.

MANAGING INVENTORY ACROSS MULTIPLE LOCATIONS AND MULTIPLE FIRMS

IV

In Sections II and III, we have been primarily concerned with the control of individual items at a single location. With just a few exceptions, which addressed the aggregate consequences of a particular decision rule, the rules and systems developed have treated the individual items and locations in isolation. In Section IV, we now explicitly consider coordinated control of both items and locations, generally in nonproduction situations. There are three important contexts where such coordination often leads to substantial cost savings. First, in Chapter 10, we deal with a single stocking point (as in Chapters 4 through 9) but now consider coordination of replenishments of a group of items to reduce costs. Such coordination makes sense if one of the following conditions occurs:

1. Several items are purchased from the same supplier.
2. Several items share the same mode of transportation.
3. Several items are produced on the same piece of equipment.

The bulk of this chapter focuses on nonproduction situations, but we do provide a substantial treatment of one particular production situation.

Chapter 11 deals with the second context where coordination can lead to appreciable cost savings—multiechelon inventory within the same firm. Material is stored at more than one geographic location, and one must account for the possibility that the supplying location is short of stock when an order is placed. It is clear that decisions concerning the same item at different locations should not be made independently.

Chapter 12 addresses the third context—managing inventories across the supply chain. This topic shares many of the complexities of the multiechelon situation, except that we are now also dealing with the coordination of multiple firms across the supply chain.

Because of these coordination issues, exact analyses are an order of magnitude more complex than those of earlier chapters. Therefore, we are frequently forced to resort to heuristic decision rules. In Chapters 13 through 16, we address coordination issues within production environments.

Chapter 10

Coordinated Replenishments at a Single Stocking Point

In this chapter, we address an important situation in which it makes sense to coordinate the control of different stock-keeping units. Specifically, the items are all stocked at the same location and they share a common supplier or mode of transportation. Alternatively, they share a common production facility. In the latter case, there is the added complexity of a finite productive capacity to be shared each period. Conceptually, the items to be coordinated could be the same item at different, parallel locations. We emphasize the word *parallel* because coordination in a multiechelon (serial) situation (one location feeding stock into another) is quite different. The latter situation will be addressed in Chapter 11.

As we will see in Section 10.1, there are several possible advantages in coordinating the replenishments or production of a group of items. The potential disadvantages will also be discussed. In Section 10.2, for the case of deterministic demand, the economic order quantity analysis of Chapter 4 (which there assumed independent control of items) is extended to the situation where there is a major fixed (setup) cost associated with a replenishment of a family of coordinated items and a minor fixed (setup) cost for each item involved in the particular replenishment. Section 10.3 extends the quantity discount arguments of Chapter 4 to the case where the discount is based on the magnitude of the total replenishment of a group of items—for example, a freight rate reduction if a carload size replenishment is achieved. Then, in Section 10.4, we discuss the complexity of the situation in which items are to be coordinated but demand is probabilistic and there are no quantity discounts. In Section 10.5, we deal with the situation where group discounts are of paramount importance so that, when a group order is placed, we normally strive to achieve a certain overall size (or dollar value) of order. Section 10.6 moves on to the production environment in which there may be a capacity constraint. In this section, we discuss both deterministic and probabilistic demand, as well as constant and time-varying demand. It should be noted that in the literature the terminology *joint replenishment* is sometimes used in lieu of *coordinated replenishment*. Finally, in Section 10.7, we briefly discuss the related problem of shipment consolidation.

10.1 Advantages and Disadvantages of Coordination

There are a number of reasons for coordinating items when making replenishment decisions. These include the following:

1. *Savings on unit purchase costs:* When a group of items is ordered from the same vendor, a quantity discount in purchase cost may be realized if the total order is greater than some breakpoint quantity. It may be uneconomical to order this much of a single item, but it could certainly make sense to coordinate several items to achieve a total order size as large as the breakpoint. For example, a distributor incurs a fixed cost for negotiating a replenishment order from her supplier. Adding a line item to the order, however, incurs a smaller fixed cost. In some cases, a vendor-imposed minimum order quantity may dictate the same sort of joint consideration. One book publisher is required by its printer to order at least four products together. And the printer produces in only three quantities: 5,000, 10,000, or 15,000.
2. *Savings on unit transportation costs:* The discussion is basically the same as above. Now, grouping individual item orders may be advisable to achieve a quantity such as a carload or a container load on a ship. Malden Mills was a textile firm that invented Polar Fleece, a unique fabric used in outdoor jackets and other items. One of our students examined shipping policies and costs as a summer project. It turns out that, because of significant discounts, shipments from their Massachusetts factory to Europe were always created to completely fill a shipping container for the ocean voyage.
3. *Savings on ordering costs:* In some cases, where the fixed (setup) cost of placing a replenishment order is high, it might make sense to put several items on a single order to reduce the annual total of these fixed costs. This is particularly relevant where the replenishment is by in-house production. In such a case, the major component of the fixed ordering cost is the manufacturing setup cost. An illustration would be bottling beer products. There are major changeover costs in converting the production line from one quality of beer to another. In contrast, the costs of changing from one container type to another are rather minor.
4. *Ease of scheduling:* Coordinated handling of a vendor group can facilitate scheduling of buyer time, receiving and inspection workload, and so forth. In fact, we have found that, by and large, managers and purchasing agents alike tend to think and deal in terms of vendors or suppliers rather than individual stock-keeping units.

On the other hand, there are possible disadvantages of using coordinated replenishment procedures. These include the following:

1. *A possible increase in the average inventory level:* When items are coordinated, some will be reordered earlier than if they were treated independently.
2. *An increase in system control costs:* By the very nature of the problem, coordinated control is more complex than independent control of individual items. Therefore, under coordinated control, review costs, computation costs, and so on are likely to be higher.
3. *Reduced flexibility:* Not being able to work with items independently reduces flexibility in dealing with unusual situations. One possible result is reduced stability of customer service on an individual-item basis.

10.2 Deterministic Case: Selection of Replenishment Quantities in a Family of Items

In this section, we consider the case where there is a family of coordinated items defined in the following manner. There is a major setup cost (A) associated with a replenishment of the family. In the procurement context, this is the fixed (or header) cost of placing an order, independent of the number of distinct items involved in the order. In the production environment, this is the changeover cost associated with converting the facility from the production of some other family to production within the family of interest. Then there is a minor setup cost (a_i) associated with including item i in a replenishment of the family. In the procurement context, a_i is often called the line cost, the cost of adding one more item or line to the requisition. From a production standpoint, a_i represents the relatively minor cost of switching to production of item i from production of some other item within the same family. In general, we are concerned with items sharing capacity, such as a container on a ship or a piece of production equipment, or with items sharing costs, such as a major fixed ordering cost. For a review of the literature on deterministic and probabilistic models, see Goyal and Satir (1989); Van Eijs (1993), and Khouja and Goyal (2008).[*]

10.2.1 Assumptions

All of the assumptions behind the derivation of the economic order quantity (Section 4.1 of Chapter 4) are retained except that now coordination of items is allowed in an effort to reduce setup costs. We review the assumptions:

1. The demand rate of each item is constant and deterministic (this deterministic assumption will be relaxed in Sections 10.4 through 10.6).[†]
2. The replenishment quantity of an item need not be an integral number of units (the extension to integral units is as discussed for a single item in Chapter 4).
3. The unit variable cost of any of the items does not depend on the quantity; in particular, there are no discounts in either the unit purchase cost or the unit transportation cost (we will relax this assumption in Section 10.3).
4. The replenishment lead time is of zero duration, but the extension to a fixed, known, nonzero duration, independent of the magnitude of the replenishment, is straightforward.
5. No shortages are allowed (Sections 10.4 through 10.6 will deal with the situation where shortages can occur).
6. The entire order quantity is delivered at the same time.

---

[*] For a related problem of finding groups of raw materials that should be ordered together, see Rosenblatt and Finger (1983); Chakravarty (1984); Rosenblatt and Kaspi (1985); Chakravarty and Martin (1988); Van Eijs et al. (1992); Hong and Hayya (1992); and Dekker et al. (1997).

[†] The situation of time-varying but deterministic demand patterns with no capacity constraint (as was the case in Chapter 5) has been analyzed by Kao (1979) See Section 10.6.2 for the case with a capacity constraint.

10.2.2 Decision Rule

In the independent EOQ analysis of Chapter 4, we showed that the EOQ expressed as a time supply was given by

$$T_{\text{EOQ}} = \sqrt{\frac{2A}{Dvr}}$$

Note that T_{EOQ} increases as the ratio A/Dv increases. Within our present context, this indicates that an item i with a high setup cost a_i and a low dollar usage rate $D_i v_i$ should probably be replenished less frequently (higher time supply) than an item j having a low a_j and a high $D_j v_j$. Because of the assumptions of deterministic demand, no shortages permitted, and instantaneous delivery, it makes sense to include an item in a replenishment only when its inventory drops to the zero level. Therefore, a reasonable type of policy to consider is the use of a time interval (T) between replenishments of the family and a set of m_i's, where m_i, an integer, is the number of T intervals that the replenishment quantity of item i will last. That is, item i will be included in every m_ith replenishment of the family.[*] For example, if $m_{17} = 3$, item 17 will only be included in every third replenishment of the family, with a replenishment quantity sufficient to last a time interval of duration $3T$. Each time it will be replenished just as its stock hits the zero level. We wish to select the values of T and the m_i's to keep the total relevant costs as low as possible.

Figure 10.1 contains a more detailed example. Items 1 and 3 are ordered every time an order is placed (i.e., m_1 and $m_3 = 1$). Item 2 is ordered every other time, so that $m_2 = 2$.

As shown in the Appendix of this chapter, the integer m_i's must be selected to minimize

$$\left(A + \sum_{i=1}^{n} \frac{a_i}{m_i} \right) \sum_{i=1}^{n} m_i D_i v_i \tag{10.1}$$

where

A = major setup cost for the family, in dollars
a_i = minor setup cost for item i, in dollars
D_i = demand rate of item i, in units/unit time
v_i = unit variable cost of item i, in \$/unit
n = number of items in the family (the items are numbered $1, 2, 3, \ldots, n-1, n$)
m_i = the integer number of T intervals that the replenishment quantity of item i will last

Also, once the best m_i's are known, the corresponding value of T is given by

$$T^*(m_1, \ldots, m_n) = \sqrt{\frac{2(A + \sum a_i/m_i)}{r \sum m_i D_i v_i}} \tag{10.2}$$

---

[*] Jackson et al. (1985) have developed an efficient procedure for the more restrictive situation where $m_i = 1, 2, 4, 8$, etc. (i.e., a multiple of 2) and T itself is also such a multiple of some basic time period (such as a day or a week). This type of restriction is commonly found in practice. Moreover, Jackson et al. (1985) have shown that their solution is, at most, 6% costlier than the "optimal" solution to the problem. We do not present the details here since their solution is more difficult conceptually than is ours. See also Roundy (1985, 1986); Atkins and Iyogun (1987); Jackson (1988); Zoller (1990); Atkins (1991); Atkins and Sun (1995), and Section 10.4 where we return to the "powers-of-two" solution. Klein and Ventura (1995) present a simple procedure for finding the optimal cycle times if T is restricted to take on one of a set of possible discrete values.

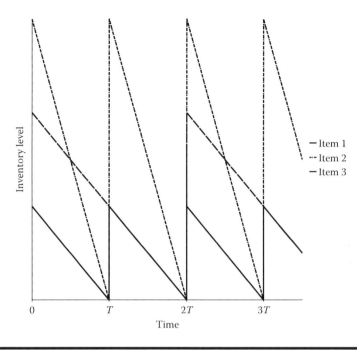

Figure 10.1 Behavior of inventory level with time.

It is worth noting that r does not appear in Equation 10.1; that is, the best values of the m_i's do not depend on the carrying charge.

Goyal (1974) has proposed a search procedure for finding the best set of m_i's. (See related research by Naddor (1975); Chakravarty (1985); Bastian (1986); Goyal (1988); Kaspi and Rosenblatt (1991); Goyal and Deshmukh (1993); Van Eijs (1993); Hariga (1994); and Viswanathan and Goyal (1997). Khouja and Goyal (2008) provides an excellent overview of this early as well as discussing more recent work.) We suggest, instead, the use of the following much simpler (noniterative) procedure which is derived in the Appendix of this chapter:

Step 1 Number the items such that

$$\frac{a_i}{D_i v_i}$$

is smallest for item 1. Set $m_1 = 1$.

Step 2 Evaluate

$$m_i = \sqrt{\frac{a_i}{D_i v_i} \frac{D_1 v_1}{A + a_1}} \tag{10.3}$$

rounded to the nearest integer greater than zero.

Step 3 Evaluate T^* using the m_i's of Step 2 in Equation 10.2.

Step 4 Determine

$$Q_i v_i = m_i D_i v_i T^* \quad i = 1, 2, \dots, n \tag{10.4}$$

In numerous tests (see Silver 1975), this procedure has produced results at or near the best possible solution. Note, however, that the results are not always optimal because the m_i values are rounded.

10.2.2.1 Numerical Illustration

A plant uses the same machinery to place three different liquids into assorted container sizes. One of these liquids (known as product XSC) is packaged in four different sized containers: 1, 5, 10, and 50 liters. The major setup cost to convert to this type of liquid is estimated to be $40. It costs $15 to switch from one container size to another. The demand rates of this family of items are shown in Table 10.1. The carrying charge has been established as 0.24 $/$/year. Moreover, there is substantial excess capacity in the packaging operation, so it is reasonable to ignore capacity considerations.

Step 2 of the decision rule gives

$$m_2 = \sqrt{\frac{15}{12,500} \frac{86,000}{55}} = 1.37 \rightarrow 1$$

$$m_3 = \sqrt{\frac{15}{1,400} \frac{86,000}{55}} = 4.09 \rightarrow 4$$

$$m_4 = \sqrt{\frac{15}{3,000} \frac{86,000}{55}} = 2.80 \rightarrow 3$$

These entries are shown in Table 10.1.

Step 3 gives

$$T^* = \sqrt{\frac{2(40 + 15/1 + 15/1 + 15/4 + 15/3)}{0.24[1(86,000) + 1(12,500) + 4(1,400) + 3(3,000)]}}$$

$$= 0.0762 \text{ year} \approx 4 \text{ weeks}$$

With Step 4, we obtain the run quantities

$Q_1 v_1 = (1)(86,000)(0.0762) = \$6,550$ (run quantity in dollars of 10-liter container)
$Q_2 v_2 = (1)(12,500)(0.0762) \approx \$950$

Table 10.1 The XSC Family of Items Problem

Item i	1[a]	2	3	4
Description	10 Liters	1 Liter	5 Liters	50 Liters
$D_i v_i$($/year)	86,000	12,500	1,400	3,000
m_i	1	1	4	3
$T^* = 0.0762$ year				
$Q_i v_i$($)	6,550	950	430	690

Note: $A = \$40$, $a_i = \$15$ (independent of i), $r = 0.24\$/$/year.

[a] As discussed in the decision logic, the items have been numbered such that item 1 has the smallest value of $a_i/D_i v_i$. (Here, the largest value of $D_i v_i$ because all the a_i's are the same.)

$$Q_3 v_3 = (4)(1{,}400)(0.0762) \approx \$430$$
$$Q_4 v_4 = (3)(3{,}000)(0.0762) \approx \$690$$

To convert these run quantities to units, we would have to divide by the v_i values.

10.2.3 A Bound on the Cost Penalty of the Heuristic Solution

By definition, a heuristic solution is not guaranteed to give a solution with a low cost penalty. Fortunately for this problem, it is possible, as shown in the Appendix of this chapter, to find a simple lower bound for the cost of the best $(T, m_i\text{'s})$ policy, namely,

$$\text{TRC}_{\text{bound}} = \sqrt{2(A + a_1)D_1 v_1 r} + \sum_{j=2}^{n} \sqrt{2a_j D_j v_j r} \tag{10.5}$$

There is an interesting interpretation of the bound when we remember that an optimal policy must have item 1 included in every replenishment of the family. The first term is the total relevant cost per unit time of an EOQ strategy (see Equation 4.5 of Chapter 4) for item 1 considered alone, if we associate the full cost $A + a_1$ with each replenishment of item 1. The second term represents a summation of the total relevant costs per unit time of an EOQ strategy of each of the other items, where a cost of only a_j is associated with a replenishment of item $j(j \neq 1)$.

For our numerical illustration, the value of this bound works out to be $2,054.15/year. As shown in the Appendix of this chapter, the cost of any $(T, m_i\text{'s})$ solution is given by

$$\text{TRC}(T, m_i's) = \frac{A + \sum_{i=1}^{n} a_i/m_i}{T} + \sum_{i=1}^{n} \frac{D_i m_i T v_i r}{2} \tag{10.6}$$

The solution developed by the heuristic has a cost of $2,067.65/year (evaluated by substituting the *integer* m_i values, found by the heuristic, into Equation 10.6). The bound clearly indicates that the heuristic solution is very close to (if not right at) the optimum for the particular example.

10.3 Deterministic Case with Group Discounts

As mentioned earlier, unit price or freight rate discounts may be offered on the total dollar value or the total volume of a replenishment made up of several items (e.g., the use of containers for shipments from Massachusetts to Europe). The inventory control manager would like to ascertain when to take advantage of such a discount. Therefore, knowing the discount structure, the major setup cost (A) of a group of items, the item characteristics (a_i's, D_i's, and v_i's), and the carrying charge (r), we want to develop decision logic for selecting the appropriate individual order quantities. The sum of the individual quantities totals a group quantity that determines whether or not a particular discount is achieved. We mention that sometimes discount structures are designed for much smaller customers than the organization under consideration. In such cases, discounts can be achieved by individual-item orders so that the more complex coordinated control is not necessarily needed.

As in the case of a single item, treated in Chapter 4, taking advantage of an offered discount reduces the replenishment costs (both the fixed and unit costs) but increases the inventory carrying costs. At first glance, it would appear that the analysis for multiple items should not be much more difficult than that for a single item. This would be the case if every item was included in every replenishment. However, we know from the preceding section that an $m_i > 1$ is likely for an item having a high value of $a_i/D_i v_i$; that is, such items should not necessarily be included in every replenishment, even to help achieve a quantity discount level. Therefore, it is conceivable that the best strategy might be one where on certain replenishments a discount was achieved, while on others it was not (even though all demand rates are assumed known and at constant levels). The analysis of such a strategy would be quite complex because the replenishment cycles would no longer all be of the same duration, T (the ones where quantity discounts were achieved would be longer than the others). Rather than attempting to explicitly model such complex possibilities, we suggest the following reasonable compromise solution (illustrated for the case of a single possible discount based on the total replenishment size, in units).

Our approach parallels that used in Section 4.5 of Chapter 4. We consider three possible solutions. The first (and if it is feasible, it is the best to use) is where a coordinated analysis, assuming a quantity discount, leads to total replenishment quantities that are always sufficient to achieve the discount. The second case is where the best result is achieved right at the breakpoint. Finally, the third possibility is the coordinated solution without a quantity discount. Therefore, first, the method of Section 10.2 is used to ascertain the m_i's and T, assuming that the discount is achieved. The *smallest*[*] replenishment quantity is computed and compared with the breakpoint quantity (Q_b) required. If it exceeds Q_b, we use the m_i's and T developed. If it is less than Q_b, then we must compare the cost of the best solution without a discount to the cost of the solution where the smallest replenishment quantity is right at the breakpoint. Whichever of these has the lower cost is then the solution to use. The steps are presented using a discount based on the total number of units ordered, while the numerical illustration that follows uses a dollar value discount. The details are as follows:

Step 1 Compute the m_i's and T as in Section 10.2 but assuming that each

$$v_i = v_{0i}(1 - d)$$

where
   v_{0i} = basic unit cost of item i without a discount
   d = fractional discount when the total replenishment equals or exceeds the breakpoint quantity Q_b

It should be noted that the set of m_i's does not depend on the size of the discount as long as the unit cost of each item is reduced by the same percentage discount (which is typically the

---

[*] What we are assuming here is that if a discount is to be achieved, it must be achieved by every replenishment; that is, we are ignoring the more complex possibility of achieving a discount on only some of the replenishments. The smallest replenishments are those where only the items with $m_i = 1$ are included. Lu (1995) develops a heuristic method for finding the order quantities when it is possible to take the discount on some replenishments and not on others. They find that this additional flexibility can yield significant savings. See also Russell and Krajewski (1992); Sadrian and Yoon (1992, 1994); Katz et al. (1994); and Xu et al. (2000).

case). This can be seen from Equation 10.3, where, if we use $v_i = v_{0i}(1 - d)$ for all i, the $(1 - d)$ terms in the numerator and denominator cancel.

Compute the size of the smallest family replenishment

$$Q_{sm} = \text{(Summation of order quantities of all items having } m_i = 1) \tag{10.7}$$

If $Q_{sm} \geq Q_b$, use the m_i's, T, and Q_i's found above. If not, proceed to Step 2.

Step 2 Scale up the family cycle time T (found in Step 1) until the smallest replenishment size equals the quantity breakpoint. This is achieved at

$$T_b = \frac{Q_b}{\text{Summation of } D_i \text{ of all items having } m_i = 1} \tag{10.8}$$

The m_i's found in Step 1 are maintained. The cost of this breakpoint solution is evaluated using the following total relevant cost expression:

$$\text{TRC}(T_b, m_i's) = (1 - d) \sum_{i=1}^{n} D_i v_{0i} + \frac{A + \sum_{i=1}^{n} \frac{a_i}{m_i}}{T_b} + \frac{r(1 - d)T_b}{2} \sum_{i=1}^{n} m_i D_i v_{0i} \tag{10.9}$$

Step 3 Use the procedure of Section 10.2 to find the m_i's, T, and Q_i's without a discount (as mentioned earlier, the best m_i's here must be the same as those found in Step 1). As shown in the Appendix of this chapter, the total relevant costs of this solution are given by a somewhat simpler expression than Equation 10.9, namely,

$$\text{TRC(best } T \text{ and } m_i's) = \sum_{i=1}^{n} D_i v_{0i} + \sqrt{2 \left(A + \sum_{i=1}^{n} \frac{a_i}{m_i} \right) r \sum_{i=1}^{n} m_i D_i v_{0i}} \tag{10.10}$$

Step 4 Compare the TRC values found in Steps 2 and 3 and use the m_i's, T, and Q_i's associated with the lower of these. If the solution right at the breakpoint is used, care must be taken (because of the likely required integer nature of the Q's) to ensure that the breakpoint is actually achieved. Adjustment of one or more Q values to the next higher integers may be necessary.

Prior to showing a numerical illustration of the above procedure, we make a few remarks related to the work of Schneider (1982), which was developed in the context of the food distribution industry. Specifically, he incorporates the idea of a modular order quantity (MOQ) for each item. Then, anytime the item is ordered, the size of the order must be an integer multiple of MOQ. Furthermore, recognizing that demand is probabilistic, he presents an approximation for the average inventory level of each item as a function of MOQ, the item's demand rate, and the vendor order quantity (the breakpoint). This permits the selection of the best MOQ for each possible breakpoint so that the various breakpoint solutions can be costed out. Other research in this area includes Chakravarty (1985).

10.3.1 Numerical Illustration

Consider three parts, EDS-031, EDS-032, and EDS-033, used in the assembly of a stabilization processor. These three products are purchased from the same domestic supplier who offers a discount of 5% if the value of the total replenishment quantity is at least $600. We use $1.50 as the basic cost of placing an order with one item involved. The inclusion of each additional item costs $0.50. It therefore follows that

$$A = \$1.50 - \$0.50 = \$1.00$$

and

$$a_i = \$0.50 \quad \text{for all } i$$

An r value of 0.24 \$/\$/year is to be used and the D_i's and v_{0i}'s are shown in columns 1 and 2 of Table 10.2.

Following the above procedure, Step 1 produces the results shown in columns 4 through 6 of Table 10.2. We can see from Equation 10.7 that

$$Q_{\text{sm}} \text{ in dollars} = 275 + 110 = \$385$$

which is below the quantity breakpoint of $600. Therefore, we proceed to Step 2. The use of Equation 10.8, suitably modified because the breakpoint is expressed in dollars, gives

$$T_b = \frac{\$600}{\$6,000/\text{year} + \$2,400/\text{year}} = 0.07143\,\text{year}$$

The cost of the breakpoint solution is from Equation 10.9:

$$
\begin{aligned}
\text{TRC}(T_b, m_i's) = {}& 0.95(8,470) + \frac{1.00 + 0.50 + 0.50 + 0.10}{0.07143} \\
& + \frac{0.24(0.95)}{2}(0.07143)(6,000 + 2,400 + 350) \\
= {}& 8,047 + 29 + 71 \\
= {}& \$8,147/\text{year}
\end{aligned}
$$

In Step 3, we know that the set of best m_i's is $m_1 = m_2 = 1$ and $m_3 = 5$ (from Step 1). Using Equation 10.10, we have

$$
\begin{aligned}
\text{TRC}(\text{best } T \text{ and } m_i's) &= 8,470 + \sqrt{2(2.10)0.24(8,750)} \\
&= 8,470 + 94 \\
&= \$8,564/\text{year} > \$8,147/\text{year}
\end{aligned}
$$

Therefore, we should use the solution of Step 2—which assures that the smallest replenishment quantity is at the breakpoint. This is achieved by using[*]

---

[*] The T and Q values have been adjusted slightly upward to ensure that $Q_1 v_{01} + Q_2 v_{02}$ is at least as large as the breakpoint value of $600.

Table 10.2 Numerical Example with Group Quantity Discount

		D_i (Units/Year) (1)	v_{0i} ($/Unit) (2)	$D_i v_{0i}$ ($/Year) (3)	m_i (4)	T (Year) (5)	$Q_i v_{0i}$ (\$) (6)	
Item i	Code							
1	EDS-001	12,000	0.5	6,000	1		275	
2	EDS-002	8,000	0.3	2,400	1	0.0459	110	}385
3	EDS-003	700	0.1	70	5		16	
				8,470				

Results of Step 1

385 < 600   Therefore go to Step 2
Final solution
$T = 0.0715$ year

Item i	m_i	Q_i (Units)	$Q_i v_i$ ($)	
1	1	858	429.00	
2	1	572	171.60	}$600.60
3	5	250	25.00	

Note: $A = \$1$, $a_i = a = \$0.50$, $r = 0.24$ \$/\$/year.

$T = 0.0715$ year
$Q_1 = 858$ units $(D_1 T)$ every group replenishment
$Q_2 = 572$ units $(D_2 T)$ every group replenishment
$Q_3 = 250$ units $(5D_3 T)$ every fifth group replenishment

We can see that a sizable saving of (8,564–8,147) or $417/year (around 4.9%) is achieved by taking the discount.

10.4 Case of Probabilistic Demand and No Quantity Discounts

Probabilistic demand greatly complicates the decision problem in a coordinated control context. Several questions must be answered:

1. How often do we review the status of items? (That is, the choice of the review interval, R.)
2. When do we reorder the group?
3. How much do we order? (In particular, whether or not to achieve a discount quantity on a particular order.)
4. How do we allocate the total order among the individual items?

Because of the coordination, it no longer follows that an item will always be at its reorder point when it is included in a replenishment. Usually, it will be above in that some other item in the family will have triggered the order. This complicates matters in two ways. First, it is now more difficult to ascertain the average inventory level of an item. Second, and more serious, the service implications of any particular reorder point are much more difficult to evaluate than in the case of individual-item control.

In the next two sections, we present an overview of two important types of coordinated control systems under probabilistic demand, particularly emphasizing the second approach.

10.4.1 (S, c, s), or Can-Order, Systems

Can-order systems are specifically geared to the situation where savings in setup costs are of primary concern (e.g., where several products are run on the same piece of equipment) as opposed to achieving a specified total replenishment size (for quantity discount purposes). Balintfy (1964) was the first to propose the use of an (S, c, s) system, a special type of continuous review system for controlling coordinated items. In such a system, whenever item i's inventory position drops to or below s_i (called its must-order point), it triggers a replenishment action that raises item i's level to its order-up-to-level S_i. At the same time, any other item j (within the associated family) with its inventory position at or below its can-order point c_j is included in the replenishment. If item j is included, a quantity is ordered sufficient to raise its level to S_j. The idea of having a can-order point is to allow an item j, whose inventory position is low enough (at c_j or lower), to be included in the order triggered by item i, thus eliminating an extra major setup cost that would likely occur in the near future when item j reaches its must-order point. On the other hand, inclusion of item j in the order is not worthwhile if its inventory position is high enough—above its can-order point (i.e., above c_j). The behavior of a typical item under such a system of control is shown in Figure 10.2.

For additional information on can-order systems, including how to establish appropriate values of the s, c, and S, see Ignall (1969); Silver (1974); Thompstone and Silver (1975); Silver (1981); Federgruen et al. (1984); Goyal and Satir (1989); Zheng (1994); Schultz and Johansen (1999); and Nielsen and Larsen (2005). Related work includes Gross et al. (1972).

10.4.2 Periodic Review System

Atkins and Iyogun (1988) suggest a procedure for this problem that outperforms can-order policies, and is easier to compute. McGee and Pyke (1996) apply this procedure at a fastener manufacturer using a spreadsheet to find the production quantities and safety stock. They modified the procedure to account for sequence-dependent setup times and a production capacity constraint. The Atkins and Iyogun (1988) procedure essentially allocates the major replenishment cost in small amounts to the products that are produced or bought most frequently, keeping the expected time to the next replenishment for these products in balance.

We now describe this procedure in some detail. First compute the EOQ and equivalent time supply for each product i using only the minor setup cost a_i:

$$\text{EOQ}_i = \sqrt{\frac{2a_i D_i}{r v_i}}$$

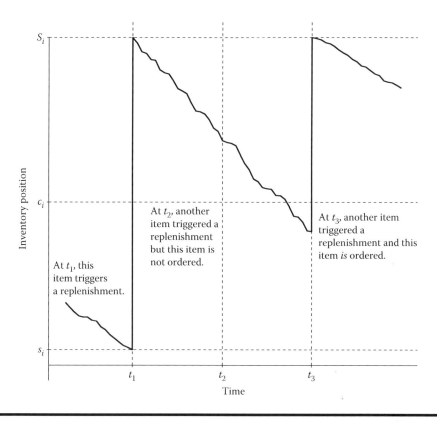

Figure 10.2   Behavior of an item under (S, c, s) control.

So the time supply, or expected *runout time*, in weeks is $EOQ_i(50)/D_i$, for product i assuming a 50-week year.[*] The result is a set of time supplies for each product in the family. Then choose the product with the smallest time supply, and denote it product 1 for now. A small portion, α_1, of the major setup cost is allocated to this first product. The idea is to eventually allocate the entire major setup cost to several products in the family. If all the costs were allocated to one product, its EOQ would increase significantly, thereby increasing its time supply. It would then be likely that this product would be purchased or produced less frequently than every cycle, and the setup cost would be incorrect. Hence, the algorithm calls for allocating only a small portion of the major setup cost to the most frequently purchased product.

Allocating more and more α_1 to the first product increases its time supply, eventually to the point that it matches the time supply of the second most frequently purchased product (based only on its a_i), denoted product 2 for now. Now allocate more of the major setup cost to the first product, and some to product 2, keeping the time supplies in balance. When these two equal time supplies increase to the time supply of the third most frequently purchased product, begin allocating α_3 to this third product. This process continues until the entire major setup is allocated—that is, $\sum_i \alpha_i = 1$. Therefore, each time a family is ordered, the entire setup cost is accounted for.

---

[*] This is essentially the same as Step 1 in Section 10.2.2.

If we express the total setup cost for an item as $(\alpha_i A + a_i)$, it is easy to solve for α_i for a given time supply:

$$\alpha_i = \frac{T_i^2 r_i v_i D_i}{2A} - \frac{a_i}{A} \tag{10.11}$$

where T_i is the time supply, EOQ_i / D_i. This expression can be calculated in a spreadsheet to quickly find all values of α_i such that the sum is 1 and the T_i values are equal for all products that have $\alpha_i > 0$. The time supply of the set of products for which $\alpha_i > 0$ is called the *base period*, or *base cycle*.

Atkins and Iyogun (1988) advocate rounding all products that are not purchased every cycle to a multiple of the base period. McGee and Pyke (1996) use powers-of-two multiples of the base period, so that each product's cycle is either the base period or $2^1, 2^2, 2^3$, etc. times the base period. Powers-of-two multiples create a greater sense of order, allowing regular preventive maintenance, communication with customers about production schedules, and easier scheduling of operators. See also the references in footnote[*].

Finally, safety stock is determined by using the periodic review (R, S) policies of Chapter 6, using the cycle time for each product as the R value.

10.4.2.1 Numerical Illustration

Four products are ordered by a distributor from a single vendor. The product data are given below, and the following information has been gathered. The fixed cost to place an order is $23, but once an order is placed, an item can be added to the order for $3. Thus, we set $A = \$20$, and $a = \$3$. The carrying charge is 24% and the delivery lead time is 1 week. Assume a 50-week year.

Product	Annual Demand	Unit Cost	Standard Deviation of Monthly Demand	EOQ	Time Supply in Weeks
1	450	$8.00	19	38	4.17
2	2,000	$12.50	50	63	1.58
3	200	$3.52	4	38	9.42
4	3,000	$33.30	88	47	0.79

The first step is to compute the EOQ and time supply based on the $3 minor ordering cost only. This calculation is shown in the table, where, for example, the EOQ for product 1 is

$$\text{EOQ}_1 = \sqrt{\frac{2a_1 D_1}{r v_1}} = \sqrt{\frac{2(3)(450)}{0.24(8)}} = 37.5 \approx 38$$

---

[*] What we are assuming here is that if a discount is to be achieved, it must be achieved by every replenishment; that is, we are ignoring the more complex possibility of achieving a discount on only some of the replenishments. The smallest replenishments are those where only the items with $m_i = 1$ are included. Lu (1995) develops a heuristic method for finding the order quantities when it is possible to take the discount on some replenishments and not on others. They find that this additional flexibility can yield significant savings. See also Russell and Krajewski (1992); Sadrian and Yoon (1992, 1994); Katz et al. (1994); and Xu et al. (2000).

Table 10.3   Solution to the Periodic Review Example

(1) Product	(2) α	(3) Time Supply in Weeks	(4) Rounded to Powers-of-Two	(5) Average Order Size
1	0.0000	4.17	4	36
2	0.1102	2.08	2	80
3	0.0000	9.42	8	32
4	0.8898	2.08	2	120

and the time supply is $37.5/450 = 0.0833$ years, or 4.17 weeks.

At this point, the major ordering cost has not been counted. So, we assign a small portion of the $20 cost to product 4, which has the smallest time supply. For example, if $\alpha_4 = 0.1$, the total setup cost is $20(0.1) + 3 = \$5$, and the new EOQ is 61, implying a time supply of 1.02 weeks. The time supply is longer, as we would expect, but it is still less than that of product 2. As we gradually increase α_4 the time supply of product 4 increases until it reaches 1.58, the time supply of product 2. This occurs at $\alpha_4 = 0.45$, which can be found from Equation 10.11, using 1.58 for T.

Now, increase α_4 slightly while increasing α_2 as well, keeping the time supplies of the two in balance. For example, if $\alpha_4 = 0.70$, and $\alpha_2 = 0.063$, the time supplies of each become 1.88 weeks. If the time supplies were to increase to 4.17, the time supply of product 1, we would increase α_1 also. This process would continue until $\sum_i \alpha_i = 1$. In this example, however, the sum of the α values reaches 1.0 before their time supplies increase to 4.17. The values are $\alpha_4 = 0.8898$, and $\alpha_2 = 0.1102$, which total to 1.0, and the time supplies of products 2 and 4 become 2.08 weeks.

Now, we have time supplies given in column 3 of Table 10.3.

The next step is to round the base period, 2.08, to an integer, or 2.0, and then round all other time supplies to powers-of-two multiples of the base period. Thus, 4.17 is rounded to $2 \times 2^1 = 4$, and 9.42 is rounded to $2 \times 2^2 = 8$.

Therefore, products 2 and 4 will be ordered every 2 weeks, while product 1 will be ordered every 4 weeks, or on every other replenishment cycle. Product 3 will be ordered every fourth cycle. On average, the amount ordered will be as given in column 5 of Table 10.3.

The final step is to use methods of Chapter 6 to find order-up-to-levels for each product, using R as given in column 4 and $L = 1$ week. This entire procedure can be built on a spreadsheet and searching for the α values. See Problem 10.9.

Tests of this procedure against the can-order policy as well as a number of other policies show that it performs very well in terms of total cost. If the major setup cost is close to zero, can-order policies perform better. However, finding optimal values of s, c, and S is not easy. See Iyogun (1991); Golany and Lev-er (1992), and Pantumsinchai (1992). For other research on this problem, see Sivazlian and Wei (1990); and Eynan and Kropp (1998). See also Song (1998) for finding the fill rate for a multi-item order.

10.5   Probabilistic Demand and Quantity Discounts

In contrast with the situation in the previous section, now it is attractive to achieve a specified group order size. One obvious example is where a discount is offered if the group order size exceeds

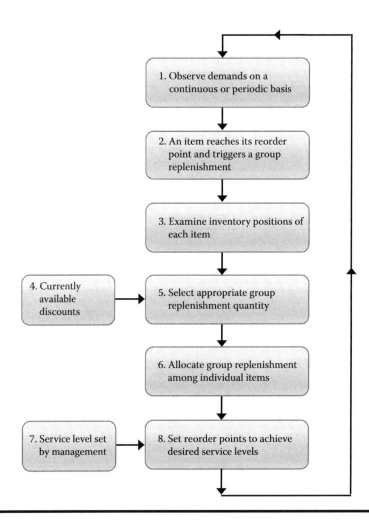

Figure 10.3 Components of a coordinated control system with quantity discounts and probabilistic demand.

some level. Another is where several products are shipped on a periodic basis in a single truck, rail car, or boat. In either case, it is important to answer the four questions posed near the start of Section 10.4.

We first provide an overview of a system developed by Miltenburg (1982) that is adapted for use on a computer. Figure 10.3 should be helpful in understanding the following discussion of Miltenburg's system.

Block 1 in Figure 10.3 indicates that either continuous or periodic review can be used. In a coordinated situation (particularly the transportation setting described earlier), a convenient periodic review interval such as a day or a week would typically be used.

As indicated in Block 2, a reorder point system is used. At a review instant, if the inventory position of any item in the family is at or below its reorder point, then a group replenishment is triggered. The setting of the reorder points will be discussed later.

The choice of an appropriate group replenishment (Block 5), of course, depends on the current inventory status of each item (Block 3) and what quantity discounts are currently available

(Block 4). In this decision, the methodology of Section 10.2 could be used; however, it is based on a steady-state model and does not take into account the current inventory situation. Thus, instead, the more appropriate decision logic of the "special opportunity to buy" situation (Section 4.9.5 of Chapter 4) is used to evaluate the various possible group replenishment sizes. We comment below on a case in which a full truckload defines the group replenishment size.

Once an overall replenishment size is selected, it must be allocated (Block 6) among the individual items. This is done to maximize the time until the next replenishment. After any given allocation, the individual stock levels are depleted by demands. The first to hit its reorder point triggers a new group order at the next review instant. Because of the probabilistic nature of the demand, the time until the next reorder is a random variable. Miltenburg models the demands for separate items as independent, diffusion processes (a diffusion process, besides being mathematically convenient, has the attractive property of implying that the total demand over any particular interval of time has a normal distribution).

Just as in Section 10.4.1, often an item will be reordered when it is above its reorder point. This is because some other item in the family will trigger the replenishment. The excess stock above the reorder point is known as residual stock. Account must be taken of this residual stock because it provides safety stock above and beyond the usual safety stock built into the reorder point. Unfortunately, the probability distribution of the residual stock of an item depends on the inventory positions of all of the items after the allocation is made. Thus, after every allocation, these distributions must be evaluated and used to establish the reorder points (Block 8) in order to provide a desired service level (Block 7).

For the case of periodic review, Miltenburg has found that a normal distribution provides a reasonable fit to the distribution of residual stock. In contrast, for the case of continuous review, a more appropriate fit is achieved by a spike at 0 and a truncated normal distribution above it (see Problem 10.8).

For other research in this area, see Low and Waddington (1967); Johnston (1980); Agnihothri et al. (1982); Miltenburg and Silver (1984a,b,c, 1988); Miltenburg (1985, 1987); Ernst and Pyke (1992); Van der Duyn Schouten et al. (1994); and Ettl et al. (2000).

10.5.1 A Full Truckload Application

In this section, we present an example of the Miltenburg system. One distributor of laminated wood products ordered extremely long beams from its Oregon supplier. The supplier required that the beams be shipped in full-truckload quantities because of the high fixed cost of sending a truck. This cost was due in part to police escorts that were required in many of the small towns the trucks had to travel through. How should the distributor determine when to order the next truckload, and how many of each beam should be ordered?

To solve this problem, we could apply the can-order system described in Section 10.4.1, but there is no guarantee that the order quantities will generate a full truckload. The periodic system of Atkins and Iyogun illustrated in Section 10.4.2 is another possibility, but again, we are faced with the problem of filling the truck exactly. The solution we describe is an application of the Miltenburg approach and is discussed by Carlson and Miltenburg (1988). It is called the service point method, and it has been used successfully in an industrial application. We shall assume that a periodic review system is employed (Block 1 of Figure 10.3). The steps are as follows:

1. Compute the acceptable shortages per replenishment cycle (ASPRC). This is a management decision, and can be computed as follows: ASPRC $= \sum_{i=1}^{n} Y_i(1 - P_2)$, where Y_i is the

expected usage of item i in the current replenishment cycle. (Y_i is simply the expected order size, Q_i, because Q_i is the average number of units used in each cycle), P_2 is the fill rate discussed in Chapter 6, and n is the number of items in the family. (This is Block 7.)

2. At every review instant, compute the expected shortages per replenishment cycle (ESPRC) if an order is not placed. Recall that ESPRC was computed as $\sigma_L G_u(k)$ in Chapter 6. In this case, however, k is different: $k = (IP - \hat{x}_{R+L})/\sigma_{R+L}$, where IP is the inventory position at the time of the review. (Now we are not concerned with s, but with the inventory position at the time of the review.) Also, because the next order, if we do not order now, will arrive $R + L$ periods later (R until the next review time, and L for delivery), we use $R + L$. (This is an application of Block 3.)

3. If ESPRC > ASPRC, place an order. If not, wait until the next review period, and check again. (This is a variation on Block 2.)

4. If the decision is to order, allocate the truck capacity to all products so as to maximize the time to the next replenishment. This last step is a bit more complicated, but can be approximated by assigning the truck capacity in proportion to the average demand of each product, adjusted for the current inventory positions, IP_i. That is, allocate (or order) an amount, Q_i, of product i, so to equate $(IP_i + Q_i)/D_i$ for all i, such that $\sum Q_i = TC$, where TC is the truck capacity.[*] (The truck capacity is the group replenishment size, or Block 5, and the allocation is Block 6.)

10.5.2 Numerical Illustration

As an illustration, assume that the distributor of laminated beams stocks just three products that are supplied by the Oregon manufacturer. The data for each product are given below. In addition, we shall simplify the problem slightly by assuming that the truck holds 850 beams of any size. In reality, we would need to insure that the truck capacity is scaled by beam sizes. Management has set a 98% fill rate target for each product.

	Product			
	1	*2*	*3*	*Total*
Annual demand, D_i	1,200	3,000	6,000	10,200
Y_i	100	250	500	850
$\hat{x}_{L+R}$	20	50	100	
σ_{L+R}	5	15	40	

We can see that, on average, a truckload composed of 100, 250, and 500 beams of products 1, 2, and 3, respectively, is ordered each month. Of course, any particular order will likely be composed of different amounts than these.

At a given review instant, assume that the inventory positions of the three products are 15, 45, and 125, respectively. We must decide whether to place an order. The ASPRC is calculated based

---

[*] See Carlson and Miltenburg (1988) for more detail.

on the product data and fill rate target:

$$\text{ASPRC} = \sum_{i=1}^{n} Y_i(1 - P_2) = (100 + 250 + 500)(1 - 0.98) = 17$$

So the k value for product 1 is computed as follows:

$$k_1 = (IP_1 - \hat{x}_{L+R})/\sigma_{L+R} = (15 - 20)/5 = -1$$

where the subscript denotes the product number. $G_u(k)$ is therefore 1.0833 (Table II.1 of Appendix II, or a spreadsheet as described in Appendix III), and

$$\text{ESPRC} = \sigma_{L+R}G_u(k) = 5(1.0322) = 5.42$$

Repeating the calculations for products 2 and 3 gives the below results.

	Product		
	1	*2*	*3*
IP	15	45	125
$\hat{x}_{L+R}$	20	50	100
σ_{L+R}	5	15	40
k	−1.000	−0.333	0.625
$G_u(k)$	1.0833	0.5876	0.1619
ESPRC	5.42	8.81	6.48

The total ESPRC is 20.71, which is higher than the acceptable number of 17. Therefore, an order should be placed.

As mentioned above, the allocation of truck capacity to the three products is somewhat complicated. For this example, we illustrate a simple approach that provides a reasonable approximation. In reality, one might want to consult the research papers mentioned above and apply an approach that could provide lower-cost solutions. For our illustration, we begin the allocation by ordering enough of each product to raise its inventory position to 80% of Y_i. (The 80% value was chosen based on work by Miltenburg (1985).) Because the inventory position of product 1 is 15, $100(0.80) - 15 = 65$ beams are initially allocated to this product. Likewise, products 2 and 3 are allocated $250(0.80) - 45 = 155$ and $500(0.80) - 125 = 275$ beams, respectively. This totals 495 beams, leaving $750 - 495 = 355$ beams to allocate. If we allocate based on average demand rates, product 1 should receive $1,200/(10,200) = 0.118$ of the total. This translates to $0.118(355) = 42$ beams. Results for all products are shown below. The truck is filled by 107 beams of product 1,259 beams of product 2, and 484 beams of product 3. Notice that $(IP_i + Q_i)/D_i$ is the same (1.22) for each product.

	Product			
	1	*2*	*3*	*Totals*
Proportion to allocate	0.118	0.294	0.588	1
Initial allocation	65	155	275	495
Remaining allocation	42	104	209	355
Total allocation, Q_i	107	259	484	850

In other research, Pantumsinchai (1992) tests a rule similar to the service point method against can-order policies and the periodic method of Atkins and Iyogun, and finds that the service point and periodic methods are quite good. For other research on the problem of finding order quantities with limits on storage space or budgets, see Roundy (1985); Hall (1988); Ventura and Klein (1988); Dagpunar (1988); Schneider and Rinks (1989); Sivazlian and Wei (1990); Mehrez and Benn-Arieh (1991); Lee (1994); Chen and Min (1994); Atkins (1991); Hariri et al. (1995); Martel et al. (1995); Güder et al. (1995); Güder and Zydiak (1999); and Haksever and Moussourakis (2005).

10.6 Production Environment

Throughout this chapter, we have noted that these coordination issues also apply in the production setting, where major and minor setups are determined by the time and cost of changing over production equipment among families and within families. Unfortunately, unless the factory is not capacity-constrained, the methods developed above cannot be applied unless adjustments are made. In this section, we discuss several dimensions of the coordination problem in a production setting. The methods we describe apply most commonly in the lower right corner of the product process matrix introduced in Chapter 2. In other words, they are applied in continuous flow processes, although they may also be applicable to batch flow processes.

10.6.1 Case of Constant Demand and Capacity: Economic Lot Scheduling Problem

A pharmaceutical manufacturer with which one of the authors consulted was faced with a bottleneck in the packaging lines. Changeover from one product to another entailed a major cleaning that could take up to 8 hours, while changes of label took only 1/2 hour. How should this firm schedule production at the packaging lines to replenish finished goods inventory in time to meet their service goals? One possibility is a can-order system. However, any system that reacts to demand occurrences is in danger of creating an idle facility for some time, followed by an overload due to multiple families reaching their reorder points at the same time. The factory simply may not have enough capacity to meet the demand. A better choice is to be proactive by creating a sequence of production that will be followed periodically, similar to the procedure described in Section 10.4.2. The solution to the economic lot scheduling problem (ELSP) is designed to do precisely that.

In the discussion that follows, we shall assume that if setup times are dependent on the sequence of production, a particular sequence of products can be developed and will be followed.[*] Thus, we will know the correct setup cost and time to use for each family and each item within the family. Membership in a family is defined by relatively low setup time or cost between members. Also, in this section we shall assume that demand is constant over time. In other words, we assume that the pharmaceutical manufacturer faces very little demand fluctuation and that there is no seasonality or trend in the demand over time. (In Sections 10.6.2 and 10.6.3, we will relax this restrictive assumption.) This constant demand case is essentially the same as the economic order quantity context discussed in Section 4.1 of Chapter 4, except there are several items with differing demand rates and finite production rates, and there is, of course, a production capacity restriction. A simple selection of independent finite-rate economic run quantities of the separate items (see Equation 4.19 of Chapter 4) can easily lead to an infeasible solution because two or more items are likely to require production at the same time.

The ELSP is to find a cycle length, a production sequence, production times, and idle times, so that the production sequence can be completed in the chosen cycle, the cycle can be repeated over time, demand can be fully met, and annual inventory and setup costs can be minimized. As it happens, this problem is notoriously difficult to solve.[†] Two factors contribute to this difficulty: the need to satisfy a production capacity constraint, and the need to have only one product in production at a time (a so-called "synchronization" constraint). A multitude of papers have been written on the ELSP. Here we provide a procedure for a simple case, and then briefly review some of the literature that extends the research into more complex situations.

Let us begin by introducing several terms used frequently in the literature. A cyclic schedule is one in which the entire system is periodic, which means that, regardless of the sequence of production, the schedule repeats itself over time. One variant of a cyclic schedule is to choose a base period and to restrict the order interval of each item to be an integer multiple of that base period. This variant includes the powers-of-two multiples mentioned above.[‡] Whatever the multiple, the batch sizes of a given item will be the same over time.[§] A special case of the base period approach is a pure rotation schedule, in which each product is produced once in each cycle.[¶] In other words, each product shares a common order interval, T. Of course, there is no guarantee that this schedule will be optimal. Therefore, when we restrict the choice of schedules to this type, we are resorting to the use of a heuristic. A second variant of a cyclic schedule is to allow items to be produced more than once in the overall period of the system, and to allow the batch sizes for each item to differ over that period.[**] We will focus here on pure rotation schedules.

---

[*] An example of sequence-dependent setups is a fastener manufacturer that has long setup times between fastener sizes, but short setup times within size. (Products that share the same size may differ in the cut of the thread, the finish, or other factors.) Another is beer production that sequences products from high quality to low quality. Within a family (or a given quality level), there are different container sizes. See Dobson (1992) and McGee and Pyke (1996).

[†] See, for example, Hsu (1983) and Gallego and Shaw (1997).

[‡] See Maxwell and Singh (1983); Jackson et al. (1985); and Roundy (1985, 1989) who have suggested the use of powers-of-two multiples of the base period for all products in the family. They show that the cost of this restriction is very low—within 6% of the optimal if a fixed base cycle is used, and within 2% if the base cycle is a variable.

[§] See Elmaghraby (1978); Hsu (1983); Schweitzer and Silver (1983); and Axsäter (1987).

[¶] See, for example, Gallego (1990).

[**] See Dobson (1987); Roundy (1989); Gallego (1992); and Zipkin (1991b).

In a pure rotation schedule, there is no chance of two products requiring production equipment at the same time because each product is scheduled only once during each cycle. Therefore, if a cycle length can be found that satisfies the capacity constraint, we will have a feasible solution to the pure rotation heuristic. In developing a pure rotation schedule, we shall focus the discussion on finding the reorder interval, rather than the batch size, partly because it is often easier to think in terms of frequency of production than in terms of number of units. The mathematical representation of the problem also becomes easier. Implementation is much easier as well because the planner can adjust batch sizes a little if demand varies, while keeping to the same reorder interval. This regularity facilitates preventive maintenance, operator scheduling, and coordination with other machines. (See Muckstadt and Roundy 1993.)

Some notation is required:

T = the common order interval, or time supply, for each product, in units of time
p_i = the production rate for product i, in units per unit time
A_i = setup cost for item i, in dollars
K_i = setup time for item i, in units of time
D_i = demand rate of item i, in units/unit time
v_i = unit variable cost of item i, in \$/unit
n = number of items in the family (the items are numbered $1, 2, 3, \ldots, n-1, n$)

The total relevant costs for item i are (following the finite production rate treatment in Section 4.8)

$$\mathrm{TRC}_i(T) = \frac{A_i}{T} + \frac{rv_i D_i(p_i - D_i)T}{2p_i}$$

Notice that $D_i T$ = the batch size for item i. Now let $H_i = (rv_i D_i(p_i - D_i))/2p_i$ to simplify the notation. The total amount of time to produce a batch of product i is $K_i + (D_i T/p_i)$, or the setup plus processing times. Therefore, we need a constraint on the total time in a cycle:

$$\sum_{i=1}^{n} \left(K_i + \frac{D_i T}{p_i} \right) \leq T \tag{10.12}$$

The problem is to minimize

$$\sum_{i=1}^{n} \mathrm{TRC}_i(T) \tag{10.13}$$

subject to

$$\sum_{i=1}^{n} \left(K_i + \frac{D_i T}{p_i} \right) \leq T \tag{10.14}$$

Equation 10.14 can be written as

$$\frac{\sum_{i=1}^{n} K_i}{1 - \sum_{i=1}^{n} \frac{D_i}{p_i}} \leq T \tag{10.15}$$

Taking the derivative of Equation 10.13 with respect to T and setting it equal to zero gives

$$T' = \sqrt{\frac{\sum_{i=1}^{n} A_i}{\sum_{i=1}^{n} H_i}} \tag{10.16}$$

If the capacity constraint Equation 10.12 is satisfied, the optimal T value is given by T'. Otherwise, T needs to be increased to insure that there is enough capacity. Then the optimal value is exactly the constraint (because $\sum_{i=1}^{n} \text{TRC}_i(T)$ is a convex function of T):

$$T'' = \frac{\sum_{i=1}^{n} K_i}{1 - \sum_{i=1}^{n} (D_i/p_i)} \tag{10.17}$$

Therefore, the optimal order interval is $\max(T', T'')$.

10.6.1.1 Numerical Illustration

A pharmaceutical manufacturer produces three bottle sizes of one product on one packaging machine. No other products are assigned to this equipment, and although other equipment is used for this product, this machine is the clear bottleneck. The data for the item are given in the following table:

Item	1	2	3
D_i (units/year)	1,850	1,150	800
v_i ($/unit)	50	350	85
p_i (units/year)	5,000	3,500	3,000
A_i ($)	125	100	110
K_i (years)	0.00068	0.00171	0.00091
H_i ($)	7,284	33,781	6,233

The carrying charge, r, is 25%. The setup times for the three items are 6, 15, and 8 hours, respectively. So, $0.00068 = 6/[(365 \text{ days})(24 \text{ hours per day})]$, because the equipment is run 24 hours per day. Using Equation 10.16, we find

$$T' = \sqrt{\frac{125 + 100 + 110}{7,284 + 33,781 + 6,233}} = 0.0842$$

From Equation 10.15, we find that

$$\frac{0.00068 + 0.00171 + 0.00091}{1 - (1,850/5,000 + 1,150/3,500 + 800/3,000)} = 0.0952 > 0.0842$$

Therefore, the order interval is too low and the capacity constraint is violated. So we set $T'' = 0.0952$ using Equation 10.17. Converting this value to weeks yields a production cycle of 0.0952(52) = 4.95, or 5 weeks. Every 5 weeks, set up and produce (5/52)(1,850) = 178 units of product 1, followed by 111 units of product 2 and 77 units of product 3. (How much time is spent for each batch?)

Gallego and Queyranne (1995) show that there is an easy way to determine how close the pure rotation schedule may be to the optimal. First compute the optimal time supply for each item taken independently using Equation 4.7 of Chapter 4 or $T_i^* = \sqrt{A_i/H_i}$. In our illustration, these values are 0.1310, 0.0544, and 0.1328, respectively. Then take the ratio of the largest to the smallest of these values, and call it R. So $R = 0.1328/0.0544 = 2.44$. The ratio of the cost of the pure rotation schedule (TRC$_{PRS}$) to the optimal (TRC$_{OPT}$) is then

$$\frac{\text{TRC}_{PRS}}{\text{TRC}_{OPT}} \leq \frac{1}{2}\left(\sqrt{R} + 1/\sqrt{R}\right)$$

In our illustration, this bound is 1.10. Thus, the cost of the pure rotation schedule is no more than 10% more than the cost of the unknown optimal solution.

We now mention a small portion of the literature on the ELSP. The earliest work in this field has been reviewed by Elmaghraby (1978). More recently, Dobson (1987) developed a formulation of the problem that allows lot sizes and therefore cycle times to vary over time. (This is the second variant of cyclic schedules mentioned above.) He was also one of the first to explicitly consider setup time, rather than using setup cost as a surrogate of setup time. See also Maxwell (1964) and Delporte and Thomas (1977). Zipkin (1991a) built on Dobson's work by taking the sequences from Dobson as givens, and then using a heuristic to find the production run times and machine idle times for each product. Taken together, Zipkin and Dobson provided a heuristic that computes feasible schedules with reasonable effort. Bourland and Yano (1997) test several heuristics and show their performance.

Additional research has addressed the ELSP with the added complication of delivery schedules when a supplier would rather run larger lot sizes than the customer desires. See Hahm and Yano (1992, 1995a,b). Silver (1995) and Viswanathan and Goyal (1997) consider the situation in which a family of items follows a cyclic schedule, but there is a limit on shelf life. The cycle length and production rate are adjusted to insure a feasible schedule. Taylor and Bolander (1996) ask a series of questions designed to help the reader understand whether MRP (Chapter 15) is the correct tool for scheduling the factory. The focus of the article is to determine whether process flow scheduling, like ELSP, is appropriate. See also Taylor and Bolander (1990, 1991, 1994) and Bolander and Taylor (1990).

In the realm of changing the givens, Gallego and Moon (1992) examine a multiple product factory that employs a cyclic schedule to minimize holding and setup costs. When setup times can be reduced, at the expense of setup costs, by externalizing setup operations, they show that dramatic savings are possible for highly utilized facilities. Allen (1990) modifies the ELSP to allow production rates to be decision variables. He then develops a graphical method for finding the production rates and cycle times for a two-product problem. The results were successfully applied in a chemical firm. See also Moon et al. (1991); Hwang et al. (1993); Gallego (1993); Gallego and Moon (1995); and Moon (1994).

Other references include Salomon (1991); Bretthauer et al. (1994); Dobson and Yano (1994); Ramudhin and Ratliff (1995); Davis (1995); Bramel et al. (2000); Gallego et al. (1996); Taylor et al. (1997); and Brahimi et al. (2006).

10.6.2 Case of Time-Varying Demand and Capacity: Capacitated Lot Sizing

The methods of Chapter 5 apply to the case of time-varying deterministic demand when there is no capacity constraint. A more difficult problem arises if it may not be possible to produce or purchase all of the desired parts in a given period. In addition, we allow production capacities to vary with time (but they are still assumed to be known with certainty). Under such circumstances, as in Chapter 5, it is most common to consider discrete time intervals such as days or weeks, each with a known capacity and known individual-item demand rates. The objective is to keep the total (across all items and all periods out to some horizon) of setup and carrying costs as low as possible, subject to (1) no backlogging of demand or loss of sales, and (2) no violations of production capacity constraints.

In Chapter 5, for the single-item, uncapacitated situation, we were able to argue that a replenishment quantity need only be placed when the inventory level is zero,[*] and also that we can restrict our attention to replenishment quantities that would cover exactly an integer number of periods of requirements. This is certainly not the case with capacitated production as evidenced by the simple, single-item example shown in Table 10.4. The only feasible schedule is to produce at capacity (i.e., 10 units) each period. Clearly, the quantity produced in period 1 lasts for 1.3 periods and, moreover, production must take place in period 2 even though the starting inventory in that period is nonzero (3 units). Thus, the nature of the solution, even for a single-item problem, is considerably more complicated when production capacities are taken into account. Not surprisingly, a mathematically optimal solution is out of the question when we are dealing with the multi-item, time-varying, capacitated case. We must again turn to the use of a heuristic solution procedure.

We outline the nature of a heuristic procedure first developed by Dixon (1979), and also presented in Dixon and Silver (1981). Subsequent industrial applications are described by Van Wassenhove and De Bodt (1983) and Van Wassenhove and Vanderhenst (1983). The latter, in particular, use the heuristic as part of an hierarchical planning framework. The procedure is forward-looking in nature; as with the basic Silver–Meal heuristic of Chapter 5, it develops the current period's production quantities using demand information for as few periods into the future as possible (a desirable characteristic when we recognize the increasing uncertainty in forecasts as we project further into the future). The heuristic makes use of the Silver–Meal heuristic's basic criterion—namely, that, if item 1 is to be produced in period 1, then we wish to select the production time supply (T_i) as the integer number of periods that produces the (first local) minimum of

Table 10.4   Single-Item, Capacitated Example

Period t	1	2	3	4	Total
Demand D_t	7	10	9	14	40
Capacity C_t	10	10	10	10	40

---

[*] The discussion here assumes a negligible lead time. If there is a known lead time L, then the order should be placed L time units before the inventory level hits 0.

the total relevant costs per unit time:

$$\text{TRCUT}(T_i) = \frac{A_i + h_i \sum_{j=1}^{T_i} (j-1) D_{ij}}{T_i}$$

where

A_i = setup cost for item i, in \$
h_i (or $v_i r$) = holding cost for item i, in \$/unit/period
D_{ij} = demand for item i in period j, in units

However, where capacity is limited, there is competition for a scarce resource and it is quite likely that not all lot sizes to be produced in a particular period can be such that the associated *TRCUT* is minimized. In order to describe how the heuristic deals with this difficulty, we need to introduce a new variable U_i, defined as follows:

$$U_i = \frac{\text{TRCUT}(T_i) - \text{TRCUT}(T_i + 1)}{k_i D_{i, T_i + 1}}$$

where k_i is the amount of production resource required per unit produced of item i.

We can see that U_i represents the marginal decrease in costs per unit time of item i per unit of capacity absorbed if the time supply produced of item i is increased from T_i to $T_i + 1$. Each item requiring production in the period under consideration has its T_i first set to unity (the minimum feasible amount that must be produced). Then, the item with the largest positive U_i has its T_i increased to $T_i + 1$. This is continued until there is insufficient capacity to increase any of the T_i's by unity or until all U_i's are negative (the latter situation would give the unconstrained Silver–Meal lot sizes for each of the items).

The above procedure clearly does not violate the capacity restriction in the particular period being scheduled. However, it does not guard against another type of infeasibility. Consider the situation where in some future period the total demand exceeds the production capacity of that period. Then, some of the requirements of that period must be satisfied by production in preceding periods. This is illustrated by our example of Table 10.4 where four of the units required in period 4 must be produced prior to that period. Thus, the heuristic also has a look-ahead feature built into it to ensure that, as the T_i's are increased (as described in the previous paragraph), enough total production is accomplished to prevent future infeasibilities.

If the heuristic is applied to a problem having a horizon with well-defined ending conditions, then the above procedure is used to establish an initial solution of all production lots out to the horizon. In such a case, the authors discuss several possible adjustments for improving on the initial solution. However, extensive testing has revealed that only one of these types of adjustments tends to be attractive in terms of potential savings without too much extra computational effort. Specifically, an attempt should be made to eliminate each scheduled lot by, if possible, allocating the production to periods where there is capacity available and the item under consideration is already being produced. The lot should be eliminated only if the increased carrying costs do not exceed the setup cost saved. In the more common case of a rolling-horizon implementation with an indefinite ending point, we do not recommend any such adjustment procedure.

There have been a number of other heuristic procedures developed for solving the capacitated, multi-item, time-varying demand problem. See reviews by De Bodt et al. (1984); Thizy and Van Wassenhove (1985); Bitran and Matsuo (1986); Bahl and Zions (1987); and Maes and

Van Wassenhove (1988). Also see, to name just a few of the many references, Florian and Klein (1971); Lambrecht and Vanderveken (1979); Baker et al. (1978); ter Haseborg (1982); Lambert and Luss (1982); Bolander and Taylor (1983); Bahl and Ritzman (1984); Blackburn and Millen (1984); Billington (1986); Maes and Van Wassenhove (1986); Taylor and Bolander (1986); Eppen and Martin (1987); Fogarty and Barringer (1987); Gunther (1987); Erenguc (1988); Prentis and Khumawala (1989); Trigeiro et al. (1989); Joneja (1990); Dixon and Poh (1990); Gilbert and Madan (1991); Maes et al. (1991); Pochet and Wolsey (1991); Salomon (1991); Thizy (1991); Campbell (1992); Diaby et al. (1992); Hill and Raturi (1992); Rajagopalan (1992); Toklu and Wilson (1992); Cattrysse et al. (1993); Mercan and Erenguc (1993); Tempelmeier and Helber (1994); Billington et al. (1994); Lotfi and Yoon (1994); Kirca (1995); Gardiner and Blackstone (1995); DeMatta and Guignard (1995); Potamianos et al. (1996); and Tempelmeier and Derstroff (1996). We make special mention of a procedure presented by Van Nunen and Wessels (1978) because it involves a conceptual approach very different from that of the Dixon and Silver heuristic. The latter builds up an initial feasible schedule with possible subsequent attempts at improvement. The van Nunen and Wessels approach is to first ignore the capacity restrictions and solve each individual item problem independently. The resulting overall solution is usually infeasible, violating one or more of the production capacities. The next step is to adjust the solution until it is feasible with as small as possible an increase in the costs. (See also Karni and Roll 1982.)

Variations on this problem are many. Eppen and Martin (1987); Pochet and Wolsey (1991); and Millar and Yang (1994) consider a related problem, but add the possibility of backordering. A number of authors have examined the case in which there is a cost to start up a facility for production, and a reservation cost that represents the cost of having the facility available regardless of whether it is producing or not. This reservation cost may represent the opportunity cost of tying up the facility or machine for the given product, and thus making it unavailable for other products. Therefore, a single item problem can be viewed as a subproblem of the more realistic multi-item case. This research applies when the facility produces a relatively small number of products, and a product may tie up the facility for more than one time period. See Karmarkar and Schrage (1985); Karmarkar et al. (1987); Sandbothe (1991); Hindi (1995); and Coleman and McKnew (1995).

10.6.3 *Probabilistic Demand: The Stochastic Economic Lot Scheduling Problem*

When demand, setup times or processing rates are not deterministic, production may not follow the production plan developed by deterministic approaches considered in the ELSP described in Section 10.6.1. If variability is high, there may be significant disruptions from using the deterministic solution in a probabilistic environment. Thus, adjustments must be made. The stochastic economic lot scheduling problem (SELSP) exactly parallels the ELSP with the added complexity of probabilistic (i.e., stochastic) demand, setup times, or processing rates.

There are two basic approaches to dealing with this problem. One is to develop a regular cyclic schedule using a solution to the deterministic problem, and then to develop a control rule that attempts to track or follow this schedule. The other is to develop a heuristic that directly decides which product to produce next and its production quantity. Therefore, the production sequence can vary over time.

Examples of the first approach include Bourland and Yano (1994) who consider the use of safety stock and idle time to account for random demand. They find that except when setup costs are very high, it is better to use safety stock rather than built-in idle time. See also Gallego (1990) who develops a real-time scheduling tool in three steps. First, replace probabilistic demands by their average values, and compute an optimal or near-optimal target cyclic schedule. Next, formulate a control policy to recover the target cyclic schedule after a disruption perturbs the inventories. Finally, find safety stocks to minimize the long run average cost of following the target schedule with the recovery policy. See also Loerch and Muckstadt (1994); Gallego (1994); and Gallego and Moon (1996).

A particularly compelling solution is by Federgruen and Katalan (1996c) who propose a simple strategy: when the facility is assigned to a given product, continue production until either a target inventory level is reached or a specific production batch is completed. The different products are produced in a given sequence, possibly with idle times inserted between the completion of an item's production batch and the setup for the next item. Federgruen and Katalan (1996c) provide a rather complex but very fast and accurate algorithm for computing the approximately optimal policy. Markowitz et al. (2000) show that setup costs should not be used as a surrogate for setup times, and that the policy of Federgruen and Katalan (1996c) should not be used when setup times are negligible but setup costs are high. They allow for switching production from one product to another based on the inventories of all products. Contrast this with the base stock policy, which relies only on the status of the product currently being produced. See also Bowman and Muckstadt (1993); Federgruen and Katalan (1994, 1996a,b, 1998); Anupindi and Tayur (1998); and Bowman and Muckstadt (1995).

The second approach, which dynamically decides which product to produce next, will generally have lower total costs, but may not be better in a real application because of disruptions to schedules of other production stages. Leachman and Gascon (1988) develop a methodology based on target cyclic schedules that are updated dynamically based on estimates of the time when inventories of various products will run out. They assume periodic review and rely on the ELSP solution for deterministic nonstationary demand to find the next item to produce and the production quantity. The schedule is modified if an item is backordered or close to a stockout. See also Leachman et al. (1991); Wein (1992); Veatch and Wein (1996); Duenyas and Van Oyen (1995, 1996); Perez and Zipkin (1997); Gascon et al. (1994); Goncalves et al. (1994); Duenyas and Van Oyen (1996); Qui and Loulou (1995); Sox and Muckstadt (1995, 1996); and Shaoxiang and Lambrecht (1996). Carr et al. (1993) consider a strategy that makes B and C items to order, while making A items to stock.

For the probabilistic time-varying demand case, see Karmarkar and Yoo (1994) and Winands et al. (2011).

10.7 Shipping Consolidation

A problem related to the coordination of multiple items is that of shipment consolidation. When a customer order arrives, the vendor must decide whether to ship the order immediately, or to wait for more orders to arrive so that the truck will be more fully loaded. Clearly, the truck entails a fixed cost regardless of the number of units shipped. In addition, costs may increase for each additional unit included. The focus here is on unit weight or volume, rather than on line items in a customer order, as above. Higginson and Bookbinder (1994) pointed out that shipment consolidation decisions include:

■ Which orders will be consolidated and which will be shipped individually?

- When will orders be released for possible shipping? Immediately, or after some time trigger or quantity trigger?
- Where will the consolidation take place? At the factory or at an off-site warehouse or terminal?
- Who will consolidate? The manufacturer, customer, or a third party?

Higginson and Bookbinder (1994) also tested three policies: a time policy that ships at a prespecified time, a quantity policy that ships when a given quantity is achieved, and a time/quantity policy that ships at the earliest of the time and quantity values. The shipper must trade off cost per unit with customer service in deciding on which policy to use. Some literature that may be helpful includes Tyworth et al. (1991); Van Eijs (1994); Bausch et al. (1995); Higginson and Bookbinder (1995); and Hall and Racer (1995).

10.8 Summary

In this chapter, we first showed the details of coordinated control under deterministic demand, both with and without quantity discounts. Then, we developed an appreciation of the complexity of the probabilistic demand situation. For the latter, we illustrated a useful model for coordinated control. We then extended the discussion to the production context and examined constant and deterministic demand, time-varying deterministic demand, and probabilistic demand.

In the next chapter, we turn to another, probably more important, type of coordination. That is, where the same item is stored at more than one location with stock flowing from one location to another.

Problems

10.1 Consider a family of four items with the following characteristics:

$$A = \$100 \quad r = 0.2/\text{year}$$

Item i	$D_i v_i$ ($/Year)	a_i ($)
1	100,000	5
2	20,000	5
3	1,000	21.5
4	300	5

a. Use the procedure of Section 10.2.2 to find appropriate values of the family cycle time T and the integer m_i's.

b. Suppose that we restricted attention to the case where every item is included in each replenishment of the family. What does this say about the m_i's? Now find the best value of T.

    c. Using Equation 10A.2 (in this Appendix), find the cost difference in the answers to parts a and b.

10.2   The Ptomaine Tavern, a famous fast-lunch spot near an institute of higher learning, procures three basic cooking ingredients from the same supplier. H. Fishman, one of the co-owners, estimates that the fixed cost associated with placing an order is $10. In addition, there is a fixed "aggravation" charge of $1 for each SKU included in the order. Usage of each of the three ingredients is relatively stable with time. The usages and basic unit values are as follows:

Item ID	Usage Rate (Units/Week)	Unit Value ($/Unit)
CO-1	300	1.00
CO-2	80	0.50
CO-3	10	0.40

    a. If a coordinated replenishment strategy is to be used, what should be the values of the m_i's?

    b. When pressed by a consulting analyst for a value of the carrying charge r (needed to compute the family cycle time T), Fishman replies: "I don't know from nothing concerning a carrying charge! All I know is that I'm satisfied with placing an order every 2 weeks." How would you use this answer to impute an approximate value of r? Discuss why it is likely to be only an approximation.

    c. What quantities of the three items should the Ptomaine Tavern order?

    d. Suppose that the supplier offers Ptomaine a 3% discount on any order totaling at least $1,350 in value (before the discount). Fishman's partner, L. Talks, likes the idea of saving on the purchase price. Fishman is not so sure about the advisability of tying up so much money in inventory. Should Ptomaine take advantage of the quantity discount offer?

10.3   Suppose you wished to estimate the annual costs of treating the members of a family independently in setting up run quantities. Under independent treatment, some major setups would be avoided, simply by the chance happening of two items of the same family being run one right after the other. How would you estimate the fraction of setups where this would happen?

10.4   A certain large manufacturing company has just hired a new member of its industrial engineering department, a Ms. V. G. Rickson. Rickson, knowing all about coordinated control, selects a family of two items with the following properties:

Item i	D_i (Units/Year)	v_i ($/Unit)
1	10,000	0.50
2	1,000	0.40

    From accounting records and discussions with operating personnel, she estimates that $A = \$5$, $a_1 = \$1$, $a_2 = \$4$, and $r = 0.2$ \$/\$/year.

    a. She computes the best values of T, m_1, and m_2. What are these values?

b. The production supervisor, Mr. C. W. Donrath, is skeptical of the value of coordination and argues that independent control is less costly, at least for this family of two items. Is he correct?

10.5 For the A, a_i's deterministic demand context, consider a special coordinated strategy—namely, where all $m_i = 1$.

a. Determine the optimal value of T.

b. Determine the associated minimum total relevant costs per unit time.

c. Find the best independent and coordinated (the latter with all $m_i = 1$ as above) strategies for each of the following two examples:

Example 1

$$A = \$10 \quad r = 0.2/\text{year}$$

Item i	D_i (Units/Year)	v_i ($/Unit)	a_i ($)
1	800	1	2
2	400	0.5	2

Example 2

$$A = \$10 \quad r = 0.2/\text{year}$$

Item i	D_i (Units/Year)	v_i ($/Unit)	a_i ($)
1	900	1	2
2	20	0.5	4

d. Determine as simple a relationship as possible that must be satisfied in order for the above special case of coordinated control (every $m_i = 1$) to be preferable to completely independent control. Ignore any system costs in your analysis. Normalize where possible to reduce the number of parameters.

10.6 The Steady–Milver Corporation produces ball bearings. It has a family of three items, which, run consecutively, do not take much time for changeovers. The characteristics of the items are as follows:

Item i	ID	D_i (Units/Year)	Raw Material ($/Unit)
1	BB1	2,000	2.50
2	BB2	1,000	2.50
3	BB3	500	1.60

Item i	Value Added ($/Unit)	Value after Production v_i ($/Unit)	a_i ($)
1	0.50	3.00	5
2	0.50	3.00	2
3	0.40	2.00	1

The initial setup cost for the family is $30. Management has agreed on an r value of 0.10 $/$/year. Production rates are substantially larger than the demand rates.

a. What are the preferred run quantities of the three items?

b. Raw material for product BB1 is acquired from a supplier distinct from that for the other two products. Suppose that the BB1 supplier offers an 8% discount on all units if an order of 700 or more is placed. Should Steady–Milver take the discount offer?

10.7 Brown (1967) has suggested a procedure for allocating a total order (of size W dollars) among a group of n items ($i = 1, 2, \ldots, n$). A particular time interval (perhaps until the next order arrives) of duration T is considered. The allocation is made so that the probability that item i runs out during T is proportional to the fraction of the group's sales that item i contributes. (Brown shows that this procedure, at least approximately, minimizes the total stock remaining at the end of the interval.) Suppose that it is reasonable to assume that demand in period T for item i is normally distributed with mean $\hat{x}_i$ and standard deviation σ_{T_i}.

a. Introducing whatever symbols are necessary, develop a routine for allocating the order among the items according to the above criterion.

b. Illustrate for the following 3 item example, in which $W = \$900$:

Item i	$\hat{x}_i$ (Units)	σ_T (Units)	v_i ($/Unit)	Initial (Before Allocation) Inventory I_i (Units)
1	100	40	1.00	50
2	300	70	2.00	100
3	250	100	1.20	250

c. Repeat part b but with

$$I_1 = 250 \quad I_2 = 150 \quad I_3 = 0$$

10.8 In Miltenburg's coordinated control system, the residual stock of a particular item is denoted by the symbol z, and in the case of periodic review, it turns out to be reasonable to approximate its distribution by a normal distribution with mean μ_z and standard deviation σ_z.

a. Why is there a spike at $z = 0$ in the continuous review case but not in the periodic review case? Also, why can z be negative in the periodic review situation?

b. Using the following equation and assuming that the demand x during $R+L$ is normally distributed with mean $\hat{x}_{R+L}$ and standard deviation σ_{R+L}, find as simple an expression as possible that the reorder point s must satisfy to ensure a probability of no stockout equal to P_1.

$$\Pr\{Stockout\} = \Pr\{x \geq s + z\}$$

$$= \int_{z_0=-\infty}^{\infty} f_z(z_0)dz_0 \int_{x=s+z}^{\infty} f_x(x_0)dx_0$$

Hint: Use the following result proved by Miltenburg (1982).

$$\int_{-\infty}^{\infty} \frac{1}{\sqrt{2\pi}\sigma_z} exp[-(z_0 - \mu_z)^2/2\sigma_z^2]$$

$$p_{u\geq} \left(\frac{s + z_0 - \hat{x}_{R+L}}{\sigma_{R+L}} \right) dz_0 = p_{u\geq} \left(\frac{s + \mu_z - \hat{x}_{R+L}}{c\sigma_{R+L}} \right)$$

where

$$c = \sqrt{1 + \sigma_z^2/\sigma_{R+L}^2}$$

c. Find s for the case where $P_1 = 0.96$, $\mu_z = 5.0$, $\sigma_z = 2.3$, $\hat{x}_{R+L} = 12.0$, and $\sigma_{R+L} = 4.2$. What s value would be used if the residual stock was ignored?

10.9 Build a spreadsheet to solve the algorithm of Section 10.4.2.

10.10 Three products are ordered by a retailer from a single vendor. The product data are given below, and the following information has been gathered. The fixed cost to place an order is $15, but once an order is placed, an item can be added to the order for $2. The carrying charge is 22% and the delivery lead time is 2 weeks. Assume a 50-week year.
   a. Use the algorithm of Section 10.4.2 to find the order frequencies for each product.
   b. If the fill rate target is 99%, use the methods of Chapter 6 to find the order-up-to-level for each product, given the results in part a.

Product	Annual Demand (Units)	Unit Cost ($)	Standard Deviation of Monthly Demand (Units)	EOQ (Units)	Time Supply (Weeks)
1	1,000	$2.00	20	95	4.77
2	500	$25.00	22	19	1.91
3	5,000	$9.45	120	98	0.98

10.11 Ten products are ordered by a distributor from a single supplier. The specific product data are given in the table below, and the following general information has been gathered. A is $25, and $a = $4. The carrying charge is 18% and the delivery lead time is 1 week.
   a. Use the algorithm of Section 10.4.2 to find the order frequencies for each product.
   b. If the fill rate target is 98%, use the methods of Chapter 6 to find the order-up-to-level for each product, given the results in part a.

Product	Monthly Demand (Units)	Unit Cost ($)	Standard Deviation of Monthly Demand (Units)
1	8	3.00	2.0
2	25	20.00	8.0
3	4	6.00	1.0
4	63	52.00	10.0
5	67	16.00	30.0
6	46	4.00	20.0
7	54	0.98	20.0
8	2	120.00	0.5
9	83	20.00	25.0
10	82	10.00	18.0

10.12 A drink bottler uses one bottling line to fill three different sized soft drink containers with the same soft drink. Other drinks are bottled on the same line as well. We shall consider only one product whose sizes are 10, 16, and 20 ounces, and whose demands are constant enough to be safely considered deterministic. The major setup cost to convert to this type of drink is estimated to be $65. It costs $22 to switch from one container size to another. The demand rates for this family of items are shown below. The carrying charge has been established as 0.24 $/$/year. Moreover, there is substantial excess capacity in the operation, so it is reasonable to ignore capacity considerations.

Item i	1	2	3
Description	10 Ounces	16 Ounces	20 Ounces
D_i (units/year)	120,000	95,000	80,000
v_i ($/unit)	0.30	0.48	0.60

a. Find the appropriate run quantities for each item.
b. Using Equation 10.5, find a lower bound on the total cost.

10.13 A retailer orders eight products from a single vendor. Assume that the demand for each product is deterministic, and is given below. The major fixed ordering cost is estimated to be $8; that is, the cost for the first item on the purchase order is $8. It costs $1 to add additional products to the purchase order. The carrying charge has been established as 0.30 $/$/year.

Item	Demand (Units/Year)	v_i ($/Unit)
1	600	2.50
2	200	12.65

Item	Demand (Units/Year)	v_i ($/Unit)
3	350	25.36
4	450	18.52
5	850	62.50
6	900	85.20
7	525	3.65
8	1,000	1.98

a. Find the appropriate run quantities for each item.
b. Using Equation 10.5, find a lower bound on the total cost.

10.14 Consider four parts used in the assembly of a particular end-item. These four parts are purchased from the same supplier who offers a discount of 4% if the value of the total replenishment quantity is at least $200. Assume that $5 is the basic cost of placing an order with one item involved. The inclusion of each additional item costs $1. $r = 0.24$ $/$/year and the D_i's and v_{0i}'s are shown below. Find the best order quantities.

Item i	D_i (Units/Year)	v_{0i} ($/Unit)
1	5,000	1.20
2	10,000	2.25
3	2,500	6.25
4	500	0.85

10.15 Consider nine parts ordered from a single supplier. The supplier offers a discount of 6% if the value of the total replenishment quantity is at least $10,000. Assume that $4.50 is the basic cost of placing an order with one item involved. The inclusion of each additional item costs $0.50. $r = 0.22$ $/$/year and the D_i's and v_{0i}'s are shown below. Find the best order quantities.

Item i	D_i (Units/Year)	v_{0i} ($/Unit)
1	2,000	11.00
2	16,000	15.00
3	1,300	18.00
4	1,000	25.00
5	3,000	32.00
6	250	65.00

Item i	D_i (Units/Year)	v_{0i} ($/Unit)
7	850	25.00
8	9,000	14.00
9	28,000	16.00

10.16 A distributor stocks six products that are supplied by the same manufacturer. The data for each product is given below. The manufacturer owns her own trucks and requires full truckload shipments. The truck can carry 400 units of the products, in any combination. Management has set a 97.25% fill rate target for each product.

Product	D_i	Y_i	$\hat{x}_{L+R}$	σ_{L+R}
1	1,000	42	83	20
2	2,000	83	167	50
3	560	23	46	10
4	850	35	70	20
5	4,200	175	350	120
6	1,000	42	83	32

At a given review instant, assume that the inventory positions of the products are 50, 175, 30, 60, 350, and 100, respectively. Should an order be placed, and if so, how much should be ordered of each product?

10.17 A manufacturer produces four products on one machine. No other products are assigned to this equipment. The data for the item are given in the following table. r is 0.24 $/$/year.

Item	1	2	3	4
D_i (units/year)	2,000	500	8,000	5,000
v_i ($/unit)	30	50	20	120
p_i (units/year)	50,000	2,000	20,000	24,000
A_i ($)	75	120	110	60
K_i (hours)	5	6	8	4
v_i ($/unit)	15	16	7.80	12.25

a. From the results of Section 10.6.1, find the optimal order interval and quantities, assuming each product is produced each cycle.
b. Briefly comment on your answer for a real environment.
c. What is the ratio of the cost of this pure rotation schedule to the optimal?

10.18 National Electronics distributes a wide variety of electronic products. National has been experiencing cash flow problems; hence it is concerned about its investment in inventories. A management consultant, Mr. P. C. Snow, has been hired to suggest ordering decision rules. As a preliminary step, he selects for study an important group of 10 products all bought from the same supplier, Harolds Corp. Based on an analysis of ordering costs, Mr. Snow estimates that the replenishment cost, A, in dollars, is given by

$$A = 46 + 5.5(V - 1,000)/1,000 \qquad (10.18)$$

where V is the total dollar value of the order and a minimum order size of $1,000 is required by Harolds. The National Electronics Controller, Mr. P. J. Schmeidler, tells Snow that the company uses an r value of 0.24 $/$/year. The sales forecast for each of the next 12 months, as well as the unit values, are shown in Exhibit 1. Snow ascertains that the requirements for any particular month must be on hand at the beginning of the month; hence he can restrict his attention to replenishments covering an integer number of months. In addition, at least initially, he decides to consider a deterministic version of the problem, namely, where lead times are known and the forecasts represent actual demands. Snow observes that because a marked seasonality exists, a simple EOQ strategy is inappropriate. He considers two possible types of decision rules:

EXHIBIT 1 Forecast (Units)

		2016				2017							
Item	Unit Value v_i $/Unit	S	O	N	D	J	F	M	A	M	J	J	A
1	70.40	4	5	5	5	5	3	2	3	2	2	2	2
2	8.40	30	30	40	40	40	30	30	20	10	10	10	10
3	49.00	15	15	15	15	15	15	11	11	9	9	10	12
4	51.40	1	1	2	2	2	1	1	1	1	1	1	1
5	15.70	24	30	30	30	30	18	12	18	12	12	12	12
6	12.40	30	30	30	30	30	20	20	20	15	15	15	15
7	5.30	31	31	35	35	35	21	21	21	15	15	10	10
8	10.10	36	50	50	50	45	25	25	18	18	15	15	15
9	18.10	33	33	44	44	44	33	33	22	22	11	11	11
10	110.50	10	10	10	10	10	10	8	8	8	6	6	6

1. Order the same fixed time supply (e.g., 2 months coverage) of all items.
2. Use a modified version of the Silver–Meal heuristic (Section 5.6 of Chapter 5), a modification is necessary because of the nature of the setup cost A which can be shared among two or more items involved in the same order.

Snow argues that any reasonable decision rule would require a group replenishment to arrive at the beginning of September, the start of the peak demand period. Hence, he assumes 0 inventories of all the items just prior to the beginning of September 2016.

a. For the fixed time supply (T months) strategy, he evaluates the order quantities, replenishment costs, and carrying costs for each of $T = 1$, 2, and 3 months. Develop these quantities and select what appears to be the best T value.

b. In an effort to appropriately modify the Silver–Meal heuristic, Snow reasons as follows. Suppose that a particular order involves n items and has a total value of V. Then, the setup cost A is given by Equation 10.18. Hence, an approximate apportioned cost for each item involved in the order is

$$\frac{46 + 5.5(V - 1,000)/1,000}{n}$$

The problem is that V depends on the order quantities, which, in turn, should depend upon the setup cost. This suggests the use of an iterative procedure. Historically, the average total value of an order involving n items has been approximately $\$600n$. Thus he starts with the Silver–Meal heuristic for each individual item using an A value of

$$A = \frac{46 + 5.5(60V - 1,000)/1,000}{n}$$
$$= 3.30 + 40.5/n \qquad (10.19)$$

The actual total value of the resulting group replenishment is determined from the individual order quantities. This implies a new value of A, the heuristic is reused, a new V is obtained, etc. until convergence takes place. Illustrate in detail these computations for the first replenishment (at the start of September), which must include all 10 items. Once a stable set of order quantities is established for a particular group replenishment, Snow moves on to the next time where at least one item requires another replenishment. The above iterative procedure involving A and V is used for that group of items. Complete, in this fashion, the 12 months of the modified heuristic computations. Determine the total replenishment and carrying costs and compare with your answer in part a.

c. Rather than iterating as above, suppose instead that for each group replenishment the individual item A was established according to Equation 10.19 and the Silver–Meal results with this A value were used. Carry through the calculations on this basis and evaluate the total replenishment and carrying costs.

d. Which approach do you think that Snow should recommend?

Appendix 10A: Derivation of Results in Section 10.2

If a time interval T between replenishments of the family and a set of m_i's are used, where m_i, an integer, is the number of T intervals that the replenishment quantity of item i will last, then this replenishment quantity (Q_i) is given by

$$Q_i = D_i m_i T \qquad (10A.1)$$

The typical sawtooth diagram of inventory is applicable so that the average inventory of item $i(\bar{I}_i)$ is

$$\bar{I}_i = \frac{Q_i}{2} = \frac{D_i m_i T}{2}$$

A group setup cost (A) is incurred every T units of time, whereas the cost a_i is incurred only once in every $m_i T$ units of time. Therefore, the total relevant costs per unit time are given by

$$\text{TRC}(T, m_i's) = \frac{A + \sum_{i=1}^{n} a_i/m_i}{T} + \sum_{i=1}^{n} \frac{D_i m_i T v_i r}{2} \tag{10A.2}$$

(Note that we do not include the purchase cost in the total relevant costs in this case. However, we do need to include it when a quantity discount is available.)

Setting

$$\frac{\partial \text{TRC}}{\partial T} = 0$$

gives the best T for the particular set of m_i's, that is,

$$-\frac{A + \sum_{i=1}^{n} a_i/m_i}{T^2} + \frac{r}{2} \sum_{i=1}^{n} m_i D_i v_i = 0$$

or

$$T^*(m_i's) = \sqrt{\frac{2(A + \sum a_i/m_i)}{r \sum m_i D_i v_i}} \tag{10A.3}$$

which is Equation 10.2.

Substitution of Equation 10A.3 back into Equation 10A.2 gives the best cost for a given set of m_i's:

$$\text{TRC}^*(m_i's) = \frac{A + \sum_{i=1}^{n} a_i/m_i}{\sqrt{\frac{2(A + \sum a_i/m_i)}{r \sum m_i D_i v_i}}} + \sqrt{\frac{2(A + \sum a_i/m_i)}{r \sum m_i D_i v_i}} \frac{r \sum m_i D_i v_i}{2}$$

which simplifies to

$$\text{TRC}^*(m_i's) = \sqrt{2 \left(A + \sum a_i/m_i \right) r \sum m_i D_i v_i} \tag{10A.4}$$

(This equation was used to directly produce Equation 10.10.)

We wish to select the m_i's to minimize $\text{TRC}^*(m_i's)$. From an inspection of Equation 10A.4, this is achieved by selecting the m_i's to minimize

$$F(m_i's) = \left(A + \sum a_i/m_i \right) \sum m_i D_i v_i \tag{10A.5}$$

The minimization of Equation 10A.5 is no simple matter because of two facts: (1) the m_i's interact (i.e., the effects of one m_i value depend on the values of the other m_i's) and (2) the m_i's must be integers (see Schweitzer and Silver 1983).

If we choose to ignore the integer constraints on the m_i's and set partial derivatives of $F(m_i's)$ equal to zero (necessary conditions for a minimum), then

$$\frac{\partial F(m_i's)}{\partial m_j} = -\frac{a_j}{m_j^2} \sum_i m_i D_i v_i + D_j v_j \left(A + \sum a_i/m_i \right) = 0$$

or

$$m_j^2 = \frac{a_j \sum m_i D_i v_i}{D_j v_j (A + \sum a_i / m_i)} \quad j = 1, 2, \ldots, n \qquad (10A.6)$$

For $j \neq k$, we have

$$m_k^2 = \frac{a_k}{D_k v_k} \frac{\sum m_i D_i v_i}{(A + \sum a_i / m_i)}$$

Dividing gives

$$\frac{m_j^2}{m_k^2} = \frac{a_j}{D_j v_j} \frac{D_k v_k}{a_k}$$

or

$$\frac{m_j}{m_k} = \sqrt{\frac{a_j}{D_j v_j} \frac{D_k v_k}{a_k}} \quad j \neq k$$

We can see that if

$$\frac{a_j}{D_j v_j} < \frac{a_k}{D_k v_k}$$

then (the continuous solution) m_j is less than (the continuous solution) m_k. Therefore, the item i having the smallest value of $a_i / D_i v_i$ should have the lowest value of m_i, namely, 1. It is reasonable to assume that this will hold even when the m_j's are restricted to being integers; of course, in this case, more than one item could have $m_j = 1$.

If the items are numbered such that item 1 has the smallest value of $a_i / D_i v_i$, then

$$m_1 = 1 \qquad (10A.7)$$

and, from Equation 10A.6,

$$m_j = \sqrt{\frac{a_j}{D_j v_j}} \sqrt{\frac{\sum m_i D_i v_i}{(A + \sum a_i / m_i)}} \quad j = 2, 3, \ldots, n \qquad (10A.8)$$

Suppose that there is a solution to these equations that results in

$$\sqrt{\frac{\sum m_i D_i v_i}{(A + \sum a_i / m_i)}} = C \qquad (10A.9)$$

Then, from Equation 10A.8, we have

$$m_j = C \sqrt{\frac{a_j}{D_j v_j}} \quad j = 2, 3, \ldots, n \qquad (10A.10)$$

Therefore,

$$\sum_{i=1}^{n} m_i D_i v_i = D_1 v_1 + \sum_{i=2}^{n} C \sqrt{\frac{a_i}{D_i v_i}} D_i v_i = D_1 v_1 + C \sum_{i=2}^{n} \sqrt{a_i D_i v_i} \qquad (10A.11)$$

Similarly,

$$\sum_{i=1}^{n} a_i/m_i = a_1 + \frac{1}{C} \sum_{i=2}^{n} \sqrt{a_i D_i v_i} \qquad (10A.12)$$

Substituting Equations 10A.11 and 10A.12 back into the left-hand side of Equation 10A.9 and squaring, we obtain

$$\frac{D_1 v_1 + C \sum_{i=2}^{n} \sqrt{a_i D_i v_i}}{A + a_1 + \frac{1}{C} \sum_{i=2}^{n} \sqrt{a_i D_i v_i}} = C^2$$

Cross-multiplication gives

$$D_1 v_1 + C \sum_{i=2}^{n} \sqrt{a_i D_i v_i} = C^2(A + a_1) + C \sum_{i=2}^{n} \sqrt{a_i D_i v_i}$$

or

$$C = \sqrt{\frac{D_1 v_1}{A + a_1}}$$

Substitution of this expression back into Equation 10A.10 gives

$$m_j = \sqrt{\frac{a_j}{D_j v_j} \frac{D_1 v_1}{A + a_1}} \quad j = 2, 3, \ldots, n \qquad (10A.13)$$

which is Equation 10.3.

   To find a lower bound on the best possible total cost, we substitute $m_1 = 1$ and the generally noninteger m_j's of Equation 10A.13 into Equation 10A.4. This leads, after considerable algebra, to the result of Equation 10.5.

References

Agnihothri, S., U. Karmarkar, and P. Kubat. 1982. Stochastic allocation rules. *Operations Research 30*(3), 545–555.

Allen, S. J. 1990. Production rate planning for two products sharing a single process facility: A real-world case study. *Production and Inventory Management Journal 31*(3), 24–29.

Anupindi, R. and S. Tayur. 1998. Managing stochastic multi-product systems: Model, measures, and analysis. *Operations Research 46*(3 supp), S98–S111.

Atkins, D. and P. Iyogun. 1988. Periodic versus "can-order" policies for coordinated multi-item inventory systems. *Management Science 34*(6), 791–796.

Atkins, D. and D. Sun. 1995. 98% Effective lot sizing for series inventory systems with backlogging. *Operations Research 43*(2), 335–345.

Atkins, D. R. 1991. The inventory joint replenishment problem with a general class of joint costs. *European Journal of Operational Research 51*, 310–312.

Atkins, D. R. and P. Iyogun. 1987. A lower bound on a class of inventory/production problems. *Operations Research Letters 6*(2), 63–67.

Axsäter, S. 1987. An extension of the extended basic period approach for economic lot scheduling problems. *Journal of Optimization Theory and Applications 52*, 179–189.

Bahl, H. and L. Ritzman. 1984. An integrated model for master scheduling, lot sizing and capacity requirements planning. *Journal of the Operational Research Society 35*(5), 389–399.

Bahl, H. C. and S. Zionts. 1987. Multi-item scheduling by Benders' decomposition. *Journal of the Operational Research Society 38*(2), 1141–1148.

Baker, K. R., P. Dixon, M. J. Magazine, and E. A. Silver. 1978. An algorithm for the dynamic lot size problem with time varying production constraints. *Management Science 24*, 1710–1720.

Balintfy, J. 1964. On a basic class of multi-item inventory problems. *Management Science 10*(2), 287–297.

Bastian, M. 1986. Joint replenishments in multi-item inventory systems. *Journal of the Operational Research Society 37*(12), 1113–1120.

Bausch, D. O., G. G. Brown, and D. Ronen. 1995. Consolidating and dispatching truck shipments of mobil heavy petroleum products. *Interfaces 25*(March–April), 1–17.

Billington, P., J. Blackburn, J. Maes, R. Millen, and L. van Wassenhove. 1994. Multi-item lotsizing in capacitated multi-stage serial systems. *IIE Transactions 26*(2), 12–17.

Billington, P. J. 1986. The capacitated multi-item dynamic lot-sizing problem. *IIE Transactions 18*(2), 217–219.

Bitran, G. R. and H. Matsuo. 1986. Approximation formulations for the single-product capacitated lot size problem. *Operations Research 34*, 63–74.

Blackburn, J. and R. Millen. 1984. Simultaneous lot-sizing and capacity planning in multi-stage assembly processes. *European Journal of Operational Research 16*, 84–93.

Bolander, S. and S. Taylor. 1983. Time phased forward scheduling: A capacity dominated scheduling technique. *Production and Inventory Management Journal 24*(1), 83–97.

Bolander, S. F. and S. G. Taylor. 1990. Process flow scheduling: Mixed-flow cases. *Production and Inventory Management Journal 31*(4), 1–6.

Bourland, K. and C. Yano. 1994. The strategic use of capacity slack in the economic lot scheduling problem with random demand. *Management Science 40*(12), 1690–1704.

Bourland, K. and C. Yano. 1997. A comparison of solutions approaches for the fixed-sequence economic lot scheduling problem. *IIE Transactions 29*, 103–108.

Bowman, R. A. and J. A. Muckstadt. 1993. Stochastic analysis of cyclic schedules. *Operations Research 41*(5), 947–958.

Bowman, R. A. and J. A. Muckstadt. 1995. Production control of cyclic schedules with demand and process variability. *Production and Operations Management 4*(2), 145–162.

Brahimi, N., S. Dauzere-Peres, N. M. Najid, and A. Nordli. 2006. Single item lot sizing problems. *European Journal of Operational Research 168*(1), 1–16.

Bramel, J., S. Goyal, and P. Zipkin. 2000. Coordination of production/distribution networks with unbalanced leadtimes. *Operations Research 48*(4), 570–577.

Bretthauer, K., B. Shetty, S. Syam, and S. White. 1994. A model for resource constrained production and inventory management. *Decision Sciences 25*(4), 561–577.

Brown, R. G. 1967. *Decision Rules for Inventory Management*. New York: Holt, Rinehart and Winston.

Campbell, G. 1992. Using short-term dedication for scheduling multiple products on parallel machines. *Production and Operations Management 1*(3), 295–307.

Carlson, M. and J. Miltenburg. 1988. Using the service point model to control large groups of items. *OMEGA 16*(5), 481–489.

Carr, S. A., A. R. Gullu, P. L. Jackson, and J. A. Muckstadt. 1993. An Exact Analysis of a Production-Inventory Strategy for Industrial Suppliers. Cornell University Operations Research and Industrial Engineering.

Cattrysse, D., M. Salomon, R. Kuik, and L. van Wassenhove. 1993. A dual ascent and column generation heuristic for the discrete lotsizing and scheduling problem with setup times. *Management Science 39*(4), 477–486.

Chakravarty, A. K. 1984. Joint inventory replenishments with group discounts based on invoice value. *Management Science 30*(9), 1105–1112.

Chakravarty, A. K. 1985. Multiproduct purchase scheduling with limited budget and/or group discounts. *Computer & Operations Research 12*(5), 493–505.

Chakravarty, A. K. and G. E. Martin. 1988. An optimal joint buyer-seller discount pricing model. *Computers and Operations Research 15*(3), 271–281.

Chen, C. and K. Min. 1994. A multi-product EOQ model with pricing consideration T.C.E. Cheng's model revisited. *Computers & Industrial Engineering 26*(4), 787–794.

Coleman, B. J. and M. A. McKnew. 1995. A quick and effective method for capacitated lot sizing with startup and reservation costs. *Computers and Operations Research 22*(6), 641–653.

Dagpunar, J. S. 1988. On the relationship between constrained and unconstrained multi-item replenishment problems. *International Journal of Operations & Production Management 9*(4), 77–81.

Davis, S. G. 1995. An improved algorithm for solving the economic lot size problem (ELSP). *International Journal of Production Research 33*(4), 1007–1026.

De Bodt, M., L. Gelders, and L. Van Wassenhove. 1984. Lot sizing under dynamic demand conditions: A review. *Engineering Costs and Production Economics 8*, 165–187.

Dekker, R., R. E. Wildeman, and F. A. van der Duyn Schouten. 1997. A review of multi-component maintenance models with economic dependence. *Mathematical Methods of Operations Research 45*(3), 411–435.

Delporte, C. and L. J. Thomas. 1977. Lot sizing and sequencing for N periods on one facility. *Management Science 23*(10), 1070–1079.

DeMatta, R. and M. Guignard. 1995. The performance of rolling production schedules in a process industry. *IIE Transactions 27*, 564–573.

Diaby, M., H. Bahl, M. Karwan, and S. Zionts. 1992. A Lagrangian relaxation approach for very-large-scale capacitated lot-sizing. *Management Science 38*(9), 1329–1340.

Dixon, P. S. 1979. *Multi-Item Lot-Sizing with Limited Capacity*. PhD thesis. University of Waterloo, Waterloo, Canada.

Dixon, P. S. and C. L. Poh. 1990. Heuristic procedures for multi-item inventory planning with limited storage. *IIE Transactions 22*(2), 112–123.

Dixon, P. S. and E. A. Silver. 1981. A heuristic solution procedure for the multi-item, single-level, limited capacity, lot-sizing problem. *Journal of Operations Management 2*(1), 23–39.

Dobson, G. 1987. The economic lot-scheduling problem: Achieving feasibility using time-varying lot sizes. *Operations Research 35*, 764–771.

Dobson, G. 1992. The cyclic lot scheduling problem with sequence dependent setups. *Operations Research 40*, 736–749.

Dobson, G. and C. A. Yano. 1994. Cyclic scheduling to minimize inventory in a batch flow line. *European Journal of Operational Research 75*, 441–461.

Duenyas, I. and M. P. Van Oyen. 1995. Stochastic scheduling of parallel queues with set-up costs. *Queueing Systems 19*(4), 421–444.

Duenyas, I. and M. P. Van Oyen. 1996. Heuristic scheduling of parallel heterogeneous queues with set-ups. *Management Science 42*(6), 814–829.

Elmaghraby, S. 1978. The economic lot scheduling problem (ELSP): Review and extensions. *Management Science 24*(6), 587–598.

Eppen, G. D. and R. K. Martin. 1987. Solving multi-item capacitated lot-sizing problems using variable redefinition. *Operations Research 35*(6), 832–848.

Erenguc, S. S. 1988. Multiproduct dynamic lot-sizing model with coordinated replenishments. *Naval Research Logistics 35*, 1–22.

Ernst, R. and D. Pyke. 1992. Component part stocking policies. *Naval Research Logistics 39*, 509–529.

Ettl, M., G. E. Feigin, G. Y. Lin, and D. D. Yao. 2000. A supply network model with base-stock control and service requirements. *Operations Research 48*(2), 216–232.

Eynan, A. and D. H. Kropp. 1998. Periodic review and joint replenishment in stochastic demand environments. *IIE Transactions 30*(11), 1025–1033.

Federgruen, A., H. Groenevelt, and H. C. Tijms. 1984. Coordinated replenishments in a multi-item inventory system with compound Poisson demand. *Management Science 30*(3), 344–357.

Federgruen, A. and Z. Katalan. 1994. Approximating queue size and waiting time distributions in general polling systems. *Queueing Systems 18*, 353–386.

Federgruen, A. and Z. Katalan. 1996a. Customer waiting time distributions under base-stock policies in single facility multi-item production systems. *Naval Research Logistics 43*, 533–548.

Federgruen, A. and Z. Katalan. 1996b. The impact of setup times on the performance of multi-class service and production systems. *Operations Research 44*(6), 989–1001.

Federgruen, A. and Z. Katalan. 1996c. The stochastic economic lot scheduling problem: Cyclical base-stock policies with idle times. *Management Science 42*(6), 783–796.

Federgruen, A. and Z. Katalan. 1998. Determining production schedules under base-stock policies in single facility multi-item production systems. *Operations Research 46*(6), 883.

Florian, M. and M. Klein. 1971. Deterministic production planning with concave costs and capacity constraints. *Management Science 18*, 12–20.

Fogarty, D. W. and R. L. Barringer. 1987. Joint order release decisions under dependent demand. *Production and Inventory Management Journal 28*(1), 55–61.

Gallego, G. 1990. An extension to the class of easy economic lot scheduling problems. *IIE Transactions 22*(2), 189–190.

Gallego, G. 1992. A minmax distribution free procedure for the (Q, R) inventory model. *Operations Research Letters 11*(1), 55–60.

Gallego, G. 1993. Reduced production rates in the economic lot scheduling problem. *International Journal of Production Research 31*(5), 1035–1046.

Gallego, G. 1994. When is a base stock policy optimal in recovering disrupted cyclic schedules? *Naval Research Logistics 41*, 317–333.

Gallego, G. and I. Moon. 1992. The effect of externalizing setups in the economic lot scheduling problem. *Operations Research 40*(3), 614–619.

Gallego, G. and I. Moon. 1995. Strategic investment to reduce setup times in the economic lot scheduling problem. *Naval Research Logistics 42*(5), 773–790.

Gallego, G. and I. Moon. 1996. How to avoid stockouts when producing several items on a single facility? What to do if you can't? *Computers and Operations Research 23*(1), 1–12.

Gallego, G. and M. Queyranne. 1995. Inventory coordination and pricing decisions: Analysis of a simple class of heuristics. In A. Sciomachen (Ed.), *Optimization in Industry 3*, New York: John Wiley & Sons.

Gallego, G., M. Queyranne, and D. Simchi-Levi. 1996. Single resource multi-item inventory systems. *Operations Research 44*(4), 580–595.

Gallego, G. and D. X. Shaw. 1997. Complexity of the ELSP with general cyclic schedules. *IIE Transactions 29*, 109–113.

Gardiner, S. and J. Blackstone. 1995. Setups and effective capacity: The impact of lot sizing techniques in an MRP environment. *Production Planning & Control 6*(1), 26–38.

Gascon, A., R. C. Leachman, and P. Lefrancois. 1994. Multi-item, single-machine scheduling problem with stochastic demands: A comparison of heuristics. *International Journal of Production Research 32*(3), 583–596.

Gilbert, K. and M. Madan. 1991. A heuristic for a class of production planning and scheduling problems. *IIE Transactions 23*(3), 282–289.

Golany, B. and A. Lev-er. 1992. Comparative analysis of multi-item joint replenishment inventory models. *International Journal of Production Research 30*(8), 1791–1801.

Goncalves, J., R. Leachman, A. Gascon, and Z. Xiong. 1994. A heuristic scheduling policy for multi-item, multi-machine production systems with time-varying, stochastic demands. *Management Science 40*(11), 1455–1468.

Goyal, S. K. 1974. A decision rule for producing Christmas greeting cards. *Operations Research 22*(4), 795–801.

Goyal, S. K. 1988. Economic ordering policy for jointly replenished items. *International Journal of Production Research 26*(7), 1237–1240.

Goyal, S. K. and S. G. Deshmukh. 1993. A note on "the economic ordering quantity for jointly replenishing items". *International Journal of Production Research 31*(12), 2959–2961.

Goyal, S. K. and A. T. Satir. 1989. Joint replenishment inventory control: Deterministic and stochastic models. *European Journal of Operational Research 38*, 2–13.

Gross, D., C. Harris, and P. Robers. 1972. Bridging the gap between mathematical inventory theory and the construction of a workable model—A case study. *International Journal of Production Research 10*(3), 201–214.

Güder, F., J. Zydiak, and S. Chaudhry. 1995. Non-stationary ordering policies for multi-item inventory systems subject to a single resource constraint. *Journal of the Operational Research Society 46*, 1145–1152.

Güder, F. and J. L. Zydiak. 1999. Ordering policies for multi-item inventory systems subject to multiple resource constraints. *Computers and Operations Research 26*(6), 583–597.

Gunther, H. O. 1987. Planning lot sizes and capacity requirements in a single stage production system. *European Journal of Operational Research 31*, 223–231.

Hahm, J. and C. Yano. 1992. The economic lot and delivery scheduling problem: The single item case. *International Journal of Production Economics 28*, 235–252.

Hahm, J. and C. A. Yano. 1995a. The economic lot and delivery scheduling problem: Models for nested schedules. *IIE Transactions 27*, 126–139.

Hahm, J. and C. A. Yano. 1995b. The economic lot and delivery scheduling problem: The common cycle case. *IIE Transactions 27*, 113–125.

Haksever, C. and J. Moussourakis. 2005. A model for optimizing multi-product inventory systems with multiple constraints. *International Journal of Production Economics 97*(1), 18–30.

Hall, N. G. 1988. A multi-item EOQ model with inventory cycle balancing. *Naval Research Logistics 35*, 319–325.

Hall, R. W. and M. Racer. 1995. Transportation with common carrier and private fleets: System assignment and shipment frequency optimization. *IIE Transactions 27*, 217–225.

Hariga, M. 1994. Two new heuristic procedures for the joint replenishment problem. *Journal of the Operational Research Society 45*(4), 463–471.

Hariri, A. M. A., M. O. Abou-El-Ata, and S. K. Goyal. 1995. The resource constrained multi-item inventory problem with price discount: A geometric programming approach. *Production Planning & Control 6*(4), 374–377.

Higginson, J. K. and J. H. Bookbinder. 1994. Policy recommendations for a shipment-consolidation program. *Journal of Business Logistics 15*(1), 87–112.

Higginson, J. K. and J. H. Bookbinder. 1995. Markovian decision processes in shipment consolidation. *Transportation Science 29*(3), 242–255.

Hill, A. V. and A. S. Raturi. 1992. A model for determining tactical parameters for materials requirements planning systems. *Journal of Operational Research Society 43*(6), 605–620.

Hindi, K. S. 1995. Efficient solution of the single-item, capacitated lot-sizing problem with start-up and reservation costs. *Journal of the Operational Research Society 46*, 1223–1236.

Hong, J. and J. C. Hayya. 1992. An optimal algorithm for integrating raw materials inventory in a single-product manufacturing system. *European Journal of Operational Research 59*, 313–318.

Hsu, W. 1983. On the general feasibility test of scheduling lot sizes for several products on one machine. *Management Science 29*, 93–105.

Hwang, H., D. B. Kim, and Y. D. Kim. 1993. Multiproduct economic lot size models with investment costs for setup, reduction and quality improvement. *International Journal of Production Research 31*(3), 691–703.

Ignall, E. 1969. Optimal continuous review policies for two product inventory systems with joint setup costs. *Management Science 15*(5), 278–283.

Iyogun, P. 1991. Heuristic methods for the multi-product dynamic lot size problem. *Journal of the Operational Research Society 42*(10), 889–894.

Jackson, P. 1988. Stock allocation in a two-echelon distribution system or "what to do until your ship comes in". *Management Science 34*(7), 880–895.

Jackson, P., W. Maxwell, and J. Muckstadt. 1985. The joint replenishment problem with a powers-of-two restriction. *IIE Transaction 17*(1), 25–32.

Johnston, F. R. 1980. An interactive stock control system with a strategic management role. *Journal of the Operational Research Society 31*(12), 1069–1084.

Joneja, D. 1990. The joint replenishment problem: New heuristics and worst case performance bounds. *Operations Research 38*(4), 711–723.

Kao, E. 1979. A multi-product dynamic lot-size model with individual and joint set-up costs. *Operations Research 27*(2), 279–289.

Karmarkar, U., S. Kekre, and S. Kekre. 1987. The dynamic lot-sizing problem with startup and reservation costs. *Operations Research 35*, 389–398.

Karmarkar, U. S. and L. Schrage. 1985. The deterministic dynamic product cycling problem. *Operations Research 33*, 326–345.

Karmarkar, U. S. and J. Yoo. 1994. The stochastic dynamic product cycling problem. *European Journal of Operational Research 73*, 360–373.

Karni, R. and Y. Roll. 1982. A heuristic algorithm for the multi-item lot-sizing problem with capacity constraints. *IIE Transactions 14*(4), 249–256.

Kaspi, M. and M. J. Rosenblatt. 1991. On the economic ordering quantity for jointly replenished items. *International Journal of Production Research 29*(1), 107–114.

Katz, P., A. Sadrian, and P. Tendick. 1994. Telephone companies analyze price quotations with Bellcore's PDSS software. *Interfaces 24*(1), 50–63.

Khouja, M. and S. Goyal. 2008. A review of the joint replenishment problem literature: 1989–2005. *European Journal of Operational Research 186*(1), 1–16.

Kirca, O. 1995. A primal-dual algorithm for the dynamic lotsizing problem with joint set-up costs. *Naval Research Logistics 42*, 791–806.

Klein, C. M. and J. A. Ventura. 1995. An optimal method for a deterministic joint replenishment inventory policy in discrete time. *Journal of the Operational Research Society 46*, 649–657.

Lambert, A. M. and H. Luss. 1982. Production planning with time-dependent capacity bounds. *European Journal of Operational Research 9*, 275–280.

Lambrecht, M. and H. Vanderveken. 1979. Heuristic procedures for the single operation, multi-item loading problem. *AIIE Transactions 11*(4), 319–326.

Leachman, R. and A. Gascon. 1988. A heuristic scheduling policy for multi-item, single-machine production systems with time-varying, stochastic demands. *Management Science 34*(3), 377–390.

Leachman, R., Z. Xiong, A. Gascon, and K. Park. 1991. An improvement to the dynamic cycle lengths heuristic for scheduling the multi-item, single-machine. *Management Science 37*(9), 1201–1205.

Lee, W. J. 1994. Optimal order quantities and prices with storage space and inventory investment limitations. *Computers in Industrial Engineering 26*(3), 481–488.

Loerch, A. G. and J. A. Muckstadt. 1994. An approach to production planning and scheduling in cyclically scheduled manufacturing systems. *International Journal of Production Research 32*(4), 851–871.

Lotfi, V. and Y. Yoon. 1994. An algorithm for the single-item capacitated lot-sizing problem with concave production and holding costs. *Journal of Operational Research Society 45*(8), 934–941.

Low, R. A. and J. F. Waddington. 1967. The determination of the optimum joint replenishment policy for a group of discount-connected stock lines. *Operational Research Quarterly 18*(4), 443–457.

Lu, L. 1995. A one-vendor multi-buyer integrated inventory model. *European Journal of Operational Research 81*, 312–323.

Maes, J., J. O. McClain, and L. N. Van Wassenhove. 1991. Multilevel capacitated lotsizing complexity and LP-based heuristics. *European Journal of Operational Research 53*, 131–148.

Maes, J. and L. Van Wassenhove. 1986. Multi item single level capacitated dynamic lotsizing heuristics: A computational comparison. *IIE Transactions 18*(2), 114–123.

Maes, J. and L. Van Wassenhove. 1988. Multi-item, single level capacitated dynamic lot-sizing heuristics: A general review. *Journal of the Operational Research Society 39*(11), 991–1004.

Markowitz, D. M., M. I. Reiman, and L. M. Wein. 2000. The stochastic economic lot scheduling problem: Heavy traffic analysis of dynamic cyclic policies. *Operations Research 48*(1), 136–154.

Martel, A., M. Diaby, and F. Boctor. 1995. Multiple items procurement under stochastic nonstationary demands. *European Journal of Operational Research 87*, 74–92.

Maxwell, W. 1964. The scheduling of economic lot sizes. *Naval Research Logistics Quarterly 11*(2–3), 89–124.

Maxwell, W. L. and H. Singh. 1983. The effect of restricting cycle times in the economic lot scheduling problem. *IIE Transactions 15*(3), 235–241.

McGee, V. E. and D. F. Pyke. 1996. Periodic production scheduling at a fastener manufacturer. *International Journal of Production Economics 46-47*, 65–87.

Mehrez, A. and D. Benn-Arieh. 1991. All-unit discounts, multi-item inventory model with stochastic demand, service level constraints and finite horizon. *International Journal of Production Research 29*(8), 1615–1628.

Mercan, H. M. and S. S. Erenguc. 1993. A multi-family dynamic lot sizing problem with coordinated replenishments: A heuristic procedure. *International Journal of Production Research 31*(1), 173–189.

Millar, H. and M. Yang. 1994. Lagrangian heuristics for the capacitated multi-item lot-sizing problem with backordering. *International Journal of Production Economics 34*, 1–15.

Miltenburg, G. J. 1982. *The Co-ordinated Control of a Family of Discount-Related Items*. PhD thesis. University of Waterloo, Waterloo, Canada.

Miltenburg, G. J. 1985. Allocating a replenishment order among a family of items. *IIE Transactions 17*(3), 261–267.

Miltenburg, G. J. 1987. Co-ordinated control of a family of discount-related items. *INFOR 25*(2), 97–116.

Miltenburg, J. G. and E. A. Silver. 1984a. Accounting for residual stock in continuous review coordinated control of a family of items. *International Journal of Production Research 22*, 607–628.

Miltenburg, J. G. and E. A. Silver. 1984b. The diffusion process and residual stock in periodic review coordinated control of families of items. *International Journal of Production Research 22*, 629–646.

Miltenburg, J. G. and E. A. Silver. 1984c. The evaluation of discounts in co-ordinated control inventory models. *Engineering Costs and Production Economics 8*, 15–25.

Miltenburg, J. G. and E. A. Silver. 1988. A microcomputer inventory control package for controlling families of items. *Engineering Costs and Production Economics 15*, 201–209.

Moon, I. 1994. Multiproduct economic lot size models with investment costs for setup reduction and quality improvement: Review and extensions. *International Journal of Production Research 32*(12), 2795–2801.

Moon, I., G. Gallego, and D. Simchi-Levi. 1991. Controllable production rates in a family production context. *International Journal of Production Research 29*(12), 2459–2470.

Muckstadt, J. A. and R. O. Roundy. 1993. Analysis of multistage production systems. In S. C. Graves, A. H. G. Rinnooy Kan, and P. H. Zipkin (Eds.), *Logistics of Production and Inventory*, Volume 4, pp. 59–131. Amsterdam: North-Holland.

Naddor, E. 1975. Optimal heuristic decisions for the s, S inventory policy. *Management Science 21*(9), 1071–1072.

Nielsen, C. and C. Larsen. 2005. An analytical study of the $q(s, s)$ policy applied to the joint replenishment problem. *European Journal of Operational Research 163*(3), 721–732.

Pantumsinchai, P. 1992. A comparison of three joint ordering inventory policies. *Decision Sciences 23*, 111–127.

Perez, A. P. and P. Zipkin. 1997. Dynamic scheduling rules for a multiproduct make-to-stock queue. *Operations Research 45*(6), 919–930.

Pochet, Y. and L. A. Wolsey. 1991. Solving multi-item lot-sizing problems using strong cutting planes. *Management Science 37*(1), 53–67.

Potamianos, J., A. J. Orman, and A. K. Shahani. 1996. Modeling for a dynamic inventory-production control system. *European Journal of Operational Research 96*, 645–658.

Prentis, E. L. and B. M. Khumawala. 1989. MRP lot sizing with variable production/purchasing costs: Formulation and solution. *International Journal of Production Research 27*(6), 965–984.

Qui, J. and R. Loulou. 1995. Multiproduct production/inventory control under random demands. *IEEE Transactions on Automatic Control 40*, 350–356.

Rajagopalan, S. 1992. A note on an efficient zero-one formulation of the multilevel lot-sizing problem. *Decision Sciences 23*(4), 1023–1025.

Ramudhin, A. and H. D. Ratliff. 1995. Generating daily production schedules in process industries. *IIE Transactions 27*, 646–656.

Rosenblatt, M. and M. Kaspi. 1985. A dynamic programming algorithm for joint replenishment under general order cost functions. *Management Science 31*(3), 369–373.

Rosenblatt, M. J. and N. Finger. 1983. An application of a grouping procedure to a multi-item production system. *International Journal of Production Research 21*(2), 223–229.

Roundy, R. 1985. 98%-effective integer-ratio lot-sizing for one-warehouse multi-retailer systems. *Management Science 31*(11), 1416–1430.

Roundy, R. 1986. A 98%-effective lot-sizing rule for a multi-product, multi-stage production/inventory system. *Mathematics of Operations Research 11*, 699–727.

Roundy, R. 1989. Rounding off to powers of two in continuous relaxations of capacitated lot sizing problems. *Management Science 35*, 1433–1442.

Russell, R. M. and L. J. Krajewski. 1992. Coordinated replenishments from a common supplier. *Decision Sciences 23*, 610–632.

Sadrian, A. and Y. Yoon. 1992. Business volume discount: A new perspective on discount pricing strategy. *Journal of Purchasing and Materials Management 28*, 43–46.

Sadrian, A. and Y. Yoon. 1994. A procurement decision support system in business volume discount environments. *Operations Research 42*, 14–23.

Salomon, M. 1991. Deterministic lotsizing models for production planning. In *Lecture Notes in Economics and Mathematical Systems*, Volume 355. Berlin: Springer-Verlag.

Sandbothe, R. A. 1991. The capacitated dynamic lot-sizing problem with startup and reservation costs: A forward algorithm solution. *Journal of Operations Management 10*(2), 255–266.

Schneider, H. and D. Rinks. 1989. Optimal policy surfaces for a multi-item inventory problem. *European Journal of Operational Research 39*, 180–191.

Schneider, M. J. 1982. Optimization of purchasing from a multi-item vendor. In *EURO V and the 25th International Conference of the Institute of Management Sciences*, Lausanne, Switzerland.

Schultz, H. and S. G. Johansen. 1999. Can-order policies for coordinated inventory replenishment with Erlang distributed times between ordering. *European Journal of Operational Research 113*(1), 30–41.

Schweitzer, P. J. and E. A. Silver. 1983. Mathematical pitfalls in the one machine multiproduct economic lot scheduling problem. *Operations Research 31*(2), 401–405.

Shaoxiang, C. and M. Lambrecht. 1996. $X - Y$ band and modified (s, S) policy. *Operations Research 44*(6), 1013–1019.

Silver, E. A. 1974. A control system for coordinated inventory replenishment. *International Journal of Production Research 12*(6), 647–671.

Silver, E. A. 1975. Modifying the economic order quantity (EOQ) to handle coordinated replenishments of two or more items. *Production and Inventory Management Journal 16*(3), 26–38.

Silver, E. A. 1981. Operations research in inventory management: A review and critique. *Operations Research 29*(4), 628–645.

Silver, E. A. 1995. Dealing with a shelf life constraint in cyclic scheduling by adjusting both cycle time and production rate. *International Journal of Production Research 33*(3), 623–629.

Sivazlian, B. D. and Y.-C. Wei. 1990. Approximation methods in the optimization of a stationary (s, S) inventory problem. *Operations Research Letters 9*(2), 105–113.

Song, J.-S. 1998. On the order fill rate in a multi-item, base-stock inventory system. *Operations Research 46*(6), 831–845.

Sox, C. R. and J. A. Muckstadt. 1995. Optimization-based planning for the stochastic lot scheduling problem. *IIE Transactions 29*(5), 349–358.

Sox, C. R. and J. A. Muckstadt. 1996. Multi-item, multi-period production planning with uncertain demand. *IIE Transactions 28*(11), 891–900.

Taylor, G. D., H. A. Taha, and K. M. Chowning. 1997. A heuristic model for the sequence-dependent lot scheduling problem. *Production Planning and Control 8*(3), 213–225.

Taylor, S. and S. Bolander. 1986. Scheduling product families. *Production and Inventory Management Journal 27*(3), 47–56.

Taylor, S. and S. Bolander. 1991. Process flow scheduling principles. *Production and Inventory Management Journal 32*(1), 67–71.

Taylor, S. G. and S. F. Bolander. 1990. Process flow scheduling: Basic cases. *Production and Inventory Management Journal 31*(3), 1–4.

Taylor, S. G. and S. F. Bolander. 1994. *Process Flow Scheduling*. Chicago: APICS.

Taylor, S. G. and S. F. Bolander. 1996. Can process flow scheduling help you? *APICS—The Performance Advantage March 6*, 44–48.

Tempelmeier, H. and M. Derstroff. 1996. A Lagrangian-based heuristic for dynamic multilevel multiitem constrained lotsizing with setup times. *Management Science 42*(5), 738–757.

Tempelmeier, H. and S. Helber. 1994. A heuristic for dynamic multi-item multi-level capacitated lotsizing for general product structures. *European Journal of Operational Research 75*(2), 296–311.

ter Haseborg, F. 1982. On the optimality of joint ordering policies in a multi-product dynamic lot size model with individual and joint set-up costs. *European Journal of Operational Research 9*, 47–55.

Thizy, J. M. 1991. Analysis of Lagrangian decomposition for the multi-item capacitated lot-sizing problem. *INFOR 29*(4), 271–283.

Thizy, J. M. and L. N. Van Wassenhove. 1985. Lagrangian relaxation for the multi-item capacitated lot-sizing problem: A heuristic implementation. *IIE Transactions 17*(4), 308–313.

Thompstone, R. M. and E. A. Silver. 1975. A coordinated inventory control system for compound Poisson demand and zero lead time. *International Journal of Production Research 13*(6), 581–602.

Toklu, B. and J. M. Wilson. 1992. A heuristic for multi-level lot-sizing problems with a bottleneck. *International Journal of Production Research 30*(4), 787–798.

Trigeiro, W. W., L. J. Thomas, and J. O. McClain. 1989. Capacitated lot sizing with setup times. *Management Science 35*(3), 353–366.

Tyworth, J. E., J. L. Cavinato, and C. J. Langley. 1991. *Traffic Management: Planning, Operations, and Control.* Reading, MA: Addison-Wesley.

Van der Duyn Schouten, F. A., M. J. G. van Eijs, and R. Heuts. 1994. The value of supplier information to improve management of a retailer's inventory. *Decision Sciences 25*(1), 1–14.

Van Eijs, M. J. G. 1993. A note on the joint replenishment problem under constant demand. *Journal of the Operational Research Society 44*(2), 185–191.

Van Eijs, M. J. G. 1994. Multi-item inventory systems with joint ordering and transportation decisions. *International Journal of Production Economics 35*, 285–292.

Van Eijs, M. J. G., R. Heuts, and J. P. C. Kleijnen. 1992. Analysis and comparison of two strategies for multi-item inventory systems with joint replenishment costs. *European Journal of Operational Research 59*, 405–412.

Van Nunen, J. A. E. E. and J. Wessels. 1978. Multi-item lot size determination and scheduling under capacity constraints. *European Journal of Operational Research 2*(1), 36–41.

Van Wassenhove, L. N. and M. A. De Bodt. 1983. Capacitated lot sizing for injection molding: A case study. *Journal of the Operational Research Society 34*(6), 489–501.

Van Wassenhove, L. N. and P. Vanderhenst. 1983. Planning production in a bottleneck department: A case study. *European Journal of Operational Research 12*, 127–137.

Veatch, M. H. and L. M. Wein. 1996. Scheduling a make-to-stock queue: Index policies and hedging points. *Operations Research 44*(4), 634–647.

Ventura, J. A. and C. M. Klein. 1988. A note on multi-item inventory systems with limited capacity. *Operations Research Letters 7*(2), 71–75.

Viswanathan, S. and S. K. Goyal. 1997. Optimal cycle time and production rate in a family production context with shelf life considerations. *International Journal of Production Research 35*(6), 1703–1712.

Wein, L. 1992. Dynamic scheduling of a multi-class make-to-stock queue. *Operations Research 40*, 724–735.

Winands, E. M., I. J. Adan, and G. Van Houtum. 2011. The stochastic economic lot scheduling problem: A survey. *European Journal of Operational Research 210*(1), 1–9.

Xu, J., L. L. Lu, and F. Glover. 2000. The deterministic multi-item dynamic lot size problem with joint business volume discount. *Annals of Operations Research 96*(1–4), 317–337.

Zheng, Y. 1994. Optimal control policy for stochastic inventory systems with Markovian discount opportunities. *Operations Research 42*(4), 721–738.

Zipkin, P. 1991a. Evaluation of base-stock policies in multiechelon inventory systems with compound-Poisson demands. *Naval Research Logistics 38*, 397–412.

Zipkin, P. H. 1991b. Computing optimal lot sizes in the economic lot scheduling problem. *Operations Research 39*(1), 56–63.

Zoller, K. 1990. Coordination of replenishment activities. *International Journal of Production Research 28*(6), 1123–1136.

Chapter 11

Multiechelon Inventory Management

In Chapters 4 through 9, we were primarily focused on single-item, single-location inventory systems. In Chapter 10, we extended the discussion to multiple items, and the issues associated with coordinating their ordering or production. In this chapter, we extend our treatment to another level of complexity—multiple locations. Specifically, we examine inventory control across the entire supply chain. As we will see, the solution approaches are somewhat more complex, but a student with solid understanding of the single-location models treated elsewhere in the book should be well prepared for this complexity. In this chapter, we focus on the multilocation setting where there is a single decision maker controlling inventory. We further investigate the important situation where decisions cross organizational boundaries in Chapter 12.

We begin in Section 11.1 by introducing supply chain management and the many issues involved with managing a supply chain. Section 11.2 discusses broader strategic issues that help to frame the discussion of the multiechelon models and insights we develop thereafter. In Section 11.3, we deal with the conceptually easiest situation of deterministic demand. Nevertheless, particularly when the demand rate varies with time, exact analysis quickly becomes quite complicated. Section 11.4 introduces the complexities of probabilistic demand in a multiechelon framework. We describe what is known as a *base stock control* system, a serial system, two arborescent (tree-structure) systems, and a "push" system. Next, in Section 11.5, for the first time, we deal with the so-called *product recovery situation*. Specifically, now, not all units demanded are consumed, but rather some units are returned, perhaps in a damaged state. As a result, both repair and purchase decisions must be made. Finally, we present additional insights in Section 11.6. As we have done throughout this book, in this chapter, we refer, sometimes very briefly, to research papers that apply to the relevant topics. These are intended for further insight and research, and can be safely ignored by many readers.

11.1 Multiechelon Inventory Management

Supply chain management (SCM) is the term used to describe the management of materials and information across the entire supply chain, from suppliers to component producers to final

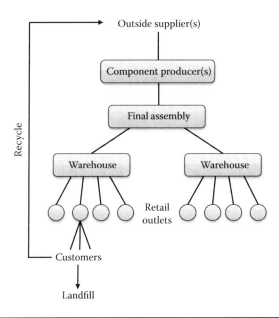

Figure 11.1 A schematic of a supply chain.

assemblers to distribution (warehouses and retailers), and ultimately to the consumer. In fact, it often includes after-sales service and returns or recycling. Figure 11.1 is a schematic of a supply chain. Supply chain management has generated much interest in recent years because of the realization that actions taken by one member of the chain can influence the performance of all others in the chain. Firms are increasingly thinking in terms of competing as part of a supply chain against other supply chains, rather than as a single firm against other individual firms. Also, as firms successfully streamline their own operations, the next opportunity for improvement is through better coordination with their suppliers and customers.

Before addressing opportunities for better coordination with suppliers and customers (in Chapter 12), we focus in this chapter on how a firm can manage inventory across multiple locations under their control. As suggested by Figure 11.1, it is often the case that a single firm makes inventory decisions at multiple locations such as their final assembly locations, warehouses, and retail outlets. Specifically, we will introduce models of multiechelon inventory systems that can be used to optimize the deployment of inventory in a supply chain and to evaluate a change in the supply chain "givens." We will look at the case of an item being stocked at more than one location with resupply being made between at least some of the locations. A three-echelon illustration is shown in Figure 11.2. Retail outlets (the first echelon) are replenished from branch warehouses (the second echelon), which are supplied from a central warehouse (the third echelon). The latter, in turn, is replenished from outside sources. Inventory management in this system is complex because demand at the central warehouse is dependent on the demand (and stocking decisions) at the branches. And demand at the branches is dependent on the demand (and stocking decisions) at the retail outlets. More generally, we refer to this as a dependent demand situation, in contrast with earlier chapters of the book where demands for different stock-keeping units were considered to be independent.

Multistage manufacturing situations (raw materials, components, subassemblies, assemblies) are conceptually very similar to multiechelon inventory systems. However, in this chapter, we

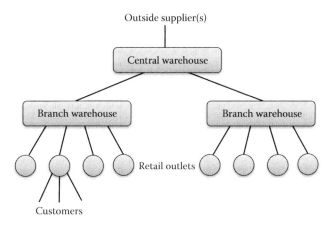

Figure 11.2   A multiechelon inventory situation.

restrict our attention to pure inventory (distribution) situations, by ignoring any finite replenishment capacities that are usually present in the production context. Nevertheless, some of the results of this chapter can be applied, with due caution, within production environments and, in fact, were developed with that type of application in mind. In addition, we will return in Chapter 15 to a detailed treatment of multistage manufacturing systems in which the concept of dependent demand will play a crucial role.

It seems intuitive that using the multiechelon inventory techniques we describe below provides significant benefit over using the methods of Chapters 6 and 7 in a multilevel setting. (See, e.g., Muckstadt and Thomas 1980.) Moreover, these multiechelon techniques have been successfully implemented. See, for example, Cohen et al. (1990) who apply multiechelon methods in a low demand, spare parts environment. Farasyn et al. (2011) describe the benefits of using inventory optimization tools, including multiechelon approaches at a large consumer goods firm. For very slow-moving items, from a single-level perspective, each retailer might hold, say, one unit, and yet sell one or fewer units per year. Because multiechelon models examine the entire system, they might recommend holding just a few parts at the warehouse level, and none at the retailers. System-wide savings can be enormous. With faster-moving parts, of course, it is likely that more units would be held at the retail level, but multiechelon methods help determine how many to hold at the retailers, and how many to hold at the warehouse.

11.2 Structure and Coordination

In this section, we provide a framework for supply chain management issues by briefly discussing structural and coordination decisions. Structural decisions are those that focus on longer-term, strategic issues and parallel the longer-term management levers described in Chapter 2. These decisions are typically made infrequently because they involve high costs to implement and change. Some structural decisions include

- Where to locate factories, warehouses, and retail sites?
- How many of these facilities to have?

- What capacity should each of these facilities have?
- When, and by how much, should capacity be expanded or contracted?
- Which facilities should produce and distribute which products?
- What modes of transportation should be used for which products, and under which circumstances?

The result of these structural decisions is a network of facilities designed to produce and distribute the products under consideration. Other parameters, such as lead times and some costs, used in our models are determined as well. The models we develop can be used to evaluate alternative structures. For example, customer service personnel often argue for having numerous stocking points close to customers. However, this leads to duplication of effort, extra carrying costs, and possibly extra transportation costs. The required inventory capacities and related costs of different options can be evaluated by the procedures of this chapter.[*]

Coordination decisions are usually taken after the structural decisions are made. Whereas structural decisions tend to be based on long-term, deterministic approximations, coordination decisions take the structure of the multiechelon network as given and focus on the short term. Therefore, they often must account for probabilistic demand and lead times. Coordination decisions include

- Should inventory stocking and replenishment decisions be made centrally or in a decentralized fashion?
- Should inventory be held at central warehouses or should these simply be used as break-bulk facilities?
- Where should inventory be deployed? In other words, should most inventory be held at a central location, or should it be pushed "forward" to the retail level?
- How should a limited and insufficient amount of stock be allocated to different locations that need it?

In addition, the four questions noted in Chapter 6 (how important is the item; periodic or continuous review; what form should the inventory policy take; and what specific cost and service goals should be set) fall in the coordination category as well. Clearly, the decisions are quite complex, and we cannot cover all of them in this chapter. However, we will present some models and insights that will help decision makers address some of the fundamental issues in supply chain management.

As a final comment, we note that costs associated with structural decisions primarily include fixed operating costs, but also some variable operating costs and certain transportation costs. Costs associated with coordination decisions include variable operating costs, inventory costs (holding and shortage), emergency transportation costs, and replenishment costs. General reviews by Clark (1972); Goyal and Gunasekaran (1990); Axsäter (1993a); Federgruen (1993); Diks et al. (1996); Thomas and Griffin (1996); and Gunasekaran et al. (2004) should be helpful for further information.

---

[*] Research relevant to structural decisions includes Arntzen et al. (1995); Cohen et al. (1995); Graves and Willems (2003b); Meixell and Gargeya (2005); and Graves and Willems (2005).

11.3 Deterministic Demand

In this section, we deal with the relatively simple situation in which the external demand rates are known with certainty. This is admittedly an idealization, but it is important to study for several reasons. First, the models will reveal the basic interactions among replenishment quantities at the different echelons. Second, as in earlier chapters, the deterministic case can serve as a basis for establishing replenishment quantities for the probabilistic case. Finally, the production environment can occasionally be approximated by deterministic demand and production rates. We also assume that replenishment lead times are known and fixed.

11.3.1 Sequential Stocking Points with Level Demand

Consider the simplest of multiechelon situations, namely, where the stocking points are serially connected—for example, one central warehouse, one retailer warehouse, and one retail outlet. Another interesting production interpretation is depicted in Figure 11.3. An item progresses through a series of operations, none of which are of an assembly or splitting nature; that is, each operation has only one predecessor and only one successor operation. For exposition purposes, we restrict ourselves to the most simple case of but two stages, denoted by a warehouse (W) and a retailer (R). Let us introduce some preliminary notation:

D = deterministic, constant demand rate at the retailer, in units/unit time (normally 1 year)
A_W = fixed (setup) cost associated with a replenishment at the warehouse, in dollars
A_R = fixed (setup) cost associated with a replenishment at the retailer, in dollars
v_W = unit variable cost or value of the item at the warehouse, in \$/unit
v_R = unit variable cost or value of the item at the retailer, in \$/unit
r = carrying charge, in \$/\$/unit time
Q_W = replenishment quantity at the warehouse, in units
Q_R = replenishment quantity at the retailer, in units

The two controllable (or decision) variables are the replenishment sizes Q_W and Q_R. Figure 11.4 shows the behavior of the two levels of inventory with the passage of time for the particular case where $Q_W = 3Q_R$. A little reflection shows that at least for the case of deterministic demand it never would make sense to have Q_W anything but an integer multiple of Q_R. Therefore, we can think of two alternative decision variables Q_R and n where

$$Q_W = nQ_R \quad n = 1, 2, \ldots \quad (11.1)$$

Note that from Figure 11.4 the inventory at the warehouse does not follow the usual sawtooth pattern even though the end usage is deterministic and constant with time. The reason is that the

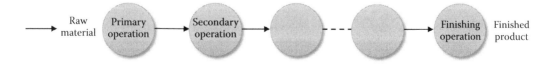

Figure 11.3 A serial production process.

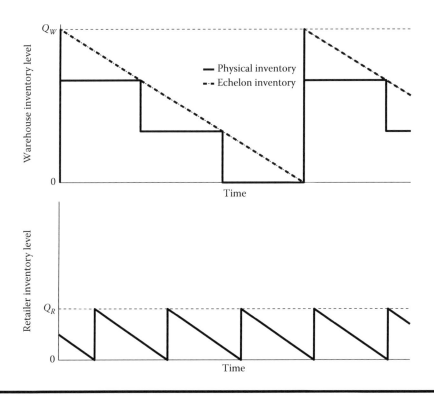

Figure 11.4 Behavior of the inventory levels in a deterministic two-stage process.

withdrawals from the warehouse inventory are of size Q_R. We could carry out an analysis using the conventional definitions of inventories, but the determination of average inventory levels becomes complicated. Instead, it is easier to use a different concept, known as *echelon stock*, introduced by Clark and Scarf (1960). They define the echelon stock of echelon j (in a general multiechelon system) as the number of units in the system that are at, or have passed through, echelon j but have as yet not been specifically committed to outside customers. When backorders are permitted, the echelon stock can be negative. With this definition and uniform end-item demand, each echelon stock has a sawtooth pattern with time, as illustrated in Figure 11.4; thus, it is simple to compute the average value of an echelon stock. However, we cannot simply multiply each average echelon stock by the standard v_r term and sum to obtain total inventory carrying costs. The reason is that the same physical units of stock can appear in more than one echelon inventory. For example, in our two-stage process, the actual finished inventory is counted in both the finished echelon inventory and the warehouse echelon inventory. The way around this dilemma is to value any specific echelon inventory at *only the value added* at that particular echelon.[*] When the decision

---

[*] In the deterministic demand case, echelon stock policies dominate installation stock policies, see Axsäter and Juntti (1996, 1997). (Installation stocks are simply the stocks at the given location, without regard for downstream stocks.) But in certain cases with probabilistic demand, the opposite might be true, particularly if the lead time to the warehouse is short relative to the lead time to the retailers. Moreover, when there is more than one retail outlet and transshipments are not possible, basing a replenishment action at the wholesale level on its echelon stock position is not likely to be optimal when the inventories at the branches are markedly unbalanced.

being made is whether to store inventory at an upstream location or at a downstream location that it supplies, the relevant holding cost is the incremental cost of moving the product to the retailer. This incremental cost is exactly the echelon holding cost. (Problem 11.2 is concerned with showing that this is equivalent to using the usual definitions of inventory levels and stock valuations.) Thus, in our two-stage example, the warehouse echelon inventory is valued at $v'_W = v_W$, while the retailer echelon inventory is valued at only $v'_R = v_R - v_W$. More generally, in a production assembly context, the echelon valuation v' at a particular stage i is given by

$$v'_i = v_i - \sum v_j \tag{11.2}$$

where the summation is over all immediate predecessors, j.

Returning to our two-stage serial situation, the total relevant (setup plus carrying) costs per unit time are given by

$$\text{TRC}(Q_W, Q_R) = \frac{A_W D}{Q_W} + \bar{I}'_W v'_W r + \frac{A_R D}{Q_R} + \bar{I}'_R v'_R r \tag{11.3}$$

where

$\bar{I}'_W$ = average value of the warehouse echelon inventory, in units
$\bar{I}'_R$ = average value of the retailer echelon inventory, in units

Substituting from Equation 11.1 and noting that the echelon stocks follow sawtooth patterns,

$$\begin{aligned}
\text{TRC}(n, Q_R) &= \frac{A_W D}{nQ_R} + n\frac{Q'_R v'_W r}{2} + \frac{A_R D}{Q_R} + \frac{Q'_R v'_R r}{2} \\
&= \frac{D}{Q_R}\left(A_R + \frac{A_W}{n}\right) + \frac{Q_R r}{2}(nv'_W + v'_R)
\end{aligned} \tag{11.4}$$

We must find the values of n (an integer) and Q_R that minimize this expression. As proved in the Appendix of this chapter, this is achieved by the following procedure:

Step 1 Compute

$$n^* = \sqrt{\frac{A_W v'_R}{A_R v'_W}} \tag{11.5}$$

If n^* is exactly an integer, go to Step 4 with $n = n^*$. Also, if $n^* < 1$, go to Step 4 with $n = 1$. Otherwise, proceed to Step 2.
Step 2 Ascertain the two integer values, n_1 and n_2, that surround n^*.
Step 3 Evaluate

$$F(n_1) = \left[A_R + \frac{A_W}{n_1}\right][n_1 v'_W + v'_R]$$

and

$$F(n_2) = \left[A_R + \frac{A_W}{n_2}\right][n_2 v'_W + v'_R] \tag{11.6}$$

If $F(n_1) \leq F(n_2)$, use $n = n_1$.
If $F(n_1) > F(n_2)$, use $n = n_2$.

Step 4 Evaluate

$$Q_R = \sqrt{\frac{2[A_R + A_W/n]D}{[nv'_W + v'_R]r}} \tag{11.7}$$

Step 5 Calculate

$$Q_W = nQ_R$$

11.3.1.1 Numerical Illustration of the Logic

Let us consider a particular liquid product, which a firm buys in bulk, then breaks down and repackages. So, in this case, the warehouse corresponds to the inventory prior to the repackaging operation, and the retailer corresponds to the inventory after the repackaging operation. The demand for this item can be assumed to be essentially deterministic and level at a rate of 1,000 liters per year. The unit value of the bulk material (v'_W or v_W) is \$1/liter, while the value added by the transforming (break and package) operation $v'_R = (v_R - v_W)$ is \$4/liter. The fixed component of the purchase charge (A_W) is \$10, while the setup cost for the break and repackage operation (A_R) is \$15. Finally, the estimated carrying charge is 0.24 \$/\$/year.

Step 1

$$n^* = \sqrt{\frac{(10)(4)}{(15)(1)}}$$
$$= 1.63$$

Step 2

$$n_1 = 1$$
$$n_2 = 2$$

Step 3

$$F(1) = \left[15 + \frac{10}{1}\right][1+4] = 125$$

$$F(2) = \left[15 + \frac{10}{2}\right][(2)(1)+4] = 120$$

that is, $F(1) > F(2)$. Thus, use $n = 2$.
Step 4

$$Q_R = \sqrt{\frac{2[15+10/2]\,1,000}{[(2)(1)+4]0.24}} \approx 167 \text{ liters}$$

Step 5

$$Q_W = (2)(167) = 334 \text{ liters}$$

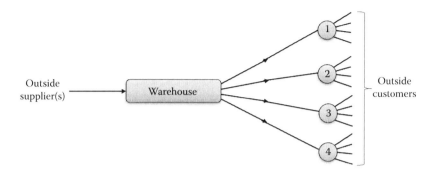

Figure 11.5   A one-warehouse, four-retailer system.

In other words, we purchase 334 liters[*] at a time; one-half of these or 167 liters are immediately broken and repackaged. When these 167 (finished) liters are depleted, a second break and repackage run of 167 liters is made. When these are depleted, we start a new cycle by again purchasing 334 liters of raw material.

11.3.1.2 Additional Research

See Roundy (1985) for research directly related to this problem. In other related research, Rosenblatt and Lee (1985) study a supplier with a single customer having level demand. The customer uses the EOQ for placing orders, and the supplier provides a discount of the form $a - bQ$, where Q is the order quantity of the customer. The supplier's order quantity is kQ, where k is an integer. Their model finds b and k so as to maximize the supplier's profit. See also Lee and Rosenblatt (1986); Moily (1986); Chakravarty and Martin (1988); Goyal and Gupta (1989); Dobson and Yano (1994); Kohli and Park (1994); Atkins and Sun (1995); Weng (1995); Aderohunmu et al. (1995); Desai (1996); Gurnani (1996); Lam and Wong (1996); and Chen and Zheng (1997b).

11.3.2 Other Results for the Case of Level Demand

For the serial situation described in the previous subsection, Szendrovits (1975) proposes a different type of control procedure—namely, where the same lot size is used at all operations but subbatches can be moved between operations. This allows an overlap between operations, and therefore reduces the manufacturing cycle time. He shows that, under certain conditions, this type of control procedure can outperform that proposed in the preceding subsection.

A number of authors have looked at systems more general in nature than the serial. Schwarz (1973) deals with a one-warehouse, *n*-retailer situation (depicted in Figure 11.5 for $n = 4$). For $n \geq 3$, he shows that the form of the optimal policy can be very complex; in particular, it requires that the order quantity at one or more of the locations vary with time even though all relevant demand and cost factors are time-invariant. He rightfully argues for restricting attention to a simpler class of strategies (where each location's order quantity does not change with time) and develops an effective heuristic for finding quite good solutions.

---

[*] Of course, it may make sense to round this quantity to a convenient shipment size such as 350 liters. In such a case, Q_R would also be adjusted (to 175 liters).

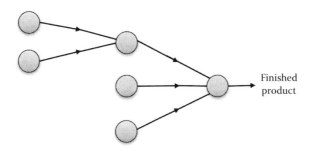

Figure 11.6   A "pure" assembly system.

Schwarz and Schrage (1975) adopt essentially the same approach for another type of deterministic situation—namely, "pure" assembly* (each node feeds into, at most, one other node as portrayed in Figure 11.6). These authors make use of a "myopic" strategy where each node and its successor are treated in isolation by much the same procedure as for the two-stage serial case discussed in the preceding section. Graves and Schwarz (1977) perform a similar analysis for arborescent systems, which are just the opposite of a "pure" assembly system—now each node has, at most, one node feeding into it.

Williams (1981, 1983) treats the general situation of an assembly network feeding into an arborescent system (assembly production followed by distribution). Maxwell and Muckstadt (1985) address the same problem but impose several restrictions (based on practical considerations) on the nature of the policy to be used, specifically:

1. The policy is nested—that is, a replenishment cannot occur at an operation unless one also occurs at immediate successor operations.
2. The policy is stationary—that is, the time between replenishments at each stage is a constant.
3. A base planning period (e.g., a shift, day, week) exists, and all reorder intervals must be an integer multiple of this base period.
4. The time between replenishments at any particular stage must be $1, 2, 4, 8, \ldots$ times the interval between replenishments at any immediately following stage. This is the "powers of 2" restriction encountered in Section 10.4.2 of Chapter 10.

The resulting mathematical formulation results in a large, nonlinear, integer programming problem, but the authors suggest an efficient, specialized solution procedure. See also Muckstadt and Roundy (1987) and Iyogun (1992).

11.3.3 Multiechelon Stocking Points with Time-Varying Demand

Here, we return to the demand situation assumed in Chapter 5 where end-item usage is known *but varies from period to period* [$D(j)$ in period j]. Also, carrying costs are incurred only on period-ending inventories. However, now we have a multiechelon assembly or distribution system. Our

---

* The term *assembly* implies a production context. However, the procedures discussed here ignore any capacity constraints.

discussion is based on the work of Blackburn and Millen (1982) specifically for an assembly structure. Moreover, for illustrative purposes, we concentrate on the simple two-stage serial process used in Section 11.3.1. (The more general case can be found in the Blackburn and Millen reference.)

One approach would be simply to use one of the procedures of Chapter 5 sequentially echelon by echelon. For example, the Silver–Meal heuristic could be utilized to schedule replenishments for the retailer (or the finished item in a production context). This would imply a pattern of requirements for the warehouse (or the primary processing stage in production), which would then be used as input to the Silver–Meal heuristic to plan the replenishments of the warehouse.[*] Although this approach is simple, it ignores the cost interdependency of the two echelons. Specifically, in choosing the replenishment strategy at the retailer the method does not take account of the cost implications at the warehouse.

Recognizing the above deficiency, we wish to develop a procedure that can still be applied sequentially (to keep implementation effort at a reasonable level) but that captures the essence of the cost interdependencies. An examination of the level demand case provides considerable insight. In particular, we repeat Equation 11.4 here:

$$\text{TRC}(Q_W, Q_R) = \frac{D}{Q_R}\left(A_R + \frac{A_W}{n}\right) + \frac{Q_R r}{2}(nv'_W + v'_R) \tag{11.8}$$

where

$$Q_W = nQ_R \tag{11.9}$$

Equation 11.8 is analogous to a *single* echelon problem (the selection of Q_R) if the adjusted fixed cost of a replenishment is

$$\hat{A}_R = A_R + \frac{A_W}{n} \tag{11.10}$$

and the adjusted unit variable cost of the item is

$$\hat{v}_R = nv'_W + v'_R \tag{11.11}$$

Note that the term A_W/n reflects that there is a warehouse setup at only every nth retailer setup. We can select Q_R (see Equation 11.7), including properly taking account of the cost impact at the warehouse, if we have a good preestimate of n. We could use Steps 1–3 of the algorithm in Section 11.3.1, but Blackburn and Millen have found that simply using Equation 11.5 and ensuring that n is at least unity works well (particularly in more complex assembly structures).

Thus we have

$$n = \max\left[\sqrt{\frac{A_W v'_R}{A_R v'_W}}, 1\right] \tag{11.12}$$

Returning to the time-varying demand situation, we compute n using Equation 11.12; then, employ the n value to obtain adjusted setup and unit variable costs for the retailer according to Equations 11.10 and 11.11. The Silver–Meal heuristic can then be applied to the retailer using $\hat{A}_R$

---

[*] Because of batching at the retailer, the resulting requirements pattern at the warehouse level is likely to contain a number of periods with no demand. As mentioned in Section 5.6.7 of Chapter 5, the unmodified Silver–Meal heuristic should be used with caution under such circumstances.

and $\hat{v}_R$. The resulting replenishments again imply a requirements pattern for the warehouse. Subsequently, the Silver–Meal, or another, lot-sizing procedure is used at the warehouse with A_W and v_W. We next show a numerical illustration. Problem 11.4 also presents a full numerical example.

11.3.3.1 Numerical Illustration

To illustrate the computation of $\hat{A}_R$ and $\hat{v}_R$, consider a two-stage process where $A_W = \$30$, $A_r = \$20$, $v_W = \$2/\text{unit}$, and $v_r = \$7/\text{unit}$. First, from Equation 11.2, we have

$$v'_R = v_R - v_W = \$5/\text{unit}$$
$$v'_W = v_W = \$2/\text{unit}$$

Then, using Equation 11.12, we obtain

$$n = \max\left[\sqrt{\frac{(30)(5)}{(20)(2)}}, 1\right]$$
$$= \max\left[1.94, 1\right]$$
$$n = 1.94$$

Finally, Equations 11.10 and 11.11 give us

$$\hat{A}_R = 20 + \frac{30}{1.94} = \$35.46$$

and

$$\hat{v}_R = 1.94(2.00) + 5.00 = \$8.88/\text{unit}$$

For related research in this area, see Zangwill (1969); Love (1972); Crowston et al. (1972, 1973a,b); Crowston and Wagner (1973); Lambrecht et al. (1981); Graves (1981); Afentakis et al. (1984); Chakravarty (1984); De Bodt et al. (1984); Blackburn and Millen (1985); Axsäter (1986); Park and Kim (1987); Afentakis (1987); Jackson et al. (1988); Joneja (1990); Rajagopalan (1992); El-Najdawi and Kleindorfer (1993); Iyogun and Atkins (1993); Lu and Posner (1994); Billington et al. (1994); Clark and Armentano (1995); and Jans and Degraeve (2007). See Simpson and Selcuk Erenguc (1996); Erengüç et al. (1999); Jans and Degraeve (2008); Buschkühl et al. (2010), and Clark et al. (2011) for reviews and discussions of this literature. Also see the references in Section 10.6.2 of Chapter 10.

11.4 Probabilistic Demand

Probabilistic demand raises several new issues and creates extreme modeling complexities in a multiechelon inventory situation. For discussion purposes, we examine the relatively simple three-echelon distribution system depicted in Figure 11.7.

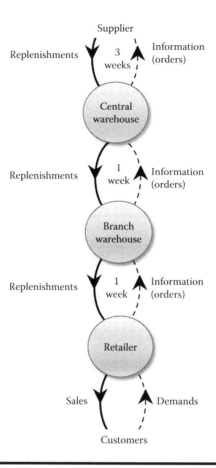

Figure 11.7   A multiechelon situation with single-stage information flow.

The replenishment lead times between adjacent levels (in each case assuming that the feeding level has adequate stock) are as follows:

Branch to retailer—1 week
Central to branch—1 week
Supplier to central—3 weeks

Straightforward use of the methods of Chapters 4 through 6 would dictate that each stocking level (retailer, branch warehouse, and central warehouse) would independently make replenishment decisions based on its own:

1. Cost factors and service considerations
2. Predicted demand—presumably forecasts based on historical demand that it has observed from the next stocking point downstream
3. Replenishment lead time from the next stocking point upstream

Such a system has three serious flaws. First, the lead time observed at, say, the retailer is dependent on whether the branch warehouse has sufficient stock to fill the order. If stock is available, the lead time is simply the transportation time plus any order processing time. If the branch is out of stock, the lead time could be as long as the branch's lead time plus the transportation time to the retailer. Second, as discussed in Section 11.3.1, it ignores the cost implications at one echelon of using certain ordering logic at another level. Third, even if end-customer demand is fairly smooth, the orders placed farther up the line become progressively larger and less frequent.

Other complicating factors include

1. How do we define service in a multiechelon situation? Normally, service is measured only at the lowest echelon. In a multiechelon system, a stockout at one of the higher echelons has only a secondary effect on service: it may lengthen the lead time for a lower echelon, which ultimately may cause customer disservice. We can see that safety stock at a particular echelon has effects on stockouts at *other* echelons. Thus, we have to be careful to avoid unnecessary duplication of safety stocks.
2. Suppose that the branch warehouse places orders of size Q_b on the central warehouse. What happens when such an order is placed and the on-hand stock at the central location is less than Q_b? Is a partial shipment made or does the system wait until the entire order can be shipped?
3. What about the possibility of an emergency shipment directly from the central warehouse to the retailer?
4. In more complicated multiechelon structures, transshipments between points at the same echelon may be possible, for example, between retailers in Figure 11.5.
5. Again, for more complicated structures when one central facility supplies several different stocking points at the next echelon, the central facility is likely to adopt a rationing policy when it faces multiple requests with insufficient stock to meet them all. Section 11.4.3 will specifically address this issue.

Researchers have developed several fundamental approaches to dealing with these complexities. One is to approximate the complex problem by simpler, and more easily solved versions. The METRIC model and extensions, to be discussed in Section 11.5.1, are examples. A second approach is to restrict the inventory policy to, say, the nested procedure discussed in Section 11.3.1. The model discussed in Section 11.4.2 employs this approach. In either case, however, the basic problem of establishing the linkages among echelons is critical. Three fundamental approaches have been developed. The first is most widely used and finds or approximates the distribution of the lead time observed by a downstream facility. See Section 11.4.2 for example. The second finds, or approximates, the distribution of the total number of orders outstanding at all retailers. These are then disaggregated to find the approximate number of orders outstanding at each retailer. The model described in Section 11.5.1 follows this approach. The third tracks each demand unit and matches it with a supply unit, calculating costs from the information obtained. Svoronos and Zipkin (1988); Axsäter (1993b); and Chen and Zheng (1997a) are papers that use this approach.

Despite all of the above complexities, Muckstadt and Thomas (1980) argue that it is indeed still worthwhile to use multiechelon control procedures. These authors show substantial benefits in a military supply context, and Lawrence (1977) does the same for a commercial distribution system. Arguing for the other side, Hausman and Erkip (1994) demonstrate that appropriately modified single echelon models lead to small cost penalties for a data set of low-volume, high-cost items. It is clear, however, that in many instances multiechelon techniques are valuable and can lead to large improvements in cost and service.

Two dimensions will be useful as we proceed: local versus global information, and centralized versus decentralized control. Local information implies that each location sees demand only in the form of orders that arrive from the locations it directly supplies. Also, it has visibility of only its own inventory status, costs, and so on. Global information implies that the decision maker has visibility of the demand, costs, and inventory status of all the locations in the system. Centralized control implies that attempts are made to jointly optimize the entire system, usually by the decisions of one individual or group. Centralized control is often identified with "push" systems, because a central decision maker pushes stock to the locations that need it most. Decentralized control implies that decisions are made independently by separate locations. Decentralized control is often identified with "pull" systems because independent decision makers pull stock from their suppliers. (See Pyke and Cohen 1990.)

The best solutions are obtained by using global information and centralized control because the decisions are made with visibility to the entire system using information for all locations. Unexpectedly large replenishment orders from downstream are eliminated. However, these solutions require cooperation and coordination across multiple parties within operations, across functions, and, in some cases, across firms. We address coordination across firms in Chapter 12.

If a firm is faced with decentralized control and local information, one good approach is for each location to use the methods of Chapters 4 through 7, probably with the probabilistic lead time formulas of Section 6.10 of Chapter 6. Benefits can be gained from sharing information even if control remains decentralized. Section 11.4.1 discusses a system that assumes decentralized control and global information.

11.4.1 Base Stock Control System

The base stock system is a response to the difficulties of each echelon deciding when to reorder based only on demand from the next lower echelon. The key change is to make end-item demand information available for decision making at all stocking points, as illustrated in Figure 11.8. This requires the use of an effective communication system that provides timely and accurate information; it also requires a high level of trust across the supply chain so that firms are willing to share potentially sensitive information. Each stocking point makes replenishments based on actual end-item customer demands rather than on replenishment orders from the next level downstream. With this modification, the procedures of Chapters 4 through 6 are now more appropriate to use.

The most common type of base stock system, in the research literature as well as in practice, is one in which an order-up-to policy is used. Order-up-to policy systems were introduced in Chapter 6 where decisions for any particular stocking point were based only on its stock position and its direct demand process. Here, we present a more general (s, S) system, more appropriate in a multiechelon situation. For each stocking point treated independently, an order quantity, Q, is established (using end-item demand forecasts) by one of the methods of Chapters 4 and 5. Next, a reorder point s is established by one of the procedures of Chapter 6, using end-item demand forecasts over the replenishment lead time appropriate to the echelon under consideration. (For example, the central warehouse would use a lead time of 3 weeks.) Then, the order-up-to-level S, also called the base stock level, is determined through the relation

$$S = s + Q \tag{11.13}$$

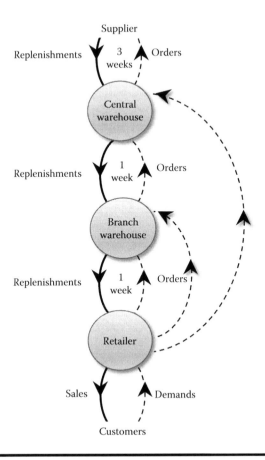

Figure 11.8   Information and stock flow in a base stock system.

In terms of physical operation, the echelon inventory position at each level is monitored according to the following relation:

$$\text{Echelon inventory position} = (\text{Echelon stock}) + (\text{On order}) \qquad (11.14)$$

where, as defined earlier, the echelon stock of echelon j is the number of units in the system that are at, or have passed through, echelon j but have as yet not been specifically committed to outside customers.

The "on-order" term in Equation 11.14 refers to an order placed by echelon j on the next *higher* echelon. To illustrate, suppose that at a point in time the physical stocks of our example (Figure 11.8) are:

Branch warehouse—50 units
Retail outlet—20 units

Furthermore, suppose that 5 units of known customer demand have not yet been satisfied. Moreover, assume that there is no order outstanding from the branch on the central warehouse and that 10 units are in transit between the branch and the retail outlet. Then, the echelon inventory position for the branch warehouse level would be

$$\text{Inventory position} = (50 + 10 + 20 - 5) + (0) = 75 \text{ units}$$

The echelon inventory position is reviewed after each transaction or on a periodic basis. (In the latter case, the replenishment lead time for safety stock calculations at each level must be increased by the review interval.) Once the echelon inventory position is known, it is compared with the reorder point s. Whenever it is at or lower than s, enough is ordered from the preceding echelon to raise the position to the base stock level S.

To repeat, ordering decisions at any stocking point in the base stock system are made as a result of end-item demand, not as a result of orders from the next level downstream. There is much less variability in the former than in the latter; hence, significantly lower safety stocks (reduced carrying costs) are achieved by the base stock system.

11.4.2 Serial Situation

For the special case of a serial system (such as in Figure 11.3), De Bodt and Graves (1985) propose a procedure for computing the replenishment quantities and reorder points that explicitly takes account of the effects of the stock at one echelon on the lead time at the next lower echelon. Their model assumes that demand information is available to all locations (global information), and that decisions are made centrally (centralized control). Again, for illustrative purposes and consistency with Section 11.3.1, we restrict our description to the case of two stages and call them *warehouse* and *retailer*. The assumptions underlying the decision rules include

1. External demand occurs only at the retailer and is a stationary process. Conceptually, the method can certainly be applied to a process that changes slowly with time where we have estimates of the mean and standard deviation of demand over suitable durations of time. We assume normally distributed forecast errors.
2. There is a deterministic replenishment lead time associated with each stage (L_W and L_R). Furthermore, the lead time L_R only begins when there is sufficient warehouse stock available to fulfill a retailer replenishment.
3. The policy used is of the (s, Q) form, that is, continuous review with four parameters:
   s_W = reorder point (based on the echelon inventory position) at the warehouse
   Q_W = order quantity at the warehouse
   s_R = reorder point at the retailer
   Q_R = order quantity at the retailer

Furthermore, for our purposes, we assume that Q_R and $Q_W = nQ_R$ have been predetermined by the procedure of Section 11.3.1. (De Bodt and Graves suggest an iterative scheme to simultaneously find the Q's and s's, similar to what was shown in Section 7.3 of Chapter 7.) Note that the restriction, $Q_W = nQ_R$, implies that the policies are nested: when the warehouse orders, the retailer also orders.

The decision rules for choosing s_R and s_W are:

1. Select s_R such that

$$s_R = \hat{x}_{L_R} + k_R \sigma_{L_R} \tag{11.15}$$

where

$\hat{x}_{L_R}$ = expected (forecast) demand over a retailer lead time

σ_{L_R} = standard deviation of forecast errors over the same interval

k_W = retailer safety factor, which satisfies

$$p_{u\geq}(k_R) = \frac{Q_R(v_R - v_W)r}{B_2 v_R D} \tag{11.16}$$

with D being the annual demand rate for the item, r the carrying charge in \$/\$/year, and B_2 the fraction of unit value charged per unit short. Note the similarity to Equation 6.22 of Chapter 6.

2. Select s_W such that

$$s_W = \hat{x}_{L_W + L_R} + k_W \sigma_{L_W + L_R} \tag{11.17}$$

where

$\hat{x}_{L_W + L_R}$ = expected demand over a warehouse lead time, plus a retailer lead time

$\sigma_{L_W + L_R}$ = standard deviation of forecast errors over the same interval

k_W = warehouse safety factor, which satisfies

$$p_{u\geq}(k_W) = \frac{Q_R[v_R + (n-1)v_W]r}{B_2 v_R D} \tag{11.18}$$

The v expression in the numerator of Equation 11.16 is equivalent with an echelon valuation. The $L_W + L_R$ combination occurs in Equation 11.17 because on every nth replenishment at the retailer, we have an associated warehouse replenishment, and the latter should be initiated when there is sufficient echelon stock to protect against shortages over a period of length $L_W + L_R$ (recall assumption 2). The derivation of Equation 11.18 (see De Bodt and Graves 1985) uses echelon stocks and hinges on distinguishing this special replenishment from the other $n-1$ retailer replenishments. Note that the right-hand side of Equation 11.16 is always smaller than that of Equation 11.18. Hence, the safety factor at the retailer is always larger than the safety factor at the warehouse level.

11.4.2.1 Numerical Illustration

We use the same example as in Section 11.3.1, the liquid product that the firm buys in bulk (the warehouse inventory), then breaks down and repackages (the retailer inventory). Suppose that the purchasing lead time is 6 weeks and the lead time for the *retailer*—that is, for repackaging—is 1 week. Also, assume that forecast errors are normally distributed with the σ for 1 week being 6.4 units, and that $\sigma_t = \sqrt{t}\sigma_1$. From Section 12.3.1, we know that the annual demand rate is 1,000 liters, $v_W = \$1/\text{liter}$, $v_r = \$5/\text{liter}$, $r = 0.24\ \$/\$/\text{year}$, $n = 2$, and $Q_R = 167$ liters. Finally, suppose that the fractional charge per unit short, B_2, is set at 0.35. From Equation 11.16, we obtain

$$p_{u\geq}(k_R) = \frac{(167)(5.00 - 1.00)(0.24)}{(0.35)(5.00)(1,000)}$$
$$= 0.0916$$

Then, Table II.1 in Appendix II gives

$$k_R = 1.33$$

Consequently, from Equation 11.15

$$s_R = \frac{1}{52}(1{,}000) + 1.33(6.4)$$
$$= 27.7, \text{ say } 28 \text{ liters}$$

From Equation 11.18

$$p_{u\geq}(k_W) = \frac{(167)[5.00 + (2-1)(1.00)](0.24)}{(0.35)(5.00)(1{,}000)}$$
$$= 0.1374$$

Thus, we have

$$k_W = 1.09$$

and, using Equation 11.17,

$$s_W = \frac{7}{52}(1{,}000) + 1.09\sqrt{7}(6.4)$$
$$= 153.1, \text{ say } 153 \text{ liters}$$

To summarize, we place a purchase order for 334 liters of bulk material when the echelon inventory position of bulk material drops to 153 liters or lower. Also, when the inventory position of the repackaged material drops to 28 liters or lower, we initiate the breaking and repackaging of 167 liters of bulk material.

11.4.2.2 Additional Research

We now comment on additional research in this area. Clark and Scarf (1960) perform one of the earliest studies in serial systems with probabilistic demand. They introduce the concept of an imputed penalty cost, wherein a shortage at a higher echelon generates an additional cost. This cost enables us to decompose the multiechelon system into a series of stages so that, assuming centralized control and the availability of global information, the ordering policies can be optimized. Clark and Scarf assume that there are no fixed costs at any stage, except the initial one. Clark and Scarf (1962) extend the analysis to allow fixed costs at every stage, and derive approximations. Lee and Whang (1999) and Chen and Samroengraja (2000) both propose performance measurement schemes for individual managers that allow for decentralized control (so that each manager makes decisions independently), and in certain instances, local information only. The result is a solution that achieves the same optimal solution as if we assumed centralized control and global information. See also Whang (1995) and Cachon and Zipkin (1999).

In general, exact, rather than approximate, analysis of these systems is quite difficult. However, Federgruen and Zipkin (1984b) derives exact performance measures, assuming the use of (s, nQ) policies. Rosling (1989) shows the equivalence of serial and assembly systems, so that in certain cases, results for serial systems can be applied to assembly systems. Chen and Zheng (1994b) study (s, nQ) policies in a serial system for both continuous review and periodic review. They find that echelon stock (s, nQ) policies are close to the optimal across any type of policy, and that installation stock policies are actually quite close to echelon stock policies. See also Lambrecht et al. (1984); Zangwill (1987); Graves (1988); Tang (1990); Van Houtum and Zijm (1991); Hodgson and Wang (1991); Axsäter and Juntti (1996); Gong et al. (1994); Glasserman and Tayur (1994, 1995); Lagodimos et al. (1995); Aderohunmu et al. (1995); Weng (1995); Zipkin (1995); and Graves and Willems (2000).

11.4.3 Arborescent Situation

In arborescent systems like the one shown in Figure 11.5, the issue of centralized control takes on a new dimension. Suppose that two retailers order 200 units each from their common supplier, and that the supplier only has 300 units available. An independent retailer would clearly choose to receive a full shipment even if it meant that the other retailer would receive only 100 units. A decision maker with centralized control, on the other hand, might choose to ship 150 units to each retailer. Zipkin (1995) shows that centralized control requires less central stock. Not all firms, however, have the capability, or the type of relationships in the supply chain, to achieve centralized control. In choosing a model of a multiechelon arborescent system, we must specify carefully the type of control system in place, as well as the information available to the decision maker.

In this section, we examine two arborescent systems, one in which the warehouse holds no stock and one in which the warehouse holds stock, and then we describe in more detail a particular push system.

11.4.3.1 Arborescent System with No Stock at the Warehouse

When the warehouse does not hold stock, it acts as a *break-bulk* facility by ordering goods in bulk, and upon receipt, breaking those quantities into smaller amounts for immediate shipment to the retailers. This differs from the numerical examples used above in that the entire amount is immediately shipped to the retailers. The warehouse decisions then are how to allocate limited inventory to multiple retailers, and how much to order each review period. Therefore, this is a push system that assumes centralized control and the availability of global information. Here we present an approach developed by Federgruen and Zipkin (1984a) for the case of periodic review.

We assume that retail demand is normally distributed, and that all retailers face identical holding and shortage costs. Additionally, we assume that demand across retailers is independent.

We use the following notation:

J = number of retailers

R = review period

S = system-wide order-up-to level; that is, the order-up-to level for the warehouse (which orders for the entire system)

B_3 = shortage cost per dollar per unit time at the retail level

r = carrying charge in \$/\$/unit time

L_{RE} = lead time from the warehouse to the retailers assuming stock is available for shipment. Note that the subscript, RE, for "retailer" differs from the use of R for retailer earlier in the chapter because of using R for *review period* in this section.

L_W = lead time to the warehouse

$\hat{x}_j$ = expected demand per period for retailer j

$\hat{x} = \sum_{j=1}^{J} \hat{x}_j$ total system-wide demand over one period

σ_j = standard deviation of demand for one period for retailer number j

$\sigma = \sum_{j=1}^{J} \sigma_j$

$\tilde{\sigma}^2$ = variance of system-wide demand $\sum_{j=1}^{J} \sigma_j^2$

The decision rule for choosing S is:

Select S such that

$$\frac{1}{R}\sum_{i=0}^{R-1}\{1 - p_{u\geq}(k_i)\} = \frac{B_3}{B_3 + r} \tag{11.19}$$

where

$$k_i = \frac{S - (L_W + L_{RE} + 1 + i)\hat{x}}{\sqrt{(L_{RE} + 1 + i)\sigma^2 + L_W\tilde{\sigma}^2}} \tag{11.20}$$

Federgruen and Zipkin also provide a number of allocation policies. We present only one: allocate stock at the warehouse to the retailers to equalize the stockout probability in the period in which they are all most likely to run out. In other words, set the following expression equal for all retailers, j:

$$\frac{IP_j + Z_j - (L_{RE} + R)\hat{x}}{\sqrt{L_{RE} + R}\sigma_j} \tag{11.21}$$

where IP_j is the inventory position for retailer j at the time of the decision and Z_j is the amount to allocate (or ship) to retailer j.

A simpler but less accurate rule for finding S is given by Eppen and Schrage (1981):

Select S such that

$$1 - p_{u\geq}\left(\frac{S - (L_W + L_{RE} + R)\hat{x}}{\sqrt{(L_{RE} + R)\sigma^2 + L_W\tilde{\sigma}^2}}\right) = \frac{B_3}{B_3 + r} \tag{11.22}$$

This equation is less accurate because it uses only the $R - 1$ term in the summation of Equation 11.19. Note the similarity to the newsvendor solution for the normal distribution given in Section 9.2.3 of Chapter 9.

11.4.3.2 Numerical Illustration

Let

$$R = 3 \text{ weeks}$$
$$L_W = 2 \text{ weeks}$$
$$L_{RE} = 1 \text{ week}$$
$$J = 7 \text{ retailers}$$
$$B_3 = 1.2$$
$$r = 0.24$$

Weekly demand:

$$\hat{x}_j = 10 \text{ units}, j = 1, \ldots, 7$$
$$\sigma_j = 2 \text{ units}, j = 1, \ldots, 7$$

Therefore, $\hat{x} = 70$ units per week, $\sigma = 14$ units, and $\tilde{\sigma}^2 = 28$. Note that for simplicity in illustration, we have used identical demand distributions at the 7 retailers.

As an illustration, using Equation 11.20, for $i = 0$

$$k_0 = \frac{S - (2 + 1 + 1)70}{\sqrt{(1 + 1)14^2 + 2(28)}} = \frac{S - 280}{\sqrt{448}}$$

The following table shows the results for $S = 300$ and $S = 500$ and the three i values ($R = 3$).

	S = 300		S = 500	
i	k_i	$1 - p_{u \geq}(k_i)$	k_i	$1 - p_{u \geq}(k_i)$
0	0.945	0.828	10.394	1.000
1	−1.970	0.024	5.911	1.000
2	−4.140	0.000	2.760	0.997
Sum		0.852		2.997

So at $S = 300$, the left side of Equation 11.19 is $(1/3)(0.852) = 0.284$ while the right side is

$$\frac{B_3}{B_3 + r} = \frac{1.2}{1.2 + 0.24} = 0.833$$

At $S = 500$, the left side is $1/3(0.2997) = 0.999$, while the right side is still 0.833.

Using trial and error, we find the optimal S to be 420. By contrast, the Eppen and Schrage formula (Equation 11.22), which does not require a search procedure, gives $S = 448$.

Now consider the allocation problem. For simplicity, assume that three retailers each have an inventory position of 40, while the remaining four have $IP = 25$. We want to allocate Z_j to retailer j so that Equation 11.21 is equal for all retailers, and so that $\sum_{j=1}^{7} Z_j = 420$. Because all retailers are equal in our example, this is identical to raising the inventory position of each retailer to the same number. We know that $Z_1 = Z_2 = Z_3$ and $Z_4 = Z_5 = Z_6 = Z_7$, and thus we wish to select Z_1 and Z_4 so that

$$\frac{40 + Z_1 - (1 + 3)(10)}{\sqrt{(1 + 3)(2)}} = \frac{25 + Z_4 - (1 + 3)(10)}{\sqrt{(1 + 3)(2)}}$$

and

$$3Z_1 + 4Z_4 = 420$$

These two linear equations can be solved for $Z_1 = 51(3/7)$ and $Z_4 = 66(3/7)$. The fractional 3 units would be allocated arbitrarily.

11.4.3.3 Additional Research

See Federgruen (1993) for an overview and review of the literature in this area. There is a vast amount of literature, and we mention just a few papers that are more recent or are particularly important. Zipkin (1984) provides a formal treatment of the idea of balanced inventories at the retailers. Jönsson and Silver (1987b) show that holding back some stock at the warehouse provides significant benefits. In an interesting paper, Graves (1996) studies a two-echelon system with order-up-to policies and deterministic delivery times. A key idea in this paper is virtual allocation, where stock at the warehouse is reserved for given retailers as demand occurs. Systems of this sort are more and more likely because of advances in information technology, such as EDI. Graves identifies three other allocation policies—ship all (where the warehouse is a break-bulk facility and holds no stock), equalize retailer inventories at regular times, and a two-interval policy that withholds some stock for a second shipment. The virtual allocation policy works well in the cases he tests. Poisson demand is assumed, and the warehouse observes demand at the retail level in real time. One additional conclusion is that stock should be held at both warehouse and retail levels, but mostly at the retail level. Therefore, the warehouse will stock out often, even while the total system is operating effectively. See Axsäter (1993c) for additional comments on allocation policies. Jackson et al. (1996) develop an interactive, two-echelon simulation game that is particularly useful from a pedagogical standpoint. See also Federgruen and Zipkin (1984a); Schwarz (1989); Jackson and Muckstadt (1989); Hill (1989); de Kok (1990); Erkip et al. (1990); Rogers and Tsubakitani (1991a); Lagodimos (1992); McGavin et al. (1993); Verrijdt and de Kok (1994, 1996); Inderfurth (1994); Forsberg (1995); Kumar et al. (1995); Diks and de Kok (1996); Van Houtum et al. (1996); Van der Heijden (1997); Van der Heijden et al. (1997); Inderfurth and Minner (1998); and Axsäter (2003).

11.4.3.4 Arborescent System with Stock at the Warehouse

When the warehouse can hold inventory, the analysis becomes yet more complicated. Now there are three types of decisions: the amount the warehouse should order from its supplier, the amount to ship from the warehouse to the retailers each period, and the amount to allocate to each retailer when stock runs short. In this section, we briefly outline a somewhat more complex model due to Matta and Sinha (1995) that is still quite practical in application.

Matta and Sinha (1995) assume that each retailer (i) orders from a single warehouse according to an (R, S_i) policy, and that the warehouse employs an (R, s_W, S_W) policy. (All locations use the same review period, R.) These policies are reasonable because many retailers order on a regular basis from their supplier. Trucks are dispatched every week, say, and the retailers must determine how much to order each week. Demand is assumed to be normally distributed and a B_3 shortage cost is applied. The cost function and algorithm developed by Matta and Sinha are approximate and are based on renewal theory and queueing theory. The power approximation discussed in Section 7.5.1 of Chapter 7 is used as well. The renewal approximations are most accurate when the order quantity at the warehouse is large.

The steps for computing the near-optimal policy are:
1. Compute the order quantity at the warehouse using a power approximation, where the order quantity equals $S_W - s_W$.
2. Compute a safety factor for each retailer using a newsvendor-type expression.
3. Compute the total cost, using an approximate total cost expression, for a given warehouse safety factor. Then search over this safety factor for the minimum total cost.
4. Find S_i, s_W, and S_W using expressions similar to those developed in Chapter 6.

11.4.3.5 Additional Research

For other research in this area, see Axsäter (1993a) for an excellent overview and review of the literature. We note several other insights here. Deuermeyer and Schwarz (1981) and Svoronos and Zipkin (1988) develop a decomposition technique for analyzing multiechelon situations. This technique begins by finding inventory policies for each retailer independently. These policies create an ordering process at the warehouse that is correlated over time—an order on a given day from a retailer implies that another order from that retailer is not likely until some time has passed. This demand structure is used to compute an inventory policy for the warehouse, which in turn generates retailer lead times. These lead times are based on the transportation time plus a delay time if the warehouse is out of stock. These new lead times can then be used to find new inventory policies at the retailers. See also Ehrhardt et al. (1981).

Zipkin (1995) shows that in some cases the average warehouse inventory level can vary widely with little change in total cost (assuming that the retailers optimize their inventory policies based on that level). Also, centralized control requires less warehouse stock than does decentralized control. Therefore, stores like Walmart, which have excellent inbound logistics and centralized control systems, need little central warehouse stock. Other firms that have decentralized control and long lead times to the warehouse, but short lead times thereafter require more warehouse stock and less stock downstream.

Hopp et al. (1999) present a model that is relatively easy to implement for the case of an arborescent system with stock at the warehouse, and illustrate how it was applied in a spare parts distribution system. Chen and Zheng (1994a) establish lower bounds on the minimum cost in serial, assembly, and one-warehouse multiretailer systems with setup costs at all levels. Schneider et al. (1995) employ the power approximation to find near-optimal values for (s, S) policies at the retailers and the warehouse. Van der Heijden et al. (1997) deal with stock allocation in a multiechelon network in which retailers have different service targets. They test various allocation rules and find that a rule called *balanced stock allocation* performs very well. Balanced stock allocation first determines the fraction of supply that is allocated from the warehouse to the retailers so that the expected amount of imbalance at the retailers is minimized. If there are more than two levels in the network, the procedure is repeated at all levels. Then all order-up-to levels are determined.

See also Rosenbaum (1981a,b); Van Beek (1981); Chakravarty and Shtub (1986); Jackson (1988); Cohen et al. (1986, 1988); Svoronos and Zipkin (1991); Zipkin (1991); Rogers and Tsubakitani (1991b); Song and Zipkin (1992, 1996); Axsäter (1993b,c); Axsater (1997); Axsäter (1995, 1997); Masters (1993); Axsäter and Zhang (1994); Banerjee and Banerjee (1994); Nahmias and Smith (1994); Shtub and Simon (1994); Aggarwal and Moinzadeh (1994); Schwarz et al. (1985); Chew and Tang (1995); Chew and Johnson (1996); Ahire and Schmidt (1996); Cachon and Lariviere (1999); Forsberg (1996); and Moinzadeh and Aggarwal (1997).

11.4.3.6 Push Control Systems

Brown (1981, 1982) has played a central role in promoting the use of a particular push system called Fair Shares. He has also developed a software system called LOGOL for implementing the ideas. It will become clear that we assume global information and centralized control.

11.4.3.6.1 Operation of a Push System

For simplicity in explanation, we deal with a two-echelon system, although the push concept can certainly be applied to systems with more echelons. The top echelon is a central warehouse supplied by a manufacturing facility or an outside vendor. The lower echelon encompasses a number of retailers meeting customer demands. The central warehouse can also have its own direct customers. The block diagram of Figure 11.9 should be helpful with respect to the following discussion. We begin in Block 1 with a review (on a periodic basis, typically daily or weekly) of the stock status of every SKU (recall that the same item at two different locations is considered as two distinct SKUs). The first point to query (Block 2) is whether or not a system replenishment from the supplier is ready for shipment. This replenishment would have been initiated a system or warehouse lead time, L_W, earlier. If indeed there is such a replenishment, then we must (Block 3) project the *net requirements* of the item at each retailer. At a particular retailer the time series (period by period) of net requirements is obtained in two steps. First, the gross requirements (ignoring the

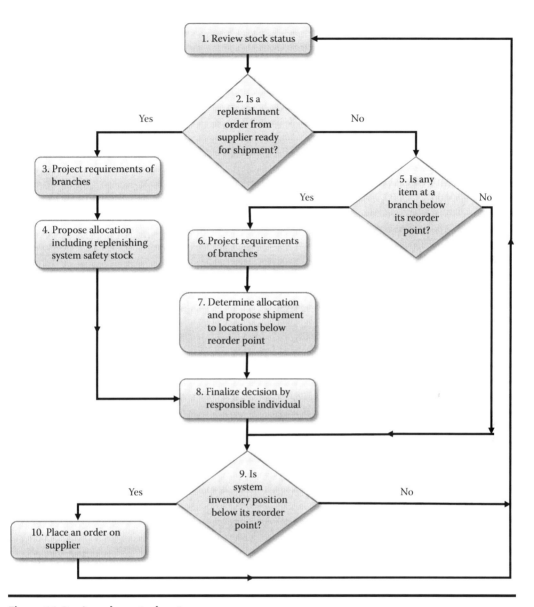

Figure 11.9   A push control system.

inventory position) are projected based on firm customer orders and forecasted demand. Then, the on-hand and on-order (in transit or firmly committed to the retailer) inventories are taken into account, reducing the gross requirements to what are called the net requirements. In addition, the requirements must be offset (moved forward) in time by the shipping time to the retailer so that the timing of the requirements as viewed by the shipping point (warehouse) is accurate. Furthermore, net requirements may be computed above a desired safety stock level. The concept of evaluating net requirements will be central to material requirements planning, a manufacturing control system that will be discussed in detail (including examples of net requirements planning) in Chapter 15.

The replenishment quantity is then allocated (Block 4). First, we ensure that adequate stock is maintained at the central warehouse; normally, this is the so-called *system safety stock*. Thus, the warehouse is not a break-bulk facility. The system safety stock is selected to trade off the costs of carrying this stock with the cost of expediting replenishments from the supplier to avoid a central warehouse stockout. The variability of *total system* demand in a warehouse lead time (L_W) is an important factor, and the decision rule of Section 6.7.5 of Chapter 6 can be used to select the safety factor. The centrally held stock is useful for responding to future needs of the retailers (prior to the arrival of the next system replenishment). The rest of the replenishment order is allocated and shipped to the retailers where it is reasonable to ensure that the time supply above the reorder point is the same for each retailer. In other words, the time to the next replenishment need will be nearly the same for each retailer. (A retailer reorder point is based on its replenishment lead time from the central warehouse, *assuming that there is adequate stock at the latter location*. The retailer safety factor can be based on, e.g., providing a desired customer service level.)

In the event that at a review time there is not a replenishment order ready for shipment, we next (Block 5) investigate the inventory position of the item at each retailer.[*] If any item at a retailer is below its reorder point, then (Block 6) the net requirements of that item at each retailer are projected (as previously discussed with respect to Block 3). An allocation (Block 7) on paper is made of the current system stock of the item (the stock is divided into what is called fair shares for the various locations). Shipments (or transshipments) are proposed for only the item–retailer combinations below their reorder points. Restrictions, such as minimum and maximum quantities to be shipped or convenient quantities (e.g., whole pallets), can easily be incorporated. Moreover, the space and weight requirements of the proposed quantities can be portrayed on a computer terminal, perhaps summarized by destination and required shipping date. This information is of prime importance for reducing freight costs.

Next (Block 8), the shipping schedule is finalized by an individual, such as a materials manager or distribution planner, who has the ultimate responsibility for the decisions. That individual may very well make changes, such as shipping some units early to fill up a freight car, ordering in an extra truck, and so forth.

The next step in the logic is to check (Block 9) the need for a system replenishment. (Specifically, is the system inventory position at or below the system reorder point?) Any such order is initiated (Block 10), possibly using coordinated control logic (such as that of Section 4.9.5 of Chapter 4). This completes the procedure at an instant of review. The process is repeated after the passage of each review interval.

This push system projects actual requirements at any particular retailer well into the future rather than just using a forecasting procedure. The methodology of Distribution Requirements Planning (DRP), to be discussed in Chapter 15, is similar to that of this push system in the sense that actual requirements are projected at each retailer. The important difference is that the lot sizes in DRP are established locally by the retailers, whereas this push system uses centralized control. Most push systems, including DRP, take a deterministic perspective and then modify it by using safety stock. Martel (2003) discusses the advantages and disadvantages of DRP, noting in particular the benefits of the level-by-level solution approach and the problems of using a static amount of safety stock in a time-varying environment. He provides analytical models for investigating other policies.

---

[*] For simplicity, we discuss the case of a single item replenishment. The logic is directly extendible to the more general case of multiple items in the same order.

11.4.3.6.2 Some Comments and Guidelines

We present a number of comments and suggested guidelines with regard to push systems in general and Fair Shares in particular.

1. As should be clear from this entire section, there is currently no simple way of analytically predicting the customer service level as a function of the retailer and system safety factors. However, Brown (1982) conducts a number of simulation experiments to test different methods of allocation of system stock among the central warehouse and the retailers. His results suggest that a firm should hold the system safety stock centrally (centrally can mean at the retailer having the highest usage of the particular item). However, this is dependent upon relatively short shipment times from the central location to the (other) retailers. (We will provide a more complete treatment of this issue in Section 11.6.2.) Where there is adequate system stock, allocate to raise retailer stocks to equal time supplies above their reorder points. When the total system stock is less than the sum of the retailer reorder points, allocate to cover the expected lead time demand plus a common fraction of each retailer's safety stock. More recent research agrees with Brown's results, but suggests more elaborate techniques for allocating shortages, such as those discussed in Section 11.4.3. Federgruen (1993) reviews some of these techniques and comments that in certain situations they are identical. See also Lagodimos (1992) who develops approximations for a periodic review arborescent system with both fair shares allocation and a first-come-first-served (FCFS) allocation policy.
2. The concept of expediting to avoid shortages of system stock works well as long as there are not too many expediting actions needed. The parameter B_1 (the implicit cost of an expediting action) can be adjusted to give a reasonable aggregate expediting workload.
3. The introduction of this push system, or any system with centralized control, will almost always require organizational adjustments even if one firm owns the entire system. The authority and responsibilities of different individuals will change. For example, retail managers will no longer be responsible for the timing and sizes of replenishments. Their primary responsibilities will now be to ensure timely transmission of accurate data on stock status and to provide forecast information. Considerable education is needed as well as the assurance that performance evaluation is properly geared to the new responsibilities. Sawchuk (1981) provides further comments on these issues based on the implementation of a push system in the Norton company. See also Hammond (1994) for a case study involving the implementation of a vendor-managed inventory (VMI) system across different firms. In this and similar cases, the interorganizational control issues become critical. Will the retailer give up control over his or her inventory decisions? If system-wide costs decrease, how will these benefits be shared?
4. Any decision rules should be viewed only as an aid to the decision maker. There are important factors, not included in the underlying models, that must be incorporated manually prior to finalizing the allocation and shipping decisions.

11.5 Remanufacturing and Product Recovery

Thus far, we have restricted our attention to consumable items. Now we turn to units that can be repaired or recovered in some way. Examples include vehicles, telephones, military equipment, computers, copying machines, glass bottles, and so on. The term *product recovery* encompasses

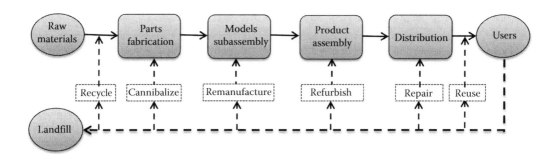

Figure 11.10 Product recovery options. (Adapted from Thierry, M. et al., 1995. *California Management Review 37*(2), 114–135.)

the handling of all used and discarded products, components, and materials. Thierry et al. (1995) note that product recovery management attempts to recover as much economic value as possible, while reducing the total amount of waste. This emerging area of research and practice is generating much interest partly due to new and proposed laws that assign responsibility to manufacturers for the ultimate disposal of their products. Because the issues are varied and complex, the field is quite broad, ranging from studies of the logistics of reusable containers to the process of designing products for disassembly. In this section, we highlight a few papers that provide a window on this field.

Thierry et al. (1995) provide a framework and a set of definitions that can help managers think about the issues in an organized way. See Figure 11.10. These authors examine the differences among various product recovery options, including repair, refurbishing, remanufacturing, cannibalization, and recycling. The whole process of manufacturing begins, of course, with product design. Today, firms are beginning to consider design for the environment (DFE) and design for disassembly (DFD) in their product development processes. See Guide and Wassenhove (2001) for further discussion on product recovery for remanufacturing activities.

The analysis of the recovery situation is considerably more complicated than that of consumables. Normally, in a recovery situation, some items cannot be recovered, so the number of units demanded is not balanced completely by the return of reusable units.[*] Thus, in addition to recovered units, we must also purchase some new units from time to time. Consequently, even at a single location, there are five decision variables: (1) how often to review the stock status, (2) when to recover returned units, (3) how many to recover at a time, (4) when to order new units, and (5) how many to order. When there are multiple locations, we must decide how many good units to deploy to a central warehouse, and how many to deploy to each retailer or field stocking location.

With consumable items, the lead time to the retailers is a transportation time from the warehouse plus a random component, depending on whether the warehouse has stock. With recoverable items, the lead time is the transportation time plus the time to recovery, if the warehouse does not have stock. So in some cases the two systems can be treated in almost the same way. However, if the recovery facility has limited capacity, or if the number of items in the system is small, the systems will differ significantly. For example, if many items have failed and are now in recovery,

---

[*] We use the term *recovered* to include any units that are returned via the repair, remanufacturing, etc., processes illustrated in Figure 11.10.

they cannot be in the field generating failures. Therefore, the demand rate at the warehouse will decline. In a consumable system, it is usually assumed that the demand rate does not depend on how many items have been consumed.

11.5.1 Multiechelon Situation with Probabilistic Usage and One-for-One Ordering

For the specific case of repairable items, see reviews by Nahmias (1981); Mabini and Gelders (1990); Diks et al. (1996); Verrijdt (1997), and Sherbrooke (2006). The Sherbrooke book provides an overview of 30 years of research on repairable item inventory in multiechelon systems. Because of the common military applications, most of the original work on this problem was done in that context. Specifically, instead of retailers, researchers studied military bases that generate demand for repairs of, say, aircraft engines. Resupply of parts comes from possible local (base) repair of failed parts and from a central repair depot that both repairs failed parts, if possible, and purchases new parts.

Some of the foundational work in this area was done by Sherbrooke (1968) and is called METRIC, for Multi-Echelon Technique for Recoverable Item Control. It in turn is based on a fundamental theorem due to Palm (1938). We now discuss these briefly before presenting a useful and relatively simple approach to this problem.

Palm's theorem states that if demand for an item is Poisson with mean m, and if the repair time for each failed unit follows *any* distribution and is independent and identically distributed with mean T, then the steady-state probability distribution for the number of units in the repair system has a Poisson distribution with mean mT. See Sherbrooke (1992b). The assumption that is required to use this simple result is that there are sufficiently many repair people and equipment so that the repair time always has the same mean. If the failure rate of the items surges so that the repair facility is overloaded, the expected time to repair will increase, and the condition of the theorem will be violated. On the other hand, a number of researchers have pointed out that this assumption is fairly realistic in real-world settings. Extensions to Palm's theorem have been developed by Feeney and Sherbrooke (1966) for compound Poisson demand, and Hillestad and Carrillo (1980) and Crawford (1981) for the case of time-varying demand and repair rates. See also Carrillo (1991).

Sherbrooke (1968), in the METRIC model, considers a two-echelon system composed of a set of bases and a central depot. Each location in the system may possess a supply of serviceable spare parts and a repair capability for converting repairable units into serviceable items. Demand at each base for each item follows a compound Poisson distribution. Because of the extensions to Palm's theorem, all that is needed is the mean repair time, if the assumption about independent repair times is made. METRIC employs this assumption to approximate the real system. The inventory policy followed for each item at each location is a continuous review $(S - 1, S)$ policy; that is, the inventory position is always kept at the same order-up-to-level S (where, of course, S varies among item–location pairs). As discussed in Section 8.1.1 of Chapter 8, an $(S - 1, S)$ policy is most appropriate for expensive, slow-moving items. This type of policy substantially simplifies the mathematical analysis, in that the net inventory of any item at any base, at a random point in time, is the difference between the particular S and the demand in a mean repair time, where the demand has a compound Poisson distribution. Furthermore, no lateral transfers are permitted between bases. Now, if there are n bases, we wish to select values of $S_i(j)$ for all bases i, and all items j, where $S_i(j)$ is the inventory position of item j at base i and $S_0(j)$ is the inventory position of item j at the depot. There is a budget constraint on the total value of the spares, and the objective is to minimize the total expected number of units backordered at the bases. Fox and Landi (1970) show that a Lagrangian minimization approach (see Appendix I at the end of the book) can substantially reduce the computational effort, and Muckstadt (1978) expands on this idea.

There is a basic problem with using the Poisson distribution for modeling military failure rates: the mean and the variance of the Poisson are equal. Therefore, it is often difficult to fit the Poisson distribution to real data. Extensions of Palm's theorem to compound Poisson demand, as in METRIC, have addressed this issue to some extent. However, these models only provide an expected value of the delay due to repair—no variance information is given. Graves (1985) develops an approach that approximates both the mean and variance of the number of items in the repair process. We now discuss this approach in more detail, returning to the setting of a single item stocked at a single warehouse and multiple retailers. First we require some notation:

N = number of retailers
$\hat{x}_i$ = expected demand per unit time at retailer i
$\hat{x}_W = \sum_i \hat{x}_i$ = expected demand per unit time at the warehouse
L_i = transportation time from the warehouse to retailer i
L_W = lead time to the warehouse
S_i = order-up-to inventory position at retailer i
S_W = order-up-to inventory position at the warehouse
I_i = inventory on hand at retailer i in steady state
I_W = inventory on hand at the warehouse in steady state
B_i = number of backorders at retailer i in steady state
B_W = number of backorders at the warehouse in steady state
W_W = time delay due to stockouts at the warehouse in steady state
O_i = number of orders outstanding at retailer i (i.e., issued to the warehouse but not yet received) in steady state
O_W = number of orders outstanding at the warehouse (i.e., issued to the outside supplier but not yet received) in steady state
h_i = holding cost per unit per unit time at retailer i
h_W = holding cost per unit per unit time at the warehouse
$B_3 v_i$ = shortage cost per unit per unit time at retailer i

Graves uses the following expression for the expected number of backorders at the warehouse:

$$E(B_W) = \sum_{j=S_W+1}^{\infty} (j - S_W) \frac{(\hat{x}_W L_W)^j}{j!} exp(-\hat{x}_W L_W)$$

$$= \sum_{j=S_W+1}^{\infty} (j - S_W) p_{p0}(j|\hat{x}_W L_W) \qquad (11.23)$$

Note that this expression follows from the definition of a Poisson random variable discussed in Section 8.1.1 of Chapter 8. The expression is the probability that a Poisson random variable with mean $\hat{x}_W L_W$ takes on the value j. Then the variance of the number of backorders at the warehouse is

$$\text{var}(B_W) = \sum_{j=S_W+1}^{\infty} (j - S_W)^2 p_{p0}(j|\hat{x}_W L_W) - [E(B_W)]^2 \qquad (11.24)$$

The following is due to a queuing theory result known as Little's law (see Heyman and Sobel 1982, pp. 399, and Chapter 16):

$$E(W_W) = E(B_W)/\hat{x}_W \tag{11.25}$$

The expression for expected inventory on hand at the warehouse is derived similarly to the expression for expected number of backorders:

$$E(I_W) = \sum_{j=0}^{S_W-1} (S_W - j)p_{p0}(j|\hat{x}_W L_W) \tag{11.26}$$

Then the average lead time at retailer i is

$$\bar{L}_i = L_i + E(W_W) \tag{11.27}$$

or the transportation time plus the expected delay due to stockouts at the warehouse.

The mean and variance of the number of orders outstanding are

$$E(O_i) = \hat{x}_i \bar{L}_i \tag{11.28}$$

and

$$\text{var}(O_i) = \left(\frac{\hat{x}_i}{\hat{x}_W}\right)^2 \text{var}(B_W) + \left(\frac{\hat{x}_i}{\hat{x}_W}\right)\left(\frac{\hat{x}_W - \hat{x}_i}{\hat{x}_W}\right) E(B_W) + \hat{x}_i L_i \tag{11.29}$$

This expression is derived from a result in Graves' paper that equates the aggregate outstanding orders at all retailers at time $t + L_i$, to the sum of the backorders at a time, t, at the warehouse plus the total demand at all retailers in the time interval between time t and time $t + L_i$. Once these values are computed, Graves fits a negative binomial distribution

$$P(O_i = j) = \left(\frac{(a_i + j - 1)!}{j!(a_i - 1)!}\right) p_i^{a_i}(1 - p_i)^j \quad j = 0, 1, 2, \ldots \tag{11.30}$$

where a_i is a positive integer and $0 < p_i < 1$ such that

$$E(O_i) = a_i(1 - p_i)/p_i \tag{11.31}$$

$$\text{var}(O_i) = a_i(1 - p_i)/p_i^2 \tag{11.32}$$

Note that we may not be able to get an (a_i, p_i) pair that exactly fits a given $E(O_i)$, $\text{var}(O_i)$. It is often necessary to round a_i to an integer, even though the mean and variance then do not exactly fit. Finally

$$\begin{aligned} E(B_i) &= \sum_{j=S_i+1}^{\infty} (j - S_i)P(O_i = j) \\ E(I_i) &= \sum_{j=0}^{\infty} (S_i - j)P(O_i = j) \end{aligned} \tag{11.33}$$

The total cost is

$$h_W E(I_W) + \sum_{i=1}^{N} h_i E(I_i) + \sum_{i=1}^{N} B_3 v_i E(B_i)$$

Graves shows that this approximation led to a wrong stocking decision in only 0.9% of the cases he examined, compared to 11.5% when using METRIC.

11.5.1.1 Numerical Illustration

Let

$$N = 2 \text{ identical retailers}$$
$$L_W = 2.4 \text{ weeks}$$
$$L_i = 2 \text{ weeks}$$
$$\hat{x}_i = 0.6 \text{ units/week}$$
$$h_W = \$2.5 \text{ per unit per week}$$
$$h_i = \$2.7 \text{ per unit per week}$$
$$B_3 v_i = \$5.1 \text{ per unit per week}$$

So $\hat{x}_W L_W = 2.88$ units. A search is required for the best triplet (S_W, S_1, S_2), where, because the retailers are identical, we can assume that $S_1 = S_2$. A spreadsheet can be used for this purpose. We illustrate the calculations for $S_W = 2$ and $S_1 = S_2 = 3$.

j	$(j - S_W)p_{p0}(j|\hat{x}_W L_W)$	$(j - S_W)^2 p_{p0}(j|\hat{x}_W L_W)$
3	0.2235	0.2235
4	0.3218	0.6437
5	0.2781	0.8342
6	0.1780	0.7118
7	0.0915	0.4576
8	0.0395	0.2372
9	0.0148	0.1033
10	0.0049	0.0389
11	0.0014	0.0129
12	0.0004	0.0038
13	0.0001	0.0010
14	0.0000	0.0003
15	0.0000	0.0001
Sum	1.1540	3.2683

So

$$E(B_W) = 1.1540 \text{ and } \text{var}(B_W) = 3.2683 - (1.1540)^2 = 1.9366$$
$$E(W_W) = 1.1540/1.2 = 0.9616$$
$$\text{and } E(I_W) = 0.2740 \text{ as shown below}$$

j	$(S_W - j)p_{p0}(j|\hat{x}_W L_W)$
0	0.1123
1	0.1617
Sum	0.2740

So

$$\bar{L}_i = 2 + 0.9616 = 2.9616 \text{ weeks}$$

$$E(O_i) = 6(2.9616) = 1.7770 \text{ units}$$

$$\text{var}(O_i) = \left(\frac{0.6}{1.2}\right)^2 (1.9366) + \left(\frac{0.6}{1.2}\right)\left(\frac{1.2 - 0.6}{1.2}\right)(1.1540) + 0.6(2)$$

$$= 1.9726 \text{ (units)}^2$$

Using Equations 11.31 and 11.32 and these values, we obtain $a_i = 17$ and $p_i = 0.900808$, after rounding a_i to an integer. Note again that this (a_i, p_i) pair does not exactly give the above $E(O_i)$ and $\text{var}(O_i)$ because a_i is forced to be an integer. Finally, using the following tables (and recalling that we are evaluating $S_i = 3$), we get the expected number of backorders and the expected inventory on hand, respectively, at retailer i.

j	$(S_i - j)P(O_i = j)$
0	0.5080
1	0.5711
2	0.2549
Sum	1.3340

j	$(j - S_i)P(O_i = j)$
4	0.0794
5	0.0662
6	0.0361
7	0.0157
8	0.0058
9	0.0019
10	0.0006
11	0.0002
Sum	0.2060

$$E(B_i) = 0.2060$$
$$E(I_i) = 1.3340$$

The total cost for these S values is

$$\text{TRC}(S_W = 2, S_1 = 3, S_2 = 3) = 2.5(0.2740) + 2(2.7)(1.3340) + 2(5.1)(0.2060) = \$9.99$$

Axsäter (1990b) provides an exact method for computing the costs directly. His method can be developed easily a programming language, but not so easily on a spreadsheet because it involves recursive calculations.

11.5.2 Some Extensions of the Multiechelon Repair Situation

In this section, we discuss several extensions of the multiechelon repair situation, and we provide a number of references that should help the interested reader pursue further investigation.

11.5.2.1 Indentured Modules

Consider a complex piece of equipment that is made up of a number of subassemblies, which in turn have components, and so forth. (The levels in this hierarchy are often called indenture levels.) An example would be an aircraft with engines, guidance system, and so forth. The equipment breaks down from time to time because of the failure of specific components. Having estimates of the failure rates of the various components, the cost of each component or subassembly, and detection and repair time distributions, an important problem, even for the case of a single location, is to ascertain how to allocate a budget among spare components, subassemblies, and complete assemblies to maximize the availability of operable complete units or, alternatively, to minimize the inventory investment to achieve a desired service level. When multiple locations are considered, this is the METRIC problem with the added complexity of indenture levels. Now, we must choose both how many of each SKU to have in stock, and where the units should be kept. Because of the indenture hierarchy, we have the further complexity that components cannot be treated independently, because any specific piece of equipment requires a particular set of components. Recognizing these complexities, it is not surprising that simple mathematical results are not forthcoming and the applications have been restricted primarily to large-scale military systems.

We next briefly review some of the key developments in this general problem area. In addition, comprehensive reviews are provided by Demmy and Presutti (1981); Nahmias (1981); and Kennedy et al. (2002).

The MOD-METRIC model due to Muckstadt (1973) addresses a two-indenture version of the above situation. The two indentures are line replaceable units (LRUs) that may be removed and replaced in the field, and shop replaceable units (SRUs) that are components or subassemblies of LRUs that must be removed or replaced in the repair shops. An $(S - 1, S)$ policy is again employed. The objective is to allocate a total budget among spares to minimize the total expected LRU base-level backorders. The heuristic solution procedure works as follows. The available budget W is divided into two components W_1 and W_2. W_1, using METRIC, is allocated among the SRUs in order to minimize the expected LRU repair delays summed over all bases. The resulting allocation determines average resupply time for each LRU at each base. Knowing these, METRIC is again used to allocate W_2 among the LRUs to minimize the expected LRU base-level backorders. These steps are repeated for various values of W_1 and W_2 (summing to W), and the best partition is used.

11.5.2.2 *Multiechelon Situation with Probabilistic Demand and Batch Ordering*

In this section, we briefly note some other relevant literature in this area. An excellent review is provided by Axsäter (1993b). The paper by Svoronos and Zipkin (1988) mentioned in Section 11.4.3 also applies in the context of repairable items. Related work includes Moinzadeh and Lee (1986) and Lee and Moinzadeh (1987a,b). Approximations by Axsäter (1993b) and Svoronos and Zipkin (1988) have been shown to be reasonably accurate. Ashayeri et al. (1996) describe an interesting case study at Olivetti that uses a deterministic model in conjunction with simulation to achieve a desired service level. Chen (2000) and Chao and Zhou (2009) also analyze the multiechelon, batch-ordering setting in a serial supply chain, presenting algorithms for determining control parameters.

11.5.2.3 *Additional Research*

There is an enormous amount of research in the repairable area. In this section, we simply highlight a few additional studies. The LMI procurement model (Logistics Management Institute 1978) essentially combines the classical series-reliability model (a complete system does not operate unless all of its component systems operate) with the basic METRIC assumptions for individual-item, two-echelon situations. The objective is to establish procurement quantities that maximize the expected number of operational pieces of equipment (complete systems), subject to a given budget constraint. A marginal analysis solution procedure is used again. Starting with no units procured, we keep increasing procurement quantities one unit at a time, always picking the unit that has the largest incremental effect on the objective function.

A number of authors explicitly consider the scheduling of the repair shop in conjunction with inventory and procurement policies. Hausman and Scudder (1982) use a simulation model to test various rules for scheduling repair operations in a shop supporting a multi-item inventory system with an indentured product structure. In contrast with most of the models discussed earlier, here, repair times need not be assumed independent. See also Pyles (1984); Pyke (1990); Gustafson (1991), and Tripp et al. (1991). Some of these references include the possibility of cannibalizing, or "borrowing" a required part from another piece of equipment that is inoperable for other reasons. See also Fisher and Brennan (1986).

Many authors have attempted to apply mathematical or queueing models to study these systems. References include some of the above as well as Hillestad (1982); Albright and Soni (1988); Ebeling (1991); and Abboud (1996). Others strive for a closer approximation to real military situations and employ simulation models. Other references include Cheung and Hausman (1995), who study the situation with more than one part failing at a time, and Verrijdt et al. (1998). See also Nahmias and Rivera (1979); Muckstadt (1979); Hollier (1980); Kaplan and Orr (1985); Sherbrooke (1986); Gross et al. (1987); Cohen et al. (1989); Cohen and Lee (1990); Gupta and Albright (1992); Cheung (1996); and Muckstadt (2005).

11.5.3 **Some Insights and Results for the More General Context of Remanufacturing and Product Recovery**

In this section, we comment on some of the many issues that arise in the more general situation of remanufacturing and product recovery. In manufacturing itself, waste disposal is a major consideration. A paper by Bloemhof-Ruwaard et al. (1996) examines the issue of designing a network with plants and waste disposal units, and coordinating the product and waste flows therein. Much of the work in this area of manufacturing is based in chemistry and other hard sciences, as firms try to redesign production processes and materials so that the process will have less environmental impact. We will not review this research here.

Most of the research in this area relevant to this book concerns products and packaging after manufacturing has been completed. For example, a large U.S. chemical company gained significant market share in water treatment chemicals by delivering its products in reusable containers. The customers (e.g., hospitals and other large institutions) need never touch the chemicals or deal with the disposal of used containers. This problem has been addressed by Goh and Varaprasad (1986); Kelle and Silver (1989); Buchanan and Abad (1995); and Castillo and Cochran (1996), among others.

Some products that are not reused as is can be disassembled so that some of the parts can be used in remanufactured products. Muckstadt and Isaac (1981) report on a model developed in connection with a manufacturer of reprographic equipment. There is a single location with two types of inventory: serviceable and repairable. See Figure 11.11. Demands for serviceable units and returns of repairable units occur probabilistically, specifically, according to independent Poisson processes with rates D and fD, respectively (where f is a fraction). In addition, repairs are done on a continuous, first-come first-served basis (e.g., at a local machine shop). Any demands for serviceable units, when none is available, are backordered at a cost per unit short per unit time. Purchases of new stock from outside involve a known lead time. With respect to purchase decisions, a continuous review (s, Q) system is used; specifically, when the inventory position drops to s or lower, a quantity Q is purchased.

Inderfurth (1997) extends the Muckstadt and Isaac model to a remanufacturing problem in which there are two decisions each period: how many returned products to remanufacture (the

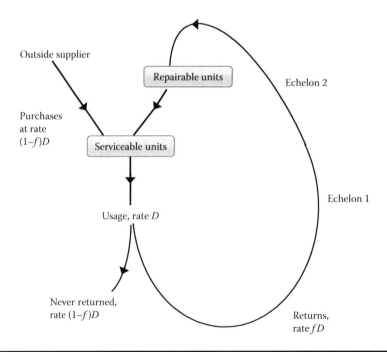

Figure 11.11 Repairable units at a single location.

remainder will be disposed of), and how many new parts to procure. In this system, returned products arrive probabilistically and are either remanufactured or thrown away. (In other words, there is no stock of returned products.) Newly procured products are stored with remanufactured products in a finished goods inventory that serves demand that arrives probabilistically. There are per-unit costs to procure, remanufacture, and dispose, and holding costs are charged against ending inventory each period. For the case of equal lead times to remanufacture and to procure, Inderfurth shows that the structure of the optimal policy is based on two parameters, L_t and U_t, in each period t. To describe the policy, we need some notation:

d_t = the number of units to be disposed of in period t

p_t = the number of units to procure in period t

IP_t = the inventory position at the beginning of period t = stock on hand (which includes products returned this period and finished goods inventory) + procurement orders outstanding + remanufacturing orders outstanding − backordered demand

The optimal policy is then

$$p_t = L_t - IP_t \quad \text{and} \quad d_t = 0 \quad \text{for } IP_t < L_t$$
$$p_t = 0 \quad \text{and} \quad d_t = 0 \quad \text{for } L_t \leq IP_t \leq U_t$$
$$p_t = 0 \quad \text{and} \quad d_t = IP_t - U_t \quad \text{for } IP_t > U_t$$

In words, if the position is lower than the lower limit, L_t, order-up-to L_t and do not dispose of any units. If the position is higher than the upper limit, U_t, dispose *down to* U_t and do not procure any units. Otherwise, do not buy or dispose. (Again, all returned units, not disposed of, are remanufactured.) Inderfurth points out that when one permits a stock of returned units waiting for disposal or remanufacturing, or when the lead times to procure and remanufacture are different, the policy is similar but more complex.

Van der Laan and Salomon (1997) propose a policy for a continuous review version of this problem. See also Heyman (1977); Penev and de Ron (1996) who study the disassembly process, and Van der Laan et al. (1996); Salomon et al. (1994); Ferrer (1995); Richter (1996); Inderfurth (1996); Guide and Spencer (1997); Taleb et al. (1997); Fleischmann et al. (1997); Guide (2000); Rudi et al. (2000); Mahadevan et al. (2003); and Savaskan et al. (2004) who study other aspects of the remanufacturing process.

11.6 Additional Insights

In this section, we comment briefly on several additional insights for multiechelon systems and supply chain management that have been developed in the literature. We begin with a well-known argument that stocks should be held centrally.

11.6.1 Economic Incentives to Centralize Stocks

In the following, we parallel the presentation of Schwarz (1981). Consider n retail outlets ($i = 1, 2, \ldots, n$). Using an independent or decentralized (s_i, EOQ$_i$) strategy for each and assuming, for simplicity, the same safety factor k at each location (equivalently, equal P_1 service levels) and demand that is independent across retailers, we have that the expected total relevant costs for the system are

$$\text{ERTC(decentralized)} = \sum_{i=1}^{n} \text{ERTC}_i$$

$$= \sum_{i=1}^{n} [\sqrt{2A_i D_i vr} + k\sigma_i vr]$$

Assuming the ordering cost A_i is the same (A) at all locations, this becomes

$$\text{ERTC(decentralized)} = \sqrt{2Avr} \sum_{i=1}^{n} \sqrt{D_i} + kvr \sum_{i=1}^{n} \sigma_i \qquad (11.34)$$

Suppose, instead, that the stocking was done at a single centralized location. Then the expected total relevant costs are

$$\text{ERTC(decentralized)} = \sqrt{2AD_c vr} + kvr\sigma_c \qquad (11.35)$$

where

$$D_c = \sum_{i=1}^{n} D_i$$

hence

$$\sqrt{D_c} \leq \sum_{i=1}^{n} \sqrt{D_i} \qquad (11.36)$$

and σ_c is the standard deviation of total demand. It can be shown that

$$\sigma_c \leq \sum_{i=1}^{n} \sigma_i \qquad (11.37)$$

This result is known, in the finance literature, as the portfolio effect. It follows from the fact that higher than average demands at some locations will be simultaneously offset by lower than average demands at some other locations. Using the conditions of Equations 11.36 and 11.37, it follows from Equations 11.34 and 11.35 that

$$\text{ETRC (centralized)} \leq \text{ETRC (decentralized)}$$

Furthermore, the difference can be quite substantial (see Problem 11.8). See also Tagaras and Cohen (1992) who show that complete pooling dominates partial pooling for the case of negligible transshipment times. Oeser and Romano (2016) empirically investigate the impact of inventory centralization based on data from German manufacturing and trading companies.

11.6.2 Where to Deploy Stock

The deterministic analysis above suggests that centralizing inventories are preferable. Some researchers have also argued for decentralizing inventories. The general question is whether the warehouse should hold back substantial inventories so that they can be allocated to retailer demands as needed, or whether most inventory should be pushed forward to the retailers? In this section, we discuss this issue.

Rosenfield (1989) describes how small "storefront" locations can hold minimal inventory and therefore keep the manufacturer close to the customer at low cost. These low-cost and low-overhead locations are cheaper to operate than full-service outlets. See also Gerchak and Gupta (1991) and Anupindi and Bassok (1999).

Now consider another factor: the product's demand rate. For very expensive low demand parts in a multi-echelon system it may be worth not storing items at field locations where demand for the item is quite low. For these locations, the product can be expedited (at additional expense) from a central location whenever a demand occurs. Field locations with sufficient demand volume may choose to stock the product. See Muckstadt and Thomas (1980).

Firms in higher demand environments will exhibit very different behavior. These firms often ship in full truckloads (FTL), and the clear choice for deployment of inventory is to push at least some of it to the retail level. The reason is that if the product does not sell today, it will almost surely sell tomorrow, and the savings from shipping FTL outweigh any small inventory savings that could be gained by holding back inventory at the warehouse level. On the other hand, some stocks, say 1 week's demand, should be held back to account for emergency requirements. The exact amount to hold back can be approximated by the models described in the previous sections of this chapter, and will depend on demand rates, transportation costs, lead times, holding costs, variability of demand, and the service objective. Related research includes Bregman et al. (1990); Ernst and Pyke (1993); Tempelmeier (1993); Ahn et al. (1994); Hall and Racer (1995); Webb and Larson (1995); and Henig et al. (1997).

In an interesting study along these lines, Barnes-Schuster et al. (2006) examine the effect of delivery lead time on the costs of a manufacturer/retailer system for three scenarios: a single retailer and a single manufacturer, multiple retailers and a single manufacturer, and a single retailer and multiple manufacturers. For each scenario, it is shown that system safety stock should not be split between a retailer and a manufacturer. In other words, the safety stock for the system should be held at either one or the other level, but not both. Note that the authors assume that the two parties cooperate and that the delivery lead time is negotiated. It is not in general true that the system safety stock should be held at only one location. From basic models and our own experiences, we can say that it becomes more attractive to retain stocks centrally when the carrying charges are appreciably lower at that stage and/or when the distribution lead times are relatively small. In this regard, downstream lead times are more crucial than the upstream ones, and effort should be devoted to reducing the downstream lead times. See also Yano (1987); Haynsworth and Price (1989); Chen and Lin (1990); Newhart et al. (1993); Inderfurth (1995); Inderfurth and Minner (1998); and Van der Heijden et al. (1997).

Minner (1997) and Graves and Willems (2000, 2003a) develop dynamic programming approaches for determining placement of safety stocks throughout a network. Graves and Willems (2000) further discuss application of the approach at Eastman Kodak. Klosterhalfen et al. (2014) addresses placement of stocks in a network when there are two sources of supply.

11.6.3 Lateral Transshipments

The most common assumption in multiechelon research is that shipments among retailers are not allowed. However, Karmarkar and Patel (1977) have shown that costs can decrease, and service can improve, if lateral transshipments are used in emergencies. If, on the other hand, transshipments are used in anticipation of stock imbalances among retailers, costs can go up due to excessive unnecessary movement of product. This problem has been studied also by Lee (1987); Robinson (1990); and Axsäter (1990a). Sherbrooke (1992a) takes the analysis a step further by allowing for lateral resupply of spare parts among retailers, and also for delayed lateral resupply if a retailer needs a part, and no other retailer has one available, but one becomes available at a later time. The benefits of this flexibility are negligible if the number of parts is very low or very high. Otherwise, there are 30%–50% reductions in backorders by using lateral resupply. Jönsson and Silver (1987a) and Diks and de Kok (1996) show that lateral transshipment is beneficial if demand variability is high, there are many retailers, and service levels are high. See also Hoadley and Heyman (1977); Pyke (1990); Evers (1996); Alfredsson and Verrijdt (1999); Mercer and Tao (1996); Archibald et al. (1997); Rudi et al. (2001); Agrawal et al. (2004); and Shao et al. 2011. See Paterson et al. (2011) for a review of the literature on lateral transshipments.

11.7 Summary

In this chapter, we have dealt with the difficult multiechelon situation in which there is a dependency of demand among SKUs. Exact analyses, particularly in the case of probabilistic demand, quickly become intractable. Thus, again, we have advocated the use of heuristic methods.

This completes our treatment of "pure" inventory situations. Beginning with Chapter 13, we address production contexts. Chapter 15 will deal with the production analog of a multiechelon situation.

Problems

11.1 Consider a two-stage serial process where the second stage is a rather minor operation adding little extra value to the product. To be more specific, suppose we have an item with these characteristics (borrowing the notation of "W" for warehouse, or primary stage and "R" for retailer, or finishing stage).

$$D = 1{,}000 \text{ units/year} \qquad v_W = \$5/\text{unit}$$
$$v_R = \$6/\text{unit} \qquad r = 0.24 \, \$/\$/\text{year}$$
$$A_W = \$20 \qquad A_R = \$10$$

Use the procedure of Section 11.3.1 to obtain Q_W, Q_R, and n. Do your results make intuitive sense?

11.2 For the situation of Section 11.3.3,

a. Verify that the average actual (not echelon) inventory at the primary stage (warehouse) is

$$I_W = \left(\frac{n-1}{2n}\right) Q_W$$

b. Show that $\bar{I}_W v_W + \bar{I}_R v_R = \bar{I}'_W v'_W + \bar{I}'_R v'_R$.

11.3 Starting with the result of Equation 11A.3 (in the Appendix of this chapter), find an expression, involving A_R, A_W, v'_W, v'_R, and n that must be satisfied for indifference between n and $n + 1$. Sketch the curves (for different n values) with suitable horizontal and vertical axes.

11.4 Consider the example that was used throughout Chapter 5. Suppose that the production process actually involves two stages (call them primary and finishing, or warehouse and retailer). The relevant cost parameters are $A_W = \$24$, $A_R = \$30$, $v_W = \$15/box$, $v_R = \$20/box$, and $r = 0.02 \ \$/\$/month$. Note that $A_W + A_R = \$54$, the A value that was used, with $v = \$20/unit$, in Chapter 5. Thus, implicitly in Chapter 5, we assumed that any replenishment at the primary stage was immediately processed through the finishing stage (with no primary inventory being retained). For simplicity, we use only the first 8 months of the requirements pattern, repeated as follows:

Month	1	2	3	4	5	6	7	8
Requirements in boxes (for finished product)	10	62	12	130	154	129	88	52

a. Use the sequential, unmodified, Silver–Meal approach to determine the patterns of replenishments. To be specific, first use the Silver–Meal at the finishing stage with $A = \$30$ and $v = \$20/box$ to get the pattern of replenishments at the finishing stage. These replenishments become the requirements pattern for the primary stage. Use these, with $A = \$24$ and $v = \$15/box$, to find the replenishments at the primary stage. Finally, cost out the overall solution.

b. Use the Blackburn–Millen approach to establish the replenishments. In particular, Equations 11.10 through 11.12 are used to establish $\hat{A}_R$ and $\hat{v}_R$ for the Silver–Meal heuristic at the finishing stage. Again, cost out the overall solution and compare with part a.

11.5 Using Equations 11.10 through 11.12, prove that

$$\frac{\hat{A}_R}{\hat{v}_R} > \frac{A_R}{v_R}$$

hence that there will tend to be more batching at the finishing stage under the Blackburn–Millen procedure than under the straight sequential approach.

11.6 A famous remote lodge in the Canadian Rocky Mountains uses fuel oil for cooking and heating purposes. The lodge is accessible by a four-wheel-drive truck on a rough forestry road. The fuel is stored in bulk in a large tank where the forestry road intersects with the main highway. On each trip, the truck can carry 3,000 liters of fuel and the main tank holds 18,000 liters. The usage of the fuel at the lodge is rather uncertain, particularly because of weather conditions. However, in a specific season of interest, data indicate that for a 1-day period, usage can be assumed to be normally distributed with $\hat{x}_1 = 450$ liters and $\sigma_1 = 125$ liters. The capacity for storage at the lodge is 4,000 liters and deliveries of 15,000 liters are made by tanker-truck to the tank on the highway. The unit cost of the fuel delivered to the main tank is $\$0.15/liter$ and the value added through the loading, trucking up the forestry road, and unloading is estimated to be $\$0.04/liter$. A lead time of

2 days is required to replenish the lodge's supply (assuming there is fuel in the main tank) and the lead time is 12 days for a replenishment of the main tank. When a stockout occurs at the lodge, propane can be used but at a cost of $0.28/liter. Assume a carrying charge of 0.26 $/$/year.

    a. What should be the reorder points for the lodge and for the main tank?

    b. Discuss some complexities ignored.

11.7 In the procedure, described in Section 11.4.1, for selecting the base stock level in a general, multiechelon situation, what implicit assumption is made about the availability of stock at the next (further removed from the customer) level? Can you suggest a correction factor that might allow relaxation of this assumption.

11.8 For the situation of Section 11.6.1, suppose that $D_i = D$ and $\sigma_i = \sigma (i = 1, 2, \ldots, n)$ and that there is no correlation among forecast errors at different locations. Under such circumstances, determine ETRC (centralized) ÷ ETRC (decentralized) as a function of n, and evaluate for $n = 2, 4,$ and 9.

11.9 Consider a warehouse that serves five retailers that face nearly identical costs and demands. The warehouse holds no stock, and therefore serves as a break-bulk facility. They order from their supplier, receive the products, and immediately distribute them to the five retailers. Find the order-up-to level for the warehouse assuming the following data:

    a. $R = 2$ weeks

             $J = 5$ retailers

           $B_3 = 0.35$ $/$/unit time

             $r = 0.24$ $/$/unit time

        $L_W = 2.5$ weeks

         $L_R = 2$ weeks

        Weekly demand

        $\hat{x}_j = 45$ units for all retailers

        $\sigma_j = 12$ units for all retailers

    b. If three retailers each have an inventory position of 160 units, while the other two have 200 units, what amount should be allocated to each, using the rule given in Section 11.4.3?

11.10 a. Repeat Problem 11.9, part a for $B_3 = 2$ $/$/unit time.

      b. Repeat Problem 11.9 part b for four retailers each having an inventory position of 220 units, while the last has only 150 units. (Assume that $B_3 = 2$ $/$/unit time.)

11.11 Using the approach of Graves in Section 11.5.1, find the total cost for $S_W = 1$ and $S_1 = 1, S_2 = 3$.

    Let

        $N = 2$

        $L_W = 3$ weeks

          $L_i = 1$ week, $i = 1, 2$

          $\hat{x}_1 = 0.2$ units/week

          $\hat{x}_2 = 0.8$ units/week

         $h_W = \$3.8$ per unit per week

          $h_i = \$5$ per unit per week

       $B_3 v_i = \$8$ per unit per week

11.12 Repeat Problem 11.11 for $S_W = 2$ and $S_1 = 1, S_2 = 2$.

11.13 A warehouse supplies a single retailer with a product. The warehouse's purchasing lead time is 3 weeks and the lead time for the retailer (from the warehouse) is 1 week. Assume that forecast errors are normally distributed with the σ for 1 week being 70 units, and that $\sigma_t = \sqrt{t}\sigma_1$. The annual demand rate is 12,000 units, and the following data have been gathered:

$v_W = \$32/\text{unit}$, $v_R = \$38/\text{unit}$, $r = 0.22$ \$/\$/year, $n = 2$, or the warehouse orders two batches of the retailer's order quantity at one time. The retailer's order quantity is $Q_R = 500$ units. Finally, suppose that the fractional charge per unit short, B_2, is set at 0.65. Find the reorder point for both locations.

11.14 A retailer supplies a small storefront outlet with a unique product. In a sense, the retailer serves as the warehouse for this outlet. The retailers' purchasing lead time is 2 weeks and the lead time for the outlet from the retailer is also 2 weeks, primarily due to order handling time. Assume that forecast errors are normally distributed with the σ for 1 week being 10 units, and that $\sigma_t = \sqrt{t}\sigma_1$. The annual demand rate is 2,000 units, and the following data have been gathered: $v_W = \$6/\text{unit}$, $v_R = \$8/\text{unit}$, $r = 0.24$ \$/\$/year, $n = 3$, or the retailer orders two batches of the outlets order quantity at one time. The outlet's order quantity is $Q_R = 200$ units. Finally, suppose that the fractional charge per unit short, B_2, is set at 0.4. Find the reorder point for both locations.

11.15 Build a spreadsheet to solve the problem in Section 11.3.1.

11.16 Build a spreadsheet to solve the problem in Section 11.4.2.

11.17 Build a spreadsheet to solve the problem in Section 11.4.3.

11.18 Build a spreadsheet to solve the problem in Section 11.5.1.

11.19 Consider a situation where a company owns a single warehouse and several retail stores. It buys a product from an outside supplier in large lot sizes, which are delivered to the warehouse after a significant lead time. Each retail store places batch orders on the warehouse, and after a substantial lead time, the order is delivered. This lead time includes order processing time and transportation time, and occasionally a delay time if the warehouse is out of stock.

 a. Is this a push or pull system?

 b. One potential problem with this system is that a retailer may request a batch of, say, Q_i units just before a second retailer requests a large batch. The supplier sends the Q_i units and then is left with very little inventory to satisfy the second request. Someone has suggested using a just-in-time (JIT) philosophy in which each unit demanded by a customer at outlet i triggers a replenishment request of $Q_i = 1$ units back to the warehouse. What would be wrong with doing this? What given(s) might you recommend changing to make this approach more attractive?

11.20 A company had a line of products with close to 100 items in it. Three different products each, in 10 different packages, contributed approximately 90% of the total dollar sales in the product line. The company worked out an arrangement with its supplier whereby the supplier held the three key items in bulk and packaged them to order in a very short lead time. In return, the company itself agreed to order the less popular items in larger replenishment quantities. Discuss the benefits from the standpoint of each of the participants in the agreement.

11.21 Consider the all-too-common situation of an organization where the purchasing department and production department have very little communication or cooperation. In particular, there is a raw material x that the purchasing department always has on hand in large quantities (the buyer receives a kickback from the vendor for paying for the material and storing it as quickly as the vendor produces it). The cost of the raw material to the company is $9.00 per dozen units. The production manager, recognizing that he always has plenty of raw material on hand, has decided to use the EOQ method of determining how big a batch of raw material he should process to produce a particular finished product each time he incurs the setup charges for the production operation. (Assume that one unit of raw material produces one unit of finished product.) These setup charges are $20 per setup and the value added in the production is $3.00 per dozen units. The annual demand for the finished product is 5,000 units/year.

   a. In the EOQ formula, what is the appropriate value of v? Discuss carefully.
   b. What run quantity, in units, should be used ($r = 0.2$ $/$/year and the replenishment rate is much higher than the demand rate)?
   c. What is the run quantity, expressed as a time supply?
   d. How might coordination between purchasing and production reduce the annual costs?

Appendix 11A: Derivation of the Logic for Computing the Best Replenishment Quantities in a Deterministic, Two-Stage Process

We wish to select n (an integer) and Q_R in order to minimize

$$\text{TRC}(n, Q_R) = \frac{D}{Q_R}\left(A_R + \frac{A_W}{n}\right) + \frac{Q_R r}{2}(n v'_W + v'_R) \tag{11A.1}$$

A convenient approach is to first set the partial derivative of TRC with respect to Q_R equal to zero and solve for the associated $Q^*_R(n)$, which is the best Q_R given the particular n value:

$$\frac{\partial \text{TRC}}{\partial Q_R} = -\frac{D}{Q^2_R}\left(A_R + \frac{A_W}{n}\right) + \frac{r}{2}(n v'_W + v'_R) = 0$$

which solves for

$$Q^*_R(n) = \sqrt{\frac{2[A_R + (A_W/n)]D}{(n v'_W + v'_R)r}} \tag{11A.2}$$

This expression is then substituted back into Equation 11A.1 to give $\text{TRC}^*(n)$, the lowest cost possible for the given value of n. The resulting equation is

$$\text{TRC}^*(n) = \sqrt{2\left[A_R + \frac{A_W}{n}\right]D(n v'_W + v'_R)r}$$

Finally, we must find the integer value of n that minimizes $\text{TRC}^*(n)$. First, we recognize that the n that minimizes the simpler expression

$$F(n) = \left[A_R + \frac{A_W}{n}\right](n v'_W + v'_R) \tag{11A.3}$$

will also minimize TRC*(n). A convenient way to find the minimizing n value is to first set

$$\frac{dF(n)}{dn} = 0$$

which gives

$$(nv'_W + v'_R)\left[-\frac{A_W}{n^2}\right] + \left[A_R + \frac{A_W}{n}\right]v'_W = 0$$

This solves for

$$n^* = \sqrt{\frac{A_W v'_R}{A_R v'_W}} \tag{11A.4}$$

which in general will not be an integer. The next step is to ascertain $F(n_1)$ and $F(n_2)$ from Equation 11A.3, where n_1 and n_2 are the two integers surrounding the n^* of Equation 11A.4. Whichever gives the lower value of F is the appropriate n to use (because the function F is convex in n). Finally, the corresponding Q_R and Q_W values are found by using this n in Equations 11A.2 and 11.1, respectively.

References

Abboud, N. E. 1996. The Markovian two-echelon repairable item provisioning problem. *Journal of the Operational Research Society 47*, 284–296.

Aderohunmu, R., A. Mobolurin, and N. Bryson. 1995. Joint vendor-buyer policy in JIT manufacturing. *Journal of the Operational Research Society 46*, 375–385.

Afentakis, P. 1987. A parallel heuristic algorithm for lot-sizing in multistage production systems. *IIE Transactions 19*(1), 34–42.

Afentakis, P., B. Gavish, and U. Karmarkar. 1984. Computationally efficient optimal solutions to the lot-sizing problem in multistage assembly systems. *Management Science 30*(2), 222–239.

Aggarwal, P. and K. Moinzadeh. 1994. Order expedition in multi-echelon production/distribution systems. *IIE Transactions 26*(2), 86–96.

Agrawal, V., X. Chao, and S. Seshadri. 2004. Dynamic balancing of inventory in supply chains. *European Journal of Operational Research 159*(2), 296–317.

Ahire, S. L. and C. P. Schmidt. 1996. A model for a mixed continuous-periodic review one-warehouse, *N*-retailer inventory system. *European Journal of Operational Research 92*, 69–82.

Ahn, B., N. Watanabe, and S. Hiraki. 1994. A mathematical model to minimize the inventory and transportation costs in the logistics system. *Computers & Industrial Engineering 27*(1–4), 229–232.

Albright, S. C. and A. Soni. 1988. Approximate steady-state distribution for a large repairable item inventory system. *European Journal of Operational Research 34*, 351–361.

Alfredsson, P. and J. Verrijdt. 1999. Modeling emergency supply flexibility in a two-echelon inventory system. *Management Science 45*(10), 1416–1431.

Anupindi, R. and Y. Bassok. 1999. Centralization of stocks: Retailers vs. manufacturer. *Management Science 45*(2), 178–191.

Archibald, T. W., S. A. Sassen, and L. C. Thomas. 1997. An optimal policy for a two depot inventory problem with stock transfer. *Management Science 43*(2), 173–183.

Arntzen, B. C., G. G. Brown, T. P. Harrison, and L. L. Trafton. 1995. Global supply chain management at digital equipment corporation. *Interfaces 25*(1), 69–93.

Ashayeri, J., R. Heuts, A. Jansen, and B. Szczerba. 1996. Inventory management of repairable service parts for personal computers: A case study. *International Journal of Operations & Production Management 16*(12), 74–97.

Atkins, D. and D. Sun. 1995. 98% Effective lot sizing for series inventory systems with backlogging. *Operations Research 43*(2), 335–345.

Axsäter, S. 1986. Evaluation of lot-sizing techniques. *International Journal of Production Research 24*(1), 51–57.

Axsäter, S. 1990a. Modelling emergency lateral transshipments in inventory systems. *Management Science 36*(11), 1329–1338.

Axsäter, S. 1990b. Simple solution procedures for a class of two-echelon inventory problems. *Operations Research 38*(1), 64–69.

Axsäter, S. 1993a. Continuous review policies for multi-level inventory systems with stochastic demand. In S. Graves, A. Rinnooy Kan, and P. Zipkin (Eds.), *Logistics for Production and Inventory*, Volume 4. Amsterdam: Elsevier (North-Holland).

Axsäter, S. 1993b. Exact and approximate evaluation of batch-ordering policies for two-level inventory systems. *Operations Research 41*, 777–785.

Axsäter, S. 1993c. Optimization of order-up-to-s policies in two-echelon inventory systems with periodic review. *Naval Research Logistics 40*, 245–253.

Axsäter, S. 1995. Approximate evaluation of batch-ordering policies for a one-warehouse, N non-identical retailer system under compound Poisson demand. *Naval Research Logistics 42*, 807–819.

Axsater, S. 1997. On deficiencies of common ordering policies for multi-level inventory control. *OR Spektrum 19*, 109–110.

Axsäter, S. 1997. Simple evaluation of echelon stock (R, Q)-policies for two-level inventory systems. *IIE Transactions 29*(8), 661–670.

Axsäter, S. 2003. Approximate optimization of a two-level distribution inventory system. *International Journal of Production Economics 81*, 545–553.

Axsäter, S. and L. Juntti. 1996. Comparison of echelon stock and installation stock policies for two-level inventory systems. *International Journal of Production Economics 45*, 305–312.

Axsäter, S. and L. Juntti. 1997. Comparison of echelon stock and installation stock policies with policy adjusted order quantities. *International Journal of Production Economics 48*, 1–6.

Axsäter, S. and W.-F. Zhang. 1994. Recursive evaluation of order-up-to-S policies for two-echelon inventory systems with compound Poisson demand. *Naval Research Logistics 43*(1), 151–157.

Banerjee, A. and S. Banerjee. 1994. A coordinated order-up-to inventory control policy for a single supplier and multiple buyers using electronic data interchange. *International Journal of Production Research 35*, 85–91.

Barnes-Schuster, D., Y. Bassok, and R. Anupindi. 2006. Optimizing delivery lead time/inventory placement in a two-stage production/distribution system. *European Journal of Operational Research 174*(3), 1664–1684.

Billington, P., J. Blackburn, J. Maes, R. Millen, and L. van Wassenhove. 1994. Multi-item lotsizing in capacitated multi-stage serial systems. *IIE Transactions 26*(2), 12–17.

Blackburn, J. and R. Millen. 1982. Improved heuristics for multi-stage requirements planning systems. *Management Science 28*(1), 44–56.

Blackburn, J. and R. Millen. 1985. An evaluation of heuristic performance in multi-stage lot-sizing systems. *International Journal of Production Research 23*(5), 857–866.

Bloemhof-Ruwaard, J. M., M. Salomon, and L. N. Van Wassenhove. 1996. The capacitated distribution and waste disposal problem. *European Journal of Operational Research 88*, 490–503.

Bregman, R. L., L. Ritzman, and L. J. Krajewski. 1990. A heuristic for the control of inventory in a multi-echelon environment with transportation costs and capacity limitations. *Journal of the Operational Research Society 41*(9), 809–820.

Brown, R. G. 1981. The new push for DRP. *Inventories & Production Magazine 1*(3), 25–27.

Brown, R. G. 1982. *Advanced Service Parts Inventory Control*. Norwich, VT: Materials Management Systems, Inc.

Buchanan, D. J. and P. L. Abad. 1998. Optimal policy for a periodic review returnable inventory system. *IIE Transactions 30*(11), 1049–1055.

Buschkühl, L., F. Sahling, S. Helber, and H. Tempelmeier. 2010. Dynamic capacitated lot-sizing problems: A classification and review of solution approaches. *OR Spectrum 32*(2), 231–261.

Cachon, G. P. and M. A. Lariviere. 1999. Capacity choice and allocation: Strategic behavior and supply chain performance. *Management Science 45*(8), 1091–1108.

Cachon, G. P. and P. H. Zipkin. 1999. Competitive and cooperative inventory policies in a two stage supply chain. *Management Science 45*(7), 936–953.

Carrillo, M. J. 1991. Note: Extensions of Palm's theorem: A review. *Management Science 37*(6), 739–744.

Castillo, E. D. and J. K. Cochran. 1996. Optimal short horizon distribution operations in reusable container systems. *Journal of the Operational Research Society 47*, 48–60.

Chakravarty, A. K. 1984. Joint inventory replenishments with group discounts based on invoice value. *Management Science 30*(9), 1105–1112.

Chakravarty, A. K. and G. E. Martin. 1988. An optimal joint buyer-seller discount pricing model. *Computers and Operations Research 15*(3), 271–281.

Chakravarty, A. K. and A. Shtub. 1986. Simulated safety stock allocation in a two-echelon distribution system. *International Journal of Production Research 24*(5), 1245–1253.

Chao, X. and S. X. Zhou. 2009. Optimal policy for a multiechelon inventory system with batch ordering and fixed replenishment intervals. *Operations Research 57*(2), 377–390.

Chen, F. 2000. Optimal policies for multi-echelon inventory problems with batch ordering. *Operations Research 48*(3), 376–389.

Chen, F. and R. Samroengraja. 2000. The stationary beer game. *Production and Operations Management 9*(1), 19–30.

Chen, F. and Y. Zheng. 1994a. Lower bounds for multi-echelon stochastic inventory systems. *Management Science 40*(11), 1426–1443.

Chen, F. and Y. Zheng. 1997a. One-warehouse multiretailer systems with centralized stock information. *Operations Research 45*(2), 275–287.

Chen, F. and Y.-S. Zheng. 1994b. Evaluating echelon stock (R, nQ) policies in serial production/inventory systems with stochastic demand. *Management Science 40*(10), 1262–1275.

Chen, F. and Y.-S. Zheng. 1997b. One-warehouse multiretailer systems with centralized stock information. *Operations Research 45*(2), 275–287.

Chen, M. S. and C. T. Lin. 1990. An example of disbenefits of centralized stocking. *Journal of the Operational Research Society 41*(3), 259–262.

Cheung, K. L. 1996. On the $(S - 1, S)$ inventory model under compound Poisson demands and i.i.d. unit resupply times. *Naval Research Logistics 43*, 563–572.

Cheung, K. L. and W. H. Hausman. 1995. Multiple failures in a multi-item spares inventory model. *IIE Transactions 27*, 171–180.

Chew, E. P. and L. A. Johnson. 1996. Service level approximations for multiechelon inventory systems. *European Journal of Operational Research 91*, 440–455.

Chew, E. P. and L. C. Tang. 1995. Warehouse-retailer system with stochastic demands—Non-identical retailer case. *European Journal of Operational Research 82*, 98–110.

Clark, A., B. Almada-Lobo, and C. Almeder. 2011. Lot sizing and scheduling: Industrial extensions and research opportunities. *International Journal of Production Research 49*(9), 2457–2461.

Clark, A. J. 1972. An informal survey of multi-echelon inventory theory. *Naval Research Logistics Quarterly 19*(4), 621–650.

Clark, A. J. and H. Scarf. 1960. Optimal policies for a multi-echelon inventory problem. *Management Science 6*(4), 475–490.

Clark, A. J. and H. Scarf. 1962. Approximate solutions to a simple multi-echelon inventory problem. In K. J. Arrow, S. Karlin, and H. Scarf (Eds.), *Studies in Applied Probability and Management Science*, pp. 88–110. Stanford, CA: Stanford University Press.

Clark, A. R. and V. A. Armentano. 1995. A heuristic for a resource-capacitated multi-stage lot-sizing problem with lead times. *Journal of the Operational Research Society 46*, 1208–1222.

Cohen, M., P. V. Kamesam, P. Kleindorfer, H. Lee, and A. Tekerian. 1990. Optimizer: IBM's multi-echelon inventory system for managing service logistics. *Interfaces 20*(1), 65–82.

Cohen, M., P. Kleindorfer, and H. Lee. 1986. Optimal stocking policies for low usage items in multi-echelon inventory systems. *Naval Research Logistics Quarterly 33*, 17–38.

Cohen, M., P. Kleindorfer, and H. Lee. 1988. Service constrained (s, S) inventory systems with priority demand classes and lost sales. *Management Science 34*(4), 482–499.

Cohen, M., P. Kleindorfer, and H. Lee. 1989. Near-optimal service constrained stocking policies for spare parts. *Operations Research 37*(1), 104–117.

Cohen, M. A., N. Agrawal, V. Agrawal, and A. Raman. 1995. Analysis of distribution strategies in the industrial paper and plastics industry. *Operations Research 43*(1), 6–18.

Cohen, M. A. and H. L. Lee. 1990. Out of touch with customer needs? Spare parts and after sales service. *Sloan Management Review 31*(2), 55–66.

Crawford, G. B. 1981. *Palm's Theorem for Nonstationary Processes*. RAND Corporation (R-2750-RC).

Crowston, W., W. Hausman, and W. Kampe II. 1973. Multistage production for stochastic seasonal demand. *Management Science 19*(8), 924–935.

Crowston, W. and M. Wagner. 1973. Dynamic lot size models for multi-stage assembly systems. *Management Science 20*(1), 14–21.

Crowston, W., M. Wagner, and A. Henshaw. 1972. A comparison of exact and heuristic routines for lot-size determination in multi-stage assembly systems. *AIIE Transactions 4*(4), 313–317.

Crowston, W. B., M. Wagner, and J. F. Williams. 1973. Economic lot size determination in multi-stage assembly systems. *Management Science 19*(5), 517–527.

De Bodt, M., L. Gelders, and L. Van Wassenhove. 1984. Lot sizing under dynamic demand conditions: A review. *Engineering Costs and Production Economics 8*, 165–187.

De Bodt, M. A. and S. C. Graves. 1985. Continuous review policies for a multi-echelon inventory problem with stochastic demand. *Management Science 31*, 1286–1295.

de Kok, A. G. 1990. Hierarchical production planning for consumer goods. *European Journal of Operational Research 45*, 55–69.

Demmy, W. S. and V. J. Presutti. 1981. Multi-echelon inventory theory in the air force logistics command. In L. B. Schwarz (Ed.), *Multi-Level Production/Inventory Control Systems: Theory and Practice*, Volume 16, pp. 279–297. Amsterdam: North-Holland Publishing Company.

Desai, V. S. 1996. Interactions between members of a marketing-production channel under seasonal demand. *European Journal of Operational Research 990*, 115–141.

Deuermeyer, B. L. and L. B. Schwarz. 1981. A model for the analysis of system service level in a warehouse-retailer distribution system. In L. B. Schwarz (Ed.), *Multi-Level Production/Inventory Control Systems: Theory and Practice*, pp. 163–193. Amsterdam: North-Holland.

Diks, E. and A. de Kok. 1996. Controlling a divergent 2-echelon network with transshipments using the consistent appropriate share rationing policy. *International Journal of Production Economics 45*(1), 369–379.

Diks, E. B., A. G. de Kok, and A. G. Lagodimos. 1996. Multi-echelon systems: A service measure perspective. *European Journal of Operational Research 95*(2), 241–263.

Dobson, G. and C. A. Yano. 1994. Cyclic scheduling to minimize inventory in a batch flow line. *European Journal of Operational Research 75*, 441–461.

Ebeling, C. E. 1991. Optimal stock levels and service channel allocations in a multi-item repairable asset inventory system. *IIE Transactions 23*(2), 115–120.

Ehrhardt, R. A., C. R. Schultz, and H. M. Wagner. 1981. (s, S) policies for a wholesale inventory system. In L. B. Schwarz (Ed.), *Multi-Level Production/Inventory Control Systems: Theory and Practice*, Volume 16, pp. 145–161. Amsterdam: North-Holland.

El-Najdawi, M. K. and P. Kleindorfer. 1993. Common cycle lot-size scheduling for multi-product, multi-stage production. *Management Science 39*(7), 872–885.

Eppen, G. and L. Schrage. 1981. Centralized ordering policies in a multi-warehouse system with lead times and random demand. In L. B. Schwarz (Ed.), *Multi-Level Production/Inventory Control Systems: Theory and Practice*, Volume 16, pp. 51–67. Amsterdam: North-Holland.

Erengüç, Ş. S., N. C. Simpson, and A. J. Vakharia. 1999. Integrated production/distribution planning in supply chains: An invited review. *European Journal of Operational Research 115*(2), 219–236.

Erkip, N., W. Hausman, and S. Nahmias. 1990. Optimal centralized ordering policies in multi-echelon inventory systems with correlated demands. *Management Science 36*(3), 381–392.

Ernst, R. and D. Pyke. 1993. Optimal base stock policies and truck capacity in a two-echelon system. *Naval Research Logistics 40*, 879–903.

Evers, P. T. 1996. The impact of transshipments on safety stock requirements. *Journal of Business Logistics 17*(1), 109–133.

Farasyn, I., S. Humair, J. I. Kahn, J. J. Neale, O. Rosen, J. Ruark, W. Tarlton, W. Van de Velde, G. Wegryn, and S. P. Willems. 2011. Inventory optimization at Procter & Gamble: Achieving real benefits through user adoption of inventory tools. *Interfaces 41*(1), 66–78.

Federgruen, A. 1993. Centralized planning models for multi-echelon inventory systems under uncertainty. In S. Graves, A. Rinnooy Kan, and P. Zipkin (Eds.), *Logistics of Production and Inventory*, Volume 4. Amsterdam: Elsevier (North-Holland).

Federgruen, A. and P. Zipkin. 1984a. Allocation policies and cost approximations for multilocation inventory systems. *Naval Research Logistics 31*, 97–129.

Federgruen, A. and P. Zipkin. 1984b. Computational issues in an infinite horizon multiechelon inventory model. *Operations Research 32*(4), 818–836.

Feeney, G. J. and C. C. Sherbrooke. 1966. The $(S-1, S)$ inventory policy under compound Poisson demand. *Management Science 12*(5), 391–411.

Ferrer, G. 1995. *Parts Recovery Problem: The Value of Information in Remanufacturing*. PhD thesis. INSEAD, Fontainebleau, France.

Fisher, W. W. and J. J. Brennan. 1986. The performance of cannibalization policies in a maintenance system with spares, repair, and resource constraints. *Naval Research Logistics 33*, 1–15.

Fleischmann, M., J. M. Bloemhof-Ruwaard, R. Dekker, E. Van der Laan, J. A. Van Nunen, and L. N. Van Wassenhove. 1997. Quantitative models for reverse logistics: A review. *European Journal of Operational Research 103*(1), 1–17.

Forsberg, R. 1995. Optimization of order-up-to-s policies for two-level inventory systems with compound Poisson demand. *European Journal of Operational Research 81*, 143–153.

Forsberg, R. 1996. Exact evaluation of (R, Q)-policies for two-level inventory systems with Poisson demand. *European Journal of Operational Research 96*, 130–138.

Fox, B. and M. Landi. 1970. Searching for the multiplier in one-constraint optimization problems. *Operations Research 18*(2), 253–262.

Gerchak, Y. and D. Gupta. 1991. On apportioning costs to customers in centralized continuous review inventory systems. *Journal of Operations Management 10*(4), 546–551.

Glasserman, P. and S. Tayur. 1994. The stability of a capacitated, multi-echelon production-inventory system under a base-stock policy. *Operations Research 42*(5), 913–925.

Glasserman, P. and S. Tayur. 1995. Sensitivity analysis for base-stock levels in multiechelon production-inventory systems. *Management Science 41*(2), 263–281.

Goh, T. N. and N. Varaprasad. 1986. A statistical methodology for the analysis of the life-cycle of reusable containers. *IIE Transactions 18*(1), 42–47.

Gong, L., A. G. de Kok, and J. Ding. 1994. Optimal leadtimes planning in a serial production system. *Management Science 40*(5), 629–632.

Goyal, S. K. and A. Gunasekaran. 1990. Multi-stage production-inventory systems. *European Journal of Operational Research 46*, 1–20.

Goyal, S. K. and Y. P. Gupta. 1989. Integrated inventory models: The buyer-vendor coordination. *European Journal of Operational Research 41*, 261–269.

Graves, S. and L. Schwarz. 1977. Single cycle continuous review policies for arborescent production/inventory systems. *Management Science 23*(5), 529–540.

Graves, S. C. 1981. A review of production scheduling. *Operations Research 29*(4), 646–675.

Graves, S. C. 1985. A multi-echelon inventory model for a reparable item with one-for-one replenishment. *Management Science 31*(10), 1247–1256.

Graves, S. C. 1988. Safety stocks in manufacturing systems. *Journal of Manufacturing and Operations Management 1*(1), 67–101.

Graves, S. C. 1996. A multiechelon inventory model with fixed replenishment intervals. *Management Science 42*(1), 1–18.

Graves, S. C. and S. P. Willems. 2000. Optimizing strategic safety stock placement in supply chains. *Manufacturing & Service Operations Management 2*(1), 68–83.

Graves, S. C. and S. P. Willems. 2003a. Erratum: Optimizing strategic safety stock placement in supply chains. *Manufacturing & Service Operations Management 5*(2), 176–177.

Graves, S. C. and S. P. Willems. 2003b. Supply chain design: Safety stock placement and supply chain configuration. *Handbooks in Operations Research and Management Science 11*, 95–132.

Graves, S. C. and S. P. Willems. 2005. Optimizing the supply chain configuration for new products. *Management Science 51*(8), 1165–1180.

Gross, D., L. C. Kioussis, and D. R. Miller. 1987. A network decomposition approach for approximating the stead-state behavior of Markovian multi-echelon reparable item inventory systems. *Management Science 33*(11), 1453–1468.

Guide, V. D. R. 2000. Production planning and control for remanufacturing: Industry practice and research needs. *Journal of Operations Management 18*(4), 467–483.

Guide, V. D. R. and M. S. Spencer. 1997. Rough-cut capacity planning for remanufacturing firms. *Production Planning and Control 8*(3), 237–244.

Guide, V. D. R. and L. N. Wassenhove. 2001. Managing product returns for remanufacturing. *Production and Operations Management 10*(2), 142–155.

Gunasekaran, A., C. Patel, and R. E. McGaughey. 2004. A framework for supply chain performance measurement. *International Journal of Production Economics 87*(3), 333–347.

Gupta, A. and S. C. Albright. 1992. Steady-state approximations for a multi-echelon multi-indentured repairable-item inventory system. *European Journal of Operational Research 62*, 340–353.

Gurnani, H. 1996. Optimal ordering policies in inventory systems with random demand and random deal offerings. *European Journal of Operational Research 95*, 299–312.

Gustafson, H. W. 1991. *Combat Support Command, Control, and Communications (CSC3)*. RAND Corporation. Santa Monica, CA.

Hall, R. W. and M. Racer. 1995. Transportation with common carrier and private fleets: System assignment and shipment frequency optimization. *IIE Transactions 27*, 217–225.

Hammond, J. H. 1994. *Barilla SpA (A)*. Boston, MA: Harvard Business School.

Hausman, W. and N. Erkip. 1994. Multi-echelon vs. single-echelon inventory control policies for low-demand items. *Management Science 40*(5), 597–602.

Hausman, W. H. and G. D. Scudder. 1982. Priority scheduling rules for repairable inventory systems. *Management Science 28*(11), 1215–1232.

Haynsworth, H. C. and B. A. Price. 1989. A model for use in the rationing of inventory during lead time. *Naval Research Logistics 36*, 491–506.

Henig, M., Y. Gerchak, R. Ernst, and D. Pyke. 1997. An inventory model embedded in a supply contract. *Management Science 43*(2), 184–189.

Heyman, D. and M. Sobel. 1982. *Stochastic Models in Operations Research*, Volume 1. New York: McGraw-Hill Company.

Heyman, D. P. 1977. Optimal disposal policies for a single-item inventory system with returns. *Naval Research Logistics 24*(3), 385–405.

Hill, R. M. 1989. Allocating warehouse stock in a retail chain. *Journal of the Operational Research Society 40*(11), 983–991.

Hillestad, R. J. 1982. Dyna-METRIC: Dynamic multi-echelon technique for recoverable item control. RAND Corporation (R-2785-AF).

Hillestad, R. J. and M. J. Carrillo. 1980. Models and techniques for recoverable item stockage when demand and the repair process are nonstationary—Part I: Performance measurement. RAND Corporation (N-1482-AF).

Hoadley, B. and D. P. Heyman. 1977. A two-echelon inventory model with purchases, dispositions, shipments, returns and transshipments. *Naval Research Logistics Quarterly 24*(1), 1–20.

Hodgson, T. J. and D. Wang. 1991. Optimal hybrid PUSH/PULL control strategies for a parallel multistage system: Part II. *International Journal of Production Research 29*(7), 1453–1460.

Hollier, R. 1980. The distribution of spare parts. *International Journal of Production Research 18*(6), 665–675.

Hopp, W. J., R. Q. Zhang, and M. L. Spearman. 1999. An easily implementable hierarchical heuristic for a two-echelon spare parts distribution system. *IIE Transactions 31*(10), 977–988.

Inderfurth, K. 1994. Safety stocks in multi-stage divergent inventory systems: A survey. *International Journal of Production Economics 35*, 321–329.

Inderfurth, K. 1995. Multistage safety stock planning with item demands correlated across products and through time. *Production and Operations Management 4*(2), 127–144.

Inderfurth, K. 1996. Modeling period review control for a stochastic product recovery problem with remanufacturing and procurement leadtimes. Preprint/Otto-von-Guericke-Univ. Magdeburg, 2, 96.

Inderfurth, K. 1997. Simple optimal replenishment and disposal policies for a product recovery system with leadtimes. *OR Spektrum 19*(2), 111–122.

Inderfurth, K. and S. Minner. 1998. Safety stocks in multi-stage inventory systems under different service measures. *European Journal of Operational Research 106*(1), 57–73.

Iyogun, P. 1992. Lot-sizing algorithm for a coordinated multi-item, multi-source distribution problem. *European Journal of Operational Research 59*, 393–404.

Iyogun, P. and D. Atkins. 1993. A lower bound and an efficient heuristic for multistage multiproduct distribution systems. *Management Science 30*(2), 204–217.

Jackson, P. 1988. Stock allocation in a two-echelon distribution system or "what to do until your ship comes in". *Management Science 34*(7), 880–895.

Jackson, P., W. Maxwell, and J. Muckstadt. 1988. Determining optimal reorder intervals in capacitated production-distribution systems. *Management Science 34*(8), 938–958.

Jackson, P. and J. A. Muckstadt. 1989. Risk pooling in a two-period, two-echelon inventory stocking and allocation problem. *Naval Research Logistics 36*(1), 1–26.

Jackson, P. L., J. A. Muckstadt, and W. Jones. 1996. https://people.orie.cornell.edu/jackson/distgame.html

Jans, R. and Z. Degraeve. 2007. Meta-heuristics for dynamic lot sizing: A review and comparison of solution approaches. *European Journal of Operational Research 177*(3), 1855–1875.

Jans, R. and Z. Degraeve. 2008. Modeling industrial lot sizing problems: A review. *International Journal of Production Research 46*(6), 1619–1643.

Joneja, D. 1990. The joint replenishment problem: New heuristics and worst case performance bounds. *Operations Research 38*(4), 711–723.

Jönsson, H. and E. A. Silver. 1987a. Analysis of a two-echelon inventory control system with complete redistribution. *Management Science 33*, 215–227.

Jönsson, H. and E. A. Silver. 1987b. Stock allocation among a central warehouse and identical regional warehouses in a particular push inventory control system. *International Journal of Production Research 25*(2), 191–205.

Kaplan, A. and D. Orr. 1985. An optimum multiechelon repair policy and stockage model. *Naval Research Logistics Quarterly 32*(4), 551–566.

Karmarkar, U. S. and N. R. Patel. 1977. The one-period N-location distribution problem. *Naval Research Logistics 24*, 559–575.

Kelle, P. and E. A. Silver. 1989. Purchasing policy of new containers considering the random returns of previously issued containers. *IIE Transactions 21*, 349–354.

Kennedy, W., J. W. Patterson, and L. D. Fredendall. 2002. An overview of recent literature on spare parts inventories. *International Journal of Production Economics 76*(2), 201–215.

Klosterhalfen, S. T., S. Minner, and S. P. Willems. 2014. Strategic safety stock placement in supply networks with static dual supply. *Manufacturing & Service Operations Management 16*(2), 204–219.

Kohli, R. and H. Park. 1994. Coordinating buyer-seller transactions across multiple products. *Management Science 40*(9), 1145–1150.

Kumar, A., L. B. Schwarz, and J. E. Ward. 1995. Risk-pooling along a fixed delivery route using a dynamic inventory-allocation policy. *Management Science 41*(2), 344–362.

Lagodimos, A. G. 1992. Multi-echelon service models for inventory systems under different rationing policies. *International Journal of Production Research 30*(4), 939–958.

Lagodimos, A. G., A. G. de Kok, and J. Verrijdt. 1995. The robustness of multi-echelon service models under autocorrelated demands. *Journal of the Operational Research Society 46*, 92–103.

Lam, S. M. and D. S. Wong. 1996. A fuzzy mathematical model for the joint economic lot size problem with multiple price breaks. *European Journal of Operational Research 95*, 611–622.

Lambrecht, M., R. Luyten, and J. Vander Eecken. 1984. Protective inventories and bottlenecks in production systems. *European Journal of Operational Research 22*, 319–328.

Lambrecht, M. R., J. Vander Eecken, and H. Vanderveken. 1981. Review of optimal and heuristic methods for a class of facilities in series dynamic lot-size problems. In L. B. Schwarz (Ed.), *Multi-Level Production/Inventory Control Systems: Theory and Practice*, Volume 16, pp. 69–94. Amsterdam: North-Holland.

Lawrence, M. 1977. An integrated inventory control system. *Interfaces 7*(2), 55–62.

Lee, H. and S. Whang. 1999. Decentralized multi-echelon supply chains: Incentives and information. *Management Science 45*(5), 633–640.

Lee, H. L. 1987. A multi-echelon inventory model for repairable items with emergency lateral transshipments. *Management Science 33*(10), 1302–1316.

Lee, H. L. and K. Moinzadeh. 1987a. Operating characteristics of a two-echelon inventory system for repairable and consumable items under batch ordering and shipment policy. *Naval Research Logistics 34*, 365–380.

Lee, H. L. and K. Moinzadeh. 1987b. Two-parameter approximations for multi-echelon repairable inventory models with batch ordering policy. *IIE Transactions 19*(2), 140–149.

Lee, H. L. and M. J. Rosenblatt. 1986. A generalized quantity discount pricing model to increase supplier's profits. *Management Science 32*(9), 1177–1185.

Logistics Management Institute. 1978. *LMI Availability System: Procurement Model*. McLean, VA.

Love, S. F. 1972. A facilities in series inventory model with nested schedules. *Management Science 18*(5), 327–338.

Lu, L. and M. Posner. 1994. Approximation procedures for the one-warehouse multi-retailer system. *Management Science 40*(10), 1305–1316.

Mabini, M. C. and L. F. Gelders. 1990. Repairable item inventory systems: A literature review. *Belgian Journal of Operations Research, Statistics and Computer Science 30*(4), 57–69.

Mahadevan, B., D. F. Pyke, and M. Fleischmann. 2003. Periodic review, push inventory policies for remanufacturing. *European Journal of Operational Research 151*(3), 536–551.

Martel, A. 2003. Policies for multi-echelon supply: DRP systems with probabilistic time-varying demands. *INFOR 41*(1), 71.

Masters, J. M. 1993. Determination of near optimal stock levels for multi-echelon distribution inventories. *Journal of Business Logistics 14*(2), 165–195.

Matta, K. F. and D. Sinha. 1995. Policy and cost approximations of two-echelon distribution systems with a procurement cost at the higher echelon. *IIE Transactions 27*, 638–645.

Maxwell, W. L. and J. A. Muckstadt. 1985. Establishing consistent and realistic reorder intervals in production-distribution systems. *Operations Research 33*(6), 1316–1341.

McGavin, E., L. Schwarz, and J. Ward. 1993. Two-interval inventory-allocation policies in a one-warehouse *N*-identical-retailer distribution system. *Management Science 39*(9), 1092–1107.

Meixell, M. J. and V. B. Gargeya. 2005. Global supply chain design: A literature review and critique. *Transportation Research Part E: Logistics and Transportation Review 41*(6), 531–550.

Mercer, A. and X. Tao. 1996. Alternative inventory and distribution policies of a food manufacturer. *Journal of the Operational Research Society 47*, 755–765.

Minner, S. 1997. Dynamic programming algorithms for multi-stage safety stock optimization. *OR Spektrum 19*(4), 261–271.

Moily, J. 1986. Optimal and heuristic procedures for component lot-splitting in multi-stage manufacturing systems. *Management Science 32*(1), 113–125.

Moinzadeh, K. and P. K. Aggarwal. 1997. An information based multi-echelon inventory system with emergency orders. *Operations Research 45*(5), 694–701.

Moinzadeh, K. and H. Lee. 1986. Batch size and stocking level in multi-echelon repairable systems. *Management Science 32*(12), 1567–1581.

Muckstadt, J. 1978. Some approximations in multi-item, multi-echelon, inventory systems for recoverable items. *Naval Research Logistics Quarterly 25*(3), 377–394.

Muckstadt, J. and L. J. Thomas. 1980. Are multi-echelon inventory methods worth implementing in systems with low-demand-rate items? *Management Science 26*(5), 483–494.

Muckstadt, J. A. 1973. A model for a multi-item, multi-echelon, multi-indenture inventory system. *Management Science 20*(4), 472–481.

Muckstadt, J. A. 1979. A three-echelon, multi-item model for recoverable items. *Naval Research Logistics Quarterly 19*(2), 199–221.

Muckstadt, J. A. 2005. *Analysis and Algorithms for Service Parts Supply Chains*. New York: Springer Science & Business Media.

Muckstadt, J. A. and M. H. Isaac. 1981. An analysis of single item inventory systems with returns. *Naval Research Logistics Quarterly 28*(2), 237–254.

Muckstadt, J. A. and R. O. Roundy. 1987. Multi-item, one-warehouse, multi-retailer distribution systems. *Management Science 33*(12), 1613–1621.

Nahmias, S. 1981. Managing repairable item inventory systems: A review. In L. B. Schwarz (Ed.), *Multi-Level Production/Inventory Control Systems: Theory and Practice*, pp. 253–277. Amsterdam: North-Holland.

Nahmias, S. and H. Rivera. 1979. A deterministic model for a repairable item inventory system with a finite repair rate. *International Journal of Production Research 17*(3), 215–221.

Nahmias, S. and S. A. Smith. 1994. Optimizing inventory levels in a two-echelon retailer system with partial lost sales. *Management Science 40*(5), 582–596.

Newhart, D. D., K. L. Stott, and F. J. Vasko. 1993. Consolidating product sizes to minimize inventory levels for a multi-stage production and distribution system. *Journal of the Operational Research Society 44*(7), 637–644.

Oeser, G. and P. Romano. 2016. An empirical examination of the assumptions of the square root law for inventory centralisation and decentralisation. *International Journal of Production Research 54*(8), 2298–2319.

Palm, C. 1938. Analysis of the Erlang traffic formula for busy-signal arrangements. *Ericsson Techniques 5*, 39–58.

Park, K. S. and D. H. Kim. 1987. Congruential inventory model for two-echelon distribution systems. *Journal of the Operational Research Society 38*(7), 643–650.

Paterson, C., G. Kiesmüller, R. Teunter, and K. Glazebrook. 2011. Inventory models with lateral transshipments: A review. *European Journal of Operational Research 210*(2), 125–136.

Penev, K. D. and A. J. de Ron. 1996. Determination of a disassembly strategy. *International Journal of Production Research 34*(2), 495–506.

Pyke, D. F. 1990. Priority repair and dispatch policies for reparable item logistics systems. *Naval Research Logistics 37*, 1–30.

Pyke, D. F. and M. A. Cohen. 1990. Push and pull in manufacturing and distribution systems. *Journal of Operations Management 9*(1), 24–43.

Pyles, R. A. 1984. *The Dyna-METRIC Readiness Assessment Model: Motivation, Capabilities, and Use*. (No. RAND/R-2886-AF). RAND Corporation, Santa Monica, CA.

Rajagopalan, S. 1992. A note on "an efficient zero-one formulation of the multilevel lot-sizing problem". *Decision Sciences 23*(4), 1023–1025.

Richter, K. 1996. The EOQ repair and waste disposal model with variable setup numbers. *European Journal of Operational Research 95*(2), 313–324.

Robinson, L. W. 1990. Optimal and approximate policies in multiperiod, multiechelon inventory models with transshipments. *Operations Research 38*(2), 278–295.

Rogers, D. F. and S. Tsubakitani. 1991a. Inventory positioning/partitioning for backorders optimization for a class of multi-echelon inventory problems. *Decision Sciences 22*, 536–558.

Rogers, D. F. and S. Tsubakitani. 1991b. Newsboy-style results for multi-echelon inventory problems: Backorders optimization with intermediate delays. *Journal of the Operational Research Society 42*(1), 57–68.

Rosenbaum, B. A. 1981a. Inventory placement in a two-echelon inventory system: An application. In L. B. Schwarz (Ed.), *Multi-Level Production/Inventory Control Systems: Theory and Practice*, Volume 16, pp. 195–207. Amsterdam: North-Holland Publishing Company.

Rosenbaum, B. A. 1981b. Service level relationships in a multi-echelon inventory system. *Management Science 27*(8), 926–945.

Rosenblatt, M. J. and H. L. Lee. 1985. Improving profitability with quantity discounts under fixed demand. *IIE Transactions 17*(4), 388–395.

Rosenfield, D. 1989. Disposal of excess inventory. *Operations Research 37*(3), 404–409.

Rosling, K. 1989. Optimal inventory policies for assembly systems under random demands. *Operations Research 37*(4), 565–579.

Roundy, R. 1985. 98%-effective integer-ratio lot-sizing for one-warehouse multi-retailer systems. *Management Science 31*(11), 1416–1430.

Rudi, N., S. Kapur, and D. F. Pyke. 2001. A two-location inventory model with transshipment and local decision making. *Management Science 47*(12), 1668–1680.

Rudi, N., D. F. Pyke, and P. O. Sporsheim. 2000. Product recovery at the Norwegian national insurance administration. *Interfaces 30*(3), 166–179.

Salomon, M., E. van der Laan, R. Dekker, M. Thierry, and A. Ridder. 1994. Product remanufacturing and its effects on production and inventory control. *ERASM Management Report Series 172*.

Savaskan, R. C., S. Bhattacharya, and L. N. Van Wassenhove. 2004. Closed-loop supply chain models with product remanufacturing. *Management Science 50*(2), 239–252.

Sawchuk, P. A. 1981. Installing a "PUSH" distribution system. In *APICS 24th Annual International Conference Proceedings, American Production and Inventory Control Society*, Atlanta, pp. 279–281.

Schneider, H., H. Rinks, and P. Kelle. 1995. Power approximations for a two-echelon inventory system using service levels. *Production and Operations Management 4*(4), 381–400.

Schwarz, L. 1981. Physical distribution: The analysis of inventory and location. *AIIE Transactions 13*(2), 138–150.

Schwarz, L. 1989. A model for assessing the value of warehouse risk-pooling: Risk-pooling over outside-supplier leadtimes. *Management Science 35*(7), 828–842.

Schwarz, L., B. Deuermeyer, and R. Badinelli. 1985. Fill-rate optimization in a one-warehouse *N*-identical retailer distribution system. *Management Science 31*(4), 488–498.

Schwarz, L. and L. Schrage. 1975. Optimal and system myopic policies for multi-echelon production/inventory assembly systems. *Management Science 21*(11), 1285–1294.

Schwarz, L. B. 1973. A simple continuous review deterministic one-warehouse *N*-retailer inventory problem. *Management Science 19*(5), 555–566.

Shao, J., H. Krishnan, and S. T. McCormick. 2011. Incentives for transshipment in a supply chain with decentralized retailers. *Manufacturing & Service Operations Management 13*(3), 361–372.

Sherbrooke, C. C. 1968. METRIC—A multi-echelon technique for recoverable item control. *Operations Research 16*(1), 103–121.

Sherbrooke, C. C. 1986. VARI-METRIC: Improved approximations for multi-indenture, multi-echelon availability models. *Operations Research 34*(2), 311–319.

Sherbrooke, C. C. 1992a. Multiechelon inventory systems with lateral supply. *Naval Research Logistics 39*, 29–40.

Sherbrooke, C. C. 1992b. *Optimal Inventory Modeling of Systems, Multi-Echelon Techniques*. New York: John Wiley & Sons.

Sherbrooke, C. C. 2006. *Optimal Inventory Modeling of Systems: Multi-Echelon Techniques*, Volume 72. New York: Kluwer Academic Publishers.

Shtub, A. and M. Simon. 1994. Determination of reorder points for spare parts in a two-echelon inventory system: The case of non identical maintenance facilities. *European Journal of Operational Research 73*, 458–464.

Simpson, N. C. and S. Selcuk Erenguc. 1996. Multiple-stage production planning research: History and opportunities. *International Journal of Operations & Production Management 16*(6), 25–40.

Song, J. and P. H. Zipkin. 1992. Evaluation of base-stock policies in multiechelon inventory systems with state-dependent demands. Part I: State-independent policies. *Naval Research Logistics 39*, 715–728.

Song, J. and P. H. Zipkin. 1996. Evaluation of base-stock policies in multiechelon inventory systems with state-dependent demands. Part II: State-dependent depot policies. *Naval Research Logistics 43*, 381–396.

Svoronos, A. and P. Zipkin. 1988. Estimating the performance of multi-level inventory systems. *Operations Research 36*(1), 57–72.

Svoronos, A. and P. Zipkin. 1991. Evaluation of one-for-one replenishment policy for multiechelon inventory systems. *Management Science 37*(1), 68–83.

Szendrovits, A. Z. 1975. Manufacturing cycle time determination for a multi-stage economic production quantity model. *Management Science 22*(3), 298–308.

Tagaras, G. and M. Cohen. 1992. Pooling in two-location inventory systems with non-negligible replenishment lead times. *Management Science 38*(8), 1067–1083.

Taleb, K. N., S. M. Gupta, and L. Brennan. 1997. Disassembly of complex product structures with parts and materials commonality. *Production Planning and Control 8*(3), 255–269.

Tang, C. S. 1990. The impact of uncertainty on a production line. *Management Science 36*(12), 1518–1531.

Tempelmeier, H. 1993. Safety stock allocation in a two-echelon distribution system. *European Journal of Operational Research 63*, 96–117.

Thierry, M., M. Salomon, J. Van Nunen, and L. N. Van Wassenhove. 1995. Strategic issues in product recovery management. *California Management Review 37*(2), 114–135.

Thomas, D. J. and P. M. Griffin. 1996. Coordinated supply chain management. *European Journal of Operational Research 94*, 1–15.

Tripp, R. S., I. K. Cohen, R. A. Pyles, R. J. Hillestad, R. W. Clarke, S. B. Limpert, and S. K. Kassicieh. 1991. A decision support system for assessing and controlling the effectiveness of multi-echelon logistics actions. *Interfaces 21*(4), 11–25.

Van Beek, P. 1981. Modelling and analysis of a multi-echelon inventory system. *European Journal of Operational Research 6*, 380–385.

Van der Heijden, M., E. B. Diks, and A. G. de Kok. 1997. Stock allocation in general multi-echelon distribution systems with (R, S) order-up-to-policies. *International Journal of Production Economics 49*(2), 157–174.

Van der Heijden, M. C. 1997. Supply rationing in multi-echelon divergent systems. *European Journal of Operational Research 101*, 532–549.

Van der Laan, E., R. Dekker, M. Salomon, and A. Ridder. 1996. An (s, Q) inventory model with remanufacturing and disposal. *International Journal of Production Economics 46-47*, 339–350.

Van der Laan, E. and M. Salomon. 1997. Production planning and inventory control with remanufacturing and disposal. *European Journal of Operational Research 102*(2), 264–278.

Van Houtum, G. J., K. Inderfurth, and W. H. M. Zijm. 1996. Materials coordination in stochastic multi-echelon systems. *European Journal of Operational Research 95*, 1–23.

Van Houtum, G. J. and W. H. M. Zijm. 1991. Computational procedures for stochastic multi-echelon production systems. *International Journal of Production Economics 23*, 223–237.

Verrijdt, J. 1997. *Design and Control of Service Part Distribution Systems.* PhD thesis. Technische Universiteit Eindhoven, the Netherlands.

Verrijdt, J., I. Adan, and A. G. de Kok. 1998. A trade off between emergency repair and inventory investment. *IIE Transactions 30*(2), 119–132.

Verrijdt, J. and A. G. de Kok. 1994. Distribution planning for a divergent N-echelon network without intermediate stocks under service restrictions. *International Journal of Production Economics 38*, 225–244.

Verrijdt, J. and A. G. de Kok. 1996. Distribution planning for a divergent depotless two-echelon network under service restrictions. *European Journal of Operational Research 89*, 341–354.

Webb, I. R. and R. C. Larson. 1995. Period and phase of customer replenishment: A new approach to the strategic inventory/routing problem. *European Journal of Operational Research 85*, 132–148.

Weng, Z. K. 1995. Channel coordination and quantity discounts. *Management Science 41*(9), 1509–1522.

Whang, S. 1995. Coordination in operations: A taxonomy. *Journal of Operations Management 12*(3–4), 413–422.

Williams, A. 1983. Decisions, decisions . . .. *Journal of the Operational Research Society 34*(4), 319–330.

Williams, J. F. 1981. Heuristic techniques for simultaneous scheduling of production and distribution in multi-echelon structures: Theory and empirical conclusions. *Management Science 27*(3), 336–352.

Yano, C. A. 1987. Stochastic leadtimes in two-level assembly systems. *IIE Transactions 19*(4), 371–378.

Zangwill, W. 1969. A backlogging model and a multi-echelon model of a dynamic economic lot size production system. *Management Science 15*(9), 506–526.

Zangwill, W. 1987. Eliminating inventory in a series facility. *Management Science 33*(9), 1150–1164.

Zipkin, P. 1984. On the imbalance of inventories in multi-echelon systems. *Math of OR 9*(3), 402–423.

Zipkin, P. 1991. Evaluation of base-stock policies in multiechelon inventory systems with compound-Poisson demands. *Naval Research Logistics 38*, 397–412.

Zipkin, P. 1995. Processing networks with planned inventories: Tandem queues with feedback. *European Journal of Operational Research 80*, 344–349.

Chapter 12

Coordinating Inventory Management in the Supply Chain

The majority of the models presented in this text examine production and inventory decisions to be made by a single firm. Suppliers were implicitly represented in those models through some production or purchase cost and perhaps some pricing schedule (e.g., quantity discount) as well as a description of some delivery process (e.g., lead time). Customers were implicitly modeled primarily through some description of demand. Of course in a supply chain, there are multiple firms making production, inventory, or other resource allocation decisions such as those modeled in this text. In this chapter, we turn our attention to explicitly consider this interaction and see how firms can better coordinate operations with their supply partners. Our experience is that there are two main (often related) areas where one can find opportunities to improve coordination: *information sharing* and *incentive alignment*. Furthermore, poor coordination can occur *externally*—among trading partners in the same supply chain—or *internally*—across organizational boundaries within the same firm.

For the most part, the initiatives presented in this chapter *change the inputs* to models discussed elsewhere in this textbook and do not require the development of new models. For example, a supplier will still use some inventory model to determine stocking quantities, but improved information sharing or incentive alignment may change the demand forecast or the desired level of service that they use for planning. This improved coordination can also happen internally. As an example, consider a firm where the sales force has additional demand information but does not communicate this information with the group responsible for inventory or production planning. (We discuss below why this seemingly odd situation might arise in practice.) If the sales force shared their improved forecast, the planning group does not need a new model but rather they simply would use the improved forecast as input into the same planning models. As we know from other chapters, using an improved forecast will result in lower safety stock, improved customer service, or both.

In Section 12.1, we discuss the causes and consequences of distorted information flow in a supply chain. Also in that section, we begin our discussion of approaches for mitigating distorted information flow and improving supply chain coordination. In Section 12.2, we discuss opportunities to improve information sharing and incentive alignment both within a firm and across multiple firms in the same supply chain. In particular, we discuss sales and operations planning (S&OP) as a process to help internal information sharing and alignment, and Collaborative Planning, Forecasting, and Replenishment (CPFR) as a process for *external* coordination across separate firms. CPFR extends information sharing across organizational boundaries. In Section 12.3, we discuss vendor-managed inventory (VMI), which is an arrangement where not only information is shared across organizational boundaries, but decision-making rights are transferred. In Section 12.4, we analyze contracts or coordinating agreements that can help align incentives across trading partners.

12.1 Information Distortion in a Supply Chain

One widely known consequence of poor coordination across partner firms in a supply chain is a phenomenon known as the "bullwhip effect." The bullwhip effect has been experienced by many students playing the "beer game" (Sterman 1995) and derives its name from the fact that relatively small variability in end-consumer demand can translate to very high variability at upstream stages in the supply chain. The typical supply chain setting for the beer game has a retailer supplied by the distributor, which is supplied by the wholesaler, which in turn is supplied by the manufacturer. Figure 12.1 shows typical ordering behavior at different stages in the supply chain for this setting. Note the relatively small variability in demand at the retailer translates to more variability at the wholesaler, even more variability at the distributer and yet even more at the manufacturer. Even when consumer demand is quite steady throughout the year, trade promotions,

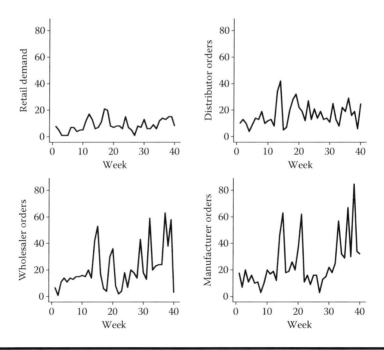

Figure 12.1　An illustration of the bullwhip effect.

volume discounts, long lead times, full-truckload discounts, and end-of-quarter sales incentives can lead to this propagation of variability as one moves further up the supply chain from the end consumer. Put in context of the inventory models discussed in this text, high demand variability at any particular stage in the supply chain leads to high cost due to inefficient use of production and warehouse resources, high transportation costs, high inventory costs, and high cost of missed sales.

The reason one typically sees such high variability in manufacturer orders in the beer game (see Figure 12.1) is that the manufacturer is making inventory decisions in reaction to the orders they see rather than the actual consumer demand. The propagation of demand variability can be somewhat mitigated by improved information sharing. In other words, if the manufacturer could make inventory decisions with full knowledge of actual consumer demand, it would be able to significantly reduce inventory and improve service.

Several studies have examined and quantified the bullwhip effect. Using industry-level data, Cachon et al. (2007) demonstrate the magnitude of the bullwhip effect across industries. Bray and Mendelson (2012) similarly report evidence of the bullwhip in practice, noting that while this effect is still present and significant, the magnitude of the bullwhip effect between 1995 and 2008 is substantially lower than the magnitude over 1974–2008. Chen et al. (2000) investigate how supply lead times and centralized demand information affect the magnitude of the bullwhip effect. The key finding is that centralized demand information (i.e., the manufacturer gets to see customer demand) can mitigate but not completely eliminate the bullwhip effect. Steckel et al. (2004) further investigate the impact of sharing demand information, also noting that, if practical, shortening supply lead times also mitigates the bullwhip effect. Other studies have expanded on these findings, examining how different supply chain configurations and different approaches for managing inventory can mitigate the bullwhip effect. See also Metters (1997) and Chen et al. (2000), and we discuss additional related work in Section 12.3.

Inventory held in central warehouses should allow factories to smooth production while meeting variable customer demand, but empirical data suggest that exactly the opposite happens. (See, e.g., Blinder 1981; Baganha and Cohen 1998; and Cachon et al. 2007.) Orders seen at the higher levels of the supply chain exhibit more variability than those at levels closer to the customer. Lee et al. (1997) and Lee and Kopczak (1997) show how four rational factors help to create the bullwhip effect: demand signal processing (if demand increases, firms order more in anticipation of further increases, thereby communicating an artificially high level of demand); the rationing game (there is, or might be, a shortage, so a firm orders more than the actual forecast in the hope of receiving a larger share of the items in short supply); order batching (fixed costs at one location lead to batching of orders); and manufacturer price variations (which encourage bulk orders). The latter two factors generate large orders that are followed by small orders, which implies increased variability at upstream locations. In addition to these rational factors, suboptimal inventory ordering behavior can exacerbate the bullwhip effect. (See Croson and Donohue 2006 and Wu and Katok 2006).

Educating supply chain managers can help reduce suboptimal inventory ordering and overreactions to, say, stockout situations. However, if the underlying reasons for the bullwhip effect are *rational* decisions, other innovations will be necessary. Indeed, innovations, such as increased communication about consumer demand via point of sale (POS) data sharing, electronic data interchange (EDI), and everyday low pricing (EDLP) (to eliminate forward buying of bulk orders), can mitigate the bullwhip effect. In fact, the number of firms ordering and receiving orders electronically is exploding. The information available to supply chain partners, and the speed with which it is available, has the potential to radically reduce inventories and increase customer service.

However, these innovations require interfirm, and often intrafirm, cooperation and coordination, which can be difficult to achieve and are discussed further below.

12.2 Collaboration and Information Sharing

12.2.1 Sales and Operations Planning

Effective information sharing is critical to improving supply chain coordination. For most firms, there are opportunities to improve information sharing within their firm even before investigating opportunities to improve coordination with their trading partners. The S&OP process can be an effective approach for improving internal collaboration. The S&OP process seeks to bring together decision makers and relevant demand and supply information across a firm, producing a coordinated demand and supply plan for several months into the future. A typical S&OP process includes at least four recurring meetings each month:

1. *Demand review.* Usually near the beginning of the month, a planning team assembles relevant demand information, potentially including statistical forecasts, input from sales, pricing or other marketing information, information from key customers, and new product plans. This part of the S&OP process should conclude with a demand review meeting where the demand plan is "approved."
2. *Supply review.* Using the approved demand plan, the second step in the S&OP process produces a supply plan to meet the demand plan if possible. This typically involves some rough-cut capacity planning as well as potential use of the production and inventory models discussed throughout this book.
3. *Pre-S&OP meeting or partnership meeting.* Many firms use different terms for this meeting, but the goal is the same: align demand and supply plans from earlier meetings and resolve any differences prior to the executive meeting (next). Companies that have successful S&OP processes will have senior leadership at these meetings and resolve gaps in the demand and supply plans. Potential gap resolution can include authorizing overtime production to meet the demand plan, and prioritizing the allocation of scarce capacity or inventory to serve key customers or markets.
4. *Executive meeting.* The final meeting in the S&OP process is the executive meeting. In this meeting, the agreed-upon demand and supply plan is presented for approval from executive leadership. The goal of this entire process is to align all business functions on the same plan based on the same information and assumptions.

Some firms will have additional meetings in their S&OP process. These meetings may include a separate meeting to review new product introductions, or an initial demand review prior to the demand review meeting. There are many articles, books, and online resources available that provide more details on this process. See, for example, Wallace (2004); Grimson and Pyke (2007); and Stahl and Wallace (2012).

The key *idea* behind S&OP is that decision makers in a firm should have access to the most accurate and up-to-date information, these decision makers should be aligned around common objectives, and they should come up with a common demand and supply plan. Like many of the coordination approaches discussed in this chapter, the idea is much simpler than the execution. Many firms struggle with effective internal collaboration through a process such as S&OP. See Oliva and Watson (2009, 2011) for a discussion of a firm that struggled to achieve alignment around their demand plan, eventually improving due in part to a combination of effective leadership and effective use of information technology (to make information more broadly available to stakeholders.)

12.2.2 Collaborative Forecasting

Collaborative forecasting is practice that has emerged and evolved over recent years to improve information sharing between trading partners. A process initially known as CFAR has evolved into CPFR (See Verity 1996). The Voluntary Interindustry Commerce Standards (VICS) committee maintained and promoted materials designed to enhance the adoption and effectiveness of CPFR. VICS has since merged with GS1 US, and materials related to CPFR can be found at `www.gslus.us.org`.

One of the main drivers for the development of a collaborative forecasting approach was poor coordination around retail promotional events. Without a process in place to share information between retailers and their suppliers, a retailer may promote or discount a product, obviously increasing demand, without informing the supplier. This of course can lead to forecasting challenges for the supplier and potentially poor product availability back to the retailer. An illustrative example is shown in Figure 12.2 where promotions occur on days 10 and 25 but the supplier is not informed. In the top chart, the supplier builds a forecast based only on the demand history available (using an exponential smoothing model such as those discussed in Chapter 3). In the

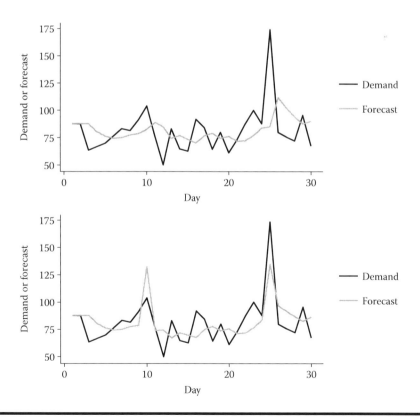

Figure 12.2   An illustration of sharing promotional information.

bottom chart, the retailer informs the supplier in advance of the promotions, so the supplier can adjust his forecast, obviously increasing forecast accuracy.

As with S&OP discussed in the previous section, the idea behind CPFR is relatively easy to understand but again the execution is difficult. Ireland and Crum (2005) provide some guidelines for the successful implementation of CPFR. Danese (2007) provides insights from seven CPFR case studies highlighting the potential for dramatic improvement but also outlining that a significant commitment of resources is required from both firms in order to achieve success.

12.3 Vendor-Managed Inventory

One of the causes of the bullwhip effect discussed above is the lack of demand information available to upstream firms in a supply chain. One business practice that has emerged to help address this effect is vendor-managed inventory (VMI). As the name implies, with VMI, the vendor assumes responsibility for determining replenishment quantities for their customer. The earliest instances of VMI involved consumer goods manufacturers assuming replenishment planning responsibilities for their large retail customers. Waller et al. (1999) provide an excellent overview of VMI, including a discussion of benefits and technology requirements. Sari (2008) jointly discusses the benefits of CPFR and VMI. Several studies develop detailed operational models for VMI. See Çetinkaya and Lee (2000); Fry et al. (2001); Dong and Xu (2002); and Disney and Towill (2003). Dejonckheere et al. (2003), among others, explicitly relate VMI to the bullwhip effect. The key idea in VMI that permits mitigation of the bullwhip effect is that when a vendor manages inventory, they can typically observe variability in actual demand at their customer, not just in orders received from their customer. Under a traditional arrangement, a retailer might operate an (s, Q) policy, as demonstrated in Figure 12.3 (with average weekly demand of five units and a reorder quantity of 25 units). Without visibility into the retailer's inventory, the vendor only sees the orders. In the case shown in Figure 12.3, the standard deviation of orders is almost three times the standard deviation of demand. With visibility into the retailer's inventory, the vendor could anticipate the inventory needed at the retailer, allowing advance scheduling of production or transportation. The precise benefits to the vendor will vary depending on the situation. Better demand information certainly can improve forecasts at the vendor. Also, VMI may enable smaller, more frequent shipments since the vendor can anticipate customer orders. The retailer also benefits by potentially receiving better product availability, in addition to shifting some of the cost of managing replenishments to the vendor.

12.4 Aligning Incentives

Numerous academic studies have addressed the topic of coordinating resource allocation decisions, such as inventory decisions, across multiple firms. Cachon (2003) provides an excellent starting point for the interested reader. In this section, we build on the single-period inventory decision presented in Chapter 9 to convey the key ideas in attempting to coordinate across multiple firms in the supply chain.

Equation 9.3 from Chapter 9 presented the expected profit for a single firm making a newsvendor quantity decision,

$$E[P_R(Q)] = (p - g)\hat{x} - (v - g)Q - (p - g + B)ES \tag{12.1}$$

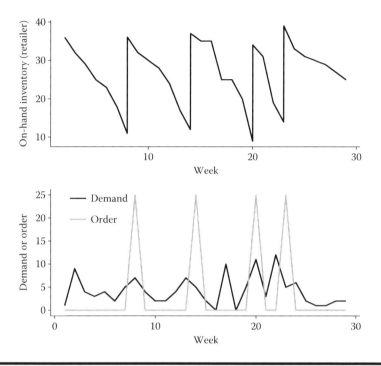

Figure 12.3   VMI can provide potential benefits by permitting the vendor to anticipate a retailer's inventory need.

with selling price p, salvage value g, unit acquisition cost v, and penalty for not satisfying demand B. ES is the expected number of units short, or expected shortages. As shown in Chapter 9, this can be rewritten as

$$E[P_R(Q)] = (p - v)\hat{x} - (v - g)EO - (p - v + B)ES$$
$$= (p - v)\hat{x} - c_o EO - c_u ES$$

where EO represents expected units left over, and c_o, c_u represent the marginal costs of overstocking or understocking, respectively (See Section 9.2.1).

In this section, we will assume that the newsvendor is a retailer, selling goods to the market at the given price p, and purchasing from a supplier at the unit acquisition cost v; thus, we add the subscript R to the above expected profit function. The existence of some supplier selling at unit cost v was implied in our earlier analysis of the newsvendor decision. Here, to examine the impact of inventory decisions across multiple firms in the supply chain, we explicitly define a profit function for the supplier. In the following sections, we analyze the impact of different contractual arrangements between the supplier and the retailer.

12.4.1 Wholesale Price Contract

The first situation that we analyze, known as the *wholesale price contract*, is one where the supplier simply chooses the wholesale price v to charge the retailer. This v of course then becomes the unit acquisition cost for the retailer. Before writing the supplier's profit function, we first define the quantity that the retailer would order as a function of the wholesale price v. That is, using the previous analysis of the newsvendor problem, we know what quantity the retailer should order for any given wholesale price v.

Equation 9.2 described the best order quantity as

$$p_{x<}(Q^*) = \frac{p - v + B}{p - g + B} \qquad (12.2)$$

This equation describes an important relationship between the wholesale price v selected by the supplier and the retailer's resulting order quantity $Q^*(v)$ for a given wholesale price. Note that as the wholesale price v increases, the ratio on the right side of Equation 12.1 decreases; thus, intuitively, the retailer's ideal order quantity decreases as the wholesale price increases.

The supplier's profit function can be written as a function of the wholesale price v and the resulting quantity chosen by the retailer $Q^*(v)$:

$$P_S(v, Q^*(v)) = (v - c)Q^*(v) \qquad (12.3)$$

where the supplier's unit production cost is c. It is not a notational accident that the expectation operator E is not present in Equation 12.3. In this wholesale price arrangement, the retailer bears all uncertainty. This will change in subsequent subsections, and in fact is the key to finding arrangements that improve profit for both the supplier and the retailer.

As the supplier increases his[*] wholesale price, the retailer chooses to purchase fewer units. Figure 12.4 shows the retailer, supplier, and total (sum of supplier and retailer) profits as the wholesale price changes. The setting shown in the figure has retail selling price $p = 10$, supplier unit cost $c = 2$, no salvage value or missed sales penalty cost ($g, B = 0$), and normally distributed demand with mean 100 and standard deviation 30.

The retailer obviously prefers a lower wholesale price, while the supplier prefers a higher wholesale price, up to a point. Eventually (around $v = 8.2$ in the figure), the supplier's profit decreases as the wholesale price increases. After this price, the retailer buys so few units that the supplier's profit decreases, even though the price per unit is very high. In general, it is complicated to find the supplier's best wholesale price for this setting, and it is not guaranteed that there is a unique optimal wholesale price.[†]

While the total profit shown in Figure 12.3 is simply the sum of the retailer's and supplier's profit functions, it is worth writing it out explicitly to emphasize one key observation. We use the subscript T for total profit.

$$E[P_T(Q)] = (p - g)\hat{x} - (v - g)Q - (p - g + B)ES + (v - c)Q \qquad (12.4)$$
$$= (p - g)\hat{x} - vQ + gQ - (p - g + B)ES + vQ - cQ \qquad (12.5)$$
$$= (p - g)\hat{x} - (c - g)Q - (p - g + B)ES \qquad (12.6)$$

---

[*] We will refer to supplier as he/his and retailer as she/her throughout this section.
[†] For many common distributions, there is a unique wholesale price. See Lariviere (2006) for a discussion of those conditions.

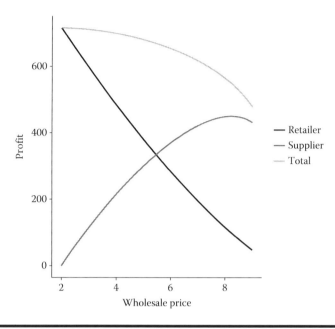

Figure 12.4 **Retailer, supplier, and total profit.**

The key observation is that the wholesale price v drops out and is replaced by the unit production cost c. This means that the order quantity that maximizes the total profit is given by

$$p_{x<}(Q^*) = \frac{p - c + B}{p - g + B} \tag{12.7}$$

and does not depend on the wholesale price v.

It is clear that the fractile in Equation 12.7 must be greater than the fractile under any wholesale price $v > c$; thus, the retailer will buy a quantity *less* than the quantity that maximizes the total profit. The system does not achieve the best overall profit in this arrangement due to something termed *double marginalization* (Spengler 1950). Since two different firms charge some positive margin on the product transaction, risk and reward are not aligned, and the system underperforms. In this case, the retailer faces all the demand uncertainty but only gets some of the profit margin. We now examine some other contractual arrangements that may help improve this situation.

12.4.2 Buyback Contract

In the previous subsection, we saw that a retailer making an inventory purchase will buy less than the quantity that maximizes overall supply chain profit because she faces all the risk of excess inventory (due to demand uncertainty) but receives only some of the overall profit margin. One way this has been addressed in practice is for the supplier to absorb some of the demand risk by agreeing to repurchase extra items at the end of the selling season.

Formally, a supplier could offer a wholesale price v_b and a buyback level b, where the supplier would pay b per unit to the retailer at the end of the selling season for any unsold inventory. For the analysis in this section, we assume $b > g$. If not, the retailer would prefer to salvage the units

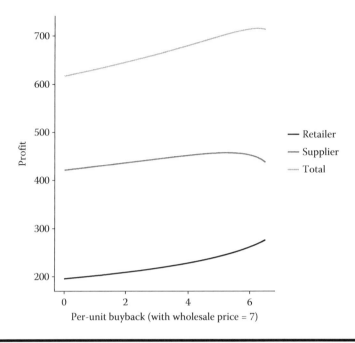

Figure 12.5 Retailer, supplier, and total profit under a buyback contract.

directly and ignore the buyback offer. In such an instance, this contractual offer would function just like the wholesale price contract above. The retailer's fractile becomes

$$p_{x<}(Q^*) = \frac{p - v_b + B}{p - b + B} \tag{12.8}$$

where v_b replaces v from above and b replaces g from above. It is clear from the fractile that as the buyback offer b is increased, the retailer's fractile, and thus her order quantity, increases. This is intuitive, since a higher buyback offer reduces the retailer's cost of overstocking. Figure 12.5 shows the effect of changing the buyback offer while keeping the wholesale price fixed at $v_b = 7$. The problem depicted in Figure 12.5 uses the same parameters as the problem from the previous section with the addition of the new parameter b.

At the same wholesale price, the retailer clearly prefers a higher buyback offer as it reduces her overstocking risk. Interestingly, and perhaps counterintuitively, the supplier also prefers higher buyback offers up to around $b = 5.2$ for the case shown in Figure 12.5. As the buyback offer increases, the supplier does take on more of the overstocking risk, but this shift of risk away from the retailer encourages the retailer to order more, thus increasing overall profitability.

Note that it is theoretically possible to find buyback and wholesale price combinations that maximize the overall supply chain profit. In fact, if one knows the values of all the cost parameters, it is straightforward to choose values of v_b and b that make the retailer's fractile (Equation 12.8) equal to the fractile that maximizes total expected profit (Equation 12.7). In our experience, it is quite challenging in practice to truly maximize supply chain profit through the use of this kind of contract for a variety of reasons we discuss below in Section 12.4.5. This should not take away from the fact that such contracts can be quite effective at improving profitability for both the supplier and the retailer.

One final note regarding these buyback contracts is that it is not required that excess inventory be physically returned to the supplier. It is often the case with these agreements that the supplier compensates the retailer in some way for excess inventory along the lines of a buyback agreement, but the units are salvaged at the retail location or sent to a third party for liquidation. The compensation to the retailer is sometimes known as "markdown money."

12.4.3 Revenue-Sharing Contract

Another coordinating contract used in practice is a revenue-sharing contract. Such a contract seeks to address the same issue from the previous section, where the retailer tends to underbuy relative to what is best for the overall supply chain. The buyback contract addresses this by shifting some of the excess inventory risk to the supplier. In a revenue-sharing agreement, the retailer agrees to share some fraction $(1 - s)$ of the revenue from products sold with the supplier in exchange for a lower wholesale price, v_s. With such an arrangement, the retailer's understocking cost becomes $sp - v_s + B$ and her overstocking cost is $v_s - g$, leading to the following fractile for the retailer:

$$p_{x<}(Q^*) = \frac{sp - v_s + B}{sp - g + B} \tag{12.9}$$

Table 12.1 presents one possible revenue-sharing agreement that improves profitability for both supplier and retailer. For comparison, the best possible profitability for the supply chain is included in Table 12.1. The best overall solution for the supply chain is found by solving the newsvendor problem as if there is a single firm with production cost c, selling price p, penalty cost B, and salvage value v. Note that both buyback and revenue-sharing agreements presented improve profitability for both parties. This is made possible by shifting risk in such a way that the retailer chooses to stock a quantity closer to the ideal quantity for the overall supply chain.

Both revenue-sharing and buyback agreements permit demand uncertainty risk to be shifted in a way that improves overall profitability. In addition to these two agreements, other coordinating contracts have been utilized in practice and studied in the academic literature. Generally, the agreements have a similar structure in that they shift risk to one party and compensate that party somehow for taking on the additional risk. The interested reader is directed to Tsay (1999); Barnes-Schuster et al. (2002); Taylor (2002); Cachon and Lariviere (2005); and Cachon (2003).

Table 12.1 Comparison of Contracts

Contract	Retailer's Quantity	Retailer Profit	Supplier Profit	Total Profit
Wholesale price, with $v = 7$	84	$195.69	$421.34	$617.03
Buyback, with $v_b = 7, b = 5.3$	111	$247.16	$457.59	$704.75
Revenue sharing, with $v_s = 2, s = 50\%$	108	$242.05	$457.25	$699.30
Best for the supply chain	125			$716.01

12.4.4 Service-Level Agreements

We analyzed buyback and revenue-sharing arrangements above in the context of a single-period ordering decision. These ideas can be extended to the case where inventory is managed in an ongoing way (e.g., the inventory models from Chapter 6), but a more common coordinating mechanism we see in these settings is that of a *service-level agreement* (SLA). As with the contracts discussed above, the objective of the SLA, at least in an inventory context,[*] is to encourage one party in the supply chain to achieve a higher level of product availability that increases overall profitability.

It is common for retailers to put customer service thresholds in an SLA. Typically, a retailer would specify a financial penalty, often called a *chargeback*[†] when a supplier fails to meet a specified service level. Chen and Thomas (2016) summarize characteristics of inventory SLAs noting that fill rate is the most common service-level measure specified. Consider one of the examples reported in Chen and Thomas (2016) from Advance Auto Parts. Advance's SLA provides for monitoring both the delivery time and the fill rate. If their supplier fails to meet either the quoted lead time or fill rate targets, there may be a disciplinary action, possibly including a fine. In addition, Advance expects 100% order fill rate delivered on the scheduled due date.

This example highlights how a retailer can influence the service they receive from a supplier by imposing a fine for failing to meet a service threshold. The inventory modeling implications of such an SLA can be complicated (see Thomas 2005; Katok et al. 2008; and Chen and Thomas 2016, for further discussion), but generally the imposition of a fine for poor service would translate to a higher missed sale penalty cost in the inventory models discussed throughout this text.

Our observation is that the imposition of chargebacks based on SLAs is a very common way of attempting to coordinate incentives. Trade articles (Harrington 2005; Cassidy 2010; Gilmore 2010) report that these chargebacks can be quite significant, suggesting that retailers are using these SLAs to try to coordinate the supply chain and improve product availability.

12.4.5 Challenges Implementing Coordinating Agreements

We conclude this section on coordinating agreements with a discussion of several challenges that arise in coordinating incentives across the supply chain.

1. *Optimal ordering decisions.* Our analysis of buyback and revenue-sharing agreements assumed that the retailer purchased the *optimal* quantity given a particular offer. Consider the buyback arrangement summarized in Table 12.1 as an example. When the supplier offers an additional buyback (of $b = 5.3$), profitability is improved for both retailer and supplier *only if* the retailer increases her order quantity. Of course, it is in her best interest to do so, but that does not always mean that she will. One possible remedy to this challenge is for the supplier and retailer to agree on a wholesale and buyback price combination *and* an order quantity.
2. *Information symmetry.* Our analysis above assumed that both supplier and retailer have the same information set. Specifically, it is assumed that the supplier knows the retail selling price and the distribution of demand. In practice of course, the retailer and supplier may have different forecasts of customer demand, and this would affect how well the two parties could identify an agreement that improved overall profitability.

---

[*] Service-level agreements are also used in other settings to achieve similar coordination objectives. For a discussion of SLA for service industries, see Hasija et al. (2008); Baron and Milner (2009); and Milner and Olsen (2008).

[†] For further discussion of chargebacks, see Craig et al. (2015) and Gilmore (2010).

3. *Compliance with the agreement.* With coordinating agreements, there is almost always some payment from one firm to the other. In the three examples discussed above, the transfer payments are: buyback (or markdown money) from the supplier to the retailer, revenue share from the retailer to the supplier, and penalty (chargeback) assessed on the supplier by the retailer. In all these cases, the transfer payment depends on information that may or may not be observable by both parties. Often, a supplier cannot see how much excess inventory is left or equivalently know what was sold. With SLAs, retailers can often issue chargebacks without any audit or verification that the specified service level was not met.[*] This can potentially be addressed by shared information systems and/or third-party audit, but of course both of those come at additional cost and effort.

4. *Multiple items, customers, locations.* The revenue-sharing and buyback agreements above were analyzed for a single item with a single supplier (and retailer). Of course in practice, a supplier could be selling hundreds if not thousands of different products to different customers. Since demand distributions, costs, prices, etc. could be quite different across these products, one might theoretically prefer different revenue shares or buybacks for each product, but this is impractical. The SLAs discussed above are often specified in a way that they can be applied to multiple items and multiple suppliers. The result is that one must often accept a blanket or standard coordinating agreement that works pretty well across multiple products and trading partners instead of trying to find the best coordinating agreement for each product and customer.

12.5 Summary

There are opportunities in all supply chains for improved coordination. Our experience suggests that in nearly all firms, the potential impact of improved coordination is quite large. In this chapter, we have presented some key ideas that are essential to improved coordination. In most cases, the ideas are simple and do not require the development of new decision-making models but rather change the input to those models by better sharing of information and alignment of incentives.

Problems

12.1 Give two examples of *rational* decision making that could cause the bullwhip effect. How might you try to mitigate this effect in practice?

12.2 What is the main objective of the S&OP process?

12.3 Why is it more challenging to implement CPFR than S&OP?

12.4 Why is it more challenging to implement VMI than CPFR?

12.5 Give a specific example of how a supplier adopting VMI might be able to reduce inventory-related costs while still maintaining the same level of customer service?

12.6 Consider the buyback agreement discussed in Section 12.4.2, and suppose you are a retailer being approached by a supplier who wants to offer a buyback. The supplier asks for you to

---

[*] For example, Gilmore (2010) asserts that at one time it was common practice for at least one large retailer to issue chargebacks automatically under the assumption that all suppliers would be in violation of the agreement.

share your demand forecast so they can prepare a buyback offer to improve profitability for both firms. Would you share your demand information with the supplier? Why or why not?

12.7 Find an example of a supply chain problem due to lack of coordination using reports or articles in the business press from the past 5 years. What specific causes discussed in this chapter are highlighted?

12.8 The Happy Toy Company (HTC) manufactures the FuzzBot, a robotic stuffed animal, at a cost of $c = \$10$ per unit. Retailers sell the FuzzBot for $40. Sales occur almost exclusively in December of each year, and the lead time is long, so retailers place one order in advance of the holiday shopping season. There is no penalty cost for missing sales ($B = 0$), and the retailers can sell excess Fuzzbots for $g = \$6$ per unit after the holiday selling season.

a. If HTC sets the wholesale price at $20, what is the ideal fractile for a retailer?

b. Suppose a retailer estimates her mean demand to be 400 and the standard deviation of her forecast error to be 80. What quantity should she order? What is her expected profit?

c. If this were the only retailer, what advice would you offer HTC regarding their wholesale price?

12.9 Consider the Happy Toy Company setting in the previous question.

a. Keeping the wholesale price equal to the previous level of $20, find the buyback value that HTC should choose to maximize expected profit.

b. Find a wholesale price and revenue share combination that improves profit for both supplier and retailer relative to the wholesale-price only contract (with $v = \$20$).

12.10 With electronic data interchange (EDI), it is possible for customers and their suppliers to have direct online communication. In some cases, the customer is shown the inventory status of the supplier prior to possibly placing an order. Briefly discuss the possible advantages and disadvantages of such a situation from the standpoint of the supplier.

12.11 Cite examples of the bullwhip effect from the local or national news or business press.

References

Baganha, M. P. and M. A. Cohen. 1998. The stabilizing effect of inventory in supply chains. *Operations Research 46*(3), S72–S83.

Barnes-Schuster, D., Y. Bassok, and R. Anupindi. 2002. Coordination and flexibility in supply contracts with options. *Manufacturing & Service Operations Management 4*(3), 171–207.

Baron, O. and J. Milner. 2009. Staffing to maximize profit for call centers with alternate service-level agreements. *Operations Research 57*(3), 685–700.

Blinder, A. S. 1981. Retail inventory investment and business fluctuations. *Brookings Papers on Economic Activity 2*, 443–505.

Bray, R. L. and H. Mendelson. 2012. Information transmission and the bullwhip effect: An empirical investigation. *Management Science 58*(5), 860–875.

Cachon, G. P. 2003. Supply chain coordination with contracts. *Handbooks in Operations Research and Management Science 11*, 227–339.

Cachon, G. P. and M. A. Lariviere. 2005. Supply chain coordination with revenue-sharing contracts: Strengths and limitations. *Management Science 51*(1), 30–44.

Cachon, G. P., T. Randall, and G. M. Schmidt. 2007. In search of the bullwhip effect. *Manufacturing & Service Operations Management 9*(4), 457–479.

Cassidy, W. B. 2010. Walmart tightens delivery deadlines. *Journal of Commerce.* https://ww.joc.com/economy-watch/ wal-mart-tightens-delivery-deadlines_20100208.html

Çetinkaya, S. and C.-Y. Lee. 2000. Stock replenishment and shipment scheduling for vendor-managed inventory systems. *Management Science 46*(2), 217–232.

Chen, C.-M. and D. J. Thomas. 2016. Inventory allocation in the presence of service level agreements. Working paper. Bucknell University, Lewisburg, PA.

Chen, F., Z. Drezner, J. K. Ryan, and D. Simchi-Levi. 2000. Quantifying the bullwhip effect in a simple supply chain: The impact of forecasting, lead times, and information. *Management Science 46*(3), 436–443.

Chen, F., J. K. Ryan, and D. Simchi-Levi. 2000. The impact of exponential smoothing forecasts on the bullwhip effect. *Naval Research Logistics 47*(4), 269–286.

Craig, N., N. DeHoratius, Y. Jiang, and D. Klabjan. 2015. Execution quality: An analysis of fulfillment errors at a retail distribution center. *Journal of Operations Management 38*, 25–40.

Croson, R. and K. Donohue. 2006. Behavioral causes of the bullwhip effect and the observed value of inventory information. *Management Science 52*(3), 323–336.

Danese, P. 2007. Designing CPFR collaborations: Insights from seven case studies. *International Journal of Operations & Production Management 27*(2), 181–204.

Dejonckheere, J., S. M. Disney, M. R. Lambrecht, and D. R. Towill. 2003. Measuring and avoiding the bullwhip effect: A control theoretic approach. *European Journal of Operational Research 147*(3), 567–590.

Disney, S. M. and D. R. Towill. 2003. The effect of vendor managed inventory (VMI) dynamics on the bullwhip effect in supply chains. *International Journal of Production Economics 85*(2), 199–215.

Dong, Y. and K. Xu. 2002. A supply chain model of vendor managed inventory. *Transportation Research Part E: Logistics and Transportation Review 38*(2), 75–95.

Fry, M. J., R. Kapuscinski, and T. L. Olsen. 2001. Coordinating production and delivery under a (z, z)-type vendor-managed inventory contract. *Manufacturing & Service Operations Management 3*(2), 151–173.

Gilmore, D. 2010. Thinking about supply chain chargebacks. *Supply Chain Digest*. http://www.scdigest.com/ASSETS/ FIRSTTHOUGHTS/10-05-14.php?cid=3464

Grimson, J. A. and D. F. Pyke. 2007. Sales and operations planning: An exploratory study and framework. *The International Journal of Logistics Management 18*(3), 322–346.

Harrington, L. 2005. From cost to profit: Service parts logistics. *Inbound Logistics*. http://www.inboundlogistics.com/cms/article/from-cost-to-profit-service-parts-logistics/

Hasija, S., E. J. Pinker, and R. A. Shumsky. 2008. Call center outsourcing contracts under information asymmetry. *Management Science 54*(4), 793–807.

Ireland, R. K. and C. Crum. 2005. *Supply Chain Collaboration: How to Implement CPFR and Other Best Collaborative Practices*. Boca Raton, FL: J. Ross Publishing.

Katok, E., D. Thomas, and A. Davis. 2008. Inventory service-level agreements as coordination mechanisms: The effect of review periods. *Manufacturing & Service Operations Management 10*(4), 609–624.

Lariviere, M. A. 2006. A note on probability distributions with increasing generalized failure rates. *Operations Research 54*(3), 602–604.

Lee, H. and L. Kopczak. 1997. Responding to the Asia-Pacific challenge. *Supply Chain Management Review 1*, 8–9.

Lee, H., P. Padmanabhan, and S. Whang. 1997. Information distortion in a supply chain: The bullwhip effect. *Management Science 43*(4), 546–558.

Metters, R. 1997. Quantifying the bullwhip effect in supply chains. *Journal of Operations Management 15*(2), 89–100.

Milner, J. M. and T. L. Olsen. 2008. Service-level agreements in call centers: Perils and prescriptions. *Management Science 54*(2), 238–252.

Oliva, R. and N. Watson. 2009. Managing functional biases in organizational forecasts: A case study of consensus forecasting in supply chain planning. *Production and Operations Management 18*(2), 138–151.

Oliva, R. and N. Watson. 2011. Cross-functional alignment in supply chain planning: A case study of sales and operations planning. *Journal of Operations Management 29*(5), 434–448.

Sari, K. 2008. On the benefits of CPFR and VMI: A comparative simulation study. *International Journal of Production Economics 113*(2), 575–586.

Spengler, J. J. 1950. Vertical integration and antitrust policy. *The Journal of Political Economy 50*(4), 347–352.

Stahl, R. A. and T. F. Wallace. 2012. S&OP principles: The foundation for success. *Foresight 29*, 30–34.

Steckel, J. H., S. Gupta, and A. Banerji. 2004. Supply chain decision making: Will shorter cycle times and shared point-of-sale information necessarily help? *Management Science 50*(4), 458–464.

Sterman, J. D. 1995. The beer distribution game. In J. Heineke and L. Meile (Eds.), *Games and Exercises for Operations Management*, pp. 101–112. Englewood Cliffs, NJ: Prentice-Hall.

Taylor, T. A. 2002. Supply chain coordination under channel rebates with sales effort effects. *Management Science 48*(8), 992–1007.

Thomas, D. J. 2005. Measuring item fill-rate performance in a finite horizon. *Manufacturing & Service Operations Management 7*(1), 74–80.

Tsay, A. A. 1999. The quantity flexibility contract and supplier-customer incentives. *Management Science 45*(10), 1339–1358.

Verity, J. W. 1996. Clearing the cobwebs from the stockroom. *Business Week October 21*, 140.

Wallace, T. F. 2004. *Sales & Operations Planning: The "How-To" Handbook*. Cincinnati, OH: TF Wallace & Co.

Waller, M., M. E. Johnson, and T. Davis. 1999. Vendor-managed inventory in the retail supply chain. *Journal of Business Logistics 20*, 183–204.

Wu, D. Y. and E. Katok. 2006. Learning, communication, and the bullwhip effect. *Journal of Operations Management 24*(6), 839–850.

PRODUCTION MANAGEMENT

V

In this section of the book, we turn to the details of decision making in a production environment where one has to face two additional complexities. First, there are capacity constraints at individual work centers, and this capacity must be shared among individual SKUs (as was discussed in portions of Chapter 10). Second, items are closely related in that raw materials are converted into final products through some production process. Thus, coordination of decisions about individual SKUs is mandatory.

In Chapter 13, we provide a broad overview of production planning and scheduling. A decision framework is presented that encompasses strategic (long-range), tactical (medium-range), and operational (short-range) concerns.

Chapter 14 addresses medium-range aggregate production planning. This encompasses the selection of work force levels and aggregate production rates, typically month by month out to a horizon of approximately 1 year. Particular emphasis is given to some of the nonmathematical aspects of aggregate planning.

In Chapter 15, we turn to assembly situations where the overriding concern is with proper coordination of materials and effective use of labor. Specifically, the details of the Material Requirements Planning system are presented.

Chapter 16 examines the popular JIT system and a related system called OPT, and introduces short-range sequencing and scheduling decisions.

The following framework that was introduced in Chapter 2 is repeated here with relevant chapter numbers included.

Product mix	Few of each; custom	Low volume; many products	High volume; several major products	Very high volume; commodity
Process pattern				
Job shop (very jumbled flow)	Sequencing rules; factory physics (Chapter 16)			
Batch flow (less jumbled)		OPT (Chapter 16)		
Worker-paced line flow				
Machine-paced line flow			MRP and JIT (Chapters 13, 15, and 16)	
Continuous, automated, rigid flow			JIT (Chapter 16)	Periodic review/ cyclic scheduling (Chapter 10)

Chapter 13

An Overall Framework for Production Planning and Scheduling

In this chapter, we present a framework that, we believe, is quite useful for managerial decision making in production situations. Although we develop a single framework, there are important differences in detail with respect to the several broad types of production contexts described in Chapter 2. Thus, Section 13.1 provides a summary statement of these types using the product-process matrix. In Section 13.2, we discuss the proposed decision-making framework that recognizes a natural hierarchy of production decisions. Section 13.3 describes options available to deal with the hierarchy of decisions, including the approach of hierarchical production planning.

13.1 Characteristics of Different Production Processes

In Chapter 2, we developed a framework for different production processes that focused on the product-process matrix, which we repeated in the introduction to Section V. In this section, again, we note a subset of the characteristics of these different processes. See Table 13.1. The important point is that management's focus in different production processes should differ. In the next section, we present a unified framework for production decision making. This framework applies across a variety of processes, but managers should emphasize different decisions depending on the process type under consideration. In continuous processes, a paramount issue should be the coordination of items at a single bottleneck operation, to achieve as high a capacity utilization as possible. In contrast, in assembly processes, the primary concern should be the appropriate coordination of raw materials, components, and so forth, across multiple stages of production. Batch flow operations may have a single bottleneck, and thus managers should focus on managing that resource. Alternatively, they may have a strong similarity to assembly processes so that material coordination should be the key management focus. Job shops tend to have moving bottlenecks, and often have few components and materials, so that the important issue is short-term scheduling.

Table 13.1 Characteristics of Production Processes

Characteristics	Job Shop	Line Flow/Batch	Assembly Line	Continuous Flow
Material requirements	Difficult to predict	More predictable	Predictable	Very predictable
Control over suppliers	Low	Moderate	High	Very high
Vertical integration	None	Very little	Some backward, often forward	Backward and forward
Raw material inventory	Small	Moderate	Varies, frequent deliveries	Large, continuous deliveries
WIP inventory	Large	Moderate	Small	Very small
Finished goods inventory	None	Varies	High	Very high
QC responsibility	Direct labor	Varies	QC specialists	Process control
Production information requirements	High	Varies	Moderate	Low
Scheduling	Uncertain, frequent changes	Frequent expediting	Process design around schedule	Inflexible, sequence dictated by technology
Operations challenges	Increasing labor and machine utilization, fast response, and breaking bottlenecks	Balancing stages, designing procedures, and responding to diverse needs	Rebalancing line, productivity improvement, and adjusting staffing levels	Avoiding downtime, timing expansions, and cost minimization

(Continued)

Table 13.1 (*Continued*)   Characteristics of Production Processes

Characteristics	Job Shop	Line Flow/Batch	Assembly Line	Continuous Flow
Process flow	No pattern	A few dominant patterns	Rigid flow pattern	Clear and inflexible
Typical size	Usually small	Moderate	Often large	Large
Type of equipment	General purpose	Combination of specialized and general purpose	Specialized, low, or high tech	Specialized, high tech
Definition of capacity	Fuzzy, often expressed in dollars	Varies	Clear, in terms of output rates	Clear, expressed in physical terms
Capacity addition	Incremental	Varies	Chunks, requires rebalancing	Mostly in chunks, requires synchronization
Bottlenecks	Shifting frequently	Shifting often, but predictable	Generally known and stationary	Known and stationary
Speed (unit/day)	Slow	Moderate	Fast	Very fast
End-of-period push for output	Much	Frequent	Infrequent	None (can't do anything)
Run length	Very short	Moderate	Long	Very long
Process changes required by new products	Incremental	Often incremental	Incremental or radical	Always radical

13.2 A Framework for Production Decision Making

In this section, we first step back from the production function and view managerial decision making from a broader perspective, identifying a hierarchy of interconnected decisions. Next, we recall another type of interrelationship among production decisions at the operational level—namely, the flow of materials between stages in a multistage process. These two types of interrelationships then lead us to a proposed framework for decision making within production contexts.

13.2.1 A Review of Anthony's Hierarchy of Managerial Decisions

Anthony (1965) proposes that managerial activities fall into three broad categories, whose names have been somewhat modified over the years to become *strategic planning, tactical planning*, and *operational control*.

In Section 1.2 of Chapter 1, we briefly discussed strategic planning within an overall corporate strategy. A key point was the need for decision making in the production/inventory area to be consistent with that in other functional areas of the organization. Strategic planning is clearly of long-range scope and is a responsibility of senior management. Tactical planning is a medium-range activity involving middle and top management, concerned with the effective use of existing resources within a given market situation. Finally, operational control involves short-range actions, typically executed by lower levels of management and nonmanagerial personnel, to efficiently carry out the day-to-day activities of the organization. We now expand the discussion of the three levels of activities, tailoring our discussion for the production context. The summary in Table 13.2 may be helpful.

Strategic (or long-range) decisions of relevance to the production area (but with important interactions with other functional areas) include which products to produce; on which of the dimensions of cost, quality, delivery, and flexibility to compete; where to locate facilities; what production equipment to use; and long-range choices concerning raw materials, energy, and labor skills. These encompass the structural decisions of supply chain management discussed in Section 11.1 of Chapter 11. However, in this book, our emphasis is on the development of production planning and scheduling decisions that have already been largely constrained by other strategic decisions made earlier. Chapter 2 included a discussion of the strategic choice between alternative decision systems for production planning and inventory management.

Tactical (medium-range) plans, with a planning horizon from 6 months to 2 years into the future, take the basic physical production capacity constraints and projected demand pattern, established by a long-range plan, and ration available resources to meet demand as effectively and as profitably as possible. Even though basic production capacity is essentially fixed by long-range considerations, production capacity can be increased or decreased within limits in the medium term. A decision can be made to vary one or more of the following: the size of the workforce, the amount of overtime worked, the number of shifts worked, the rate of production, the amount of inventory, the shipping modes used, and possibly the amount of subcontracting utilized by the company. These plans, in turn, constrain but provide stability to what can be done at the operational level.

Operational (short-term) activities provide the day-to-day flexibility needed to meet customer requirements on a daily basis within the guidelines established by the more aggregate plans discussed above. Short-range operating schedules take the orders directly from customers, or as generated by the inventory system, and plan in detail how the products should be processed through a plant. In most cases, detailed schedules are drawn up for 1 week, then 1 day, and finally

Table 13.2   Summary of Anthony's Hierarchy Applied to the Production Function

Category of Activity	Strategic	Tactical	Operational
General types of decisions	Plans for acquisition of resources	Plans for utilization of resources	Detailed execution of schedules
Managerial level	Top	Middle	Low
Time horizon	Long (2+ years)	6–24 months	Short range
Level of detail	Very aggregated	Aggregated	Very detailed
Degree of uncertainty	High	Medium	Low
Examples of variables under control of management	Products to sell; on which dimensions to compete; sizes and locations of facilities; nature of equipment (e.g., general purpose vs. specialized); long-term raw material and energy contracts; labor skills needed; and nature of production planning and inventory management decision systems	Operation hours of plants; work force sizes; inventory levels; subcontracting levels; output rates; and transportation modes used	What to produce (procure), when, on what machine (from which vendor), in what quantity, and in what order; order processing and follow-up; and material control

one shift in advance. The schedules involve the assignment of products to machines, the sequencing and routing of orders through the plant, the determination of replenishment quantities for each SKU, and so on.

13.2.2 Integration at the Operational Level

There is a second type of integration of decisions that is important in the context of multistage production operations. Specifically, at the operational level, we must ensure that there is proper coordination of the input streams of the various raw materials, components, subassemblies, and so forth. In addition, as already evidenced in Chapter 10, there should be a coordination of actions back through all the stages of distribution from the ultimate customer to the interface with production.

13.2.3 The Framework

We now present a single general framework for planning and scheduling within a production environment. It embraces Anthony's hierarchy of decisions as well as integration at the operational level. As noted above, the differences evident in the product-process matrix among various types of processes will lead to a different importance being assigned to specific components of the decision

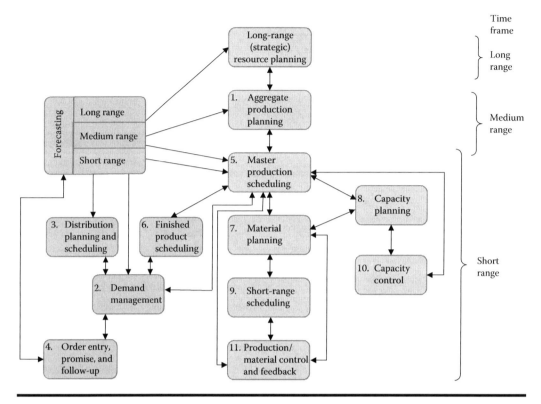

Figure 13.1 A production decision-making framework.

structure. The framework essentially encompasses what is called Closed Loop, Materials Requirements Planning (see, e.g., Wight 1995; Vollmann et al. 2005), a system developed primarily for assembly industries, but the framework is flexible enough to accommodate other process types (see Taylor et al. 1981). A more recent development over the last decade or two is the wide adoption of Enterprise Resource Planning (ERP) Systems that follow a similar structure as that described below, but provide real-time visibility of the status of the factory for all departments, including marketing, operations, accounting, and so on. A key feature of these systems is that they operate from a single database, so that any changes to the schedule are immediately available to all departments. More will be said about ERP systems, and their strengths and shortcomings, in Chapter 15. Frequent references will be made to Figure 13.1 in the following discussion. We restrict our attention to the tactical (medium-range) and operational (short-range) levels. Furthermore, where a chapter number is indicated in the text in connection with a particular block of the diagram, further coverage of that block will be provided (or has been provided) in the designated chapter. In Figure 13.1, bidirectional arrows indicate that the information flow goes both ways.

13.2.3.1 Aggregate Planning

We begin with medium-range aggregate production planning (Block 1 and Chapter 14). Here, we are primarily concerned with establishing production rates, work force sizes, and inventory levels for something on the order of 6–24 months into the future. The time frame used is generally 1 month, and the planning is done on an aggregate basis for families of items (produced on the same equipment)—that is, not on an individual-item basis. This stage of planning assumes extreme importance for continuous-process industries for two reasons. First, it is designed to properly utilize capacity. Second, because of the typical similarity of the products and the flow nature of the production line, it is relatively easy to aggregate and subsequently disaggregate in the shorter-range planning. This is usually not the case in other production environments. Finally, in continuous-process situations, the aggregate planning must properly take into account planned maintenance downtime. Note that the arrow to long-range planning is bidirectional. This indicates that long-range plans can influence the set of choices available in the aggregate production-planning process, and that the aggregate plan can influence long-range strategic plans.

13.2.3.2 Demand Management

We next turn to demand management (Block 2). This function coordinates demand requirements and supply information emanating from five different sources. Obviously, one source is the forecasting module, which tends to be of greater importance in industries that produce to stock. Forecasting is less useful in, say, job shops because of the great uncertainty in make-to-order environments. Distribution planning and scheduling (Block 3 and Chapter 11) represent a second component of the framework that interacts with demand management. They include possible interplant transfers and international distribution. Another input to demand management emanates from order entry, promise, and follow-up (Block 4). In order entry, actual customer orders are received and short-range forecasts of remaining demand in the current period must be modified accordingly. Order promise involves making a firm delivery commitment, based on material and capacity availability. These two ideas are illustrated in Figure 13.2. Follow-up involves activities such as customer notification, lot identification, and so forth.

13.2.3.3 Master Production Scheduling

The central component of the whole framework is master production scheduling (Block 5 and Chapter 15). It serves as the primary interface between marketing and production, and essentially drives all of the shorter-run operations. In essence, the master production schedule takes the aggregate production plan (with its implied constraints) and disaggregates it into a production schedule of specific products to be produced in particular time periods at each manufacturing facility. Normally the time periods are smaller (weeks or days) than those used in the aggregate plan. The master schedule must be realistic because, among other things, it is used to provide customer delivery dates. Note, however, that it is a statement of production, and not of market demand.

Note that the overriding concern in a continuous-process situation is to use up the available capacity at the bottleneck, or pacing, operation. Thus, the master scheduling would normally be done at that stage. In other environments, the choice of where to master schedule is less obvious. Sometimes, it is convenient to master schedule the finished products themselves. However, when there are many combinations of options in a finished product, it may be preferable to master schedule at the subassembly or other intermediate product levels. The same idea applies when myriad packages are used for the same basic product. This issue is directly related to establishing a *push–pull* point in the process. That is, components are pushed to an inventory stocking location according to a master schedule based on forecasts, and then final products are customized to order (pulled) after that point using the component inventory. In some cases, the push–pull point is as far downstream as the retailer, where some final customization is performed. In other cases, it is before the packaging operation in the factory, or even at the start of the production process. The

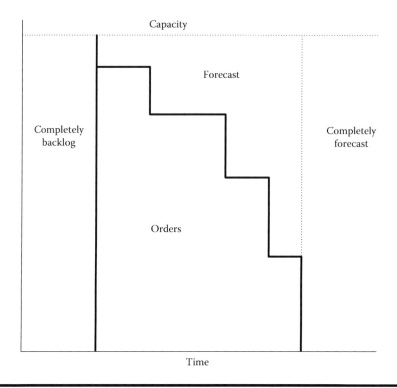

Figure 13.2   Order backlog and forecasts. (Adapted from the Boise Cascade control system described in Bolander, S.F. et al. 1981. *Manufacturing Planning and Control in Process Industries*, Falls Church, Virginia: American Production and Inventory Control Society.)

JIT ideal is to "pull" products through the entire factory based on final customer demand. Another important issue is how far into the future to prepare a master schedule. Each item that can appear on the master schedule will have a cumulative lead time (to purchase its raw materials, process them into components, etc.). The longest of these cumulative lead times is the minimum planning horizon for the master schedule.

Ideally, we would like to establish the master schedule while still keeping the total relevant costs as low as possible. This is often a feasible goal where there is a single known bottleneck operation. However, in contexts such as assembly and job shops, the bottleneck and the flow patterns often change with the product mix. Furthermore, there are likely to be many more short-range changes because of alterations in customer orders, machine breakdowns, material supply difficulties, and so on. Thus, in these situations, we strive, essentially by trial and error, for a feasible solution that is flexible and reasonable from a cost standpoint (Chapter 15). Moreover, in such an environment, the master schedule is likely to be more heavily influenced by negotiating among marketing, production, and finance than is the case in the process industries. If there are high setup times, which is often true in continuous-process industries and job shops, special care is required when coordinating the aggregate production plan with the master schedule, so that the planned production can actually be accomplished.

Lest we leave the impression that the master-scheduling problem is easy in process industries, note that firms in these industries sometimes use very simplistic decision rules at the master-scheduling stage—they keep the factory running regardless of demand and inventory levels. We are aware of some firms, facing slack demand, that should have incurred the high cost of shutting down the factory for a time so that inventory costs could be held down. The incentives for managers, however, implied that the factory would be kept running no matter how much its capacity exceeded current demand.

As mentioned above, frequent updates of input information are likely. To prevent extreme instability of plans, organizations often use the concept of *time fences*. For example, no order can be changed, without senior management approval, within the last 2 days prior to its production time; changes within 1 week of scheduled production require sign-off by the vice president of manufacturing; and any reasonable change beyond a production lead time can be instituted by the master scheduler in charge. These time fences "freeze" the master schedule for the given time periods, helping to resolve the common conflict between sales and manufacturing. Some enterprise systems (ERP) claim to reduce the time fences to zero—essentially creating a real-time scheduling system. However, even these systems, at this writing, must run the bill of materials explosion and lead time offsetting (see Chapter 15) in batch mode on a periodic basis. Thus, there is visibility in real time, but time fences are still required.

The master production schedule can be used effectively in a "what-if" mode to determine the consequences of proposed tactical actions. For example, it can determine the production/inventory impact of a special sales campaign proposed by marketing. Also, the "what-if" mode can be used to provide input for longer-range resources planning. There are a number of excellent references on master production scheduling, in particular Vollmann et al. (2005) and Wight (1995).

13.2.3.4 Finished Product Scheduling

Finished product scheduling (Block 6) is relevant when the master scheduling is not done at the finished product stage. In this case, items must be scheduled for processing from the master schedule level to the finished products stage. For example, the master scheduling may be done to stock with bulk products. That is, they are "pushed" to the inventory location of bulk products. Then, final packaging is scheduled to firm customer orders as they develop. Finished product scheduling, in turn, is likely to activate shipping schedules (an indirect link to Block 3 is shown in Figure 13.1 via demand management). Finished product scheduling is sometimes called *final assembly schedul-ing*, but, as seen above, the finishing operations need not involve assembly. As noted above, finished product scheduling may often be initiated by customer request. That is, final products are "pulled" by customers from their previous stocking location.

13.2.3.5 Material Planning

Material planning (Block 7 and Chapter 15) explodes the master production schedule into implied, detailed production/procurement schedules (timing and quantities) of all components and raw materials. This is a complicated, crucial task in assembly contexts where there can be literally thousands of SKUs and components involved. In contrast, in most process and job shop situations material planning is relatively straightforward.

13.2.3.6 Capacity Planning

Capacity planning (Block 8 and Chapter 15) really occurs at two times in assembly environments. First, there is what is called a rough-cut capacity check that is done during preparation of the master production schedule. The purpose is to make a rough check on the feasibility of the master schedule (against labor availability, raw materials, and limiting pieces of equipment) before carrying out the details of material planning. If the master schedule is not feasible at this stage, it should be adjusted. This is highlighted by the bidirectional arrow between Blocks 5 and 8. Then, once the material planning is undertaken, detailed calculations afford a more accurate check on the feasibility of the plan. In continuous-process situations, feasibility constraints may be directly built into a mathematical model that permits scheduling of the bottleneck operation. When capacity planning reveals serious capacity problems, this type of information should be relayed back to the higher level aggregate production planning (this direct link is not shown in Figure 13.1). Capacity planning can also be of considerable assistance to suppliers in that a projection of order workload can help reduce a supplier's lead time and production cost.

13.2.3.7 Short-Range Scheduling

Once a feasible master production schedule is established, we turn to short-range scheduling (Block 9 and Chapter 16). Again, this is a relatively simple task in most process industry settings, whereas it involves much more detail in job shops, batch flow, and assembly operations. Here, we are concerned with the final scheduling of production, usually narrowed down to a finer time frame than was used in the master schedule. Orders are released to production (in-house) and suppliers (purchasing) and indicate, *based on the master production schedule and the associated material plan*, which items should be replenished and their associated due dates. This can include the rescheduling of open (tentative) orders. Within house, dispatch lists are typically used on a daily basis first thing in the morning. These *suggest* the sequence in which jobs are to be run on each machine on the shop floor. We emphasize the word *suggest*, because the shop foreman knows the hidden agendas that can never be factored into a priority-setting model; for example, Frank simply dislikes running certain types of jobs, particularly on a Monday morning. Material planning and short-range scheduling involve the use of lead times (including queueing time at work centers) that must be reestimated as shop conditions change (Chapter 16). There are a number of techniques for short-term scheduling. An important one is called finite loading (or finite scheduling), which sets a detailed schedule for each job through each work center, and explicitly accounts for limited capacity at each work center. Capacity planning (Block 8), on the other hand, uses *infinite loading*. In other words, rather than scheduling jobs based on capacity, it loads jobs on work centers assuming infinite capacity and then reports any capacity overloads.

13.2.3.8 Capacity Control

Capacity control (Block 10) monitors the level of output, compares it to planned levels, and executes short-term corrective actions if substantial (defined by control limits) deviations are observed. An output control report is a useful device. It shows the actual output, the planned output, and the actual input for each work center. The input is needed because it may reveal that a center's output is down simply because it was starved for input. The possible short-term actions include the use of overtime, transfer of people between work centers, alternative routings of jobs, and so forth.

13.2.3.9 Production/Material Control and Feedback

Finally, the whole system will not operate effectively without the detailed production/material control and feedback function (Block 11). First, there must be well-defined responsibility (involving signing authority) for key short-range actions such as the accepting of dispatch lists. Production status and inventory records must be punctually and accurately updated. Prompt feedback on deviations from the plan must be provided. Possible deviations include changes in customer orders and forecasts, manufacturing problems (machine breakdowns, quality difficulties with a particular batch, lower-than-average yield, etc.), vendor problems (quality, can't meet promise dates, etc.), inventory inaccuracies (identified by stock counts), engineering changes, and so on. More generally, feedback and corrective actions on any perceived problems are to be actively encouraged, which is precisely the idea behind quality circles. Other short-run control functions include the measurement of yield, scrap, energy utilization, and productivity, as well as possible records for product traceability.

We cannot overemphasize this control and feedback function. One firm we consulted with seemed like a war zone because of the battles between the manufacturing planning department, which generated the production plans, and the manufacturing department, which made the products. The manufacturing department did not believe that the plans were feasible, and made daily, even hourly, adjustments as they saw fit. The planning department grew increasingly frustrated, and rarely, if ever, verified that the information used in their plans was correct. Needless to say, any attempts to improve production plans required better communication and cooperation between these two departments.

13.3 Options in Dealing with the Hierarchy of Decisions

In Section 13.2.1, a hierarchy of three levels (strategic, tactical, and operational) of managerial decision making was presented. The purpose of this section is to identify three different possible approaches for coping with this hierarchy, one of which will be treated in detail in Section 13.3.4. In discussing these approaches, we keep in mind the following managerial objectives with respect to production planning, scheduling, and control (partly adapted from Meal et al. 1987):

1. Effective determination of required production resources.
2. Efficient *planned* use of these resources at the tactical level of aggregate planning.
3. Efficient *actual* use of the resources at the operational level.
4. Plans at the operational level that stay within the aggregate constraints implied by the output of the tactical level—that is, short-term behavior that, *in the aggregate*, is consistent with tactical plans.
5. A planning system for the organization that is realistic—specifically, having a planning procedure that fits the organizational structure and is practical to implement.

13.3.1 Monolithic Modeling Approach

Several papers (see, e.g., Dzielinski and Gomory 1965; Lasdon and Terjung 1971; Newsom 1975), in response to objectives 2, 3, and 4 above, propose one large (monolithic) mathematical programming model of the combined tactical production planning and operational scheduling problem. The idea is to input a vast amount of data on demands, production requirements, capacities, bills of materials, and so on, and then optimize the entire system at one time using a large-scale mathematical program. Quite aside from severe computational and input data requirements (e.g., detailed individual-product demand information projected out to the reasonably long horizon needed for aggregate planning), this approach does not normally provide an appropriate response to the fifth objective, except perhaps in a very simple, capacity-oriented flow situation with highly centralized control. One of the authors worked with such a model, and its replacement, at a large chemical company. Apparently, the model had been used effectively for several years. Over time, however, fewer and fewer people understood the details, until only two or three people in the entire company could work with it. The input data requirements were just too demanding, and the users were not comfortable with the mathematical program. It was eventually replaced by a much simpler spreadsheet model.

13.3.2 Implicit Hierarchical Planning

Many organizations, with structures that permit hierarchical planning, do their planning with very little, if any, formal interconnections (of a costing or constraint nature) between the tactical and operational levels. Thus, hierarchical planning is done only on an implicit basis. The MRP system is often based on deterministic models of production lead times and capacity. Problems with capacity constraints are handled on a trial-and-error basis without much prescriptive advice. The rationale is that in most assembly situations, it is very difficult to define aggregate capacity in the first place, and that the capacity is flexible, in any event, in the short run. Thus, modeling efforts, involving precise capacities, at the tactical level are questionable exercises in this environment. Queueing models of the factory have addressed this issue and have been successfully applied in many firms. See Chapter 16 for more detail on these models.

13.3.3 Explicit Hierarchical Planning

A third approach that explicitly develops a sequence of models that account for the interconnections among levels was first proposed by Hax and Meal (1975) and is known as *hierarchical production planning*. (See also Holt et al. 1960; Axsäter 1981; Bitran et al. 1982; Meal 1984; Oliff and Burch 1985; Seward et al. 1985; Meal et al. 1987; Singhal and Singhal 1996. For a thorough review and discussion, see Bitran et al. 1993.) Specifically, separate decision stages are adopted for the strategic, tactical, and operational phases, recognizing the way that most organizations are structured.

Consistent with the framework of Figure 13.1, once strategic decisions are made, an aggregated model is used to assist middle and senior management with the tactical production planning. Then, *explicitly* subject to the constraints imposed by the aggregate plan, a lower level of management, with the aid of a more detailed model, develops the shorter-range production and procurement schedules for individual items. Thus, such a so-called hierarchical decision system permits effective coordination throughout the organizational structure, establishing consistent subgoals at each level that become progressively more specific as they approach the actual day of manufacture. Moreover, appropriate managerial levels now address only questions of prime interest to them. For example, individual-item scheduling details should not be of concern to senior managers who need more of an aggregate perspective. Clearly, this framework directly addresses the objective of having a system that is consistent with the structure of most organizations. The trade-off is that *tangible* costs may be increased somewhat, as compared to the monolithic model approach, because of the sequential nature of the decision making. However, consistent with our earlier findings on individual-item decision rules, it would appear that effectively developed hierarchical (sequential) procedures can do almost as well, from a tangible cost standpoint, as do monolithic approaches, particularly when there is a well-defined bottleneck operation. We expand on the Hax and Meal hierarchical planning system in the next section. Examples of other hierarchical planning systems, some of which report on successful implementations, are provided by Axsäter (1976), Van Beek (1977), McClain and Thomas (1980), Dempster et al. (1981), Axsäter and Jönsson (1984), Edwards et al. (1985), Liberatore and Miller (1985), Oliff (1987), Bitran and Tirupati (1988), Leong et al. (1989), Ari and Axsäter (1988), de Kok (1990), Aardal and Larsson (1990), Bowers and Jarvis (1992); and Venkataraman and Smith (1996).

13.3.4 The Hax–Meal Hierarchical Planning System

In this section, we present some details of an explicit hierarchical system developed by Hax and Meal (1975), and then, we show some of the subsequent modifications that have appeared in the literature. In deciding on the distinct decision levels to use within a hierarchical planning system, Meal et al. (1987) identify three principles:

1. *Lead time and planning horizon.* Decisions are more likely to be placed at different hierarchical levels if they have significantly different implementation lead times or required planning horizons.
2. *Similarity.* If a number of resources or products are similar with respect to certain decisions, it may be convenient to make decisions about the aggregate resource or product at a higher level in the hierarchy and then some, more detailed, decisions at a lower level.
3. *Natural locus of decision making.* Some decisions, such as plant investments, are decided by the board of directors while others, such as lot sizing, are made by production control personnel.

Based on the above principles, the original Hax–Meal system incorporated four hierarchical levels. The first was the strategic level, involving capacity provision decisions and the assignment of products to specific manufacturing facilities. We do not discuss this level here but will instead concentrate on the other three levels of the system. That is, we assume that the capacity of each manufacturing facility is given, as is the set of products that are produced there. The three further levels are primarily a consequence of recognizing three different levels of detail in the product structure:

1. *Individual items.* This is the finest level of detail for scheduling and control purposes (what we have denoted earlier as SKUs). A given product may generate a large number of items differing in certain characteristics such as size, color, packaging, etc.
2. *Families.* As mentioned in Chapter 10, a family of items, in a production context, shares a common major manufacturing setup cost. There are benefits from coordinating the replenishments of such items. In chemical processes, families are often called *process train units.* Members of a family are often found using a set of techniques that have been termed *group technology.*
3. *Types.* These are groups of families whose production quantities are to be determined *simultaneously* by an aggregate production plan. As will be discussed in Chapter 14, the choice of a group of families relates to the selection of the aggregate unit of measure. (Families in a group should have similar seasonal demand patterns and approximately the same production rate as measured by inventory investment produced per unit time.)

These three product levels, considered in reverse order, give the three hierarchical decision levels:

Level 1 Aggregate production planning subsystem (involves types).
Level 2 Family scheduling subsystem (allocates production quantities for each type among families belonging to that type).

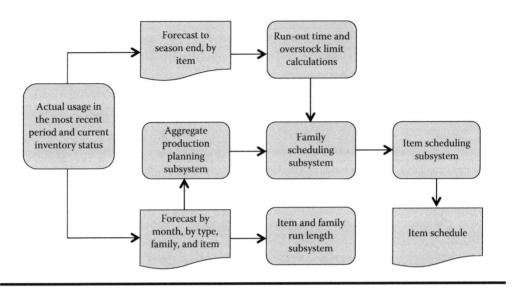

Figure 13.3   A summary of the hierarchical decision system.

Level 3 Individual-item scheduling subsystem (allocates production quantity of each family among its individual-item members).

A summary of the resulting hierarchical decision system is shown in Figure 13.3.

13.3.4.1 The Aggregate Production-Planning System

The aggregate production-planning system addresses the tactical issues of selecting workforce levels, production rates, and inventories for each product type. We will address this topic in detail in Chapter 14. Hax and Meal use a linear programming model, but a simple spreadsheet model is easy and may be even more practical. Demand (requirements) for each product type must be estimated, along with the initial inventory, the planned workforce levels, and the maximum and minimum inventory levels specified by management. Minimum inventory levels are based on the degree of uncertainty per monthly forecast, using techniques such as those discussed in Chapter 6. Production-smoothing considerations also play a role in the determination of maximum/minimum levels. The result of the aggregate planning system is a master production plan for a period, say, 12 months into the future, for each product type. However, normally the implementation would be on a rolling-horizon basis; that is, only the first month results of the aggregate model are implemented. In other words, the only data conveyed to the next level of planning are the aggregated type production quantities for the first month. (Further discussion of rolling horizons will be provided in Section 14.3.)

13.3.4.2 The Family Scheduling Subsystem

The family scheduling subsystem attempts to set the family run quantities correctly and to preserve the aggregate schedule determined at the aggregate planning level. The decision system at this level must schedule just enough production in each of the families in the product type to use up the aggregate production time scheduled in the previous level. Therefore, the schedule should not incur excess major setup costs associated with the production of the multiple families in each product type. At the same time, customer-service standards should be met without violating the upper limits (or the so-called "overstock" limits) on stock levels of individual SKUs. Within these constraints, any family belonging to a type can be produced. To achieve these ends, a four-step procedure is recommended by Hax and Meal:

1. Determine the families in each type that must be run in the scheduling interval to meet SKU service requirements.
2. Determine the initial family run quantities to minimize average cycle inventory and changeover cost.
3. Adjust the initial family run quantities so that all the product time scheduled to a particular type is used up.
4. At each of the above steps, check individual SKU overstock limits.

To determine which families must be scheduled for production in the current month, Hax and Meal calculate the run-out time for each family. All families with a run-out time less than, say, one month will be scheduled. This is equivalent to assuming that families are scheduled once a month and that the production lead time is negligible. Next, a trial-run quantity is computed, perhaps using the methods of Chapters 4, 5, or 10, and the resulting total production time is compared with available capacity. If there is excess capacity, the trial family run quantities must be increased so that this excess capacity is used up and thereby the aggregate production-smoothing requirements are met. Otherwise, the trial-run quantities are decreased so that the required production will be possible within the available capacity.

13.3.4.3 The Individual-Item Scheduling Subsystem

What remains to be done is the scheduling of individual SKUs within each family so that the total run quantities are effectively utilized. Hax and Meal, in their original system, recommend the scheduling of individual SKUs by calculating quantities that equate the expected run-out times for all items in a family. Effectively, this keeps average inventories as low as possible in that most items will be at reasonably low levels when the family is next scheduled. Recall that the concept of equal time supplies was central in the coordinated control decision rules of Chapter 10. See Bolander et al. (1981) and Rhodes (1977) for specific applications of this concept.

13.3.4.4 Extensions to the Original System

We briefly discuss three types of modifications that have been suggested to improve the overall performance of the Hax–Meal hierarchical planning system.

13.3.4.4.1 Avoiding Infeasibilities in Disaggregation

In the original Hax–Meal system, because of the aggregation at the production-planning stage, it is possible to obtain a feasible solution to the aggregate problem that will not permit any feasible solution to the subsequent family and individual-item scheduling problems. This may happen because of an imbalance among the initial inventories of the various items that make up the product type of the aggregate problem. Bitran and Hax (1977) suggest the computation of the so-called *effective individual-item demands* that circumvent this difficulty. The effective demands are determined for each item by subtracting its current inventory level from the sum of its forecasted demand and SS. The sum of all individual-item effective demands belonging to a product type gives the effective demand for that type.

13.3.4.4.2 An Alternative Procedure for Selecting Family Run Quantities

Family run quantities are often based on first trying to use individual-item EOQs. From Chapter 4, we know that the EOQ is an order quantity that strikes a balance between setup and carrying costs. However, total inventory costs for a product type are predetermined by the results of the aggregate planning stage. Thus, Bitran and Hax (1977) suggest a different family scheduling routine that allocates the aggregate production of a product type to families in an effort to minimize only the total of family setup costs.

13.3.4.4.3 The Case of Very High Setup Costs

In Hax and Meal's particular selection of hierarchical levels, setup costs are ignored in the aggregate planning decision. Thus, an *implicit* assumption is that the choice of the aggregate plan will have a relatively minor (as a percentage of total relevant costs) impact on setup costs. Consequently, a modification is suggested by Bitran et al. (1981) when setup costs are relatively large. Operationally, we can define "relatively large" as when the unmodified procedure leads to a solution (through to and including the family stage) where setup costs are larger than 10% of the total costs (holding, overtime, and setup). The modification proceeds essentially as follows:

Step 1. The aggregate planning problem for the product type is solved in the usual fashion (i.e., ignoring setup costs).
Step 2. The family scheduling solution is obtained for the period under consideration, using up the aggregate production time determined by Step 1. At the same time, the marginal benefit (in terms of reduced setup costs) of having a larger aggregate production rate is obtained. If this is positive and there are regular production hours unallocated in the current (Step 1) solution to the aggregate problem, then, additional regular hours are added to the solution until the marginal benefit is no longer positive or until all regular production time is used up.

In essence, Step 2 provides a type of feedback between the two levels of the hierarchical system to compensate for having ignored setup costs at the aggregate planning stage. Graves (1982) discusses a more formalized feedback linkage by means of a Lagrange multiplier formulation.

The need for a modification when setup costs are relatively high points out a fact mentioned earlier—namely, that the appropriate hierarchical decomposition is dependent on the cost structure of the particular firm as well as on its organizational structure.

13.4 Summary

In this chapter, we have provided a general framework for decision making within a production environment. In Chapter 14, we elaborate on the tactical, aggregate production-planning phase. Then, in Chapters 15 and 16, we address three planning and scheduling approaches that are applicable in different contexts. It is critical to emphasize that no single approach to production planning, scheduling, and control is appropriate for all situations and that what we present are not the only possibilities. For instance, in an assembly environment, other approaches, involving mathematical models that more explicitly deal with costs, are suggested by Brown et al. (1981), Maxwell and Muckstadt (1981), and others.

Problems

13.1 Select a local process industry and illustrate the production planning, scheduling, and control activities through the use of the individual blocks of Figure 13.1.

13.2 Repeat Problem 13.1 but for an assembly situation.

13.3 Using Figure 13.1, indicate major interactions with other functional areas of the firm and identify in which blocks they take place.

13.4 Consider the by-products created in process situations.
  a. Give an example of one such by-product.
  b. Discuss why lot sizing and inventory management are complicated by by-products. Illustrate for the case of a single major product, a single by-product, and deterministic level demand.

13.5 From the literature, select an approach, different from Anthony's, to grouping management activities. What are the main differences between your selection and Anthony's structure? Do they matter?

13.6 Formulate a linear programming model of the aggregate planning problem discussed in Section 13.3.4.

13.7 When the market turns down, the usual North American response is often to layoff workers, thus reducing capacity. Suppose instead that the same number of workers were retained. What could be done, in terms of setups and inventory levels, with the associated excess capacity? What about the same questions related to an upturn in the market?

13.8 Formulate a linear programming model that encompasses aggregate planning, master scheduling, material planning, and short-range scheduling.

References

Aardal, K. and T. Larsson. 1990. A benders decomposition-based heuristic for the hierarchical production planning problem. *European Journal of Operational Research 45*, 4–14.

Anthony, R. N. 1965. *Planning and Control Systems: A Framework for Analysis*. Boston, MA: Harvard University Graduate School of Business.

Ari, E. A. and S. Axsäter. 1988. Disaggregation under uncertainty in hierarchical production planning. *European Journal of Operational Research 35*, 182–188.

Axsäter, S. 1976. Coordinating control of production-inventory. *International Journal of Production Research 14*(6), 669–688.

Axsäter, S. 1981. Aggregation of product data for hierarchical production planning. *Operations Research 29*(4), 744–756.

Axsäter, S. and H. Jönsson. 1984. Aggregation and disaggregation in hierarchical production planning. *European Journal of Operational Research 17*(2), 338–350.

Bitran, G., E. Haas, and A. Hax. 1981. Hierarchical production planning: A single stage system. *Operations Research 29*(4), 717–743.

Bitran, G., E. Haas, and A. Hax. 1982. Hierarchical production planning: A two stage system. *Operations Research 30*(2), 232–251.

Bitran, G. and A. Hax. 1977. On the design of hierarchical production planning systems. *Decision Sciences 8*(1), 28–55.

Bitran, G. R. and D. Tirupati. 1988. Planning and scheduling for epitaxial wafer production facilities. *Operations Research 36*(1), 34–49.

Bitran, G. R., D. Tirupati, G. L. Nemhauser, and A. H. G. Rinnooy Kan. 1993. Hierarchical production planning. In S. C. Graves, A. H. G. Rinnooy Kan, and P. H. Zipkin (Eds.), *Logistics of Production and Inventory*, Volume 4, pp. Chapter 10. Amsterdam: North-Holland.

Bolander, S. F., R. C. Heard, S. M. Seward, and S. G. Taylor. 1981. *Manufacturing Planning and Control in Process Industries*. Falls Church, VA: American Production and Inventory Control Society.

Bowers, M. R. and J. P. Jarvis. 1992. A hierarchical production planning and scheduling model. *Decision Sciences 23*(1), 144–158.

Brown, G. G., A. Geoffrion, and G. Bradley. 1981. Production and sales planning with limited shared tooling at the key operation. *Management Science 27*(3), 247–259.

de Kok, A. G. 1990. Hierarchical production planning for consumer goods. *European Journal of Operational Research 45*, 55–69.

Dempster, M. A. H., M. L. Fisher, L. Jansen, B. J. Lageweg, J. K. Lenstra, and A. H. G. Rinnooy Kan. 1981. Analytical evaluations of hierarchical planning systems. *Operations Research 29*, 707–716.

Dzielinski, B. P. and R. E. Gomory. 1965. Optimal programming of lot sizes, inventory and labor allocations. *Management Science 11*(9), 874–890.

Edwards, J. R., H. M. Wagner, and W. P. Wood. 1985. Blue bell trims its inventory. *Interfaces 15*(1), 34–52.

Graves, S. 1982. Using Lagrangian techniques to solve hierarchical production planning problems. *Management Science 28*(3), 260–275.

Hax, A. C. and H. C. Meal. 1975. Hierarchical integration of production planning and scheduling. In M. A. Geisler (Ed.), *Logistics*, Volume 1, pp. 53–69. Amsterdam: North-Holland.

Holt, C. C., F. Modigliani, J. F. Muth, and H. A. Simon. 1960. *Planning Production, Inventories and Work Force*. Englewood Cliffs, NJ: Prentice-Hall.

Lasdon, L. and R. Terjung. 1971. An efficient algorithm for multi-item scheduling. *Operations Research 19*(4), 946–969.

Leong, G. K., M. D. Oliff, and R. E. Markland. 1989. Improved hierarchical production planning. *Journal of Operations Management 8*(2), 90–114.

Liberatore, M. J. and T. Miller. 1985. A hierarchical production planning system. *Interfaces 15*(4), 1–11.

Maxwell, W. and J. A. Muckstadt. 1981. Coordination of production schedules with shipping schedules. In L. B. Schwarz (Ed.), *Multi-Level Production/Inventory Control Systems: Theory and Practice*, pp. 127–143. Amsterdam, The Netherlands: North-Holland.

McClain, J. O. and L. J. Thomas. 1980. *Operations Management: Production of Goods and Services*. Englewood Cliffs, NJ: Prentice-Hall.

Meal, H. C. 1984. Putting production decisions where they belong. *Harvard Business Review 62*(2), 102–111.

Meal, H. C., M. H. Wachter, and D. C.Whybark. 1987. Material requirements planning in hierarchical production planning systems. *International Journal of Production Research 25*(7), 947–956.

Newsom, E. F. P. 1975. Multi-item lot size scheduling by heuristic. Part I: With fixed resources. *Management Science 21*(10), 1186–1193.

Oliff, M. D. 1987. Disaggregate planning for parallel processors. *IIE Transactions 19*(2), 215–221.

Oliff, M. D. and E. E. Burch. 1985. Multiproduct production scheduling at Owens-Corning Fiberglas. *Interfaces 15*(5), 25–34.

Rhodes, P. 1977. A paint industry production planning and smoothing system. *Production and Inventory Management Journal 18*(4), 17–29.

Seward, S. M., S. G. Taylor, and S. F. Bolander. 1985. Progress in integrating and optimizing production plans and scheduling. *International Journal of Production Research 23*(3), 609–624.

Singhal, J. and K. Singhal. 1996. Alternative approaches to solving the Holt et al. model and to performing sensitivity analysis. *European Journal of Operational Research 91*, 89–98.

Taylor, S., S. Seward, and S. Bolander. 1981. Why the process industries are different. *Production and Inventory Management Journal 22*(4), 9–24.

Van Beek, P. 1977. An application of dynamic programming and the HMMS rule on two-level production control. *Zeitschrift für Operations Research Band 21*, B133–B141.

Venkataraman, R. and S. B. Smith. 1996. Disaggregation to a rolling horizon master production schedule with minimum batch-size production restrictions. *International Journal of Production Research 34*(6), 1517–1537.

Vollmann, T. E., W. L. Berry, D. C. Whybark, and F. R. Jacobs. 2005. *Manufacturing Planning and Control for Supply Chain Management.* New York: McGraw-Hill/Irwin.

Wight, O. 1995. *Manufacturing Resource Planning: MRP II: Unlocking America's Productivity Potential.* New York: John Wiley & Sons.

Chapter 14

Medium-Range Aggregate Production Planning

Within the framework of Figure 13.1 of Chapter 13, we now deal with the medium-range, tactical problem of establishing aggregate production rates, workforce sizes, inventory levels, and, possibly, shipping rates. We present an evaluative overview of a selection of approaches to dealing with this problem.

While a rich selection of decision systems that could be used in corporate planning exists, modeling has progressed much faster than application in practice. More careful consideration needs to be given to the *process and strategy* of introducing aggregate planning technology. There is a need in most organizations to demonstrate to management that the aggregate planning problem can be structured rationally with the assistance of relatively simple, yet powerful, concepts and models. Managers need to understand that these models can add rigor to their decision-making process. Many analysts, on the other hand, do not seem to be aware of the current low level of implemented aggregate planning theory, or of the qualitative aspects of the decisions. They need to understand that the "solution" to the aggregate planning model is only one piece in a large and complex puzzle. This is why we will emphasize in our overview the implementability and practicality of existing methodology.

In Section 14.1, we elaborate on the aggregate planning problem. Next, in Section 14.2, we discuss relevant costs. Section 14.3 deals with the choice of the planning horizon and appropriate ending conditions. In Section 14.4, we discuss two pure strategies, a level workforce strategy and a "chase demand" strategy. Then, in Sections 14.5 through 14.8, we describe various methods for coping with the aggregate production-planning problem (see Table 14.1). Section 14.9 addresses dealing with uncertainty in the aggregate planning problem.

14.1 The Aggregate Planning Problem

The aggregate production-planning problem, in its most general form, can be stated as follows: Given a set of (usually monthly) forecasts of demand for a single product, or for some measure

Table 14.1   Classification of Selected Aggregate Production-Planning Methods and Models

Classification	Type of Model/Method	Type of Cost Structure	Discussed in
A. Feasible solution methods	1. Barter 2. Graphical/tabular (spreadsheet)	General/not explicit linear/discrete	14.5.1 14.5.2
B. Mathematically optimal methods	3. LP models 4. Transportation models 5. Linear decision rules	Linear/continuous linear/quadratic/ continuous	14.6 14.10
C. Heuristic decision procedures	6. Simulation search procedures	General/explicit	14.7
	7. Management coefficients	Not explicit	14.8.1
	8. Projected capacity utilization ratios	Not explicit	14.8.2

of output that is common across several products, what should be specified for each period in terms of[*]

1. Size of the workforce W_t
2. Rate of production P_t
3. Quantity shipped[†] S_t

The problem is usually resolved analytically by minimizing the expected total costs over a given planning interval (6–24 months). The cost components include:

1. Cost of regular payroll and overtime
2. Cost of changing the production rate from one period to the next (including items such as costs of layoffs, hiring, training, learning, etc.)
3. Cost of carrying inventory
4. Cost of insufficient capacity in the short run

In the above definition, only three decision variables need to be determined: the required workforce size W_t (number of workers in month t); the required production rate per period P_t (units of product produced in month t); and the number of units S_t, to be shipped in month t. The

---

[*] One other decision variable in some situations is the amount of subcontracting utilized, which we are implicitly ignoring here.

[†] Note that we use the shipments S_t as opposed to the demand D_t, because, in general, it is not profitable to always try to meet all demands on time.

resulting inventory at the end of month t, I_t, can then be determined as follows:

$$I_t = I_{t-1} + P_t - S_t \qquad (14.1)$$

The workforce variable W_t does not appear directly in Equation 14.1 because any set level of W_t in turn determines the possible range of values that P_t can assume. That is, the size of the workforce is always *one* of the important determinants of the feasible rate of production per period.

Equation 14.1 is for the case of production to stock. An analogous result for the situation of production to order is the following (assuming that shipments are sent out as soon as possible):

$$B_t = B_{t-1} - P_t + D_t \qquad (14.2)$$

where

B_t = backlog at the end of period t
D_t = actual (or forecasted) demand in period t

It would appear at first glance that production smoothing (i.e., stabilizing the work load) is not possible in a make-to-order situation. However, see Problem 14.2, which is based on the work of Cruickshanks et al. (1984).

Returning to the make-to-stock situation, note that the solution to the problem posed is greatly simplified *if average demand over the planning interval is expected to be constant.* Under such a circumstance, there would be no need to consider changing the level of the production rate, nor the size of the workforce, nor the planned quantity to be shipped from period to period.[*] The firm would need to only carry sufficient inventory (SS) to balance the risk of running out using one of the many methods described in Chapter 6. The appropriate size of regular payroll would be given by

$$W_t = \frac{\text{Average demand rate per period}}{\text{Productivity per worker}} \qquad (14.3)$$

We would have to consider using overtime only to replenish SS on occasion, or if Equation 14.3 resulted in a noninteger number of workers so that it might be worthwhile to round down and make up the shortfall through working overtime.

The complexity in the aggregate production-planning problem, therefore, arises from the fact that in most situations, demand per period is not constant but varies from month to month according to some known (seasonal) pattern. Only then do the following questions need to be answered:[†]

1. Should inventory investment be used to absorb the fluctuation in demand over the planning period by accumulating inventories during slack periods to meet demand in peak periods?
2. Why not absorb the fluctuations in demand by varying the size of the workforce by hiring or laying off workers?

---

[*] We are also assuming implicitly that there are no problems in receiving a constant supply of raw materials and labor at a fixed wage rate. If the supply of raw materials or labor is seasonal, then these statements need to be modified.
[†] See footnote *.

3. Why not keep the size of the workforce constant and absorb fluctuations in demand by changing the rate of production per period by working shorter or longer hours as necessary, including the payment of overtime?

4. In process industries and where capacity exceeds average demand over a long period, should periodic shutdowns be used or should the plant be throttled (run at less than full capacity)? Taylor (1980) discusses this situation (see also Problem 14.3).

5. Why not keep the size of the workforce constant and meet the fluctuation in demand through planned backlogs or by subcontracting excess demand?

6. Is it always profitable to meet all fluctuations in demand or should some orders not be accepted?

To develop decision models to analyze these questions, a number of logical conditions and data requirements have to be met. First, as is obvious from the above discussion, management must be able to forecast reasonably accurately the *fluctuating* demand over an appropriately long planning interval (horizon). The degree to which the different models that we will discuss are affected by forecast errors will vary, but they will all require that a reasonable forecast of demand be made (e.g., by using the regression methods referred to in Chapter 3).

Second, to apply any of the *analytical* approaches, management must be able to develop a single overall *surrogate* measure of output and sales for all the different products to be scheduled by an aggregate production plan. For example, instead of using every size and color of house paint for this level of planning, the surrogate measure could be gallons of paint (including all sizes and all colors). This can prove to be quite difficult in practice, particularly when one realizes that disaggregation of the aggregate results is subsequently needed for the shorter-range, operational, scheduling actions. For example, from the aggregate number of gallons of paint to be produced in a month, schedulers must decide the number of quart cans of sky-blue paint to be produced each week. (See, e.g., Axsäter 1981; Bitran et al. 1981; Boskma 1982; Axsäter et al. 1983.) Some examples of units used in past applications include gallons of beer, man-hours of assembly labor, machine hours in some job shop situations, cases of cans packed by a cannery, and dollars of sales, where each dollar of sales represented approximately the same amount of productive effort to be allocated.

Third, to use an *analytical* approach, management must be able to identify and measure the costs associated with the various options raised above (these costs are discussed in greater detail in the next section). In actual fact, there are important intangibles that must also be included in deciding on whether to use a particular decision system (e.g., the willingness of management to use a specific model as a decision aid, the commitment of the firm to the community and the workers, the effects on public image or public relations of frequent layoffs of employees, etc.).

Prior to discussing costs, we mention the importance of taking a broad look at the aggregate planning problem, particularly questioning some of the "givens." For example, if there is a highly seasonal demand pattern, can inducements or promotions be given to customers to reduce the seasonal variations? Is there a possibility of expanding the product line to include items with a counterbalancing seasonal pattern (e.g., snowblowers and lawnmowers)? Also, can the capacity be expanded by sharing equipment with other organizations whose needs occur during a different period of the year? An interesting reference on these kinds of issues is Galbraith (1969). Finally, Tersine et al. (1986) discuss the possibility of varying lot sizes as an alternative to overtime and undertime in aggregate scheduling. Additional references on aggregate planning include Nam and Logendran (1992), and Thomas and McClain (1993) who provide an overview and survey; Oliff and Burch (1985), Akine and Roodman (1986), Stadtler (1988), de Kok (1990),

Thompson et al. (1993), Heath and Jackson (1994), Behnezhad and Khoshnevis (1996), and Singhal and Singhal (1996).

14.2 The Costs Involved

There are essentially six categories of costs involved in the aggregate planning problem:

1. Regular-time costs. The cost of producing a unit of output during regular working hours, including direct and indirect labor, materials, and manufacturing expenses.
2. Overtime costs. The costs associated with using manpower beyond normal working hours.
3. Production-rate change costs. The expenses incurred in altering the production rate substantially from one period to the next.
4. Inventory associated costs. The costs (both out of pocket and lost opportunity) associated with carrying inventory.
5. Costs of insufficient capacity in the short run. The costs incurred when demands occur while an item is out of stock. These costs can be a result of backordering, lost sales revenue, and loss of goodwill. Also included in this category are the costs of actions initiated to prevent shortages—for example, split lots and expediting.
6. Control system costs. The costs associated with operating the control system. These include the costs of acquiring the data required by analytical decision rules, the costs of computational effort exerted per time period, and any implementation costs that depend on the particular control system chosen.

As Vergin (1980) points out, percentage cost comparisons among alternative procedures should only include controllable variable costs and not fixed costs that are independent of the policy used.

14.2.1 Costs of Regular-Time Production

A typical relation of cost versus production rate is shown in Figure 14.1. Usually some form of monotonic increasing behavior is assumed (i.e., the costs associated with any production rate must be at least as high as those associated with any lower production rate). The near-vertical rises in Figure 14.1 occur because additional pieces of equipment are required at increasingly higher rates of production. The shape of the cost curve may also depend on the particular time period under consideration.

A major portion of regular-time production costs in Figure 14.1 is the regular-time wage bill paid to the full-time workforce. When such costs are identified separately, they are usually assumed to increase linearly with the size of the workforce, as shown in Figure 14.2. It is becoming more difficult in practice to reduce the size of the workforce from 1 month to the next because of commitment to employees, social pressures, public opinion, and union contracts. For instance, many countries have labor laws that make firing workers extremely expensive. In addition, some firms have discovered (the hard way) that layoffs can result in a loss of organizational knowledge. As a result, W_t, the size of the workforce, in many situations, is in effect a constant and not a decision variable that can be altered at will. Under such a circumstance, the regular wage bill becomes a fixed cost, as shown by the dotted line in Figure 14.2, and the size of the intercept "C" in Figure 14.1 would have to be adjusted upward accordingly.

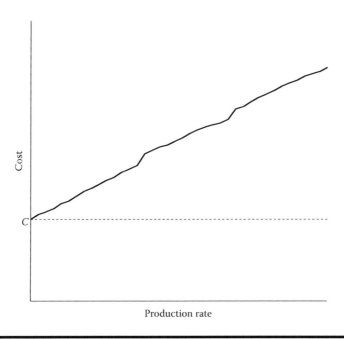

Figure 14.1   Typical costs of regular-time production.

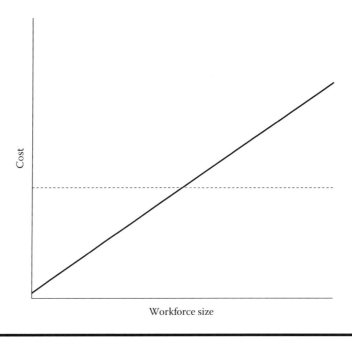

Figure 14.2   Typical costs of regular-time labor.

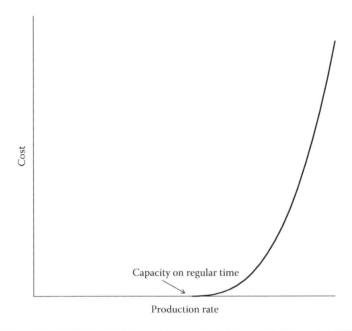

Figure 14.3 Typical shape of overtime costs for a given work force size.

14.2.2 Overtime Costs

The general graphical form of overtime costs is illustrated in Figure 14.3. The production rate is at first increased beyond the regular-time capacity at little extra cost. This is because only the bottleneck operations need to run on overtime. However, with continued increases in the production rate, more and more of the operation must be run at overtime premium rates. The overtime cost curve rises sharply at higher levels of overtime because the efficiency with which workers produce, when asked to work longer and longer hours, starts to decrease. It is clear that a single overtime cost curve versus production rate is appropriate only for a fixed size of workforce; that is, there really is a whole set of curves for different workforce sizes. There is also a different curve for every t (when workforce is held constant) if workers are asked to work overtime day after day. Such *cumulative tiredness* results in curves with increasing slopes for every consecutive t worked on overtime.[*] The curve in Figure 14.3 is not always as smooth as shown. Discontinuities are common.

For process industries, the costs of regular time and overtime can be as illustrated in Figure 14.4. If production exceeds the rated capacity, costs increase due to lower-quality, problems with workforce skill, and so on. Likewise, if production is throttled so that the equipment runs less than the rated capacity, quality may degrade, materials may not be used efficiently, and so on.

14.2.3 Costs of Changing the Production Rate

These costs consist primarily of production rate changes brought on by changes in the size of the workforce, although they may also include shutdown and startup costs. Workforce changes incur

---

[*] Employees of some General Motors factories have been known to complain because they were being asked to work overtime for long periods of time without a break.

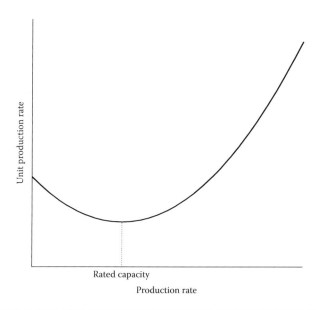

Figure 14.4 **Typical shape of production costs for process industries.**

costs of training, reorganization, terminal pay, loss of morale, and decreased interim productivity. Decreased productivity results in part from the learning process when new labor or equipment must be started up. Another source of changeover costs is a common union requirement that specifies that, when an individual is laid off, the person's job can revert to any other union member according to seniority. As a result, the layoff of a few workers can result in *bumping*, a series of shifts in job responsibilities.

A typical behavior of the costs as a function of the change in rates is shown in Figure 14.5. Two points are worth mentioning. First, the curve is usually asymmetric about the vertical axis; that is, a unit increase does not necessarily cost the same as a unit decrease. Second, ideally there should be a separate curve for each starting rate P_{t-1}. This is because a change from 90% to 95% of production capacity has an appreciably different cost than going from 30% to 35%.

Some decision models do not include a cost for changing production rates. Instead, they may specify that the change from one period to the next can be no greater than a certain value, or simply that the rate itself must stay within certain bounds. A special case is where no change is permitted from period to period. This level workforce policy will be discussed in Section 14.4.

As we will see, the inclusion/exclusion of an explicit cost of changing the production rate becomes an important issue in the evaluation of alternative analytical models. In particular, this type of cost is very difficult to determine explicitly (see Welam 1978).

14.2.4 Inventory Associated Costs

As discussed in Chapter 1, there is a cost associated with tying up funds invested in the inventory. The standard approach is to say that the costs for T periods are given by

$$\text{Costs for } T \text{ period} = \bar{I}_1 vr + \bar{I}_2 vr + \cdots + \bar{I}_T vr = vr \sum_{t=1}^{T} \bar{I}_t \qquad (14.4)$$

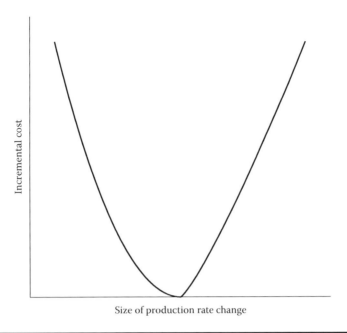

Figure 14.5   **Costs of production rate changes.**

where

$\bar{I}_t$ = average inventory (in aggregate units) in period t, often approximated by the starting or ending inventory

v = unit variable cost of the aggregate unit used[*]

r = carrying charge in \$/\$/period (possibly determined implicitly through the use of exchange curves)

14.2.5 *Costs of Insufficient Capacity in the Short Run*

When an unplanned shortage actually occurs, the costs can be expressed in a number of different ways. The method to use is a function of the environment of the particular stocking point under consideration. Planners should note that the costs of actions taken to avoid shortages are also relevant. These types of costs, discussed in more detail in Chapter 6, are generally quite difficult to estimate.

Some aggregate planning models look at the combined costs of inventory and insufficient short-run capacity in a slightly different and more aggregate fashion. From fundamental economic arguments, which we presented in Chapter 4, the best inventory level in period t can be shown to be proportional to the square root of D_t, where D_t is the forecasted demand for period t. Over a narrow range, any square root function can be approximated reasonably well by a straight line;

---

[*] At any point, the actual value of the inventory will depend on the mix of individual items present. The variable v represents the unit value of an average mix. This suggests that, if possible, the aggregate unit should be selected so that it represents roughly the same dollar value of each of the underlying individual SKUs.

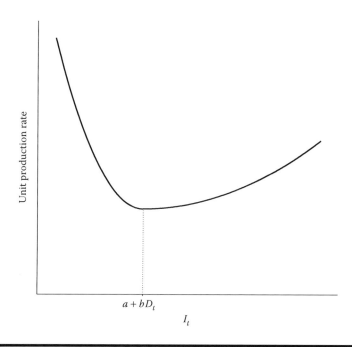

Figure 14.6 Inventory and shortage costs.

that is,

$$\text{Optimal inventory in period } t \approx a + bD_t \qquad (14.5)$$

where *a* and *b* are constants. Deviations from the optimal inventory result in extra costs. Too high an inventory results in excessive carrying costs; too low an inventory leads to inordinate shortage costs. For a given value of D_t, the typical curve of costs versus inventory is shown in Figure 14.6. There is actually a whole set of such curves, one for each value of D_t, the current aggregate rate of demand. See Holt et al. (1960) for a complete discussion.

14.3 The Planning Horizon

The planning horizon refers to how far into the future to use information (particularly demand forecasts) in making the current decisions. This concept was first introduced in Chapter 5 with respect to individual item lot sizing, but it is equally applicable in aggregate production planning. *If the demand pattern (and other factors) were known with certainty*, then, the quality of the decision making would never decrease as the planning horizon was increased. However, demand forecasts become less accurate the further into the future we make projections. Considerable testing has revealed that, in a seasonal demand situation, the horizon should be chosen to extend beyond the next seasonal peak. This point is particularly important when using the pure level or chase strategies discussed in the next section. To keep the number of decision variables manageable, an effective device is to vary the length of the individual periods within the horizon—for example, by using a 12-month horizon composed of six 1-month periods, followed by two 3-month intervals (see Boskma 1982). Perhaps more important is the use of a reasonable desired ending inventory level to reflect the on-going nature of the business beyond the artificial horizon (see Baker and Peterson 1979; McClain and Thomas 1977).

As in Chapter 5, it is also important to recognize that implementation, because of future uncertainties, is almost always on a rolling-horizon basis. Specifically, if the planning horizon is of length T periods, only the first period results of the aggregate planning model are implemented. Then, at the end of the first period, a new (rolling) horizon of T periods is used to establish new results, and so on.

The notion of a frozen period is often also relevant. In particular, if planning is done at the beginning of January, the first period where changes can be made may be February (or even later). That is, the previous decisions for January (or even longer) may be frozen. This is usually the case where scarce, highly skilled labor is involved or where the production process itself involves long delays in bringing about a change in the output rate. See Venkataraman and Smith (1996).

14.4 Two Pure Strategies: Level and Chase

Some firms are completely constrained with regard to their ability to hire or fire workers. For instance, if skilled labor is scarce and the process requires highly skilled people, managers may decide to pursue a strategy that maintains a constant workforce size. Inventory builds up in periods of slack demand and is drawn down in periods of peak demand. This strategy is called *level* because of the level, or constant, workforce size. Other firms face extremely high inventory costs, but are able to hire and fire somewhat freely. Managers in these firms may pursue a strategy that keeps inventory levels at a minimum. Seasonal fluctuations are met with changes in the workforce size. This strategy is often called *chase* because production chases, or exactly meets, demand requirements. Most strategies, of course, are a mixture of these two, with limited changes in workforce so that the total of the costs of changing the number of workers and the costs of inventory are kept to a minimum. In this section, we briefly discuss these two pure strategies. See also Johansen and Riis (1995).

If it is possible to specify costs for hiring and firing, it may not be necessary to restrict attention to a pure level strategy, even if the skilled workers are scarce. Presumably the costs to find new workers, and the costs to train them, would be included in the hiring cost. However, managers may decide that these costs are too difficult to specify accurately, and therefore may require that hiring and firing be kept to a minimum. Other reasons for pursuing a level strategy include the potential for loss of organizational knowledge, a commitment to the community and the workers, fewer disruptions to the production process, and better employee relations and morale. A level strategy is common in plants that are the major employer in a town, and where inventory carrying and backorder costs are relatively low. The disadvantages include high inventory investment, potentially high shortage costs, and the possibility of employees working less than full time. Examples, as reported by Bolander et al. (1981), are the Clairol Products Division and the Champion Spark Plug Company. Edwards et al. (1985) describe a successful use of aggregate planning at an apparel manufacturer who wanted a level production schedule throughout the year, so that inventory buildups were required. Tadei et al. (1995) describe the successful use of aggregate planning in the food industry.

A chase strategy, in its pure form, meets demand with minimal inventory investment by hiring and firing each month. Of course, there are many ways of chasing demand that do not require frequent hiring and firing. For example, the firm may use overtime and undertime, subcontracting, and backorders to try to limit the number of changes to the workforce size. Chase strategies are possible if the firm has a monopoly on the labor market, so that workers have few other places to work. They are common when inventory costs are high, the required skill level is low, or there is a significant service component to the operation. (In services, it is typically not possible to build inventory in advance of demand because the customer participates in the service. Think of a haircut; the service provider must be available when the demand arrives.) The disadvantages of a chase strategy include poor morale and employee relations, high hiring and firing costs, disruptions to operations, and high costs of overtime and subcontracting, if these are permitted.

Clearly the choice among level, chase, or some mixture of the two is based on both quantitative and qualitative factors. As noted above, managers should not ignore models that compute the total costs of various options. To base decisions purely on gut feel and qualitative information runs the risk of ignoring significant costs and benefits. On the other hand, to base decisions purely on the output of a quantitative model runs the risk of ignoring extremely important nonquantitative issues such as employee morale. In the remaining sections of this chapter, we primarily discuss quantitative models that should enhance the rigor of the decision process, while recognizing that they serve as only one input to that process.

14.5 Feasible Solution Methods

14.5.1 General Comments

Most *feasible solution methods* make only a limited effort toward achieving some form of explicit trade-off among the competing costs discussed earlier. The apparent overriding objective for these approaches is the achievement of *any* feasible allocation of available resources that guarantees that daily orders will be met.

Each group within an organization approaches the determination of appropriate inventory investment (and thereby production smoothing and workforce balancing) with somewhat different motives. Top managers, along with financial managers, are pleased if inventory investment can be successfully kept at a "minimum." Operating managers prefer long production runs, smooth production and workforce levels, and thereby fluctuating inventory investment. Marketing managers usually elect for a larger inventory investment because they want high availability of finished goods. They are less concerned with the length of production runs. Buxey (2005) surveyed plant managers and discovered that at JIT ethos often led to chase strategies being used, unless skilled labor or facilities constraints required a more level approach.

Feasible solution methods attempt to strike a compromise position among the desires of the various parties involved. Many companies have a scheduling committee, consisting of all the senior managers affected by an aggregate production plan, that considers various alternatives and agrees on a feasible set of trade-offs. Frequently, a compromise is achieved through a bargaining or bartering process. "You support my production plan for our division in England and I will support your request in Argentina."[*] Often, these kinds of compromises are struck as part of the S&OP process. See Section 12.2 for further discussion of S&OP.

An approach often used in industry is to take last year's plan and adjust it slightly to meet this year's conditions. The danger in doing this lies in the implicit assumption that any previous plan

---

[*] Donald B. McCaskill, former Executive Vice President of Warner-Lambert Company, as quoted in "Management Science in Canada," Management Science, Vol. 20, No. 4, December 1973, p. 571.

Table 14.2   Cumulative Forecasts (in Production Hours)

Month	Forecast Demand D_t	Cumulative Demand
1. September	30	30
2. October	30	60
3. November	120	180
4. December	90	270
5. January	60	330
6. February	30	360

was close to optimal. By pursuing such a course, management takes the risk of getting locked into a series of poor plans.

In many situations, a graphic or tabular approach, such as that to be discussed in the next section, can be used to focus attention on the key trade-offs to be made in aggregate production planning.

14.5.2 An Example of a Graphic–Tabular Method

The most useful and rudimentary analytical extensions of bartering are of a graphic–tabular nature.[*] Such procedures, while not mathematically optimal, always guarantee a feasible solution.[†] Such methods are surprisingly effective and difficult to improve on, and are quite practical given the ease of developing them on a spreadsheet. See for example Shafer (1991) and Penlesky and Srivastava (1994).

Consider the forecasted demand pattern in Table 14.2 and its plot as cumulative demand in Figure 14.7. In Figure 14.7, we have assumed that 30 production hours were in inventory at the beginning of period zero. The same 30 production hours are to be made available at the end of the 12-month planning horizon. The straight line labeled as the level strategy represents a constant or level workforce production plan using 60 hours of production per month, which meets the expected cumulative forecast of 360 production hours for the 12-month period.[‡] The pure chase production strategy is also drawn in Figure 14.7, assuming a buffer stock of 30 production hours each month. (The data for Figure 14.7 are contained in Tables 14.3 and 14.4.) Because we assume a buffer stock of 30 production hours for the chase strategy, its curve in Figure 14.7 is always 30 production hours above the cumulative demand curve. Any other plans that represent a mixture of chase and level can also be drawn.

Note from Figure 14.7 and Table 14.3 that if sales materialize as expected, then a level production plan is feasible, although, during months 4 and 5, inventory would be drawn down to zero. This highlights the fact that the timing of the aggregate planning process is important. Note that

---

[*] See for example Close (1968).

[†] In Section 14.10, we note that feasibility at this stage of production planning may not guarantee a feasible solution at the master scheduling or materials requirements planning stage.

[‡] The single surrogate measure of output in which all variables are measured is assumed to be production hours. A single hour of production may require several man-hours to achieve.

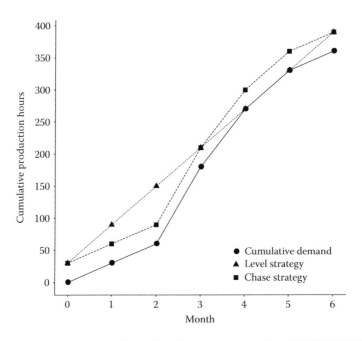

Figure 14.7   Graphs of cumulative forecasts and alternative production plans.

the highest demand occurs in months 3 and 4, and that there is sufficient time to build up inventory in advance. If the planning decisions were made closer to the peak demand months, it would not be possible to have a level workforce without incurring backorders until demand slackens. To avoid backorders, the firm must hire people immediately, and then try to maintain a level workforce for the remainder of the horizon. If they pursue this strategy, however, there will be excess productive capacity through the latter part of the horizon. Backorders, on the other hand, may be

Table 14.3   Level Production Plan (60 Hours of Production/Month)

End of Month	Production	Cumulative Available	Expected Cumulative Demand	Expected Inventory
0				30
1	60	90	30	60
2	60	150	60	90
3	60	210	180	30
4	60	270	270	0
5	60	330	330	0
6	60	390	360	30

Table 14.4 Chase Production Plan

End of Month	Production	Cumulative Avail	Cumulative Demand	Expected Inventory
0				30
1	30	60	30	30
2	30	90	60	30
3	120	210	180	30
4	90	300	270	30
5	60	360	330	30
6	30	390	360	30

unacceptable. This discussion is consistent with a comment made in Section 14.3 that in a seasonal demand situation, the horizon should be chosen to extend past the next seasonal peak.

Any cumulative production plan that ends up at the level of the cumulative demand or higher in period 6 in Figure 14.7 without violating any exogenous constraints would be feasible. This is a very powerful and useful result, quite easily achieved graphically or on a spreadsheet. In fact, these tables and this figure were developed on a spreadsheet in a matter of minutes. The selection of a low-cost, feasible production plan in this way, from among the infinite numbers possible, will be somewhat more difficult, as we will see.

In Figure 14.7, any vertical distance between a cumulative demand line and a cumulative production line represents the amount of inventory OH at that time. For example, at the end of month 2, we expect to have 30 hours of production OH with the chase strategy and 90 hours with the level strategy. The possibility of running out of inventory in months 4 and 5 could render the level plan infeasible if company policy dictates that such a possibility should not be tolerated in any production plan considered. In addition, there is usually a maximum level of inventory that is allowed by management policy. This could represent either the maximum space available in the warehouse for storage or, alternatively, the maximum investment risk that management is willing to take. In Figure 14.7, warehouse capacity is assumed to be 120 production hours. The vertical distance between any cumulative production plan and any cumulative sales forecast must be greater than zero and less than 120 production hours. Such constraints are called exogenous; that is, they are imposed externally on the decision problem.

Other exogenous constraints are possible. For example, we may include the fact that the manufacturing facility can support a maximum regular production rate of 90 production hours/month and that management is unwilling to take the chance of losing key personnel by producing less than 30 production hours/month. The graphical approach to these constraints would be to draw a line with a slope of 90 hours/month and another line with a slope of 30 hours/month. Any production plan that exceeded the higher slope or was less than the lower slope would be infeasible. A simpler approach is to use an **IF** function in the spreadsheet to highlight an infeasibility if these constraints are violated.

Throughout this discussion, we have relied on the expected or forecast demand, without regard to variability. This focus is because the typical approach is to consider variability only when making shorter-term operational plans. In addition, aggregate forecasts are generally more accurate than

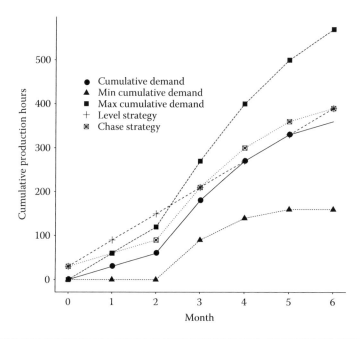

Figure 14.8 Graphs of min and max cumulative forecasts and alternative production plans.

individual-item forecasts. Occasionally, however, managers want to consider demand variability at this stage of planning. If demand may take on significantly larger values than expected, managers may want to consider an aggregate plan that contains flexibility for higher levels of demand. There are several ways to accommodate this problem. One is to plot the minimum and maximum demand on a graph similar to Figure 14.7. This is illustrated in Figure 14.8. Note that, due to the fact that forecasts are less reliable further in the future, the minimum and maximum demand curves spread wider over the horizon. If managers are concerned with never stocking out, it will be necessary to plan for higher production rates starting in month 2 (for the chase strategy) or month 3 (for the level strategy). Another approach is to build the minimum and maximum forecasts directly into the table, using, say, the maximum, to calculate the aggregate plan. Yet another approach is to establish the required SSs period by period (as in Chapter 6), recognizing the uncertainty of forecasts and the costs of carrying SS and of having stockouts. Finally, we could use a simulation in a spreadsheet, to generate numerous possible realizations of aggregate demand. We repeat, however, that all these approaches are most appropriate at the short-term planning and scheduling stage.

To this point, we have not considered costs in our discussion. In practice, including costs would be the next level of sophistication attempted. By assuming the data in Table 14.5, the expected cost of the chase production plan can be computed as in Table 14.6. Again, this table can be developed on a spreadsheet in minutes.

Having found the expected cost for one feasible production plan, there remains the task of trying to find a better feasible production plan that is lower in cost. One option is a level plan as in Table 14.7. Other possibilities would involve an ad hoc trial-and-error search. Alternatively, an analytic technique could be introduced as discussed in the next section. In the two plans tested, we can see that the overall cost is higher in the chase plan because of higher hiring and firing costs and the overtime worked in month 3. This overtime is due to the maximum regular production

Table 14.5 Cost Structure and Initial Conditions

Cost of regular time (R/T) labor = \$200/production hour
Cost of overtime (O/T) labor = \$300/production hour
Cost of hiring = \$120/production hour
Cost of firing = \$70/production hour
Cost of carrying inventory = \$40/production hour/month
Initial inventory = 30 production hours
Initial workforce = 25 production hours/month

Table 14.6 Expected Cost of Chase Production Plan

t	Work Force				Production		Expected	
	W_{t-1}	Hired	Fired	W_t	R/T	O/T	D_t	I_t
0	—	—	—	25	—	—	—	30
1	25	5	0	30	30	0	30	30
2	30	0	0	30	30	0	30	30
3	30	90	0	120	90	30	120	30
4	120	0	30	90	90	0	90	30
5	90	0	30	60	60	0	60	30
6	60	0	30	30	30	0	30	30
Totals		95	90	360	330	30	360	180
Costs		\$11,400	\$6,300		\$66,000	\$9,000		\$7,200
Total cost =		\$99,900						

rate of 90 hours. Sustained regular production at a higher rate is considered infeasible, so overtime or subcontracting is required in month 3. The chase plan, on the other hand, has somewhat lower inventory costs. Moreover, the inventory costs of the chase plan are solely due to deliberately specifying a buffer stock of 30 production hours.

Before proceeding, let us summarize the steps that are required in using the spreadsheet procedure we have been discussing:

1. Plot the cumulative expected forecasts on a graph or list them in a table (as in Figure 14.7 or Table 14.3).
2. Consider first a plan involving a constant production rate that involves minimal hiring/firing costs. Check to see if this level production plan is feasible, given exogenous constraints. Determine the total cost of this plan.

Table 14.7   Expected Cost of Chase Production Plan

t	Work Force				Production		Expected	
	W_{t-1}	Hired	Fired	W_t	R/T	O/T	D_t	I_t
0	—	—	—	25	—	—	—	30
1	25	35	0	60	60	0	30	60
2	60	0	0	60	60	0	30	90
3	60	0	0	60	60	0	120	30
4	60	0	0	60	60	0	90	0
5	60	0	0	60	60	0	60	0
6	60	0	0	60	60	0	30	30
Totals		35	0	360	360	0	360	210
Costs		$4,200	$0		$72,000	$0		$8,400
Total cost =		$84,600						

3. Consider next a plan tailored to match the forecast fluctuations as exactly as feasible (i.e., a chase plan). This will result in a plan with minimum inventory holding costs. Determine the total cost of this plan.
4. After examining the plans derived in Steps 2 and 3, investigate plans intermediate in position.[*] These plans should attempt to trade off inventory holding (and shortage) costs versus hiring/firing, regular-time, and overtime costs. For each plan devised, determine the total cost.
5. Select the feasible plan found by the steps above that is considered to be most desirable by a scheduling committee or its equivalent.

Note that the best we can hope for is to design a mathematically near-optimal plan by this procedure. One shortcoming of our heuristic method is that it does not usually yield mathematically optimal trade-offs between the costs of regular-time and overtime production, carrying costs, and hiring/firing costs. What is worse, we can never be quite sure how near or far from a mathematical optimum any particular plan is. Close (1968) claims that experienced production planners can, by trial and error, design feasible production plans that are surprisingly close to the mathematically optimal trade-offs possible. This appears to be the case because (as we saw in Chapter 4) the total cost equation of most inventory decision models tends to be U-shaped, with a shallow bottom around the optimal point. It is, of course, quite possible that mathematical optimality is an irrelevant criterion to the managers on a scheduling committee who strive to "optimize" far more complex objectives.

---

[*] Note that the level and chase plans in Figure 14.7 do not bound the large number of alternatives possible. That is, the optimal plan does not necessarily lie on a line drawn between them. Nevertheless, the two graphs from Steps 2 and 3 usually greatly delimit the number of alternatives we would want to consider evaluating.

14.6 Linear Programming Models

Linear programming (LP) models can be tailored to suit the particular cost and decision structure faced by most companies, which accounts in part for the relatively large number of formulations in existence. Many relatively complex mathematical programming models, including both linear and integer (fixed) cost components, have been proposed. For example, see those by Hanssmann and Hess (1960) and Haehling von Lanzenaur (1970).

Because of the large variety of possible formulations, we start by listing the common assumptions made by most formulations and then present a relatively simple specific model, one which is similar to those that have been implemented in practice.

Most practical linear programs are structured as follows:

1. Demand is taken as deterministic and known

$$D_t \text{ in period } t \ (t = 1, 2, \ldots, T)$$

2. The costs of regular-time production in period t are usually described by piecewise linear or convex functions as shown in Figure 14.9 (convex merely means that the slopes of the linear segments become larger and larger; i.e., the marginal cost is increasing).
3. The cost of changes in the production rate is usually taken to be piecewise linear as shown in Figure 14.10. Note that the cost function need not be symmetrical about the vertical axis. An alternate approach in some LP formulations is not to assign costs for changes in production rates at all, but rather to limit the size of the change or, alternatively, to simply put upper and lower bounds on the allowable production rates in any particular period.
4. Upper and lower bounds on production rates and inventory levels are usually specified.

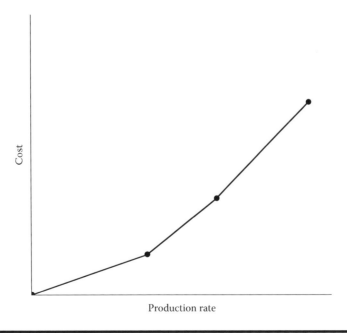

Figure 14.9 The cost of regular-time production in general LP models.

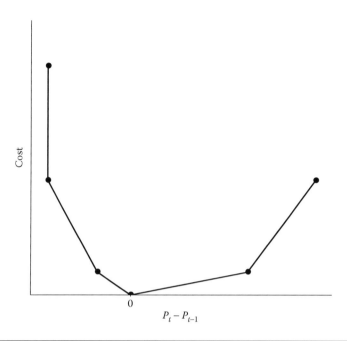

Figure 14.10 Costs of changes in the production rate in general LP models.

5. The inventory carrying cost can be different for each period.
6. It is usually assumed that there is a single production facility serving a given market.
7. In most formulations, backorders or lost sales are not permitted.

One of the most important characteristics of LP models is that many constraints can be directly incorporated in the problem formulation. Also, many product categories can be included in the formulation rather than having to define a single, overall surrogate measure of production.

These characteristics are illustrated in the following LP model:

Minimize

$$C_{\mathrm{TOT}} = \sum_{t=1}^{T} \sum_{i=1}^{n} [c_{i1} I_{it} + c_{i2} W_{it} + c_{i3} W'_{it} + c_{i4} H_{it} + c_{i5} F_{it}]$$

subject to

$$
\left.
\begin{aligned}
I_{it} &= I_{i,t-1} + W_{it} + W'_{it} - D_{it} \\
W_{it} &= W_{i,t-1} + H_{it} - F_{it} \\
&\quad\quad W'_{it} \le a_{it} W_{it}
\end{aligned}
\right\}
\begin{aligned}
i &= 1, \ldots, n \\
t &= 1, \ldots, T
\end{aligned}
\tag{14.6}
$$

and

$$I_{it}, W_{it}, W'_{it}, H_{it}, F_{it} \ge 0 \quad \text{for all } i, t$$

where

n = number of product groups

T = length of the planning horizon

I_{it} = inventory of product group i at the end of t

W_{it} = workforce level (expressed as a regular-time production rate) of product group i at the end of t

W'_{it} = amount of overtime of product group i during t

a_{it} = maximum fraction overtime allowed for group i in t

H_{it} = amount of hiring for product group i in t

F_{it} = amount of firing for product group i in t

D_{it} = forecasted requirements (demand) for group i in t

c = unit cost coefficients

While this may look complicated, the basic ideas expressed in the total cost objective function are quite simple. The total cost is the sum of the costs of carrying inventory, regular workforce, overtime, and hiring and firing for each product group i during each period t. These decision variables are constrained by three equations. The inventory at the end of t (I_{it}) is equal to the inventory in the previous period, plus the amount of product produced on regular time (W_{it}), plus the amount of product produced on overtime (W'_{it}), less the amount of product shipped to meet requirements (D_{it}). The second constraint specifies that the regular-time workforce in this period must be equal to that in the previous period plus the amount hired less the amount fired. (Both H_{it} and F_{it} will never be greater than zero at the same time.) The third constraint sets available overtime in any period equal to or less than a fraction a_{it} of the regular-time workforce capacity. The last line of Equation 14.6 states that all decision variables must be positive or zero.

Note the similarity of the objective function in Equation 14.6 to the way we evaluated costs in the graphic–tabular method in Section 14.5. But compared to the graphic–tabular methods, the LP model in Equation 14.6 yields an optimal rather than near-optimal solution. It also determines the size of the workforce W_{it} directly by including costs associated with hiring and firing in the objective function rather than requiring a manager to specify them exogenously.

In Equation 14.6, no constraints are placed on the maximum or minimum levels of inventory, workforce, or production, nor are backorders allowed. Note that these could be introduced without difficulty by defining additional constraint equations. However, each additional constraint specified, as well as each additional product group chosen for individual attention, increases the computational effort required.

In Section 14.1, we discussed the difficulties in choosing appropriate groups of items to aggregate. The above LP formulation now provides some insight. Members of the same aggregate product group should share similar seasonal patterns (D_{it}), production costs (c_{i2}, c_{i3}), carrying costs (c_{i1}), and hiring/firing costs (c_{i4}, c_{i5}). Obviously, to reduce the number of variables and constraints needed, some severe compromises have to be made in defining product families.

14.6.1 Strengths and Weaknesses

One of the basic weaknesses of LP approaches (and most other aggregate planning techniques as noted above) is the assumption of deterministic demand. However, tests by Dzielinski et al. (1963), using a deterministic model under stochastic conditions, indicate that under many situations, the deterministic model can perform favorably, particularly when we recognize that the solution is adapted as new forecast information becomes available (i.e., a rolling-horizon implementation is used).

Another shortcoming of LP models is the requirement of linear cost functions. However, the possibility of piecewise linearity improves the validity (at a cost of additional computational effort).

An important benefit of an LP model is the potential use of the dual solution to obtain the implicit costs of constraints such as maximum allowable inventory levels. Also, parametric methods allow a simple determination of the production plan for conditions somewhat different from those for which the primary solution is obtained. This last property is useful for two purposes: first, for sensitivity analysis on the estimated cost parameters; second, as a means of measuring the effects of slight changes in one or more of the uncertain conditions.

One could argue that another restriction of LP is that there must be a single measure of effectiveness (normally total costs). In actual fact, a generalization, known as goal programming, permits the use of multiple objectives, which, however, must be prioritized. A chemical industry example is provided by Sanderson (1978) where the objectives include:

1. Minimize the quantity of by-products burnt as fuel.
2. Minimize the unplanned purchases of materials.
3. Minimize the intersite movements of materials.
4. Minimize the variation in production levels between time periods.
5. Minimize the deviation of stocks from preferred levels.
6. Minimize the deviation of production from preferred levels.

See also Goodman (1974), Lockett and Muhlemann (1978), Akine and Roodman (1986), Thomas and McClain (1993), and Wagner et al. (1993).

14.6.2 The Inclusion of Integer Variables in LP Formulations

One or more integer variables are required to include a setup cost or any other type of concave cost in an LP model. (Concave means that the marginal cost is decreasing.) The reason for the integer variable in the case of the setup cost is because of the discontinuity in the cost function. When production is zero, the cost is zero; however, a very small production quantity incurs the entire setup cost.

The need for the integer variable in the case of the concave function is not as obvious. To illustrate the need, let us look at a simple example.

Consider a product whose concave production costs are as follows:

Rate Range (Pieces/Period)	Cost per Additional Piece Produced
0–10	2
11 and higher	1

The basic LP model would select production units only from the 11 and higher range because they are less expensive than those in the 0–10 range. Of course, this is physically impossible. An integer variable is needed to force the use of all of the 0–10 range before any production can be taken from the next higher range. The integer variable takes on only the value 0 or 1. The 11 and higher range can only be used when the integer variable is 1. The logic is arranged such that it is 1 only when the entire 0–10 range is used.

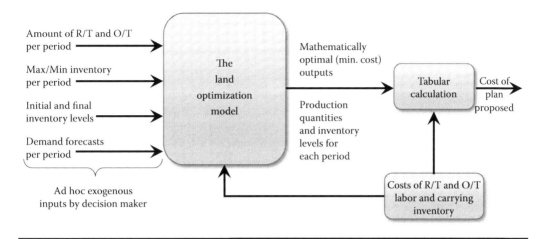

Figure 14.11   Aggregate planning involving the Land algorithm.

The presence of integer variables drastically increases the computational time. While thousands of continuous decision variables can be handled in a regular linear program, only problems with considerably fewer integer variables can be solved in a reasonable length of time with existing solution algorithms.

14.6.3 The Land Algorithm

In many cases, it is extremely difficult to specify hiring and firing costs. There are simply too many qualitative issues that must be included, and managers are unwilling to specify a dollar figure. It is possible in these situations to modify the LP formulation so that managers can specify a set of exogenous inputs, such as the amount of regular-time production available per period. See Figure 14.11. Other inputs may include the maximum amount of overtime per period, the amount of subcontracting allowed, the minimum and maximum allowable inventory levels per period, and the final inventory level at the end of the horizon. The linear program therefore has more constraints, but a simpler objective function—because there are no explicit costs associated with hiring and firing, the cost function includes only production and carrying costs. Thus, for a set of specific inputs, the optimization model finds the minimum cost production quantities and inventory levels.

Management can then specify different resource patterns (in particular, patterns of workforce size with time) and, for each, the algorithm gives the minimum total of regular-time, overtime, and inventory carrying costs.

A simple optimization technique for this problem was originally proposed by Bowman (1956), and was subsequently extended by Beale and Morton (1958) and Land (1958). The model proposed by Land is suitable for manual computations, and is a special form of LP called a transportation model.

14.7 Simulation Search Procedures

Vergin (1966) observes that in many cases, it is difficult to find an analytical solution to the mathematical decision model that is representative of the situation faced in practice by a manager.

Therefore, he argued, it is better to model the actual cost functions very accurately in mathematical or tabular form, so that functions more complex than those allowed in approaches such as LP could be solved.[*] For a horizon of T months, in general terms, in the case of the aggregate scheduling decision:

$$C_{\text{TOT}} = f(W_1, P_1, W_2, P_2, \ldots, W_T, P_T) \qquad (14.7)$$

As in any simulation, the approach is to vary the variables (e.g., the workforce sizes and production rates) systematically until a reasonable (and, we hope, near optimal) solution is obtained.

A comprehensive computerized search methodology, known as *Search Decision Rules (SDR)*, has been developed by Taubert (1968). He defines C_{TOT} as a function of $(W_t, P_t, W_{t-1}, I_{t-1}, D_t; t = 1, 2, \ldots, 12)$ and then identifies the factors within C_{TOT} by the following vectors:

$$
\begin{aligned}
\{D\} &= \text{Decision vector} = P_t, W_t; t = 1, 2, \ldots, 12 \\
\{S\} &= \text{Stage vector} = W_{t-1}, I_{t-1}; t = 1, 2, \ldots, 12 \qquad (14.8)\\
\{P\} &= \text{Parameter vector} = \text{Cost coefficients at time } t \text{ and } D_t; t = 1, 2, \ldots, 12
\end{aligned}
$$

His decision vector contains 24 (or $2T$) independent decision variables, which are the desired output of SDR methodology. The stage vector transmits information about the *state* of the system from time $t-1$ to t. The parameter vector contains the forecasts and the cost coefficients *that could be different from stage to stage*.

SDR searches directly for decision vectors that reduce C_{TOT}. Computer search routines attempt to optimize all stages simultaneously generating 24 trial decisions per iteration. The search procedure terminates when successive iterations result in small reductions in C_{TOT}.

The appropriate procedure for generating alternative values of the decision vector $\{D\}$ is not obvious. The generation technique chosen determines the total time to arrive at a near-optimal C_{TOT}, or whether a near-optimal C_{TOT} is even achieved. Refer to the book by Baker (1995) and to references at the end of the paper by Taubert (1968), for detailed discussions of computer search techniques. See also Gilgeous (1989).

Taubert, using SDR, was able to achieve results that were very close to those realized by other mathematically optimal methods. Similar results have been obtained by a number of other researchers. (See, e.g., Lee and Khumawala 1974; Mellichamp and Love 1978). In their book, Buffa and Miller (1979) summarize the advantages and disadvantages of SDR methodology as follows:

SDR Advantages

1. Permits realistic modeling free from many restrictive assumptions, such as closed-form mathematical expressions, linear/quadratic cost functions, and so on.
2. Permits a variation in mathematical structure from stage to stage (heterogeneous stages), so that anticipated system changes such as the introduction of new products or production equipment, reorganizations, wage increases, etc. can be considered.
3. Provides the operating manager with a set of current and projected decisions.

---

[*] Although computing power has mitigated some of these concerns, it is still true that managers do not respond favorably to huge mathematical models that are yet approximations of reality. As mentioned in Chapter 13, one of the authors worked with a chemical company that had a massive linear program for planning production at one of its organic chemicals plants. As it happened, only two or three people in the entire corporation really understood the model. Needless to say, its use was resisted by a number of people.

4. Permits optimized disaggregate decision making.

5. Lends itself to evolutionary cost model development and provides solutions at desired points in the iterative process.

6. Facilitates sensitivity analysis and provides sensitivity data while the search routine is converging on a solution.

7. Easily handles cash flow discounting, nonlinear utility functions, multiple objective functions, and complex constraints.

8. Offers the potential of solving many otherwise impossible operations planning problems.

9. The methodology is general and can be applied to single or multistage decision problems that are not related to aggregate planning—for example, determining the optimal capital structure of a firm given a forecast of interest rates, stock performance, etc., or determining a least cost allocation of manpower to activities defined by a critical path network.

SDR Disadvantages

1. Optimization using computer search routines is an art and it is currently impossible to state, *a priori*, which search routine will give the best performance on a particular objective function.

2. Decisions made by this methodology may not, and in general will not, represent the absolute global optimum.

3. Response surface dimensionality appears to be a limiting factor (in terms of computational effort required).[*]

14.8 Modeling the Behavior of Managers

In Sections 14.6 and 14.7, we examined models that attempted to capture the important trade-offs required by an aggregate scheduling decision through explicit specification of relatively complex mathematical formulations. These modeling strategies confront analysts and managers with other difficulties that result from their limited ability to measure the more qualitative costs required and the difficulties that result from having to derive solutions to more complex mathematical formulations. In the next section, we demonstrate a somewhat simpler type of strategy that does not necessarily require the *explicit* specification of cost parameters. In this section, we discuss a different philosophy of modeling based on implicit cost measurement; in particular, we identify key costs and other factors by interviewing managers and observing their actual decision-making behavior. Specifically, we comment on two distinct approaches of this general nature.

14.8.1 Management Coefficients Models

Based on his consulting experiences, Bowman (1963) suggests:

> Managerial decisions might be improved more by making them more consistent from one time to another than by approaches purporting to give "optimal solutions" to

---

[*] A potential tool for a drastic reduction in computational time is the branch-and-bound technique (see, e.g., Wagner 1975). Moreover, sophisticated metaheuristics, such as tabu search (see, e.g., Glover and Laguna 1993; Glover et al. 1993), simulated annealing (see, e.g., Kirkpatrick et al. 1983; Eglese 1990), and genetic algorithms (see, e.g., Goldberg 1989; Reeves 1993; Gen and Cheng 1997), can be very effective in dealing with this type of situation. See also Chapter 4 in Baker (1995).

explicit models ... especially for problems where intangibles (run-out costs, delay penalties) must otherwise be estimated or assumed.

Bowman justifies this philosophy by observing that experienced managers are generally quite aware of the criteria and cost factors that influence decisions they must make. Over many repetitive decisions, their decision behavior is unbiased; that is, on the average, they make the correct response to the decision environment they face. However, their behavior is probably more erratic than it should be. Spurious influences such as emergency phone calls from superiors, suppliers, and customers can produce deviations from normal (average) behavior when not warranted. Therefore, Bowman argues, why not adopt an approach to modeling that tries to keep a manager closer to his or her average decision behavior by dampening out most of these erratic reactions. Bowman names his methodology/philosophy the *Management Coefficients Model*. Gordon (1966) and Kunreuther (1969), among others, report applications of this approach.

The Management Coefficients approach utilizes standard multiple regression methodology to fit decision rules to historical data on actual managerial decision behavior. By fitting a regression line, the analyst attempts to capture the average historical relationship between cues in the environment and management's responses. Erratic behavior (the residuals) is "dampened" by minimizing least squares. The first step is to select a form of relationship between the decision variable(s) and the historical data on environmental variables. For example, in the production smoothing and workforce-balancing situation, Bowman chooses

$$P_t = aP_{t-1} + b_1 D_t + b_2 D_{t+1} + c \tag{14.9}$$

That is, the production rate (P_t) in period t is assumed to be linearly related to the production rate in the previous period (P_{t-1}) and the estimated orders in periods t (D_t) and $t + 1(D_t + 1)$. There should, of course, be no other reason, except the logic of the relationships involved, to limit such regressions (in general) to linear forms.

Having fit a management coefficients regression equation, the analyst can now describe to management how they would have normally (on the average) reacted in the past to any forecasted pattern of future orders, given the existing OH inventories and prevailing production rates. Presumably this would cause decision makers to consider more carefully and justify any contemplated deviation from a conventional response to cues in the environment based on their actions in the past.

Tests with actual company data have shown that costs can be significantly reduced through the use of the Management Coefficients approach.[*] A serious drawback of the procedure is the essentially subjective selection of the form of the rule. It very easily can be selected incorrectly. Implicit in Bowman's approach is the assumption that past decisions are a good basis for future actions. A historical orientation could prevent a manager from quickly adapting to new conditions in a rapidly changing competitive environment.

Kunreuther (1969) proposes general criteria that must be satisfied if a Management Coefficients approach is to have validity in an actual application. The criteria are similar to conditions that must hold for any least-squares regression model. Kunreuther makes two recommendations.

---

[*] Strictly speaking, we cannot really compare actual company decision behavior with a Management Coefficients model, or with any of the other explicit models discussed in this chapter. The objective function that management may have tried to minimize with their past decisions may have been far different from any of the ones we have formulated mathematically.

When managers have limited information regarding future sales, as when they develop their initial production estimate, then, a plan based on average (past) decision behavior is very useful. However, when environmental cues provide reliable information on future sales of specific items (as when subsequent revisions are made in the Master Plan), then, actual decisions are clearly superior to those suggested by an averaging rule. In such cases, variance in managerial action is beneficial rather than costly to a firm and should not be averaged out.

14.8.2 Manpower Decision Framework

Colley et al. (1977) interview a number of managers responsible for aggregate planning decisions and observe that these managers feel that capacity utilization in both the short and long term is of paramount importance in their decision making. Thus, two ratio factors are proposed, specifically,

$$\text{Current period ratio (CPR)} = \frac{\text{Demand for current period production}}{\text{Current period productive capacity}} \qquad (14.10)$$

$$\text{Planning period ratio (PPR)} = \frac{\substack{\text{Average monthly order backlog and additional} \\ \text{anticipated orders in planning period}}}{\text{Monthly period productive capacity}} \qquad (14.11)$$

where the planning period is the time interval out to the planning horizon.

Ideally, we would like to have both of these ratios close to unity. Nine possible states are defined by three possible values of each of the two ratios. These values are (1) significantly below unity, (2) near unity, and (3) significantly above unity. The boundaries (e.g., between "significantly below" and "near" unity) are under management control. In consultation with management, plausible courses of action are defined for each of the states. For example, if we were in the state

$$\text{CPR} \gg 1 \quad \text{and} \quad \text{PPR} \cong 1$$

then, the use of some overtime would likely be appropriate. On the other hand, if both

$$\text{CPR} \gg 1 \quad \text{and} \quad \text{PPR} \gg 1$$

then, one would probably wish to increase the size of the workforce.

14.9 Planning for Adjustments Recognizing Uncertainty

Earlier, we have argued that one way to cope with uncertainty is to use a rolling-horizon implementation of a deterministic model. An alternative methodology has been proposed by Hanssmann (1962) and Magee and Boodman (1967). Specifically, an *initial* solution is developed (by some procedure such as LP) for a horizon of T periods. This initial solution implies target (or planned) inventory and production levels through time. Since actual demands differ from the forecasted values, we may wish to adjust the individual production rates from their original planned levels rather than reusing the original algorithm at each period with a rolling horizon of T periods.

Let P_t be the production rate actually used in period t and P_t^* be the originally planned (or target) production rate for period t. Suppose that there is a lead time of L periods to initiate a change in production rate (i.e., any change in the production rate for period $t + L + 1$ must be initiated at the end of period t). Then, a simple, plausible decision rule is to schedule a production rate P_{t+L+1} equal to the originally planned rate, P^*_{t+L+1}, plus an adjustment equal to a fraction f of the projected inventory discrepancy at the end of period $t + L$. That is,

$$P_{t+L+1} = P^*_{t+L+1} + f \left[I^*_t - I_t - \sum_{j=1}^{L}(P_{t+j} - P^*_{t+j}) \right] \qquad (14.12)$$

where
I_t = actual inventory level at the end of period t
I^*_t = target inventory level at the end of period t resulting from the original plan

The summation term in Equation 14.12 represents the additional change in the inventory level that will occur during the lead time L as a result of prior adjustments in $P_{t+1}, P_{t+2}, \ldots, P_{t+L}$.

Obviously, the behavior of the decision rule depends on the choice of f. As f increases, the changes in the production rate will tend to be larger, but the buffer inventory required to protect against the demand uncertainty will decrease. *Under the restrictive assumption of stationary demand* (i.e., an average demand level that does not change with time) and *independent forecast errors* from period to period, we can show (see Johnson and Montgomery 1974; Magee and Boodman 1967) that

$$E[|\Delta P|] \approx 0.8 \sqrt{\frac{2f}{2 - f}} \sigma_\varepsilon \qquad (14.13)$$

$$\sigma_{ID} \approx \sqrt{\frac{1 + L(2f - f^2)}{2f - f^2}} \sigma_\varepsilon \qquad (14.14)$$

where
$\Delta P = P_t - P_{t-1}$ = change in the production rate from period to period (with stationary demand ΔP would be 0 if there were no forecast errors)
$E[|\Delta P|]$ = expected (or average) absolute change in the production rate
σ_{ID} = standard deviation of the inventory discrepancy $I_t - I^*_t$
σ_ε = standard deviation of errors of forecasts over a unit time period

14.9.1 The Production-Switching Heuristic

The above analysis assumes that any (continuous) production rate can be used. In fact, there may only be a few discrete possible production rates—for example, high (H), normal (N), and low (L). In such a case, a plausible strategy (first suggested by Orr (1962), and subsequently developed by Elmaleh and Eilon (1974), and Mellichamp and Love (1978)), still under a stationary demand pattern, is to use three inventory levels $a > c > b$ and to set the production rate in period $t + 1$ as follows:

$$P_{t+1} = \begin{cases} H & \text{if } I_t \text{ drops below } b \\ N & \text{if } I_t \text{ crosses } c \\ L & \text{if } I_t \text{ rises above } a \\ P_t & \text{otherwise} \end{cases} \qquad (14.15)$$

The values of a, b, and c are found by some form of search procedure. Lambrecht et al. (1982) show that the decision rule of Equation 14.15 can lead to sizable cost penalties when the demand pattern is no longer stationary (i.e., the average rate changes appreciably with time). Furthermore, they suggest a heuristic modification to cope with this more realistic situation, namely, recomputing the a, b, and c values when the average demand over the next n periods deviates by more than $x\%$ from the average demand over the interval used to last compute the parameter values. The quantities n and x now become two additional controllable parameters of the policy. See also Oliff and Leong (1987) and Hwang and Cha (1995).

14.10 Summary

In this chapter, we have presented a number of approaches to dealing with the aggregate production-planning problem. The major insight we have attempted to convey is the need to always select decision models for aggregate planning whose sophistication matches the complexity demanded by the specific situation faced. Shycon (1974) points out the fact that management scientists can get carried away in their search for generality and optimality. In Shycon's words, they "tend, at times, to over-agonize about the entire analytical process. In short, sometimes they make more of a big deal of the routines of analysis than is justified by the type of result required and the management need." We believe that aggregate planning theory is one area where management scientists have gotten somewhat carried away.

Interesting insights on some of the issues with implementation are found in DuBois and Oliff (1991). In a survey of a number of companies in the southeast United States, more than 50% of the firms responded that they used a formal computer-supported decision procedure for aggregate planning, 20% relied on the use of past decisions extrapolated into the future, and 11% used rule-of-thumb procedures. Many managers noted that they lacked the cost information needed to use formal models. And many indicated that they did not have the option of making continuous changes in workforce levels. Rather, they added or subtracted workers in discrete groups.

The sparse number of applications in this area is therefore in large part the result of two related reasons. First, some management scientists have tried to construct more and more complex models, which are not easy to adapt to changing real-world conditions, and which rely on data that are not easily gathered. It is for this reason that we have not discussed certain other mathematical models and associated decision rules, such as the Linear Decision Rule (see, e.g., Holt et al. 1955; Schwarz and Johnson 1978), which, although of conceptual value, have had little practical implementation. Second, the manager is too often isolated from the process of achieving a "solution." It is possible that every time we claim we have invented a more complex model that includes more decision variables than ever before, or when we claim that some computerized algorithms can achieve an "optimal solution" faster than ever before, we are actually progressively alienating more and more managers from ever considering using decision models for aggregate planning. Managers do not necessarily want quick and easy answers. Nor do they want to delegate too much of the analysis to an inanimate process. They may well want to agonize, to make deals and compromises, and to derive "personalized" alternative plans—even at the expense of mathematical optimality. Plans may be personalized in the sense that they are partly the result of qualitative factors, data availability, historical accidents, and organizational compromises in which the decision makers have a stake, or in the sense that every feasible schedule in practice will include details such as John Doe, the best tool and die maker in your plant, being on vacation in August. Thus, in summary, the aggregate plan is too important a decision to a manager for that individual not to spend a considerable amount of time developing it.

In the next chapter, we move to the next levels of production planning—master scheduling and requirements planning. It is important to note at this point that disaggregating the aggregate plan is not always an easy task. When setup times are significant or dependent on the sequence of production, a plan that appears feasible at the aggregate planning level may not be feasible when disaggregated. It may be necessary to revise the aggregate plan as more detailed information is used. See Axsäter and Jönsson (1984), Aucamp (1987), and Venkataraman and Smith (1996).

Problems

14.1 Foot, Inc. is in the process of setting their aggregate production rate for skis for each of the next 4 quarters. The forecasts are given below. There will be 20 units (hundreds of skis) OH at the beginning of the next quarter (Q1). Currently they have 30 fixed-pay employees, each of whom can produce 5 units per quarter.

Quarter	Forecast (Hundreds of Skis)
1	140
2	160
3	200
4	120

    a. Find a level strategy assuming that only 30 employees are available.
    b. Find a chase strategy that seeks to maintain zero inventory.
    c. Use LP to find the optimal solution if the cost of hiring is $1,500 per employee, the cost of firing is $4,000 per employee, the inventory holding cost is $800 per hundred skis, and the cost of shortage is $5,000 per hundred skis.
    d. Find the cost of the level and chase plans using the costs in part (c).

14.2 In a production-to-stock situation where demand varies with time, an aggregate planning strategy is to produce to stock, using this stock to level out the required production rate through time. In a make-to-order situation, this strategy cannot directly be used because of the custom nature of the demands. However, smoothing is still available through the introduction of the so-called planning window, which represents the amount by which the promise time exceeds the production lead time. Within such a planning window, we can smooth requirements without causing any backorders.

    a. By what means could we increase the planning window?
    b. Discuss the costs relevant in choosing the appropriate size of the window.

14.3 Consider a plant whose capacity, on the average, exceeds the demand rate. There are two options for balancing the capacity used with the demand rate: Option 1—periodic shutdown and Option 2—throttling (running at less than full capacity).
Let

    c_{sf} = semifixed operating cost, in $/year, incurred whenever the plant is operating (independent of the operating level)
    h = inventory carrying cost, in $/unit/year

C_s = cost of shutting down and starting up, in $

P_{max} = production rate when the plant is running at capacity, in units/year

For the case where the demand rate is level at D units/year, answer the following questions:

a. For Option 1, write an expression for the total relevant cost per year as a function of the time T between shutdowns and the production rate used P units/year.

b. Find the best value of T for a given value of P.

c. What P value(s) need(s) to be considered in the shutdown strategy? What are the total relevant costs for this(these) value(s)?

d. What is the decision rule for choosing, based on costs, between shutdown and throttling?

14.4 Clarke Manufacturing Co. makes 30 types of transformers. Sales forecasts for the next year for type CT-OIM are as below:

Month	Demand (Units)
January	600
February	500
March	700
April	900
May	1,000
June	1,200
July	900
August	700
September	700
October	600
November	600
December	600

A maximum of 24 transformers of this type can be produced at one time. The company works one shift of 8 hours/day. Each worker can handle only one machine at a time and produces, on the average, 2 units per 8-hour shift.

Regular-time pay is $4/hour and the overtime rate is $6/hour. Cost of carrying inventory is set at 0.02 $/$/month. Costs of hiring and firing a worker are estimated at $360 and $180, respectively. Union contract stipulates that at least 14 workers have to be employed at all times. On December 31 of this year, there are 15 workers on the payroll and inventory OH was 300 units. The desired minimum level of inventory is set at 300 units with each unit costed at $50. Normally a worker can work a maximum of 56 hours in a week. Assume that each month is 20 working days in length.

Formulate the least-cost production plan for the next year using a graphic or spreadsheet approach. (No backordering is allowed, and inventory level should not go too far below 300 units.)

14.5 Starship Ltd. manufactures boats and distributes them across the United States. Part C-212 (a motor) is the most expensive single part. The table below shows the sales forecast for this part for the next 12 months.

Month	Forecast (Units)
1	300
2	600
3	300
4	400
5	800
6	900
7	1,600
8	1,800
9	1,600
10	800
11	500
12	400

Labor cost is $40 per unit at regular time and $60 per unit at overtime. Carrying cost is $2 per month per unit.

A maximum of 20 people can work at any time. The monthly output on regular time per person is 50 units. Overtime production can be up to 40% of regular-time production. Maximum monthly inventory is set at 1,000 units. There is no constraint on the minimum number of workers per month.

a. Develop an optimal production schedule for the next 12 months. Indicate
    i. Total cost of production (i.e., labor and carrying costs)
    ii. Total production every month
    iii. Total number of workers required in each of the 12 months
b. Draw the cumulative production and demand curves.

14.6 In Problem 14.5, if the maximum inventory limit is raised to 1,300 units, how much is the increased storage capacity worth?

14.7 Refer to Problem 14.5. If at least 12 workers have to be employed at all times, how does this constraint affect the total cost?

14.8 A version of the LP model of the aggregate scheduling problem that has been used in practice is

Minimize

$$C_{\text{TOT}} = \sum_{i=1}^{N} \sum_{t=1}^{T} (v_{it} P_{it} + I_{it} v_i r_t) + \sum_{t=1}^{T} (c_w W_t + c_0 W_t')$$

subject to

$$P_{it} + I_{i,t-1} - I_{it} = D_{it} \begin{cases} t = 1, \ldots, T \\ i = 1, \ldots, N \end{cases}$$

$$\sum_{i=1}^{N} k_i P_{it} - W_t - W'_t = 0 \quad t = 1, \ldots, T$$

$$0 \leq W_t \leq W_{t,\max} \quad t = 1, \ldots, T$$

$$0 \leq W'_t \leq W'_{t,\max} \quad t = 1, \ldots, T$$

$$P_{it}, I_{it} \geq 0 \begin{cases} t = 1, \ldots, T \\ i = 1, \ldots, N \end{cases}$$

    a. Define all symbols, giving the units of all variables as well as the point in time when they must be measured.

    b. Which data must be collected or estimated?

    c. Which variables must be assigned values and which variables are decision variables whose values will be determined by an LP algorithm?

14.9   a. Using Equations 14.13 and 14.14, prove that

        i. $E[|\Delta P|]$ increases as f increases

        ii. σ_{ID} decreases as f increases

    b. Outline an analysis to aid in the choice of the value of f, based on cost considerations.

14.10 For the historical data given below, fit by linear regression, suitable management coefficients equations similar to those in Equation 14.9. Explain your reasons for selecting the number of variables in your model. Is it a good enough fit that you would recommend it for future decisions? Discuss.

Period t	D_t	S_t	W_t	P_t	I_t
1	—	—	84	515	302
2	445	445	79	447	303
3	438	426	75	418	295
4	321	356	72	390	329
5	396	388	69	380	321
6	376	373	68	371	319
7	292	331	67	365	353
8	455	416	67	377	314
9	400	386	67	376	308
10	355	363	69	376	316
11	289	338	71	386	364
12	430	427	74	386	358
13	395	475	77	455	337
14	513	496	81	482	323
15	505	503	85	504	325

14.11 Hefty, Inc. is in the process of setting their aggregate production rate for soccer balls for each of the next 12 months. The forecasts are given below. There will be no inventory at the beginning of the year. Currently they have 50 fixed-pay employees, each of whom can produce 9 units (100 soccer balls) per month.
a. Find a level strategy assuming that only 50 employees are available.
b. Find a chase strategy that seeks to maintain zero inventory.

Period t	Forecast Demand D_t
1	300
2	300
3	350
4	400
5	450
6	500
7	650
8	600
9	475
10	475
11	450
12	450

14.12 Consolidated Steel manufactures a product for which the monthly demands are expected to be 800, 1,200, 1,000, 1,600, 1,600, 1,400, 1,200, 1,600, 1,200, 1,400, 1,200, and 1,600 units respectively. Both regular production and overtime production are possible. The costs per unit of regular and overtime production are $8 and $12, respectively, for the first 6 months. During the second 6 months, these costs each increase by $1, to $9 and $13. Inventory carrying charges are $1 per unit per month. Initial inventory is 100 units, and it is desired to have 100 units in inventory at the end of the year. Given that regular-time production capacity is 1,000 units per month and overtime production capacity is 400 units per month, how should production be scheduled to minimize costs? Formulate an appropriate LP without taking into account the time value of money.

14.13 For the Clarke Manufacturing Co. example in Problem 14.4, assume that in addition to the data already given:
1. Cumulative tiredness reduces daily production by 5% per day for each 10 hours of overtime worked in a month
2. Sales forecasts are expected to be within 25% of actual sales
3. Backlogging is allowed at a penalty of $30 per unit
   Instead of the data given in Problem 14.4, assume that the company has only a choice of working either one or two shifts throughout the year (i.e., they cannot hire and fire during their fiscal year). A maximum of 16 transformers can be produced at

a time on either shift, although the productivity on the second shift tends to be about 30% lower. Recommend a production plan for Clarke Manufacturing under these revised assumptions.

14.14 The Galt Gadget Company is a manufacturer of automotive parts. For a particular line of products, the demand in the next 6 periods (each of 2-months duration) is estimated to be

Period, j	1	2	3	4	5	6
Demand, $D(j)$ (in shifts of output)	30	60	60	60	160	50

The company does not permit deliberate backlogging. However, the requirements shown for a period can be satisfied (if necessary) by production in that period. With a prespecified manpower schedule, the production available is as follows:

Period	Type of Production	Cost (Hundreds of Dollars) per Shift of Output	Capacity (Shifts)
1	Regular	8	46
2	Regular	8	44
	2nd shift	11	22
3	Regular	8	44
	2nd shift	11	44
4	Regular	8	43
	2nd shift	11	43
5	Regular	8	43
	2nd shift	11	43
	3rd shift	17	20
6	Regular	8	42
	2nd shift	11	42

The initial inventory (at the beginning of period 1) is 20 shifts of output. An inventory carrying charge of $200 per shift of output is incurred when material is carried forward from one period to the next (no charge is incurred during a period). Warehouse limitations restrict the inventory at the end of any period to be no larger than 70 shifts of output.

a. Determine the production plan for Galt Gadget that minimizes the total of production and carrying costs over the 6-month period if an ending inventory of 15 shifts of output is desired. In particular, show the actual production by period-shift and the ending inventory of each period.

b. From your solution, can you infer the value (in terms of reduced costs) of increasing the warehouse capacity to 72 shifts of output? 75 shifts of output?

14.15 Build a spreadsheet to calculate the cost of a pure level strategy. Use the data from Problem 14.4 to test it.

14.16 Build a spreadsheet to calculate the cost of a pure chase strategy. Use the data from Problem 14.4 to test it.

14.17 Build a spreadsheet to calculate the cost of any strategy. In other words, allow for input by management. Use the data from Problem 14.4 to test it.

14.18 How would you modify the LP model if the cost of regular-time capacity of, say, x, labor-hours is incurred even if the x hours are not completely used? Is this a realistic scenario?

14.19 What factors would encourage the management to take a chase strategy? Level?

14.20 Table 14.8 shows sales of a washing machine in successive 5-week periods. (The last period in each year is 6 weeks plus a couple of days—assume it is actually 5 weeks). It is required to develop methods of planning production when the only information on which to base the decisions is current inventory and current production rate. There are four possible production rates that can be used: 7,500, 15,000, 22,500, and 30,000 per period. The costs of changing production rates are increased by a step of 7,500 per period—$70,000 reduction of 7,500 per period—$130,000. The lead time to affect a production rate change is one period.

   a. Over the period concerned, the rule the company used was to change the production rate whenever inventory exceeded 70,000 or fell below 20,000. The inventory at the beginning of period 6, Year 1, was 33,000 and the production rate in that period was 15,000. Evaluate the number of changes and the cost of changes if the company's policy is used.

   b. Suppose shortages can be estimated at $100 per unit demand not met when it occurred (demand is backlogged). The value of a washing machine is $200; thus it is possible to estimate inventory and shortage costs. Do so (no great accuracy is required—assume a carrying charge of 15% per year).

Table 14.8   Demand for Washing Machines

Period	Year 1	Year 2	Year 3	Year 4	Year 5	Year 6	Year 7
1		17,103	15,876	33,059	17,409	13,110	17,877
2		18,494	19,727	18,093	26,633	21,647	15,741
3		15,474	18,571	16,078	19,303	19,349	19,207
4		26,978	34,800	24,542	24,994	23,893	
5		24,021	20,042	24,705	17,251	17,162	
6	13,974	23,488	9,369	24,301	22,672	16,220	
7	14,342	22,984	12,460	19,199	22,014	25,078	
8	17,350	24,931	33,746	20,049	27,144	24,342	
9	21,612	27,837	26,861	21,518	26,579	20,866	
10	22,439	24,416	23,247	16,982	32,407	20,150	
Total	89,717	225,726	214,699	218,526	236,406	201,817	52,825

Table 14.9   Decision Rule for Setting Production Rate

Inventory	Present Production Rate			
	7,500	15,000	22,500	30,000
<0	30,000	30,000	30,000	30,000
0–10,000	22,500	22,500	30,000	30,000
10–20,000	22,500	22,500	22,500	30,000
20–30,000	15,000	22,500	22,500	22,500
30–40,000	15,000	15,000	22,500	22,500
40–50,000	15,000	15,000	22,500	22,500
50–60,000	15,000	15,000	15,000	15,000
60–70,000	7,500	15,000	15,000	15,000
70–80,000	7,500	7,500	7,500	15,000
>80,000	7,500	7,500	7,500	7,500

    c. Compare the rule in part (a) with a rule described by Table 14.9. The new production rate is given by the intersection of the present inventory row and the present production rate column.

    d. Develop some other reasonable rule that could be used and compare it with (a) and (c).

14.21   An organization faces a significantly seasonal demand pattern for a number of items. The production capacity during the peak season is not able to meet the peak demand rate. The organization is considering offering discounts to customers who order/take significant quantities of items ahead of the peak season. What is the cost trade-off involved in the discount offer? Outline an economic analysis that might be undertaken.

References

Akine, U. and G. Roodman. 1986. A new approach to aggregate production planning. *IIE Transactions 18*(1), 88–94.

Aucamp, D. C. 1987. A lot-sizing policy for production planning with applications in MRP. *International Journal of Production Research 25*(8), 1099–1108.

Axsäter, H., H. Jönsson, and A. Thorstenson. 1983. Approximate aggregation of product data. *Engineering Costs and Production Economics 7*(2), 119–126.

Axsäter, S. 1981. Aggregation of product data for hierarchical production planning. *Operations Research 29*(4), 744–756.

Axsäter, S. and H. Jönsson. 1984. Aggregation and disaggregation in hierarchical production planning. *European Journal of Operational Research 17*(2), 338–350.

Baker, K. and D. Peterson. 1979. An analytic framework for evaluating rolling schedules. *Management Science 25*(4), 341–351.

Baker, K. R. 1995. *Elements of Sequencing and Scheduling.* Hanover, New Hampshire: The Amos Tuck School, Dartmouth College.

Beale, E. M. L. and G. Morton. 1958. Solution of a purchase-storage programme—Part 1. *Operational Research Quarterly 9*(3), 174–187.

Behnezhad, A. R. and B. Khoshnevis. 1996. Integration of machine requirements planning and aggregate production planning. *Production Planning and Control 7*(3), 292–298.

Bitran, G., E. Haas, and A. Hax. 1981. Hierarchical production planning: A single stage system. *Operations Research 29*(4), 717–743.

Bolander, S. F., R. C. Heard, S. M. Seward, and S. G. Taylor. 1981. *Manufacturing Planning and Control in Process Industries.* Falls Church, VA: American Production and Inventory Control Society.

Boskma, K. 1982. Aggregation and the design of models for medium-term planning of production. *European Journal of Operational Research 10*, 244–249.

Bowman, E. H. 1956. Production scheduling by the transportation method of linear programming. *Operations Research 4*(1), 100–103.

Bowman, E. H. 1963. Consistency and optimality in managerial decision making. *Management Science 9*(2), 310–321.

Buffa, E. S. and J. G. Miller. 1979. *Production Inventory Systems: Planning and Control* (3rd ed.). Homewood, IL: Richard D. Irwin, Inc.

Buxey, G. 2005. Aggregate planning for seasonal demand: Reconciling theory with practice. *International Journal of Operations and Production Management 25*(11), 1083–1100.

Close, J. 1968. A simplified planning scheme for the manufacturer with seasonal demand. *Journal of Industrial Engineering 19*(9), 454–462.

Colley, J. L., R. D. Landel, and R. R. Fair. 1977. *Production Operations Planning and Control.* San Francisco, CA: Holden-Day.

Cruickshanks, A., R. Drescher, and S. Graves. 1984. A study of production smoothing in a job shop environment. *Management Science 30*(3), 368–380.

de Kok, A. G. 1990. Hierarchical production planning for consumer goods. *European Journal of Operational Research 45*, 55–69.

DuBois, F. L. and M. D. Oliff. 1991. Aggregate production planning in practice. *Production and Inventory Management Journal 32*(3), 26–30.

Dzielinski, B., C. Baker, and A. Manne. 1963. Simulation tests of lot size programming. *Management Science 9*(2), 229–258.

Edwards, J. R., H. M. Wagner, and W. P. Wood. 1985. Blue bell trims its inventory. *Interfaces 15*(1), 34–52.

Eglese, R. W. 1990. Simulated annealing: A tool for operational research. *European Journal of Operational Research 46*, 271–281.

Elmaleh, J. and S. Eilon. 1974. A new approach to production smoothing. *International Journal of Production Research 12*(6), 673–681.

Galbraith, J. 1969. Solving production smoothing problems. *Management Science 15*(12), 665–674.

Gen, M. and R. Cheng. 1997. *Genetic Algorithms and Engineering Design.* New York: John Wiley & Sons.

Gilgeous, V. 1989. Modelling realism in aggregate planning: A goal search approach. *International Journal of Production Research 27*(7), 1179–1183.

Glover, F. and M. Laguna. 1993. Tabu search. In C. R. Reeves (Ed.), *Modern Heuristic Techniques for Combinatorial Problems,* pp. 70–141. Oxford: Blackwell Scientific Publishing.

Glover, F., E. Taillard, and D. de Werra. 1993. A user's guide to tabu search. *Annals of Operations Research 41*, 3–38.

Goldberg, D. 1989. *Genetic Algorithms in Search, Optimization, and Machine Learning.* Reading, MA: Addison-Wesley.

Goodman, D. 1974. A goal programming approach to aggregate planning of production and work force. *Management Science 20*(12), 1569–1575.

Gordon, J. R. M. 1966. *A Multi-Model Analysis of an Aggregate Scheduling Decision.* PhD thesis. Massachusetts Institute of Technology, Cambridge, MA.

Haehling von Lanzenaur, C. 1970. Production and employment scheduling in multistage production systems. *Naval Research Logistics 17*(2), 193–198.

Hanssmann, F. 1962. *Operations Research in Production and Inventory Control*. New York: John Wiley & Sons.

Hanssmann, F. and S. W. Hess. 1960. A linear programming approach to production and employment scheduling. *Management Technology 1*(1), 46–51.

Heath, D. and P. Jackson. 1994. Modeling the evolution of demand forecasts with application to safety stock analysis in production/distribution systems. *IIE Transactions 26*(3), 17–30.

Holt, C., F. Modigliani, and H. Simon. 1955. A linear decision rule for production and employment scheduling. *Management Science 2*(1), 1–30.

Holt, C. C., F. Modigliani, J. F. Muth, and H. A. Simon. 1960. *Planning Production, Inventories and Work Force*. Englewood Cliffs, NJ: Prentice-Hall.

Hwang, H. and C. N. Cha. 1995. An improved version of the production switching heuristic for the aggregate production planning problem. *International Journal of Production Research 33*, 2567–2577.

Johansen, J. and J. O. Riis. 1995. Managing seasonal fluctuations in demand. Practice and experience of selected industrial enterprises. *Production Planning and Control 6*(5), 461–468.

Johnson, L. A. and D. C. Montgomery. 1974. *Operations Research in Production Planning, Scheduling and Inventory Control*, pp. 236–242. New York: John Wiley & Sons.

Kirkpatrick, S., C. D. Gelatt, and M. P. Vecchi. 1983. Optimization by simulated annealing. *Science 220*, 671–680.

Kunreuther, H. 1969. Extensions of Bowman's theory on managerial decision making. *Management Science 15*(8), B415–B439.

Lambrecht, M. R., R. Luyten, and J. Vander Eecken. 1982. The production switching heuristic under non-stationary demand. *Engineering Costs and Production Economics 7*, 55–61.

Land, A. H. 1958. Solution of a purchase-storage programme—Part 2. *Operational Research Quarterly 9*(3), 188–197.

Lee, W. and B. Khumawala. 1974. Simulation testing of aggregate production models in an implementation methodology. *Management Science 20*(6), 903–911.

Lockett, A. G. and A. P. Muhlemann. 1978. A problem of aggregate scheduling—An application of goal programming. *International Journal of Production Research 16*(2), 127–135.

Magee, J. F. and D. M. Boodman. 1967. *Production Planning and Inventory Control*, pp. 199–208, 361–364. New York: McGraw-Hill.

McClain, J. and J. Thomas. 1977. Horizon effects in aggregate production planning with seasonal demand. *Management Science 23*(7), 728–736.

Mellichamp, J. and R. M. Love. 1978. Production switching heuristics for the aggregate planning problem. *Management Science 24*(12), 1242–1251.

Nam, S.-J. and R. Logendran. 1992. Aggregate production planning: A survey of models and methodologies. *European Journal of Operational Research 61*, 255–272.

Oliff, M. D. and E. E. Burch. 1985. Multiproduct production scheduling at Owens-Corning Fiberglas. *Interfaces 15*(5), 25–34.

Oliff, M. D. and G. K. Leong. 1987. A discrete production switching rule for aggregate planning. *Decision Sciences 18*(4), 582–597.

Orr, D. 1962. A random walk production-inventory policy. *Management Science 9*(1), 108–122.

Penlesky, R. and R. Srivastava. 1994. Aggregate production planning using spreadsheet software. *Production Planning and Control 5*(6), 524–532.

Reeves, C. R. 1993. Improving the efficiency of tabu search for machine sequencing problems. *Journal of the Operational Research Society 44*, 375–382.

Sanderson, I. W. 1978. An interactive production planning system in the chemical industry. *Journal of the Operational Research Society 29*(8), 731–739.

Schwarz, L. and R. E. Johnson. 1978. An appraisal of the empirical performance of the linear decision rule for aggregate planning. *Management Science 24*(8), 844–849.

Shafer, S. 1991. A spreadsheet approach to aggregate scheduling. *Production and Inventory Management Journal 32*(4), 4–10.

Shycon, H. N. 1974. Perspectives on management science applications. *Interfaces 4*(3), 23.

Singhal, J. and K. Singhal. 1996. Alternative approaches to solving the Holt et al. model and to performing sensitivity analysis. *European Journal of Operational Research 91*, 89–98.

Stadtler, H. 1988. Medium term production planning with minimum lot sizes. *International Journal of Production Research 26*(4), 553–566.

Tadei, R., M. Trubian, J. Avendano, F. Della Croce, and G. Menga. 1995. Aggregate planning and scheduling in the food industry: A case study. *European Journal of Operational Research 87*(3), 564–573.

Taubert, W. H. 1968. A search decision rule for the aggregate scheduling problem. *Management Science 14*(6), 343–359.

Taylor, S. 1980. Optimal aggregate production strategies for plants with semifixed operating costs. *AIIE Transactions 12*(3), 253–257.

Tersine, R., W. Fisher, and J. Morris. 1986. Varying lot sizes as an alternative to undertime and layoffs in aggregate scheduling. *International Journal of Production Research 24*(1), 97–106.

Thomas, L. J. and J. O. McClain. 1993. An overview of production planning. In S. Graves, A. Rinnooy Kan, and P. Zipkin (Eds.), *Logistics of Production and Inventory*, Volume 4, pp. Chapter 7. Amsterdam: Elsevier (North-Holland).

Thompson, S. D., D. T. Watanabe, and W. J. Davis. 1993. A comparative study of aggregate production planning strategies under conditions of uncertainty and cyclic product demands. *International Journal of Production Research 31*(8), 1957–1979.

Venkataraman, R. and S. B. Smith. 1996. Disaggregation to a rolling horizon master production schedule with minimum batch-size production restrictions. *International Journal of Production Research 34*(6), 1517–1537.

Vergin, R. C. 1966. Production scheduling under seasonal demand. *Journal of Industrial Engineering 17*(5), 260.

Vergin, R. C. 1980. On "a new look at production switching heuristics for the aggregate planning problem." *Management Science 26*(11), 1185–1186.

Wagner, H. M. 1975. *Principles of Operations Research* (2nd ed.), pp. 484–493. Englewood Cliffs, NJ: Prentice-Hall.

Wagner, H. M., V. A. Vargas, and N. N. Kathuria. 1993. The accuracy of aggregate LP production planning models. In R. Sarin (Ed.), *Perspectives in Operations Management: Essays in Honor Elwood S. Buffa*, pp. 359–387. Hingham, MA: Kluwer.

Welam, U. P. 1978. An HM management science type interactive model for aggregate planning. *Management Science 24*(5), 564–575.

Chapter 15

Material Requirements Planning and Its Extensions

In the previous chapter, we discussed the aggregate planning problem of setting workforce and other resource requirements for the medium term. Now, we turn to the shorter-term problem of determining and meeting requirements for all parts and components that are assembled into each final product. Here, the overriding concern is with the coordination of materials at the various stages of the production process. The concept of dependent demand, introduced in the multi-echelon context of Chapter 11, is vital in this chapter as well. Demand for components and raw materials is, to a large extent, determined when production schedules are established for parent items in which these materials are used. For example, in automobile assembly the requirements of a certain type of engine are known accurately when the assembly schedules of automobiles, in which the engine is used, are specified. The concepts discussed in this chapter are most applicable in situations where there are many stages of production and many parts and components that must be coordinated. It should become clear as the material requirements planning (MRP) system is described that it would not be useful in a continuous flow process industry or in a job shop. We shall see why in Section 15.7.

In Section 15.1, the complexity of multistage assembly manufacturing is discussed. Historically, in such contexts, many organizations have attempted to use the types of replenishment systems discussed earlier in the book (e.g., in Chapters 4 through 6). Section 15.2 addresses the weaknesses of such an approach. Next, in Section 15.3, we review Closed-Loop MRP, a planning and control system encompassed in the decision framework introduced in Chapter 13. The details of the material planning, a central component of this system, are laid out in Section 15.4. Another key element, capacity requirements planning (CRP), is presented in Section 15.5. Section 15.6 is concerned with the conceptual extension of MRP to distribution. In Section 15.7, we discuss some of the weaknesses of MRP, and in Section 15.8 we introduce an extension to MRP called enterprise resource planning (ERP).

15.1 The Complexity of Multistage Assembly Manufacturing

Let us consider a manufacturing facility with a number of different work centers or stations. To achieve each finished product (in the form of an assembly of several components), processing through several of these centers is required. Inventories can exist in the following forms:

1. Raw materials
2. WIP raw materials to component parts
3. Component parts
4. WIP component parts to subassemblies
5. Subassemblies
6. WIP subassemblies to assemblies
7. Assemblies

As mentioned earlier, the requirements through time of a particular component are primarily dictated by the production schedules of the next (closer-to-the-end-items) level of components in which this element is used. That is, there are complicated interactions between the production schedules (and associated inventories) of the various level items. Furthermore, relatively smooth demand for end products can produce erratic requirements through time for a particular component because of the batching of assemblies, subassemblies, etc. This is illustrated in the example of Table 15.1 and Figure 15.1, where the same subassembly A is used in three different assemblies. Notice that the end usage of each assembly is uniform with time, yet the requirements for subassembly A are highly erratic with time because the assemblies (for setup cost reasons)

Table 15.1   Illustration of Erratic Requirements for a Subassembly[a]

Time Period	1	2	3	4	5	6	7	8	9	10
Assembly X										
Demand	10	10	10	10	10	10	10	10	10	10
Production	30	–	–	30	–	–	30	–	–	30
Assembly Y										
Demand	5	5	5	5	5	5	5	5	5	5
Production	20	–	–	–	20	–	–	–	20	–
Assembly Z										
Demand	7	7	7	7	7	7	7	7	7	7
Production	14	–	14	–	14	–	14	–	14	–
Subassembly A										
Requirements	64	–	14	30	34	–	44	–	34	30

[a] Each of assemblies *X*, *Y*, *Z* requires one unit of subassembly *A*. For simplicity, we assume negligible assembly times.

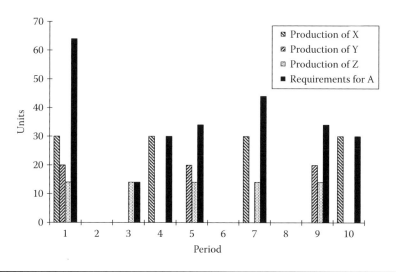

Figure 15.1 Graphical representation of erratic requirements for subassembly when demand for the end-items is constant.

are made in lots covering more than a single period of demand. (Recall that in a multiechelon production–distribution system, this was part of the bullwhip effect discussed in Chapter 12.)

The complexity is magnified in a manufacturing environment when many components are required to assemble a single finished item. Inadequate supplies of any of the components will lead to delays as well as excess in-process stock of the other components. To make matters worse, there are capacity constraints at the work centers (only so many machine-hours of production can be achieved at a particular work center).

The cost components include setup and value-added costs at machine centers (or replenishment costs for raw materials), inventory carrying costs, production overtime costs, shortage costs, and system control costs. Ideally, given the forecasts of usage of end-items by time period, it would be desirable to establish production run (purchase) quantities through time at the various levels of manufacturing. The goal is to keep the total relevant costs as low as possible, while not violating any of the constraints. The problem is not static in nature; there are likely to be many short-run changes, such as alterations in customer orders, machine breakdowns, quality rejections, and so forth. With the current state of the art, an "optimal" solution to this complex problem is out of the question. Instead, the best we can hope for is a feasible solution that generates reasonable costs. This is the philosophy underlying the Closed-Loop, MRP system to be discussed later in this chapter. However, before that discussion, we now emphasize the weaknesses of using traditional replenishment systems in a manufacturing environment.

15.2 The Weaknesses of Traditional Replenishment Systems in a Manufacturing Setting

By "traditional replenishment" we mean decision rules, discussed earlier in this book, whereby the timing and sizes of orders of each item are determined independently, in particular under the assumption of statistically independent demand for each item. The weaknesses of such an approach in an assembly environment include the following:

1. There is no need to statistically forecast the requirements of a component. Once the production plans for all items in which it is used have been established, then the requirements of the component follow, as dependent demand, by simple arithmetic. (This is illustrated by the example of Table 15.1.)
2. The procedures for establishing SSs are usually based on reasonably smooth demand. As discussed above (and illustrated in Table 15.1), this is usually unrealistic in the case of component items.
3. Traditional replenishment systems are geared to replenish stocks immediately following large demands that drive inventories to low levels. Again, in an erratic demand situation, a large demand may be followed by several known periods of inactivity. In such a situation, it makes no sense to immediately replenish the stock because unnecessary carrying costs would be incurred.
4. Where several components are needed for a single assembly, the inventories of these individual components should not be treated in isolation. To illustrate, consider the case where 20 different components are required for a particular assembly. Suppose, under independent control of the components, that for each component there is a 95% chance that it is in stock. Then, the probability of being able to build a complete assembly is only $(0.95)^{20}$ or 0.36; that is, 64% of the time at least one of the components would be unavailable, thus delaying the completion of an assembly.[*]

15.3 Closed-Loop MRP

MRP was initially developed without any capacity checks or input from other departments; thus, the production plan often was not believable to anyone outside of the production function. Closed-Loop MRP is an enhancement that includes capacity checks (to be discussed in Section 15.5), which are used iteratively with the master production schedule (MPS) and the component production plans (from MRP), to generate feasible schedules. Manufacturing Resources Planning, or MRP II, is an additional enhancement that converts a number of the outputs of production planning and control into financial terms—for example, inventories in dollars, labor budget, shipping budget, standard hours of output in dollars, vendor dollar commitments, and so forth. It also draws on forecasts from other departments, such as marketing, so that its schedules are more acceptable across the firm.[†]

Closed-Loop MRP was already covered to a large extent in the production decision-making framework of Section 13.2.3 of Chapter 13. For convenience, Figure 15.2 is a repetition of the major portion of Figure 13.1 that is normally considered as Closed-Loop MRP.

Aggregate production planning (Block 1) was treated in detail in Chapter 14, but there is one added difficulty in an assembly context. The problem is deciding for which work center the planning should be done, since, in the shorter run, capacity difficulties could crop up at other work centers. A similar dilemma will be mentioned with respect to master production scheduling. In

---

[*] See Ernst and Pyke (1992) for a more thorough discussion of this problem.

[†] Another version of MRP that explicitly addresses capacity constraints, and does so in a rigorous way, is MRP-C. See Tardif (1995) and Hopp and Spearman (2011).

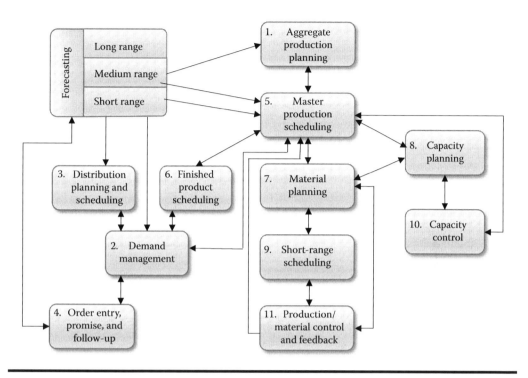

Figure 15.2   Closed-loop materials requirements planning.

this chapter (Sections 15.4 and 15.5), we concentrate on material planning (Block 7) and capacity planning (Block 8). Other than the following brief comments on master production scheduling (Block 5) as it particularly relates to Closed-Loop MRP, we do not say anything further about the other components of Figure 15.2. To review these components, see Section 13.2.3 of Chapter 13.

Master production scheduling is a complex task in an MRP assembly setting. For one thing, the bottleneck work center can shift depending on the changing nature of the workload and the labor force available in the short run. Thus, what appears on the surface to be a feasible master schedule may not be so when the detailed, implied needs at other production (and procurement) stages are worked out. Thus, although some bottleneck scheduling decision rules (to be discussed in Chapter 16) may be of assistance in master scheduling, a trial-and-error component, as well as some negotiation, will almost certainly be required.

There are a number of excellent writings on MRP (see, e.g., New 1974; Orlicky 1975; Landvater and Gray 1989; Baker et al. 1993; Wight 1995; Vollmann et al. 2005). The APICS has several study guides on the topic. In addition, numerous software packages are commercially available. For a description of how to implement MRP on a spreadsheet, see Lambrecht and Van den Wijngaert (1985), Sounderpandian (1994), and Friend and Ghobbar (1999).

In summary, Closed-Loop MRP is a system for generating reasonable feasible solutions to the complex problem of planning, scheduling, and controlling production/procurement in a dynamically changing, assembly situation. As noted by Zolnick (1996), Closed-Loop MRP tries to answer the following questions:

1. What are we going to make? (using the forecast)
2. What does it take to make it? (using resource requirements and bills of materials)

3. What do we have? (using inventory records)
4. What do we need, and when? (using manufacturing schedules)

15.4 Material Requirements Planning

To review from Chapter 13, MRP (or simply materials planning) takes the MPS and explodes it into implied, detailed production/procurement schedules (timing and quantities) of all components and raw materials. This includes scheduling modifications (cancellations, adjustments in delivery dates, etc.) caused by changes in various conditions such as customer order sizes, quality difficulties, and so on. As was discussed in Chapter 13, master scheduling need not be done at the end-item level. If, however, master scheduling is done at a component-item level, in addition to MRP, finished product scheduling must be used for the processing stages from the master schedule level to the finished products level.

15.4.1 Some Important Terminology

Before describing the MRP approach, it is necessary first to discuss some concepts that we will use in our description.

15.4.1.1 Bill of Materials

To properly take into account of the dependent nature of demand, we must have a means of projecting the needs, in terms of components, for a production lot of a particular assembly or subassembly. The bill of materials helps us achieve this goal. In the so-called modular form, a bill of materials for a particular inventory item (termed the parent) shows all of its *immediate* components and their numbers per unit of the parent. This is illustrated in Figure 15.3 for a red plastic toy car. Because this car snaps together, there are no fasteners listed in the bill of materials.

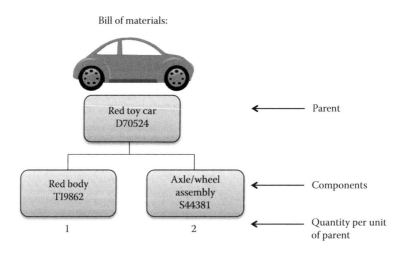

Figure 15.3   A red plastic toy car.

The meaning of the word *immediate* becomes clear when we see that in Figure 15.3, we do not subdivide the axle/wheel assembly, S44381, into its components, an axle and two wheels. This would be done on a separate modular bill of materials for that item. See Danese and Romano (2005) for a discussion of an application of modular bills of assembly.

Where there is a wide diversity of end-items (because of a number of optional choices available to the customer), we would not bother to develop a bill of materials for each end-item; instead, the first level would be that of the major subassemblies. This is completely analogous with master scheduling the major subassemblies instead of the end-items.

It will be convenient to discuss a second common form of the bill of materials after we introduce the concept of level coding. There are also other, less common types of bills of materials that are used for special purposes, such as phasing in an engineering replacement for an existing item.

15.4.1.2 Level Coding

To provide a systematic framework for exploding back the implications on all components of a given MPS, it is convenient to use a particular method of coding the individual SKUs. Each item (or, equivalently, its bill of materials) is assigned a level code according to the following logic:

Level 0 A finished product (or end-item) not used as a component of any other product.

Level 1 A direct component of a level 0 item. At the same time, the level 1 item could also be a finished product in itself. To illustrate, consider the example of a real (not toy) automobile and tires. A particular type of tire could be sold as a finished product in its own right. However, if it was used as a direct component in the manufacture of one or more types of level 0 automobiles, it would be classified as a level 1 item.

Level 2 A direct component of a level 1 item. Again, as shown by the dotted lines in Figure 15.4, a level 2 item could be used as a direct component of a level 0 item or could even itself be a finished product.

$$\vdots$$

Level n A direct component of a level $(n-1)$ item, that is, a component on a bill of materials with level code $(n-1)$.

Thus, an item may be sold as a finished product, and be used as a component of, say, a level 2 item. As a finished product, the item could be coded as level 0, but as a component, it could be coded as level 3. MRP resolves this conflict by always choosing the higher number (in this case, level 3) to code the item.

This coding process is continued all the way back to raw materials, which are themselves given appropriate level codes. We can see that the level coding is equivalent with the concept of indentures used in Chapter 11 when describing multiechelon inventory systems involving repairable items.

Figure 15.5 illustrates a portion of the coding for the red toy car. In this case, the car goes into a gift pack that is composed of this car and a red truck. The example of Figure 15.5 provides an opportunity to show the so-called *indented* form of a bill of materials, illustrated in Table 15.2. This table lists all the components, and shows that there is a one-to-one correspondence between the indentation and the level code. Note that the car is sold in three forms, a car and truck gift pack, a multicar pack, and as an individual item (regular external demand).

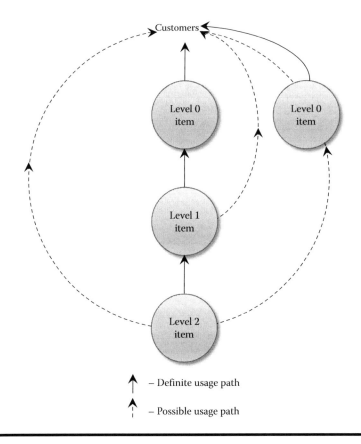

Figure 15.4   Level coding.

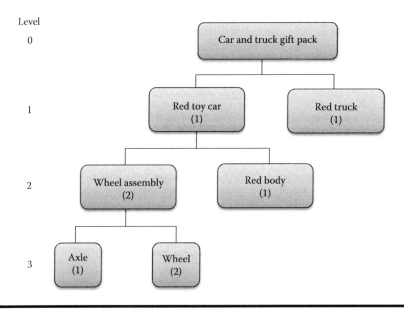

Figure 15.5   Level-by-level coding for the red toy car.

Table 15.2   Indented Bill of Materials for the Toy Car

Part Number	Description			
	Car			
	Truck Gift Pack			
	Multi-Car Pack			
	Red Car External Sales			
D70524		Red Toy Car		
S44381			Axle/Wheel Assembly	
S11844				Axle
R21174				Wheel
T19862			Red Body	

15.4.1.3 Lead Times (Offsetting)

This concept was already introduced in Section 11.4.3 of Chapter 11 when we dealt with multiechelon inventory systems. The manufacturing operation represented by a particular bill of materials, for example, combining components B and C in a complicated machining activity to produce subassembly A, may require a considerable length of time. This is particularly so because such operations tend to be performed on batches of items, and these batches compete with other orders for the use of a particular machine. In this case, if certain requirements of item A are specified for a particular date (the due date), then we must properly offset (or phase in time) the order release date for the machining operation. In other words, the corresponding units of parts B and C must be available at a suitably earlier time, recognizing the waiting and processing time at the particular operation. The waiting time varies with the shop load, and is quite difficult to estimate. (We will comment on this issue in Section 15.7. See also Karmarkar 1993.) To illustrate, suppose that the toy car (Figure 15.5) assembly operation is performed in a batch size such that approximately 1 week (under current shop conditions) is required from order release to completion date. Therefore, if 100 toy car assemblies were required by, for example, June 26, then 100 of the components—namely, the body and axle/wheel assembly—would be needed by June 19. (See Figure 15.6.) The components, in turn, would have to be started in production in time to be available by June 19. In this example, suppose that the lead time to produce bodies is 1 week, and the lead time to produce axle/wheel assemblies is 2 weeks. Then, the correct number of bodies must be initiated into production by June 12, and the correct number of axle/wheel assemblies by June 5. In the case of a raw material, the lead time (offset) is the time that elapses from when we decide to send the purchase order until the moment when the material is physically present for the first processing operation.

15.4.1.4 Routing

For each item that could appear on the master schedule, the routing shows the sequence of production operations (and associated work centers) and the standard hours for each operation. The routings are essential for the capacity-planning activity to be discussed in Section 15.5.

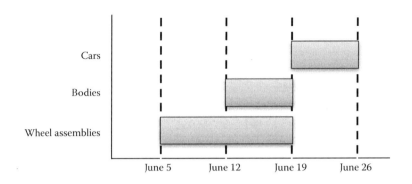

Figure 15.6   Illustration of lead time offsetting.

15.4.2 Information Required for MRP

To carry out MRP, the following input information is essential:

1. *The MPS projected out to the planning horizon.* The MPS is carried out at the end-item level and contains the production schedule for all end-items for each time period out to the planning horizon.
2. *The inventory status of each item (including possible backorders).* Accurate stock status information is essential because MRP, in contrast with traditional replenishment systems, establishes the timing of replenishments to keep inventories as low as possible. Thus, errors in stock status can cause severe problems. In many companies utilizing MRP, it has been found appropriate to physically hold stocks in limited access stores. Also, regular physical counts to verify records are necessary.
3. The timing and quantities involved in any outstanding or planned replenishment orders.
4. Forecasts (which can be partially or entirely firm customer orders) of demand for each component, *subject to direct customer demand*, by time period out to the planning horizon.
5. All relevant bills of materials and associated level codes.
6. All routings.[*]
7. Production or procurement lead times (offsets) for each operation.
8. Possible scrap (or yield) allowances for some operations (e.g., to convert item B into item A, we require, because of losses, on the average, 105 units of B to obtain 100 good units of A).

As will be discussed later, additional information may be needed in order to determine the replenishment quantities (lot sizes) for any specific item.

15.4.3 The General Approach of MRP

In this section, to provide an overview, we present only an outline of the general approach used in MRP, purposely omitting details that we discuss in the numerical illustration in the following section.

---

[*] Routings are only needed in CRP (to be treated in Section 15.5).

MRP seeks to overcome the weaknesses of traditional replenishment systems in a manufacturing environment by making specific use of the dependent demand for components, and by accounting for the time-varying (erratic) nature of the requirements for components. Moreover, the inventories of different components needed for the same operation are coordinated to avoid a shortage of one element delaying the operation and tying up the other components in inventory.

MRP begins with an MPS that provides the timing (order release dates) and quantities of production of all end-items (level 0) on a discrete time basis (normally a 1-week period or *bucket* is used).[*] The product files (bills of materials) indicate the immediate component items and their quantities per unit of each parent item. Thus, a time series of requirements (at each order release date of the level 0 items) is generated for the level 1 items. For each level 1 item, one must add to this time series of *dependent* requirements any requirements for externally generated direct independent demand (e.g., as service parts). The result is a new series of requirements by time period, known as the gross requirements of the item.[†]

Next, for each level 1 item, the existing inventory position (quantities OH and already on order) is allocated against the gross requirements to produce a modified series of requirements by time period, known as the net requirements of the item. Figure 15.7 shows the graphic relationship among the inventory position, cumulative gross requirements, and cumulative net requirements for a situation in which there are nonzero gross requirements at the start of periods 1, 2, 4, and 5, and there is an initial OH inventory with a replenishment due at the start of period 3. When the dotted line is above the solid line in Figure 15.7, there are no positive net requirements; the distance between the lines represents the projected OH inventory.[‡] However, when the solid line is above the dotted one, the distance between them gives the cumulative net requirements.

The next step is to provide appropriate coverage for the net requirements of the level 1 item under consideration by adjusting previously scheduled replenishment actions, or by initiating new replenishment actions. In other words, net requirements are covered by planned receipts of replenishment lots. When these lots are backed off (offset) over the lead time, such replenishment actions are then known as planned order releases. The possible adjustments include:

1. Increasing (or decreasing) a replenishment quantity
2. Advancing (or delaying) an order due date
3. Canceling an order

A new action involves specification of the item involved, the order quantity, the order release date, and the order receipt date. In selecting the timing and sizes of replenishments, we are faced with a primarily deterministic, time-varying pattern of demands, which is precisely the situation encountered in Chapters 5 and 10. Therefore, the Silver–Meal heuristic (the details are in Section 5.6.1 of Chapter 5) or the Dixon–Silver heuristic (Section 10.6.2 of Chapter 10) could certainly be used for selecting the lot sizes. However, quite often, a very simple solution

---

[*] In its usual form, MRP requires the handling of scheduling on a discrete time basis, the normal period being 1 week. That is, requirements and replenishment quantities are shown no finer than on a weekly time basis. Conceptually, one could certainly handle MRP on an event (transaction) basis rather than by the somewhat artificial basis of discrete time slots. Some software firms that provide add-on software for ERP systems operate MRP in a real-time mode. See Section 15.8.1.

[†] We have already used the concepts of gross and net requirements in Section 11.4.3 of Chapter 11 where we dealt with a nonproduction, multiechelon, dependent demand situation.

[‡] Equivalently, one could think of this situation as representing negative cumulative net requirements.

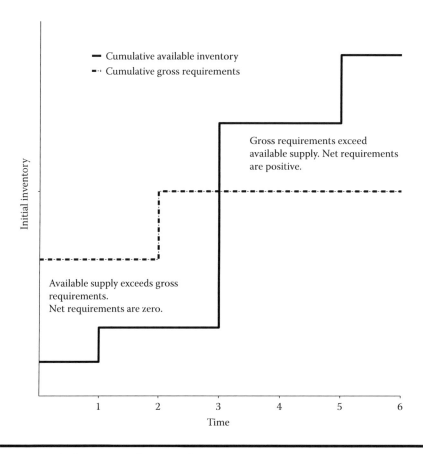

Figure 15.7 Cumulative gross and net requirements.

is appropriate—namely, to cover each net requirement with a separate replenishment quantity. This type of solution is known as a lot-for-lot (LFL) strategy. It is appropriate when

1. The requirements pattern is very erratic; that is, the only requirements are large occasional quantities, or
2. The production operation involved has a very low setup cost, which is usually the case in an assembly operation.

The PPB heuristic (discussed in Section 5.6.5 of Chapter 5) has also been used in MRP systems. Tests by De Bodt and Van Wassenhove (1983) have revealed that the basic *fixed* EOQ can perform surprisingly well when there is considerable uncertainty in the demand pattern. (See also the references in Chapters 5 and 10.) If an analytic procedure such as the EOQ, the Silver–Meal heuristic, or PPB heuristic is used, provision must be made for manual overrides. The scheduler may possess information not included in the model; for example, "Georges Roy, the foreman at a particular work center, will never accept a small lot size late in a month." Finally, it should be emphasized that lot-sizing decisions at the level under consideration have ramifications back through all the component levels. However, the usual lot-sizing procedures consider carrying and setup costs only at the level being scheduled. Thus, at best, a suboptimization is achieved and one should really use multiechelon adjustments in carrying and setup costs, as discussed in Section 11.3 of Chapter 11.

Once coverage is completed for the level 1 item, the associated bill of materials indicates which level 2 items are used as components. The order release dates and quantities of the level 1 item thus imply requirements through time of level 2 items. This is done for all level 1 items. Again, any direct external demand for a level 2 item, as well as any dependent usage as a direct component of any level 0 items, must be included to obtain the gross requirements of that item. These requirements are then netted, covered, and so on.

The above process is continued all the way back to all raw materials, which, in turn, have their requirements properly covered by purchasing actions. In summary, for each item, we determine gross requirements, net requirements, planned order receipts, and planned order releases. The latter, in turn, contribute to the gross requirements of the item's immediate components. Note that all items having a given level code are processed before any items on the next (higher numbered) level. It is clear that a computer is indispensable for the step-by-step explosion of requirements at the next level implied by each individual-item coverage pattern.

15.4.4 A Numerical Illustration of the MRP Procedure

We use the toy car to illustrate the MRP procedure. Reference is made to the product structure depicted in Figure 15.3. Recall that the car is sold in three forms, a car and truck gift pack, a multicar gift pack, and as an individual item (regular external demand). Suppose that through the first stages of the application of MRP, we have arrived at the order release patterns for the gift pack and the multicar pack, as shown in Table 15.3. These patterns, together with direct external demand, imply the gross requirements pattern for the car, shown at the bottom of the table. The gross requirement in a period is the total number of units in the order releases of the two parents planned for that period, plus any anticipated direct external demand in the period. (We can simply take the total here because precisely 1 unit of the component is needed for each unit of either parent.)

Next, the net requirements for item D70524 are determined by taking into account of any OH inventory and any released or planned orders. Suppose that

1. The offset (production lead time), including queue and move time, for this item is 1 period.
2. The current OH inventory is 10 units and there are no released or planned orders.

Table 15.3 Derivation of Gross Requirements for the Red Toy Car[a]

Source	Time Period	1	2	3	4	5	6	7	8
Parent	Car								
	Truck Gift Pack, D63321, planned order releases	–	50	–	20	–	40	–	60
Parent	Multi-Car Pack, D63322, planned order releases	–	–	30	–	–	30	–	–
External	Regular Pack, D70524, direct external demand for Red Toy Car	–	–	10	–	–	–	10	–
Total	Red Toy Car, D70524, gross requirements	–	50	40	20	–	70	10	60

[a] 1 unit of D70524 is used in each of D63321 and D63322.

Table 15.4 Material Requirements Plan for the Toy Car, D70524

Lead time = 1 period									
Order quantity = not fixed[a]									
Time Period	*0*	*1*	*2*	*3*	*4*	*5*	*6*	*7*	*8*
1. Projected gross requirements	0	0	50	40	20	0	70	10	60
2. Scheduled receipts	0	0	0	0	0	0	0	0	0
3. Planned order receipts									
4. Projected net inventory[b] at the end of period	10	10	−40	−80	−100	−100	−170	−180	−240
5. Net requirements	0	0	40	40	20	0	70	10	60
6. Planned order releases									

[a] Provision can be made to have a preestablished, fixed order quantity.
[b] Recall that Net Inventory = (On-hand)− (Backorders).

Table 15.4 shows a partial material requirements plan for item D70524. Row 4, the projected net inventory, gives the cumulative net requirements when the projected net inventory is negative. This is completely consistent with the earlier discussion of Figure 15.7. When there are cumulative requirements in any period (i.e., the net inventory is negative), the current plan does not adequately provide coverage for the projected needs. Thus, some change is in order. Row 5 gives the net requirements in each period. Note that

$$N_t = \max\{0, G_t + SS_t - I_{t-1}^+\} \qquad (15.1)$$

where
N_t = net requirements in period t
G_t = gross requirements for period t
SS_t = safety stock level in period t
I_{t-1} = ending inventory in period $t - 1$
I_{t-1}^+ = max $\{0, I_{t-1}\}$

Note that the SS level can change over time.

Scheduled receipts (row 2) differ from planned receipts (row 3) in that scheduled receipts have been released to the shop floor or to vendors. Planned receipts have yet to be released, and therefore can still be changed. Note from Table 15.4 that we have no scheduled receipts, and we have, for now, left the planned receipts row blank. Because there are net requirements for 40 (40 = 50 + 0 − 10, assuming no SS) cars in period 2, an order should be scheduled for receipt in that period. This order should be released in period 1 because the lead time is 1 period. It might be attractive to order enough to also cover the requirements in later periods (particularly periods 3 and 4). In any event, at least tentative coverage should now be made for all periods out to the horizon (here period 8). These will appear in the planned order releases row that is blank in Table 15.4. If the LFL lot-sizing rule is used, a lot is ordered each time a requirement appears. Table 15.5 gives

Table 15.5   Revised Material Requirements Plan (LFL) for the Red Toy Car, D70524

Lead time = 1 period									
Order quantity = not fixed (use LFL)									
Time Period	*0*	*1*	*2*	*3*	*4*	*5*	*6*	*7*	*8*
1. Projected gross requirements	0	0	50	40	20	0	70	10	60
2. Scheduled receipts	0	0	0	0	0	0	0	0	0
3. Planned order receipts	0	0	40	40	20	0	70	10	60
4. Projected net inventory at the end of period	10	10	0	0	0	0	0	0	0
5. Net requirements	0	0	40	40	20	0	70	10	60
6. Planned order releases	0	40	40	20	0	70	10	60	0

the resulting schedule, which is the revised plan for item D70524. Note how the planned order receipts equal the projected gross requirements from periods 3 through 8. Also note that scheduled receipts are still zero because no orders have been released yet. Next period, if the order for 40 is actually released, it will appear as a scheduled receipt in period 2, while the planned order receipt for period 2 will be zero.

Rather than necessarily using the LFL rule, let us consider the following. Suppose that the setup cost for the car assembly operation is $7.50, the added variable cost at this operation is $8/unit, and the inventory carrying charge has been set at 0.005 $/$/period (if the period was 1 week, this would correspond to an *r* value of 0.26 $/$/year). If the Silver–Meal heuristic was used, then the logic of Section 5.6.1 of Chapter 5 gives the production schedule (in terms of planned order receipts of toy cars) shown in Row 3 of Table 15.6. With the 1-period lead time, the corresponding schedule of order releases is shown in Row 6. Only the order in period 1 would be released, and the later quantities are tentative for planning purposes. Neglecting any need for SS, we now see from Row 4 that adequate coverage is provided.

Now, we show all the remaining levels of the MRP computations for the car's components, assuming that the Silver–Meal heuristic (as opposed to LFL) is used for the car itself. The gross requirements for the axle/wheel assembly used in the red toy car are derived in part from planned order releases of the red car. Note that because there are two axle/wheel assemblies per red car, the gross requirements are twice the planned order releases for each period. This calculation illustrates that the order release period of a parent item (the red car, in this case) is exactly the gross requirement period of the component (the axle/wheel assembly). Thus, the first row of Table 15.7 is 2 times the last row of Table 15.6. Now, the axle/wheel assembly, item R21174, is used in cars of all colors. MRP would first be used to determine the order release patterns of these other cars. Suppose this has already been done and the combined order releases, expressed in terms of numbers of axle/wheel assemblies required, are as shown in the second row of Table 15.7. Assuming that there is no other usage of the axle/wheel assemblies, then the last row of Table 15.7 provides the gross requirements. These entries become the gross requirements in Row 1 of the axle/wheel assembly MRP table (the second one in Table 15.8), which is an initial material requirements plan for item R21174.

Table 15.6 Revised Material Requirements Plan (Silver–Meal) for the Toy Car, D70524

	0	1	2	3	4	5	6	7	8
Lead time = 1 period									
Order quantity = not fixed (use Silver–Meal)									
Time Period	*0*	*1*	*2*	*3*	*4*	*5*	*6*	*7*	*8*
1. Projected gross requirements	0	0	50	40	20	0	70	10	60
2. Scheduled receipts	0	0	0	0	0	0	0	0	0
3. Planned order receipts	0	0	100	0	0	0	80	0	60
4. Projected net inventory at the end of period	10	10	60	20	0	0	10	0	0
5. Net requirements	0	0	40	40	20	0	70	10	60
6. Planned order releases	0	100	0	0	0	80	0	60	0

Table 15.7 Determination of Gross Requirements for Wheel Assembly, R21174

Source	*Time Period*	*1*	*2*	*3*	*4*	*5*	*6*	*7*	*8*
Parent	Red Car, D70524, planned order releases	200	0	0	0	160	0	120	0
Parent	Other Colors Toy Cars, planned order releases	700	500	300	700	300	200	600	100
Total	Gross requirements of the wheel assembly	900	500	300	700	460	200	720	100

There are a number of points that should be emphasized from Table 15.8. First, note that the material requirements plan for the red car body, item B71298, indicates that an order of 150 units is due to arrive in period 1. This is fortunate because of the 100 bodies that are required for the production of red cars, as shown in the last row of Table 15.6 and in the first row of the first MRP plan in Table 15.8. Then, 50 units should be left in the inventory until period 5, when 80 more red bodies are needed, creating a net requirement of 30. Note that the lead time for red bodies is 1 period, and the lot-sizing rule is LFL. Therefore, a planned order release of 30 units is needed in period 4 to combine with the 50 in inventory to satisfy the requirement for 80 units in period 5. Then, 60 units are planned for release in period 6 to satisfy the needed 60 units in period 7.

In the MRP plan for axle/wheel assemblies (the second plan in Table 15.8), we can see that 1,200 axle/wheel assemblies are in inventory at the beginning of period 1, and that 200 more will be received in period 1. The lead time is 2 periods and the lot-sizing rule is LFL, which together yield the planned order releases of 300, 700, 460, and so on. These are listed in the last row of the axle/wheel assembly plan.

Now, note how these planned order releases become the projected gross requirements for the axle (shown in the third MRP plan of Table 15.8). Because one axle is needed for each axle/wheel assembly, the gross requirements for axles are identical to the planned order releases of axle/wheel assemblies. To begin production of 300 axle/wheel assemblies in period 1, 300 axles need to be

Table 15.8   Material Requirements Plan for the Remaining Components

Material Requirements Plan for the Red Car Body, B71298									
Lead time = 1 period									
Order quantity = LFL									
Time Period	0	1	2	3	4	5	6	7	8
1. Projected gross requirements	0	100	0	0	0	80	0	60	0
2. Scheduled receipts	0	150[a]	0	0	0	0	0	0	0
3. Planned order receipts	0	0	0	0	0	30	0	60	0
4. Projected net inventory at the end of period	0	50	50	50	50	0	0	0	0
5. Net requirements	0	0	0	0	0	30	0	60	0
6. Planned order releases	0	0	0	0	30	0	60	0	0

[a] This order was released in period 0.

Material Requirements Plan for the Axle/Wheel Assembly, R21174									
Lead time = 2 periods									
Order quantity = LFL									
Time Period	0	1	2	3	4	5	6	7	8
1. Projected gross requirements	0	900	500	300	700	460	200	720	100
2. Scheduled receipts	0	200[a]	0	0	0	0	0	0	0
3. Planned order receipts	0		0	300	700	460	200	720	100
4. Projected net inventory at the end of period	1,200	500	0	0	0	0	0	0	0
5. Net requirements	0	0	0	300	700	460	200	720	100
6. Planned order releases	0	300	700	460	200	720	100	0	0

[a] This order was released in period −1.

Material Requirements Plan for the Axle, A31456									
Lead time = 3 periods									
Order quantity = 500									
Time Period	0	1	2	3	4	5	6	7	8
1. Projected gross requirements	0	300	700	460	200	720	100	0	0

(Continued)

Table 15.8 (*Continued*)   Material Requirements Plan for the Remaining Components

Time Period	0	1	2	3	4	5	6	7	8
2. Scheduled receipts	0	0	0	0	0	0	0	0	0
3. Planned order receipts	0	0	0	0	500	500	0	0	0
4. Projected net inventory at the end of period	1,480	1,180	480	20	320	100	0	0	0
5. Net requirements	0	0	0	0	180	720	100	0	0
6. Planned order releases	0	500	500	0	0	0	0	0	0

Material Requirements Plan for the Wheel, W54376									
Lead time = 1 period									
Order quantity = LFL									
Time Period	0	1	2	3	4	5	6	7	8
1. Projected gross requirements	0	600	1,400	920	400	1,440	200	0	0
2. Scheduled receipts	0	0	0	0	0	0	0	0	0
3. Planned order receipts	0	0	1,280	920	400	1,440	200	0	0
4. Projected net inventory at the end of period	720	120	0	0	0	0	0	0	0
5. Net requirements	0	0	1,280	920	400	1,440	200	0	0
6. Planned order releases	0	1,280	920	400	1,440	200	0	0	0

ready in period 1. The other important item to note in the MRP plan for axles is the fixed order quantity of 500 units. These axles are purchased from a vendor who offers a significant quantity discount on orders of 500, 1,000, 1,500, etc. The purchasing department has determined that the best order quantity is 500 (perhaps using methods of Section 4.5 of Chapter 4). The three-period lead time, in conjunction with this order quantity, gives rise to the plan shown.

Finally, consider the plan for wheels (item W54376). The projected gross requirements of 600, 1,400, 920, and so on are exactly twice the planned order releases of axle/wheel assemblies, because there are two wheels per axle/wheel assembly. The plan for wheels follows directly from the one-period lead time and the LFL lot-sizing rule.

In this example, there are no problems evident in the plans for the red toy car and its components. If, however, we were not expecting 150 red car bodies to arrive in period 1, there would be an immediate problem with the 100 bodies due in period 1. If a problem of this sort were to arise, we would need to expedite an outstanding order or release a high priority new order. Alternatively, the entire plan would have to be adjusted. If problems are evident far enough in the future, the master schedule usually can be modified. On the other hand, when problems arise unexpectedly—due to bad quality, late deliveries, machine breakdowns, etc.—we may need to expedite as well as modify the plan. We will discuss other courses of action in Section 15.5.

15.4.5 *The Material Requirements Plan and Its Uses*

The material requirements plan provides several types of information of use to management, particularly in an environment that changes due to alterations in customer demands, scrap output, equipment failures, and so forth. The information includes:

1. Actual and projected inventory status of every item.
2. Listing of released and planned orders by time period. This document is useful for two purposes. First, in a summary form, it is fed back to the aggregate planning stage (discussed in Chapter 14) of a hierarchical planning system. Second, it is a necessary input for detailed CRP (to be discussed in Section 15.5).
3. Rescheduling and cancellation notices. These are particularly helpful in establishing and adjusting order priorities for both in-house production and outside procurement (Block 9 of Figure 15.2; see also Section 13.2.3 of Chapter 13 for further discussion).

Because of the typically very large number of SKUs in an MRP system, the material requirements plan must display information on an exception basis; that is, only for those items where an immediate action is likely necessary. Examples include:

1. An open order exists this period, but will not cover the existing backorders.
2. Net requirements exist this period, but the next open order is not due until a future period.
3. An open order exists this period, but the net requirements can be met from the already-existing, OH inventory.

15.4.6 *Low-Value, Common-Usage Items*

There is one exception to the rule of using MRP level-by-level explosion to ascertain the requirements of all items—specifically, low-value items having high-usage rates, which are the basic components of many items (e.g., bolts, washers, nuts, etc.). The costs of precise physical control are likely to be prohibitive in view of the low cost of carrying SSs of such items. Therefore, for such items, it is preferable to use one of the control systems suggested in Chapter 6, such as a continuous-review, order-point, order-quantity system (i.e., a two-bin system) or a periodic-review, order-up-to-level system. The methods suggested in Chapter 8 should be considered as well. One of the authors of this book has worked with a fastener manufacturer that was required to hold 2 weeks of projected demand in SS for a major U.S. automotive assembler. The automotive firm held very little SS, and apparently applied a simple rule for these low-value parts.

15.4.7 *Pegging*

Consider an item that is used as a component of several other items. The straightforward use of MRP, as discussed above, leads to gross requirements on this item that are generated from a number of sources. In some circumstances, it may be important to know which items generated which amounts of these requirements. In particular, if a shortage of the item is imminent, it would be helpful to know which subassemblies, assemblies, finished products, and, ultimately, customer orders would be affected. The solution is as follows. When the production (procurement) schedule of an item is exploded to generate gross requirements on the next (higher numbered) level items, these requirements are "pegged" with an identification of the item generating them.

It is clear that additional processing effort is required in connection with pegging. Therefore, this procedure should be used only when the information generated is of paramount importance. Even though computing costs are quite low now, the costs (including human effort) of maintaining accurate data and reporting may be high.

15.4.8 Handling Requirements Updates

Our discussions so far have implicitly dealt with MRP as a process carried out once per basic time period. However, changes in various inputs are certainly not restricted to occurring only once per period. These possible changes include:

1. Changes in the master schedule or in direct external demand for components
2. Identified discrepancies in inventory records
3. Changes in machine availability (e.g., due to breakdowns)
4. Actual completion time or quantity different from planned
5. Engineering changes in product structure (bill of materials)
6. Changes in costs, lead times, etc.

MRP must be able to effectively cope with such changes. There are two very different options available: regeneration and net change.

In the regeneration method, the entire MRP process, as discussed in Sections 15.4.3 and 15.4.4, is carried out once per period (typically the period is 1 week), known as batch processing. All changes that have taken place since the previous regeneration are incorporated in the new run. In the net change approach, one does not wait until the next period to incorporate a change and replan coverage. Replanning takes place on essentially a continuous time basis. For a particular change, the possible effects are limited to components of the item causing the change. Therefore, the modification of the previous schedule tends to be much more limited than under regeneration. In effect, only a partial explosion of requirements is undertaken each time a change is processed.

In a regenerative system, between regenerations, only the OH and on-order inventory levels of each item are updated. This updating uses the standard inventory transactions such as demands, receipts, quality losses, and so forth. The possible associated changes in requirements for component items are not updated until the next regeneration. Hence, these requirements become less reliable as the period between regenerations progresses. In contrast, these requirements (and associated priorities, etc.) are kept up to date in a net change system as changes are incorporated on a frequent basis.

Obviously the up-to-date, rapid response capability of a net change system is desirable. Furthermore, the computational effort is more evenly spread through time than in a regenerative arrangement. Primarily for the first of these reasons, it appears that most users will eventually switch to a net change system. However, there are potential problems that should not be overlooked. First, net change tends to promulgate any earlier errors. Therefore, on an occasional basis, a regeneration is necessary to purge the system of errors. Second, some judgment must be used in deciding on how quickly to process the different changes; too frequent processing can lead to unnecessary instability. For example, the size of an order may jump up and down several times prior to actual production, causing factory workers to disbelieve the schedule. Finally, net change is somewhat less efficient from a data-processing standpoint. This last reason is less important today than it was in the 1980s; but with increased computing power comes the responsibility for increased discipline in making changes to avoid instability.

Instability in MRP systems is often called "nervousness." Nervousness specifically refers to frequent rescheduling actions, such as expediting, delaying or canceling an order, or to changing the size of an order. In Section 5.6.9 of Chapter 5, we discussed some research that focuses on reducing, or dampening, system nervousness. Various dampening methods include freezing the master schedule for some periods in the future; SS; safety time; safety capacity; freezing certain orders so that they cannot be changed; and expediting.[*]

A major benefit of freezing the master schedule is to mediate the conflict between sales and manufacturing. The sales force often wants frequent changes to the production schedule as they try to be responsive to customers. Manufacturing would prefer infrequent changes so that less time is spent in the setup, and production plans can be followed. A compromise can be reached if the master schedule is frozen for, say, 3 weeks, so that no changes can be made within the frozen period.

15.4.9 *Coping with Uncertainty in MRP*

To this point, our discussions on MRP have essentially ignored the effects of uncertainty. In Chapter 6, we coped with uncertainty by the introduction of SSs. This is not the predominant approach in many MRP applications in industry. The usual argument is that SSs are not really appropriate in a dependent demand situation. Instead, one can more effectively avoid shortages and excess inventories through the adjustment of lead times, by expediting or, more generally, by shifting priorities of shop and vendor orders.

In general, however, to avoid frequent expediting, it is necessary to consider SS or safety time (scheduling orders for completion slightly ahead of the required time) for dependent demand items when there is uncertainty. Clearly, inventories increase as a result. The effect is pronounced when there are multiple components that must be assembled into a single final product, because most components arrive early. Whybark and Williams (1976) provide considerable qualitative insight concerning four general sources of uncertainty in an MRP setting. The sources of uncertainty are (1) supply timing, (2) supply quantity (e.g., variable yield), (3) demand timing, and (4) demand quantity. Sources (2) and (4), involving quantity uncertainty, are best handled by SSs, whereas safety times are more appropriate for sources (1) and (3). However, a quantitative analysis of exactly how much SS or safety time is appropriate for each item is very complicated because of the erratic, time-varying, dependent nature of the demand patterns. Summary guidelines include the possible use of the following:

1. SS in items with direct external usage
2. SS in items produced by a process with a significantly variable yield
3. SS in items produced at a bottleneck operation
4. SS in certain semifinished items used for a myriad of end-items
5. Safety time in raw materials

---

[*] We list here some of the references from Chapter 5, along with additional references: Carlson et al. (1979), Kropp et al. (1983), Kropp and Carlson (1984), Blackburn et al. (1986), Sridharan et al. (1987), Sridharan and Berry (1990a,b), Metters (1993), DeMatta and Guignard (1995), Gupta and Brennan (1995), Toklu and Wilson (1995), Ho et al. (1995), Zhao et al. (1995), Ho and Carter (1996), Koh et al. (2002), and D'Avino et al. (2014). Some of these papers (such as Carlson et al. 1979; Kropp et al. 1983) suggest cost-based rules for determining whether or not to reschedule. Others (such as Gupta and Brennan 1995; Zhao et al. 1995) examine the interactions between lot-sizing rules and the dampening procedure.

Perhaps, the most common approach is to provide SS primarily for end-items using the methods of Chapter 6. However, these methods also apply to SS of semifinished items and at a bottleneck. Methods of Chapter 16 can be useful in setting SS at a bottleneck and in setting safety time for raw materials.

Further research on dealing with uncertainty in MRP includes a paper by Buzacott and Shanthikumar (1994b). They conclude that safety time is usually only preferable to SS when it is possible to make accurate forecasts of future required shipments over the lead time. Otherwise, SS is more robust in coping with changes in customer requirements in the lead time or with fluctuations in forecasts of lead time demand. See also Miller (1979), De Bodt et al. (1982), De Bodt and Van Wassenhove (1983), Lambrecht et al. (1984), Lowerre Jr. (1985), Yano (1987a,b), Kurtulus and Pentico (1988), Karmarkar (1993), Vargas (1994), Shore (1995), Vargas and Metters (1996), Guide Jr and Srivastava (2000), and Bangash et al. (2004).

15.5 Capacity Requirements Planning

This section deals with Block 8 of Figure 15.2. CRP determines the needed capacity through time at each work center as implied by a given or proposed MPS. This is required because, left alone, MRP assumes that infinite capacity is available.[*] Recall that there were no capacity checks in our example in Section 15.4.4, so that a plan to produce, say 1,280 wheels in period 1, is assumed to be feasible. As mentioned in Chapter 13, CRP takes place at two stages in time. First, a rough-cut capacity plan (RCCP) is sometimes used to evaluate a *tentative* MPS.[†] In this regard, a useful construct to employ is the *load profile*. It gives the approximate needs, in terms of various resources by time period, that are associated with 1 unit of a particular end-item being put in the master schedule in a base period. Such load profiles are developed for each item that could be master scheduled, but need only be done once in a lifetime for a product (except, of course, if engineering or process changes take place). The MRP system itself can be used to develop these load profiles (for details, see Ptak and Smith 2011). The profiles are then used to project the likely resource requirements by time period of the particular trial master schedule.

The second and more common type of CRP is more detailed in nature. A (possibly revised) master schedule is exploded through the material requirements plan (Section 15.4), producing a set of released and planned shop (and supplier) orders. Using the routings of the various items, these orders are converted into requirements (e.g., in machine hours) period by period at the various work centers. In this way, the required capacity for each center in each time period is determined. In the usual approach, known as *infinite loading*, capacity constraints are ignored in developing the capacity profile. In *finite loading*, a master schedule is built up that stays within capacity constraints at all work centers. (Finite loading will be discussed in Chapter 16.) Figure 15.8 shows a schematic representation of the required profile, obtained by infinite loading, for a hypothetical center.

Next, the actual capacity through time is estimated. This is typically not equal to the gross capacity scheduled. Instead, an account must be taken of machine breakdowns, time for minor

---

[*] For integrating production scheduling with MRP, see Section 15.8.1, Hastings et al. (1982), and Maxwell et al. (1983).

[†] See for example Schmitt et al. (1984).

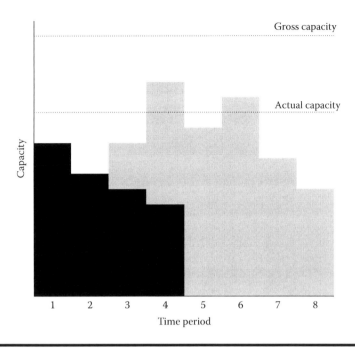

Figure 15.8   Required and actual capacity profiles.

maintenance, delays because of missing parts, and so on. An efficiency factor expressed as

$$\text{Efficiency factor} = \frac{\text{Actual capacity}}{\text{Gross capacity}}$$

can be estimated (and updated) based on recent actual operating experience. Figure 15.8 shows an estimated actual capacity that is level through time.

The next step is to compare required with actual capacity. As an example, consider Figure 15.9 that illustrates the load on the wheel department due to the red car orders. No other end-item orders have been included here; so, the planned orders are directly from Table 15.8 for wheels: 1,280, 920, 400, 1,440, and 200, in weeks 1 through 5. Let us assume for the moment that this department only makes wheels for red cars. The wheel machine, rather than labor, is the constraint in the wheel department, and this machine can make 15 wheels per hour and is operated for two 8-hour shifts, which is 5 days per week (i.e., 80 hours per week). All maintenance is done by a separate maintenance team on the third shift. Therefore, for instance, the 1,280 wheels due to be made in week 1 require 1,280/15 = 85.3 hours while only 2(8)(5) = 80 hours are available. It is clear that in weeks 1 and 4 there will be capacity problems.

For week 4, the solution is quite easy: simply shift some production ahead to week 3. The load on the department is therefore more balanced, and the orders due in week 4 will be available on time. Note, however, that if this component is made from other components, their schedules will have to be modified. For week 1 production cannot be shifted earlier. Some other action is needed, such as working overtime in week 1, delivering a few units at the beginning of week 2, or drawing on SS if any is available. Once the reconciling decisions have been made, the master schedule may be revised and MRP may run again.

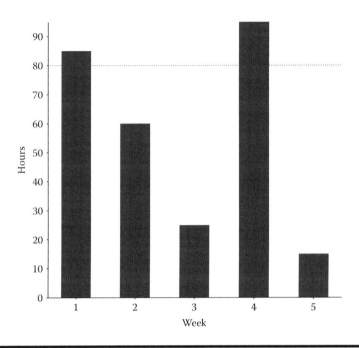

Figure 15.9   Machine hours for red cars for the wheel department.

Other possible courses of action include:

1. Use subcontracting
2. Purchase some processed parts instead of raw materials
3. Transfer personnel or machine time from underutilized areas
4. Use alternate (less desirable) routings for certain orders
5. If all else fails, modify the master schedule itself

15.6 Distribution Requirements Planning

In Chapter 11, we examined multiechelon inventory systems and pointed out the drawbacks of using independent control of the same item at different locations, particularly without information flow between the different echelons. A very natural extension of MRP, that addresses this problem, is known as distribution requirements planning (DRP). See for example Stenger and Cavinato (1979) and Martin (1995).

To describe DRP, let us consider a relatively simple, two-echelon system with a central warehouse (perhaps adjacent to a production facility) and several branch warehouses. Also, to facilitate understanding, our discussion will concentrate on the control of a single product. Rather than using an order-point system, each branch warehouse develops a master schedule—that is, a projected replenishment pattern to satisfy the net requirements at that location. Almost certainly, forecasts of demand will first be needed to establish the gross requirements. The item located at a branch warehouse has a level code of 0 and its bill of materials is extremely simple. Its only component is the same item (level code 1) at the central warehouse. Note that an SKU is defined by both the product and the location, so that the same product at two different locations is recorded as two different SKUs. The offset is the replenishment lead time between the central and the branch warehouse. Once all of the master schedules are established at the branches, the dependent demand at the central warehouse can be calculated. Any projected direct customer demand on the central warehouse can be added to the dependent demand, resulting in gross requirements, and so forth. If there is a linkage back to a manufacturing facility (plant), then the same item at the plant has a level code of 2. Thus, we see that DRP is an extension of MRP into the distribution phase of operations. A company only involved in distribution (i.e., no manufacturing) can still make use of DRP, including a possible linkage back to a (manufacturing) supplier's MRP system.

When one recognizes that replenishment lot sizing at the separate branches should not be done independently, but rather by taking into account of the system-wide stock status, then it is a very natural further step from DRP to the push control system of Section 11.4.3 of Chapter 11 (also, see Brown et al. 1981). Recall that the push system uses information about stock status at all locations in making replenishment decisions.

15.7 Weaknesses of MRP

MRP has been a remarkably powerful and pervasive tool in the industry for several decades. From the previous discussion, it should be clear that firms, facing chaos because of many items with complex bills of materials, can benefit tremendously from the control and order provided by MRP. When demand exhibits significant seasonality, or when demand can surge or decline, MRP's ability to plan ahead provides clear benefits. Successful implementations of MRP are seen in thousands of manufacturing firms. Nevertheless, MRP has some significant drawbacks that should be mentioned. In this section, we highlight four primary weaknesses, and then briefly discuss a number of others.

1. *Lead times.* For the purposes of lead time offsetting, the MRP system is given a lead time by the user for each component and part. MRP takes this lead time to be deterministic. As noted by Karmarkar (1993), from the perspective of MRP, the lead time is an attribute of the part, rather than of the condition of the shop. We know, however, that the lead time for, say, wheels will certainly not always be 1 week, but will vary depending on how busy the people and machines are at that department. If multiple parts are required by that department, or if the batch size is unusually large, the lead time will be longer than 1 week. If the department is not busy, or the batch size is small, the lead time will be shorter than 1 week. How do users adjust for variable lead times? Clearly, the most common solution is to inflate the lead time given to MRP so that orders are rarely delivered late. This implies, of course, that orders are most often delivered early, which increases WIP and finished goods inventory.[*]

2. *Lot sizes.* As we discovered in Section 10.6.2 of Chapter 10, multi-item and multilevel lot sizing is an extremely difficult problem for which optimal solutions are typically not available. Therefore, users must rely on heuristics that may not apply to their situations. In an informal survey of MRP vendors, one of the authors discovered that most MRP systems do not provide extensive support to any lot-sizing rules other than EOQ and LFL. Even if the support is available, most users seem to rely on simple rules that may generate higher cost than is possible by using other rules.

---

[*] The reference by Karmarkar (1993) is an excellent one for understanding lead times. See also Chapter 16; Yano (1987a,b).

3. *SS*. MRP systems, again, do not typically support SS calculations based on formulas such as those in Chapter 6 or 7 of this book. In fact, the user is required to input the desired SS values for each item and component at each stage of production. Because there is little known about the appropriate levels of SS at the component level, users must guess at good values. As in the case of lead times, users may protect against costly stockouts by unnecessarily inflating SS levels.*

4. *Incentives for improvement*. The last primary weakness of MRP is directly related to the previous three. Because of the significant effort required to gather and input data such as SS levels, lot sizes, and lead times, people are reluctant to make regular changes to these values. In fact, as MRP is being installed, it is often desirable to inflate all these values to avoid startup problems. The pitfall arises when MRP is running smoothly, and the firm has gained control over its schedules. There is now little incentive for digging into the system to change values that are working well. The fact that these values reside on a computer only aggravates the problem, because people often assume that numbers given by a computer must be correct!

We continue the list with weaknesses that are somewhat minor.

5. *Data*. Data consistency is a common problem with MRP. Many firms do not have "closed" warehouses, so that nearly any employee can remove the components. One of the authors recently worked with a manufacturer in Germany who faced enormous problems because the inventory levels recorded on the computer did not match the actual levels in the facility. This manufacturer is now putting strict controls on who moves the inventory into and out of stocking locations. If the data are not accurate, MRP's schedules will be of questionable value. Of course, this issue has often been cited as a major benefit of MRP, because the firm must take control of inventories, schedules, and data accuracy.

6. *Design changes*. Often, a new product is changed in small ways once it is in production. Unfortunately, most design changes must be accounted for in the MRP system. This is an added burden, but if neglected, the MRP system will be inaccurate.

7. *Data input*. Every change that must be recorded in MRP, and every new product that must be added to the system, puts a load on data input personnel. In fact, one of the authors worked with a firm in which shop floor workers frequently neglected to input data to the MRP system, creating high levels of frustration in the department responsible for maintaining the system.

8. *Data output*. MRP can generate reams of paper when printing reports. Handling the large amounts of data can be a significant task.

9. *Completed work*. MRP does not recognize completed work until an inventory transaction is made. Gilman (1995) notes that if a customer cancels an order, it is possible that finished units will be available, but will not appear in the system. Along the same lines, when a large batch is partially completed, some units may be available to satisfy some customers' needs. Left alone, MRP does not recognize this availability.

---

* For research on SSs in MRP systems, see Section 15.4.9 and Section 9.7.4 of Chapter 9 as well as New (1975), Miller (1979), Mehta (1980), De Bodt and Van Wassenhove (1983), Lambrecht et al. (1984), Schmidt and Nahmias (1985), Wijngaard and Wortmann (1985), Carlson and Yano (1986), De Bodt et al. (1982), Guerrero et al. (1986), McClelland and Wagner (1988), Yano and Carlson (1988), Graves (1988), Benton (1991), Baker et al. (1993), Legodimos and Anderson (1993), Buzacott and Shanthikumar (1994a), Campbell (1995), Inderfurth and Minner (1998), and Dellaert and Jeunet (2005).

10. *Implementation.* Ptak (1991) notes that implementing Closed-Loop MRP is a complex task that requires political savvy and technical expertise. The results, and the process itself, can be beneficial, but managers should not ignore the difficulties of the implementation process.

11. *Where MRP applies.* Too often, consultants and software developers sell a system that is not applicable to their client's production process. MRP has certainly fallen into this trap. As noted in Chapter 13, MRP applies well in situations of multiple items with complex bills of materials. However, it is not particularly useful, and can even be detrimental, in the upper left and bottom right of the product-process matrix. In the bottom right are continuous processes with relatively simple bills of materials, and tightly linked stages of production. The entire factory operates almost like a single machine. MRP clearly provides excessive capability for exploding the bill of materials and offsetting the lead times. The important issues in this environment have to do with scheduling production where there are significant, sequence-dependent, setups. Methods of Sections 10.6.1 and 10.6.3 of Chapter 10 would be more appropriate. In the upper left of the product-process matrix are job shops in which queueing time is 80%–95% of the time a batch spends in the shop. Lead times are extremely variable, and bottlenecks are difficult to identify because they shift frequently. The bill of materials is often very simple. For example, a single piece of cast metal is processed through multiple metal-cutting stages to create a final product. Therefore, the bill of materials explosion capability of MRP is not necessary; and because MRP uses fixed lead times (weakness #1 noted above), it is inappropriate in these environments. Methods to be discussed in Chapter 16 are more suited to job shops.

Some of these weaknesses, of course, can be mitigated by careful implementation and maintenance of the MRP system, assuming that the production process matches the requirements as discussed in weakness #11. Several advances in production planning and scheduling systems specifically answer some of these weaknesses. For example, the JIT system, to be discussed in the next chapter, focuses on continuous improvement, and therefore is ideal for dealing with weakness #4.[*] In the next section, we introduce a system that also addresses some of these weaknesses. For additional comments on MRP implementation issues, see Anderson and Schroeder (1984) and Petroni (2002).

15.8 ERP Systems

An ERP system can be viewed as a direct extension of MRP. The fundamental difference is that, while MRP is primarily a tool for the production department, under ERP, the entire firm operates from the same data. At this writing, it is still common for ERP systems to run on a single database in a client–server environment although "cloud" implementations are gaining ground. Without going into the details of computing technology, the single database simply means that all functions within the firm draw on, and add to, the same data; the client–server environment means that users have personal computers with functional modules on their desks, while the large database is kept

---

[*] Note however that Axsäter and Rosling (1994) show that MRP systems dominate Kanban systems (the control systems associated with JIT) because MRP systems can replicate Kanban systems. More will be said about this in the next chapter.

centrally. Client–server systems can be expanded reasonably easily at low cost. In this section, we discuss ERP systems and some of the advances in production planning that accompany them.

Because all functions operate with the same data, coordination is facilitated.[*] Imagine that a sales person receives a call from a customer requesting delivery of a batch of a certain product if a given price and delivery schedule can be met. With ERP, the sales person can query the system to check the feasibility of delivering the product at the required time and price. Because this is the same data used in production planning, it should be very accurate for quoting a delivery date; and because it is the data used by the sales department, the price should reflect the firm's current pricing schedule. If the order is accepted, the sales person, or an order entry clerk, can input the order into the system. The system then automatically verifies the delivery schedule, the price, and the credit performance of the customer.

Sales forecasting is supported in ERP by simple models resembling those discussed in Chapter 3, and then these data are used as input to the production-planning process. Production planning, capacity planning, and inventory control in ERP are identical to these functions in MRP (with RCCP and CRP), because ERP employs standard MRP logic. Production and purchase orders are initiated by the system, just as in an MRP system. Some ERP systems have simulation capability so that alternative plans can be examined when capacity limits are exceeded.

As batches are processed in the factory, the ERP system tracks their progress so that production and sales personnel can update the projected completion date. Therefore, if a customer calls a sales representative, the system can be queried and an accurate delivery date can be communicated. Some systems have modules that report many performance evaluation measures, including lead times, capacity utilization, and costs.

Once the batch is ready for shipment, ERP supports shipping activities, including routing, choosing the destinations for the truck, and setting shipment dates. As in all the functions, the logistics person responsible can override the ERP choices if a better alternative arises. ERP even generates a pick-list within the context of its warehouse management system. Financial information is generated by the system as well, so that billing, accounts receivable, and financial planning are all supported by the same data. Unlike an MRP system, therefore, ERP systems are used by managers at all levels, and in all departments.

Our discussion raises some important points about ERP systems. First, they have many of the benefits of MRP systems. They provide a new level of planning and control, this time to multiple functions or processes across the entire firm. MRP brought control to chaotic factories, but its data are isolated from sales, finance, accounting, and other functions. ERP facilitates communication among all these functions because they now operate with the same data. Second, ERP is explicitly a multilocation system, so that global firms do not need to bind together multiple MRP and DRP systems. In fact, some ERP systems accommodate exchange rate adjustments, multiple modes, production planning across multiple sites, and other functions that are critical for global firms.

There are, however, some drawbacks to these systems. First, they face many of the same weaknesses that MRP faces because the production modules are simply composed of standard MRP logic. Second, in spite of the success stories on ERP implementation,[†] the costs of implementation are typically very high, and firms often become so dependent on consultants for implementation,

---

[*] See Whang et al. (1995) for a more complete description.

[†] Edmondson et al. (1996), for example, note how an ERP system by the German company SAP helped in inventory reduction efforts at Nokia, the Finnish cellular phone maker. Stewart (1997) describes the transformation at Owens Corning due, in part, to SAP.

that the payback takes much longer than anticipated. Consulting fees for implementation are typically at least as costly as the software itself, and often are two to three times as high. For example, the cost of software, hardware, consulting, and training at Owens Corning was $110 million (Stewart 1997). For further discussion on factors affecting the success of ERP implementations, see Bingi et al. (1999), Holland and Light (1999), Hong and Kim (2002), Aladwani (2001), and Hendricks et al. (2007). Third, because the databases of these systems are somewhat open, there are potential security problems that must be carefully managed. Finally, although many companies have used an ERP implementation as a catalyst for reengineering, they have found that their reengineering efforts are tied to the technology. Suppose a firm wants to reorganize departments so that manufacturing and sales now operate together. No longer will there be a single manufacturing department and a separate sales department serving multiple customer types. Rather, there will be a series of departments that include both manufacturing and sales personnel, each of which is devoted to a single customer type. ERP implementations can facilitate this reorganization because people, machines, and reporting relationships can all be adjusted while the software is being installed. However, the software is not completely flexible, and many firms find that they are limited in their reorganization options by the software. Other firms try to install the software without addressing fundamental problems in their business processes and find that they rarely achieve the promised benefits. Therefore, it is important to view the implementation as a combined software/reengineering project.

One final note on implementation: When a global firm pursues a large ERP project, the implementation is often done in phases. For example, Gillette's European subsidiary, Braun, implemented SAP before Gillette in the United States did. Because the software requires specifying product data and business processes consistently across the entire corporation, Braun's specifications limited the choices Gillette could make. Fortunately, Gillette was involved in the Braun implementation so that their own implementation went reasonably smoothly.

15.8.1 Enhancements to ERP Systems

Because basic ERP systems typically rely on simple MRP logic, a number of software firms have developed advanced production planning and scheduling programs that run with ERP. In recent years, some of these firms have been acquired and integrated into ERP systems. The two largest ERP providers, SAP and Oracle, each offer advanced planning capabilities with their ERP systems. These systems often employ finite scheduling tools that ensure that capacity limits are not exceeded. These tools can schedule backward from an order's due date, after first netting out the finished goods, WIP, and raw material inventory, to determine the latest possible start time for the order. They can also schedule forward from the current date (or the availability of raw materials) to determine the earliest possible start time. This latter time is called a "hard" constraint, because nothing can be built without materials or started before the current time. The latest possible start time, on the other hand, is not a hard constraint because orders can be delivered late. Rather than simply identifying constraints, as MRP does, an ERP system with advanced planning capabilities will suggest solutions. Again, because this happens in real time, the user can see what ripple effects will occur if an order is inserted into the schedule, or if some other disruption occurs. Perhaps, a proposed order is from a preferred customer, and therefore is vital to the firm. Managers can see immediately what other orders will be delayed, and by how much, if this order is accepted. Chapter 16 addresses these techniques in more detail.

These systems address some of the weaknesses of ERP, but they are subject to problems of their own. A challenge that carries over from our discussion in Section 15.7 is that of maintaining data integrity. Dreibelbis et al. (2008) and Loshin (2010) provide thorough treatments of issues and challenges associated with maintaining master data in ERP systems. Vosburg and Kumar (2001) and Knolmayer and Röthlin (2006) present master data management case studies.

In addition to master data challenges, the capability of making real-time changes to the schedule can create a whole new level of nervousness in the production plan; firms must enforce discipline in making changes so that new orders are not inserted too frequently. Otherwise, it will be difficult for production personnel, not accurately knowing what orders are coming next, to plan ahead effectively.

15.9 Summary

In this chapter, we have concentrated on production/inventory planning and control in assembly situations. The emphasis has been on Closed-Loop MRP. Such a system is not optimal in a mathematical sense, but, when compared with the more traditional replenishment systems, it typically provides substantial benefits to the management, in terms of excellent summarized information, in a complex multistage, manufacturing environment. Perhaps, the major benefit of MRP in a complex manufacturing setting is the ability to more quickly adapt to a rapidly changing environment. This is partly accomplished through the use of the logic in a simulation mode to provide answers to "what-if" questions.

Implementation difficulties must be recognized. In particular, organizations have not given enough emphasis to the input of reliable information nor to the maintenance of a stable MPS. This is probably the result, in large part, of inadequate education and training of management, supervisors, and other staff, concerning the nature and proper operation of MRP systems. This challenge extends on a broader scale to ERP implementations.

Problems

15.1 For the product structure sketched in Figure 15.10, which of the items could be subject to independent demand? Which to dependent demand?

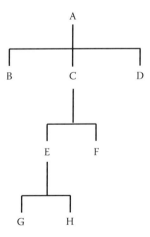

Figure 15.10 Product structure for Problem 15.1.

15.2 A company assembles three distinct finished SKUs, items Fl, F2, and F3. The bills of materials for these items, their components, and quantities are as follows:

Item F1: Composed of 1 unit of A1, A2, and A3 and 2 units of A4
Item F2: Composed of 1 unit of Al, A2, and A4
Item F3: Composed of 1 unit of A1 and A4
Item A1: Composed of 1 unit of A2 and B1
Item A2: Composed of 1 unit of A4, B1, and B2
Item A3: Composed of 1 unit of B1 and B2 and 3 units of B3
Item A4: Composed of 2 units of B3
Item B1: Composed of 1 unit of B3 and C1
Item B2: Composed of 1 unit of C1 and C2
Item B3: Composed of 1 unit of C1, C2, and C3

Items C1, C2, and C3 are purchased parts. In addition, items A1 and A2 are sold directly to customers as spare parts.

Develop a level-by-level coding for the 13 items.

15.3 Suppose that the offsets for the operations involved in Problem 15.2 are as follows:

Assembly, Fabrication, or Purchase of Item	Offset (Weeks)
F1	1
F2	2
F3	1
A1	1
A2	2
A3	2
A4	3
B1	1
B2	3
B3	1
C1	2
C2	2
C3	4

Considering only F1 and its components,
a. What minimum horizon should be used for the MPS?
b. If the only customer order is for 10 units of item F1 in week 15 and there is no initial OH or on-order stock for any items, establish the procurement and production schedule for item F1 and all its components.

15.4 If an MPS is overstated (i.e., not realizable), briefly discuss which of the following will or will not result:

a. A material plan inconsistent with the MPS will be developed.

b. Inappropriate priorities will result on shop/vendor orders.

c. Overstated capacity requirements will be set.

d. Excessive component part inventories will be developed.

e. It will be easier to cope with large, unexpected orders.

15.5 Eastern Telecom produces switching units for telecommunication systems. They use MRP and an analyst, Maxine Schultz, is responsible for the lot sizing of a class of SKUs including item C703. Based on earlier material planning, there is an OH inventory of 230 units of item C703 at the start of week 1, a firm order of 200 units to be delivered in week 3, and a tentative order of 290 units scheduled for delivery in week 6. This and other relevant information is portrayed in Table 15.9.

a. Compute the projected net inventory at the end of each week as well as the cumulative net requirements.

b. What is the impact on net inventory and cumulative net requirements of moving the delivery of the tentative order into week 4?

c. How does the discussion in Section 5.6.9 of Chapter 5 relate to this type of action?

15.6 Consider a particular work center A in a manufacturing environment. The gross capacity per week is 120 hours with an efficiency factor of 0.85. The projected workload for weeks 6, 7, and 8 is as shown in Table 15.10.

a. How would the entries in the "Standard Hours Required" column be obtained?

b. Sketch a graphic representation of the workload at the center.

c. What change(s) would you recommend in the schedule of orders?

15.7 An inventory control manager at a manufacturing plant described his method of separating parts for a particular end-item into two physically separate storage areas: one for assembly parts, the other for spare parts. His last comment was "I've often wondered whether there was any way of combining the inventory for both types of parts so that there would be enough safety stock to meet emergencies from either of the two sources." What suggestions would you have in this regard? Include a discussion of some of the less tangible factors.

Table 15.9 Material Plan for Item C703

Lead time = 1 period								
Order quantity = not fixed								
Time Period	*0*	*1*	*2*	*3*	*4*	*5*	*6*	*7*
Projected gross requirements	–	150	120	130	180	160	90	100
Planned order receipts	–	–	–	200	–	–	290	–
Projected net inventory at the end of period	230							

Table 15.10   Workload at Center A

Week	Order No.	Standard Hours Required
6	315	41.2
	317	18.8
	314	22.5
7	322	51.5
	318	47.6
	313	27.7
	320	15.8
8	327	43.5
	325	16.7
	326	15.4

15.8   Review the operations of a local manufacturing firm and briefly discuss the applicability of MRP.

15.9   Develop a spreadsheet to compute the MRP records illustrated in Table 15.8.

15.10   Consider a company that produces a number of items where each end-item is made up of a number of components.

   a.   In such a situation, what are the advantages of keeping SS in raw materials, as opposed to finished goods?

   b.   What conditions are necessary to permit keeping the SS only in raw materials?

15.11   In an MRP context, suppose that the following are properties of the system:

   a.   Production of product i is done in batches. Production of a batch is commenced only at the beginning of a time period. The time to complete a batch of product i is k_i time periods where k_i is always a positive integer.

   b.   The requirements of an item i in terms of its immediate component items are given by a bill of materials vector

$$\underline{b}_i = [b_{i1}, b_{i2}, \ldots, b_{iN}]$$

where b_{im} = number of units of item m that are required as direct components of item i, and N is the total number of items in the inventory (i.e., many of the b_{im}'s will be zero).

   c.   Production of any item (component or finished) is done only to exactly match demand, that is, no attempt is made to save on setup costs or achieve economies of long runs.

   Consider the following numerical example with four items:

   Item 1—used only as a finished product
   Item 2—finished product and two components of item 1 (i.e., each unit of item 1 requires two units of item 2 as components).

Item 3—single component of item 1 and three components of item 2
Item 4—single component of item 2 and two components of item 3

Item	Initial Inventory (at Time 0)	Production (or Assembly) Time for Batch (Periods)	Finished Product Requirements in Period			
			8	9	10	11
1	2	2	10	9	7	10
2	4	1	5	15	0	6
3	3	3	–	–	–	–
4	5	1	–	–	–	–

a. Develop the production schedule assuming that the company wishes to have no inventory OH at the end of period 11.
b. Attempt to use the mathematical notation shown earlier to write your procedure in algorithmic form (i.e., in a form that could be easily coded for computer usage).
c. What real-life complexities have been neglected in this problem description?

15.12 Consider a company that is producing two types of style goods items, call them A and B. Each is composed of components as follows:

A—made up of 2 units of C, 1 unit of D, and 1 unit of E.
B—made up of 1 unit of C, 3 units of D, and 1 unit of E.

The final assembly operation of A or B is irreversible, that is, an A cannot be disassembled to use the components to help make up a B unit or vice versa. The assembly operation takes a negligible amount of time. Such is not the case with the preparation of the component units (the C's, D's, and E's). Thus, the company's strategy is to prepare all of the C's, D's, and E's prior to the selling period but to not assemble them until later. The selling period is short—all demands are backordered during it and sales (when demand can be satisfied) are made at the end of the period. It is company policy to fill as many A requests as possible (by assembling available components) and then use the remaining components to satisfy as much B demand as possible. Unused components are sold as scrap.
a. Indicate, introducing whatever symbols you feel are appropriate, how the company should establish the quantities of C, D, and E to stock.
b. Suppose, rather than being style goods, A and B had continuing demands from period to period. Why would this appreciably complicate the analysis?

15.13 Consider the situation of a company manufacturing a seasonal item with one optional component. There are n different possible components but only one is used in each finished item. Before a particular season, the company knows, with certainty, that the total number of finished items that will be demanded is N; however, they don't know precisely which options will be required. From historical data, the probability that optional component i (with unit variable cost, v_i) is needed in any particular item is known at the value p_i where $\sum_{i=1}^{n} p_i$.

a. What is the probability distribution of the total number of units of component i needed for the N finished items? What are its mean and standard deviation?

b. Using a continuous approximation to the distribution of part 15.13a, ascertain a decision rule for allocating a total stock of W among the n optional components so as to minimize the total expected number of finished units (out of the N demanded) that can't be satisfied because of a missing optional component.

c. Illustrate for the following example:

$$W = \$14{,}000 \quad N = 300$$

Option, i	v_i ($/Unit)	p_i
1	30	0.5
2	40	0.2
3	60	0.3

15.14 Consider the part explosion diagram shown in Figure 15.11. Four distinct parts exist and are separately numbered in the upper half of each diagram node. The diagram has three levels. The lower half of each part consists of two pieces. The lower-left portion indicates the number of parts required at the next higher assembly level; the lower-right portion indicates whether the part is to be made (M) or purchased (P). Required lead times are shown on the links of the diagram. For example, orders for Part D must be initiated 5 weeks before they are required for use in building Part C. Orders for Part A must be

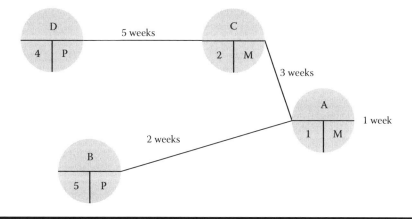

Figure 15.11   Part explosion diagram for Problem 15.14.

initiated 1 week before they are required by the master schedule. The following data apply for each of the parts:

Part	A	B	C	D
Setup cost/order	$50	–	$50	–
Purchase cost/order	–	$50	–	$50
Carrying cost/unit-year	20	15	10	5

The annual demand for Part A is 1,500 units. The current planning horizon is 20 weeks and has resulted in the master schedule below:

Period	1	2	3	4	5	6	7	8	9	10	11	12	13	14	15	16	17	18	19	20
Demand	–	–	–	–	–	–	–	–	–	–	–	–	–	10	20	30	40	30	20	10

Time periods are in weeks, and at the present time, it is the beginning of week 1. There are 50 work weeks per year. All present stock balances are zero.

a. Develop an MRP schedule for Part A using LFL. Show gross requirements, scheduled receipts, planned order receipts, inventory at the end of the period, net requirements, and planned order releases.

b. Assuming LFL orders in part 15.14a, develop an MRP schedule for Part B, using EOQ lot sizing.

15.15 For item F2 in Problem 15.2, there are customer orders for 20 units in week 18. Considering this item only, develop the purchase and production schedule for F2 and all its components, assuming that no inventory OH is available.

15.16 For item F3 in Problem 15.2, there are customer orders for 25 units in week 17. Considering this item only, develop the purchase and production schedule for F3 and all its components, assuming that 8 units of inventory of F3, 3 units of A4, and 5 units of B3 are available.

15.17 Now combine Problems 15.2, 15.15, and 15.16; that is, assume that the demands listed in those problems arise, and the inventory listed in Problem 15.16 is available. Develop the purchase and production schedule for F1, F2, and F3, and all their components.

References

Aladwani, A. M. 2001. Change management strategies for successful ERP implementation. *Business Process Management Journal* 7(3), 266–275.

Anderson, J. C. and R. G. Schroeder. 1984. Getting results from your MRP system. *Business Horizons* 27(3), 57–64.

Axsäter, S. and K. Rosling. 1994. Multi-level production-inventory control: Material requirements planning or reorder point policies? *European Journal of Operational Research* 75, 405–412.

Baker, K. R., G. L. Nemhauser, and A. H. G. Rinnooy Kan. 1993. Requirements planning. In S. C. Graves, A. H. G. Rinnooy Kan, and P. H. Zipkin (Eds.), *Logistics of Production and Inventory*, Volume 4, pp. Chapter 11. Amsterdam: North-Holland.

Bangash, A., R. Bollapragada, R. Klein, N. Raman, H. B. Shulman, and D. R. Smith. 2004. Inventory requirements planning at lucent technologies. *Interfaces 34*(5), 342–352.

Benton, W. C. 1991. Safety stock and service levels in periodic review inventory systems. *Journal of the Operational Research Society 42*(12), 1087–1095.

Bingi, P., M. K. Sharma, and J. K. Godla. 1999. Critical issues affecting an ERP implementation. *IS Management 16*(3), 7–14.

Blackburn, J., D. Kropp, and R. Millen. 1986. Comparison of strategies to dampen nervousness in MRP systems. *Management Science 32*(4), 413–429.

Brown, G. G., A. Geoffrion, and G. Bradley. 1981. Production and sales planning with limited shared tooling at the key operation. *Management Science 27*(3), 247–259.

Buzacott, J. and J. Shanthikumar. 1994a. Safety stock versus safety time in MRP controlled production systems. *Management Science 40*(12), 1678–1689.

Buzacott, J. A. and J. G. Shanthikumar. 1994b. Safety stock versus safety time in MRP controlled production systems. *Management Science 40*(12), 1678–1689.

Campbell, G. M. 1995. Establishing safety stocks for master production schedules. *Production Planning and Control 6*(5), 404–412.

Carlson, R., J. Jucker, and D. Kropp. 1979. Less nervous MRP systems: A dynamic economic lot-sizing approach. *Management Science 25*(8), 754–761.

Carlson, R. C. and C. A. Yano. 1986. Safety stocks in MRP-systems with emergency setups for components. *Management Science 32*(4), 403–412.

Danese, P. and P. Romano. 2005. Finn-Power Italia develops and implements a method to cope with high product variety and frequent modifications. *Interfaces 35*(6), 449–459.

D'Avino, M., V. De Simone, and M. M. Schiraldi. 2014. Revised MRP for reducing inventory level and smoothing order releases: A case in manufacturing industry. *Production Planning and Control 25*(10), 814–820.

De Bodt, M. and L. Van Wassenhove. 1983. Lot sizes and safety stocks in MRP: A case study. *Production and Inventory Management Journal 24*(1), 1–16.

De Bodt, M. A., L. N. Van Wassenhove, and L. F. Gelders. 1982. Lot sizing and safety stock decisions in an MRP-system with demand uncertainty. *Engineering Costs and Production Economics 6*, 67–75.

Dellaert, N. and J. Jeunet. 2005. An alternative to safety stock policies for multi-level rolling schedule MRP problems. *European Journal of Operational Research 163*(3), 751–768.

DeMatta, R. and M. Guignard. 1995. The performance of rolling production schedules in a process industry. *IIE Transactions 27*, 564–573.

Dreibelbis, A., E. Hechler, I. Milman, M. Oberhofer, P. van Run, and D. Wolfson. 2008. *Enterprise Master Data Management: An SOA Approach to Managing Core Information*. Boston: Pearson Education.

Edmondson, G., P. Elstrom, and P. Burrows. 1996. At Nokia, a comeback—And then some. *Business Week December 2*, 106.

Ernst, R. and D. Pyke. 1992. Component part stocking policies. *Naval Research Logistics 39*, 509–529.

Friend, C. H. and A. A. Ghobbar. 1999. Extending visual basic for applications to MRP: Low budget spreadsheet alternatives in aircraft maintenance. *Production and Inventory Management Journal 40*(4), 9.

Gilman, A. 1995. MRP-4U? *APICS—The Performance Advantage March 5*, 40–42.

Graves, S. C. 1988. Safety stocks in manufacturing systems. *Journal of Manufacturing and Operations Management 1*(1), 67–101.

Guerrero, H. H., K. R. Baker, and M. H. Southard. 1986. The dynamics of hedging the master schedule. *International Journal of Production Research 24*(6), 1475–1483.

Guide Jr, V. and R. Srivastava. 2000. A review of techniques for buffering against uncertainty with MRP systems. *Production Planning and Control 11*(3), 223–233.

Gupta, S. M. and L. Brennan. 1995. MRP systems under supply and process uncertainty in an integrated shop floor control environment. *International Journal of Production Research 33*(1), 205–220.

Hastings, N., P. Marshall, and R. Willis. 1982. Schedule based MRP: An integrated approach to production scheduling and material requirements planning. *Journal of the Operational Research Society 33*(11), 1021–1029.

Hendricks, K. B., V. R. Singhal, and J. K. Stratman. 2007. The impact of enterprise systems on corporate performance: A study of ERP, SCM, and CRM system implementations. *Journal of Operations Management 25*(1), 65–82.

Ho, C. J. and P. L. Carter. 1996. An investigation of alternative dampening procedures to cope with MRP system nervousness. *International Journal of Production Research 34*(1), 137–156.

Ho, C.-J., W.-K. Law, and R. Rampal. 1995. Uncertainty-dampening methods for reducing MRP system nervousness. *International Journal of Production Research 33*(2), 483–496.

Holland, C. P. and B. Light. 1999. A critical success factors model for ERP implementation. *IEEE Software 16*(3), 30–36.

Hong, K.-K. and Y.-G. Kim. 2002. The critical success factors for ERP implementation: An organizational fit perspective. *Information and Management 40*(1), 25–40.

Hopp, W. J. and M. L. Spearman. 2011. *Factory Physics*. Long Grove, IL: Waveland Press.

Inderfurth, K. and S. Minner. 1998. Safety stocks in multi-stage inventory systems under different service measures. *European Journal of Operational Research 106*(1), 57–73.

Karmarkar, U. S. 1993. Manufacturing lead times, order release and capacity loading. In S. Graves, A. Rinnooy Kan, and P. Zipkin (Eds.), *Logistics of Production and Inventory*, Volume 4, pp. Chapter 6. Amsterdam: Elsevier (North-Holland).

Knolmayer, G. F. and M. Röthlin. 2006. Quality of material master data and its effect on the usefulness of distributed ERP systems. In *International Conference on Conceptual Modeling*, pp. 362–371. Berlin Heidelberg: Springer.

Koh, S., S. Saad, and M. Jones. 2002. Uncertainty under MRP-planned manufacture: Review and categorization. *International Journal of Production Research 40*(10), 2399–2421.

Kropp, D., R. Carlson, and J. Jucker. 1983. Heuristic lot-sizing approaches for dealing with MRP system nervousness. *Decision Sciences 14*(2), 156–169.

Kropp, D. and R. C. Carlson. 1984. A lot-sizing algorithm for reducing nervousness in MRP systems. *Management Science 20*(2), 240–244.

Kurtulus, I. and D. W. Pentico. 1988. Materials requirements planning when there is scrap loss. *Production and Inventory Management Journal 29*(2), 18–21.

Lambrecht, M., R. Luyten, and J. Vander Eecken. 1984. Protective inventories and bottlenecks in production systems. *European Journal of Operational Research 22*, 319–328.

Lambrecht, M., J. Muckstadt, and R. Luyten. 1984. Protective stocks in multi-stage production systems. *International Journal of Production Research 22*(6), 1001–1025.

Lambrecht, M. R. and F. Van den Wijngaert. 1985. A microcomputer system for MRP and scheduling in Essochem. *Production and Inventory Management Journal 26*(4), 59–69.

Landvater, D. and C. Gray. 1989. *MRP II Standard System*. Essex Junction, VT: Oliver Wight Limited Publications, Inc.

Legodimos, A. G. and E. J. Anderson. 1993. Optimal positioning of safety stocks in MRP. *International Journal of Production Research 31*(8), 1797–1813.

Loshin, D. 2010. *Master Data Management*. Amsterdam: Morgan Kaufmann.

Lowerre Jr., W. M. 1985. Protective scheduling smoothes jittery MRP plans: Buffer forecast error the key. *Production and Inventory Management Journal 26*(1), 1–21.

Martin, A. J. 1995. *DRP: Distribution Resource Planning: The Gateway to True Quick Response and Continuous Replenishment*. New York: John Wiley & Sons.

Maxwell, W., J. Muckstadt, L. J. Thomas, and J. VanderEecken. 1983. A modeling framework for planning and control of production in discrete parts manufacturing and assembly systems. *Interfaces 13*(6), 92–104.

McClelland, M. K. and H. M. Wagner. 1988. Location of inventories in an MRP environment. *Decision Sciences 19*(3), 535–553.

Mehta, N. 1980. How to handle safety stock in an MRP system. *Production and Inventory Management Journal 21*(3), 16–21.

Metters, R. D. 1993. A method for achieving better customer service, lower costs, and less instability in master production schedules. *Production and Inventory Management Journal 34*(4), 61–65.

Miller, J. G. 1979. Hedging the master schedule. In L. P. Ritzman, L. J. Krajewski, W. L. Berry, S. H. Goodman, S. T. Hardy, and L. D. Vitt (Eds.), *Disaggregation Problems in Manufacturing and Service Organizations*, pp. 237–256. Hingham, MA: Martinas Nijhoff.

New, C. 1974. Lot-sizing in multi-level requirements planning systems. *Production and Inventory Management Journal 15*(4), 57–71.

New, C. 1975. Safety stocks for requirements planning. *Production and Inventory Management Journal 16*(2), 1–18.

Orlicky, J. A. 1975. *Material Requirements Planning*. New York: McGraw-Hill.

Petroni, A. 2002. Critical factors of MRP implementation in small and medium-sized firms. *International Journal of Operations and Production Management 22*(3), 329–348.

Ptak, C. and C. Smith. 2011. *Orlicky's Material Requirements Planning 3/E*. New York: McGraw Hill Professional.

Ptak, C. A. 1991. MRP, MRP II, OPT, JIT, and CIM—Succession, evolution, or necessary combination. *Production and Inventory Management Journal 32*(2), 7–11.

Schmidt, C. P. and S. Nahmias. 1985. Optimal policy for a two-stage assembly system under random demand. *Operations Research 33*(5), 1130–1145.

Schmitt, T. G., W. L. Berry, and T. E. Vollmann. 1984. An analysis of capacity planning procedures for a material requirements planning system. *Decision Sciences 15*, 522–541.

Shore, H. 1995. Fitting a distribution by the first two moments (partial and complete). *Computational Statistics and Data Analysis 19*, 563–577.

Sounderpandian, J. 1994. MRP on spreadsheets: An update. *Production and Inventory Management Journal 35*(3), 60–64.

Sridharan, V. and L. Berry. 1990a. Freezing the master production schedule under demand uncertainty. *Decision Sciences 21*(1), 97–120.

Sridharan, V. and W. L. Berry. 1990b. Master production scheduling make-to-stock products: A framework for analysis. *International Journal of Production Research 28*(3), 541–558.

Sridharan, V., W. L. Berry, and V. Udayabhanu. 1987. Freezing the master production schedule under rolling planning horizons. *Management Science 33*(9), 1137–1149.

Stenger, A. and J. Cavinato. 1979. Adapting MRP to the outbound side-distribution requirements planning. *Production and Inventory Management Journal 20*(4), 1–14.

Stewart, T. A. 1997. Owens Corning: Back from the dead. *Fortune May 26*, 118–126.

Tardif, V. 1995. *Detecting Scheduling Infeasibilities in Multi-Stage, Finite Capacity, Production Environments*. PhD thesis. Northwestern University, Evanston, IL.

Toklu, B. and J. M. Wilson. 1995. An analysis of multi-level lot-sizing problems with a bottleneck under a rolling schedule environment. *International Journal of Production Research 33*(7), 1835–1847.

Vargas, V. A. 1994. *The Stochastic Version of the Wagner–Whitin Production Lot-Size Model*. PhD thesis. Emory University, Atlanta, GA.

Vargas, V. A. and R. Metters. 1996. Adapting lot-sizing techniques to stochastic demand through production scheduling policy. *IIE Transactions 28*, 141–148.

Vollmann, T. E., W. L. Berry, D. C. Whybark, and F. R. Jacobs. 2005. *Manufacturing Planning and Control for Supply Chain Management*. New York: McGraw-Hill/Irwin.

Vosburg, J. and A. Kumar. 2001. Managing dirty data in organizations using ERP: Lessons from a case study. *Industrial Management and Data Systems 101*(1), 21–31.

Whang, S., W. Gilland, and H. Lee. 1995. *Information Flows in Manufacturing under SAP R/3*. PhD thesis, Stanford, CA.

Whybark, D. C. and J. C. Williams. 1976. Material requirements planning under uncertainty. *Decision Sciences 7*(4), 595–606.

Wight, O. 1995. *Manufacturing Resource Planning: MRP II: Unlocking America's Productivity Potential*. New York: John Wiley & Sons.

Wijngaard, J. and J. C. Wortmann. 1985. MRP and inventories. *European Journal of Operational Research 20*, 281–293.

Yano, C. A. 1987a. Setting planned lead times in serial production systems with tardiness costs. *Management Science 33*, 95–106.

Yano, C. A. 1987b. Stochastic leadtimes in two-level assembly systems. *IIE Transactions 19*(4), 371–378.

Yano, C. A. and R. C. Carlson. 1988. Safety stocks for assembly systems with fixed production intervals. *Journal of Manufacturing and Operations Management 1*(2), 182–201.

Zhao, X., J. C. Goodale, and T. S. Lee. 1995. Lot-sizing rules and freezing the master production schedule in material requirements planning systems under demand uncertainty. *International Journal of Production Research 33*(8), 2241–2276.

Zolnick, S. 1996. The manufacturing game: Thingamajigs teach MRP. *APICS—The Performance Advantage 6*, October, 64–70.

Chapter 16

Just-in-Time, Optimized Production Technology and Short-Range Production Scheduling

In the last chapter, we introduced MRP as a production planning system appropriate for assembly environments in which dependent demand may be erratic. We argued that MRP is overkill for continuous process industries because of the relatively simple bills of materials and the closely linked production stages. Even in discrete parts manufacturing, when demand is high and relatively stable, MRP is not appropriate because it is specifically designed to look ahead and plan for periods of high or low demand. Likewise, MRP does not explicitly account for bottlenecks in production because it assumes infinite capacity. In this chapter, we discuss two production planning and scheduling systems, Just-in-Time (JIT) and Optimized Production Technology (OPT®), that answer these limitations of MRP. We then introduce tools for decisions made at the level of the individual operator or machine. These are called short-range production scheduling tools and are particularly valuable for use in conjunction with OPT and in job shops.

JIT is a very popular system that is most appropriate in high-volume repetitive manufacturing. OPT was designed for manufacturing environments that have one or more severe bottlenecks. In both cases, however, the systems go far beyond production planning and scheduling to include broad statements of the philosophy of manufacturing, and even of business strategy.[*] Short-range scheduling applies at the level of the individual work center, where operators and supervisors need to know which of a set of jobs in queue, or soon to be in queue, to process next.

In Section 16.1, we discuss the JIT production system, including the philosophy and the production scheduling scheme, and we examine its strengths and weaknesses. Then, in Section 16.2,

---

[*] See Golhar and Stamm (1991) for a bibliography of more than 200 articles on JIT and OPT and White et al. (1999) for a survey and discussion of JIT adoption among manufacturers. See also Ohno (1988).

we focus on OPT and the management of bottlenecks. We discuss a series of rules for bottleneck management, and we introduce a related system that strives for constant work-in-process inventory (CONWIP). In Section 16.3, we discuss a number of tools and insights for short-range production scheduling.

16.1 Production Planning and Scheduling in Repetitive Situations: Just-in-Time

The JIT system was first introduced by the Toyota Motor Corporation. JIT is one of the primary pillars of a total manufacturing system often called the Toyota Production System (TPS), "lean manufacturing," or more simply "lean" (see Womack and Jones 2010). The lean system rigorously focuses on the elimination of waste in manufacturing. As such, it encompasses product design, equipment selection, materials management, quality assurance, line layout, job design, and productivity improvement. Inventories are reduced as much as possible to increase productivity, to improve quality, and to reduce production lead times (and hence, customer response times). The TPS was developed over a period of 20 years and led to higher quality, lower cost, and substantially less labor time per vehicle than was achieved by Toyota's international competitors. This success defied conventional wisdom, which assumed that the much larger plants of, say, the American auto assemblers, should have exhibited the cost benefits of economies of scale. By the mid-1980s, large numbers of major international manufacturing firms had developed a strong interest in understanding and replicating Toyota's system.[*]

The production/material control system associated with JIT is called the Kanban system. In this section, we discuss the broader philosophical aspects of JIT and then describe the Kanban system in detail. We then examine the strengths and weaknesses of JIT. For other general descriptions of JIT, see Bowman (1991), Groenevelt et al. (1993), and Vollmann et al. (2005).

16.1.1 Philosophy of JIT

The JIT philosophy directly addresses some of the fundamental weaknesses of MRP. Specifically, the goal of JIT is to *remove all waste* from the manufacturing environment, so that the right quantity of products is produced with the highest quality, at exactly the right time (neither late nor early), with zero inventory, zero lead time, and no queues. "Waste" means anything—inventory, disruptions, or poor quality, for example—that impairs the firm from attaining these goals. In fact, any activity that disrupts the flow of products and does not contribute to making or selling them is waste. Common examples are counting, scheduling, moving, and sorting. In addition to eliminating waste, JIT seeks to eliminate all uncertainty, including machine breakdowns. Manufacturing is to be flexible, but in a flow process. For these ambitious goals to be achieved, the firm must pursue continuous improvement, or *kaizen*, as it is called in Japan (see Imai 1997). Therefore,

---

[*] In an interesting study, Lieberman and Demeester (1999b) trace the diffusion of JIT production in the Japanese automotive sector, as reflected by inventory reductions in a sample of 52 suppliers and assemblers. They show that most inventory reductions occurred during a remarkable burst of activity starting in the late 1960s. Companies affiliated with Toyota were the early adopters but were followed very quickly by others in Japan. By the late 1970s, nearly all of the firms in the sample had made drastic reductions in inventory. Work-in-process and suppliers' finished goods fell by nearly two-thirds on average.

JIT is *dynamic*, rather than *static*.[*] We argued in Chapter 15 that MRP tends to be static because once certain numbers are entered into the computer, no one has responsibility to change them. JIT works in exactly the opposite way. As soon as improvements are made, new improvements are sought.

In working toward these goals, particularly that of zero inventory, an organization follows a set of systematic steps (see also Bielecki and Kumar 1988 and Piper and McLachlin 1990):

1. The need for high levels of quality (both in-house and with suppliers) is given extreme emphasis. (See Mefford 1989.)
2. Setup or changeover times are reduced as much as possible.[†]
3. Lead times are reduced as much as possible.
4. Lot sizes are reduced as much as possible and standardized. Very low setup times permit the economical production of small lot sizes, and increase the utilization of equipment.[‡]
5. Work-in-process (WIP) inventory is removed from the stockroom and put on the factory floor, where it is visible. Therefore, WIP levels are immediately evident, and can be counted quickly.
6. Once the plant is in reasonable balance (in terms of stable, roughly equal workloads at the different work centers), inventory is systematically reduced.[§] Each reduction usually leads to the identification of a problem (bottleneck) area. When the problem is discovered, inventory is temporarily increased so that production will continue smoothly until the problem is solved.
7. The problem is resolved in a cost-effective fashion (e.g., by procedural changes and equipment adjustments).

Steps 6 and 7 are repeated over and over until no further improvements can be realized or until zero inventory is achieved between pairs of work centers. In the latter case, the possibility of automated piece-by-piece transfer between the two centers is considered.

A useful analogy to the above philosophy is to equate inventories to water in a river. When the water is at a high level, dangerous rocks (obstacles to higher productivity) below the surface are concealed. As the water (inventory) level is lowered, the rocks are identified and removed, thus keeping the waterway open (improving productivity) at a lower water (inventory) level. Contrast this to MRP, which often has been implemented with high inventories to mask problems. Then there is little incentive to change an effective operating system. Recall from Figure 6.9 that it is often more beneficial to "change the givens" of the inventory system than to optimize the amount of inventory given certain parameters. In other words, if a firm can reduce, say, the fixed order cost (A) in the economic order quantity (EOQ), it will lower the entire total relevant cost curve. Even significant departures from the optimal order quantity with this new A value are less expensive than the optimal order quantity with the original A value. See Silver (1992) for several numerical examples.

---

[*] See Jaikumer and Bohn (1992) for an interesting perspective on the dynamic approach.
[†] See Porteus (1986), Hay (1987), Spence and Porteus (1987), Severson (1988), Fine and Porteus (1989), Freeland et al. (1990), MacDuffie et al. (1996), Jain et al. (1996), Leschke (1996), Samaddar and Kaul (1995), and the references in Section 4.1.
[‡] For an interesting discussion on the relationship of setup times to lot sizes in a JIT environment, see Zangwill (1987, 1992), Gerchak and Wang (1994), Duenyas (1994), and Samaddar (1996).
[§] See Sarker and Harris (1988) and Miltenburg and Sinnamon (1989) for research on balanced workload.

There are impressive benefits to be gained through the adoption of the JIT philosophy, which we will discuss in Section 16.1.3. It is important to note, however, that, as with any successful development in manufacturing, there is a tendency to inappropriately ascribe all sorts of benefits to JIT. There is no reason, for instance, that this same philosophy cannot apply in an MRP environment. Certainly, the fact that stock is on the factory floor makes it easier to assign responsibility for inventory reduction, but excellent firms have been able to achieve the same goals in other environments. We will have more to say about this issue in Section 16.1.3.[*]

16.1.2 Kanban Control System

JIT production is appropriate in a high-volume, repetitive manufacturing environment. The different stages of production are tightly linked with very little in-process inventories. Final assembly needs dictate the inflow of subassemblies, triggering the production of new subassemblies, and so on. Each feeding work center produces only what its following (or consuming) work center uses to satisfy the assembly schedule. The associated manual information system is known as Kanban (the Japanese word for a *card* or *signboard*). The amount of in-process inventory between any two work centers is strictly (and easily) controllable by the number of cards assigned to that particular pair of centers. Kanban therefore operates in a pull mode, whereas MRP operates in a push mode. "Pull" implies that production is initiated at a given work center only when its output is needed (pulled) by the next stage of production. If the next stage is slow for some reason, the upstream center does not produce. "Push," on the other hand, implies that the work center produces to a forecast, regardless of whether the parts are needed immediately or not. (See Section 11.4 for further discussion of push and pull systems.)

We next outline the requirements for Kanban control, and then we provide a more detailed description of the Kanban system. For other comments on Kanban systems, see Sugimori et al. (1977), Kimura and Terada (1981), Huang et al. (1983), Monden (1984), Seidmann (1988), Karmarkar and Kekre (1989), and Karmarkar (1993).

16.1.2.1 Requirements for Kanban Control

Kanban control is not appropriate in many environments, even if the JIT philosophy is. For Kanban to be truly effective, there are several conditions that must first exist or be developed. These include

1. Employee motivation, and mutual trust between workers and management. These qualities are extremely difficult to develop, and can take years if there has been a culture of mistrust. New and heavy responsibilities may be put on the individual workers (see Kamata 1982, for a scathing attack on the system).
2. A multiskilled work force, to provide flexibility in scheduling.[†]

---

[*] Sutton (1991) discusses changes in cost accounting that must accompany JIT. See also Richmond and Blackstone (1988).

[†] See Zavadlav et al. (1996) for insights about production lines in which workers help one another. In general, if the rules for when to move and help another worker are established correctly, this flexibility provides substantial improvement over lines in which workers stay at their own stations. See Benjaafar (1996) for research on machine flexibility.

3. Good relationships with suppliers, so that joint improvement efforts are possible. Suppliers have significant impact on cost, quality, and delivery performance of end products. See Chapman (1989).

4. Extremely high levels of quality, so that production will not be interrupted due to parts with poor quality.[*] One method of achieving high levels of product quality is called *poka-yoke*, which essentially means foolproof operations. Poka-yoke tries to ensure that simple errors are not made. For example, some parts feeders are composed of a chute through which parts have to fall before they are used. Appropriately sizing the chute ensures that no part is too big or too small.

5. Low setup times, and therefore small batch sizes. Without low setup times, any effort to reduce batch sizes will be doomed because too much capacity will be lost in setups. Small batch sizes reduce average cycle stock, and provide faster feedback on quality problems. The latter point holds because often a work center discovers quality problems created by its immediate predecessor. If a large batch is run before the problem is found, many parts have to be repaired or discarded. In some firms, 8-hour setup times have been decreased to less than 10 minutes. The goal of setups under 10 minutes has been called single-minute exchange of dies, or SMED. (See, e.g., Shingo 1985, 1989.)

6. Highly reliable equipment, so that a low inventory system will not be forced to shut down while repairs are made.[†] One way to achieve high levels of equipment quality is through total preventive maintenance (TPM). One of the authors visited a new JIT line at a major international automobile assembler. The manager of industrial relations cited reliable equipment as a significant advantage in management's relationship with the line workers. The high level of frustration associated with trying to build good cars with poor equipment has been replaced by satisfaction with reliable equipment.

7. A stable master production schedule, so that a daily production rate can be set and followed. (See Chapman 1990 and Spencer and Cox 1994.)

8. High-volume, repetitive manufacturing because this environment is likely to be conducive to the previous seven requirements.

9. Some excess capacity so that variability can be met without constantly injecting more Kanbans (inventory) into the system. This also permits time for the workers to experiment with ways to eliminate waste. (See Crawford and Cox 1991.)

16.1.2.2 Detailed Description of the Kanban Control System

The cards, tags, or tickets represent the information system for JIT manufacturing. Henceforth, we simply refer to them as cards. They have two major functions in a JIT system: first, to provide the mechanism for the short-range implementation of the system *given a prescribed number of cards*; and second, to facilitate reduction in the WIP inventory through a systematic removal of cards from the system. (See Schonberger 1983.)

Two types of cards are used:

1. *Move cards.* These authorize the transfer of one standard container[‡] of a specific part from the outbound stockpoint of the work center where it is produced to the inbound stockpoint

---

[*] Zipkin (1995) shows that poor quality degrades performance primarily by wasting productive capacity.

[†] See Abdulnour et al. (1995) for research on machine maintenance policies in a JIT system.

[‡] As discussed earlier, in JIT as low lot sizes as possible are used. Thus, a standard container could contain but a single unit of a specific item.

of the center where it is to be used. A set of move cards is issued for the exclusive use of a single SKU between a specific pair of work centers. Move cards are often called "withdrawal cards."

2. *Production cards.* These authorize the production of one standard container of a specific part at a particular work center in order to replace a container just taken from the outbound stockpoint of that center. A set of production cards is issued exclusively for the production of a specific SKU at a particular work center.

A card typically contains the following information:

1. The Kanban number (the identification of the specific card)
2. The part number
3. The name and description of the part
4. The place where the card is used (the two associated centers in the case of a move card; a single center for a production card)
5. The number of units in the standard container

Every container at an inbound stockpoint must have a move card attached. When such a container is selected for use in production at the specific center (A in Figure 16.1), an employee detaches the move card and attaches it to an empty container (call it C1), which is then moved to the outbound stockpoint of the supplying work center (B in Figure 16.1). There the move card is detached and placed on a full container (call it C2), thus leaving the empty container C1 without any card. The production card is removed from container C2 and placed in a collection box. The full replacement container C2 (now with only a move card attached) is transferred to the consuming

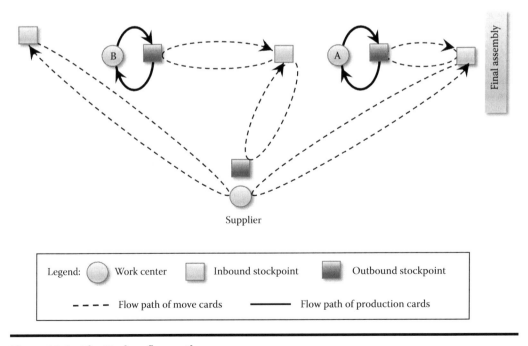

Figure 16.1 The Kanban flow paths.

center (A). On a frequent basis, an employee at the supplying center (B) picks up the released production cards from the collection box. For each such card, a container of parts is produced. Empty containers (such as C1) are used to receive these parts. Each such filled container has a production card attached to it and is then placed in the outbound stockpoint of the center (B). The making of each container of parts requires a withdrawal of parts from the inbound stockpoint of the supplying center (B), which initiates further upstream actions. Figure 16.1 (adapted from Hall 1983) also shows the flow paths of the different types of cards, including move cards to and from a supplier of raw materials. Supplier shipments of containers, each holding a small number of units, necessitate a shipping procedure different from the method that has been conventionally used. Instead of infrequent large shipments involving only one or a few SKUs, now many items are delivered in smaller quantities in each shipment. To reduce the implied high transportation cost, suppliers may employ a mixed loading system in which multiple suppliers share the same vehicle. Moreover, where a supplier is geographically remote from the production facility, the lead time can obviously be reduced through electronic transmission of a supply need as opposed to the physical shipment of a container and associated Kanban. See Kim (1985).

The Kanban system is completely manual. For it to operate properly, strict adherence to three rules is required:

1. A standard container must always be filled with the prescribed number of parts.
2. A container must not be moved forward until authorized by the receipt of a move card.
3. A container of parts must not be produced until authorized by a detached production card.

It should be clear that if a work center is short of materials, it stops working. Therefore, it will not have parts available for the next operation downstream, and eventually the entire line will shut down. In this way, any major disruption causes a line stoppage, and problems are dramatically visible.

16.1.2.3 JIT, MRP, and Reorder Point Systems

In this subsection, we compare JIT with MRP and a reorder point-order quantity system. From the details of the Kanban system, we can see that it is simply an (s, Q) system. (See, e.g., Liberopoulos and Dallery 2003.) The reorder point s is the number of containers or Kanbans for a particular part at a particular stage of production. Just as with a reorder point, the number of production Kanbans should cover the average time to produce plus a safety factor. This, in the language of Chapter 6, is equivalent to the average lead time demand plus safety stock, which is the reorder point in an (s, Q) system. The order quantity Q is the production container size for the part. (In some Kanban implementations, the work center produces multiple containers at once to save setup time. Therefore, the order quantity is actually several container sizes.)

Now note that MRP can exactly replicate an (s, Q) system as well. Imagine the MRP records in Table 15.8, except with stable demand for the end item. If safety stock is built into the MRP requirements for a part, the reorder point is simply the lead time demand plus safety stock. MRP will initiate a production order when the projected net requirements fall below the safety stock level. The order will be released one lead time prior to when it is required. The order quantity is completely flexible in MRP, within capacity constraints, so it can be set at the Kanban container size if we choose. Therefore, Kanban is an (s, Q) system, and MRP can replicate both (s, Q) systems and Kanban. In fact, Axsäter and Rosling (1994) show that MRP dominates both systems because it is more general, and can imitate either. Why then is JIT such an improvement on MRP, which is an improvement on (s, Q) systems, as discussed in Section 15.2.

The answer is twofold. First, JIT is not better for certain environments. For example, in multistage systems where end-item demand fluctuates widely, the Kanban system does not work well. Moreover, even when end-item demand is relatively level, fluctuations in component requirements can be caused by batching decisions, which are made because of high setup times/costs. So, if there are significant setup times, and parts are therefore batched for production, dependent demand will fluctuate widely, and Kanban and (s, Q) systems will not be appropriate. A related reason is that if there are multiple items, and high changeover times between items, batching will be necessary. Therefore, we see why Kanban applies in high-volume lines where equipment is more likely to be dedicated, and not as well in certain other cases. If setup times are low, and small lot sizes are used with a lot-for-lot replenishment strategy, variability in demand is not amplified back through the system. And, if the assembly schedule is relatively level and lead times are short, there is no need for substantial safety stocks. Then Kanban is entirely appropriate.

The second reason JIT is considered better than MRP is that JIT answers one of the primary problems with MRP—the lack of incentives for improvement (weakness #4 listed in Section 15.7). One tenet of the philosophy of JIT is to continuously improve, a facet that is missing from MRP. As noted in Chapter 15, MRP could be made to fit a continuous improvement environment, and it often has. But JIT makes that improvement explicit. Related to this point is the fact that MRP requires computer control, while JIT relies on manual control. Changes due to improvements are far easier to implement because of the decentralized nature of the manual control. In general, of course, computers make many things easier to do. But in this case, the centralized control implied by MRP and the consequent removal of the planning process from the factory floor suggest that the manual, decentralized Kanban system is better for implementing changes due to incremental improvement.

In many cases, a hybrid of MRP and JIT is desirable. Karmarkar (1989) describes how and when to blend MRP and JIT. One appealing approach is to use MRP for planning and Kanban for scheduling. A production plan is developed by the MRP system in the form of a master production schedule, using forecast information. Purchases may also be initiated by MRP. Then Kanban is used to control the movement of materials between work centers. This way, the production plan can account for nonstationary demand, but improvements can be pursued at the individual work center level. Responsibility and control are given to work center supervisors, leading to a greater sense of ownership. The factory may do the master production schedule at a component stage, and then do final assembly in response to customer orders. This then would be the push–pull point: push to component inventory, and pull thereafter, using Kanban for material movement. See Monden (1993) for an application of a hybrid MRP/JIT system at Nissan. See also Hall (1983), Groenevelt and Karmarkar (1988), Olhager and Östlund (1990), Flapper et al. (1991), Hodgson and Wang (1991), Pyke and Cohen (1990, 1993, 1994), Ptak (1991), Deleersnyder et al. (1992), Wang et al. (1996), Kern and Wei (1996), and Wang and Xu (1997).

MRP has cost advantages over JIT (excluding system control costs) when demand varies highly with time and setup times are relatively large. On the other hand, JIT does very well, from a cost standpoint, for stable (repetitive) demand and relatively low setup times. Moreover, achieving reductions in setup times may be more important than the choice between the two scheduling systems. In related work, Krajewski et al. (1987) undertake a massive simulation study of 36 factors that influence the performance of production systems. These factors include setup times, lot sizes, employee flexibility, yield rates, and so on. The authors show which factors are most important to the performance of both MRP and Kanban, and they discover that changing these factors is more important than which scheduling system is used. For example, if lot sizes are decreased as setup times are reduced, operating performance improves regardless of the production planning and scheduling system used. Thus, the performance of Kanban can deteriorate significantly if setup times or lot sizes are high. See also Goddard (1982) and Vendemia et al. (1995).

16.1.3 Benefits and Weaknesses of JIT

There are many benefits to a successful implementation of JIT. These include

- Reduced WIP inventory, and therefore less space and cost.
- Higher quality, due to continuous improvement efforts and to small lot sizes.
- Higher productivity.[*]
- Short lead times. If lead times are reduced using continuous improvement, firms that make-to-stock can sometimes change to make-to-order, and therefore become more responsive to customers. (See Blackburn 1992.) Also, if lead times are short, there is less need to track inventory as it progresses through the factory.
- Low control costs due to the decentralized nature of the system. Production planning is much simpler if schedules are level, because the planning problem is simply to find the best level schedule. Shop floor control is easier because schedules are predictable and constant. Like-wise, scheduling of workers is not necessary. They have assigned tasks and they are trained to help if an adjacent work station falls behind.
- Less paperwork.
- Higher reliability of production because problems are visible.[†]
- Visible, predictable amounts of inventory. Inventory records of WIP and materials can be updated only when finished goods have been recorded into stock. Therefore, there is no need for complex WIP inventory tracking systems.

There are, however, some weaknesses and warnings that must be mentioned. Probably the most compelling warning is not really a weakness of JIT at all. Rather, it is the tendency for production managers to implement the system where it does not fit. For instance, one of the authors worked with a firm that was struggling with high setup times and very low throughput rates. A student intern from a local university suggested implementing JIT because "JIT means high quality, low setup times, and low inventory." Fortunately, managers realized that these are prerequisites to JIT, not the certain result of it. In a less encouraging case, JIT was implemented in a large job shop without first reducing long setup times. This factory soon developed huge backlogs due to capacity lost to frequent setups. (For other comments on this topic see, e.g., Inman 1993.)

It should be clear from the discussion in this chapter that JIT is not appropriate in job shops where products are made to order, variability is high, and demand is extremely nonstationary.

---

[*] Lieberman and Demeester (1999a) use historical data for 52 Japanese automotive companies to evaluate the inventory–productivity relationship. They find that inventory reductions stimulated gains in productivity, rather than vice versa. On average, each 10% reduction in inventory led to about 1% gain in labor productivity, with a lag of about 1 year. See also Huson and Nanda (1995).

[†] See Moinzadeh and Aggarwal (1997) who find that cost reduction efforts are more effective when the production system is more reliable. See also Schonberger (1984).

Production is not smooth because bottlenecks shift continually. The high levels of variability imply high levels of inventory, but it is difficult to know exactly what inventory to put into the system when products are all made to order. How do managers know how many Kanban cards to use? Also, it is possible that some Kanbans will be inserted into the system for a product that will not be produced again for a long time, and WIP will be held unnecessarily.[*]

JIT is also not appropriate in continuous process industries where stages of production are tightly linked. There is no need for Kanbans to control movement of materials because the entire facility operates as a single machine. Finally, because JIT is a reactive pull system, it is not appropriate for environments in which demand can fluctuate widely, but can be forecast. Karmarkar (1989) notes that a reactive system would be a poor choice for a McDonald's restaurant located outside a football stadium. If the kitchen staff waits to begin cooking until the game ends and customers' order, the lines will be extremely long. It is far better to "push" production according to a forecast. In other words, begin cooking long enough before the game ends so that most customers can be served from inventory. See also Koenigsberg (1985).

Other weaknesses include

- JIT is vulnerable to plant shutdowns, demand surges, and other uncertain events, primarily because of the low levels of inventory.[†] There is no method of adjusting capacity within a given time frame—usually 1–6 months—except within narrow limits. When a plant shuts down, for whatever reason, many workers can be left idle. This has happened in dramatic fashion some years ago when a General Motors brake plant went out on strike, and within about 2 weeks, all of GM's North American factories were closed, not to mention dozens of supplier plants. Likewise, a supply chain disruption due to a volcano in Iceland shut down factories as far away as Germany and the United States.
- JIT cannot accommodate frequent new product introductions because of the need to introduce Kanban cards for all the components and parts.
- Frequent deliveries of small lots can generate highway congestion, a phenomenon that has been observed in crowded cities in Japan. See Miller (1991) and Moinzadeh et al. (1997).
- Some writers point to enormous improvements from JIT at a number of firms. Zipkin (1991) notes, however, that much of the research is quite shoddy in that it attempts to assign all improvements to JIT programs. No attempt is made to discern which benefits are due to JIT and which are due to, say, high levels of capital investment. The true causes of the improvements are not clear. See also Funk (1989) and Rees et al. (1989).

There are also a number of myths about JIT that should be discussed. One is that JIT requires suppliers to locate close to their customers. But many firms use JIT within the factory, and hold enough raw materials inventory to buffer for materials arriving from distant suppliers. Another myth is that JIT firms use sole sourcing for all components. The fact is that most Japanese firms use two or three suppliers for nearly every purchased component. They foster competition amongst the suppliers so that price, quality, and delivery performance will continuously improve. Needless to say, suppliers are not always excited about the competition. (See McMillan 1990; Helper and Sako 1995.)

---

[*] However, Gravel and Price (1988) for an implementation of Kanban in a small factory that is somewhat like a job shop. See also Ashton and Cook Jr. (1989).

[†] Moinzadeh and Aggarwal (1997) show that determining proper safety stock levels is very important in JIT systems that experience disruptions.

A third myth is that workers enjoy working in JIT plants. Although it is true that a worker without a job will be very motivated to work in a JIT plant, it is not clear that the same worker would prefer a JIT plant to another. Much research has been done on this problem, and it seems that the unfortunate fact is that repetitive work tends to be boring, regardless of whether or not the JIT philosophy is in place. This is true, over time, even if workers rotate among different jobs. In JIT plants, workers often feel the pressure to work faster and faster, and to achieve higher and higher quality. In this light, it is interesting that Japanese automotive firms have had trouble finding assembly workers in Japan, and have begun moving to more a flexible, less connected, workflow. These new production lines are less pressured than traditional JIT lines. See also Klein (1989) and Womack et al. (1991). Other references on JIT include Hall (1988); Billesbach and Schniederjans (1989); and Blackburn (1991).

16.2 Planning and Scheduling in Situations with Bottlenecks: Optimized Production Technology

As we have seen in Chapter 15 and in Section 16.1, MRP and Kanban apply in very different production environments, and both are very effective when applied and used properly. Neither, however, is effective when there is a significant bottleneck. The OPT system was popularized in a book called *The Goal* (Goldratt and Cox 1986a), and applies specifically to bottleneck situations.[*] The OPT philosophy follows a set of principles called the theory of constraints (TOC), and has also been called synchronous manufacturing (SM). In this section, we discuss the philosophy of OPT and TOC, and then discuss the rules for managing bottlenecks. Finally, we introduce a related system called CONWIP, and we comment on some research that relates to one of the rules of OPT.

16.2.1 Philosophy of OPT

As its name implies, TOC focuses on constraints, as opposed to JIT, which attempts to eliminate waste at all operations. The typical constraint in a manufacturing setting is a bottleneck in production. However, as we shall soon see, OPT broadens the definition. In fact, OPT attempts to focus the efforts of the firm on a single goal—making money. But, in *The Goal* and other materials, this single-minded focus seems to expand to include such things as continuous improvement, and the welfare of the workers and of the surrounding community. Nevertheless, most observers suggest that in OPT, making money is the one goal for the firm. In pursuing this goal, three key financial performance measures are stressed: net profit, return on investment, and cash flow. And three operating performance measures—throughput, inventory, and operating expense—determine the level of these financial measures.

Throughput, in OPT, is the rate at which the manufacturing firm sells finished goods. Normally, we think of throughput as the production rate. One valuable contribution of OPT is the recognition that a high production rate is worthless unless the firm can sell its products. Marketing activities therefore contribute to increasing throughput, and the constraint might well be the marketplace itself. This simple insight encourages manufacturing and marketing personnel to work toward the same goal.

---

[*] See also Jacobs (1984) and Goldratt (1988).

Inventory is defined by OPT to be "the money the firm has invested in purchasing things which it intends to sell" (Goldratt and Cox 1986a, p. 59). These include, of course, raw materials, components, and finished goods that have been bought by the firm, but not yet sold. In a departure from standard accounting practice, labor and overhead are not included in inventory.

Operating expense is the cost of converting inventory into throughput. It includes direct and indirect labor, electricity, and so on.

If throughput increases, while inventory and operating expenses remain constant, net profit, return on investment, and cash flow all will increase because revenues increase with no increase in cost. Likewise, if operating expenses decrease, while throughput and inventory remain constant, all three financial measures will increase because the cost of production decreases with no loss in revenue. In summary, if a firm can increase throughput, while decreasing operating expense and inventory, its financial performance will improve.

The problem is that constraints can impede performance. Constraints may be internal resources, like a bottleneck machine; market related, like the demand level; or policy related, like a policy that allows no work on weekends. The TOC suggests a procedure for dealing with constraints:

1. Identify the primary constraint.
2. Find out how to exploit the constraint.
3. Subordinate everything else to the decisions made in step 2.
4. Elevate the constraint so that a higher level of performance can be achieved.
5. If the constraint is eliminated, go back to step 1.

The OPT system is composed of 10 rules that help carry out this procedure and that focus on improving the three operating measures so that financial performance is enhanced. We discuss the 10 rules in the next subsection. But first we present a simple example to illustrate the procedure.

Two products, A and B, are produced at a factory. The price, cost of raw materials, and market demand for each product are given in Table 16.1. Also given is the time on the bottleneck machine. We can see from the table that the contribution margin for product A is $80 − $20 = $60, and for B is $100 − $25 = $75. (We are assuming for simplicity that labor cost is identical for the two products, and therefore is ignored.) Because only 60 hours per week are available on the bottleneck machine, the firm cannot produce the entire market demand for both products (which would require 95 hours per week). The usual choice in traditional manufacturing and accounting is to produce as much as possible of the product with the higher unit profit, or product B. Table 16.2

Table 16.1 Data for OPT Example

Product	A	B
Price ($/unit)	80	100
Cost of materials ($/unit)	20	25
Market demand per week (units)	200	110
Time on the bottleneck machine (hours/unit)	0.2	0.5
Contribution margin ($/unit)	60	75

Table 16.2   Traditional Solution to the Production Quantities Problem

Product	A	B
Production quantity	25	110
Hours on the bottleneck	5	55
Contribution per week ($)	1,500	8,250
Total	$9,750	

Table 16.3   OPT Solution to the Production Quantities Problem

Product	A	B
Production quantity	200	40
Hours on the bottleneck	40	20
Contribution per week ($)	12,000	3,000
Total	$15,000	

shows this solution. Produce the entire market demand of product B, using 55 hours on the bottleneck. Then, use the remaining bottleneck capacity by producing 25 units of A. Total contribution margin per week is $9,750.

Now, TOC suggests the following five-step procedure in this situation. First, identify the constraint, which is easy in this example. The machine is the clear bottleneck. Second, use this constraint to its fullest. In this case, we should determine the contribution per bottleneck hour used, rather than the contribution per unit. For product A, this is

$$\frac{\$60}{0.2 \text{ hour}} = \$300/\text{hour}$$

whereas for product B, this is

$$\frac{\$75}{0.5 \text{ hour}} = \$150/\text{hour}$$

Therefore, the firm should produce as many of A as possible. Third, subordinate everything else to the decision made in step 2, or use the remainder of the capacity for B. The results in Table 16.3 show the clear superiority of the OPT solution.

Of course, the OPT solution can be replicated by a simple linear program and run quickly a spreadsheet. The linear program is as follows. Let X = the number of units of product A to produce, and $Y =$ the number of units of product B to produce. Then,

Maximize $60X + 75Y$

subject to:

$$
\begin{aligned}
X &\leq 200 \quad \text{(Market demand limit for } A) \\
Y &\leq 110 \quad \text{(Market demand limit for } B) \\
0.2X + 0.5Y &\leq 60 \quad \text{(Available hours on the bottleneck machine)} \\
X, Y &\geq 0
\end{aligned}
$$

See also Luebbe and Finch (1992) and Simons Jr. et al. (1996).

This example shows that applying the first three steps of the above procedure can be powerful. Step 4, elevating the constraint, can be pursued in a multitude of ways, some of which are discussed below in the 10 rules of OPT. Others include

- Reducing variability at and surrounding the constraint
- Increasing the capacity of the constraint
- Eliminating all idle time at the constraint
- Shifting some of the constraint's workload to other resources

16.2.1.1 The 10 Rules of OPT: Managing Bottlenecks

A bottleneck is a stage in the process of producing, storing, distributing, or selling that restricts throughput. Thus, some production stage may be the bottleneck. For instance, in the manufacture of bicycle helmets, there is a curing process that can take up to 20 hours. In one such factory, this operation was a clear bottleneck, and expanding its capacity would increase throughput significantly. Of course, operating expenses would increase as well. As noted above, limited demand, relative to production capacity, may also be the key constraint. Most of the 10 rules of OPT focus on managing bottleneck and nonbottleneck resources.

1. *Utilization and activation of a resource are not synonymous.* Activating a machine means, say, using it to process parts, whereas utilizing the machine means processing only those parts that can be turned into throughput. There is no gain from running a nonbottleneck machine if its output will only build up inventory in front of a bottleneck.
2. *The level of utilization of a nonbottleneck is not determined by its own potential, but by some other constraint in the system.* The utilization of a nonbottleneck is limited by the rate of the bottleneck machine. We have seen this principle violated at a paper company, which ran its equipment 24 hours per day, 7 days per week. Unfortunately, there was limited market demand, and the company had to rent new warehouse space to store the excess finished goods inventory. Operating expense and inventory increased, with no gain in throughput. OPT would recommend slowing down the equipment, or shutting it down for a time, so that the output of the plant equals market demand. However, very high shutdown and startup costs might rule out a complete shutdown. (See the discussion in Sections 14.1 and 14.2.) An alternative would be to spend more time in changeover, thereby reducing output of the plant and decreasing average cycle stock.
3. *An hour lost at the bottleneck is an hour lost for the total system.* This rule parallels and extends rule #1, and helps managers focus on all activities at the bottleneck. For example, is it possible

to keep the bottleneck running while its operators take a lunch break? If not, the firm should train other workers to run the bottleneck machine so that it is never needlessly idled. One factory we worked with had a bottleneck at the packaging operation. Although other departments ran two shifts, this operation ran three shifts. As it happened, however, the number of employees in packaging was greater than any other department, and the parking lot could not accommodate the cars of two shifts of packaging employees at the same time! The line had to be shut down for 1/2 hour between shifts to allow time for the parking lot to clear and refill. The simple, inexpensive solution was a parking lot expansion that added 1.5 hours of bottleneck time every day.

4. *An hour saved at a nonbottleneck is a mirage.* Throughput will not increase with savings at a nonbottleneck. Therefore, managers should focus improvement efforts elsewhere. The time spent by a job at a bottleneck is composed of setup and processing time, while the time spent at a nonbottleneck includes setup, processing, and idle time. Reducing the setup time at a bottleneck saves time for the entire system. On the other hand, reducing setup time at a nonbottleneck may simply increase idle time. Of course, if setup time is reduced, batch sizes and average cycle stock at the nonbottleneck can be reduced. But the gain from setup reduction at the bottleneck is clearly greater.

5. *The bottleneck governs the throughput and inventory in the system.* Inventory should be used carefully so that the bottleneck is never starved for parts to process. We will discuss the details of scheduling nonbottlenecks and bottlenecks in Section 16.2.2 when we introduce "drum-buffer-rope" (DBR) scheduling.

6. *The transfer batch size should not necessarily equal the production batch size.* Imagine a large batch being run on a nonbottleneck just prior to a bottleneck. If the bottleneck is starved for parts at the moment, it would be desirable to get it started on part of the next batch, even though the nonbottleneck is still processing the remainder. We introduced this topic in Section 11.3.2.

7. *The production batch size should not be the same from stage to stage in the process.* Lot sizes at bottlenecks should, in general, be larger than at nonbottlenecks, so that less time is lost to setups. Of course, the small batches from the nonbottleneck need to arrive at the bottleneck in time to be rejoined into a large batch. Determining the best lot sizes and scheduling them are very difficult tasks.[*]

8. *Capacity and priority should be considered simultaneously.* Because the lead time for a given batch depends on the priority given to it at a machine, and on the capacity of the machine, priority rules should be determined in conjunction with the capacity of the machine. In fact, the capacities at all constrained resources should be considered. We noted in Chapter 15 that the assumption of fixed lead times is a fundamental flaw in MRP. OPT explicitly recognizes the relationship among priority sequencing rules, capacities of equipment, and the lead time of a batch. It is less clear about which sequencing rules to choose. We will discuss priority sequencing rules in the next section of this chapter.

9. *Balance flow, not capacity.* The flow through the plant should equal market demand. Too often, managers focus on the "rated capacity" of individual machines, and ignore disruptions due to variability. Imagine a series of three machines each of which can produce, on average, one part per hour. Because of variability in processing times, however, the machines may lose time to blocking (nowhere to put a completed part) and starving (no part to work on),

[*] For related research, see Luss and Rosenwein (1990).

especially if there is little WIP inventory to buffer for the variability. The flow through this system will clearly be less than one part per hour. It is the flow, not the rated capacity, that should be balanced with the market demand. There is much research on this issue. See, for example, Hillier and Boling (1966), Hillier and Boling (1979), Conway et al. (1988), Hopp and Simon (1989), Baker et al. (1990), Hillier and So (1991), Dallery and Gershwin (1992), Buzacott and Shantikumar (1993), Baker et al. (1993), Powell and Pyke (1996), Hillier and So (1996), and the references therein.

10. *The sum of local optima is not equal to the optimum of the whole.* This principle is clear in the theory of mathematical optimization, and we saw it in Chapter 7 when we discussed sequential versus simultaneous selection of inventory policies. Therefore, although it is not particularly new, managers need to be reminded that problems develop when supervisors at bottlenecks, supervisors at nonbottlenecks, and marketing personnel, all optimize for their own goals. Many supervisors try to run their equipment at full capacity, while many marketing personnel try to earn larger bonuses by selling more at the end of the quarter. The previous nine rules make clear that the financial performance of the firm will suffer with this type of arrangement.

In applying these 10 rules, one should recognize several types of resource sequences in a facility. We describe four:

1. *A bottleneck is prior to a nonbottleneck.* The nonbottleneck should run only when parts are available from the bottleneck. There is probably little other choice because of the constraint just prior to it.
2. *A nonbottleneck is prior to a bottleneck.* The nonbottleneck should be paced to match the rate of the bottleneck. If it runs slower, the bottleneck will be starved (see Rule #5); and if it runs faster, inventory will build up.
3. *A nonbottleneck and a bottleneck are just prior to an assembly stage.* The nonbottleneck should be paced at the rate of the bottleneck. Otherwise, inventory of its parts will build up at the assembly point.
4. *Both a nonbottleneck and a bottleneck supply independent market demand.* Both should produce as close to the rate of market demand for their separate products as possible.

The next subsection elaborates on these points.

16.2.2 Drum-Buffer-Rope Scheduling

To operationalize the 10 rules under different sequences of resources, Goldratt and Cox (1986b) suggest following a scheduling scheme called drum-buffer-rope. The analogy given is of Boy Scouts on a hike. The distance from the fastest to the slowest boy in line is equivalent to inventory in the system. If faster boys (nonbottlenecks) run ahead, the distance to the end of the line (or inventory) increases. Of course, the slowest boy (the bottleneck) is trailing along at the end of the line. Now how should the Scout leader deal with this situation? He could put the slowest boy at the beginning of the line, and therefore force all faster boys to slow down. Alternatively, he could require each boy to grab hold of a rope so that, even if the slowest boy is not in the lead, the faster boys cannot run too far ahead. The length of the rope is analogous to the amount of inventory (buffer) in the system. Having a flexible rope allows the faster boys to stop and tie their shoes, for instance, without stopping the slowest boy from continuing. If the rope is too short, the slowest boy may have to stop when another boy ties his shoes. If it is too long, there is too much distance between the slowest and the fastest. In the absence of a real rope, the Scout leader could beat a drum at the pace of the slowest boy, requiring all others to listen and walk at that pace, and to stay within a specified distance. Other aspects of the 10 rules apply here as well. For example, if it is possible to take the heaviest things out of the slowest boy's backpack, he can move faster.

In the factory, the drumbeat is used to pace all operations upstream from the bottleneck at the bottleneck rate. In other words, all upstream nonbottleneck resources operate at the bottleneck's pace. A buffer is set before the bottleneck to insure that it can always maintain the pace of the drum, even if a prior nonbottleneck slows down for some reason. Nonbottleneck operations are scheduled to keep the buffer at the appropriate level using a hypothetical rope. Jobs are released only at the rate of the bottleneck, but far enough in advance so that a time buffer is maintained before the bottleneck. Inventory does not increase unnecessarily, and the bottleneck is never starved. Jobs, therefore, are pulled from the first production stage by the bottleneck. After the bottleneck, jobs are pushed to the end of the line as quickly as possible.

16.2.3 A Related System: CONWIP

In this subsection, we discuss a system that is quite similar to DBR, but that answers some of the deficiencies of the DBR system. Specifically, the algorithm the DBR scheduling software uses is unpublished; thus, users do not know how the software generates schedules. Managers face a "black box," where the details of the scheduling algorithm are not made public. Consequently, managers often do not trust the schedule generated by the OPT software. As a result, the acceptance and use of this software in industry has been limited. The CONWIP system is very similar to DBR, but has been developed from clear and rigorous analysis.

The CONWIP system was developed by Spearman and Hopp, and has been successfully applied in numerous firms. (See, e.g., Spearman et al. 1990.) In CONWIP, the firm specifies, and holds constant, the total amount of WIP in the factory. When a job leaves the factory to the finished goods warehouse, another job is released into production. CONWIP defines production quantities in terms of *standard parts*, which are based on the time a part takes at the bottleneck. Say part X requires 1 hour at the bottleneck and part Y requires 2 hours. When a job of 10 units of part Y leaves the factory, a job of 20 units of part X is released, assuming that X is next in queue. Therefore, the total WIP, adjusted for the bottleneck processing rate, is held constant. The stage-to-stage movement of materials is quite flexible, because unlike Kanban, batches are pushed from station to station once the job is released to the line. There is no requirement to maintain constant WIP at each station. (Recall that in the Kanban system, the WIP or the number of Kanbans must be specified for each station.) Now, CONWIP can specify that new jobs be released to the factory either when a job leaves the bottleneck, or when a job leaves the factory. If releases are linked to completions at the bottleneck, CONWIP and DBR are nearly identical. However, DBR requires that frequent decisions be made regarding releases to the shop, while in CONWIP those decisions are automatic.

CONWIP employs an order backlog, or a list of jobs waiting to begin processing. Jobs can be proactively sequenced, thereby reducing sequence-dependent setups. (Kanban, on the other hand, is generally regarded as inappropriate for lines that have significant setup times.) The research associated with CONWIP helps managers specify the optimal amount of inventory, and the best timing of releases.

To set the "length of the rope" or the amount of inventory in a CONWIP system, we can employ a fundamental law from queueing theory called Little's law (Little 1961). We will have more to say about this topic in the next section, but for now we simply note that in most circumstances the average amount of WIP equals the product of the average lead time and the average output rate (jobs completed per unit time) from the factory. Increased amounts of WIP will generate increased output, because the bottleneck is starved less often. However, as WIP continues to increase, output levels off to the rate of output of the bottleneck. Also, as WIP increases, lead time also increases. These relationships seem quite simple, but they are not linear, so they should be used with care. In complex environments, it is best to develop a simple simulation to test for the best CONWIP level.

Relevant research includes Spearman and Zazanis (1992) who develop mathematical models to show that pull systems, like Kanban, DBR, and CONWIP, are inherently easier to control than push systems. There is less congestion in pull systems because the variability of WIP is reduced. They also show that stabilizing the WIP level is more important to improving effectiveness than the fact of using a pull control system. They show that CONWIP has higher throughput than Kanban.[*] CONWIP also accommodates multiple part types on a single line better than Kanban, which would require inventory of each part type at each station. In CONWIP, WIP inventory tends to build up at the bottleneck stations, which increases the utilization of the bottleneck and therefore the throughput of the line. This last feature, of course, parallels DBR.

Spearman et al. (1990) also note that push systems schedule throughput and measure the amount of WIP, while pull systems set the amount of WIP and measure throughput. Recall that MRP can be used to plan production, while Kanban is used to control materials movement on the floor. MRP can be used with CONWIP in an identical way. See, for example, Betz (1996). Other related research includes Bertrand and Wortmann (1981), Bertrand (1983), Bemelmans (1985), Weindahl (1987), Weindahl and Springer (1988), Spearman (1990), Hopp and Spearman (1991), Graham (1992), Donahue and Spearman (1993), Duenyas et al. (1993), Karmarkar (1993), Muckstadt and Tayur (1995a,b), and Hopp and Spearman (2011).

16.2.4 Benefits and Weaknesses of OPT

In this section, we describe some of the benefits and weaknesses of OPT. Probably the biggest benefit of OPT is its simplicity, and the clear presentation of the OPT principles in *The Goal*. Countless students and managers have read *The Goal*, and subsequently have made dramatic progress in identifying and managing bottlenecks. No longer must they attack the entire production system at once, as in MRP and JIT. Rather, OPT recommends focusing energies on a particular problem area, called a bottleneck. When a bottleneck is relieved, a new bottleneck appears, and the process starts over again. This continuous improvement approach resembles the *kaizen* of JIT, but in a prioritized fashion, and is a major departure from the static approach of MRP. The rules for managing bottlenecks provide useful, but largely qualitative guidelines. In contrast, analytical research on CONWIP systems has provided managers with some good tools for scheduling production in factories that have one or more identifiable bottlenecks. There have been a number of successful implementations of OPT and CONWIP, although to date their impact has been more muted than either MRP or JIT. Nevertheless, the aforementioned and the following references suggest that it is quite possible to successfully implement OPT and CONWIP in many environments. See Feather

---

[*] Gstettner and Kuhn (1996) show that a Kanban system can have lower WIP inventory than CONWIP for the same production rate. However, their Kanban system has been modified from the traditional one. See also Sipper and Shapira (1989).

and Cross (1988), Wheatley (1989), Fawcett and Pearson (1991), Wu et al. (1994), Spencer and Cox (1995), and Guide et al. (1996) and the references therein.

On the negative side, we again note that OPT is appropriate for environments in which there are identifiable bottlenecks. Consequently, it is not suitable for continuous processes in which the entire factory operates as a single machine. In fact, in continuous processes, managers usually speak of *gating* or *pacing* operations, rather than of bottlenecks. Identifying the pacing operation is important, but because all operations are linked, OPT's rules for managing bottlenecks are not especially useful. There is nothing gained by rules that recommend how to pace bottlenecks and nonbottlenecks when all machines have to operate at the same pace. In this context, production scheduling requires another set of tools, which we discussed in Chapter 10.

In a similar vein, OPT is not helpful in complex job shops because there are multiple moving bottlenecks. In other words, as new orders arrive and are released into production, the bottlenecks shift from operation to operation. Pacing the factory by the bottleneck is impossible when managers do not know where the bottleneck will be an hour from now. Finally, OPT is not particularly useful in a shop that is suitable for Kanban control because bottlenecks have already been eliminated.

Many researchers and practitioners alike have complained that the OPT software was a black box, because its details were never published, and as noted in Section 16.2.3, managers have been unwilling to rely on a black box. A final weakness applies to nearly any methodology used in manufacturing, but is particularly dangerous with OPT because of its remarkable simplicity. With such a simple set of rules, and such an appealing presentation of them, students or practitioners (in a workshop or seminar setting) learn a tremendous amount of useful information in a short period of time. The problem arises when they try to apply these rules in environments in which they are not appropriate, or that are very complex. For instance, in the absence of the CONWIP research, the rules for the size of the time buffer before the bottleneck are not specified. Severe problems can develop if DBR scheduling is implemented with a "rope" that is too short. Likewise, when complexities such as multiple bottlenecks are present, it is not clear how one should schedule production. The CONWIP research does not address all such complexities. Students and practitioners must be encouraged to resort to simulation to test various scheduling rules, rather than only relying on qualitative insights.

16.3 Short-Range Production Scheduling

In Chapters 13 and 15, we briefly introduced the short-range production scheduling problem and noted that it is concerned with the final scheduling of production. Short-range scheduling applies at the level of the individual work center, where operators and supervisors need to know which of a set of jobs in queue, or soon to be in queue, to process next. The answer is contained in dispatch lists that are typically developed once a day, but with more automated systems, they may be updated more frequently. These lists suggest the sequence in which jobs are to be run on each machine. Sometimes, the list will indicate that a work center should remain idle until an important job arrives, even if there are other jobs waiting in queue. We will describe methods for developing these lists later in this section. We shall first consider where these methods apply and where they do not. Specifically, we begin by discussing environments in which these short-term scheduling aids are not particularly helpful.

As noted in Chapter 13, the task of determining which job to perform next on every machine in the shop is relatively simple in most continuous process industry settings because the entire factory operates as a single machine. Therefore, there are no decisions to be made at individual work stations. The optimal sequence of production is often driven by sequence-dependent setup times, and it is best to use the methods discussed in Chapter 10. Also, continuous process industries generally use repetitive production, whereas most of this chapter applies best in nonrepetitive environments. In a similar vein, as noted above, the Kanban system applies best in repetitive manufacturing environments, and employs a very simple job release mechanism. Thus, the approaches we introduce here are not particularly useful in a firm that uses the Kanban system.

The information in this chapter is relevant, however, in factories that employ MRP, that have severe bottlenecks, or that are job shops. In MRP environments, a master schedule and a material requirements plan have been developed from the MRP system, and now the focus is on the sequence of individual jobs at each work station. In fact, as noted in Chapter 15, many MRP systems include finite loading software, a topic we discuss in Section 16.3.2.2. In situations with a severe bottleneck, factory performance is driven by one machine. The OPT philosophy can (and should) draw on the insights and procedures from this section in determining the best sequence of jobs at the bottleneck. Finally, as we have noted throughout Chapters 10 through 15, most of the techniques discussed in those chapters do not apply well in job shops. In fact, many job shops approach the scheduling problem directly with the approaches described throughout this section, without the use of a master schedule or a material requirements plan. These methods are extremely useful in these complex environments, so much so that another name for this section could well be Job Shop Scheduling.

In Section 16.3.1, we introduce some of the key issues in short-range scheduling, including some important definitions. Then, in Section 16.3.2, we discuss many useful techniques for this problem, including Gantt charts, priority sequencing rules, and finite scheduling. In Section 16.3.3, we introduce the one machine, deterministic sequencing problem. Finally, in Section 16.3.4, we discuss the general job shop scheduling problem. We introduce some formulas and intuition that are helpful in such a complex environment, and we give some insight into which procedures are most applicable.

16.3.1 Issues in Short-Term Scheduling

Throughout this section, we will use the terms *work station*, *machine*, and *work center* interchangeably. A work center may be composed of multiple machines, and the term *work station* is often used to indicate a machine and its operator. For our purposes, however, each of these entities will be considered a single resource on which jobs must be scheduled. Therefore, if the machines in a work center are scheduled individually, the scheduling problem is focused at the machine level. On the other hand, if the work center can be treated as a single resource for scheduling purposes, we shall treat it as such. One final note on terminology: we will use the terms *factory* and *shop* interchangeably.

Other activities at the short-term scheduling level can include

- ◼ Tracking WIP inventory. This task involves finding the location of each part in the factory. In many shops, this is a trivial task, but in large factories, it can be very time consuming.
- ◼ Monitoring the status of machines and people. For example, has a particular computer numerically controlled (CNC) milling machine been repaired yet? Or, is the setup of a certain machine finished? If the shop uses bar codes to scan jobs in and out of a work center, or to record when a setup is started and finished, the information system can more easily track WIP and machine status. However, it is essential that operators maintain discipline and never neglect scanning these activities.

- Tracking throughput. Throughput is defined as the number of jobs (or individual items) completed per unit time. A manager might be interested in a report of last month's throughput compared to an established standard for the shop.
- Production/material control and feedback. This is Block 11 in Figure 13.1. It interacts with short-range scheduling, and provides feedback on deviations from due dates, quality standards, and so on.

For firms that have the financial resources to implement large computerized systems, and the complexity to require them, manufacturing execution systems (MES) may be considered. MES are computerized systems that track and document the activities that transform raw materials into finished goods. MES are designed to work in real time and to control machines, materials, personnel, and even support services (see Meyer et al. 2009). As such, they incorporate management of product definitions, resources, scheduling, and dispatching, as well as collection of production data, among other tasks.

16.3.1.1 Some Important Definitions

In this subsection, we define some important terms. We begin with the structure of the shop, and then discuss several scheduling terms. Finally, we introduce a number of performance measures used in this context. First, we note one important distinction: *static* versus *dynamic* scheduling. Static scheduling assumes that a set of jobs is available for processing on one or more machines, and that no additional jobs will arrive before these jobs are completed. Dynamic scheduling allows for the possibility of new job arrivals over time.

16.3.1.2 Shop Structure

Researchers and practitioners have developed specialized terms that indicate the layout of machines, and the flow of jobs through a factory. A *flow shop* is a shop design in which machines are arranged in series. Jobs begin processing on an initial machine, proceed through several intermediary machines, and conclude on a final machine. The series arrangement implies a linear structure to the shop as illustrated in Figure 16.2. This figure shows a pure flow shop in which jobs must be processed on each machine in exactly the same order. A general flow shop is somewhat different, in that a job may skip a particular machine. For instance, although every job must proceed from left to right in Figure 16.2, some jobs may go from machine 1 to, say, machine 3 and then machine 4.

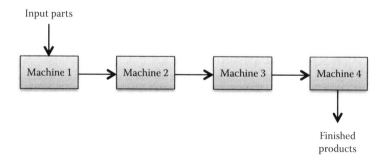

Figure 16.2   A pure flow shop.

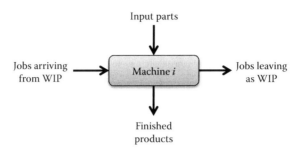

Figure 16.3 One machine in a job shop.

A job shop does not have the same restriction on workflow as a flow shop. In a job shop, jobs can be processed on machines in any order. The typical job shop, from a research standpoint, is one in which there are *m* machines and *n* jobs to be processed. Each job requires *m* operations, one on each machine, in a specific order, but the order can be different for each job. Real job shops are more complicated. Jobs may not require all *m* machines, and yet they may have to visit some machines more than once. Clearly, workflow is not unidirectional in a job shop. Any given machine may observe new jobs arriving from outside the shop (as new inputs), and from other machines within the shop (as WIP). The same machine may be the last machine for a particular job, or it may be an intermediate processing step. Thus, the workflow can be illustrated as in Figure 16.3.

One final complication in the structure of the shop is that of parallel machines. Thus far, we have described situations in which there is only one machine that can perform a given operation. In practice, there are often multiple copies of the same machine. Therefore, a job arriving at a work center can be scheduled on any one of a number of machines. This of course adds more flexibility, but it complicates the scheduling problem further. It is important to determine whether jobs can really run on multiple machines. One of the authors worked with a factory that had multiple "identical" machines, purchased from the same manufacturer. Yet, some parts ran with higher quality on one machine than on any other, and the higher-quality machine was different for different parts.

16.3.1.3 Schedules

The objective of short-term scheduling is to provide a schedule for each job and each machine. We now discuss important terminology pertaining to scheduling. In precise terms, a *schedule* provides the order in which jobs are to be done, and it projects the start time of each job at each work center. In other words, it allocates time intervals on one or more machines to each of one or more jobs. A *sequence* only lists the order in which jobs are to be done. A *dispatch list* is a sequence, but only for those jobs currently at the work center. In other words, a sequence can include jobs that are yet to arrive at a work center.

In general terms, a schedule requires a number of inputs. These include a list for each job of the operations that need to be performed, and the amount of time each operation takes. Also required is a list of any precedence constraints that describe which operations need to precede others. If workers are to be scheduled at the same time, the scheduler needs a list of all required resources for each job and each operation.

Consider for a moment a single machine that must process four jobs. We shall assume that setup times are included in the specification of the total processing time, and that setup times are not dependent on the sequence of production. We shall also assume that a job cannot be interrupted once it has begun processing. To establish a schedule for the four jobs, we must decide which job to process first, second, and so on. There are four choices for the job to process first. Once this job has been chosen, there are three jobs left to choose from for the job to process second. Then, there are two jobs to choose from for the third job in order; and finally, the last remaining job is scheduled to run fourth. In other words, there are $4! = 24$ distinct possibilities.* Each possibility is called a permutation of the four jobs. If a schedule, as in this case, can be characterized completely by a permutation of the number of jobs, it is called a permutation schedule. Notice what happens as the number of jobs increases. If there are six jobs, the number of possible schedules is $6! = 720$, and if there are 20 jobs, the number is 2.433×10^{18}. Clearly, it would be impossible to search for the best schedule out of so many possibilities. Much of scheduling theory is devoted to providing simple optimal, or near-optimal, solutions to problems such as these.

Our assumptions for the permutation schedule example included one pertaining to preemption, specifically that no preemption or interruption can occur. Of course, preemption may be allowed in some environments. When a job is preempted, another job interrupts its processing. Once the preempting job is finished on this machine, the original job can resume processing. If the discipline is *preempt-resume*, the job begins at the point at which it was interrupted. Metal cutting operations would be an example. The metal already removed does not need to be removed again. If, however, the job must start over, the discipline is called *preempt-repeat*. An example would be a printing operation in which the setup time is quite long. Imagine an operator setting up equipment by adjusting colors, say, when the work is preempted. The preempting job may require an entirely new color scheme, and the operator will have to begin the original setup from scratch when the preempting job is finished.

16.3.1.4 *Performance Measures*

Think back to the 24 different schedules for the four job problem mentioned above. How can we evaluate which of these is best? The answer is to identify a number of performance measures that can be used to evaluate different schedules. Some have to do with the time a job spends in the shop; others pertain to performance relative to due dates; and yet others concern utilization of production resources. Note that these three general goals may conflict with one another. For example, as we shall see in later in this section, high utilization (which is generally considered a positive thing) increases the time spent in the shop, and may degrade due date performance. Therefore, in spite of the very short-term nature of the scheduling problem, much larger strategic concerns are at issue. Managers should be very concerned with the trade-offs implied in the choice of which performance measure to use. For example, this choice may implicitly trade off WIP inventory cost, with customer satisfaction regarding meeting due dates, and with amortization of investment in equipment. The three general goals can be specified in more precise performance measures. We discuss each in turn.

One performance measure that is a primary focus of managers is the *average WIP level*. High WIP levels mean that more money is invested in inventory, and therefore is not available for other purposes. Average WIP is directly related to the time jobs spend in the shop. *Flowtime* is the

---

* $n! = n(n - 1) \cdots (1)$ and is stated as "n factorial."

amount of time a job spends from the moment it is ready for processing until its completion, and includes any waiting time prior to processing. The *makespan* is the total time for all jobs to finish processing. In a static problem, it is also the flowtime of the last job to be processed. For a single machine problem, the makespan is the same regardless of the schedule, assuming we do not allow any idle time between jobs. It is simply the sum of all processing times of all jobs to be scheduled. For more than one machine, the makespan will change with the schedule.

Some performance measures have to do with performance relative to each job's due date. These include *lateness, earliness,* and *tardiness.* Lateness is the amount of time a job is past its due date. Lateness is a negative number if a job is early. Tardiness equals lateness if the job is late, or zero if it is on time or early. Earliness is the amount of time prior to its due date at which a job's processing is complete. It is the negative of lateness if the job is early, or zero if it is on time or late.

Managers concerned with amortizing investment in equipment would like to see the equipment continually processing jobs that bring in revenue. *Machine utilization* and *labor utilization* are the primary performance measures of shop utilization. Utilization is the fraction of available time spent processing jobs. These measures focus on the cost of production, rather than on due date performance.

With so many choices, it is important to understand the effect that the scheduling rule has on shop performance. And it is important to understand the firm's manufacturing strategy so that shop performance supports the strategy. In this light, we may be interested in a number of different functions of one or more performance measures. For example, it may be important to know the average flowtime, the maximum flowtime, or the variance of flowtime, as a result of the scheduling rule chosen. And the choice of the scheduling rule may be based on an objective of minimizing the makespan, minimizing the average flowtime, minimizing the maximum or average tardiness, and so on. We will show which sequencing rules are known to be optimal for a number of different objectives in Sections 16.3.3 and 16.3.4.

16.3.2 Techniques for Short-Term Scheduling

In this section, we describe two useful procedures for the short-term scheduling problem. We focus on deterministic problems, although, as we shall see, the methods can be applied in probabilistic environments as well. We begin by introducing a general-purpose approach for visualizing schedules, and then proceed to a discussion of finite loading.

16.3.2.1 Gantt Charts

The Gantt chart was developed by Henry L. Gantt in the early 1900s as a simple visual technique for sequencing jobs on machines. It is best understood by an example. Assume that four jobs must be processed on two machines (A and B). The times (in days) required to process the jobs on each machine are given in Table 16.4. The Gantt chart for processing the jobs in numerical order is given in Figure 16.4, with job numbers listed in each bar. Note how the start and completion times of each job are immediately visible. One of the authors worked with a firm that produced several thousand products, and used a large Gantt chart for scheduling. The chart was placed on a huge magnetic board in the planning department so that schedulers and shop floor supervisors could see what work had to be done and what new jobs could be inserted into the schedule. See also Duersch and Wheeler (1981) and Adelsberger and Kanet (1991) for computerized versions of the Gantt chart.

Table 16.4 Requirements (Days) for Gantt Chart Illustration

Job	Machine	
	A	*B*
1	7	2
2	1	5
3	2	4
4	5	3

Gantt charts have several drawbacks. The first relates to the firm with several thousand parts—it is very difficult to manipulate a Gantt chart for so many parts even if the chart is computerized. And the advantage of simple visibility can be lost. The large magnetic Gantt chart mentioned in the previous paragraph had provided a remarkable sense of order in the midst of a chaotic environment. However, as the firm grew and the number of parts surpassed 3,000, this tool was no longer effective. The scheduling problem was just too complex. The second drawback is that a Gantt chart is purely descriptive. The planner has no way of knowing how good the solution is. For this reason, it is important to understand simple rules that generate optimal, or near-optimal, solutions, a topic we investigate shortly.

16.3.2.2 Finite Loading Systems

One application of Gantt charts is in finite loading systems, although these systems do not require the visibility of Gantt charts. Finite loading systems are specifically designed to address the issue of finite capacity, and, as noted in Chapter 15, have been applied in conjunction with MRP systems. They often employ a material requirements plan from MRP and then schedule work centers based on work center capacities and job priorities.

Imagine that we want to run the finite loading system on Monday morning. Assume that at a given machine, a job is partly finished, there is a queue of jobs waiting for processing, and some new jobs have arrived at the shop. The finite loading system will account for the amount of processing time remaining for the job currently running on the machine, and then sequence all remaining jobs, including the new ones, so that the available capacity of the machine is not exceeded. And it

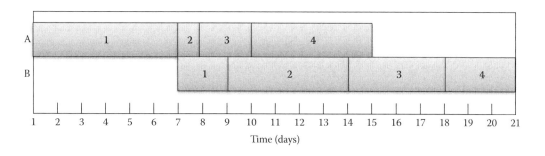

Figure 16.4 Illustration of a Gantt chart.

Table 16.5 Setup and Processing Times for Finite Loading Examples

Job	Priority	Machine A	B	C
1	Second	2	5	4
2	First	6	3	7
3	Third		4	7

will do this for all machines in the shop. The output of a finite loading system is a set of start times and completion times for each job at each work center. These systems often have the capability to look ahead to jobs that have yet to arrive at the work center. So these systems might, for instance, leave a machine idle for a short time waiting for an important job to arrive. Two types of loading are commonly used. The first, horizontal loading, begins with the highest-priority job and schedules it through all work centers. Then, the system schedules the next job through all work centers. This process repeats until all jobs have been scheduled. If a lower-priority job can be scheduled ahead of a higher-priority job at some machine, without delaying the scheduled start time of the higher-priority job, the system will do so. As an example, we shall use the data in Table 16.5. Three jobs are to be scheduled on three machines, although job 3 only requires machines B and C. Horizontal loading first takes the highest-priority job, job 2, and schedules it on all three machines. The second-priority job, job 1, is then scheduled, followed by job 3. See Figure 16.5. Note that job 3 can be assigned to machine B at time zero because this machine is idle, and inserting job 3 will not cause a delay in jobs 1 or 2. Job 3 cannot be assigned to machine C, however, until the higher-priority jobs are finished, because it cannot be processed earlier without delaying jobs 1 and 2. Job 3 requires 7 hours on machine C, but only 5 hours are available (after it is completed on machine B). If it happens, that job 2 is delayed on machine A or B, it may be possible to insert job 3 on machine C next time the system is run. The makespan in this example is 27. Horizontal schedules can suffer from excess idle time because the focus is on jobs rather than on machines. In other words, jobs are put into the schedule in sequence without regard for gaps in the machine schedule.

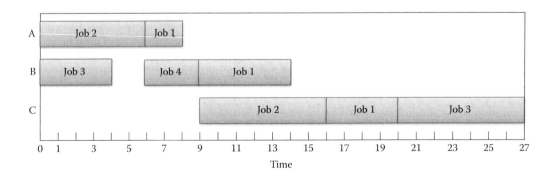

Figure 16.5 An example of horizontal loading.

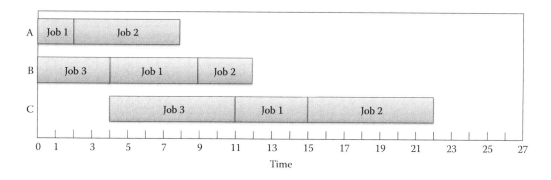

Figure 16.6   An example of vertical loading.

The second type of loading, called vertical loading, focuses on each work center in turn, and determines which job to load next. In our example, the system would begin, say, at machine A, and load on it all jobs that are available. Because only jobs 1 and 2 require machine A, we must decide which to load first based on some priority rule. As an example, assume now that the chosen priority rule is to schedule in increasing order of processing time on the machine. (We shall call this rule shortest processing time [SPT].) Therefore, job 1 would be loaded first, followed by job 2. See Figure 16.6. Now focus on machine B. The only job available for processing at time 0 on machine B is job 3. So it would be scheduled first. When it finishes at time 4, only job 1 can be scheduled. Likewise, on machine C, the first job that can be scheduled is job 3 at time 4. When it completes, job 1 is the only remaining job that can be scheduled. The makespan is now 22.

Enns (1996) calls horizontal loading the "blocked-time" approach and compares it with an "event-driven" approach. The latter is similar to vertical loading in that it focuses on each machine and chooses a sequence based on all available jobs. The system reschedules each time an "event" occurs. Events are job completions at work centers, arrivals of new jobs, and so on. The "event-driven" approach attempts to never idle a machine if any jobs are ready for processing. Enns notes that event-driven schedules perform better in terms of flowtime, flowtime variability, and mean tardiness.

Another distinction, as described briefly in Chapter 15, is that of forward (or front) loading versus backward (or back) loading. Forward loading starts from the current time and builds the schedule forward in time. The user observes the completion time of each batch and can therefore determine if due dates will be met. If not, the schedule is infeasible. Backward loading starts from the due date of each job and schedules backward in time. The user then observes the start time of each job, and if it is prior to the current time, the schedule is infeasible. In either type of loading, if the schedule is infeasible, the user must adjust due dates, schedule overtime, employ subcontracting, or take other actions. The numerical illustration we have used in this section is an example of forward loading.

Sharma and Wilson (1995) argue that finite loading systems have a fundamental flaw because they typically take an MRP requirements plan as given, and then try to schedule orders. It is likely, however, that the MRP plan will be infeasible, and so the finite loading system will actually be working with a bad schedule. The alternative is to perform both planning and scheduling functions at the same time. Sharma and Wilson also suggest loading the bottleneck first, and then loading upstream and downstream from there. Advanced planning and scheduling (APS) systems, to be discussed shortly, do exactly this.

In related research, Enns (1996) notes that event-driven schedules are most compatible with forward loading. Guerrero and Kern (1988) discuss the use of forward loading and backward loading in the process of determining which customer orders to accept. See also Akkan (1996) who considers the possible use of overtime in determining whether to insert a new job into the schedule. Taylor and Bolander (1990) and Bolander and Taylor (1990) discuss forward loading and backward loading in process industries. See also Hastings et al. (1982).

Recall from Section 15.5 that capacity requirements planning (CRP) checks for violations of capacity, but that it does not actually schedule jobs to stay within available capacity. Finite loading and APS, used in conjunction with MRP and CRP, can simulate the load on each work center reasonably accurately because, unlike CRP, they do not rely on average queue times. If setup and processing times are not too variable, they should predict start and completion times fairly precisely.

Finally, we note that finite loading does not claim to find the best schedules. Rather, it attempts to generate good, feasible schedules. Near-optimization techniques do exist in APS and finite scheduling software, as noted in Section 15.8.1. APS systems use complex solution procedures, such as branch and bound, tabu, or other search algorithms, and with enough computing power can find near optimal solutions quite quickly. See Van Wassenhove and Gelders (1978), Kirk-patrick et al. (1983), Glover (1990), Conway and Maxwell (1993), Glover et al. (1993), Glover and Laguna (1993), and Chapter 4 of Baker (1995) for a description of some of these techniques.

16.3.3 Deterministic Scheduling of a Single Machine: Priority Sequencing Rules

In this section, we describe several priority sequencing rules that are designed to provide good schedules for a number of different objectives. We focus on a single machine to keep the discussion relatively straightforward and to provide fundamental insights. Several points should be made at the start. First, these sequencing rules do not explicitly consider the issue of capacity. They simply sequence jobs in priority order, without regard for capacity constraints. Second, they are generally designed to require very little computational effort and time, even for a large number of jobs. In fact, most require only sorting through all jobs based on one parameter, such as due date. Recall the number of possible permutation schedules for a set of 20 jobs. If it can be shown that sorting on one parameter provides the optimal schedule for the relevant performance objective, the computational effort (compared with evaluating all permutations) has been drastically reduced. Before discussing several specific cases, we first describe different sequencing rules in general terms. The issue at hand is which of a set of available jobs to process next.

Random Choose the next job at random. This rule is sometimes used in chaotic factories, but we certainly do not advocate using it.

FCFS First-come first-served. Choose the jobs in the order in which they arrive. This again is a common rule. It is viewed as being *fair*, and many service operations, such as call centers, use it for that reason. We shall see below, however, that it is far from optimal for many objectives.

SPT Choose the job that has the shortest processing time. Recall that, in this context, processing time includes setup time. (This appears to be the rule many students use in determining which assignment to work on next. It certainly was ours when we were students!)

SWPT Choose the job that has the shortest weighted processing time. A weight is assigned to each job based on the job's value (holding cost) or on its cost of delay.

EDD Choose the job that has the earliest due date. The previous two rules focus on the time a job spends in the shop, and they ignore any due date information. This rule specifically focuses on due dates.

CR Critical ratio. Compute the ratio (processing time remaining until completion)/(due date–current time), and choose the job with the highest ratio (provided it is positive). The ratio will be large for longer-remaining processing times, and for smaller amounts of time from the current time until the due date. However, if a job is late, the ratio will be negative, or the denominator will be zero, and the job should be given the highest priority. In the event that there is more than one late job, schedule the late jobs in SPT order.

There are dozens of priority sequencing rules that have been proposed in the literature for a multitude of objectives. We shall focus on just a few to provide insights. Valuable references for further information include Conway et al. (1967), Baker (1974), Coffman Jr. (1976), Graves (1981), French (1982), Lawler et al. (1982), Lawler et al. (1993), and Baker (1995). See Woolsey (1982) for an entertaining perspective on several of these rules in real shops.

16.3.3.1 Preliminary Comments and Notation

We focus on the simplest case of n jobs with deterministic processing times to be scheduled on one machine. This simple case will provide insights that are fundamental for more complex environment of a job shop with random processing times and dynamic arrivals. However, it is also useful for factories that have a single bottleneck. Recall that DBR and CONWIP, discussed earlier in this chapter, provide some insight into when jobs should be released into the shop. As jobs arrive at the bottleneck, however, they need to be prioritized. CONWIP explicitly allows for sequencing jobs based on sequence-dependent setup times, but if setup times are sequence independent, some sequencing rule is needed.

Two insights are immediately evident for the single machine case. First, the optimal schedule for most objectives will have no inserted idle time. Idle time simply delays jobs, leading to worse due date performance and longer flowtimes. The exception is when there is a penalty for jobs being early. Second, the optimal schedule most often will have no preemption. Preemption delays the preempted job, and advances the preempting job. But if the latter job should be advanced, it should have been scheduled prior to the preempted job in the first place.

We need some notation for the examples that follow.

$$C_j = \text{completion time for job } j$$

$$d_j = \text{due date for job } j$$

$$E_j = \text{earliness for job } j = \max(0, -L_j)$$

$$F_j = \text{flowtime for job } j = C_j - r_j$$

$$F = \text{total flowtime for all jobs to be scheduled} = \sum_{j=1}^{n} F_j$$

$$L_j = \text{lateness for job } j = C_j - d_j$$

$$n = \text{number of jobs to be scheduled}$$

$$p_j = \text{processing time for job } j$$

Table 16.6 Data for Single Machine Illustrations

Job	Processing Time, p_j (in Days)	Due Date, d_j (Day)
1	7	8
2	1	12
3	5	6
4	2	4
5	6	18

r_j = ready time for job j, or the time at which job j is available for processing

T_j = tardiness for job j = $\max(0, L_j)$

Note that lateness can be negative. One final piece of notation: $[k] = i$ means that the kth job in the sequence is job i.

16.3.3.2 Results and Insights

We now give examples and results for several sequencing rules. We shall use one numerical illustration, with data given in Table 16.6, for most of this subsection. The current time is time zero, so that job 1 is due 8 days from now.

16.3.3.3 FCFS

The FCFS schedule is quite easy to develop: simply take the jobs in the order in which they arrive. The results are given in Table 16.7, where the jobs have been numbered in the order of their arrival and all are available at time zero (i.e., each $r_j = 0$). To illustrate one job, consider job 3. Its completion time is 13, which is its start time plus its 5 day processing time. Job 3 begins in day 8 because job 2 completes on that day. Its flowtime is 13 days because it is ready at time 0, and it completes on day 13. Because its due date is 6, it is late by 7 days. The makespan is 21 days.

Table 16.7 FCFS Schedule

Job j	p_j	d_j	C_j	F_j	L_j	E_j	T_j
1	7	8	7	7	−1	1	0
2	1	12	8	8	−4	4	0
3	5	6	13	13	7	0	7
4	2	4	15	15	11	0	11
5	6	18	21	21	3	0	3
Average				12.8	3.2	1	4.2
Maximum				21	11	4	11

Table 16.8   SPT Schedule

Job j	p_j	d_j	C_j	F_j	L_j	E_j	T_j
2	1	12	1	1	−11	11	0
4	2	4	3	3	−1	1	0
3	5	6	8	8	2	0	2
5	6	18	14	14	−4	4	0
1	7	8	21	21	13	0	13
Average				9.4	−0.2	3.2	3
Maximum				21	13	11	13

Also note that the completion times and the flowtimes are identical for each job. Jobs spend, on average, 12.8 days in the shop, and on average they are tardy by 4.2 days. Now let us see what happens when SPT is employed.

16.3.3.4 SPT

In this case, we arrange jobs in increasing order of processing times, as shown in Table 16.8. Note that [2] = 4, or the second job in sequence is job 4. The mean flowtime is lower than that for the FCFS case, as are the average tardiness and lateness. The maximum lateness and tardiness are higher than in the FCFS case, however. This latter result illustrates a common complaint with SPT schedules, that they can cause long jobs to be very late. (Many students will recognize this problem. The long job is that history paper that never seems to get done.) There has been a significant amount of research that proves that SPT schedules are optimal for a number of objectives. We prove one case here, and then give the results for several others.

Total flowtime is minimized by SPT sequencing. The proof uses a simple argument that appears in the appendix to this chapter. Many proofs in simple sequencing problems follow a similar logic. Begin with a sequence that is *not* ordered according to the rule in question. Then interchange two jobs, so their ordering follows the rule. If the performance measure always improves, the sequencing rule is optimal for that performance measure. Similar proofs establish that SPT sequences are optimal for minimizing: (1) total flowtime, (2) mean flowtime, (3) mean waiting time, (4) mean lateness, and (5) total lateness.

16.3.3.5 EDD

It can be shown (by the same method illustrated in the appendix for SPT) that earliest due date sequencing minimizes *maximum lateness* and *maximum tardiness*. It is intuitive that on these performance measures, EDD outperforms SPT, and other rules based only on processing times, because the latter ignore due date information. We shall see, however, that our intuition is not always correct. EDD sequences are extremely easy to develop so long as each job has a due date attached. The EDD schedule for our numerical example is shown in Table 16.9. Note that, as expected, the average flowtime has now increased (compared with the SPT rule).

Table 16.10 provides a summary of key performance measures for these three sequencing rules.

692 ■ *Inventory and Production Management in Supply Chains*

Table 16.9 EDD Schedule

Job j	p_j	d_j	C_j	F_j	L_j	E_j	T_j
4	2	4	2	2	−2	2	0
3	5	6	7	7	1	0	1
1	7	8	14	14	6	0	6
2	1	12	15	15	3	0	3
5	6	18	21	21	3	0	3
Average				11.8	2.2	0.4	2.6
Maximum				21	6	2	6

Table 16.10 Summary of Results for Numerical Example

Rule	Average Flowtime	Average Lateness	Average Earliness	Maximum Earliness	Average Tardiness	Maximum Tardiness	Number of Tardy Jobs
FCFS	12.8	3.2	1.0	4	4.2	11	3
SPT	9.4	−0.2	3.2	11	3.0	13	2
EDD	11.8	2.2	0.4	2	2.6	6	4

16.3.4 General Job Shop Scheduling

It is clear from the previous section that many of the research results in scheduling are very specialized. If one happens to manage a shop that has a single significant bottleneck, the single machine research can be helpful. On the other hand, how should a manager approach a job shop that is more complex than the stylized cases we have been discussing? Real job shops have many complicating factors. For instance, there may be multiple machines that can process jobs at a given work center. There may be numerous different routings through the shop depending on the characteristics of the job. Each job will have a set of precedence constraints that specify the order in which operations must be performed. In addition, processing times may be probabilistic rather than deterministic. Do the results for the simple cases apply to more complex and realistic settings?

Complex job shops do not lend themselves to analytical research, and, therefore, most of the research has been based on simulation experiments. The problem with simulation is that it is difficult to know whether the results of the experiment apply in environments different from those tested. For example, we might simulate a job shop composed of six highly utilized machines and find that using SPT at each machine provides good performance for, say, mean flowtime. The question then is whether that result will generalize to shops of 10 machines, or to shops with some parallel machines, or to shops with lower utilization rates. In this section, we provide some insights into the general job shop scheduling problem.

16.3.4.1 Dynamic, Probabilistic Job Shop

In the dynamic probabilistic job shop, jobs arrive randomly over time, and processing times are probabilistic. This is the most realistic job shop situation, and, of course, the most difficult to model. Nevertheless, results from simpler contexts may provide insights. In the dynamic job shop, each machine may have a queue of jobs awaiting processing at any point in time. One primary decision is which job the machine operator should process next. We shall see, for example, that in many cases, shop performance improves if each machine operator chooses the job in his or her queue that has the shortest processing time. In other words, each operator uses SPT. First, however, we discuss some ideas that provide intuition about shop performance in general. We borrow the term *factory physics* from Hopp and Spearman (2011) for this discussion. Then we address the dynamic probabilistic job shop scheduling problem.

16.3.4.2 Factory Physics

Imagine a job shop that has just five machines, as illustrated in Figure 16.7. We begin with a very simple case. For the moment, let us assume that many jobs arrive at the shop at the same time, and that all jobs follow the same routing through the shop. In fact, let us assume that they proceed in order of machine letter, so that machine A is first, and so on. For now, assume that there is no space that can hold inventory waiting for processing. In Figure 16.7, the triangles that represent buffers (space for inventory) are always empty and can hold no jobs. (Note that WIP includes the contents of these buffers—zero for now—*and* any jobs that are being processed.) Finally, assume that processing times are deterministic, and are 1 hour for each job on each machine. It is relatively easy to understand the dynamics of this shop. Machine A is never starved for work because of the large number of jobs to be released to the shop. Once each machine has a job to work on, each machine will process a job for exactly 1 hour, and then pass it on to the next machine. The average throughput (TH = the number of completed jobs per unit time) of the shop will be 1 job per hour.

Now relax just one of the assumptions, that of deterministic processing times. There is still no buffer space, there is still a large number of jobs waiting to be released to the shop, and the routings remain the same. The processing times, however, now average 1 hour, but may vary from 0.5 to 1.5 hours, according to a uniform distribution. What will be the throughput of the shop? It seems

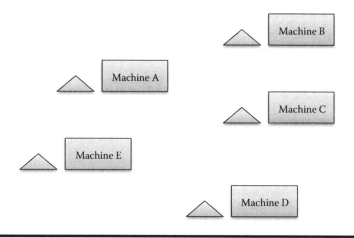

Figure 16.7   A five machine job shop.

clear that the average throughput should decrease, because of the variability in processing times. Any time a machine works slower than 1 hour, while the machine just upstream works faster, the upstream machine is *blocked*. It has nowhere to put the job, and must wait for the slower machine to finish before it can get started on the next job. Likewise, any time a machine works faster, while the machine just upstream works slower, the machine is *starved*. It has no job to work on, even though it is available. In this case, simulation results show that the average throughput is 0.78 jobs per hour, or a loss of 22% of the theoretical capacity of the shop. The key insight is that *variability decreases average throughput.*

How, then, can one increase the throughput of this shop? Because much of the loss in throughput is due to blocking and starving, we could eliminate these effects by increasing the buffers from their current size of zero—in other words, increasing WIP. Imagine now that we increase the provision for WIP, so that over time, the buffers, represented by triangles in the figure, fill up with jobs. When a machine works faster than average, it can put its completed job in the input buffer of the next machine in line, and it can begin processing the next job in its own input buffer. Blocking and starving will be eliminated if there is enough inventory in the buffers. However, if there is room for only a small amount of inventory in each buffer, there may still be occasional blocking and starving. The effect of increasing WIP on throughput is shown in Figure 16.8. Note that there is no throughput if there is no WIP (i.e., there is never a job in the shop), and that small amounts of WIP dramatically increase average throughput. At some point, however, average throughput levels off. In fact, the maximum throughput for this line is exactly 1 job per hour. If there were an infinite amount of inventory in front of each machine, so blocking and starving were completely eliminated, each machine would complete an average of 1 job per hour. In a shop with a bottleneck, the average throughput can never exceed the average throughput of the bottleneck. The key insight is that *WIP increases average throughput, but only to a point. That point is the average throughput of the bottleneck.*

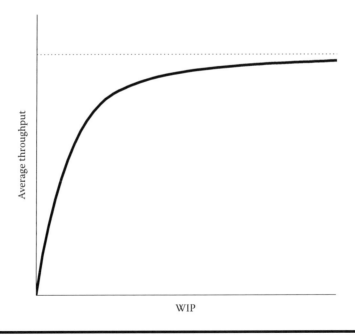

Figure 16.8 Relationship of WIP to average throughput.

Many production managers have seen the benefits of WIP reduction (discussed in this chapter in the context of JIT). We can see, however, that inventory reduction, without reduction of variability, serves only to decrease throughput. Well, then, why not increase WIP in every job shop? First of all, from the last point, we know that the throughput benefits disappear after WIP increases somewhat. In addition, consider the time it takes a job, 1, to proceed through the shop, or the cycle time (CT). If there is no other job in the shop for job 1's entire cycle time, it never has to wait in queue. Every machine is available when it arrives, and on average, the CT will be 5 hours, or 1 hour for each of the 5 machines. However, when there are other jobs in the shop, that is, there is WIP, the job may have to wait in queue for some time before processing, and the CT will increase. Imagine that we observe a job going through the shop, and that there is just one other job in the shop at that time. The probability of our job having to wait for the other job is quite low. Hence, WIP inventory must increase more for there to be a significant effect on cycle time. The relationship is shown in Figure 16.9. Notice how the cycle time increases very slowly as WIP first increases. Then, it increases more rapidly, and finally levels off to increase at a linear rate. The straight line portion of the curve has slope equal to the average processing time at the bottleneck. One more unit of WIP simply means that jobs must wait at the bottleneck for one more job to be processed. The key insight is that *cycle time increases with WIP*.

These insights apply to any complex job shop as well as to the simple case we have been using to develop intuition. In fact, we have used these insights in a consulting engagement to help managers examine whether they should inject more WIP into the shop. Perhaps the shop is operating with too little WIP, so that a small increase would generate a large increase in throughput with little cost in cycle time. In other words, the shop is operating on the steep part of the WIP–TH curve and on the shallow part of the WIP–CT curve. On the other hand, if the shop is really further up on the curves, increases in WIP will generate little gain in throughput, at the expense of a large increase in CT.

A fundamental law from queueing theory, due to Little (1961), captures some of these relationships.

$$WIP = TH(CT) \tag{16.1}$$

If there are 100 jobs in WIP on average, and the average throughput of the shop is 20 jobs per day, it is intuitive that the average time a job spends in the shop will be 5 days. This relationship, known as Little's law, holds for any factory, not just job shops. It can be extremely valuable for predicting delivery times, if the manager has a sense for average WIP and average throughput.[*] But it should be used in conjunction with the insights from Figures 16.8 and 16.9 when determining how much WIP to hold on average.

One final insight is illustrated in Figure 16.10. Recall that utilization is the proportion of available time that a machine is occupied. As utilization increases, average cycle time increases, at first gradually, and then very rapidly. This relationship is evident from the insights we have discussed. To increase utilization, machines must be starved less often, and this is accomplished by increasing WIP. When utilization approaches 1.0, the machine is busy constantly, and cycle times get longer and longer. If there is variability in job arrivals or in processing times, the problem is magnified because machine (idle) time, lost due to variability, can never be recaptured. Throughout this book, we have emphasized "changing the givens," and this is no exception. If managers can reduce the variability of the machine, by training or process improvements, or if they can reduce

---

[*] For other insights on predicting delivery times, see the papers by Enns listed in the references at the end of this chapter, and Cheng (1985).

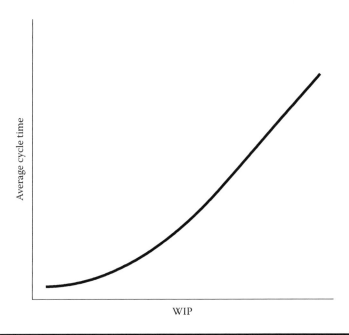

Figure 16.9 Relationship of WIP to average cycle time.

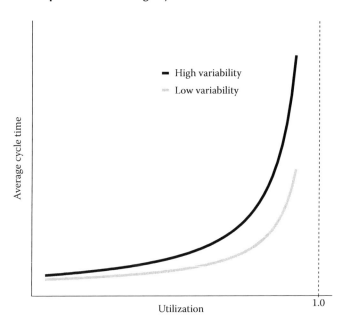

Figure 16.10 Variability and utilization.

the variability of job arrivals, by demand management, the entire curve shifts downward. So, we present these graphs and relationships as valuable tools for understanding the physics of factories, but we stress the importance of process improvements that shift the tradeoff curves in a favorable direction.

We note that the discussion in this subsection did not address sequencing. In the next subsection, we provide insights on sequencing in dynamic probabilistic job shops.

16.3.4.3 Results for Dynamic, Probabilistic Job Shop Scheduling

Most of this chapter has assumed deterministic processing times. When we allowed for probabilistic processing times, we only presented general insights. Now we present some specific insights and tools for more complex probabilistic job shops. Because much of queueing theory is unable to accommodate priority sequencing rules, researchers have relied on simulation to gain insight into sequencing in complex job shops. The discussion that follows is based on a vast amount of research that is summarized in Panwalkar and Iskander (1977), who provide a survey of over 100 dispatching rules; Conway (1965a,b), Conway et al. (1967), Blackstone et al. (1982), Chapter 5 of Hax and Candea (1984), Lawler et al. (1993), MacCarthy and Liu (1993), and Chapter 12 of Baker (1995).

We begin with some general comments on sequencing rules in this environment. *Local* rules require information only about the queue at the current machine. SPT is a local rule because to use it only requires knowledge of the processing times of each job in the current queue. *Global* rules require information about conditions elsewhere in the shop as well. Work in next queue (WINQ) is an example. This rule assigns the highest priority to the job that will join the queue (after this operation is completed) with the smallest workload, where workload is the sum of processing times waiting in that queue. *Static* rules have priorities that do not change over time. EDD is a static rule because the due date does not change as the job progresses through the shop. SPT, on the other hand, is static at a particular machine, but a job's priority can change as the job moves through the shop. *Dynamic* rules generate priorities that can change over time. Minimum slack time (MST) is a dynamic rule. Jobs are given priority based on slack time, which is the difference between the job's due date and the earliest possible completion time. The earliest possible completion time is the sum of the processing times of all remaining operations of this job, and it changes as operations are completed. Sequencing rules are generally divided into two groups: rules for relieving shop congestion and rules for meeting due dates. We discuss each in turn.

Shop congestion is exhibited by high levels of WIP, or by long flowtimes. A number of simulation experiments have found that SPT is the best rule for many performance measures related to shop congestion, including mean flowtime, mean lateness, and WIP level. SPT, however, may cause long jobs to experience excessive delays because short jobs keep arriving and moving to the front of the queue. Therefore, the variance of job flowtimes may be quite high with SPT. One possible solution would be to use truncated SPT (or TSPT). TSPT imposes a time limit on jobs in the queue. Any jobs exceeding the limit are sequenced according to FCFS. If no jobs in queue exceed the limit, use SPT. TSPT, therefore, moves jobs to the front of the queue if they have been waiting too long. As expected, flowtime variance decreases if we use TSPT in place of SPT, but WIP and average flowtimes increase. Another solution is to use relief SPT (RSPT) which employs FCFS until the queue length hits a value Q. Then we switch the rule to SPT. The results are roughly the same as with TSPT: flowtime variance decreases at the expense of higher average flowtime and WIP.

Our intuition might suggest that performance will improve if global rules are used in place of the local rules described in the last paragraph. Two possibilities are WINQ, which was mentioned above, and expected work in next queue (XWINQ). XWINQ is identical to WINQ except that it accounts for jobs that are expected to arrive to the subsequent queue. In fact, neither rule outperforms SPT when the performance measure is the mean number of jobs in queue. It seems that SPT is consistently the best, or near best, rule for relieving shop congestion.

There are many possible rules designed to meet due dates (assuming due dates are preassigned). These include EDD and CR, which were introduced above. A number of other rules have been proposed by researchers, including

MST Minimum slack time. Slack time is the difference between the time until the due date and the remaining processing time. Choose the job with the minimum slack time.

ODD Operation due date. For this rule, we create operation due dates that serve as intermediate deadlines prior to the real due date. Operation due dates may be spaced evenly between the job arrival and the final due date, by dividing the time interval by the number of operations to be performed. However, several other ways to set ODDs have been investigated. One is to use a proportion of the job's total work. Research has shown that this method performs better than dividing the time interval by the number of operations. In either case, the operation due dates are used to assign priority to jobs exactly as EDD does. It is worth noting that the operation due dates outperform job-based due dates (like EDD) in certain experiments. See Kanet and Hayya (1982) and Baker (1984).

S/OPN Slack time per operation. The job with the smallest ratio between slack time (as described above) and the number of operations remaining is given highest priority.

A/OPN Allowance per operation. The remaining allowance is the time between the current date and the due date. The job with the smallest ratio between remaining allowance and the number of remaining operations is given the highest priority.

MOD Modified operation due date. A modified operation's due date is the larger of its original operation due date and its earliest possible finish time.

COVERT COVERT stands for "C over T," where C represents the delay cost for the job, and T represents the processing time for this operation. Jobs are sequenced according to the ratio of delay cost to processing time.

Simulation experiments have shown that S/OPN performs best when the performance measure is the fraction of jobs tardy or the variance of job lateness. SPT, however, actually is very close on fraction of jobs tardy, and is even better than S/OPN for mean job lateness and mean flowtime. Some of these results are surprising because SPT ignores due date information. SPT even outperforms EDD on average job lateness! For variance of job lateness, however, EDD performs better than SPT. COVERT has been shown to outperform other rules on mean job tardiness. But, in general, the studies show mixed results for mean job tardiness. For the performance measure of mean tardiness, computed only for the set of tardy jobs, CR and similar rules have been shown to perform best. In addition to the surveys mentioned above, see Conway (1965b), and Carroll (1965) for some of these results.

To summarize, research would indicate that for most performance measures, except variance of job lateness and mean job tardiness, managers should direct each machine operator to sequence jobs in queue according to the SPT rule. There may be other exceptions, however, and it is best to simulate the specific job shop environment to be scheduled.

In a complex shop with the financial resources to employ a more elaborate scheduling system, it would be wise to consider APS systems, which we briefly introduced earlier in this chapter and in Chapter 15. These systems use advanced search techniques such as genetic algorithms, tabu search, and simulated annealing, among others. See the review by Blazewicz et al. (1996) for an excellent introduction to some of this literature, and Fleischmann and Meyr (2003) for a more detailed discussion. In very general terms, these techniques begin with a trial solution (an SPT sequence at the bottleneck, for instance), and then make some adjustment. The adjustment might be a pairwise interchange of two jobs in the sequence at a given machine. If the new sequence looks promising, retain that sequence and continue making interchanges, eventually including interchanges at all machines in the shop. Stop when the improvements are small. For large shops with multiple machines and jobs in progress, this process clearly involves a huge number of calculations, which is why sophisticated search algorithms are necessary. Fortunately, researchers have made significant progress on these algorithms, not to mention the dramatic increases in computing speed. For further information, see the references at the end of Section 16.3.2.

For other research on job shop scheduling, see Ragatz and Mabert (1984), Scudder and Hoffman (1987), Wein (1988), Vepsalainen and Morton (1988), Enns (1994), Brah (1996), and Randhawa and Zeng (1996). Hopp and Spearman (2011) provide details on the application of CONWIP in job shops. Treleven (1989) and Gargeya and Deane (1996) examine job shops that are constrained by multiple resources, including machines, labor, and tools.

16.4 Summary

In this chapter, we have discussed two powerful systems, JIT and OPT, that have made intellectual and practical contributions to operations management in the past several decades. JIT has had a significant impact on practicing managers, having been implemented in numerous plants. One of its primary contributions is the paradigm shift from the static view of reorder point systems or MRP to one of continuous improvement. OPT has had less of an impact as a production scheduling system, but it has made an important contribution by encouraging managers to focus improvement efforts on bottlenecks. We then reviewed a multitude of sequencing rules that apply to simple job shops. We used the rules and insights developed for the simple cases to establish effective procedures for more complex job shops. We discovered that the SPT rule is remarkably robust, providing the best, or near best, results for a number of different performance measures in a number of different environments. Managers should be wary of using this single rule in every environment, however, because other rules outperform it in some cases. Simulating the shop is a useful exercise to find the best strategy for each particular case. As always, managers should try to change the givens by controlling due dates, employing lot streaming, and eliminating bottlenecks.

Problems

16.1   For a local manufacturer, discuss the production process and describe why MRP, JIT, or OPT would be most suitable.

16.2   Describe the difference between a process batch and a transfer batch.

16.3   What could be the primary contribution of JIT for a manufacturer using MRP?

16.4   What could be the primary contribution of OPT for a manufacturer using MRP?

16.5   Which is better, MRP, JIT, or OPT?

16.6   Describe the difference between push and pull systems for production. Why is MRP often called a push system, while JIT is called a pull system?

16.7   Design a simple production line using a real or hypothetical product and simple materials. Develop Kanbans for this line, and with your group, operate until it runs smoothly.

16.8   Describe drum-buffer-rope scheduling and apply it to a production line of a local organization.

16.9 For the following data, assume that there are 80 hours of time available on the bottleneck machine. Find the number of each product to produce using the conventional approach based on contribution margin. Then find the optimal numbers using the OPT principles.

Product	A	B
Price ($/unit)	600	400
Cost of materials ($/unit)	250	200
Market demand per week (units)	50	30
Time on the bottleneck machine (hours/unit)	2	1
Contribution margin per unit ($/unit)	350	200

16.10 Why is short-term scheduling not as appropriate for continuous process industries? Give an example of how some of the methods discussed in this chapter nevertheless might be helpful.

16.11 Why is short-term scheduling not as appropriate for Kanban systems? Give an example of how some of the techniques discussed in this chapter nevertheless might be helpful.

16.12 Briefly discuss how a local firm uses, or might use, Gantt charts and finite scheduling.

16.13 Briefly discuss how a local firm uses, or might use, sequencing rules such as SPT or EDD.

16.14 What insights from single machine deterministic problems are useful for more complex job shops? What warnings would you give a manager who is planning to use these insights?

16.15 Give an intuitive explanation of why SPT performs so well for (a) mean flowtime and (b) mean lateness.

16.16 Explain how short-term scheduling tools work together with MRP. Explain how short-term scheduling tools work together with ERP.

16.17 Find and diagram the process flows for a local flow shop.

16.18 Find and diagram the process flows for a local job shop.

16.19 The choice of sequencing rule should be consistent with the operations strategy of the firm. What strategies are consistent with SPT? With EDD? With FCFS? With SWPT? With S/OPN? With MOD?

16.20 Draw a Gantt chart for the following data, assuming that jobs are sequenced in order of job number.

	Machine	
Job	A	B
1	3	9
2	5	1
3	6	8
4	4	7

16.21 Draw a Gantt chart for the following data, assuming that jobs are sequenced in order of job number.

	Machine	
Job	A	B
1	30	45
2	20	15
3	60	25
4	80	10
5	50	50
6	20	70

16.22 Use the following data to forward load the three machines.

		Machine		
Job	Priority	A	B	C
1	First	5	3	7
2	Second		7	3
3	Third	4	5	6

16.23 Use the data in the previous example to backward load the three machines, assuming the three jobs have due dates of 20, 12, and 21, respectively.

16.24 Assuming a priority sequence of 1, 2, 3, 4, schedule the following jobs on a Gantt chart. The numbers in parentheses indicate the order in which operations are to be done. So job 1 requires 3 hours on A, followed by 4 hours on B. Job 2 begins on machine C for 2 hours, and then proceeds to machine B for 3 hours, and so on.

		Machine	
Job	A	B	C
1	3(1)	4(2)	
2	4(3)	3(2)	2(1)
3		1(1)	3(2)
4	2(1)	1(2)	2(3)

16.25 Assuming the data in the previous example, improve the makespan by changing the schedule.

16.26 Backward load for the following data:

Job	Due Date	Machine		
		A	B	C
1	4	2(1)	3(2)	4(3)
2	3	4(2)		6(1)
3	4	1(3)	3(1)	2(2)
4	4	3(3)	3(2)	4(1)
5	2	5(1)	3(2)	

16.27   For the following data (assuming jobs are numbered in the order of their arrival), find the mean flowtime, mean lateness, mean tardiness, and maximum tardiness.
   a.   Using FCFS
   b.   Using SPT
   c.   Using EDD
   d.   Using CR
   e.   Using SWPT, assuming weights of 3, 4, 6, 10, 1, respectively

Job	Processing Time, p_j (Days)	Due Date, d_j (Day)
1	5	10
2	3	8
3	9	11
4	7	16
5	15	22

16.28   For the following data (assuming jobs are numbered in the order of their arrival), find the mean flowtime, mean lateness, mean tardiness, and maximum tardiness.
   a.   Using FCFS
   b.   Using SPT
   c.   Using EDD
   d.   Using CR
   e.   Using SWPT, assuming weights of 2, 2, 3, 5, 1, 6, 1, respectively

Job	Processing Time, p_j (Days)	Due Date, d_j (Day)
1	8	19
2	6	10
3	1	5

Job	Processing Time, p_j (Days)	Due Date, d_j (Day)
4	9	22
5	12	30
6	15	50
7	2	6

16.29 A job shop manager has been asked what the average lead times in her shop have been. Unfortunately, no one has been tracking these values. She has, however, been able to determine that approximately 14 jobs are completed per week, and that there are currently 65 jobs in the shop. What should she quote as the average lead time? Should she tell marketing that the next job can be delivered in this time?

16.30 The president of a small job shop is frustrated by the amount of money he is spending on WIP inventory, and by the low revenues the shop has been earning. The sales manager tells the president that if she could quote lead times of 1 week, sales could increase significantly. The average job generates $2,000 in revenue, and the shop has been earning about $16,000 per week. A count of WIP reveals that the average number of jobs in WIP is about 20. What should you tell the president?

Appendix 16A: Proof that SPT Minimizes Total Flowtime

Total flowtime is minimized by SPT sequencing. The proof uses the following simple argument. The flowtime of the kth job in sequence is

$$F_{[k]} = \sum_{i=1}^{k} p_{[i]} \tag{16A.1}$$

So the flowtime for job 3, for instance, is $1 + 2 + 5 = 8$, or $p_{[1]} + p_{[2]} + p_{[3]} = p_2 + p_4 + p_3$. The total flowtime, then, is

$$\sum_{k=1}^{n} F_{[k]} = \sum_{k=1}^{n} \sum_{i=1}^{k} p_{[i]} \tag{16A.2}$$

In our example, the total flowtime is

$$\sum_{k=1}^{n} \sum_{i=1}^{k} p_{[i]} = p_{[1]} + (p_{[1]} + p_{[2]}) + (p_{[1]} + p_{[2]} + p_{[3]}) + (p_{[1]} + p_{[2]} + p_{[3]} + p_{[4]})$$

$$+ (p_{[1]} + p_{[2]} + p_{[3]} + p_{[4]} + p_{[5]})$$

$$= 5p_{[1]} + 4p_{[2]} + 3p_{[3]} + 2p_{[4]} + p_{[5]}$$

$$= 1 + (1 + 2) + (1 + 2 + 5) + (1 + 2 + 5 + 6) + (1 + 2 + 5 + 6 + 7) = 47$$

and the mean flowtime is $47/5 = 9.4$. Now consider a sequence that is not SPT, and call this sequence S. Because it is not SPT, there must be two jobs, i and j, with i before j such that $p_i > p_j$. Now interchange jobs i and j in the sequence and call the new sequence S'. See Figure 16A.1. All

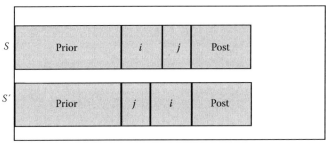

Time

Figure 16A.1   Illustration of the SPT optimality for total flowtime.

jobs prior to jobs i and j (labeled "Prior" in Figure 16A.1) will finish processing at the same time in both S and S' because they are unchanged. Likewise, all jobs after i and j (labeled "Post" in Figure 16A.1) will have identical completion times in both sequences. But the flowtimes of i and j will clearly differ. In sequence S, the flowtime of job i will equal the flowtime of the last job in the set of jobs prior to job $i + p_i$. And the flowtime of job j will equal the flowtime of the last job in the set of jobs prior to job $i + p_i + p_j$. Compare these with sequence S'. The flowtime of job j will equal the flowtime of the last job in the set of jobs prior to job $j + p_j$. And the flowtime of job i will equal the flowtime of the last job in the set of jobs prior to job $j + p_j + p_i$. The difference between total flowtime of S and S' is

$$p_i + (p_i + p_j) - (p_j + p_j + p_i) = p_i - p_j > 0$$

The latter inequality is because $p_i > p_j$. Therefore, the total flowtime of S' is less. Now if any further jobs are not ordered in SPT sequence, repeat the same interchange and the total flowtime will decrease. Thus, total flowtime is minimized using SPT sequences.

References

Abdulnour, G., R. A. Dudek, and M. L. Smith. 1995. Effect of maintenance policies on the Just-in-Time production system. *International Journal of Production Research 33*(2), 565–583.

Adelsberger, H. and J. Kanet. 1991. The Leitstand—A new tool for computer-integrated manufacturing. *Production and Inventory Management Journal 32*(1), 43–47.

Akkan, C. 1996. Overtime scheduling: An application of finite-capacity real-time scheduling. *Journal of the Operational Research Society 47*, 1137–1149.

Ashton, J. E. and F. X. Cook Jr. 1989. Time to reform job shop manufacturing. *Harvard Business Review 67*(2), 106–111.

Axsäter, S. and K. Rosling. 1994. Multi-level production-inventory control: Material requirements planning or reorder point policies? *European Journal of Operational Research 75*, 405–412.

Baker, K. R. 1974. *Introduction to Sequencing and Scheduling.* New York: John Wiley & Sons.

Baker, K. R. 1984. Sequencing rules and due date assignments in a job shop. *Management Science 30*, 1093–1104.

Baker, K. R. 1995. *Elements of Sequencing and Scheduling.* Hanover, NH: The Amos Tuck School, Dartmouth College.

Baker, K. R., S. G. Powell, and D. F. Pyke. 1990. Buffered and unbuffered assembly systems with variable processing times. *Journal of Manufacturing and Operations Management 3*, 200–223.

Baker, K. R., S. G. Powell, and D. F. Pyke. 1993. Optimal allocation of work in assembly systems. *Management Science* 39(1), 101–106.

Bemelmans, R. 1985. Capacity oriented production planning in case of a single bottleneck. *Engineering Costs and Production Economics* 9, 135–140.

Benjaafar, S. 1996. Modeling and analysis of machine sharing in manufacturing systems. *European Journal of Operational Research* 91(1), 56–73.

Bertrand, J. W. M. 1983. The use of workload information to control job lateness in controlled and uncontrolled release production systems. *Journal of Operations Management* 3(2), 79–92.

Bertrand, J. W. M. and J. Wortmann. 1981. *Production Control and Information Systems for Component Manufacturing.* Amsterdam: Elsevier.

Betz Jr., H. J. 1996. Common sense manufacturing, a method of production control. *Production and Inventory Management Journal* 37(1), 77–81.

Bielecki, T. and P. R. Kumar. 1988. Optimality of zero-inventory policies for unreliable manufacturing systems. *Operations Research* 36(4), 532–541.

Billesbach, T. J. and N. J. Schniederjans. 1989. Applicability of Just-in-Time techniques in administration. *Production and Inventory Management Journal* 30(3), 40–45.

Blackburn, J. 1991. Time based competition: JIT as a weapon. *APICS—The Performance Advantage* 1(1), 30–31, 34.

Blackburn, J. D. 1992. Time-based competition: White-collar activities. *Business Horizons* 35(4), 96–101.

Blackstone, J. J. H., D. T. Phillips, and G. L. Hogg. 1982. A state-of-the-art survey of dispatching rules for manufacturing job shop operations. *International Journal of Production Research* 20(1), 27–45.

Blazewicz, J., W. Domschke, and E. Pesch. 1996. The job shop scheduling problem: Conventional and new solution techniques. *European Journal of Operational Research* 93, 1–33.

Bolander, S. F. and S. G. Taylor. 1990. Process flow scheduling: Mixed-flow cases. *Production and Inventory Management Journal* 31(4), 1–6.

Bowman, D. J. 1991. If you don't understand JIT how can you implement it? *Industrial Engineering* 23(2), 38–39.

Brah, S. A. 1996. A comparative analysis of due date based job sequencing rules in a flow shop with multiple processors. *Production Planning and Control* 7(4), 362–373.

Buzacott, J. A. and J. G. Shantikumar. 1993. *Stochastic Models of Manufacturing Systems.* Englewood Cliffs, NJ: Prentice-Hall, Inc.

Carroll, D. C. 1965. *Heuristic Sequencing of Single and Multiple Component Jobs.* PhD thesis. MIT, Cambridge, MA.

Chapman, S. 1989. Just-in-Time supplier inventory: An empirical implementation model. *International Journal of Production Research* 27(12), 1993–2007.

Chapman, S. N. 1990. Schedule stability and the implementation of Just-In-Time. *Production and Inventory Management Journal* 31(3), 66–70.

Cheng, T. C. E. 1985. Analysis of flow-time in a job shop. *Journal of the Operational Research Society* 36, 225–230.

Coffman Jr., E. G. 1976. *Computer & Job/Shop Scheduling Theory.* New York: John Wiley & Sons.

Conway, R. W. 1965a. Priority dispatching and job lateness in a job shop. *Journal of Industrial Engineering* 16(4), 228–237.

Conway, R. W. 1965b. Priority dispatching and work-in-process inventory in a job shop. *Journal of Industrial Engineering* 16(2), 123–130.

Conway, R. W. and W. L. Maxwell. 1993. *PRS: Production Reservation System: User's Guide.* New York: C-Way Associates.

Conway, R. W., W. L. Maxwell, J. O. McClain, and L. J. Thomas. 1988. The role of work-in-process inventories in serial production lines. *Operations Research* 36, 229–241.

Conway, R. W., W. L. Maxwell, and L. W. Miller. 1967. *Theory of Scheduling.* Reading, MA: Addison-Wesley.

Crawford, K. M. and J. F. Cox. 1991. Addressing manufacturing problems through the implementation of just-in-time. *Production and Inventory Management Journal* 32(1), 33–36.

Dallery, Y. and S. B. Gershwin. 1992. Manufacturing flow line systems: A review of models and analytical results. *Queueing Systems* 12, 3–94.

Deleersnyder, J. L., T. Hodgson, R. King, P. O'Grady, and A. Savva. 1992. Integrating Kanban type pull systems and MRP type push systems: Insights from a Markovian model. *IIE Transactions 24*(3), 43–56.

Donahue, K. and M. Spearman. 1993. Improving the design of stochastic production lines: An approach using perturbation analysis. *International Journal of Production Research 31*(12), 2789–2806.

Duenyas, I. 1994. The limitations of suboptimal policies. *Interfaces 24*(5), 77–84.

Duenyas, I., W. J. Hopp, and M. L. Spearman. 1993. Characterizing the output process of a CONWIP line with deterministic processing and random outages. *Management Science 39*, 975–988.

Duersch, R. and D. B. Wheeler. 1981. An interactive scheduling model for assembly-line manufacturing. *International Journal of Modelling & Simulation 1*(3), 241–245.

Enns, S. T. 1994. Job shop leadtime requirements under conditions of controlled delivery performance. *European Journal of Operational Research 77*, 429–439.

Enns, S. T. 1996. Finite capacity scheduling systems: Performance issues and comparisons. *Computers and Industrial Engineering 30*(4), 727–739.

Fawcett, S. and J. Pearson. 1991. Understanding and applying constraint management in today's manufacturing environments. *Production and Inventory Management Journal 32*(3), 46–55.

Feather, J. and K. Cross. 1988. Workflow analysis, Just-in-Time technique simplify administrative process in paperwork operation. *Industrial Engineering 20*(1), 32–40.

Fine, C. H. and E. L. Porteus. 1989. Dynamic process improvement. *Operations Research 37*(4), 580–591.

Flapper, S. D. P., G. J. Miltenburg, and J. Wijngaard. 1991. Embedding JIT into MRP. *International Journal of Production Research 29*(2), 329–341.

Fleischmann, B. and H. Meyr. 2003. Planning hierarchy, modeling and advanced planning systems. *Handbooks in Operations Research and Management Science 11*, 455–523.

Freeland, J. R., J. P. Leschke, and E. N. Weiss. 1990. Guidelines for setup-cost reduction programs to achieve zero inventory. *Journal of Operations Management 9*(1), 85–100.

French, S. 1982. *Sequencing and Scheduling: An Introduction to the Mathematics of the Job-Shop*. Chichester: Horwood.

Funk, J. L. 1989. A comparison of inventory cost reduction strategies in a JIT manufacturing system. *International Journal of Production Research 27*(7), 1065–1080.

Gargeya, V. B. and R. H. Deane. 1996. Scheduling research in multiple resource constrained job shops: A review and critique. *International Journal of Production Research 34*(8), 2077–2097.

Gerchak, Y. and Y. Wang. 1994. Periodic review inventory models with inventory-level-dependent demand. *Naval Research Logistics 41*(1), 99–116.

Glover, F. 1990. Tabu search: A tutorial. *Interfaces 20*(4), 74–94.

Glover, F. and M. Laguna. 1993. Tabu search. In C. R. Reeves (Ed.), *Modern Heuristic Techniques for Combinatorial Problems*, pp. 70–141. Oxford: Blackwell Scientific Publishing.

Glover, F., E. Taillard, and D. de Werra. 1993. A user's guide to tabu search. *Annals of Operations Research 41*, 3–38.

Goddard, W. E. 1982. Kanban versus MRP II—Which is best for you? *Modern Materials Handling* (5), 40–48.

Goldratt, E. M. 1988. Computerized shop floor scheduling. *International Journal of Production Research 26*(3), 443–455.

Goldratt, E. M. and J. Cox. 1986a. *The Goal* (Revised ed.). Croton-on-Hudson, NY: North River Press.

Goldratt, E. M. and J. Cox. 1986b. *The Race*. Croton-on-Hudson, NY: North River Press.

Golhar, D. Y. and C. L. Stamm. 1991. The Just-in-Time philosophy: A literature review. *International Journal of Production Research 29*(4), 657–676.

Graham, I. 1992. Comparing trigger and Kanban control of flow-line manufacture. *International Journal of Production Research 30*(10), 2351–2362.

Gravel, M. and W. L. Price. 1988. Using Kanban in a job shop environment. *International Journal of Production Research 26*(6), 1105–1118.

Graves, S. C. 1981. A review of production scheduling. *Operations Research 29*(4), 646–675.

Groenevelt, H. and U. S. Karmarkar. 1988. A dynamic Kanban system case study. *Production and Inventory Management Journal 29*, 46–50.

Groenevelt, H., G. L. Nemhauser, and A. H. G. Rinnooy Kan. 1993. The Just-in-Time system. In S. C. Graves, A. H. G. Rinnooy Kan, and P. H. Zipkin (Eds.), *Logistics of Production and Inventory*, Volume 4, pp. Chapter 12. Amsterdam: North-Holland.

Gstettner, S. and H. Kuhn. 1996. Analysis of production control systems Kanban and CONWIP. *International Journal of Production Research 34*(11), 3253–3273.

Guerrero, H. H. and G. M. Kern. 1988. How to more effectively accept and refuse orders. *Production and Inventory Management Journal 29*(4), 59–63.

Guide Jr., V. D. R., R. Srivastava, and M. S. Spencer. 1996. Are production systems ready for the green revolution? *Production and Inventory Management Journal 37*(4), 70–76.

Hall, R. W. 1983. *Zero Inventories*. Homewood, IL: Dow Jones-Irwin.

Hall, R. W. 1988. Cyclic scheduling for improvement. *International Journal of Production Research 26*(3), 457–472.

Hastings, N., P. Marshall, and R. Willis. 1982. Schedule based MRP: An integrated approach to production scheduling and material requirements planning. *Journal of the Operational Research Society 33*(11), 1021–1029.

Hax, A. C. and D. Candea. 1984. *Production and Inventory Management*. Englewood Cliffs, NJ: Prentice-Hall.

Hay, E. J. 1987. Any machine setup time can be reduced by 75%. *Industrial Engineering 19*(8), 62–67.

Helper, S. R. and M. Sako. 1995. Supplier relations in Japan and the United States: Are they converging? *Sloan Management Review 36*(3), 77–84.

Hillier, F. and R. Boling. 1966. The effect of some design factors on the efficiency of production lines with variable operation times. *Journal of Industrial Engineering 17*(12), 651–658.

Hillier, F. and R. Boling. 1979. On the optimal allocation of work in symmetrically unbalanced production line systems with variable operation times. *Management Science 25*, 721–728.

Hillier, F. S. and K. C. So. 1991. The effect of the coefficient of variation of operation times on the allocation of storage space in production line systems. *IIE Transactions 23*(2), 198–206.

Hillier, F. S. and K. C. So. 1996. On the robustness of the bowl phenomenon. *International Journal of Production Research 89*, 496–515.

Hodgson, T. J. and D. Wang. 1991. Optimal hybrid PUSH/PULL control strategies for a parallel multistage system: Part II. *International Journal of Production Research 29*(7), 1453–1460.

Hopp, W. J. and J. T. Simon. 1989. Bounds and heuristics for assembly-like queues. *Queueing Systems 4*, 137–156.

Hopp, W. J. and M. L. Spearman. 1991. Throughput of a constant work in process manufacturing line subject to failures. *International Journal of Production Research 29*(3), 635–655.

Hopp, W. J. and M. L. Spearman. 2011. *Factory Physics*. Long Grove, IL: Waveland Press.

Huang, P. Y., L. P. Rees, and B. W. Taylor. 1983. A simulation analysis of the Japanese Just-in-Time technique (with Kanbans) for a multiline, multistage production system. *Decision Sciences 14*, 326–344.

Huson, M. and D. Nanda. 1995. The impact of Just-in-Time manufacturing on firm performance in the US. *Journal of Operations Management 12*(3–4), 297–310.

Imai, M. 1997. *Gemba Kaizen: A Commonsense, Low-Cost Approach to Management*. New York: McGraw-Hill Professional.

Inman, R. R. 1993. Inventory is the flower of all evil. *Production and Inventory Management Journal 34*(4), 41–45.

Jacobs, F. R. 1984. OPT uncovered: Many production planning and scheduling concepts can be applied with or without the software. *Industrial Engineering 16*(10), 32–41.

Jaikumer, R. and R. Bohn. 1992. A dynamic approach to operations management: An alternative to static optimization. *International Journal of Production Economics 27*, 265–282.

Jain, S., M. E. Johnson, and F. Safai. 1996. Implementing setup optimization on the shop floor. *Operations Research 43*(6), 843–851.

Kamata, S. 1982. *Japan in the Passing Lane: An Insider's Account of Life in a Japanese Auto Factory*. New York: Pantheon Books.

Kanet, J. J. and J. C. Hayya. 1982. Priority dispatching with operation due dates in a job shop. *Journal of Operations Management 2*(3), 167–175.

Karmarkar, U. 1989. Getting control of Just-in-Time. *Harvard Business Review 67*(5), 122–131.

Karmarkar, U. S. 1993. Manufacturing lead times, order release and capacity loading. *Handbooks in Operations Research and Management Science 4*, 287–329.

Karmarkar, U. S. and S. Kekre. 1989. Batching policy in Kanban systems. *Journal of Manufacturing Systems 8*, 317–328.

Kern, G. M. and J. C. Wei. 1996. Master production rescheduling policy in capacity-constrained Just-in-Time Make-to-Stock environments. *Decision Sciences 27*(2), 365–387.

Kim, T. M. 1985. Just-in-time manufacturing system: A periodic pull system. *International Journal of Production Research 23*(3), 553–562.

Kimura, O. and H. Terada. 1981. Design and analysis of pull systems, a method of multi-stage production control. *International Journal of Production Research 19*(3), 241–253.

Kirkpatrick, S., C. D. Gelatt, and M. P. Vecchi. 1983. Optimization by simulated annealing. *Science 220*, 671–680.

Klein, J. A. 1989. The human costs of manufacturing reform. *Harvard Business Review 67*(2), 60–66.

Koenigsberg, E. 1985. Seppuku in the stockroom. *Interfaces 15*(4), 86–88.

Krajewski, L., B. King, L. Ritzman, and D. Wong. 1987. Kanban, MRP and shaping the manufacturing environment. *Management Science 33*(1), 39–57.

Lawler, E. L., J. K. Lenstra, and A. H. G. Rinnooy Kan. 1982. Recent developments in deterministic sequencing and scheduling: A survey. In M. A. H. Dempster, J. K. Lenstra, and A. H. G. Rinnooy Kan (Eds.), *Deterministic and Stochastic Scheduling*, pp. 35–73. Dordrecht: Reidel.

Lawler, E. L., J. K. Lenstra, A. H. G. Rinnooy Kan, and D. B. Shmoys. 1993. Sequencing and scheduling: Algorithms and complexity. In S. Graves, A. Rinnooy Kan, and P. Zipkin (Eds.), *Logistics of Production and Inventory*, Volume 4, pp. Chapter 9. Amsterdam: Elsevier (North-Holland).

Leschke, J. P. 1996. An empirical study of the setup-reduction process. *Production and Operations Management 5*(2), 121–131.

Liberopoulos, G. and Y. Dallery. 2003. Comparative modelling of multi-stage production-inventory control policies with lot sizing. *International Journal of Production Research 41*(6), 1273–1298.

Lieberman, M. and L. Demeester. 1999a. Inventory reduction and productivity growth: Linkages in the Japanese automotive industry. *Management Science 45*(4), 466–485.

Lieberman, M. B. and L. Demeester. 1999b. Inventory reduction and productivity growth: Linkages in the Japanese automotive industry. *Management Science 45*(4), 466–485.

Little, J. D. C. 1961. A proof for the queueing formula: $L = \lambda W$. *Operations Research 9*, 383–387.

Luebbe, R. and B. Finch. 1992. Theory of constraints and linear programming: A comparison. *International Journal of Production Research 30*(6), 1471–1478.

Luss, H. and M. Rosenwein. 1990. A lot-sizing model for Just-in-Time manufacturing. *Journal of the Operational Research Society 41*(3), 201–209.

MacCarthy, B. L. and J. Liu. 1993. Addressing the gap in scheduling research: A review of optimization and heuristic methods in production scheduling. *International Journal of Production Research 31*(1), 59–79.

MacDuffie, J. P., K. Sethuraman, and M. L. Fisher. 1996. Product variety and manufacturing performance: Evidence from the international automotive assembly plant study. *Management Science 42*(3), 350–369.

McMillan, J. 1990. Managing suppliers: Incentive systems in Japanese and U.S. industry. *California Management Review 32*(4), 38–55.

Mefford, R. N. 1989. The productivity nexus of new inventory and quality control techniques. *Engineering Costs and Production Economics 17*, 21–28.

Meyer, H., F. Fuchs, and K. Thiel. 2009. *Manufacturing Execution Systems (MES): Optimal Design, Planning, and Deployment: Optimal Design, Planning, and Deployment*. New York: McGraw-Hill Professional.

Miller, K. L. 1991. "Just-in-Time" is becoming just a pain. *Business Week June 17*, 100H.

Miltenburg, J. and G. Sinnamon. 1989. Scheduling mixed-model multi-level Just-in-Time production systems. *International Journal of Production Research 27*(9), 1487–1509.

Moinzadeh, K. and P. K. Aggarwal. 1997. Analysis of a production/inventory system subject to random disruptions. *Management Science 43*(11), 1577–1588.

Moinzadeh, K., T. D. Klastorin, and E. Berk. 1997. The impact of small lot ordering on traffic congestion in a physical distribution system. *IIE Transactions 29*(8), 671–680.

Monden. 1984. A simulation analysis of the Japanese Just-in-Time technique (with Kanbans) for a multiline, multistage production system: A comment. *Decision Sciences 15*, 445–447.

Monden, Y. 1993. *Toyota Production System* (Second Edition ed.). Atlanta, GA: Industrial Engineering Press.

Muckstadt, J. A. and S. R. Tayur. 1995a. A comparison of alternative Kanban control mechanisms. I. background and structural results. *IIE Transactions 27*, 140–150.

Muckstadt, J. A. and S. R. Tayur. 1995b. A comparison of alternative Kanban control mechanisms. II. experimental results. *IIE Transactions 27*, 151–161.

Ohno, T. 1988. *Toyota Production System: Beyond Large Scale Production*, Volume (Original Japanese edition published in 1978.). Portland, OR: Productivity Press.

Olhager, J. and B. Östlund. 1990. An integrated push-pull manufacturing strategy. *European Journal of Operational Research 45*, 135–142.

Panwalkar, S. S. and W. Iskander. 1977. A survey of scheduling rules. *Operations Research 25*(1), 45–61.

Piper, C. J. and R. McLachlin. 1990. Just-in-Time production: Eleven achievable dimensions. *Operations Management Review 7*(3), 1–8.

Porteus, E. L. 1986. Investing in new parameter values in the discounted EOQ model. *Naval Research Logistics Quarterly 33*(1), 39–48.

Powell, S. G. and D. F. Pyke. 1996. Allocation of buffers to serial production lines with bottlenecks. *IIE Transactions 28*(1), 18–29.

Ptak, C. A. 1991. MRP, MRP II, OPT, JIT, and CIM—Succession, evolution, or necessary combination. *Production and Inventory Management Journal 32*(2), 7–11.

Pyke, D. F. and M. A. Cohen. 1990. Push and pull in manufacturing and distribution systems. *Journal of Operations Management 9*(1), 24–43.

Pyke, D. F. and M. A. Cohen. 1993. Performance characteristics of stochastic integrated production-distribution systems. *European Journal of Operational Research 68*(1), 23–48.

Pyke, D. F. and M. A. Cohen. 1994. Multiproduct integrated production-distribution systems. *European Journal of Operational Research 74*, 18–49.

Ragatz, G. L. and V. A. Mabert. 1984. A simulation analysis of due date assignment rules. *Journal of Operations Management 5*(1), 27–39.

Randhawa, S. U. and Y. Zeng. 1996. Job shop scheduling: An experimental investigation of the performance of alternative scheduling rules. *Production Planning and Control 7*(1), 47–56.

Rees, L. P., P. Y. Huang, and B. W. Taylor. 1989. A comparative analysis of an MRP lot-for-lot system and a Kanban system for a multistage production operation. *International Journal of Production Research 27*(8), 1427–1443.

Richmond, L. E. and J. H. Blackstone. 1988. Just-in-Time in the plastics processing industry. *International Journal of Production Research 26*(1), 27–34.

Samaddar, S. 1996. The limits of Japanese production systems. *Interfaces 26*(4), 66–68.

Samaddar, S. and T. Kaul. 1995. Effects of setup and processing time reductions on WIP in the JIT production systems. *Management Science 41*(7), 1263–1265.

Sarker, B. R. and R. D. Harris. 1988. The effect of imbalance in a Just-in-Time production system: A simulation study. *International Journal of Production Research 26*(1), 1–18.

Schonberger, R. J. 1983. Applications of single-card and dual-card Kanban. *Interfaces 13*(4), 56–67.

Schonberger, R. J. 1984. Just-in-Time production systems: Replacing complexity with simplicity in manufacturing management. *Industrial Engineering 16*(10), 52–63.

Scudder, G. D. and T. R. Hoffman. 1987. The use of cost-based priorities in random and flow shops. *Journal of Operations Management 7*(1–2), 217–232.

Seidmann, A. 1988. Regenerative pull (Kanban) production control policies. *European Journal of Operational Research 35*, 401–413.

Severson, D. 1988. The SMED system for reducing change over times: An exciting catalyst for companywide improvement and profits. *Production and Inventory Management 8*(10), 10–16.

Sharma, K. and J. Wilson. 1995. What's wrong with finite capacity scheduling. *APICS—The Performance Advantage 5*, March, 37–39.

Shingo, S. 1985. *A Revolution in Manufacturing: The SMED System*. Cambridge, MA: Productivity Press.

Shingo, S. 1989. *Study of the Toyota Production System from an Industrial Engineering Viewpoint*. Cambridge, MA: Productivity Press.

Silver, E. A. 1992. Changing the givens in modelling inventory problems: The example of Just-in-Time systems. *International Journal of Production Economics 26*, 347–351.

Simons Jr., J. V., W. P. Simpson III, B. J. Carlson, S. W. James, C. A. Lettiere, and B. A. Mediate Jr. 1996. Formulation and solution of the drum-buffer-rope constraint scheduling problem (DBRCSP). *International Journal of Production Research 34*(9), 2405–2420.

Sipper, D. and R. Shapira. 1989. JIT vs. WIP—A tradeoff analysis. *International Journal of Production Research 27*(6), 903–914.

Spearman, M. and M. Zazanis. 1992. Push and pull production systems: Issues and comparisons. *Operations Research 40*(3), 521–532.

Spearman, M. L. 1990. Customer service in pull production systems. *Operations Research 40*, 949–957.

Spearman, M. L., D. L. Woodruff, and W. J. Hopp. 1990. CONWIP: A pull alternative to Kanban. *International Journal of Production Research 28*(5), 879–894.

Spence, A. M. and E. L. Porteus. 1987. Setup reduction and increased effective capacity. *Management Science 33*(10), 1291–1301.

Spencer, M. and J. Cox. 1994. Sales and manufacturing coordination in repetitive manufacturing: Characteristics and problems. *International Journal of Production Economics 37*, 73–81.

Spencer, M. S. and J. F. Cox. 1995. Optimum production technology (OPT) and the theory of constraints (TOC): Analysis and genealogy. *International Journal of Production Research 33*(6), 1495–1504.

Sugimori, Y., K. Kusunoki, F. Cho, and S. Uchikawa. 1977. Toyota production system and Kanban system materialization of Just-in-Time and respect-for-human system. *International Journal of Production Research 15*, 553–564.

Sutton, J. R. 1991. New cost management tools offer competitive advantage. *Industrial Engineering 23*(9), 20–23.

Taylor, S. G. and S. F. Bolander. 1990. Process flow scheduling: Basic cases. *Production and Inventory Management Journal 31*(3), 1–4.

Treleven, M. D. 1989. A review of the dual resource constrained system research. *IIE Transactions 21*, 279–287.

Van Wassenhove, L. N. and L. Gelders. 1978. Four solution techniques for a general one-machine scheduling problem. *European Journal of Operational Research 2*, 281–290.

Vendemia, W. G., B. E. Patuwo, and M. S. Hung. 1995. Evaluation of lead time in production/inventory systems with non-stationary stochastic demand. *Journal of the Operational Research Society 46*, 221–233.

Vepsalainen, A. P. J. and T. E. Morton. 1988. Improving local priority rules with global lead-time estimates. *Journal of Manufacturing and Operations Management 1*, 102–118.

Vollmann, T. E., W. L. Berry, D. C. Whybark, and F. R. Jacobs. 2005. *Manufacturing Planning and Control for Supply Chain Management*. New York: McGraw-Hill/Irwin.

Wang, D., X. Z. Chen, and Y. Li. 1996. Experimental push/pull production planning and control system. *Production Planning and Control 7*(3), 236–241.

Wang, D. and C. Xu. 1997. Hybrid push/pull production control strategy simulation and its applications. *Production Planning and Control 8*(2), 142–151.

Wein, L. M. 1988. Scheduling semiconductor wafer fabrication. *IIE Transactions on Semiconductor Manufacturing 1*, 15–130.

Weindahl, H. P. 1987. *Load-Oriented Production Control*. Vienna: Hanser Verlag.

Weindahl, H. P. and G. Springer. 1988. Manufacturing process control on the basis of the input/output chart—A chance to improve the operational behaviour of automatic production systems. *International Journal of Advanced Manufacturing Technology 3*, 55–69.

Wheatley, M. 1989. OPTimising production's potential. *International Journal of Operations & Production Management 9*(2), 38–44.

White, R. E., J. N. Pearson, and J. R. Wilson. 1999. JIT manufacturing: A survey of implementations in small and large us manufacturers. *Management Science 45*(1), 1–15.

Womack, J. P. and D. T. Jones. 2010. *Lean Thinking: Banish Waste and Create Wealth in Your Corporation.* New York: Simon and Schuster.

Womack, J. P., D. T. Jones, and D. Roos. 1991. *The Machine That Changed the World: The Story of Lean Production.* New York: Harper Perennial.

Woolsey, G. 1982. The fifth column: Production scheduling as it really is. *Interfaces 12*(6), 115–118.

Wu, S., J. S. Morris, and T. M. Gordon. 1994. A simulation analysis of the effectiveness of drum-buffer-rope scheduling in furniture manufacturing. *Computers & Industrial Engineering 26*(4), 757–764.

Zangwill, W. 1987. Eliminating inventory in a series facility. *Management Science 33*(9), 1150–1164.

Zangwill, W. 1992. The limits of Japanese production theory. *Interfaces 22*(5), 14–25.

Zavadlav, E., J. O. McClain, and L. J. Thomas. 1996. Self-buffering, self-balancing, self-flushing production lines. *Management Science 42*(8), 1151–1164.

Zipkin, P. 1995. Processing networks with planned inventories: Tandem queues with feedback. *European Journal of Operational Research 80*, 344–349.

Zipkin, P. H. 1991. Does manufacturing need a JIT revolution? *Harvard Business Review 69*(1), 40–50.

Chapter 17

Summary

With this chapter we come to the end of a long journey that has included broad qualitative discussion and detailed quantitative analysis. In this brief chapter, we look back at the process of improving operations, focusing on operations strategy and changing the givens, and we look ahead to possible future developments.

17.1 Operations Strategy

Early in the book (Chapter 1), we outlined a framework for operations strategy that highlighted the role of inventory management and production planning and scheduling in the larger strategic direction of the firm. Because we have delved deeply into formulas and procedures for the remainder of the book, we hasten to reemphasize the need to pursue inventory and production improvements with the operations and business strategies in mind. It is possible that the most important improvements may not be in the inventory or production areas at all. Although we have rarely seen a firm that could not reduce inventories or improve production policies, these same firms may have much bigger problems in another area, such as human resources or facilities location. One large automotive supplier bought a $475 million division of a competitor giving them a total of 99 plants worldwide. The first item on the priority list after the deal closed was to consolidate and rationalize these plants. Ultimately, they closed approximately 25 plants, moving their production to the remaining facilities. This issue of structuring the supply chain had longer-term implications, and represented much more significant benefits, than immediate improvements in inventory management or production planning and scheduling policies.

A similar case occurred when a large manufacturer of garment labels merged with its primary competitor. After the merger, the new firm had two factories that produced essentially the same products, so the smaller plant was closed and its production shifted to the larger one. This decision about facility focus and location was important, but relatively straightforward. The trouble began when the production control manager at the larger plant tried to schedule production. It seems that although the new products represented only a 10% increase in volume, the manager had been in a firefighting mode for several years, and this additional volume completely disrupted his somewhat haphazard policies. New procedures, such as those described in Chapter 16, were desperately needed and were ultimately implemented after a very painful process.

In other words, consultants and managers whose aim is to improve inventory or production should be certain that these areas are high on the firm's priority list. If they are not, other actions should be taken first.

A second key point is that the actions taken in these areas should be consistent with the operations objectives (cost, quality, delivery, flexibility), and with the remainder of the management levers (as described in Chapter 1). For instance, Shackleton Furniture sells customized furniture, offering literally thousands of SKUs when considering all styles, types of wood, and stains and other finishes. Now, assembled and finished furniture requires large amounts of storage space and can be easily scratched or damaged. This fact, along with the large number of SKUs, suggests that Shackleton generally has very slow customer lead times. In other words, it has chosen to emphasize flexibility and quality as its primary operations objectives, at the expense of delivery. However, Shackleton sometimes will store wood components that have been cut to shape, but not finished. Thus, workers can draw on these components when a customer order arrives, thereby decreasing delivery times.

An alternate strategy is exhibited by IKEA, which offers relatively limited variety in most furniture styles. Finished products are stocked at the store in boxes. IKEA will not customize, and in fact they require the customer to assemble the product at home. They are clearly emphasizing cost and delivery over flexibility. IKEA's inventory policies are thus quite different from Shackleton's. It would be a mistake for a well-intentioned consultant to apply the same inventory policies to both firms. Each may be able to cut inventories without sacrificing their objectives, but the consultant should understand the firms and their objectives before applying any inventory formulas. Indeed, the best approach will likely be to examine finished goods inventories at IKEA, and partially finished component inventories at Shackleton. Forecasting demand at these levels, and then applying appropriate inventory formulas could be extremely valuable.

17.2 Changing the Givens

Throughout this book we have argued for the dramatic benefits that come from *changing the givens* of an operation. In other words, if it is possible to decrease the replenishment lead time or the underlying variability, for example, the benefits are often far greater than those achieved by optimizing inventory or production without these changes. Consultants and managers, therefore, should always be thinking on two levels. First, what are the key drivers of high cost or poor service, and can these be changed? Second, how can the inventory or production policies be improved *within* the givens?

We discovered in the chapters on inventory management that demand and supply variability, lead time, and inventory ordering and holding costs are key drivers of inventory system costs and customer service. In the production context, key drivers are lead time, variability, setup and holding costs, setup time, and bottlenecks. The consultant or manager should investigate if there is some way to reduce, say, variability. Perhaps the firm observes variability that is simply the result of poor communication about ultimate consumer demand (recall the bullwhip effect described in Chapter 12). Instead of spending time and energy finding inventory and production policies that respond well to that variability, it is better to find ways to lower it. For example, will the firm's immediate customers provide information about their demand as it occurs? If so, artificial fluctuations due to the customer's ordering decisions can be eliminated. Likewise, are there delays in the transmission of information that are causing lead times to be longer than they should? If so, there may be ways to expedite information transmission and therefore decrease inventories. In summary, we advocate first examining the givens and making a serious attempt at changing them.

At the same time, however, because sometimes it is impossible to make these major changes in the short term, it is critical to consider the possibility of improving inventory and production within the givens. For instance, if the firm wishes to modernize its communications with its supply chain partners, it may be necessary to develop new relationships with them. If the relationships have been characterized by mistrust and miscommunication, it may take some time to win their trust so that a more open sharing of data is acceptable to all parties. In some cases, a firm is so small that its suppliers or customers are unwilling to make significant adjustments to ordering procedures. Of course, as we have noted throughout the book, internal resistance may be strong as well. In all these cases, the consultant or manager should work toward changing the givens while attempting to improve the existing inventory or production policies. The procedures and insights presented in this book provide ways to improve operations in the short term, often without lengthy external negotiations, and they can also be used to quantify the potential benefits from changing the givens. And as longer-term improvements are completed, then the tools presented in this book can be used to achieve even greater benefits.

17.3 Future Developments

Many academics and consultants make a practice of predicting the future with apparent confidence. While this may be good for generating business, it is a dangerous activity that we will try to avoid. It has been said that "the problem with our times is that the future is not what it used to be." The world is changing fast, and it is a challenge just to keep up, let alone predict what will happen next. Nevertheless, current trends suggest that there are certain likely developments in the inventory and production planning and scheduling areas. In this section, we briefly cite several of these.

One of the most influential developments in inventory and production over the past 20 years is the widespread adoption of enterprise resource planning (ERP) systems such as SAP (as discussed in Chapter 15). These systems often have their own sophisticated inventory and production modules, but they can also be augmented with similar modules from third-party software firms. As more companies in supply chains implement these systems, it becomes possible to efficiently share data up and down the chain. Furthermore, it will be easier to jointly optimize inventory and production policies across the chain. This clearly requires enormous trust and communication, but leading supply chains are already showing excellent results. We anticipate that this trend will expand beyond the largest firms and supply chains to include smaller- and medium-sized firms.

Another trend is the advent of extremely rapid home delivery of consumer goods. Amazon and other Internet retailers regularly offer next day, or 2 day, delivery at low cost. At this writing, Amazon is experimenting with home delivery by drone within 30 minutes of an order. Note that inventory policies must be adjusted to accommodate such delivery promises. The formulas are the same, but the inputs must be updated so that appropriate policies can be applied.

A third trend is a drive to increase the accuracy of data. It is well known that the amount of inventory recorded in the information system rarely matches the actual inventory on the shelves. One solution to this, and to a host of other issues, is the use of radio frequency identification (RFID) tags. To date, RFID has been successfully applied in many environments, but the cost of the tags is usually too high to be used on individual SKUs. As RFID cost decreases, we anticipate their increasing deployment in industry, and therefore improved information accuracy. The net effect will be to reduce variability. Firms can thus decrease inventories, but only if they update the inputs to their inventory formulas. Another trend has to do with shifting global supply chains. At one time, there was a veritable tidal wave of companies moving operations or suppliers to

China. Lead times therefore increased, but unit costs decreased dramatically as well. At this writing, many firms are moving operations to other low-cost countries, or back to North America. As fuel costs, wages, and political circumstances change, these shifts will surely continue. Indeed, we suspect that over time, other countries will emerge as global manufacturing hubs. Once again, our inventory formulas are applicable in all these situations, but the inputs must be adjusted for the given environment.

A related trend is the explosion of emerging markets. At one time, China was thought to be solely a source for inexpensive production. As the Chinese middle class has grown, however, firms are finding ready markets for their products. This trend is evident across the globe, from Vietnam to Brazil. Hence, global demand patterns are shifting, adding complexity to the issue of where to locate manufacturing.

As mentioned in Chapter 3, one thing is certain—whatever projections we make, they will be at least somewhat wrong. The world *is* changing, and will continue to change. Consultants and managers who want to prosper in the coming years should work toward improving operations within the givens. But they should also continually keep an eye to emerging trends, technologies, and conceptual developments that will allow for, and require, new ways of changing the givens.

Appendix I: Elements of Lagrangian Optimization

We consider the case where we wish to maximize (or minimize) a function $f(x, y)$ of two variables x and y, but subject to a constraint or side condition that x and y must satisfy, denoted by

$$g(x, y) = 0 \tag{I.1}$$

x and y are not independent; selection of a value of one of them implies a value (or values) of the other through Equation I.1. Therefore, we cannot find the maximum (or minimum) of $f(x, y)$ by simply equating the partial derivatives of f (with respect to x and y) to zero. Instead, we could solve Equation I.1 for y in terms of x, then substitute this expression into $f(x, y)$ to obtain a new function, call it $h(x)$, which depends only on the single variable x. Then a maximum (or minimum) of $h(x)$ could be found by setting $dh(x)/dx = 0$, and so on. The associated value of y would be found from Equation I.1

I.1 Illustration

Suppose we wish to minimize

$$f(x, y) = \frac{x}{2} + 2y$$

subject to both variables being positive and

$$xy = 4$$

This constraint can be rewritten as

$$xy - 4 = 0$$

which, from Equation I.1, shows us that

$$g(x, y) = xy - 4$$

The constraint can also be written as

$$y = \frac{4}{x} \tag{I.2}$$

717

By substituting into $f(x, y)$, we obtain

$$\frac{x}{2} + 2\frac{4}{x} = \frac{x}{2} + \frac{8}{x}$$

which we call $h(x)$. Then,

$$\frac{dh(x)}{dx} = \frac{1}{2} - \frac{8}{x^2}$$

Also,

$$\frac{d^2h(x)}{dx^2} = \frac{16}{x^2}$$

At a minimum, we require

$$\frac{dh(x)}{dx} = 0$$

In other words,

$$\frac{1}{2} - \frac{8}{x^2} = 0$$

or

$$x^2 = 16$$

Because x must be positive, the only valid solution is $x = 4$. Because the second derivative is positive, we have a minimum at $x = 4$. The y value corresponding to $x = 4$ is found from Equation I.2 to be $y = 1$. Finally, the minimum value of $f(x, y)$ is

$$f(4, 1) = \frac{4}{2} + 2(1) = 4$$

We now state, without proof, an alternative method for finding the maximum (or minimum) value of $f(x, y)$ subject to the constraint

$$g(x, y) = 0 \qquad (I.3)$$

We set up the new function

$$L(x, y, M) = f(x, y) - Mg(x, y)$$

where M is called a *Lagrange multiplier*. This new function is now treated as a function of the three *independent* variables x, y, and M. Therefore, we set

$$\frac{\partial L}{\partial x} = 0, \quad \text{that is,} \quad \frac{\partial f}{\partial x} - M\frac{\partial g}{\partial x} = 0 \qquad (I.4)$$

$$\frac{\partial L}{\partial y} = 0, \quad \text{that is,} \quad \frac{\partial f}{\partial y} - M\frac{\partial g}{\partial y} = 0 \qquad (I.5)$$

$$\frac{\partial L}{\partial M} = 0, \quad \text{that is,} \quad g(x, y) = 0 \qquad (I.6)$$

The last equation simply regenerates the constraint of Equation I.3. For any given value of M, Equations I.4 and I.5 can be solved for the two unknowns x and y. Of course, this (x, y) pair will likely not satisfy the condition of Equation I.3. Therefore, we keep trying different M values until this method leads to an (x, y) pair satisfying Equation I.3. In some cases the appropriate value of M can be found analytically by substituting expressions for x and y, found from Equations I.4 and I.5, into Equation I.3.

I.2 Illustration

Consider the same problem treated earlier, namely, for positive variables x and y

$$\text{Minimize} \quad f(x, y) = \frac{x}{2} + 2y$$
$$\text{Subject to} \quad g(x, y) = xy - 4 = 0 \tag{I.7}$$

Our new function is

$$L(x, y, M) = \frac{x}{2} + 2y - M(xy - 4)$$

Equations I.4 and I.5 give

$$\frac{1}{2} - My = 0 \quad \text{or} \quad y = \frac{1}{2M} \tag{I.8}$$

and

$$2 - Mx = 0 \quad \text{or} \quad x = \frac{2}{M} \tag{I.9}$$

The following table illustrates values of x, y, and $g(x, y)$ for various values of M.

M	$x = \dfrac{2}{M}$	$y = \dfrac{1}{2M}$	$g(x, y) = xy - 4$
2	1	1/4	$-15/4$
1	2	1/2	-3
1/2	4	1	$0 \leftarrow$
1/4	8	2	12
1/5	10	5/2	21

The behavior of $g(x, y)$ as a function of M is also sketched in Figure I.1. It is clear that there is but a single value of M, namely $1/2$, where $g(x, y) = 0$; that is, the constraint of Equation I.7 is satisfied. (In this example the value of M could be found more directly, without trial and error, by substituting Equations I.8 and I.9 into Equation I.7.)

From this table, we see that the corresponding (x, y) pair is

$$x = 4 \quad y = 1$$

precisely the same result we obtained by the earlier more direct method involving substitution, and so on.

At this stage, the reader may be asking, "Who needs all this aggravation when the direct substitution method seems so much simpler?" For the example analyzed, we certainly do not need the

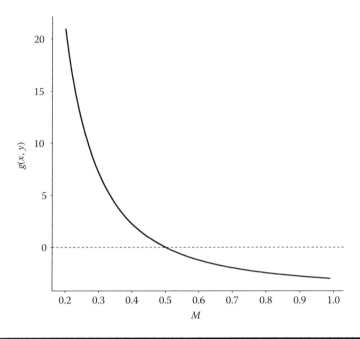

Figure I.1   Illustration of $g(x, y)$ as a function of the Lagrange multiplier M.

Lagrange multiplier approach. However, the method is directly extendable to the case where there are several, instead of just two, variables involved. Also, in the substitution method we can quickly run into problems if the constraint is of such a form that one variable cannot be easily expressed in terms of the other variables. This would have been the case in this illustration if the constraint had been instead, for example,

$$x^3 y^2 - 3xy^4 = 6$$

This type of complexity does not hamper the Lagrange approach.

We now present, without proof,[*] the general Lagrange approach for maximizing (or minimizing) a function $f(x_1, x_2, \ldots, x_n)$ of n variables $x_1, x_2, \ldots, x_n$ subject to a single equality constraint

$$g(x_1, x_2, \ldots, x_n) = 0 \qquad (\text{I.10})$$

Step 1 Form the function

$$L(x_1, x_2, \ldots, x_n) = f(x_1, x_2, \ldots, x_n) - Mg(x_1, x_2, \ldots, x_n)$$

---

[*] A proof is provided in Sokolnikoff et al. (1958).

Step 2 Evaluate the n partial derivatives and set each equal to zero.

$$\frac{\partial L}{\partial x_1} = 0$$

$$\frac{\partial L}{\partial x_2} = 0$$

$$\vdots$$

$$\frac{\partial L}{\partial x_n} = 0$$

This gives n equations in the $n+1$ variables $x_1, x_2, \ldots, x_n$, and M.

Step 3 For a particular value of M, solve the set of equations (obtained in Step 2) for the x's. Then evaluate $g(x_1, x_2, \ldots, x_n)$.

Step 4 If $g(x_1, x_2, \ldots, x_n) = 0$, we have found a set of x's which are candidates for maximizing (or minimizing) the function f subject to satisfying the condition of Equation I.10. Otherwise, we return to step 2 with a different value of M. There are different methods available for searching for the M values where the constraint is satisfied. At the very least we can develop graphs, similar to Figure I.1, as we proceed. This certainly would indicate in what direction to explore M values. In some cases, one can obtain analytic solutions for the required M values.

Reference

Sokolnikoff, I. S., R. M. Redheffer, and J. Avents. 1958. Mathematics of physics and modern engineering. *Journal of the Electrochemical Society 105*(9), 249–251.

Appendix II: The Normal Probability Distribution

The normal is undoubtedly the most important single probability distribution in decision rules of production planning and inventory management, as well as in general usage of probability (particularly in the area of applied statistics). This appendix is devoted to a discussion of the properties of the normal distribution, particularly those needed for the decision rules of the main text.

II.1 The Probability Density Function

The probability density function (p.d.f.) of a normal variable x, with mean $\hat{x}$ and standard deviation σ_x, is denoted by

$$f_x(x_0) = \frac{1}{\sigma_x\sqrt{2\pi}}e^{-(x_0-\hat{x})^2/(2\sigma_x^2)} \tag{II.1}$$

A typical graph of the normal p.d.f. is shown in Figure II.1.

Using the p.d.f., one can verify that the mean and standard deviation are, indeed, the $\hat{x}$ and σ_x in Equation II.1.

II.2 The Unit (or Standard) Normal Distribution

An important special case of the normal distribution is the one where the mean value is 0 and the standard deviation is 1. This is known as the unit normal or standard normal distribution. We denote such a variable by u. Thus,

$$f_u(u_0) = \frac{1}{\sqrt{2\pi}}e^{-u_0^2/2} \tag{II.2}$$

with $E(u) = 0$ (or $\hat{u} = 0$) and $\sigma_u = 1$. This p.d.f. is shown graphically in Figure II.2.

723

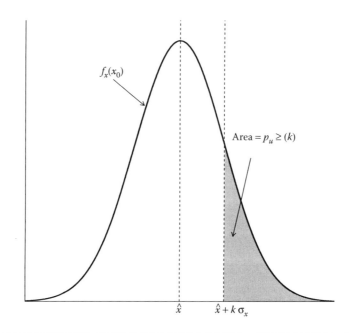

Figure II.1   The normal p.d.f.

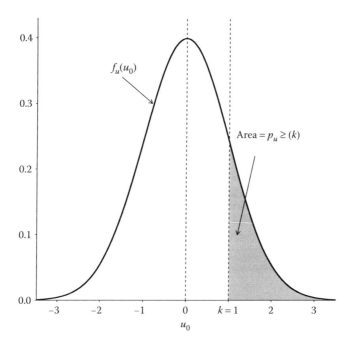

Figure II.2   The unit (or standard) normal p.d.f.

A quantity of frequent interest is the probability that u is at least as large as a certain value k.

$$p_{u\geq}(k) = \text{prob}(u \geq k) = \int_k^\infty f_u(u_0)du_0$$

$$= \int_k^\infty \frac{1}{\sqrt{2\pi}}e^{-\frac{u_0^2}{2}}du_0 \qquad \text{(II.3)}$$

There is no indefinite integral for

$$e^{\frac{-x^2}{2}}dx$$

Therefore, Equation II.3 has to be numerically integrated. (The result represents the shaded area in Figure II.2.) This has been done and $p_{u\geq}(k)$ has been tabulated as in Table II.1, for a range of k values. Only positive values of k are shown in the table. If $p_{u\geq}(k)$ is needed for a negative argument, it is clear from the symmetry of Figure II.2 that

$$p_{u\geq}(-k) = 1 - p_{u\geq}(k) \qquad \text{(II.4)}$$

A second quantity of interest relative to production/inventory decision rules, denoted by $G_u(k)$, is given by

$$G_u(k) = \int_k^\infty (u_0 - k)f_u(u_0)du_0 \qquad \text{(II.5)}$$

Note that the derivative of the unit normal p.d.f. can be expressed as follows:

$$\frac{df_u(u_0)}{du_0} = \frac{d}{du_0}\left(\frac{1}{\sqrt{2\pi}}e^{-u_0^2/2}\right) = -u_0\left(\frac{1}{\sqrt{2\pi}}e^{-u_0^2/2}\right) = -u_0 f_u(u_0) \qquad \text{(II.6)}$$

Using this relationship, we can write $G_u(k)$ as follows:

$$G_u(k) = \int_k^\infty (u_0 - k)f_u(u_0)du_0$$

$$= \int_k^\infty u_0 f_u(u_0)du_0 - \int_k^\infty kf_u(u_0)du_0$$

$$= \int_k^\infty -\frac{df_u(u_0)}{du_0}du_0 - \int_k^\infty kf_u(u_0)du_0$$

$$= f_u(k) - kp_{u\geq}(k) \qquad \text{(II.7)}$$

Table II.1 also shows $G_u(k)$ as a function of k. Note that it can be shown that

$$G_u(-k) = G_u(k) + k \qquad \text{(II.8)}$$

Table II.1 Some Functions of the Unit Normal Distribution

k	$f_u(k)$	$p_{u \geq}(k)$	$G_u(k)$	$J_u(k)$	$G_u(-k)$	$J_u(-k)$	k
0.00	0.3989	0.5000	0.3989	0.5000	0.3989	0.5000	0.00
0.01	0.3989	0.4960	0.3940	0.4921	0.4040	0.5080	0.01
0.02	0.3989	0.4920	0.3890	0.4842	0.4090	0.5162	0.02
0.03	0.3988	0.4880	0.3841	0.4765	0.4141	0.5244	0.03
0.04	0.3986	0.4840	0.3793	0.4689	0.4193	0.5327	0.04
0.05	0.3984	0.4801	0.3744	0.4613	0.4244	0.5412	0.05
0.06	0.3982	0.4761	0.3697	0.4539	0.4297	0.5497	0.06
0.07	0.3980	0.4721	0.3649	0.4466	0.4349	0.5583	0.07
0.08	0.3977	0.4681	0.3602	0.4393	0.4402	0.5671	0.08
0.09	0.3973	0.4641	0.3556	0.4321	0.4456	0.5760	0.09
0.10	0.3970	0.4602	0.3509	0.4251	0.4509	0.5849	0.10
0.11	0.3965	0.4562	0.3464	0.4181	0.4564	0.5940	0.11
0.12	0.3961	0.4522	0.3418	0.4112	0.4618	0.6032	0.12
0.13	0.3956	0.4483	0.3373	0.4044	0.4673	0.6125	0.13
0.14	0.3951	0.4443	0.3328	0.3977	0.4728	0.6219	0.14
0.15	0.3945	0.4404	0.3284	0.3911	0.4784	0.6314	0.15
0.16	0.3939	0.4364	0.3240	0.3846	0.4840	0.6410	0.16
0.17	0.3932	0.4325	0.3197	0.3782	0.4897	0.6507	0.17
0.18	0.3925	0.4286	0.3154	0.3718	0.4954	0.6606	0.18
0.19	0.3918	0.4247	0.3111	0.3655	0.5011	0.6706	0.19
0.20	0.3910	0.4207	0.3069	0.3594	0.5069	0.6806	0.20
0.21	0.3902	0.4168	0.3027	0.3533	0.5127	0.6908	0.21
0.22	0.3894	0.4129	0.2986	0.3473	0.5186	0.7011	0.22
0.23	0.3885	0.4090	0.2944	0.3413	0.5244	0.7116	0.23
0.24	0.3876	0.4052	0.2904	0.3355	0.5304	0.7221	0.24
0.25	0.3867	0.4013	0.2863	0.3297	0.5363	0.7328	0.25
0.26	0.3857	0.3974	0.2824	0.3240	0.5424	0.7436	0.26
0.27	0.3847	0.3936	0.2784	0.3184	0.5484	0.7545	0.27
0.28	0.3836	0.3897	0.2745	0.3129	0.5545	0.7655	0.28

(Continued)

Table II.1 (*Continued*)   Some Functions of the Unit Normal Distribution

k	$f_u(k)$	$p_{u \geq}(k)$	$G_u(k)$	$J_u(k)$	$G_u(-k)$	$J_u(-k)$	k
0.29	0.3825	0.3859	0.2706	0.3074	0.5606	0.7767	0.29
0.30	0.3814	0.3821	0.2668	0.3021	0.5668	0.7879	0.30
0.31	0.3802	0.3783	0.2630	0.2968	0.5730	0.7993	0.31
0.32	0.3790	0.3745	0.2592	0.2915	0.5792	0.8109	0.32
0.33	0.3778	0.3707	0.2555	0.2864	0.5855	0.8225	0.33
0.34	0.3765	0.3669	0.2518	0.2813	0.5918	0.8343	0.34
0.35	0.3752	0.3632	0.2481	0.2763	0.5981	0.8462	0.35
0.36	0.3739	0.3594	0.2445	0.2714	0.6045	0.8582	0.36
0.37	0.3725	0.3557	0.2409	0.2665	0.6109	0.8704	0.37
0.38	0.3712	0.3520	0.2374	0.2618	0.6174	0.8826	0.38
0.39	0.3697	0.3483	0.2339	0.2570	0.6239	0.8951	0.39
0.40	0.3683	0.3446	0.2304	0.2524	0.6304	0.9076	0.40
0.41	0.3668	0.3409	0.2270	0.2478	0.6370	0.9203	0.41
0.42	0.3653	0.3372	0.2236	0.2433	0.6436	0.9331	0.42
0.43	0.3637	0.3336	0.2203	0.2389	0.6503	0.9460	0.43
0.44	0.3621	0.3300	0.2169	0.2345	0.6569	0.9591	0.44
0.45	0.3605	0.3264	0.2137	0.2302	0.6637	0.9723	0.45
0.46	0.3589	0.3228	0.2104	0.2260	0.6704	0.9856	0.46
0.47	0.3572	0.3192	0.2072	0.2218	0.6772	0.9991	0.47
0.48	0.3555	0.3156	0.2040	0.2177	0.6840	1.0127	0.48
0.49	0.3538	0.3121	0.2009	0.2136	0.6909	1.0265	0.49
0.50	0.3521	0.3085	0.1978	0.2096	0.6978	1.0404	0.50
0.51	0.3503	0.3050	0.1947	0.2057	0.7047	1.0544	0.51
0.52	0.3485	0.3015	0.1917	0.2018	0.7117	1.0686	0.52
0.53	0.3467	0.2981	0.1887	0.1980	0.7187	1.0829	0.53
0.54	0.3448	0.2946	0.1857	0.1943	0.7257	1.0973	0.54
0.55	0.3429	0.2912	0.1828	0.1906	0.7328	1.1119	0.55
0.56	0.3410	0.2877	0.1799	0.1870	0.7399	1.1266	0.56
0.57	0.3391	0.2843	0.1771	0.1834	0.7471	1.1415	0.57

(Continued)

Table II.1 (*Continued*)   Some Functions of the Unit Normal Distribution

k	$f_u(k)$	$p_{u \geq}(k)$	$G_u(k)$	$J_u(k)$	$G_u(-k)$	$J_u(-k)$	k
0.58	0.3372	0.2810	0.1742	0.1799	0.7542	1.1565	0.58
0.59	0.3352	0.2776	0.1714	0.1765	0.7614	1.1716	0.59
0.60	0.3332	0.2743	0.1687	0.1730	0.7687	1.1870	0.60
0.61	0.3312	0.2709	0.1659	0.1697	0.7759	1.2024	0.61
0.62	0.3292	0.2676	0.1633	0.1664	0.7833	1.2180	0.62
0.63	0.3271	0.2643	0.1606	0.1632	0.7906	1.2337	0.63
0.64	0.3251	0.2611	0.1580	0.1600	0.7980	1.2496	0.64
0.65	0.3230	0.2578	0.1554	0.1569	0.8054	1.2656	0.65
0.66	0.3209	0.2546	0.1528	0.1538	0.8128	1.2818	0.66
0.67	0.3187	0.2514	0.1503	0.1507	0.8203	1.2982	0.67
0.68	0.3166	0.2483	0.1478	0.1478	0.8278	1.3146	0.68
0.69	0.3144	0.2451	0.1453	0.1448	0.8353	1.3313	0.69
0.70	0.3123	0.2420	0.1429	0.1419	0.8429	1.3481	0.70
0.71	0.3101	0.2389	0.1405	0.1391	0.8505	1.3650	0.71
0.72	0.3079	0.2358	0.1381	0.1363	0.8581	1.3821	0.72
0.73	0.3056	0.2327	0.1358	0.1336	0.8658	1.3993	0.73
0.74	0.3034	0.2297	0.1334	0.1311	0.8734	1.4165	0.74
0.75	0.3011	0.2266	0.1312	0.1283	0.8812	1.4342	0.75
0.76	0.2989	0.2236	0.1289	0.1257	0.8889	1.4519	0.76
0.77	0.2966	0.2206	0.1267	0.1231	0.8967	1.4698	0.77
0.78	0.2943	0.2177	0.1245	0.1206	0.9045	1.4878	0.78
0.79	0.2920	0.2148	0.1223	0.1181	0.9123	1.5060	0.79
0.80	0.2897	0.2119	0.1202	0.1157	0.9202	1.5243	0.80
0.81	0.2874	0.2090	0.1181	0.1133	0.9281	1.5428	0.81
0.82	0.2850	0.2061	0.1160	0.1110	0.9360	1.5614	0.82
0.83	0.2827	0.2033	0.1140	0.1087	0.9440	1.5802	0.83
0.84	0.2803	0.2005	0.1120	0.1064	0.9520	1.5992	0.84
0.85	0.2780	0.1977	0.1100	0.1042	0.9600	1.6183	0.85
0.86	0.2756	0.1949	0.1080	0.1020	0.9680	1.6376	0.86

(*Continued*)

Table II.1 (*Continued*)   Some Functions of the Unit Normal Distribution

k	$f_u(k)$	$p_{u \geq}(k)$	$G_u(k)$	$J_u(k)$	$G_u(-k)$	$J_u(-k)$	k
0.87	0.2732	0.1922	0.1061	0.09987	0.9761	1.6570	0.87
0.88	0.2709	0.1894	0.1042	0.09776	0.9842	1.6766	0.88
0.89	0.2685	0.1867	0.1023	0.09570	0.9923	1.6964	0.89
0.90	0.2661	0.1841	0.1004	0.09367	1.0004	1.7163	0.90
0.91	0.2637	0.1814	0.09860	0.09168	1.0086	1.7364	0.91
0.92	0.2613	0.1788	0.09680	0.08973	1.0168	1.7567	0.92
0.93	0.2589	0.1762	0.09503	0.08781	1.0250	1.7771	0.93
0.94	0.2565	0.1736	0.09328	0.08593	1.0333	1.7977	0.94
0.95	0.2541	0.1711	0.09156	0.08408	1.0416	1.8184	0.95
0.96	0.2516	0.1685	0.08986	0.08226	1.0499	1.8393	0.96
0.97	0.2492	0.1660	0.08819	0.08048	1.0582	1.8604	0.97
0.98	0.2468	0.1635	0.08654	0.07874	1.0665	1.8817	0.98
0.99	0.2444	0.1611	0.08491	0.07702	1.0749	1.9031	0.99
1.00	0.2420	0.1587	0.08332	0.07534	1.0833	1.9247	1.00
1.01	0.2396	0.1562	0.08174	0.07369	1.0917	1.9464	1.01
1.02	0.2371	0.1539	0.08019	0.07207	1.1002	1.9683	1.02
1.03	0.2347	0.1515	0.07866	0.07048	1.1087	1.9904	1.03
1.04	0.2323	0.1492	0.07716	0.06892	1.1172	2.0127	1.04
1.05	0.2299	0.1469	0.07568	0.06739	1.1257	2.0351	1.05
1.06	0.2275	0.1446	0.07422	0.06590	1.1342	2.0577	1.06
1.07	0.2251	0.1423	0.07279	0.06443	1.1428	2.0805	1.07
1.08	0.2227	0.1401	0.07138	0.06298	1.1514	2.1034	1.08
1.09	0.2203	0.1379	0.06999	0.06157	1.1600	2.1265	1.09
1.10	0.2179	0.1357	0.06862	0.06018	1.1686	2.1498	1.10
1.11	0.2155	0.1335	0.06727	0.05883	1.1773	2.1733	1.11
1.12	0.2131	0.1314	0.06595	0.05749	1.1859	2.1969	1.12
1.13	0.2107	0.1292	0.06465	0.05619	1.1946	2.2207	1.13
1.14	0.2083	0.1271	0.06336	0.05491	1.2034	2.2447	1.14
1.15	0.2059	0.1251	0.06210	0.05365	1.2121	2.2688	1.15

(Continued)

Table II.1 (*Continued*)   Some Functions of the Unit Normal Distribution

k	$f_u(k)$	$p_{u\geq}(k)$	$G_u(k)$	$J_u(k)$	$G_u(-k)$	$J_u(-k)$	k
1.16	0.2036	0.1230	0.06086	0.05242	1.2209	2.2932	1.16
1.17	0.2012	0.1210	0.05964	0.05122	1.2296	2.3177	1.17
1.18	0.1989	0.1190	0.05844	0.05004	1.2384	2.3424	1.18
1.19	0.1965	0.1170	0.05726	0.04888	1.2473	2.3672	1.19
1.20	0.1942	0.1151	0.05610	0.04775	1.2561	2.3923	1.20
1.21	0.1919	0.1131	0.05496	0.04664	1.2650	2.4175	1.21
1.22	0.1895	0.1112	0.05384	0.04555	1.2738	2.4429	1.22
1.23	0.1872	0.1093	0.05274	0.04448	1.2827	2.4684	1.23
1.24	0.1849	0.1075	0.05165	0.04344	1.2917	2.4942	1.24
1.25	0.1826	0.1056	0.05059	0.04242	1.3006	2.5201	1.25
1.26	0.1804	0.1038	0.04954	0.04141	1.3095	2.5462	1.26
1.27	0.1781	0.1020	0.04851	0.04043	1.3185	2.5725	1.27
1.28	0.1758	0.1003	0.04750	0.03947	1.3275	2.5989	1.28
1.29	0.1736	0.09853	0.04650	0.03853	1.3365	2.6256	1.29
1.30	0.1714	0.09680	0.04553	0.03761	1.3455	2.6524	1.30
1.31	0.1691	0.09510	0.04457	0.03671	1.3546	2.6794	1.31
1.32	0.1669	0.09342	0.04363	0.03583	1.3636	2.7066	1.32
1.33	0.1647	0.09176	0.04270	0.03497	1.3727	2.7339	1.33
1.34	0.1626	0.09012	0.04179	0.03412	1.3818	2.7615	1.34
1.35	0.1604	0.08851	0.04090	0.03330	1.3909	2.7892	1.35
1.36	0.1582	0.08691	0.04002	0.03249	1.4000	2.8171	1.36
1.37	0.1561	0.08534	0.03916	0.03170	1.4092	2.8452	1.37
1.38	0.1539	0.08379	0.03831	0.03092	1.4183	2.8735	1.38
1.39	0.1518	0.08226	0.03748	0.03016	1.4275	2.9019	1.39
1.40	0.1497	0.08076	0.03667	0.02942	1.4367	2.9306	1.40
1.41	0.1476	0.07927	0.03587	0.02870	1.4459	2.9594	1.41
1.42	0.1456	0.07780	0.03508	0.02799	1.4551	2.9884	1.42
1.43	0.1435	0.07636	0.03431	0.02729	1.4643	3.0176	1.43
1.44	0.1415	0.07493	0.03356	0.02661	1.4736	3.0470	1.44

(Continued)

Table II.1 (*Continued*)   Some Functions of the Unit Normal Distribution

k	$f_u(k)$	$p_{u\geq}(k)$	$G_u(k)$	$J_u(k)$	$G_u(-k)$	$J_u(-k)$	k
1.45	0.1394	0.07353	0.03281	0.02595	1.4828	3.0765	1.45
1.46	0.1374	0.07215	0.03208	0.02530	1.4921	3.1063	1.46
1.47	0.1354	0.07078	0.03137	0.02467	1.5014	3.1362	1.47
1.48	0.1334	0.06944	0.03067	0.02405	1.5107	3.1664	1.48
1.49	0.1315	0.06811	0.02998	0.02344	1.5200	3.1967	1.49
1.50	0.1295	0.06681	0.02931	0.02285	1.5293	3.2272	1.50
1.51	0.1276	0.06552	0.02865	0.02227	1.5386	3.2578	1.51
1.52	0.1257	0.06426	0.02800	0.02170	1.5480	3.2887	1.52
1.53	0.1238	0.06301	0.02736	0.02115	1.5574	3.3198	1.53
1.54	0.1219	0.06178	0.02674	0.02061	1.5667	3.3510	1.54
1.55	0.1200	0.06057	0.02612	0.02008	1.5761	3.3824	1.55
1.56	0.1182	0.05938	0.02552	0.01956	1.5855	3.4140	1.56
1.57	0.1163	0.05821	0.02494	0.01906	1.5949	3.4458	1.57
1.58	0.1145	0.05705	0.02436	0.01856	1.6044	3.4778	1.58
1.59	0.1127	0.05592	0.02380	0.01808	1.6138	3.5100	1.59
1.60	0.1109	0.05480	0.02324	0.01761	1.6232	3.5424	1.60
1.61	0.1092	0.05370	0.02270	0.01715	1.6327	3.5749	1.61
1.62	0.1074	0.05262	0.02217	0.01670	1.6422	3.6077	1.62
1.63	0.1057	0.05155	0.02165	0.01627	1.6516	3.6406	1.63
1.64	0.1040	0.05050	0.02114	0.01584	1.6611	3.6738	1.64
1.65	0.1023	0.04947	0.02064	0.01542	1.6706	3.7071	1.65
1.66	0.1006	0.04846	0.02015	0.01501	1.6801	3.7406	1.66
1.67	0.0989	0.04746	0.01967	0.01461	1.6897	3.7743	1.67
1.68	0.0973	0.04648	0.01920	0.01423	1.6992	3.8082	1.68
1.69	0.0957	0.04551	0.01874	0.01385	1.7087	3.8423	1.69
1.70	0.0940	0.04457	0.01829	0.01348	1.7183	3.8765	1.70
1.71	0.0925	0.04363	0.01785	0.01311	1.7278	3.9110	1.71
1.72	0.0909	0.04272	0.01742	0.01276	1.7374	3.9456	1.72
1.73	0.0893	0.04182	0.01699	0.01242	1.7470	3.9805	1.73

(Continued)

Table II.1 (*Continued*)   Some Functions of the Unit Normal Distribution

k	$f_u(k)$	$p_{u\geq}(k)$	$G_u(k)$	$J_u(k)$	$G_u(-k)$	$J_u(-k)$	k
1.74	0.0878	0.04093	0.01658	0.01208	1.7566	4.0155	1.74
1.75	0.0863	0.04006	0.01617	0.01176	1.7662	4.0507	1.75
1.76	0.0848	0.03920	0.01578	0.01144	1.7758	4.0862	1.76
1.77	0.0833	0.03836	0.01539	0.01112	1.7854	4.1218	1.77
1.78	0.0818	0.03754	0.01501	0.01082	1.7950	4.1576	1.78
1.79	0.0804	0.03673	0.01464	0.01052	1.8046	4.1936	1.79
1.80	0.0790	0.03593	0.01428	0.01023	1.8143	4.2298	1.80
1.81	0.0775	0.03515	0.01392	0.009952	1.8239	4.2661	1.81
1.82	0.0761	0.03438	0.01357	0.009677	1.8336	4.3027	1.82
1.83	0.0748	0.03362	0.01323	0.009409	1.8432	4.3395	1.83
1.84	0.0734	0.03288	0.01290	0.009148	1.8529	4.3765	1.84
1.85	0.0721	0.03216	0.01257	0.008893	1.8626	4.4136	1.85
1.86	0.0707	0.03144	0.01226	0.008645	1.8723	4.4510	1.86
1.87	0.0694	0.03074	0.01195	0.008403	1.8819	4.4885	1.87
1.88	0.0681	0.03005	0.01164	0.008167	1.8916	4.5262	1.88
1.89	0.0669	0.02938	0.01134	0.007937	1.9013	4.5642	1.89
1.90	0.0656	0.02872	0.01105	0.007713	1.9111	4.6023	1.90
1.91	0.0644	0.02807	0.01077	0.007495	1.9208	4.6406	1.91
1.92	0.0632	0.02743	0.01049	0.007282	1.9305	4.6791	1.92
1.93	0.0620	0.02680	0.01022	0.007075	1.9402	4.7178	1.93
1.94	0.0608	0.02619	0.009957	0.006874	1.9500	4.7567	1.94
1.95	0.0596	0.02559	0.009698	0.006677	1.9597	4.7958	1.95
1.96	0.0584	0.02500	0.009445	0.006486	1.9694	4.8351	1.96
1.97	0.0573	0.02442	0.009198	0.006299	1.9792	4.8746	1.97
1.98	0.0562	0.02385	0.008957	0.006118	1.9890	4.9143	1.98
1.99	0.0551	0.02330	0.008721	0.005941	1.9987	4.9542	1.99
2.00	0.0540	0.02275	0.008491	0.005769	2.0085	4.9942	2.00
2.01	0.0529	0.02222	0.008266	0.005601	2.0183	5.0345	2.01
2.02	0.0519	0.02169	0.008046	0.005438	2.0280	5.0750	2.02

(Continued)

Table II.1 (*Continued*)   Some Functions of the Unit Normal Distribution

k	$f_u(k)$	$p_{u\geq}(k)$	$G_u(k)$	$J_u(k)$	$G_u(-k)$	$J_u(-k)$	k
2.03	0.0508	0.02118	0.007832	0.005279	2.0378	5.1156	2.03
2.04	0.0498	0.02068	0.007623	0.005125	2.0476	5.1565	2.04
2.05	0.0488	0.02018	0.007418	0.004974	2.0574	5.1975	2.05
2.06	0.0478	0.01970	0.007219	0.004828	2.0672	5.2388	2.06
2.07	0.0468	0.01923	0.007024	0.004686	2.0770	5.2802	2.07
2.08	0.0459	0.01876	0.006835	0.004547	2.0868	5.3219	2.08
2.09	0.0449	0.01831	0.006649	0.004412	2.0966	5.3637	2.09
2.10	0.0440	0.01786	0.006468	0.004281	2.1065	5.4057	2.10
2.11	0.0431	0.01743	0.006292	0.004153	2.1163	5.4479	2.11
2.12	0.0422	0.01700	0.006120	0.004029	2.1261	5.4904	2.12
2.13	0.0413	0.01659	0.005952	0.003909	2.1360	5.5330	2.13
2.14	0.0404	0.01618	0.005788	0.003791	2.1458	5.5758	2.14
2.15	0.0396	0.01578	0.005628	0.003677	2.1556	5.6188	2.15
2.16	0.0387	0.01539	0.005472	0.003566	2.1655	5.6620	2.16
2.17	0.0379	0.01500	0.005320	0.003458	2.1753	5.7054	2.17
2.18	0.0371	0.01463	0.005172	0.003353	2.1852	5.7490	2.18
2.19	0.0363	0.01426	0.005028	0.003251	2.1950	5.7928	2.19
2.20	0.0355	0.01390	0.004887	0.003152	2.2049	5.8368	2.20
2.21	0.0347	0.01355	0.004750	0.003056	2.2147	5.8810	2.21
2.22	0.0339	0.01321	0.004616	0.002962	2.2246	5.9254	2.22
2.23	0.0332	0.01287	0.004486	0.002871	2.2345	5.9700	2.23
2.24	0.0325	0.01255	0.004358	0.002783	2.2444	6.0148	2.24
2.25	0.0317	0.01222	0.004235	0.002697	2.2542	6.0598	2.25
2.26	0.0310	0.01191	0.004114	0.002613	2.2641	6.1050	2.26
2.27	0.0303	0.01160	0.003996	0.002532	2.2740	6.1504	2.27
2.28	0.0297	0.01130	0.003882	0.002453	2.2839	6.1959	2.28
2.29	0.0290	0.01101	0.003770	0.002377	2.2938	6.2417	2.29
2.30	0.0283	0.01072	0.003662	0.002302	2.3037	6.2877	2.30
2.31	0.0277	0.01044	0.003556	0.002230	2.3136	6.3339	2.31

(Continued)

Table II.1 (*Continued*)   Some Functions of the Unit Normal Distribution

k	$f_u(k)$	$p_{u\geq}(k)$	$G_u(k)$	$J_u(k)$	$G_u(-k)$	$J_u(-k)$	k
2.32	0.0270	0.01017	0.003453	0.002160	2.3235	6.3802	2.32
2.33	0.0264	0.009903	0.003352	0.002092	2.3334	6.4268	2.33
2.34	0.0258	0.009642	0.003255	0.002026	2.3433	6.4736	2.34
2.35	0.0252	0.009387	0.003159	0.001962	2.3532	6.5205	2.35
2.36	0.0246	0.009137	0.003067	0.001900	2.3631	6.5677	2.36
2.37	0.0241	0.008894	0.002977	0.001839	2.3730	6.6151	2.37
2.38	0.0235	0.008656	0.002889	0.001781	2.3829	6.6626	2.38
2.39	0.0229	0.008424	0.002804	0.001724	2.3928	6.7104	2.39
2.40	0.0224	0.008198	0.002720	0.001668	2.4027	6.7583	2.40
2.41	0.0219	0.007976	0.002640	0.001615	2.4126	6.8065	2.41
2.42	0.0213	0.007760	0.002561	0.001563	2.4226	6.8548	2.42
2.43	0.0208	0.007549	0.002484	0.001512	2.4325	6.9034	2.43
2.44	0.0203	0.007344	0.002410	0.001463	2.4424	6.9521	2.44
2.45	0.0198	0.007143	0.002337	0.001416	2.4523	7.0011	2.45
2.46	0.0194	0.006947	0.002267	0.001370	2.4623	7.0502	2.46
2.47	0.0189	0.006756	0.002199	0.001325	2.4722	7.0996	2.47
2.48	0.0184	0.006569	0.002132	0.001282	2.4821	7.1491	2.48
2.49	0.0180	0.006387	0.002067	0.001240	2.4921	7.1989	2.49
2.50	0.0175	0.006210	0.002004	0.001199	2.5020	7.2488	2.50
2.51	0.0171	0.006037	0.001943	0.001160	2.5119	7.2989	2.51
2.52	0.0167	0.005868	0.001883	0.001122	2.5219	7.3493	2.52
2.53	0.0163	0.005703	0.001826	0.001085	2.5318	7.3998	2.53
2.54	0.0158	0.005543	0.001769	0.001049	2.5418	7.4506	2.54
2.55	0.0154	0.005386	0.001715	0.001014	2.5517	7.5015	2.55
2.56	0.0151	0.005234	0.001662	0.0009800	2.5617	7.5526	2.56
2.57	0.0147	0.005085	0.001610	0.0009473	2.5716	7.6040	2.57
2.58	0.0143	0.004940	0.001560	0.0009156	2.5816	7.6555	2.58
2.59	0.0139	0.004799	0.001511	0.0008848	2.5915	7.7072	2.59
2.60	0.0136	0.004661	0.001464	0.0008551	2.6015	7.7591	2.60

(Continued)

Table II.1 (*Continued*)   Some Functions of the Unit Normal Distribution

k	$f_u(k)$	$p_{u\geq}(k)$	$G_u(k)$	$J_u(k)$	$G_u(-k)$	$J_u(-k)$	k
2.61	0.0132	0.004527	0.001418	0.0008263	2.6114	7.8113	2.61
2.62	0.0129	0.004396	0.001373	0.0007984	2.6214	7.8636	2.62
2.63	0.0126	0.004269	0.001330	0.0007713	2.6313	7.9161	2.63
2.64	0.0122	0.004145	0.001288	0.0007452	2.6413	7.9689	2.64
2.65	0.0119	0.004025	0.001247	0.0007198	2.6512	8.0218	2.65
2.66	0.0116	0.003907	0.001207	0.0006953	2.6612	8.0749	2.66
2.67	0.0113	0.003793	0.001169	0.0006715	2.6712	8.1282	2.67
2.68	0.0110	0.003681	0.001132	0.0006485	2.6811	8.1818	2.68
2.69	0.0107	0.003573	0.001095	0.0006262	2.6911	8.2355	2.69
2.70	0.0104	0.003467	0.001060	0.0006047	2.7011	8.2894	2.70
2.71	0.0101	0.003364	0.001026	0.0005838	2.7110	8.3435	2.71
2.72	0.0099	0.003264	0.0009928	0.0005636	2.7210	8.3978	2.72
2.73	0.0096	0.003167	0.0009607	0.0005441	2.7310	8.4524	2.73
2.74	0.0093	0.003072	0.0009295	0.0005252	2.7409	8.5071	2.74
2.75	0.0091	0.002980	0.0008992	0.0005069	2.7509	8.5620	2.75
2.76	0.0088	0.002890	0.0008699	0.0004892	2.7609	8.6171	2.76
2.77	0.0086	0.002803	0.0008414	0.0004721	2.7708	8.6724	2.77
2.78	0.0084	0.002718	0.0008138	0.0004556	2.7808	8.7279	2.78
2.79	0.0081	0.002635	0.0007870	0.0004396	2.7908	8.7837	2.79
2.80	0.0079	0.002555	0.0007611	0.0004241	2.8008	8.8396	2.80
2.81	0.0077	0.002477	0.0007359	0.0004091	2.8107	8.8957	2.81
2.82	0.0075	0.002401	0.0007115	0.0003946	2.8207	8.9520	2.82
2.83	0.0073	0.002327	0.0006879	0.0003807	2.8307	9.0085	2.83
2.84	0.0071	0.002256	0.0006650	0.0003671	2.8407	9.0652	2.84
2.85	0.0069	0.002186	0.0006428	0.0003540	2.8506	9.1221	2.85
2.86	0.0067	0.002118	0.0006213	0.0003414	2.8606	9.1793	2.86
2.87	0.0065	0.002052	0.0006004	0.0003292	2.8706	9.2366	2.87
2.88	0.0063	0.001988	0.0005802	0.0003174	2.8806	9.2941	2.88
2.89	0.0061	0.001926	0.0005606	0.0003060	2.8906	9.3518	2.89

(*Continued*)

Table II.1 (*Continued*) **Some Functions of the Unit Normal Distribution**

k	$f_u(k)$	$p_{u\geq}(k)$	$G_u(k)$	$J_u(k)$	$G_u(-k)$	$J_u(-k)$	k
2.90	0.0060	0.001866	0.0005417	0.0002950	2.9005	9.4097	2.90
2.91	0.0058	0.001807	0.0005233	0.0002843	2.9105	9.4678	2.91
2.92	0.0056	0.001750	0.0005055	0.0002740	2.9205	9.5261	2.92
2.93	0.0055	0.001695	0.0004883	0.0002641	2.9305	9.5846	2.93
2.94	0.0053	0.001641	0.0004716	0.0002545	2.9405	9.6433	2.94
2.95	0.0051	0.001589	0.0004555	0.0002452	2.9505	9.7023	2.95
2.96	0.0050	0.001538	0.0004396	0.0002363	2.9604	9.7614	2.96
2.97	0.0048	0.001489	0.0004247	0.0002276	2.9704	9.8207	2.97
2.98	0.0047	0.001441	0.0004101	0.0002193	2.9804	9.8802	2.98
2.99	0.0046	0.001395	0.0003959	0.0002112	2.9904	9.9399	2.99
3.00	0.0044	0.001350	0.0003822	0.0002034	3.0004	9.9998	3.00
3.01	0.0043	0.001306	0.0003689	0.0001959	3.0104	10.0599	3.01
3.02	0.0042	0.001264	0.0003560	0.0001887	3.0204	10.1202	3.02
3.03	0.0040	0.001223	0.0003436	0.0001817	3.0303	10.1807	3.03
3.04	0.0039	0.001183	0.0003316	0.0001749	3.0403	10.2414	3.04
3.05	0.0038	0.001144	0.0003199	0.0001684	3.0503	10.3023	3.05
3.06	0.0037	0.001107	0.0003087	0.0001621	3.0603	10.3634	3.06
3.07	0.0036	0.001070	0.0002978	0.0001561	3.0703	10.4247	3.07
3.08	0.0035	0.001035	0.0002873	0.0001502	3.0803	10.4862	3.08
3.09	0.0034	0.001001	0.0002771	0.0001446	3.0903	10.5480	3.09
3.10	0.0033	0.0009676	0.0002672	0.0001391	3.1003	10.6099	3.10
3.11	0.0032	0.0009354	0.0002577	0.0001335	3.1103	10.6720	3.11
3.12	0.0031	0.0009043	0.0002485	0.0001288	3.1202	10.7343	3.12
3.13	0.0030	0.0008740	0.0002396	0.0001236	3.1302	10.7968	3.13
3.14	0.0029	0.0008447	0.0002311	0.0001188	3.1402	10.8595	3.14
3.15	0.0028	0.0008164	0.0002227	0.0001152	3.1502	10.9224	3.15
3.16	0.0027	0.0007888	0.0002147	0.0001098	3.1602	10.9855	3.16
3.17	0.0026	0.0007622	0.0002070	0.0001062	3.1702	11.0488	3.17
3.18	0.0025	0.0007364	0.0001995	0.0001028	3.1802	11.1123	3.18

(Continued)

Table II.1 (*Continued*)   Some Functions of the Unit Normal Distribution

k	$f_u(k)$	$p_{u\geq}(k)$	$G_u(k)$	$J_u(k)$	$G_u(-k)$	$J_u(-k)$	k
3.19	0.0025	0.0007114	0.0001922	0.00009884	3.1902	11.1760	3.19
3.20	0.0024	0.0006871	0.0001852	0.00009392	3.2002	11.2399	3.20
3.21	0.0023	0.0006637	0.0001785	0.00009143	3.2102	11.3040	3.21
3.22	0.0022	0.0006410	0.0001720	0.00008792	3.2202	11.3683	3.22
3.23	0.0022	0.0006190	0.0001657	0.00008455	3.2302	11.4328	3.23
3.24	0.0021	0.0005976	0.0001596	0.00008002	3.2402	11.4975	3.24
3.25	0.0020	0.0005770	0.0001537	0.00007716	3.2502	11.5624	3.25
3.26	0.0020	0.0005571	0.0001480	0.00007514	3.2601	11.6275	3.26
3.27	0.0019	0.0005377	0.0001426	0.00007109	3.2701	11.6928	3.27
3.28	0.0018	0.0005190	0.0001373	0.00006831	3.2801	11.7583	3.28
3.29	0.0018	0.0005009	0.0001322	0.00006559	3.2901	11.8240	3.29
3.30	0.0017	0.0004834	0.0001273	0.00006315	3.3001	11.8899	3.30
3.31	0.0017	0.0004665	0.0001225	0.00006163	3.3101	11.9560	3.31
3.32	0.0016	0.0004501	0.0001179	0.00005922	3.3201	12.0223	3.32
3.33	0.0016	0.0004342	0.0001135	0.00005586	3.3301	12.0888	3.33
3.34	0.0015	0.0004189	0.0001093	0.00005467	3.3401	12.1555	3.34
3.35	0.0015	0.0004041	0.0001051	0.00005252	3.3501	12.2224	3.35
3.36	0.0014	0.0003897	0.0001012	0.00004963	3.3601	12.2896	3.36
3.37	0.0014	0.0003758	0.00009734	0.00004729	3.3701	12.3569	3.37
3.38	0.0013	0.0003624	0.00009365	0.00004553	3.3801	12.4244	3.38
3.39	0.0013	0.0003495	0.00009009	0.00004471	3.3901	12.4921	3.39
3.40	0.0012	0.0003369	0.00008666	0.00004192	3.4001	12.5600	3.40
3.41	0.0012	0.0003248	0.00008335	0.00004040	3.4101	12.6281	3.41
3.42	0.0012	0.0003131	0.00008016	0.00003888	3.4201	12.6964	3.42
3.43	0.0011	0.0003018	0.00007709	0.00003801	3.4301	12.7649	3.43
3.44	0.0011	0.0002909	0.00007413	0.00003649	3.4401	12.8336	3.44
3.45	0.0010	0.0002803	0.00007127	0.00003503	3.4501	12.9025	3.45
3.46	0.0010	0.0002701	0.00006852	0.00003363	3.4601	12.9716	3.46
3.47	0.0010	0.0002602	0.00006587	0.00003129	3.4701	13.0409	3.47

(Continued)

Table II.1 (*Continued*)   **Some Functions of the Unit Normal Distribution**

k	$f_u(k)$	$p_{u \geq}(k)$	$G_u(k)$	$J_u(k)$	$G_u(-k)$	$J_u(-k)$	k
3.48	0.0009	0.0002507	0.00006331	0.00003029	3.4801	13.1104	3.48
3.49	0.0009	0.0002415	0.00006085	0.00002900	3.4901	13.1801	3.49
3.50	0.0009	0.0002326	0.00005848	0.00002756	3.5001	13.2500	3.50
3.51	0.0008	0.0002241	0.00005620	0.00002738	3.5101	13.3201	3.51
3.52	0.0008	0.0002158	0.00005400	0.00002627	3.5201	13.3904	3.52
3.53	0.0008	0.0002078	0.00005188	0.00002521	3.5301	13.4609	3.53
3.54	0.0008	0.0002001	0.00004984	0.00002418	3.5400	13.5316	3.54
3.55	0.0007	0.0001926	0.00004788	0.00002243	3.5500	13.6025	3.55
3.56	0.0007	0.0001854	0.00004599	0.00002133	3.5600	13.6736	3.56
3.57	0.0007	0.0001785	0.00004417	0.00002134	3.5700	13.7449	3.57
3.58	0.0007	0.0001718	0.00004242	0.00002047	3.5800	13.8164	3.58
3.59	0.0006	0.0001653	0.00004073	0.00001857	3.5900	13.8881	3.59
3.60	0.0006	0.0001591	0.00003911	0.00001883	3.6000	13.9600	3.60
3.61	0.0006	0.0001531	0.00003755	0.00001805	3.6100	14.0321	3.61
3.62	0.0006	0.0001473	0.00003605	0.00001731	3.6200	14.1044	3.62
3.63	0.0005	0.0001417	0.00003460	0.00001660	3.6300	14.1769	3.63
3.64	0.0005	0.0001363	0.00003321	0.00001515	3.6400	14.2496	3.64
3.65	0.0005	0.0001311	0.00003188	0.00001448	3.6500	14.3225	3.65
3.66	0.0005	0.0001261	0.00003059	0.00001462	3.6600	14.3956	3.66
3.67	0.0005	0.0001213	0.00002935	0.00001402	3.6700	14.4689	3.67
3.68	0.0005	0.0001166	0.00002816	0.00001343	3.6800	14.5424	3.68
3.69	0.0004	0.0001121	0.00002702	0.00001203	3.6900	14.6161	3.69
3.70	0.0004	0.0001078	0.00002592	0.00001234	3.7000	14.6900	3.70
3.71	0.0004	0.0001036	0.00002486	0.00001095	3.7100	14.7641	3.71
3.72	0.0004	0.00009962	0.00002385	0.00001102	3.7200	14.8384	3.72
3.73	0.0004	0.00009574	0.00002287	0.00001043	3.7300	14.9129	3.73
3.74	0.0004	0.00009201	0.00002193	0.000009979	3.7400	14.9876	3.74
3.75	0.0004	0.00008842	0.00002103	0.000009592	3.7500	15.0625	3.75
3.76	0.0003	0.00008496	0.00002016	0.000009190	3.7600	15.1376	3.76

(*Continued*)

Table II.1 (*Continued*)   Some Functions of the Unit Normal Distribution

k	$f_u(k)$	$p_{u\geq}(k)$	$G_u(k)$	$J_u(k)$	$G_u(-k)$	$J_u(-k)$	k
3.77	0.0003	0.00008162	0.00001933	0.000008687	3.7700	15.2129	3.77
3.78	0.0003	0.00007841	0.00001853	0.000008302	3.7800	15.2884	3.78
3.79	0.0003	0.00007532	0.00001776	0.000007947	3.7900	15.3641	3.79
3.80	0.0003	0.00007235	0.00001702	0.000007686	3.8000	15.4400	3.80
3.81	0.0003	0.00006948	0.00001632	0.000007269	3.8100	15.5161	3.81
3.82	0.0003	0.00006673	0.00001563	0.000007068	3.8200	15.5924	3.82
3.83	0.0003	0.00006407	0.00001498	0.000006671	3.8300	15.6689	3.83
3.84	0.0003	0.00006152	0.00001435	0.000006448	3.8400	15.7456	3.84
3.85	0.0002	0.00005906	0.00001376	0.000006140	3.8500	15.8225	3.85
3.86	0.0002	0.00005669	0.00001317	0.000005797	3.8600	15.8996	3.86
3.87	0.0002	0.00005442	0.00001262	0.000005632	3.8700	15.9769	3.87
3.88	0.0002	0.00005223	0.00001208	0.000005377	3.8800	16.0544	3.88
3.89	0.0002	0.00005012	0.00001157	0.000005078	3.8900	16.1321	3.89
3.90	0.0002	0.00004810	0.00001108	0.000004945	3.9000	16.2100	3.90
3.91	0.0002	0.00004615	0.00001061	0.000004700	3.9100	16.2881	3.91
3.92	0.0002	0.00004427	0.00001016	0.000004388	3.9200	16.3664	3.92
3.93	0.0002	0.00004247	0.000009723	0.000004214	3.9300	16.4449	3.93
3.94	0.0002	0.00004074	0.000009307	0.000004059	3.9400	16.5236	3.94
3.95	0.0002	0.00003908	0.000008908	0.000003963	3.9500	16.6025	3.95
3.96	0.0002	0.00003748	0.000008525	0.000003801	3.9600	16.6816	3.96
3.97	0.0002	0.00003594	0.000008158	0.000003610	3.9700	16.7609	3.97
3.98	0.0001	0.00003446	0.000007806	0.000003429	3.9800	16.8404	3.98
3.99	0.0001	0.00003304	0.000007469	0.000003293	3.9900	16.9201	3.99
4.00	0.0001	0.00003167	0.000007145	0.000003069	4.0000	17.0000	4.00

Another special function of the unit normal, $J_u(k)$, was introduced in Section 7.5.2. Using the relationship in Equation II.6, and integration by parts, the following relationship can be established:

$$J_u(k) = \int_k^{\infty} (u_0 - k)^2 f_u(u_0)\, du_0$$
$$= (1 + k^2) p_{u\geq}(k) - k f_u(k) \tag{II.9}$$

Values of $J_u(k)$ have been tabulated in Table II.1.

Two other results, potentially useful in some advanced modeling contexts, are that

$$\int_{-\infty}^{\infty} f_u(u_0) p_{u \geq}(au_0 + b)\, du_0 = p_{u \geq}\left(\frac{b}{\sqrt{1 + a^2}}\right) \tag{II.10}$$

and

$$\int_{-\infty}^{\infty} f_u(u_0) G_u(au_0 + b)\, du_0 = \sqrt{1 + a^2}\, G_u\left(\frac{b}{\sqrt{1 + a^2}}\right) \tag{II.11}$$

The proof, involving expressing $p_{u \geq}(k)$ or $G_u(k)$ in integral form and then changing the order of integration, was presented by Silver and Smith (1981).

II.3 Relating Any Normal Distribution to the Unit Normal

Consider a normally distributed variable x with mean $\hat{x}$ and standard deviation σ_x. One quantity of direct interest in measuring customer service is the probability of A where A is the event that x takes on a value greater than or equal to $\hat{x} + k\sigma_x$. Now,

$$\text{prob}(A) = \Pr\{x \geq \hat{x} + k\sigma_x\}$$

$$= \int_{\hat{x}+k\sigma_x}^{\infty} f_x(x_0)\, dx_0$$

$$= \int_{\hat{x}+k\sigma_x}^{\infty} \frac{1}{\sigma_x \sqrt{2\pi}} exp[-(x_0 - \hat{x})^2 / 2\sigma_x^2]\, dx_0 f_x(x_0)\, dx_0 \tag{II.12}$$

Consider

$$u_0 = (x_0 - \hat{x})/\sigma_x$$

We have that $du_0/dx_0 = 1/\sigma_x$. Also, when $x_0 = \hat{x} + k\sigma_x$, we have that $u_0 = k$; and when $x_0 \to \infty$, $u_0 \to \infty$.

Therefore, substituting these expressions into Equation II.12 results in

$$\text{prob}(A) = \int_k^{\infty} \frac{1}{\sqrt{2\pi}} exp(-u_0^2/2)\, du_0$$

Using Equation II.3 and the definition of event A, we have

$$\text{prob}\{x \geq \hat{x} + k\sigma_x\} = p_{u \geq}(k) \tag{II.13}$$

This is represented by the shaded area in Figure II.1.

Another quantity of interest is the expected shortage per replenishment cycle.

$$\text{ESPRC} = \int_{\hat{x}+k\sigma_x}^{\infty} (x_0 - \hat{x} - k\sigma_x) f_x(x_0) dx_0$$

$$= \int_{\hat{x}+k\sigma_x}^{\infty} (x_0 - \hat{x} - k\sigma_x) \frac{1}{\sigma_x\sqrt{2\pi}} exp[-(x_0 - \hat{x})^2/2\sigma_x^2] dx_0 \qquad (\text{II.14})$$

The key again is to substitute

$$u_0 = (x_0 - \hat{x})/\sigma_x$$

The result, after considerable simplification, is

$$\text{ESPRC} = \sigma_x \int_k^{\infty} (u_0 - k) f_u(u_0) du_0$$

From Equation II.5, this is seen to be

$$\text{ESPRC} = \sigma_x G_u(k) \qquad (\text{II.15})$$

II.4 Further Properties of the Normal Distribution

In Chapter 9 we have a normally distributed variable x with mean $\hat{x}$ and standard deviation σ_x, and we let

$$Q = \hat{x} + k\sigma_x$$

The first quantity desired is $p_{x<}(Q)$. Clearly,

$$p_{x<}(Q) = \text{prob}(x < Q) = 1 - \text{prob}(x \geq Q)$$
$$= 1 - \text{prob}(x \geq \hat{x} + k\sigma_x)$$

Using Equation II.10, we have

$$p_{x<}(Q) = 1 - p_{u\geq}(k) = p_{u<}(k)$$

The function $p_{u<}(k)$ is sometimes denoted $\Phi(k)$ in the literature. We may show several other properties:

$$\int_Q^{\infty} x_0 f_x(x_0) dx_0 = \int_Q^{\infty} (x_0 - Q) f_x(x_0) dx_0 + \int_Q^{\infty} Q f_x(x_0) dx_0$$

$$= \int_{\hat{x}+k\sigma_x}^{\infty} (x_0 - \hat{x} - k\sigma_x) f_x(x_0) dx_0 + \int_{\hat{x}+k\sigma_x}^{\infty} Q f_x(x_0) dx_0$$

Use of Equations II.10 through II.12 gives us

$$\int_{Q}^{\infty} x_0 f_x(x_0) dx_0 = \sigma_x G_u(k) + Q p_{u \geq}(k) \tag{II.16}$$

Also, if the chance of a negative value of x is very small, then

$$\int_{0}^{Q} x_0 f_x(x_0) dx_0 \approx \int_{-\infty}^{Q} x_0 f_x(x_0) dx_0$$

$$= \int_{-\infty}^{\infty} x_0 f_x(x_0) dx_0 - \int_{Q}^{\infty} x_0 f_x(x_0) dx_0$$

$$= \hat{x} - \int_{Q}^{\infty} x_0 f_x(x_0) dx_0$$

This result applies to any p.d.f. Then the use of Equation II.13 leads to

$$\int_{0}^{Q} x_0 f_x(x_0) dx_0 \approx \hat{x} - \sigma_x G_u(k) - Q p_{u \geq}(k)$$

Reference

Silver, E. A. and D. Smith. 1981. Setting individual item production-rates under significant lead time conditions. *INFOR 19*(1), 1–19.

Appendix III: Approximations and Excel Functions

In this appendix we provide approaches to calculating key values for the normal, Gamma, and Poisson distributions on a spreadsheet. As discussed in Appendix II and throughout the book, the normal distribution is widely used in inventory and production planning models. As we will see in Section III.1, there are concise and effective approaches for calculating all the values necessary for models presented in this book for the normal distribution. For Gamma and Poisson distributions, we present approaches using spreadsheet functions for calculating expected units short per cycle and probability of stocking out in a cycle. These expressions can be used along with search procedures in spreadsheets to apply the models in this text.

III.1 Approximations and Excel Functions for the Normal Distribution

The approximations are extremely accurate—*much* more so than the typical error in other input data (such as r and A). The functions listed in upper case are Excel functions.

Given a k value:

$$f_u(k) = \frac{1}{\sqrt{2\pi}} exp(-k^2/2)$$

The Excel function NORMDIST can be used to return this value

$$f_u(k) = \text{NORMDIST}(k, 0, 1, \text{FALSE})$$

where NORMDIST is an Excel function returning the p.d.f. of the normal distribution at the value k. The second and third arguments (0 and 1 in this case) are the mean and standard deviation. The fourth argument is a switch which returns the p.d.f. when set to FALSE and the c.d.f. when set to TRUE.

Given $f_u(k)$

$$k = \sqrt{2\ln\left(\frac{1}{f_u(k)\sqrt{2\pi}}\right)}$$

We can also use this NORMDIST Excel function to find $p_{u\geq}(k)$.

$$p_{u\geq}(k) = 1 - \text{NORMDIST}(k, 0, 1, \text{TRUE})$$

Alternatively, Waissi and Rossin (1996) provide a very accurate approximation

$$p_{u\geq}(k) = 1 - \frac{1}{1 + e^{-\sqrt{\pi}(\beta_1 k^5 + \beta_2 k^3 + \beta_3 k)}}$$

where $\beta_1 = -0.0004406$, $\beta_2 = +0.0418198$, $\beta_3 = +0.9000000$.

As noted in Equation 19.7 in Appendix II,

$$G_u(k) = f_u(k) - k p_{u\geq}(k)$$

This can be calculated in Excel for a given k as

$$G_u(k) = \text{NORMDIST}(k, 0, 1, \text{FALSE}) - k[1 - \text{NORMDIST}(k, 0, 1, \text{TRUE})] \qquad \text{(III.1)}$$

Given $G_u(k)$

$$k = \frac{a_0 + a_1 z + a_2 z^2 + a_3 z^3}{b_0 + b_1 z + b_2 z^2 + b_3 z^3 + b_4 z^4}$$

where

$$z = \sqrt{\ln(25/G_u(k))^2} \qquad b_0 = 1$$
$$a_0 = -5.3925569 \qquad b_1 = -7.2496485 \times 10^{-1}$$
$$a_1 = 5.6211054 \qquad b_2 = 5.07326622 \times 10^{-1}$$
$$a_2 = -3.8836830 \qquad b_3 = 6.69136868 \times 10^{-2}$$
$$a_3 = 1.0897299 \qquad b_4 = -3.29129114 \times 10^{-3}$$

This rational approximation looks somewhat messy, but on a spreadsheet it is very easy to implement. Furthermore, one can verify the spreadsheet using Table II.1 in Appendix II. These formulas allow one to find reorder points for an item on a single line of a spreadsheet, implying that a manager can perform computations for even several thousand items quite easily. Note that the inverse functions (k, given a value of $G_u(k)$, for instance) can be used to find implicit performance measures, as discussed in Sections 6.8 and 6.11. See Tijms (1994) and Shore (1982) for other approximations, particularly if working with computer code that does not have the normal distribution built in. There are excellent rational approximations for all of the above functions.

From Equation II.9 in Appendix II

$$J_u(k) = (1 + k^2) p_{u\geq}(k) - k f_u(k)$$

Using Excel functions, we can express $J_u(k)$ as follows:

$$J_u(k) = (1 + k^2)[1 - \text{NORMDIST}(k, 0, 1, \text{TRUE})] - k \, \text{NORMDIST}(k, 0, 1, \text{FALSE})$$

For a given value of $J_u(k)$ we can find k using the following rational approximation, which can be built on a spreadsheet very quickly. See Schneider (1981).

$$k = \frac{a_0 + a_1 z + a_2 z^2 + a_3 z^3}{b_0 + b_1 z + b_2 z^2 + b_3 z^3}$$

where for $0 \leq J_u(k) \leq 0.5$

$$z = \sqrt{\ln(1/J_u(k))^2}$$

$$a_0 = -4.18884136 \times 10^{-1} \qquad b_0 = 1$$

$$a_1 = -2.5546970 \times 10^{-1} \qquad b_1 = 2.1340807 \times 10^{-1}$$

$$a_2 = 5.1891032 \times 10^{-1} \qquad b_2 = 4.4399342 \times 10^{-2}$$

$$a_3 = 0 \qquad b_3 = -2.6397875 \times 10^{-3}$$

while for $J_u(k) > 0.5$

$$z = J_u(k)$$

$$a_0 = 1.1259464 \qquad b_0 = 1$$

$$a_1 = -1.3190021 \qquad b_1 = 2.8367383$$

$$a_2 = -1.8096435 \qquad b_2 = 6.5593780 \times 10^{-1}$$

$$a_3 = -1.1650097 \times 10^{-1} \qquad b_3 = 8.2204352 \times 10^{-3}$$

III.2 Excel Functions for the Gamma Distribution

As mentioned in Section 6.7.14 of Chapter 6 the Gamma distribution may be a good choice when the total lead time demand distribution is skewed to the right, or when the variability is higher than is reasonable for the normal distribution. Some of the approximations for the Gamma are somewhat more complex, so we present only those easily implemented on a spreadsheet. (Refer to Burgin (1975); Burgin and Norman (1976) for approximations.) Tyworth et al. (1996) describe a spreadsheet implementation using the Gamma distribution.

Suppose demand follows a Gamma distribution described by shape parameter α and scale parameter β, implying a mean of $\alpha\beta$ and standard deviation of $\sqrt{\alpha}\beta$. With a reorder point s and order quantity Q,

$$\text{Probability of a stockout} = 1 - P_1 = 1 - \text{GAMMADIST}(s, \alpha, \beta, \text{TRUE})$$

where TRUE is a switch to indicate that the cumulative distribution function is desired. (FALSE indicates the p.d.f.)

Given a probability of not stocking out P_1,

$$\text{Reorder point } s = \text{GAMMAINV}(P_1, \alpha, \beta)$$

As shown in Section 6A.7 of the Appendix to Chapter 6, the expected units short per replenishment cycle for Gamma distributed demand is

$$\text{ESPRC} = \alpha\beta[1 - F(s; \alpha + 1, \beta)] - s[1 - F(s; \alpha, \beta)]$$

where $f(x; \alpha, \beta)$ and $F(x; \alpha, \beta)$ represent the p.d.f. and the cumulative distribution function, respectively, of the Gamma distribution with parameters α and β.

We can calculate this expected units short per cycle expression for the Gamma distribution using Excel functions as follows:

$$\text{ESPRC} = \alpha\beta[1 - \text{GAMMADIST}(s, \alpha + 1, \beta, \text{TRUE})] - s[1 - \text{GAMMADIST}(s, \alpha, \beta, \text{TRUE})]$$

Using this value for ESPRC, we can find the fill rate, P_2, for a given reorder point s and order quantity Q

$$\text{Fill rate} = P_2 = 1 - \frac{\text{ESPRC}}{Q}$$

Using this expression, one can search for the value s that achieves the desired fill rate for a given value of Q.

III.3 Excel Functions for the Poisson Distribution

As mentioned in Chapter 8, the Poisson distribution may be useful for modeling demand for slow-moving items. Following the definitions from Equation 8.29 in the Appendix of Chapter 8, the probability mass function (p.m.f.) is given by

$$p_x(x_0) = \Pr\{x = x_0\} = \frac{a^{x_0} e^{-a}}{x_0!} x_0 = 0, 1, 2, \ldots \tag{III.2}$$

where a is the single parameter of the distribution. The cumulative distribution function is

$$P_x(s) = \sum_{x_0=0}^{s} p_x(x_0)$$

For a given reorder point s, P_1 can be calculated with Excel functions

$$P_1 = \text{POISSON}(s, a, \text{TRUE})$$

where TRUE is a switch to indicate that the cumulative distribution function is desired. (FALSE indicates the p.m.f.) Since the Poisson distribution is a discrete distribution, there is not a convenient way to find the inverse as there is with Normal and Gamma. It is possible to calculate P_1 values over a range of values of s and select the desired s. Typically the desired s will be the smallest value that achieves *at least* the desired value for P_1.

To calculate the fill rate for a given reorder point s and order quantity Q, we first calculate ESPRC as specified in Equation 8.30 from the Appendix to Chapter 8 which we restate here.

$$\text{ESPRC} = a[1 - P_x(s - 1)] - s[1 - P_x(s)]$$

Using Excel functions,

$$\text{ESPRC} = a[1 - \text{POISSON}(s - 1, a, \text{TRUE})] - s[1 - \text{POISSON}(s, a, \text{TRUE})]$$

Then, as above, fill rate can be calculated as

$$\text{Fill rate} = P_2 = 1 - \frac{\text{ESPRC}}{Q}$$

References

Burgin, T. A. 1975. The gamma distribution and inventory control. *Operational Research Quarterly 26*(3i), 507–525.

Burgin, T. A. and J. M. Norman. 1976. A table for determining the probability of a stockout and potential lost sales for a gamma distributed demand. *Operational Research Quarterly 27*(3i), 621–631.

Schneider, H. 1981. Effect of service-levels on order-points or order-levels in inventory models. *International Journal of Production Research 19*(6), 615–631.

Shore, H. 1982. Simple approximations for the inverse cumulative function, the density function and the loss integral of the normal distribution. *Applied Statistician 31*(2), 108–114.

Tijms, H. C. 1994. *Stochastic Models: An Algorithmic Approach.* New York: John Wiley & Sons.

Tyworth, J. E., Y. Guo, and R. Ganeshan. 1996. Inventory control under gamma demand and random lead time. *Journal of Business Logistics 17*(1), 291–304.

Waissi, G. R. and D. R. Rossin. 1996. A sigmoid approximation of the standard normal integral. *Applied Mathematics and Computation 77*, 91–95.

Author Index

Subject Index